# Immune Regulation and Immunotherapy in Autoimmune Disease

# Immune Regulation and Immunotherapy in Autoimmune Disease

Edited by

## Jingwu Zhang, M.D., Ph.D.
*Shanghai JiaoTong University School of Medicine*
*Shanghai, China*

and

*Baylor College of Medicine*
*Houston, Texas, USA*

 Springer

Jingwu Zhang, M.D., Ph.D.
Institute of Health Sciences
Shanghai Institutes for Biological
    Sciences
Chinese Academy of Sciences
Shanghai JiaoTong University
    School of Medicine
Shanghai, China
and
Department of Neurology
Baylor College of Medicine
Houston, TX, USA

Library of Congress Control Number: 2006928326

ISBN-10: 0-387-36002-6          e-ISBN-10: 0-387-36003-4
ISBN-13: 978-0-387-36002-7          e-ISBN-13: 978-0-387-36003-4

Printed on acid-free paper.

9 8 7 6 5 4 3 2 1

springer.com

# Preface

This book critically reviews both clinical and immunological aspects of autoimmune disease with a strong emphasis on multiple sclerosis (MS). Research in MS is one of the fastest-developing areas in modern medicine. It employs some of the newest concepts in autoimmune mechanisms and an array of new treatments that would have been considered science fiction only two decades ago. It is an area in which research findings are being actively translated into treatment strategies. Advances in this area have both clinical and scientific implications for other autoimmune conditions that share many similarities with MS. The book, which comprises 24 chapters contributed by experts and thought leaders in the field, is designed to provide new insights into two arenas: our current understanding of autoimmune mechanisms and immune regulation and the latest developments in immunotherapy.

This book is intended for both researchers and clinicians. Its purpose is not only to provide a comprehensive review on the recent advances in the two arenas mentioned above but also to reflect the current opinions or concepts that influence our thinking about the disease mechanisms and our way of treating MS patients. Many of these issues have emerged from recent studies and are somewhat contradictory to traditional thinking. For example, recent studies have indicated that MS is more heterogeneous in many aspects than traditionally thought. Pathologically, in addition to demyelination and inflammation, there is axonal loss or damage detectable in the central nervous system (CNS) lesions. There is a neurodegenerative component of the disease. It is currently debated as to how significant this neurodegenerative component is and whether the neurodegenerative process found in MS is secondary to inflammation. Even with the CNS demyelination and inflammation characteristic of MS, there are distinct patterns characterized by a differential presence of heterogeneous inflammatory cells (T cells, B cells, macrophages) or antibodies in association with varying degrees of demyelination and inflammation. In addition to pathological heterogeneity, there are well known genetic, immunological, and clinical variations in MS. The heterogeneous nature of MS has important therapeutic implications. A particular drug may be more efficacious in one subset of MS than another because of the different disease processes involved. In this regard, it is important to determine suitable treatments for different subgroups of MS patients using biomarkers characteristic of each subgroup. With this knowledge, combination therapy can be developed to capture multiple attack points involved in the various MS subgroups. Our oncology colleagues have been utilizing this strategy elegantly for years to treat cancer patients. A number of chapters in this book are devoted to these issues.

Another interesting aspect reviewed extensively here is related to how significantly the immunological research has contributed to our understanding of the disease and has translated to new treatments for MS. Much of the immunological research in MS centers on the hypothesis that myelin autoreactive T cells play an important role in CNS inflammation and demyelination of MS. These autoreactive T cells are abnormally activated—perhaps by microbial infection through molecular mimicry mechanisms—and undergo clonal expansion in conjunction with aberrant regulatory mechanisms that normally keep them in check. Based on this hypothesis, numerous specific immune therapeutic strategies have been developed through immunological research and have been proven effective in experimental autoimmune encephalomyelitis (EAE), an animal model for MS. To name a few, these approaches include altered peptide ligands, myelin-induced oral tolerance, T cell receptor (TCR) peptides, DNA vaccination and T cell vaccination (for targeting autoreactive T cells), and finally monoclonal antibodies directed at a variety of cytokines/integrins or their receptors (reducing inflammation or blocking T cell or other inflammatory cells from entering the CNS). Unfortunately, although we can cure EAE by precisely targeting a component of the TCR–peptide–major histocompatibility complex required for T cell activation, many of these approaches have failed in pilot studies or controlled human clinical trials.

There are several levels of complexity in this regard. First, the true myelin autoantigen(s) is unknown. The candidate myelin antigens used in all immunological studies are extrapolated from EAE in which the disease is commonly induced against myelin basic protein (MBP), proteolipid protein, and myelin oligodendrocyte glycoprotein. The best evidence indicative of the potential involvement of myelin antigens in MS perhaps comes from a recent clinical trial in which an altered peptide of MBP was tested to inactivate circulating T cells recognizing the immune dominant epitope (residues 83-99) of MBP. Although the approach works well in inbred rodents in which the TCR repertoire, including the contact residues involved in recognition of the MBP peptide, is highly restricted, the TCR of human MBP-reactive T cells is considerably degenerated. Thus, alanine substitution at the key TCR contact residue is able to render these T cells inactive in one MS patient but may be ineffective or may even activate the same autoreactive T cells in another. Indeed, some MS patients who received injections of the altered peptide experienced clinical exacerbation and increased lesion load, as indicated by magnetic resonance imaging. It was evident that T cells recognizing the immunodominant MBP peptide were activated by the treatment in these patients. Even if these myelin antigens are involved in the autoimmune mechanism of MS as autoantigens, there are other unresolved issues, such as "epitope spreading."

The T cell repertoire and TCR makeup are much more heterogeneous and complex in humans than in inbred rodents. This complexity is one of the key problems preventing us from extrapolating what works effectively for EAE to the treatment of MS. For example, it is known that, unlike highly restricted TCR V gene usage in myelin-autoreactive T cells seen in EAE mice, myelin-autoreactive T cells in MS display a highly diverse TCR V gene distribution pattern even in the context of the DRB1*1501 genetic background, making TCR-based immunotherapy difficult. Altered peptide ligand of MBP is another example in which a high degree of TCR degeneracy in human MBP-reactive T cells makes a critical difference, as described above. By the same token, it is debatable as to whether EAE is an adequate research model for MS. To say the least, human MS involves highly complex genetic and immunological processes, making EAE at best an incomplete match for MS.

In this regard, a "humanized mouse model" would be more suited for immunological research. Such a mouse model has been successfully generated in NOD-SCID or Rag-deficient immune compromised mice by implanting human stem cells and thymus to reconstitute an entire human immune system.

Furthermore, when a therapeutic strategy is too specific, it may carry with it significant pitfalls for the reasons described above. Many strategies are now designed to target more "downstream" processes by blocking T cell entrance into the CNS or reducing the CNS inflammation seen in MS. Interferon-β and humanized antibody to integrin-α4 are good examples. Immunological research during the last 15 to 20 years has provided important lessons, as described above, and has produced exciting results. We now have interferon-β, glatiramer acetate, perhaps natalizumab, and many other immunotherapies that are currently being tested in clinical trials at various stages. Moreover, forward-looking research, including stem cell approaches, has brought new promise that damaged myelin or neuronal tissue may be repaired or regenerated by stem cells when inflammation and demyelination are under therapeutic control.

In conclusion, we are entering an exciting time in history—witnessing rapid development of new treatments for MS and learning how to treat the disease effectively. Several chapters in this book review some of the highlights in the field and provide expert opinions on what the future holds for our patients who suffer daily from this devastating disease. I am grateful to our contributors, many of whom are long-time collaborators and friends, for bringing together this unique and timely book.

Jingwu Zhang

# Contents

**Part II:   Novel Immunotherapeutic Strategies and Emerging Treatments**

# Contributors

**Ana C. Anderson**
Center for Neurologic Diseases
Brigham and Women's Hospital
Harvard Medical School
Boston, MA, USA

**Jean-François Bach**
Université René Descartes -
    Paris V - INSERM U580
Hôpital Necker-Enfants Malades
Paris, France

**Richard K. Burt**
Division of Immunotherapy
Northwestern University
Feinberg School of Medicine
Chicago, IL, USA

**Sophie Candon**
Université René Descartes -
    Paris V - INSERM U580
Hôpital Necker-Enfants Malades
Paris, France

**Lucienne Chatenoud**
Inserm U580
Universite René Descartes, Paris 5
Hospital Necker-Enfants Malades
Paris, France

**Sunil V. Cherry**
Laboratory of Molecular Immunology
Harvard Medical School
Center for Neurologic Disease
Department of Neurology
Brigham and Women's Hospital
77 Avenue Louis Pasteur
Boston, MA, USA

**Leonard Chess**
Department of Medicine and
    Pathology
Rheumatology Division
Columbia University
College of Physicians and Surgeons
New York, New York, USA

**Irun R. Cohen**
Department of Immunology
The Weizmann Institute of Science
Rehovot, Israel

**Nicole E. Culbertson**
Medical Center
Portland VA medical Center
Neuroimmunology Research
Portland, OR, USA

**Chen Dong**
Department of Immunology
MD Anderson Cancer Center
Houston, TX, USA

**Manuel A. Friese**
MRC Human Immunology Unit and
  Department of Clinical Neurology
Weatherall Institute of Molecular
  Medicine
John Radcliffe Hospital
University of Oxford
Oxford, UK

**Lars Fugger**
MRC Human Immunology Unit and
Department of Clinical Neurology
Weatherall Institute of Molecular
  Medicine
John Radcliffe Hospital
University of Oxford
Oxford, UK

**David A. Hafler**
Neurology (Neuroscience)
Harvard Medical School
Brigham and Women's Hospital,
  NRB 641
77 Avenue Louis Pasteur
Boston, MA, USA

**Hans-Peter Hartung**
Department of Neurology
Heinrich Heine University
Düsseldorf, Germany

**Bernhard Hemmer**
Department of Neurology
Heinrich Heine University
Düsseldorf, Germany

**Reinhard Hohlfeld**
Department of Neuroimmunology
Max Planck-Institute for
  Neurobiology
Martinsried, Germany

**Jian Hong**
Department of Neurology
Baylor Multiple Sclerosis Center
Baylor College of Medicine
Houston, Texas, USA

**Steven Jacobson**
Viral Immunology Section,
  NINDS/NIH
Bethesda, MD, USA

**Hong Jiang**
Department of Medicine
College of Physicians and Surgeons
Columbia University
New York, New York, USA

**Bernd C. Kieseier**
Department of Neurology
Heinrich Heine University
Düsseldorf, Germany

**Jeffery D. Kocsis**
Department of Neurology
Yale University School of Medicine
VA Connecticut Healthcare System
Neuroscience Research Center
West Haven, CT, USA

**Adam P. Kohm**
Department of Microbiology-
  Immunology
Feinberg School of Medicine
Northwestern University
Chicago, IL, USA

**Vijay K. Kuchroo**
Center for Neurologic Diseases
Brigham and Women's Hospital
Harvard Medical School
Boston, MA, USA

**Vipin Kumar**
Torrey Pines Institute for
   Molecular Studies
La Jolla, CA, USA

**Peter E. Lipsky**
National Institute of Arthritis and
   Musculoskeletal and Skin Diseases
National Institutes of Health
Bethesda, MD, USA

**Yvonne Loh**
Division of Immunotherapy
750 N. Lakeshore Drive
Room 649
Chicago, IL, USA

**Claudia F. Lucchinetti**
Department of Neurology
Mayo Clinic College of Medicine
Rochester, MN, USA

**Natalia Martin-Orozco**
Department of Immunology
MD Anderson Cancer Center
Houston, TX, USA

**Til Menge**
Department of Neurology
Heinrich Heine University
Düsseldorf, Germany

**Stephen D. Miller**
Department of Microbiology-
   Immunology
Feinberg School of Medicine
Northwestern University
Chicago, IL, USA

**Paolo A. Muraro**
Clinical Reader in Neuroimmunology
Department of Cellular and Molecular
   Neuroscience
Division of Neuroscience and Mental
   Health

Faculty of Medicine,
   Imperial College
Charing Cross Campus
St. Dunstan's Road, London W6 8RP,
   UK

**Stefan Nessler**
Department of Neurology
Heinrich Heine University
Düsseldorf, Germany

**Roza Nurieva**
Department of Immunology
MD Anderson Cancer Center
Houston, TX, USA

**Kevin C. O'Connor**
Laboratory of Molecular
   Immunology
Harvard Medical School
Center for Neurologic Disease
Department of Neurology
Brigham and Women's Hospital
77 Avenue Louis Pasteur
Boston, MA, USA

**Francisco J. Quintana**
Center for Neurologic Diseases
Harvard Medical School
Boston, MA, USA

**Christine Radtke**
Department of Neurology
Yale University School of
   Medicine
VA Connecticut Healthcare System
Neuroscience Research Center
West Haven, CT, USA

**Hendrik Schulze-Koops**
Nikolaus Fiebiger Center for
   Molecular Medicine
Clinical Research Group III
University of Erlangen-Nuremberg
Erlangen, Germany

**Sheri M. Skinner**
Department of Neurology
Baylor Multiple Sclerosis Center
Baylor College of Medicine
Houston, Texas, USA

**Trevor R.F. Smith**
Torrey Pines Institute for Molecular
    Studies
La Jolla, CA, USA

**Xiaolei Tang**
Torrey Pines Institute for Molecular
    Studies
La Jolla, CA, USA

**Arthur A. Vandenbark**
Portland VA Medical Center
Neuroimmunology Research
Portland, OR, USA

**Larissa Verda**
Division of Immunotherapy
Northwestern University
Feinberg School of Medicine
Chicago, IL, USA

**Peter M. Vogt**
Department of Neurology
Yale University School of Medicine
VA Connecticut Healthcare System
Neuroscience Research Center
West Haven, CT, USA

**Rhonda R. Voskuhl**
Department of Neurology
University of California
Los Angeles, CA, USA

**Hartmut Wekerle**
Department of Neuroimmunology
Max Planck Institute for
    Neurobiology
Martinsried, Germany

**Heinz Wiendl**
Department of Neurology
University of Wuerzburg
Wuerzburg, Germany

**Elizabeth L. Williams**
Viral Immunology Section,
    NINDS/NIH
Bethesda, MD, USA

**Takashi Yamamura**
Department of Immunology
National Institute of Neuroscience
Tokyo, Japan

**Ying C.Q. Zang**
Department of Neurology
Baylor Multiple Sclerosis Center
Baylor College of Medicine
Houston, Texas, USA

**Jingwu Zhang**
Institute of Health Sciences
Shanghai Institutes for Biological
    Sciences
Chinese Academy of Sciences
Shanghai JiaoTong University
    School of Medicine
Shanghai, China
Department of Neurology
Baylor College of Medicine
Houston, TX, USA

**Dun Zhou**
Department of Neurology
Heinrich Heine University
Düsseldorf, Germany

# Part I
## Immune Regulation and Autoimmune Disease

# 1
# Autoimmune Response and Immune Tolerance

Ana C. Anderson and Vijay K. Kuchroo

## 1   Introduction

Central mechanisms of tolerance are responsible for purging highly self-reactive cells from the developing T cell repertoire. However, central tolerance is not complete, and potentially deleterious self-reactive T cells are seeded to the peripheral immune compartment. Activation of these cells results in autoimmunity. Thus, peripheral mechanisms of tolerance exist to provide an added layer of protection against the activation of such cells. The relative importance of central versus peripheral mechanisms of tolerance has been debated over the years, with each being favored at different times. Ultimately, it is clear that both are needed given that detects in either of these mechanisms favors the development of autoimmunity. During the last few years, the discovery of the expression of "tissue-specific" antigens in the thymus and keen interest in regulatory T cells has put both of these mechanisms of tolerance at center stage.

## 2   Thymic Selection of the Self-Reactive Repertoire

The ability to discriminate between self and non-self antigens is imposed on developing T cells in the thymus through the processes of positive and negative selection. In the thymus, T cells that can recognize peptide fragments derived from protein antigens presented in the context of self major histocompatibility complex (MHC) molecules receive important survival signals that allow them to continue through development. This process is known as positive selection. The antigens expressed in the thymus are predominantly of self origin. Consequently, all T cells are inherently self-reactive. Thus, a second process known as negative selection serves to eliminate T cells that react too strongly with self-antigen.

Recent work has focused on the impact of thymic expression of "tissue-specific" antigens on the positive and negative selection of the self-reactive repertoire. The transcription factor AIRE (autoimmune regulator) has emerged as an important regulator of "tissue-specific" antigen expression in medullary thymic epithelial cells (mTECs), specialized cells that play a role in deletion of self-reactive cells and/or the generation of regulatory T cells. Both of these are discussed below.

## 2.1    Impact of Self-Antigen

For many years, it was hypothesized that a major factor in organ-specific autoimmunity was lack of negative selection against self-antigens sequestered behind anatomical barriers such as the eye, testis, and brain. However, several studies have now shown that many "tissue-specific" antigens are indeed expressed in the thymus (reviewed in Kyewski et al., 2002). The expression of such antigens in the thymus can shape the self-reactive T cell repertoire by two mechanisms: deletion of self-reactive cells and induction of regulatory T cells.

The first evidence that expression of a "tissue-specific" antigen in the thymus affected the selection of high-affinity T cells came from studies of the T cell repertoire to myelin basic protein (MBP), a known target antigen in experimental autoimmune encephalomyelitis (EAE), an animal model of autoimmunity of the central nervous system (CNS). For several years, it was known that MBP was expressed in the thymus (Fritz and Zhao, 1996; Mathisen et al., 1993; Pribyl et al., 1993, 1996). However, evidence that this expression affected the selection of the T cell repertoire to MBP came later from studies in MBP-deficient mice. One study found that the T cell repertoire to MBP that develops in MBP-deficient C3H (H-$2^k$) mice recognizes MBP 79-87 with at least three logs higher avidity relative to the repertoire generated in wild-type mice (Targoni and Lehman, 1998). A similar study of MBP-deficient mice on the Balb/c (H-$2^d$) background found that MBP-deficient mice respond strongly to both MBP 59-76 and 89-101, whereas wild-type mice fail to respond (Yoshizawa et al., 1998). Although not formally demonstrated in either case, the data suggest that expression of MBP in the thymus of wild-type mice results in deletion of T cells that recognize MBP with high avidity.

Myelin proteolipid protein (PLP), another CNS autoantigen, is also expressed in the thymus. Interestingly, the form of PLP that predominates in the thymus is the splice variant DM20, which lacks residues 116–150 (Anderson et al., 2000; Klein et al., 2000). Consequently, only cells that recognize epitopes in DM20 would be expected to undergo central tolerance. Indeed, analysis of the T cell repertoire to PLP in PLP-deficient C57BL/6 (H-2$^b$) mice suggests that T cells reactive to epitopes contained with DM20 undergo tolerance (Klein et al., 2000). In SJL (H-2$^s$) mice, which are highly susceptible to EAE induced with PLP 139-151, the predominant expression of DM20 in the thymus appears to be responsible for the lack of negative selection of cells reactive to the PLP 139-151 epitope. This partly explains the high frequency (0.2% of CD4$^+$ T cells) of PLP 139-151 reactive T cells in the peripheral repertoire of SJL mice and the immunodominance of the PLP 139-151 epitope in this strain (Anderson et al., 2000; Reddy et al., 2004). Indeed, expression of PLP 139-151 in utero in SJL embryos results in a reduced frequency of PLP 139-151-reactive cells in the peripheral T cell repertoire and the loss of immunodominance of the PLP 139-151 epitope (Anderson et al., 2000).

Perhaps the most striking illustration of the impact of self-antigen expression on the developing T cell repertoire is the autoimmune disorder known as APECED (autoimmune polyendocrinopathy-candidiasis-ectodermaldystrophy), a disorder characterized by autoimmunity of the endocrine glands. APECED is a result of deficiency in AIRE, a transcription factor that regulates the expression of many "tissue-specific" antigens and is expressed predominantly in mTECs Zuklys et al., 2000). To gain insight into how AIRE controls the expression of "tissue-specific" antigens in mTECs, two groups have performed large-scale analyses of the genes regulated by AIRE (Derbinski et al., 2005; Johnnidis et al., 2005). Both of these groups found that AIRE-dependent genes are clustered in the genome with both AIRE-dependent and AIRE-independent genes interspersed within a cluster. The fact that AIRE-dependent and AIRE-independent genes are contiguous in these clusters suggests that AIRE does not function solely via an effect on chromatin remodeling. A working hypothesis is that epigenetic factors expressed in mTECs open the DNA, allowing specific transcription factors, such as AIRE, to initiate gene transcription. Aside from its function as a transcription factor, AIRE has recently been shown to have E3 ligase activity, raising the possibility that this enzymatic activity of AIRE may also be important in its role in self-tolerance (Uchida et al., 2004).

Although it is clear that AIRE is an important regulator of the expression of "tissue-specific" antigens in the thymus, it should be noted that the autoimmune syndrome that results from AIRE deficiency is restricted to endocrine glands. Thus, there must be other factors that regulate the expression of other "tissue-specific" antigens involved in other autoimmune diseases. Indeed, it has recently been estimated that at least 21% of the "tissue-specific" antigens expressed in mTECs are AIRE-independent (Derbinski et al., 2005).

mTEC expression of "tissue-specific" antigens could affect immune tolerance by promoting the negative selection of self-reactive T cells or by promoting the selection of regulatory T cells. Two independent studies have employed a double transgenic model to examine these possibilities (Anderson et al., 2005; Liston et al., 2003). In these studies, the thymic selection of transgenic TCRs is analyzed in mice that express the cognate antigen under the control of the rat insulin promoter. The rat insulin promoter was chosen specifically because insulin expression is controlled by AIRE (Anderson et al., 2002). Both of these studies found that AIRE-deficient mice failed to delete transgenic T cells under conditions that normally result in deletion. In addition, both of these studies found no defects in either the generation or function of regulatory T cells. This is in agreement with the results from another group (Kuroda et al., 2005). Thus it appears that AIRE primarily affects the negative selection of self-reactive thymocytes rather than the selection/generation of regulatory T cells in the thymus.

Interestingly, one study of AIRE deficient mice found that they developed a Sjorgens-like syndrome concomitant with autoreactivity directed against α-fodrin despite normal α-fodrin expression in the thymus (Kuroda et al., 2005). This observation led the authors to propose that in addition to its function in regulating the transcription of "tissue-specific" antigens in the thymus AIRE may influence antigen processing and/or presentation in mTECs (Kuroda et al., 2005). Indeed, several studies have now found that AIRE may regulate several genes involved in antigen processing and presentation, such as MHC, H2-O and H2-M, cathepsins (E, H, K, S, Z), CD80, tap1, several chemokines (ccl17, ccl22, cxc19, ccl19, cxcl10, ccl25) and some cytokines (Anderson et al., 2005; Derbinski et al., 2005; Johnnidis et al., 2005). Furthermore, AIRE-deficient mTECs have been shown to be less efficient at presenting antigen to T cells relative to wild-type mTECs (Anderson et al., 2005). However, one study has shown that at least some of the genes involved in antigen presentation (cathepsin S, MHC class II, CD80, H2-M) are upregulated in AIRE-deficient mTECs, suggesting that the upregulation of at least some of these genes may reflect an AIRE-independent gene program initiated by maturation in mTECs (Derbinski et al., 2005).

Although the examination of AIRE and the regulation of "tissue-specific" antigen expression in the thymus highlights the role of the thymus in deleting self-reactive cells, an equally important function of the thymus is to generate regulatory T cells. Neonatal thymectomy, recently shown to preclude the generation of regulatory T cells, has long been known to result in multiorgan autoimmunity (reviewed in Sakaguchi, 2004)). Studies of influenza hemagglutinin (HA) transgenic mice bearing various T cell receptors (TCRs) specific for HA have now provided direct evidence that regulatory cells are generated in the thymus and that their affinity for antigen may teeter on the edge of the range of affinities that would be expected to lead to negative selection (Jordan et al., 2001). These studies examined the thymic development of regulatory cells specific for a neo-self antigen. However, there is evidence that expression of a bona fide self-antigen in the thymus is important for the generation of regulatory T cells. The first study to demonstrate this utilized mice deficient in interphotoreceptor binding protein (IRBP), a retinal antigen that is a target in experimental autoimmune uveitis (EAU). Examination of IRBP expression in the thymus using conventional techniques had shown previously that thymic expression of IRBP correlates with disease susceptibility and resistance, with resistant strains exhibiting expression and susceptible strains lacking expression (Egwuagu et al., 1997). Not surprisingly, the EAU-susceptible B10.RIII (H-2$^r$) strain was shown to lack expression of IRBP in the thymus. However, it was later shown that the T cell repertoire to

IRBP that develops in IRBP-deficient B10.RIII mice differs significantly from that of wild-type B10.RIII mice (Avichezer et al., 2003). This observation prompted reexamination of IRBP expression in the thymus of this strain. Indeed, ultrasensitive immunohistochemistry has now shown that IRBP is expressed in very small clusters of cells in the thymic medulla and corticomedullary junction (Avichezer et al., 2003). Not only does this thymic expression of IRBP affect the negative selection of the T cell repertoire to IRBP, it also may affect the generation of regulatory T cells. Depletion of regulatory T cells in wild-type B10.RIII mice increases susceptibility to EAU. Although this alone is not surprising, it has also been shown that regulatory cells that protect against EAU do not develop in thymectomized wild-type mice; and thymectomized mice grafted with an IRBP-deficient thymus exhibit more aggressive disease (Avichezer et al., 2003). Taken together, these data support the hypothesis that thymic expression of IRBP may affect the selection of both the effector T cell repertoire and the regulatory T cell repertoire to IRBP.

While the aforementioned study implicates thymic expression of IRBP in the generation of regulatory T cells that protect against EAU, the authors did not directly demonstrate that these regulatory cells are specific for IRBP. The first demonstration of the antigen specificity of CD4$^+$CD25$^+$ regulatory T cells (discussed below) came from the examination of PLP 139-151 specific T cells in the EAE model (Reddy et al., 2004). This study utilized PLP 139-151/IA$^s$ tetramers to determine the frequency of PLP 139-151-reactive T cells in the CD4$^+$CD25$^+$ and CD4$^+$CD25$^-$ populations in the EAE-susceptible SJL and EAE-resistant B10.S strains and found that most of the tetramer-positive T cells in the SJL strain are in the CD4$^+$CD25$^-$ population, whereas in the resistant B10.S strain there was a significantly higher proportion of tetramer-positive T cells in the CD4$^+$CD25$^+$ population relative to the CD4$^+$CD25$^-$ population. Depletion of CD4$^+$CD25$^+$ T cells by anti-CD25 treatment resulted in development of autoimmunity in the normally resistant B10.S mice. Thymic expression of PLP may underlie the difference in tetramer-positive CD4$^+$CD25$^+$ T cells in these two strains as B10.S mice express PLP earlier during development in the thymus than SJL mice (Z. Illes and V.K.K., unpublished data).

Alas, thymic expression of self-antigen is not the sine qua non for the establishment of immune tolerance. This is exemplified by examining the repertoire generated to MBP 121-150 in B10.PL (H-2$^u$) mice (Huseby et al., 2001). In wild-type mice, the immunodominant epitope of MBP is 1-11, with MBP-121-150 being a subdominant epitope. In contrast, in MBP-deficient B10.PL mice, the immunodominant eptiope is MBP121-150, with MBP 1-11 being a subdominant T cell epitope (Harrington et al., 1998). These data suggested that the T cell repertoires to these two epitopes undergo differential tolerance induction in the presence of MBP. Indeed, TCR transgenic mice expressing a TCR specific for MBP 121-150 undergo deletion in the thymus (Huseby et al., 2001). However, MBP 121-150 is not included in golli-MBP, the embryonic form of MBP that is expressed in the thymus (Fritz and Kalvakolanu, 1995; Mathisen et al., 1993; Pribyl et al., 1993; Zelenika et al., 1993). This conundrum was resolved when experiments with bone marrow chimeras demonstrated that bone marrow-derived antigen-presenting cells acquire MBP exogenously and mediate negative selection in the thymus (Huseby et al., 2001).

# 3    Peripheral Mechanisms of Tolerance

The self-reactive T cells that manage to escape the various mechanisms of central tolerance discussed above are seeded to the periphery where they can be kept in check by mechanisms of peripheral tolerance. Among them are the molecules that regulate T cell anergy and the various populations of naturally occurring and induced regulatory T cells. These are discussed below.

## 3.1    Role of Regulatory T Cells

Regulatory function has been ascribed to many different T cell populations. They include $CD4^+CD25^+$, Tr1, and Th3 cells. $CD4^+CD25^+$ cells, often referred to as Tregs, are generated in the thymus and mediate suppression by direct cell contact. In contrast, Tr1 and Th3 cells are induced in the periphery and mediate suppression via the production of the immunoregulatory cytokines interleukin-10 (IL-10) and transforming growth factor-$\beta$ (TGF$\beta$), respectively.

Tregs are most often identified by their constitutive expression of CD25 (IL-2 receptor alpha chain) in addition to IL-2R$\beta$ and the common cytokine receptor-$\gamma$-chain ($\gamma$c). Consequently, Tregs constitutively express a high-affinity IL-2 receptor. Not surprisingly, it has been demonstrated that Tregs are crucially dependent on IL-2 for their maintenance and survival (reviewed in Malek and Bayer, 2004). Although the body of literature examining the biology of $CD4^+CD25^+$ Tregs is large, the use of CD25 as marker for Tregs is problematic, as activated CD4 and CD8 T cells also express CD25. Recently, the forkhead transcription factor FoxP3 has been identified as a key player in the generation and function of Tregs (Fontenot et al., 2003; Hori et al., 2003; Khattri et al., 2003). Scurfy mice, which fail to produce functional FoxP3 due to a frameshift muation (Brunkow et al., 2001), exhibit a CD4 T cell-mediated lymphoproliferative disease characterized by wasting and lymphocytic infiltration in multiple organs (Blair et al., 1994; Godfrey et al., 1991, 1994; Kanangat et al., 1996; Lyon et al., 1990). This phenotype, which closely resembles that of CTLA-4-deficient (Tivol et al., 1995; Waterhouse et al., 1995) and TGF$\beta$–deficient (Kulkarni et al., 1993; Shull et al., 1992) mice, has now been shown to be due to a lack of $CD4^+CD25^+$ T cells with regulatory function.

(Fontenot et al., 2003; Hori et al., 2003; Khattri et al., 2003). Several lines of evidence further underscore the crucial role of FoxP3 in the generation and function of $CD4^+CD25^+$ regulatory cells. $CD4^+CD25^+$ regulatory T cells express high levels of FoxP3 (Fontenot et al., 2003; Hori et al., 2003; Khattri et al., 2003) and do not develop from FoxP3-deficient bone marrow (Fontenot et al., 2003). Moreover, expression of FoxP3 confers regulatory function to $CD25^-$ and CD8 cells (Fontenot et al., 2003; Hori et al., 2003; Khattri et al., 2003).

Little is known regarding the mechanism by which FoxP3 instructs regulatory T cell development and function. A recent study suggests that NF-κB and NFAT may be involved, as FoxP3 interacts with both of these transcription factors and blocks their ability to induce transcription of known target genes, including cytokine genes (Bettelli et al., 2005). More important clues may come from the examination of reporter mice in which GFP has been fused to FoxP3. Examination of the "green" cells in these mice has already shown that not all FoxP3-positive cells express CD25, again highlighting the limited use of CD25 as a marker for regulatory cells (Fontenot et al., 2005).

The most actively studied subsets of induced regulatory cells are Th3 and Tr1. Th3 cells were first discovered while studying the mechanism of suppression of EAE induced by oral feeding of MBP (Chen et al., 1994). Th3 cells produce high levels of TGFβ, promote class switching to immunoglobulin A (IgA), and suppress both Th1 and Th2 cells (Weiner, 2001). Th3 cells act via TGFβ which also appears to act as a differentiation factor for these cells (Seder et al., 1998). In this regard, dendritic cells (DCs) from the gut have been shown to produce TGFβ after oral feeding and thus may be key to the induction of Th3 cells in the gut (Akbari et al., 2001).

Tr1 cells were first generated in vitro by chronic stimulation of CD4$^+$ T cells in the presence of IL-10. Tr1 cells produce high levels of IL-10 and low levels of IL-2 and have been shown to prevent colitis (Groux et al., 1997). DCs may also promote the generation of Tr1 cells, as pulmonary DCs isolated from mice after intranasal administration of antigen produce IL-10 and induce T cells with a Tr1-like phenotype (Akbari et al., 2001).

The existence of Tr1 and Th3 cells has raised the question of what the relation of these induced cells is to the naturally occurring CD4$^+$CD25$^+$ cells. Several lines of evidence have fueled this debate. First, CD4$^+$CD25$^+$ cells are induced by oral feeding (Thorstenson and Khoruts, 2001; Zhang et al., 2001). Second, TGFβ can drive the expansion of CD4$^+$CD25$^+$ cells in vivo (Peng et al., 2004) and more importantly appears to be a key cytokine involved in the peripheral homeostasis of CD4$^+$CD25$^+$ cells and the maintenance of FoxP3 expression (Marie et al., 2005). Third, several groups have implicated TGFβ in the regulatory function of CD4$^+$CD25$^+$ cells. CD4$^+$CD25$^+$ cells express surface-bound TGFβ, and anti-TGFβ abrogates their suppressor function (Nakamura et al., 2001). Moreover, CD4$^+$CD25$^+$ cells from TGFβ–deficient mice fail to suppress colitis (Nakamura et al., 2004) and TGFβ signaling in CD4$^+$CD25$^+$ cells appears to be required for suppressor function (Marie et al., 2005). Lastly, CD4$^+$CD25$^+$ cells from IL-10-deficient mice fail to protect immunodeficient mice from a T cell-mediated wasting disease (Annacker et al., 2001). Thus it appears that both TGFβ and IL-10 may be necessary for the suppressive action of CD4$^+$CD25$^+$ cells in vivo.

## 3.2   Role of Co-stimulation

It is well accepted that T cells need to receive two distinct signals to become activated (Bretscher and Cohn, 1970). The first signal is ligation of the T cell receptor by peptide/MHC and provides specificity to the interaction. The second signal is provided by the interaction of co-stimulatory molecules on the surface of antigen-presenting cells (APCs) such as B cells, DCs, and macrophages with their counterreceptors on the surface of T cells. The most well characterized family of co-stimulatory molecules is the B7 family. The B7 family has expanded in recent years and now includes B7-1/B7-2, ICOSL (B7h), and PD-L1/PDL-2, which are expressed on the surface of APCs and interact with their counterreceptors CD28/CTLA-4, ICOS, and PD-1 on the surface of T cells.

Ligation of the TCR in the absence of B7-mediated co-stimulation results in a state of long-term T cell nonresponsiveness, or T cell anergy, in vitro (Jenkins and Schwartz, 1987). In vivo, co-stimulatory molecules are not expressed on most tissues unless there is inflammation. Thus, T cells that recognize self-antigen in a particular tissue are rendered anergic owing to the lack of co-stimulation. Evidence for this mechanism of tolerance comes from studies in which mice that are normally tolerant to the expression of an allogeneic MHC in the pancreas develop autoimmune diabetes when B7-1 is co-expressed in the pancreas (Guerder et al., 1994).

The importance of the B7:CD28/CTLA-4 co-stimulatory pathway in maintaining peripheral tolerance is further emphasized by the phenotype of mice deficient in CTLA-4. These mice develop a lymphoproliferative disorder characterized by infiltration of various organs and death a few weeks after birth (Tivol et al., 1995; Waterhouse et al., 1995). The phenotype of these mice supports the hypothesis that the interaction of B7 molecules with CTLA-4 is critical for the regulation of peripheral T cell responses. Indeed, CTLA-4 appears to be required for anergy induction, as blocking antibodies against CTLA-4 but not CD28 prevent anergy induction (Perez et al., 1997), and CTLA-4 deficient T cells fail to become tolerized in response to tolerogenic stimuli in vivo (Greenwald et al., 2001).

The recent identification of alternate splice variants of CTLA-4 adds a layer of complexity to our understanding of how CTLA-4 regulates peripheral T cell responses (Fig. 1.1). CTLA-4 is found on mouse chromosome 1 in a genetic interval that also contains CD28 and ICOS. Genetic studies have shown that this interval (called Idd5.1), when replaced with the interval from the diabetes-

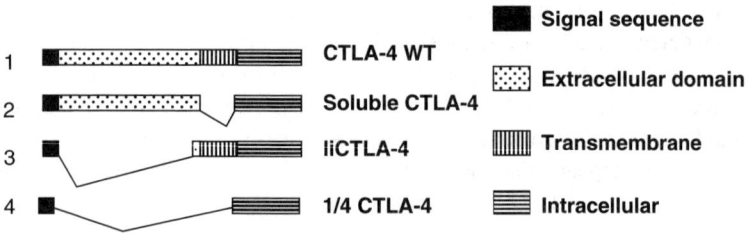

FIGURE 1.1. Four splice variants of CTLA-4. CTLA-4 WT is the full-length protein. Soluble CTLA-4 lacks the transmembrane domain. liCTLA-4 lacks the extracellular domain including the B7 binding motif. 1/4 CTLA-4 lacks both the extracellular and transmembrane domains.

resistant C57BL/10 strain, confers resistance to diabetes in susceptible NOD mice (Colucci et al., 1997; Hill et al., 2000; Lamhamedi-Cherradi et al., 2001). Although CTLA-4 is an attractive candidate gene in the Idd5.1 locus, no polymorphisms in the coding sequence of CTLA-4 have been identified. However, differential expression of a recently identified CTLA-4 splice variant that lacks the extracellular B7 binding domain called liCTLA-4 (ligand-independent CTLA-4) correlates with disease susceptibility, as liCTLA-4 is expressed at low levels in the autoimmune-susceptible strains (NOD and SJL) and at higher levels in their resistant counterparts (C57BL/6.H2-$^{g7}$ and B10.S) (Ueda et al. 2003; Vijayakrishnan et al., 2004). Furthermore, LiCTLA-4 appears to signal negatively in T cells (Vijayakrishnan et al., 2004). At present, little is known about the expression pattern and functional effects of the other CTLA-4 splice variants. To what extent liCTLA-4 or any of the other CTLA-4 splice variants are important for controlling peripheral T cell responses and whether the absence of one or more of these forms is responsible for the phenotype of CTLA-4-deficient mice remains to be determined.

In addition to their role in regulating the outcome of T cell receptor ligation, B7/CD28 family members may also affect the development, maintenance, and function of CD4$^+$CD25$^+$ regulatory T cells. Almost all B7/CD28 family members are expressed in the thymus. Given the data suggesting that Tregs recognize their cognate antigen with high affinity (Jordan et al., 2001) and the fact that co-stimulation affects the affinity of T cell/APC interactions, it is possible that B7/CD28 family members participate in the selection of Tregs in the thymus. That CD28-deficient mice exhibit an 80% reduction in the percentage of CD4 single positive CD25$^+$ T cells in the thymus supports this notion (Tang et al., 2003).

In the periphery, Tregs are reduced in B7-1/B7-2-deficient mice as well as in CD28-deficient mice, indicating a role for co-stimulation in the maintenance of Tregs (Salomon et al., 2000). In addition, B7/CD28 family members may be important in Treg function. Tregs express CD28, CTLA-4, PD-1, PDL-l, and ICOS. The role of CTLA-4 in Treg function is controversial, as two studies with blocking antibodies suggested that CTLA-4 is required for suppression (Read et al., 2000; Takahashi et al., 2000), whereas the results of another, similar study suggest that CTLA-4 has no role (Chai et al., 2002). Lastly, B7-1/2-deficient T cells are resistant to suppression (Paust et al., 2004), raising the possibility that the B7 on responder T cells interacts with CTLA-4 on Tregs and this T–T interaction is important for downregulating immune responses.

Other B7/CD28 family members have also been implicated in Treg function. ICOS antibody treatment precipitates diabetes onset in NOD mice by disrupting the balance of Tregs and effector T cells (Herman et al., 2004). Furthermore, ICOS has been shown to be crucial for the development of mucosal tolerance as both oral and nasal tolerance cannot be induced in ICOS$^{-/-}$ mice (Miyamoto et al., 2005). There are also data suggesting that PD-1/PDL-1 interactions may also play a role in Treg function (Baecher-Allan et al., 2001), but this notion requires further investigation.

B7/CD28 family members may also be important determinants of DC-induced immunity versus tolerance. For many years, DCs were thought of as the professional

APCs responsible for initiating T cell responses. However, it is now clear that immature DCs are tolerogenic. Maturation of DCs in response to inflammatory stimuli results in the upregulation of both B7-1 and B7-2 (De Smedt et al., 1996), consequently tipping the balance in favor of immunity. Interestingly, ligation of B7-1 and B7-2 on DCs by either CTLA-4/Ig or membrane-bound CTLA-4 has been shown to induce expression of indoleamine 2,3-dioxygenase (IDO), an intracellular enzyme that catabolizes tryptophan to by-products that inhibit T cell proliferation (Grohmann et al., 2002). Furthermore, pharmacological inhibition of IDO abrogates the protective effect of CTLA-4/Ig treatment in a mouse allotransplant model (Grohmann et al., 2002). These observations point to complex model of bidirectional signaling involving B7, CD28, and CTLA-4 where B7/CD28 signals promote T cell activation and B7/CTLA-4 signals downregulate T cell activation directly by inhibiting IL-2 production and cell cycle progression in T cells (Brunner et al., 1999; Greenwald et al., 2001; Walunas et al., 1996) and indirectly via expression of IDO by APCs.

Other B7/CD28 family members are also expressed on DCs. PDL-1 is expressed on immature DCs and is upregulated during maturation (Yamazaki et al., 2002). PDL-1 deficiency renders the normally resistant 129S4/SvJae strain susceptible to EAE and stimulation of naive T cells with PDL-1-deficient DCs results in increased interferon-$\gamma$ (IFN$\gamma$) production (Latchman et al., 2004). Both PD1 and PDL-1 are increased in the CNS of mice with EAE, and blockade of either PD-1 or PDL-1 augments disease (Salama et al., 2003). Similarly, in the NOD mouse, PDL-1 is expressed on inflamed islets; and blockade of either PD-1 or PDL-1 precipitates diabetes onset (Ansari et al., 2003). Taken together, these data support a negative regulatory role for PD-1/PDL-1 interactions. PDL-2 is also upregulated on DCs in response to inflammatory stimuli. In contrast to PDL-1, ligation of PDL-2 appears to enhance the antigen presentation function of DCs (Nguyen et al., 2002). Thus, as in the case with B7/CD28/CTLA-4, bidirectional signaling may also take place between PD-1 and PDL-1 and PDL-2. Lastly, ICOSL (B7h) is also expressed in immature DCs as well as Langerhans cells in the skin (Witsch et al., 2002). Unlike the other B7/CD28 family members, ICOSL ligand is not upregulated in response to inflammation and thus may serve primarily in regulating the induction of tolerance by immature DCs. Indeed, upregulation of ICOSL by lung DCs after intranasal administration of antigen has been associated with IL-10 production and tolerance in a model of airway hyperreactivity (Akbari et al., 2002).

## 3.3   Role of E3 Ligases

As mentioned above, TCR ligation in the absence of co-stimulation results in T cell anergy. TCR ligation alone induces the mobilization of intracellular calcium and activation of the calcium-dependent phosphatase calcineurin (Jenkins et al., 1987). Calcineurin dephosphorylates the transcription factor NFAT, which then translocates to the nucleus and initiates a gene program associated with anergy (Macian et al., 2002). Among the genes upregulated by calcium signaling in anergic T cells are several E3 ligases, which function to transfer ubiquitin from an E2 ubiquitin donor to a specific protein substrate. Ubiquitination not only targets proteins for proteasomal degradation; it can also alter protein function and/or trafficking between subcellular compartments (Haglund et al., 2003). Of the three families of E3 ubiquitin ligases discovered thus far, only the RING and HECT family ligases have been demonstrated to have an effect in the immune system.

A RING family E3 ligase, c-cbl was the first E3 ligase to be implicated in T cell anergy (Boussiotis et al., 1997). c-cbl is phosphorylated early after the induction of anergy in T cells; it can associate with Zap70 and phosphorylated CD3 ζ and can promote the ubiquitination of CD3 ζ. This ubiquitination may function in the ligand-induced downregulation of TCRs. In support of this is the observation that c-cbl-deficient thymocytes have abnormally high levels of TCRs (Murphy et al., 1998; Naramura et al., 1998). However, the fact that c-cbl-deficient mice do not develop spontaneous autoimmunity calls into question its importance in the maintenance of peripheral tolerance.

Cbl-b is another RING family E3 ligase that is implicated in peripheral T cell anergy. Cbl-b is upregulated in response to calcium mobilization and activation of calcineurin during the induction of anergy (Heissmeyer et al., 2004). It may be a more important determinant of the anergic phenotype than c-cbl, as cbl-b-deficient T cells are resistant to anergy induction (Heissmeyer et al., 2004) and cbl-b deficient mice develop systemic autoimmunity (Bachmaier et al., 2000). In addition, cbl-b has been shown to regulate signals negatively through CD28 (Chiang et al., 2000; Fang and Liu, 2001), and co-stimulation through CD28 promotes the degradation of cbl-b (Zhang et al., 2002).

The gene related to anergy in lymphocytes (GRAIL) is a third RING family E3 ligase implicated in T cell anergy. GRAIL was identified as an anergy-induced gene in a screen of transcripts differentially induced after TCR stimulation of T cell clones with and without co-stimulation (Anandasabapathy et al., 2003). GRAIL is also induced upon calcium mobilization and calcineurin activation (Heissmeyer et al., 2004). Expression of GRAIL in T cell hybridomas reduces both IL-2 and IL-4 production (Anandasabapathy et al., 2003), whereas expression of otubain, a protein that binds GRAIL and increases its turnover, enhances IL-2 production (Soares et al., 2004). Thus, the effect of GRAIL in regulating T cell anergy may be via its regulation of IL-2.

The two HECT family E3 ligases that may participate in anergy are Itch and Nedd4. The extent of upregulation of Itch in response to calcium signaling is similar to that of cbl-b and GRAIL; and restimulation of anergic T cells induces the translocation of both Itch and Nedd4 from the cytosol to the plasma membrane (Heissmeyer et al., 2004). Itch is an attractive candidate in the regulation of T cell anergy because it participates in the endocytic pathway (Angers et al., 2004; Qiu et al., 2000). Moreover, Itch-deficient mice develop a spontaneous fatal systemic lymphoproliferative disease (Fang et al., 2002). Although Nedd4 may also be important in T cell anergy, much less is known about it. Nedd4 is known to interact with and regulate the membrane expression of ion channels through ubiquitination and lysosomal degradation (Staub et al., 1997). Whether this involvement in lysosomal degradation relates to its potential function in T cell anergy requires further investigation.

# 4 Conclusion

Examination of "tissue-specific" antigen expression in the thymus and its regulation by AIRE and the identification of FoxP3 as a critical determinant of regulatory T cell development and function represent significant advances in our understanding of the selection and regulation of self-reactive T cell repertoire, but there is much that remains unknown. Elucidation of the factors that regulate AIRE-independent "tissue-specific" gene expression in the thymus as well as the epigenetic factors that allow for transcription factors such as AIRE to bind DNA represent the next major advance in our attempts to fully understand central tolerance. As for the mechanisms of peripheral tolerance, the fury over regulatory T cells is likely to continue, and the next wave will likely include identification of the genetic targets of FoxP3. Similarly, analysis of the biochemistry underlying T cell anergy will continue. In all of these areas, we will likely see advancing technologies, such as RNAi and two-photon imaging of live tissue, employed to address these important questions. At the same time, the debate over the relative importance of central versus peripheral mechanisms of tolerance will likely continue. In the end, however, one truth remains: Despite the immune system's best efforts, autoimmunity persists and multiple mechanisms are needed to keep it under control.

## References

Akbari, O., DeKruyff, R. H., Umetsu, D. T. (2001) Pulmonary dendritic cells producing IL-10 mediate tolerance induced by respiratory exposure to antigen. *Nat. Immunol.* 2:725-731.

Akbari, O., Freeman, G. J., Meyer, E. H., et al. (2002) Antigen-specific regulatory T cells develop via the ICOS-ICOS-ligand pathway and inhibit allergen-induced airway hyperreactivity. *Nat. Med.* 8:1024-1032.

Anandasabapathy, N., Ford, G. S., Bloom, D., et al. (2003) GRAIL: an E3 ubiquitin ligase that inhibits cytokine gene transcription is expressed in anergic CD4+ T cells. *Immunity* 18:535-547.

Anderson, A. C., Nicholson, L. B., Legge, K. L., et al. (2000) High frequency of autoreactive myelin proteolipid protein-specific T cells in the periphery of naive mice: mechanisms of selection of the self-reactive repertoire. *J. Exp. Med.* 191:761-770.

Anderson, M. S., Venanzi, E. S., Klein, L., et al. (2002) Projection of an immunological self shadow within the thymus by the aire protein. *Science* 298:1395-1401.

Anderson, M. S., Venanzi, E. S., Chen, Z., et al. (2005) The cellular mechanism of Aire control of T cell tolerance. *Immunity* 23:227-239.

Angers, A., Ramjaun, A. R., McPherson, P. S. (2004) The HECT domain ligase itch ubiquitinates endophilin and localizes to the trans-Golgi network and endosomal system. *J. Biol. Chem.* 279:11471-11479.

Annacker, O., Pimenta-Araujo, R., Burlen-Defranoux, O., et al. (2001) CD25+CD4+ T cells regulate the expansion of peripheral CD4 T cells through the production of IL-10. *J. Immunol.* 166:3008-3018.

Ansari, M. J., Salama, A. D., Chitnis, T., et al. (2003) The programmed death-1 (PD-1) pathway regulates autoimmune diabetes in nonobese diabetic (NOD) mice. *J. Exp. Med.* 198:63-69.

Avichezer, D., Grajewski, R. S., Chan, C., et al. (2003). An Immunologically privileged retinal antigen elicits tolerance: major role for central selection mechanisms. *J. Exp. Med.* 198:1665-1676.

Bachmaier, K., Krawczyk, C., Kozieradzki, I., et al. (2000) Negative regulation of lymphocyte activation and autoimmunity by the molecular adaptor Cbl-b. *Nature* 403:211-216.

Baecher-Allan, C., Brown, J. A., Freeman, G. J., Hafler, D. A. (2001) CD4+CD25+ regulatory cells in human peripheral blood. *J. Immunol.* 167:1245-1253.

Bettelli, E., Dastrange, M., Oukka, M. (2005) FoxP3 interacts with nuclear factor of activated T cells and NF-kB to repress cytokine gene expression and effector functions of T helper cells. *Proc. Natl. Acad. Sci. USA* 102:5138-5143.

Blair, P. J., Bultman, S. J., Haas, J. C., et al. (1994) CD4+CD8− T cells are the effector cells in disease pathogenesis in the scurfy (sf) mouse. *J. Immunol.* 153:3764-3774.

Boussiotis, V. A., Freeman, G. J., Berezovskaya, A., et al. (1997) Maintenance of human T cell anergy: blocking of IL-2 gene transcription by activated Rap1. *Science* 278: 124-128.

Bretscher, P., Cohn, M. (1970) A theory of self-nonself discrimination. *Science* 169: 1042-1049.

Brunkow, M. E., Jeffery, E. W., Hjerrild, K. A., et al. (2001) Disruption of a new forkead/winged-helix protein, scurfin, results in the fatal lymphoproliferative disorder of the scurfy mouse. *Nat. Genet.* 27:68-73.

Brunner, M. C., Chambers, C. A., Chan, F. K., et al. (1999) CTLA-4-mediated inhibition of early events of T cell proliferation. *J. Immunol.* 162:5813-5820.

Chai, J. G., Tsang, J. Y., Lechler, R., et al. (2002) CD4+CD25+ T cells as immunoregulatory T cells in vitro. *Eur. J. Immunol.* 32:2365-2375.

Chen, Y., Kuchroo, V. K., Inobe, J., et al. (1994) Regulatory T cell clones induced by oral tolerance: suppression of autoimmune encephalomyelitis. *Science* 265:1237-1240.

Chiang, Y. J., Kole, H. K., Brown, K., et al. (2000) Cbl-b regulates the CD28 dependence of T-cell activation. *Nature* 403:216-220.

Colucci, F., Bergman, M. L., Penha-Goncalves, C., et al. (1997) Apoptosis resistance of nonobese diabetic peripheral lymphocytes linked to the Idd5 diabetes susceptibility region. *Proc. Natl. Acad. Sci. USA* 94:8670-8674.

De Smedt, T., Pajak, B., Muraille, E., et al. (1996) Regulation of dendritic cell numbers and maturation by lipopolysaccharide in vivo. *J. Exp. Med.* 184:1413-1424.

Derbinski, J., Schulte, A., Kyewski, B., Klein, L. (2001) Promiscuous gene expression in medullary thymic epithelial cells mirrors the peripheral self. *Nat. Immunol.* 2:1032-1039.

Derbinski, J., Gabler, J., Brors, B., et al. (2005) Promiscuous gene expression in thymic epithelial cells is regulated at multiple levels. *J. Exp. Med.* 202:33-45.

Egwuagu, C. E., Charukamnoetkanok, P., Gery, I. (1997) Thymic expression of autoantigens correlates with resistance to autoimmune disease. *J. Immunol.* 159:3109-3112.

Fang, D., Liu, Y. C. (2001) Proteolysis-independent regulation of PI3K by Cbl-b-mediated ubiquitination in T cells. *Nat. Immunol.* 2:870-875.

Fang, D., Elly, C., Gao, B., et al. (2002) Dysregulation of T lymphocyte function in itchy mice: a role for Itch in Th2 differentiation. *Nat. Immunol.* 3:281-287.

Fontenot, J. D., Gavin, M. A., Rudensky, A. Y. (2003) Foxp3 programs the development and function of CD4+CD25+ regulatory T cells. *Nat. Immunol.* 4:330-336.

Fontenot, J. D., Rasmussen, J. P., Williams, L. M., et al. (2005) Regulatory T cell lineage specification by the forkhead transcription factor FoxP3. *Immunity* 22:329-341.

Fritz, R. B., Kalvakolanu, I. (1995) Thymic expression of the golli-myelin basic protein gene in the SJL/J mouse. *J. Neuroimmunol.* 57:93-99.

Fritz, R. B., Zhao, M.-L. (1996) Thymic expression of myelin basic protein (MBP). *J. Immunol.* 157:5429-5253.

Godfrey, V. L., Wilkinson, J. E., Russell, L. B. (1991) X-linked lymphoreticular disease in the scurfy (sf) mutant mouse. *Am. J. Pathol.* 138:1379-1387.

Godfrey, V. L., Rouse, B. T., Wilkinson, J. E. (1994) Transplantation of T cell-mediated, lymphoreticular disease from the scurfy (sf) mouse. *Am. J. Pathol.* 145:281-286.

Greenwald, R. J., Boussiotis, V. A., Lorsbach, R. B., et al. (2001) CTLA-4 regulates induction of anergy in vivo. *Immunity* 14:145-155.

Grohmann, U., Orabona, C., Fallarino, F., et al. (2002) CTLA-4-Ig regulates tryptophan catabolism in vivo. *Nat. Immunol.* 3:1097-1101.

Groux, H., O'Garra, A., Bigler, M., et al. (1997) A CD4$^+$ T-cell subset inhibits antigen-specific T-cell responses and prevents colitis. *Nature* 389:737-742.

Guerder, S., Meyerhoff, J., Flavell, R. (1994) The role of the T cell costimulator B7-1 in autoimmunity and the induction and maintenance of tolerance to peripheral antigen. *Immunity* 1:155-166.

Haglund, K., Sigismund, S., Polo, S., et al. (2003) Multiple monoubiquitination of RTKs is sufficient for their endocytosis and degradation. *Nat. Cell. Biol.* 5:461-466.

Harrington, C. J., Paez, A., Hunkapiller, T., et al. (1998) Differential tolerance is induced in T cells recognizing distinct epitopes of myelin basic protein. *Immunity* 8:571-580.

Heissmeyer, V., Macian, F., Im, S. H., et al. (2004) Calcineurin imposes T cell unresponsiveness through targeted proteolysis of signaling proteins. *Nat. Immunol.* 5:255-265.

Herman, A. E., Freeman, G. J., Mathis, D., Benoist, C. (2004) CD4+CD25+ T regulatory cells dependent on ICOS promote regulation of effector cells in the prediabetic lesion. *J. Exp. Med.* 199:1479-1489.

Hill, N. J., Lyons, P. A., Todd, J. A., et al. (2000) Nod Idd5 locus controls insulitis and diabetes and overlaps the orthologous CTLA4/IDDMI2 and NRAMP1 loci in humans. *Diabetes* 49:1744-1747.

Hori, S., Nomura, T., Sakaguchi, S. (2003) Control of regulatory T cell development by the transcription factor Foxp3. *Science* 299:1057-1061.

Huseby, E. S., Sather, B., Huseby, P. G., Goverman, J. (2001) Age-dependent T cell tolerance and autoimmunity to myelin basic protein. *Immunity* 14:471-481.

Jenkins, M. K., Schwartz, R. H. (1987) Antigen presentation by chemically modified splenocytes induces antigen-specific T cell unresponsiveness in vitro and in vivo. *J. Exp. Med.* 165:302-319.

Jenkins, M. K., Pardoll, D. M., Mizuguchi, J., et al. (1987) Molecular events in the induction of a nonresponsive state in interleukin 2-producing helper T-lymphocyte clones. *Proc. Natl. Acad. Sci. USA* 84:5409-5413.

Johnnidis, J. B., Venanzi, E. S., Taxman, D. J., et al. (2005) Chromsomal clustering of genes controlled by the aire transcription factor. *Proc. Natl. Acad. Sci. USA* 102:7233-7238.

Jordan, M. S., Boesteanu, A., Reed, A. J., et al. (2001) Thymic selection of CD4+CD25+ regulatory T cells induced by an agonist self-peptide. *Nat. Immunol.* 2:301-306.

Kanangat, S., Blair, P., Reddy, R., et al. (1996) Disease in the scurfy (sf) mouse is associated with overexpression of cytokine genes. *Eur. J. Immunol.* 26:161-165.

Khattri, R., Cox, T., Yasako, S. A., Ramsdell, F. (2003) An essential role for Scurfin in CD4+CD25+ T regulatory cells. *Nat. Immunol.* 4:337-342.

Klein, L., Klugmann, L., Nave, K. A., et al. (2000) Shaping of the autoreactive T-cell repertoire by a splice variant of self protein expressed in thymic epithelial cells. *Nat. Med.* 6:56-61.

Kulkarni, A. B., Huh, C. G., Becker, D., et al. (1993) Transforming growth factor beta 1 null mutation in mice causes excessive inflammatory response and early death. *Proc. Natl. Acad. Sci. USA* 90:770-774.

Kuroda, N., Mitani, T., Takeda, N., et al. (2005) Development of autoimmunity against transcriptionally unrepressed target antigen in the thymus of aire-deficient mice. *J. Immunol.* 174:1862-1870.

Kyewski, B., Derbinski, J., Gotter, J., Klein, L. (2002) Promiscuous gene expression and central T-cell tolerance: more than meets the eye. *Trends Immunol.* 23:364-371.

Lamhamedi-Cherradi, S. E., Boulard, O., Gonzalez, C., et al. (2001) Further mapping of the Idd5.1 locus for autoimmune diabetes in NOD mice. *Diabetes* 50:2874-2878.

Latchman, Y. E., Liang, S. C., Wu, Y., et al. (2004) PD-L1-deficient mice show that PD-L1 on T cells, antigen presenting cells, and host tissues negatively regulates T cells. *Proc. Natl. Acad. Sci. USA* 101:10691-10696.

Liston, A., Lesage, S., Wilson, J., et al. (2003) Aire regulates negative selection of organ-specific T cells. *Nat. Immunol.* 4:350-354.

Lyon, M. F., Peters, J., Glenister, P. H., et al. (1990) The scurfy mouse mutant has previously unrecognized hematological abnormalities and resembles Wiskott-Aldrich syndrome. *Proc. Natl. Acad. Sci. USA* 87:2433-2437.

Macian, F., Garcia-Cozar, F., Im, S. H., et al. (2002) Transcriptional mechanisms underlying lymphocyte tolerance. *Cell* 109:719-731.

Malek, T. R., Bayer, A. L. (2004) Tolerance, not immunity crucially depends on IL-2. *Nat. Rev. Immunol.* 4:665-674.

Marie, J. C., Letterio, J. J., Gavin, M., Rudensky, A. Y. (2005) TGF-beta 1 maintains suppressor function and FoxP3 expression in CD4+CD25+ regulatory T cells. *J. Exp. Med.* 201:1061-1067.

Mathisen, P. M., Pease, S., Garvey, J., et al. (1993) Identification of an embryonic isoform of myelin basic protein that is expressed widely in the mouse embryo. *Proc. Natl .Acad. Sci. USA* 90:10125-10129.

Miyamoto, K., Kingsley, C. I., Zhang, X., et al. (2005) The ICOS molecule plays a crucial role in the development of mucosal tolerance. *J. Immunol.* 175:7341-7347.

Murphy, M. A., Schnall, R. G., Venter, D. J., et al. (1998) Tissue hyperplasia and enhanced T-cell signaling via Zap-70 in c-Cbl-deficient mice. *Mol. Cell. Biol.* 18:4872-4882.

Nakamura, K., Kitani, A., Strober, W. (2001) Cell contact-dependent immunosuppression by CD4(+)CD25(+) regulatory T cells is mediated by cell surface-bound transforming growth factor beta. *J. Exp. Med.* 194:629-644.

Nakamura, K., Kitani, A., Fuss, I., et al. (2004) TGF-beta 1 plays an important role in the mechanism of CD4+CD25+ regulatory T cell activity in both humans and mice. *J. Immunol.* 172:834-842.

Naramura, M., Kole, H. K., Hu, R. J., Gu, H. (1998) Altered thymic positive selection and intracellular signals in Cbl-deficient mice. *Proc. Natl. Acad. Sci. USA* 95:15547-15552.

Nguyen, L. T., Radhakrishnan, S., Ciric, B., et al. (2002) Cross-linking the B7 family molecule B7-DC directly activates immune functions of dendritic cells. *J. Exp. Med.* 196:1393-1398.

Paust, S., Lu, L., McCarty, N., Cantor, H. (2004) Engagement of B7 on effector T cells by regulatory T cells prevents autoimmune disease. *Proc. Natl. Acad. Sci. USA* 101:10398-10403.

Peng, Y., Laouar, Y., Li, M. O., et al. (2004) TGF-beta regulates in vivo expansion of FoxP3-expressing CD4+CD25+ regulatory T cells responsible for protection against diabetes. *Proc. Natl. Acad. Sci. USA* 101:4572-4577.

Perez, V. L., Van Parijs, L., Biuckians, A., et al. (1997) Induction of peripheral T cell tolerance in vivo requires CTLA-4 engagement. *Immunity* 6:411-417.

Pribyl, T. M., Campagnoni, C. W., Kampf, K., et al. (1993) The human myelin basic protein gene is included within a 179-kilobase transcription unit: expression in the immune and central nervous systems. *Proc. Natl. Acad. Sci. USA* 90:10695-10699.

Pribyl, T. M., Campagnoni, C. W., Kampf, K., et al. (1996) Expression of the myelin proteolipid protein gene in the human fetal thymus. *J. Neuroimmunol.* 67:125-130.

Qiu, L., Joazeiro, C., Fang, N., et al. (2000) Recognition and ubiquitination of Notch by Itch, a hect-type E3 ubiquitin ligase. *J. Biol. Chem.* 275:35734-35737.

Read, S., Malmstrom, V., Powrie, F. (2000) Cytotoxic T lymphocyte-associated antigen 4 plays an essential role in the function of CD25(+) CD4(+) regulatory cells that control intestinal inflammation. *J. Exp. Med.* 192:295-302.

Reddy, J., Illes, Z., Zhang, X., et al. (2004) Myelin proteolipid protein-specific CD4+CD25+ regulatory cells mediate genetic resistance of experimental autoimmune encephalomyelitis. *Proc. Natl. Acad. Sci. USA* 101:15434-15439.

Sakaguchi, S. (2004) Naturally arising CD4+ regulatory T cells for immunologic self-tolerance and negative control of immune responses. *Annu. Rev. Immunol.* 22:531-562.

Salama, A. D., Chitnis, T., Imitola, J., et al. (2003) Critical role of the programmed death-1 (PD-1) pathway in regulation of experimental autoimmune encephalomyelitis. *J. Exp. Med.* 198:71-78.

Salomon, B., Lenschow, D. J., Rhee, L., et al. (2000) B7/CD28 costimulation is essential for the homeostasis of the CD4+CD25+ immunoregulatory T cells that control autoimmune diabetes. *Immunity* 12:431-440.

Seder, R. A., Marth, T., Sieve, M. C., et al. (1998) Factors involved in the differentiation of TGF-beta-producing cells from naive CD4+ T cells: IL-4 and IFN-gamma have opposing effects, while TGF-beta positively regulates its own production. *J. Immunol.* 160:5719-5728.

Shull, M. M., Ormsby, I., Kier, A. B., et al. (1992) Targeted disruption of the mouse transforming growth factor-gene results in multifocal inflammatory disease. *Nature* 359:693-699.

Soares, L., Seroogy, C., Skrenta, H., et al. (2004) Two isoforms of otubain 1 regulate T cell anergy via GRAIL. *Nat. Immunol.* 5:45-54.

Staub, O., Gautschi, I., Ishikawa, T., et al. (1997) Regulation of stability and function of the epithelial $Na^+$ channel (ENaC) by ubiquitination. *EMBO J.* 16:6325-6336.

Takahashi, T., Tagami, T., Yamazaki, S., et al. (2000) Immunologic self-tolerance maintained by CD25(+)CD4(+) regulatory T cells constitutively expressing cytotoxic T lymphocyte-associated antigen 4. *J. Exp. Med.* 192:303-310.

Tang, Q., Henriksen, K. J., Boden, E. K., et al. (2003) Cutting edge: CD28 controls peripheral homeostasis of CD4+CD25+ regulatory T cells. *J. Immunol.* 171:3348-3352.

Targoni, O. S., Lehman, P. V. (1998) Endogenous myelin basic protein inactivates the high avidity T cell repertoire. *J. Exp. Med.* 187:2055-2063.

Thorstenson, K. M., Khoruts, A. (2001) Generation of anergic and potentially immunoregulatory CD25+CD4 T cells in vivo after induction of peripheral tolerance with intravenous or oral antigen. *J. Immunol.* 167:188-195.

Tivol, E. A., Borriello, F., Schweitzer, A. N., et al. (1995) CTLA-4 deficient mice exhibit massive lymphoproliferation and multi-organ lymphatic infiltration: a critical negative immunoregulatory role of CTLA-4. *Immunity* 3:541-547.

Uchida, D., Hatakeyama, S., Matsushima, A., et al. (2004) Aire functions as an E3 ubiquitin ligase. *J. Exp. Med.* 199:167-172.

Ueda, H., Howson, J. M., Esposito, L., et al. (2003) Association of the T-cell regulatory gene CTLA4 with susceptibility to autoimmune disease. *Nature* 423:506-511.

Vijayakrishnan, L., Slavik, J. M., Illes, Z., et al. (2004) An autoimmune disease-associated CTLA-4 splice variant lacking the B7 binding domain signals negatively in T cells. *Immunity* 20:563-575.

Walunas, T. L., Bakker, C. Y., Bluestone, J. A. (1996) CTLA-4 ligation blocks CD28-dependent T cell activation. *J. Exp. Med.* 183:2541-2550.

Waterhouse, P. W., Penninger, J. M., Timms, E., et al. (1995) Lymphoproliferative disorders with early lethality in mice deficient in CTLA-4. *Science* 270:985-988.

Weiner, H. L. (2001) Induction and mechanism of action of transforming growth factor-b-secreting Th3 regulatory cells. *Immunol. Rev.* 182:207-214.

Witsch, E. J., Peiser, M., Hutloff, A., et al. (2002) ICOS and CD28 reversely regulate IL-10 on reactivation of human effector T cells with mature dendritic cells. *Eur. J. Immunol.* 32:2680-2686.

Yamazaki, T., Akiba, H., Iwai, H., et al. (2002) Expression of programmed death 1 ligands by murine T cells and APC. *J. Immunol.* 169:5538-5545.

Yoshizawa, I., Bronson, R., Dorf, M. E., Abromson-Leeman, S. (1998) T-cell responses to myelin basic protein in normal and MBP-deficient mice. *J. Neuroimmunol.* 84:131-138.

Zelenika, D., Grima, B., Pessac, B. (1993) A new family of transcripts of the myelin basic protein gene: expression in brain and in immune system. *J. Neurochem.* 60:1574-1577.

Zhang, J., Bardos, T., Li, D., et al. (2002) Cutting edge: regulation of T cell activation threshold by CD28 costimulation through targeting Cbl-b for ubiquitination. *J. Immunol.* 169:2236-2240.

Zhang, X., Izikson, L., Liu, L., Weiner, H. L. (2001) Activation of CD4+CD25+ regulatory T cells by oral antigen administration. *J. Immunol.* 167:4245-4253.

Zuklys, S., Balciunaite, G., Agarwal, A., et al. (2000) Normal thymic architecture and negative selection are associated with AIRE expression, the gene defective in the autoimmune-polyendocrinopathy-candidiasis-ectodermal dystrophy (APECED). *J. Immunol.* 165:1976-1983.

# 2

# Priming Regulatory T Cells and Antigen-Specific Suppression of Autoimmune Disease

Trevor R.F. Smith, Xiaolei Tang, and Vipin Kumar

## 1   Summary

It is now widely recognized that regulatory T cells (Treg) play an important role in protecting the body from autoimmune diseases. T cells with suppressor activity have been identified in both CD4+ and CD8+ T cell populations in humans and animal models of inflammatory disease. Here we provide a brief review of the field of T cell suppression with special emphasis on CD8+ T cell-mediated regulation of immune responses to self. We focus on the role of CD8+ Treg in the control of myelin basic protein-reactive T cells in experimental autoimmune encephalomyelitis, a model for human multiple sclerosis. We address how Treg can be specifically induced to downregulate an immune response.

## 2   Introduction

Central tolerance in the thymus is the immune system's primary mechanism of purging the body of self-reactive T cells. However, thymic tolerance is not absolute, and potentially pathogenic T cells with specificity to self-antigens are present in the periphery (Bouneaud et al., 2000). Once activated such T cells are capable of mediating inflammatory reactions against one's own tissues, and these responses may manifest as autoimmune disease (e.g., multiple sclerosis, type 1 diabetes, Crohn's disease). The immune system has thus developed peripheral mechanisms to control aberrant autoimmune reactions. In addition to T cell anergy (Schwartz, 2003) and activation-induced cell death (AICD) (Kabelitz et al., 1993), active suppression by regulatory T cells has now been recognized as a major peripheral mechanism to protect against detrimental autoimmune reactions (Kumar, 2004; Sakaguchi, 2005; Shevach, 2000). In experimental systems where Treg has been depleted or disabled, autoimmune disease has been frequently observed (Piccirillo and Shevach, 2004). In turn, experimental strategies targeting Treg are likely to prove effective in the treatment of autoimmunity (Hori et al., 2003).

Here we discuss the importance of Treg, describing how they can be targeted to help protect against or treat autoimmune disease. We focus on the studies performed in our laboratory that have deciphered the role of CD4 and CD8 Treg in mediating antigen-specific protection from experimental autoimmune encephalomyelitis (EAE), a surrogate mouse model for multiple sclerosis in humans.

# 3  Identification of T Cells With Suppressor Function

The existence of a T cell population with a regulatory role in autoimmune disease was proposed after the experiments performed by Nishizuka and Sakakura in 1969. They reported that thymectomy in a neonatal (day 3) rodent resulted in the development of autoimmune disease. Further experiments demonstrated that reconstitution of the day 3 thymectomized rodent's T cell population with thymocytes from an adult mouse protected against autoimmune disease. T cells with suppressor function were first described by Gershon and Kondo during the 1970s. It was demonstrated that adoptive transfer of T cells derived from an animal tolerant to a given antigen suppressed antibody responses to that antigen in the recipient (Gershon and Kondo, 1970). In similar systems the downregulation of type I hypersensitivity or cell-mediated delayed hypersensitivity reactions, after the transfer of T cells derived from tolerized animals, was reported (Askenase et al., 1975; Takatsu and Ishizaka, 1975). The suppressor T cell activity was found to reside in the Lyt2+ (CD8) T cell population (Cantor et al., 1976). However, further characterization of this T suppressor cell population was hindered by the inability to characterize these cells molecularly in a well defined antigenic system. Furthermore, realization of T cells with different type 1 and type 2 cytokine secretion during the mid-1980s led many immunologists to believe that T cell-mediated suppression could be explained by conventional T cell subsets secreting counterinhibitory cytokines, not by a population of T cells with dedicated suppressor function (Green and Webb, 1993; Salgame et al., 1991).

Despite skepticism, several laboratories continued to study T cell-mediated suppression (Kumar and Sercarz, 1993; Sakaguchi et al., 1985, 1995). In 1995, Sakaguchi and colleagues published the seminal paper identifying a subpopulation of T-helper cells constitutively expressing the CD25 cell surface marker with the ability to protect against autoimmune disease. Athymic BALB/c mice inoculated with T cell suspensions depleted of CD4+CD25+ T cells developed spontaneous autoimmune disease. Disease could be prevented if mice were reconstituted with CD4+CD25+ T cells shortly after inoculation with the CD4+CD25− T cells (Sakaguchi et al., 1995). Discovery of a cell surface phenotype to identify a population of T cells with suppressor function rekindled the interest of many immunologists into T cells with dedicated suppressor function. CD4+CD25+ Tregs have been characterized as a naturally occurring anergic population that is FoxP3+, CTLA-4+ in both humans and animals (Piccirillo and Shevach, 2004). To date the mechanism of suppression by CD4+CD25+ T cells has yet to be fully defined. However, it is generally agreed that the mechanism is contact-dependent, and it involves different molecules, including interleukin-10 (IL-10) or trans-forming growth factor-β (TGFβ) in some models. It is likely that within the CD4+CD25+ T cell population there are multiple subsets of T cells utilizing dif-ferent mechanisms of suppression, as most studies are based on the isolation of polyclonal populations.

Other experimentally induced CD4+ Treg subsets have been described. These include Th3 cells, which can be induced by oral tolerance protocols (Weiner, 2001). Such cells mediate their suppressor function through secretion of TGFβ. Tr1 cells mediate suppression by secretion of IL-10 and can be induced by various immu-nization protocols that may be dependent on naive T cell interactions with antigen-presenting cells (APCs) in a non-activated/immature state (Jonuleit et al., 2000).

CD8 Treg populations have been described in many systems. They include CD8+CD28− Treg (Filaci and Suciu-Foca, 2002), CD8+CD75+ Treg (Zimring and Kapp, 2004), CD8+CD25+ Treg (Cosmi et al., 2003), and anti-T cell receptor (TCR) CD8αα+ Treg (Kumar, 2004; Tang et al., unpublished data). CD8+CD28− Treg can be generated in vitro from naive T cells in an allogeneic mixed leukocyte reaction (MLR) or by the direct use of IL-10 (Filaci and Suciu-Foca, 2002). CD8+CD28− Tregs have been shown to mediate allosuppression by inhibiting APC maturation. Analogous to CD4+CD25+ Treg cells, CD8+ T cells co-expressing the CD25 molecule with suppressive ability have been identified. Akin to their CD4+ counterparts, CD8+CD25+ Tregs appear to be a naturally occurring popu-lation, FoxP3+ and CTLA4+; and their regulatory function is cell–cell contact-dependent (Cosmi et al., 2003). Peripheral CD8+CD25+ T cells with in vitro suppressor ability have also been shown to be a naturally occurring population in major histocompatibility complex II (MHC II)-deficient mice (Bienvenu et al., 2005). Although similar to CD4+CD25+ T cells with respect to CTLA-4 and Foxp3 expression, murine CD8+CD25+ T cells strongly proliferate and produce interferon-γ (IFNγ) upon in vitro stimulation in the absence of exogenous IL-2. It has yet to be determined if CD8+CD25+ Treg cells function in vivo to protect against autoimmune disease or whether their mechanism of action is IFNγ-dependent. However, naturally occurring rat CD8+CD45RC$^{low}$ have been shown in vivo to protect against CD4+ T cell-mediated graft-versus-host-disease by a mechanism that appears to be cell–cell contact-dependent (Xystrakis et al., 2004).

## 3.1 Studying Treg Populations at the Clonal Level Using the EAE Model

Experimental autoimmune encephalomyelitis can be induced in susceptible animals by adoptive transfer of T cell clones or lines reactive to myelin antigens or by immunization with myelin proteins, such as myelin basic protein (MBP) emulsified in complete Freund's adjuvant (CFA). PL/J and B10.PL mice spontaneously recover from and become refractory to subsequent induction of EAE (Kumar, 2004). Evidence suggests that CD8 Tregs arise naturally to protect from EAE (Madakamutil et al., 2003; Tang et al., in press). In 1992, it was demonstrated that after CD8+ T cell depletion prior to induction of EAE the mice were no longer protected from reinduction of EAE (Jiang et al., 1992). Additionally, a more chronic form of EAE was seen in CD8-/-PL/J H-$2^u$ mice (Koh et al., 1992). Our laboratory has elucidated a feedback inhibition regulatory mechanism that incorporates both anti-TCR CD4+ and CD8+ Tregs that control MBP-reactive CD4+ T cells and thereby protect animals from EAE (Kumar, 2004). The induction of anti-TCR Treg has also been demonstrated in Lewis rats and B10.PL or PL/J mice during the recovery from EAE (Vandenbark et al., 1996; Jiang and Chess, 2000). The EAE model in the B10.PL or PL/J mouse provides the ideal system to study the induction and function of Treg. First, immunization with MBP emulsified in CFA induces a clonotypic CD4+ T cell response to MBP directed toward a single determinant (Kibler et al., 1977; Zamvil et al., 1986); second, most of the pathogenic CD4+ T cells responding to this determinant express the Vβ8.2 TCR chain (Acha-Orbea et al., 1988; Kumar et al., 1989; Urban et al., 1988). These characteristics provide a specific target for regulation. Third, disease is generally monophasic; most of the animals recover spontaneously and are refractory to disease reinduction. The observation that animals became resistant to disease indicates the induction of an active regulatory mechanism to downregulate the pathogenic CD4+ T cell population. Further investigation demonstrated that immunization with peptides derived from the Vβ8.2+ TCR chain in both mice and rats could protect the animal from EAE, indicating that a population of Tregs specific for TCR determinants on the pathogenic T cell population could regulate disease (Howell et al., 1989; Kumar and Sercarz, 1993; Kumar et al., 1995; Vandenbark et al., 1989). Importantly, these TCR-derived peptides are also immunogenic in the human population and have been used to halt progression of multiple sclerosis (Saruhan-Direskeneli et al., 1993; Vandenbark et al., 1996).

## 3.2 Role of Regulatory T Cells in Human Autoimmune Disease

Studies confirming a role for Treg in human autoimmune disease are limited and fraught with practical difficulties. Major hindrances include identification of disease-associated autoantigens, heterogeneity of the human T cell repertoire, sample collection, and a lack of in vivo experimental techniques. Despite such limitations, studies have detected defects in functional Treg populations in

TABLE 2.1. Treg in Autoimmune Diseases.

| Regulatory cell | Arthritis | | Multiple sclerosis | | Diabetes | | Inflammatory bowel disease | |
|---|---|---|---|---|---|---|---|---|
| | Human | Rodent | Human | Rodent | Human | Rodent | Human | Rodent |
| CD4+, CD25+, Treg | +[1] | +[2] | +[3] | +[4] | ±[5/6] | +[7] | +[8] | +[9] |
| Tr1 | ± | +[10] | ±[11] | +[12] | ± | +[13] | ± | +[14] |
| Th3 | ND | ND | +[5] | +[16] | ND | ND | ND | + |
| Anti-TCR, Treg/ CD8, Treg | ND | +[17] | +[18] | +[19] | +[20] | +[21] | +[22] | ND |

Table supplies the evidence for a protective role of different subsets of Treg in humans and the experimental rodent model of various autoimmune diseases. Rodent model for arthritis, collagen-induced arthritis; for multiple sclerosis, EAE; for diabetes, nonobese diabetic (NOD) mouse; for inflammatory bowel disease, experimental colitis.

+, evidence for; ±, evidence for and against; ND, not determined.

[1]Ehrenstein et al., 2004; [2]Morgan et al., 2005; [3]Vigiletta et al., 2004; [4]Kohm et al., 2002; [5]Putnam et al., 2005; [6]Lindley et al., 2005; [7]Chen et al., 2005; [8]Kelsen et al., 2005; [9]Martin et al., 2004; [10]Quattrocchi et al., 2001; [11]Aharoni et al., 2005; [12]Cua et al., 2001; [13]You et al., 2004; [14]Groux et al., 1997; [15]Fukaura et al., 1996; [16]Weiner, 2001; [17]Honda et al., 2004; [18]Zhang et al., 1995; [19]Madakumatil et al., 2003; [20]Bisikiriska et al., 2005; [21]Panoutsakopoulou et al., 2004; [22]Brimnes et al., 2005.

autoimmune disease and allergy, and immunotherapeutic studies suggest that targeting Treg populations may be critical for efficacy (Bisikirska et al., 2005; Hong et al., 2005; Kumar et al., 2001; Lan et al., 2005).

Both CD4+ and CD8+ T cells with regulatory function have been described in many autoimmune settings and represent a target for disease intervention. Table 2.1 summarizes studies in the autoimmune disease setting in which regulatory T cells have been shown to be associated.

# 4    Immunotherapeutic Strategies to Treat Autoimmune Disease

The prevalence of autoimmune disease in the human population is steadily rising, with approximately 5% of the U.S. population currently afflicted (Jacobson et al., 1997). The mainstay immunosuppressive treatments include corticosteroid and nonsteroidal antiinflammatory or cytotoxic drugs. These treatments are palliative aimed at nonspecifically reducing inflammation or killing dividing cells. The usage of such drugs is hindered by their limited selectivity and significant toxicity. However, novel therapeutic strategies based on immunomodulation of inflammatory cytokine networks are now available, and therapies targeting Treg are currently being developed (see below).

Inflammatory cytokine networks can be modulated to curtail autoimmune manifestations. For example, an anti-tumor necrosis factor-$\alpha$ (TNF$\alpha$) monoclonal antibody (mAb) has been approved by the Food and Drug Administration (FDA) and has shown efficacy in the treatment of rheumatoid arthritis (RA) and Crohn's disease (Feldman and Steinman, 2005). IFN$\beta$ has demonstrated efficacy in the treatment of relapsing remitting multiple sclerosis (MS) (Revel, 2003). However, both of these treatment regimens still suffer from a lack of specificity, which hinders both their efficacy and safety. For example, IFN$\beta$ treatment is effective in only 30% of relapsing remitting MS patients, and its mechanism of action is still unclear (Revel, 2003). Anti-TNF$\alpha$ treatment has been associated with exacerbation of latent tuberculosis in some patients (Keane et al., 2001).

## 4.1   Induction of Treg (Nonspecific)

As discussed above, Treg populations can downregulate pathogenic immune responses and represent a valid target for immune intervention. However, the techniques employed to induce these cells have yet to be optimized. In experimental TCR-transgenic models disease-regulating CD4+CD25+ Treg can be generated and/or activated by antigen vaccination, (Thorstenson and Khoruts, 2001; Zhang et al., 2001). The generation of regulatory cells has also been demonstrated by the use of general immunosuppressive drugs. Barrat and colleagues demonstrated that a cocktail of vitamin D$_3$ and dexamethasone could induce IL-10-producing CD4+ T cells with in vivo suppressor ability (Barrat et al., 2002). Recently, Bisirikia and colleagues demonstrated that OKT3 (a modified anti-CD3 mAb) activated a population of CD8+CD25+ Treg cells that could suppress the in vitro responses of CD4+ T cells derived from type 1 diabetes patients (Bisirikia et al., 2005). Furthermore, treatment with copolymer-I, a polymer consisting of four randomly joined amino acids that has proven efficacy in the treatment of MS, increased the number of Foxp3+ CD4+CD25+ T cells in the peripheral blood of MS patients. In vitro analysis showed copolymer-I generated CD4+CD25+ T cells expressing high levels of FoxP3 and exhibiting increased regulatory ability (Hong et al., 2005). Karandikar and colleagues have also published data suggesting that copolymer-I treatment targets CD8+ Treg (Karandiker et al., 2002). Initially CD8+ T cells from MS patients were found to respond poorly to copolymer-I. However, after copolymer-I treatment the patients' CD8+ T cell responses were restored. Interestingly, significantly enhanced IFN$\gamma$ expression by the CD8 T cells was recorded, and it was proposed that copolymer-I treatment restored the IFN$\gamma$-dependent suppressor ability of CD8 T cells. The above studies indicate that the augmentation of Treg ability correlates with the efficacy of immunotherapy.

## 4.2   Induction of Treg (Antigen-Specific)

The ideal immunosuppressive therapeutic strategy would be one that specifically targets the disease driving immune cells (i.e., the cells reacting directly to the body's self-antigens). By blocking or modulating the function of these cells one would be able to manipulate the underlying cause of disease. Such treatment strategies are dependent on identification of the autoantigen or antigens involved in disease pathogenesis and the cells reacting to these antigens. Strategies based on inducing tolerance to defined antigens have shown much promise in experimental systems where the inflammatory response is directed toward a single autoantigen on a defined genetic background (Bielekova and Martin, 2001). In such systems the animal could be vaccinated in a tolerogenic manner with a peptide (e.g., p87-99 epitope of MBP) or whole antigen (e.g., MBP) to modulate the disease course (Brocke et al., 1996). However, on translation onto heterogeneous human backgrounds, the efficacy of such treatments has been unpredictable (Bielokova et al., 2000; Kappos et al., 2000). One hindrance to therapies based on targeting tissue antigens is the phenomenon of epitope spreading (Sercarz et al., 1993). Although the initial immunological response may be toward a single determinant, during the period between the initial insult and the manifestation of clinical disease the focus of autoimmune reaction may have spread to other tissue antigens. Thus peptide therapy targeting only one or a few autoantigens may be insufficient at the time of clinical disease.

Another therapeutic strategy with antigen specificity targets receptors expressed on pathogenic disease-mediating T cell populations. When treating autoimmune disease, the aim of T cell vaccination (TCV) is to induce a population of T cells that react specifically against and inhibit the disease causing the pathogenic T cell population. Evidence suggests that TCV induces a population of Tregs that recognize clonotypic antigens on pathogenic T cells (Sun et al., 1988; Zhang et al., 1995). Experimental evidence suggests that these antigens are expressed within the variable region of the TCR on the pathogenic T cell (Jiang et al., 1998). Studies have demonstrated that vaccination with TCR-derived peptides can protect mice from EAE (Kumar and Sercarz, 1993; Kumar et al., 1995; Vandenbark et al., 1989). This indicates that within the bodies' immune system there is a naturally occurring subset of Treg that recognize conserved TCR regions with the ability once primed to downregulate cells expressing a particular TCR. Tregs have been shown to reside in both CD4+ and CD8+ T cell populations (Kumar and Sercarz, 1993; Tang et al., unpublished data). The induction of Treg can explain how the B10.PL mouse spontaneously recovers from EAE and becomes resistant to further disease induction. In this model we have shown that MBP immunization plus adjuvant leads to a TCRVβ8.2+ CD4+ T cell-mediated inflammatory disease, and recovery is associated with the activation of CD4+ and CD8+ T cells reactive against epitopes within the TCRVβ8.2+ (Kumar, 2004).

Over the last decade our laboratory has put together the pieces and deciphered the regulatory network engaged in suppressing the EAE-mediating TCRVβ8.2+ CD4+ T cell responses in the B10.PL mouse. We have now characterized two Treg (CD4+ and CD8+) populations that work together specifically to downregulate the pathogenic T cells. Injection of a peptide from the Vβ8.2+ TCR framework 3 region was demonstrated to protect mice from EAE, and characterization of the reacting T cells revealed that this MHC II-restricted peptide induces a Vβ14+ TCR CD4+ Treg population (Kumar and Sercarz, 1993). Additionally, CD8 depletion revealed that CD8+ T cells are also essential for disease protection and downregulation of the Vβ8.2+ TCR T cell population (Kumar et al., 1997). Recent studies have characterized a novel population of CD8+ Tregs. The CD8+ Treg population recognizes a different determinant in the CDR2 region of the Vβ8.2+ TCR in the context of the non-classical MHC I molecule Qa-1(Tang et al., in press). The data suggest that CD4+ Tregs help in the recruitment and activation of the CD8 Tregs. It is the CD8 Tregs that ultimately kill the pathogenic CD4+ T cell population by apoptosis induction (Madakamutil et al., 2003). This mechanism of clonotypic feedback regulation is depicted in Figure 2.1.

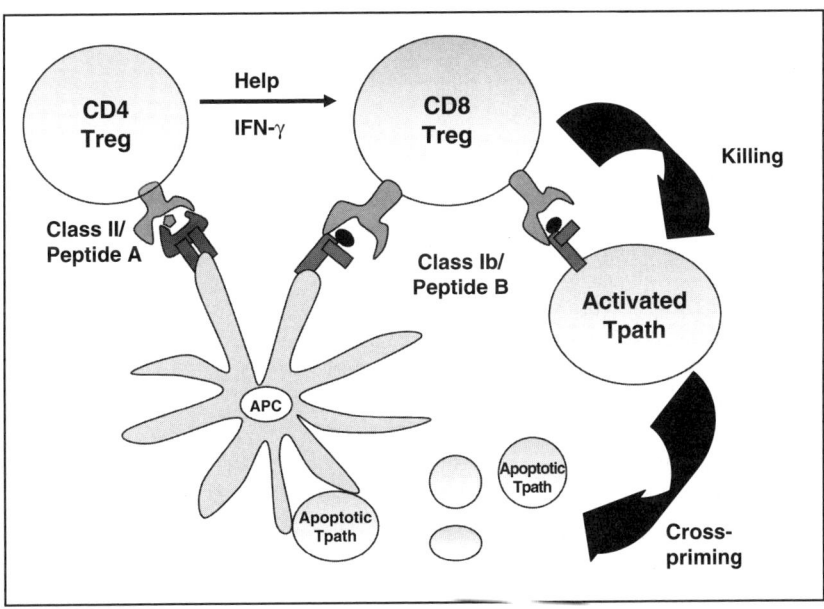

FIGURE 2.1. Negative-feedback mechanism targeting T cell receptors (TCRs). The working model of negative feedback TCR-based regulation depicts Treg-mediated killing of activated Tpath cells in a TCR-specific manner. CD4+ Treg recognizes a TCR-derived peptide in the context of a major histocompatibility complex II (MHC II) molecule and provides help for the CD8+ Treg. CD8+ Treg, recognizing another distinct peptide from the same TCR in the context of the nonclassic MHC class Ib molecule Qa-1, kills the activated pathogenic T cell population by inducing apoptosis. To complete the negative feedback loop, the apoptotic T cells are captured by antigen-presenting cells (APCs), and their TCR-derived peptides are processed and presented to cross-prime Treg.

## 5    Antigen-Presenting Cells and Priming of Treg

Antigen-specific immunotherapy is dependent on Tregs recognizing a specific peptide determinant in the context of an MHC molecule displayed on the surface of an APC. Therefore, the APC is central to the immunotherapeutic mechanism. The most effective APC in the priming of T cells is the dendritic cell (DC). The ability of a DC to prime an immune response is dependent on its maturation state. Immature DCs are highly adept at capturing antigens but have low cell surface levels of co-stimulatory molecules hindering their ability to prime T cells strongly. On the other hand, under certain conditions such as inflammation, DCs receive signals to become activated. For example, DCs can be innately stimulated through their Toll-like receptors by products derived from microbes such as lipopolysaccharide (LPS) or unmethylated CG dinucleotides (CpGs) (Pasare and Medzhitov, 2004). Activation of the DC results in upregulation of co-stimulatory molecules and secretion of soluble mediators, resulting in an enhanced ability to stimulate T cells. The activation state thus determines the type of immune response a DC can mediate. In the "nonactivated," immature state DCs may present antigens to naive T cells in such a context that induces peripheral tolerance (Hawiger et al., 2001), although mature DCs can mediate strong "productive" immune responses to fight infection.

Various types of Treg may be primed by activated and nonactivated DCs. Immature DCs have been demonstrated to be essential in the development of Tr1-like cells. Jonuleit and colleagues demonstrated the in vitro generation of IL-10-producing T cells after stimulating naive CD4+ T cells with immature DCs (Jonuleit et al., 2000). The in vivo generation of CD8 Treg has been demonstrated after immunization with immature DCs pulsed with influenza matrix protein (IMP). Here it was shown that CD8 T cells isolated 7 days after immunization could suppress T effector cell responses to IMP in vitro (Dhodapkar and Steinman, 2002). Faunce et al. demonstrated that DCs rendered tolerogenic by exposure to TGFβ and MBP could induce CD8+ Tregs that could protect against EAE (Faunce et al., 2004). In turn, CD8+ Treg cells have been shown to mediate the APC function. Chang and colleagues demonstrated that CD8+CD28− T suppressor cells acted on immature DCs, inducing upregulation of the inhibitory receptors ILT3 and ILT4, rendering the APCs capable of anergizing T-helper cells (Chang et al., 2002). Furthermore, the group showed that human CD8+CD28− T suppressor cells from heart allograft recipients acted on both professional APCs and endothelial cells carrying donor MHC I antigens, converting them to a tolerogenic state (Chang et al., 2002; Manavalan et al., 2004). The above studies are examples of how CD8 Tregs can both mediate and be mediated by APCs. The CD8 Treg function is absolutely dependent on interactions with APCs.

The mechanisms involved in the priming of anti-TCR CD8+ Tregs are currently under investigation. Anti-TCR CD8 Treg cells recognize their cognate antigens in the context of the nonclassical MHC I molecule Qa-1. Evidence that Qa-1 molecules may be involved in this regulation came from studies that demonstrated that the activity of CD8+ T cell hybridomas generated from T cell-vaccinated mice appeared to be Qa-1b-dependent (Jiang and Chess, 2000; Jiang et al., 1995, 1998). Our laboratory has now generated functional CD8+ Treg clones reactive to a specific TCR peptide. We have shown that upon adoptive transfer these Qa-1-restricted CD8+ Treg clones protect against Vβ8.2+ TCR T cell-mediated autoimmune disease. Furthermore, Tregs react to the p42-50 peptide derived from the Vβ8.2+ TCR (Tang et al., in press). Functional Qa-1 molecules are upregulated on the surface of activated DCs and T and B lymphocytes (Sarantopoulis et al., 2004; Tang et al., in press) and, as depicted in Figure 2.1, are targets for regulation in our model. Currently, we are investigating how the Qa-1 molecules acquire their cognate antigen p42-50 derived from the Vβ8.2+ TCR in the EAE model. We know that in the B10.PL mouse 20% to 30% of the peripheral TCR repertoire is Vβ8+; and during an immune response a significant proportion of these cells undergo apoptotic death (in press)). The uptake of apoptotic cells and the cross-presentation of antigen determinants derived from these cells has been widely described (Albert et al., 1998; Mougneau et al., 2002).

Our data suggest that DCs can capture apoptotic CD4 T cells and process and present Vβ8.2+ TCR-derived peptides to prime CD4+ and CD8+ Tregs. Such a mechanism explains how EAE can be naturally downregulated in the B10-PL mouse without the need for exogenous antigens to stimulate Treg. This mechanism may also be applicable in the downregulation of an antiviral immune response (Kumar, 2004). For example, cells expressing a unique TCR-Vβ chain can account for up to 50% of the peripheral T cell population during an antiviral immune response (Murali-Krishna et al., 1998). We would predict specific TCR-derived epitopes, captured from apoptotic T cells generated during an antiviral response, to be presented by APCs to prime Treg. The primed Treg cells would orchestrate the TCR-specific downregulation of the antiviral response.

One caveat to this proposed mechanism was that APCs may present TCR-derived peptides from apoptotic T cells during normal peripheral turnover. If Tregs were primed under such circumstances, the body's ability to launch productive immune responses to evading pathogens would be hindered. However, our preliminary data indicate that the effector function of the Tregs is under strict control of the APCs. Using immature/nonactivated DCs pulsed with apoptotic Vβ8.2TCR+ T cells, we have been able to demonstrate only weak CD4+ Treg priming in in vitro co-cultures. However, if after pulsing with apoptotic Vβ8.2TCR+ T cells the DCs were activated, strong CD4+ and CD8+ Treg priming was detectable. These observations suggest that Treg may be functionally primed only under conditions when they encounter activated DCs. Such DCs would be constantly sampling apoptotic T cells but functionally priming Tregs only when they receive activation signals. Thus CD8+ Treg would function to downregulate TCR-specific responses only during inflammation and not under steady-state conditions (Fig. 2.2). Furthermore, we have found that CD11c+ DCs isolated from mice during EAE could stimulate CD4Treg in vitro, whereas DCs isolated from naive mice could not (unpublished data). We have additional data indicating that only the activated Vβ8.2TCR+ T cells are regulated. Nonactivated T cells have low-level Qa-1 cell surface expression and are not targeted by the

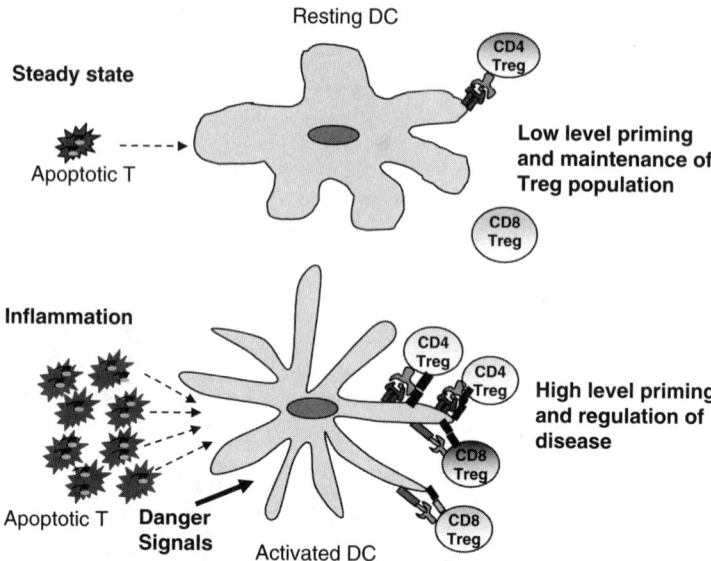

FIGURE 2.2. Dendritic cells (DCs) present TCR-derived peptides from apoptotic T cells to Treg. For priming of TCR-reactive Treg, we predict that apoptotic TCRVβ8.2+ T cells are captured by the APCs, and the TCR-derived peptides are processed and presented. In the steady state, this may occur during natural apoptotic turnover of T cells. During inflammatory disease, when there is a high level of apoptosis, capture may be increased. The apoptotic T cell's TCR-derived peptides may then be cross-presented via the MHC II pathway to prime CD4+ Treg and via the nonclassic MHC class Ib pathway to prime CD8+ Treg. In an inflammatory milieu, activated DCs would provide stronger Treg priming and thus, predictably, boost efficient immune regulation.

Qa-1-restricted Treg. Therefore, only the MBP-reactive T cells that have been activated are targeted, and most of the TCRVβ8.2+ T cells are not subjected to Treg-mediated suppression.

## 6   Conclusion

Here we have discussed the involvement of Tregs in both human and experimental animal autoimmune disease settings. There is now overwhelming evidence that Tregs play an important role in suppressing aberrant immune responses to self. Several mechanisms of suppression have been proposed, and identifying the appropriate regulatory T cell to treat specific disease processes may be essential for effective treatment protocols. We have focused on a TCR-based regulatory mechanism that specifically operates on activated T cells. We believe that therapies targeting this mechanism will downregulate and confer protection against aberrant immune reactions involving T cell responses.

*Acknowledgments.* This work was supported by grants from the National Institutes of Health, National Multiple Sclerosis Society, Multiple Sclerosis National Research Institute, and the Arthritis Foundation to V.K.

# References

Acha-Orbea, H., Mitchell, D. J., Timmermann L., et al. (1988) Limited heterogeneity of T cell receptors from lymphocytes mediating autoimmune encephalomyelitis allows specific immune intervention. *Cell* 54:263-273.

Aharoni, R., Eilam R., Domev, H., et al. (2005) The immunomodulator glatiramer acetate augments the expression of neurotrophic factors in brains of experimental autoimmune encephalomyelitis mice. *Proc. Natl. Acad. Sci. U S A* 102:19045-19050.

Albert, M. L., Pearce, S. F., Francisco, L. M., et al. (1998) Immature dendritic cells phagocytose apoptotic cells via alphavbeta5 and CD36, and cross-present antigens to cytotoxic T lymphocytes. *J. Exp. Med.* 188:1359-1368.

Askenase, P. W., Hayden, B. J., Gershon, R. K. (1975) Augmentation of delayed-type hypersensitivity by doses of cyclophosphamide which do not affect antibody responses. *J. Exp. Med.* 141:697-702.

Barrat, F. J., Cua, D. J., Boonstra, A., et al. (2002) In vitro generation of interleukin 10-producing regulatory CD4(+) T cells is induced by immunosuppressive drugs and inhibited by T helper type 1 (Th1)- and Th2-inducing cytokines. *J. Exp. Med.* 195: 603-616.

Bielekova, B., Martin, R. (2001) Antigen-specific immunomodulation via altered peptide ligands. *J. Mol. Med.* 79:552-565.

Bielekova, B., Goodwin, B., Richert, N., et al. (2000) Encephalitogenic potential of the myelin basic protein peptide (amino acids 83-99) in multiple sclerosis: results of a phase II clinical trial with an altered peptide ligand. *Nat. Med.* 6:1167-1175.

Bienvenu, B., Martin, B., Auffray, C., et al. (2005) Peripheral CD8+CD25+ T lymphocytes from MHC class II-deficient mice exhibit regulatory activity. *J. Immunol.* 175:246-253.

Bisikirska, B., Colgan, J., Luban, J., et al. (2005) TCR stimulation with modified anti-CD3 mAb expands CD8+ T cell population and induces CD8+CD25+ Tregs. *J. Clin. Invest.* 115:2904-2913.

Bouneaud, C., Kourilsky, P., Bousso, P. (2000) Impact of negative selection on the T cell repertoire reactive to a self-peptide: a large fraction of T cell clones escapes clonal deletion. *Immunity* 13:829-840.

Brimnes, J., Allez, M., Dotan, I., et al. (2005) Defects in CD8+ regulatory T cells in the lamina propria of patients with inflammatory bowel disease. *J. Immunol.* 174:5814-5822.

Brocke, S., Gijbels, K., Allegretta, M., et al. (1996) Treatment of experimental encephalomyelitis with a peptide analogue of myelin basic protein. *Nature* 379: 343-346.

Cantor, H., Shen, F. W., Boyse, E. A. (1976) Separation of helper T cells from suppressor T cells expressing different Ly components. II. Activation by antigen: after immunization, antigen-specific suppressor and helper activities are mediated by distinct T-cell subclasses. *J. Exp. Med.* 143:1391-1340.

Chang, C. C., Ciubotariu, R., Manavalan, J. S., et al. (2002) Tolerization of dendritic cells by T(S) cells: the crucial role of inhibitory receptors ILT3 and ILT4. *Nat. Immunol.* 3:237-243.

Chen, Z., Herman, A. E., Matos, M., et al. (2005) Where CD4+CD25+ T reg cells impinge on autoimmune diabetes. *J. Exp. Med.* 202:1387-1397.

Cosmi, L., Liotta, F., Lazzeri, E., et al. (2003) Human CD8+CD25+ thymocytes share phenotypic and functional features with CD4+CD25+ regulatory thymocytes. *Blood.* 102:4107-4114.

Cua, D. J., Hutchins, B., LaFace, D. M., et al. (2001) Central nervous system expression of IL-10 inhibits autoimmune encephalomyelitis. *J. Immunol.* 166:602-608.

Dhodapkar, M. V., Steinman, R. M. (2002) Antigen-bearing immature dendritic cells induce peptide-specific CD8(+) regulatory T cells in vivo in humans. *Blood* 100:174-177.

Ehrenstein, M. R., Evans, J. G., Singh, A., et al. (2004) Compromised function of regulatory T cells in rheumatoid arthritis and reversal by anti-TNFalpha therapy. *J. Exp. Med.* 200:277-285.

Faunce, D. E., Terajewicz, A., Stein-Streilein, J. (2004) Cutting edge: in vitro-generated tolerogenic APC induce CD8+ T regulatory cells that can suppress ongoing experimental autoimmune encephalomyelitis. *J. Immunol.* 172:1991-1995.

Feldmann, M., Steinman, L. (2005) Design of effective immunotherapy for human autoimmunity. *Nature* 435:612-619.

Filaci, G., Suciu-Foca, N. (2002) CD8+ T suppressor cells are back to the game: are they players in autoimmunity? *Autoimmun. Rev.* 1:279-283.

Fukaura, H., Kent, S. C., Pietrusewicz, M. J., et al. (1996) Induction of circulating myelin basic protein and proteolipid protein-specific transforming growth factor-beta1-secreting Th3 T cells by oral administration of myelin in multiple sclerosis patients. *J. Clin. Invest.* 98:70-77.

Gershon, R. K., Kondo, K. (1970) Cell interactions in the induction of tolerance: the role of thymic lymphocytes. *Immunology* 18:723-737.

Green, D. R., Webb, D. R. (1993) Saying the 'S' word in public. *Immunol. Today* 14:523-525.

Groux, H., O'Garra, A., Bigler, M., et al. 1997. A CD4+ T-cell subset inhibits antigen-specific T-cell responses and prevents colitis. *Nature* 389:737-742.

Hawiger, D., Inaba, K., Dorsett, Y., et al. (2001) Dendritic cells induce peripheral T cell unresponsiveness under steady state conditions in vivo. *J. Exp. Med.* 194:769-779.

Honda, A., Ametani, A., Matsumoto, T., et al. (2004) Vaccination with an immunodominant peptide of bovine type II collagen induces an anti-TCR response, and modulates the onset and severity of collagen-induced arthritis. *Int. Immunol.* 16:737-745.

Hong, J., Li, N., Zhang, X., et al. (2005) Induction of CD4+CD25+ regulatory T cells by copolymer-I through activation of transcription factor Foxp3. *Proc. Natl. Acad. Sci. U S A* 102:6449-6454.

Hori, S., Takahashi, T., Sakaguchi, S. (2003) Control of autoimmunity by naturally arising regulatory CD4+ T cells. *Adv. Immunol.* 81:331-371.

Howell, M. D., Winters, S. T., Olee, T., et al. (1989) Vaccination against experimental allergic encephalomyelitis with T cell receptor peptides. *Science* 246:668-670.

Jacobson, D. L., Gange, S. J., Rose, N. R., Graham, N. M. (1997) Epidemiology and estimated population burden of selected autoimmune diseases in the United States. *Clin Immunol Immunopathol* 84:223-243.

Jiang, H., Chess, L. (2000) The specific regulation of immune responses by CD8+ T cells restricted by the MHC class Ib molecule, Qa-1. *Annu. Rev. Immunol.* 18:185-216.

Jiang, H., Zhang, S. I., Pernis, B. (1992) Role of CD8+ T cells in murine experimental allergic encephalomyelitis. *Science* 256:1213-1215.

Jiang, H., Ware, R., Stall, A., et al. (1995) Murine CD8+ T cells that specifically delete autologous CD4+ T cells expressing V beta 8 TCR: a role of the Qa-1 molecule. *Immunity* 2:185-194.

Jiang, H., Kashleva, H., Xu, L. X., et al. (1998) T cell vaccination induces T cell receptor Vbeta-specific Qa-1-restricted regulatory CD8(+) T cells. *Proc. Natl. Acad. Sc.i U S A* 95:4533-4537.

Jonuleit, H., Schmitt, E., Schuler, G., et al. (2000) Induction of interleukin 10-producing, nonproliferating CD4(+) T cells with regulatory properties by repetitive stimulation with allogeneic immature human dendritic cells. *J. Exp. Med.* 192:1213-1222.

Kabelitz, D., Pohl, T., Pechhold, K. (1993) Activation-induced cell death (apoptosis) of mature peripheral T lymphocytes. *Immunol. Today* 14:338-339.

Kappos, L., Comi, G., Panitch, H., et al. (2000) Induction of a non-encephalitogenic type 2 T helper-cell autoimmune response in multiple sclerosis after administration of an altered peptide ligand in a placebo-controlled, randomized phase II trial: the Altered Peptide Ligand in Relapsing MS Study Group. *Nat. Med.* 6:1176-1182.

Karandikar, N. J., Crawford, M. P., Yan, X., et al. (2002) Glatiramer acetate (Copaxone) therapy induces CD8(+) T cell responses in patients with multiple sclerosis. *J. Clin. Invest.* 109:641-649.

Keane, J., Gershon, S., Wise, R. P., et al. (2001) Tuberculosis associated with infliximab, a tumor necrosis factor alpha-neutralizing agent. *N. Engl. J. Med.* 345:1098-1104.

Kelsen, J., Agnholt, J., Hoffmann, H. J., et al. (2005) FoxP3(+)CD4(+)CD25(+) T cells with regulatory properties can be cultured from colonic mucosa of patients with Crohn's disease. *Clin. Exp. Immunol.* 141:549-557.

Kibler, R. F., Fritz, R. B., Chou, F., et al. (1977) Immune response of Lewis rats to peptide C1 (residues 68-88) of guinea pig and rat myelin basic proteins. *J. Exp. Med.* 146:1323-1331.

Koh, D. R., Fung-Leung, W. P., Ho, A., et al. (1992) Less mortality but more relapses in experimental allergic encephalomyelitis in CD8−/− mice. *Science* 256:1210-1213.

Kohm, A. P., Carpentier, P. A., Anger, H. A., Miller, S. D. (2002) Cutting edge: CD4+CD25+ regulatory T cells suppress antigen-specific autoreactive immune responses and central nervous system inflammation during active experimental autoimmune encephalomyelitis. *J. Immunol.* 169:4712-4716.

Kumar, V. (2004) Homeostatic control of immunity by TCR peptide-specific Tregs. *J. Clin. Invest.* 114:1222-1226.

Kumar, V., Sercarz, E. E. (1993) The involvement of T cell receptor peptide-specific regulatory CD4+ T cells in recovery from antigen-induced autoimmune disease. *J. Exp. Med.* 178:909-916.

Kumar, V., Kono, D. H., Urban, J. L., Hood, L. (1989) The T-cell receptor repertoire and autoimmune diseases. *Annu. Rev. Immunol.* 7:657-682.

Kumar, V., Tabibiazar, R., Geysen, H. M., Sercarz, E. (1995) Immunodominant framework region 3 peptide from TCR V beta 8.2 chain controls murine experimental autoimmune encephalomyelitis. *J. Immunol.* 154:1941-1950.

Kumar, V., Coulsell, E., Ober, B., et al. (1997) Recombinant T cell receptor molecules can prevent and reverse experimental autoimmune encephalomyelitis: dose effects and involvement of both CD4 and CD8 T cells. *J. Immunol.* 159:5150-5156.

Kumar, V., Sercarz, E., Zhang, J., Cohen, I. (2001) T-cell vaccination: from basics to the clinic. *Trends Immunol.* 22:539-540.

Lan, R. Y., Ansari, A. A., Lian, Z. X., Gershwin, M. E. (2005) Regulatory T cells: development, function and role in autoimmunity. *Autoimmun. Rev.* 4:351-363.

Lindley, S. C., Dayan, M., Bishop, A., et al. (2005) Defective suppressor function in CD4(+)CD25(+) T-cells from patients with type 1 diabetes. *Diabetes* 54:92-99.

Madakamutil, L. T., Maricic, I., Sercarz, E., Kumar, V. (2003) Regulatory T cells control autoimmunity in vivo by inducing apoptotic depletion of activated pathogenic lymphocytes. *J. Immunol.* 170:2985-2992.

Manavalan, J. S., Kim-Schulze, S., Scotto, L., et al. (2004) Alloantigen specific CD8+CD28- FOXP3+ T suppressor cells induce ILT3+ ILT4+ tolerogenic endothelial cells, inhibiting alloreactivity. *Int. Immunol.* 16:1055-1068.

Martin, B., Banz, A., Bienvenu, B., et al. (2004) Suppression of CD4+ T lymphocyte effector functions by CD4+CD25+ cells in vivo. *J. Immunol.* 172:3391-3398.

Morgan, M. E., Flierman, R., van Duivenvoorde, L. M., et al. (2005) Effective treatment of collagen-induced arthritis by adoptive transfer of CD25+ regulatory T cells. *Arthritis Rheum.* 52:2212-2221.

Mougneau, E., Hugues, S., Glaichenhaus, N. (2002) Antigen presentation by dendritic cells in vivo. *J. Exp. Med.* 196:1013-1016.

Murali-Krishna, K., Altman, J. D., Suresh M., et al. (1998) Counting antigen-specific CD8 T cells: a reevaluation of bystander activation during viral infection. *Immunity* 8:177-187.

Nishizuka, Y., Sakakura, T. (1969) Thymus and reproduction: sex-linked dysgenesis of the gonad after neonatal thymectomy in mice. *Science* 166:753-755.

Panoutsakopoulou, V., Huster, K. M., McCarty, N., et al. (2004) Suppression of autoimmune disease after vaccination with autoreactive T cells that express Qa-1 peptide complexes. *J. Clin. Invest.* 113:1218-1224.

Pasare, C., Medzhitov, R. (2004) Toll-like receptors: linking innate and adaptive immunity. *Microbes Infect.* 6:1382-1387.

Piccirillo, C. A., Shevach, E. M. (2004) Naturally-occurring CD4+CD25+ immunoregulatory T cells: central players in the arena of peripheral tolerance. *Semin. Immunol.* 16: 81-88.

Putnam, A. L., Vendrame, F., Dotta, F., Gottlieb, P. A. (2005) CD4+CD25 high regulatory T cells in human autoimmune diabetes. *J. Autoimmun.* 24:55-62.

Quattrocchi, E., Dallman, M. J., Dhillon, A. P., et al. (2001) Murine IL-10 gene transfer inhibits established collagen-induced arthritis and reduces adenovirus-mediated inflammatory responses in mouse liver. *J. Immunol.* 166:5970-5978.

Revel, M. (2003) Interferon-beta in the treatment of relapsing-remitting multiple sclerosis. *Pharmacol. Ther.* 100:49-62.

Sakaguchi, S. (2005) Naturally arising Foxp3-expressing CD25+CD4+ regulatory T cells in immunological tolerance to self and non-self. *Nat. Immunol.* 6:345-352.

Sakaguchi, S., Fukuma, K., Kuribayashi, K., Masuda, T. (1985) Organ-specific autoimmune diseases induced in mice by elimination of T cell subset. I. Evidence for the active participation of T cells in natural self-tolerance; deficit of a T cell subset as a possible cause of autoimmune disease. *J. Exp. Med.* 161:72-87.

Sakaguchi, S., Sakaguchi, N., Asano, M., et al. (1995) Immunologic self-tolerance maintained by activated T cells expressing IL-2 receptor alpha-chains (CD25): breakdown of a single mechanism of self-tolerance causes various autoimmune diseases. *J. Immunol.* 155:1151-1164.

Salgame, P., Abrams, J. S., Clayberger, C., et al. (1991) Differing lymphokine profiles of functional subsets of human CD4 and CD8 T cell clones. *Science* 254:279-282.

Sarantopoulos, S., Lu, L., Cantor, H. (2004) Qa-1 restriction of CD8+ suppressor T cells. *J. Clin. Invest.* 114:1218-1221.

Saruhan-Direskeneli, G., Weber, F., Meinl, E., et al. (1993) Human T cell autoimmunity against myelin basic protein: CD4+ cells recognizing epitopes of the T cell receptor beta chain from a myelin basic protein-specific T cell clone. *Eur. J. Immunol.* 23:530-536.

Schwartz, R. H. (2003) T cell anergy. *Annu. Rev. Immunol.* 21:305-334.

Sercarz, E. E., Lehmann, P. V., Ametani, A., et al. (1993) Dominance and crypticity of T cell antigenic determinants. *Annu. Rev. Immunol.* 11:729-766.

Shevach, E. M. (2000) Regulatory T cells in autoimmmunity. *Annu. Rev. Immunol.* 18:423-449.

Sun, D., Ben Nun, A., Wekerle, H. (1988) Regulatory circuits in autoimmunity: recruitment of counter-regulatory CD8+ T cells by encephalitogenic CD4+ T line cells. *Eur. J. Immunol.* 18:1993-1999.

Tang, X., Maricic, I., et al., Regulation of immunity by a novel population of Qa-1 restricted CD8 alpha +TCR alpha beta + T cells. *J. Immunol.* (in press).

Takatsu, K., Ishizaka, K. (1975) Reaginic antibody formation in the mouse. VI. Suppression of IgE and IgG antibody responses to ovalbumin following the administration of high dose urea-denatured antigen. *Cell. Immunol.* 20:276-289.

Thorstenson, K. M., Khoruts, A. (2001) Generation of anergic and potentially immunoregulatory CD25+CD4 T cells in vivo after induction of peripheral tolerance with intravenous or oral antigen. *J. Immunol.* 167:188-195.

Urban, J. L., Kumar, V., Kono, D. H., et al. (1988) Restricted use of T cell receptor V genes in murine autoimmune encephalomyelitis raises possibilities for antibody therapy. *Cell* 54:577-592.

Vandenbark, A. A., Hashim, G., Offner, H. (1989) Immunization with a synthetic T-cell receptor V-region peptide protects against experimental autoimmune encephalomyelitis. *Nature* 341:541-544.

Vandenbark, A. A., Chou, Y. K., Whitham, R., et al. (1996) Treatment of multiple sclerosis with T-cell receptor peptides: results of a double-blind pilot trial. *Nat. Med.* 2:1109-1115.

Viglietta, V., Baecher-Allan, C., Weiner, H. L., Hafler, D. A. (2004) Loss of functional suppression by CD4+CD25+ regulatory T cells in patients with multiple sclerosis. *J. Exp. Med.* 199:971-979.

Weiner, H. L. (2001) Induction and mechanism of action of transforming growth factor-beta-secreting Th3 regulatory cells. *Immunol. Rev.* 182:207-214.

Xystrakis, E., Dejean, A. S., Bernard, I., et al. (2004) Identification of a novel natural regulatory CD8 T-cell subset and analysis of its mechanism of regulation. *Blood* 104:3294-3301.

You, S., Chen, C., Lee, W. H., et al. (2004) Presence of diabetes-inhibiting, glutamic acid decarboxylase-specific, IL-10-dependent, regulatory T cells in naive nonobese diabetic mice. *J. Immunol.* 173:6777-6785.

Zamvil, S. S., Mitchell, D. J., Moore, A. C., et al. (1986) T-cell epitope of the autoantigen myelin basic protein that induces encephalomyelitis. *Nature* 324:258-260.

Zhang, J., Vandevyver, C., Stinissen, P., Raus J. (1995) In vivo clonotypic regulation of human myelin basic protein-reactive T cells by T cell vaccination. *J. Immunol.* 155:5868-5877.

Zhang, X., Izikson, L., Liu, L., Weiner, H. L. (2001) Activation of CD25(+)CD4(+) regulatory T cells by oral antigen administration. *J. Immunol.* 167:4245-4253.

Zimring, J. C., Kapp, J. A. (2004) Identification and characterization of CD8+ suppressor T cells. *Immunol. Res.* 29:303-312.

# 3
# Peripheral T Cell Regulation and Autoimmunity

Hong Jiang and Leonard Chess

## 1   Introduction

The immune system is capable of mounting effective immune responses to a virtually infinite variety of foreign pathogens or tumor cells while avoiding harmful immune responses to self. This is achieved by central thymic selection and peripheral immune regulation. It is clear that thymocytes expressing T cell receptors (TCRs) with high affinity/avidity for self-peptide/major histocompatibility complex (MHC) complexes undergo apoptosis and are deleted centrally in the thymus (Kappler et al., 1987). During thymic development immature T cells migrate from the bone marrow to the thymus, where they express virgin αβ TCRs and encounter self-peptides derived from endogenous proteins bound to MHC molecules. Selection of the T cell repertoire in the thymus involves two steps. First, positive selection permits the survival of cells binding thymic self-MHC/peptide complexes with sufficient affinity/avidity. Those thymocytes with affinity/avidity too low to be positively selected die of neglect in the thymus. The thymocytes that "pass" positive selection are therefore, from the onset, MHC-restricted (2). Second, negative selection deletes thymocytes whose TCRs interact with MHC/self-peptide complexes with high affinity/avidity by a process involving apoptotic cell death (Bevan, 1977; Bevan and Fink, 1978; Hengartner et al., 1988; von Boehmer and Kisielow, 1990; Waldmann, 1978; Zinkernagel, 1978). Undoubtedly, negative selection must have been preserved during evolution to prevent pathogenic autoimmunity by deleting the high affinity/avidity and therefore potentially pathogenic self-reactive T cells (Nossal, 1994; Sprent and Webb, 1995). Overall, only 3% of the cells that differentiate in the thymus survive thymic positive and negative selection (Shortman et al., 1990). These T cells then migrate to and populate the peripheral lymphoid organs (lymph nodes and spleen), where they function to induce, effect, and regulate cellular immune responses. Although selected on self-peptides these T cells form the peripheral repertoire capable of reacting to foreign antigens. This is possible because it is known that a single TCR cross-reacts with multiple peptides and the T cell cross-reactivity is necessary to maintain an immune system with sufficient flexibility to adapt to a continuously changing antigenic environment (Garcia et al., 1998; Mason, 1998).

Despite the fact that approximately 97% of self-reactive T cells are deleted by central thymic positive and negative selection, some T cells with intermediate affinity/avidity to self-antigen escape thymic negative selection and are released into the periphery (Bouneaud et al., 2000; Jiang et al., 2003; Sandberg et al., 2000). Although these self-reactive T cell clones display lower affinity/avidity to MHC/self-peptide complexes than do the truly high affinity/avidity self-reactive T cell clones that are deleted intrathymically, they are capable of self-peptide-driven proliferation, and some may differentiate into pathogenic effector cells (Jiang et al., 2003, 2005; Kuchroo et al., 2002; Mason, 1998). To avoid pathogenic autoimmunity, various peripheral regulatory mechanisms have evolved to fine-tune the self-reactive TCR repertoire and suppress clonal expansion of those self-reactive clones with TCRs of affinity/avidity that are not sufficiently high to be eliminated intrathymically but high enough to induce pathogenic autoimmunity. Thus, under normal circumstances, despite the abundance of self-reactive clones in the periphery, clinical autoimmunity is usually well controlled.

## 2   Biological and Clinical Impact of Immunoregulation

The idea that immunoregulation plays a central role in all immune responses was originally postulated more than three decades ago by the late Richard Gershon on the basis of two fundamental concepts. The first is the concept of homeostasis, initially coined by the American physiologist Walter Cannon in 1932, who was impressed by the "wisdom of the body" in being capable of guaranteeing with efficiency the control of physiological equilibrium. The immune system is indeed a homeostatic system that maintains its basic structure and its goal to create the correct tempo, magnitude, duration, and quality of immune responses to avoid disease secondary to excessive or too little immune response, which is either unnecessary or inadequate. Many of these homeostatic mechanisms are indeed carried out by the intrinsic mechanisms of immunoregulation, particularly secretion of cytokines by Th1, Th2, and the Tr1 and Tr3 subsets. The second concept implicit in the notion of immunoregulation is that because all T cells are selected on the basis of reactivity to self-antigens in the thymus the potential for pathogenic autoimmunity is great. As a consequence, immunoregulation requires that regulatory mechanisms, particularly "suppressor cells," discriminate self from non-self. In the absence of this cognitive property, the suppression of self-reactive cells would be necessarily accompanied by general nonspecific immunosuppression, leading to an increased incidence of infections or the outgrowth of tumors. Indeed, many therapeutic pharmacological agents currently employed to treat immunological disease, whether they be autoimmune disorders or allograft rejection, are plagued by these complications of indiscriminate immunosuppression. Implicit in these notions is the concept that the various physiological mechanisms operating in the periphery to regulate immune responses, to both self and foreign antigens, fall into two general categories. They must function to control the magnitude and class of immune responses and/or discriminate self from non-self (Cohn, 2004).

## 2.1 Regulation by Controlling the Magnitude and Class of Immune Responses

One level of controlling the magnitude and class of immune responses resides at the initial clonal activation of the TCR by MHC/peptide complexes. T cell signaling is dependent on the affinity and duration of binding of the TCRs with MHC/peptide complexes (Davis et al., 1998; Savage et al., 1999) and subsequently influences many events during the evolution of immune responses. In this regard, TCR triggering induces not only activation and differentiation events but also apoptotic pathways of antigen-induced cell death (Lenardo et al., 1999). In addition, other receptor/ligand interactions become pivotal in immunoregulation. For example, one of the earliest antigen activation-induced cell surface molecules expressed by T cells is CD40L (Lederman et al., 1992). A critical consequence of the interaction of CD40L with CD40 expressed on antigen-presenting cells (APCs) is the upregulation of other key co-stimulatory molecules, including CD80 and CD86 (Caux et al., 1994; Klaus et al., 1994). These molecules interact with CD28 or CTLA-4 molecules on T cells to determine whether the outcome of antigen triggering is functional T cell activation (via CD28) or, alternatively, inactivation (anergy) and tolerance induction (via CTLA-4) (Durie et al., 1994; Klaus et al., 1994; Koulova et al., 1991 Lenschow et al., 1996). Thus, blockade of CD80 or CD86 triggering is known to induce T cell anergy (Boussiotis et al., 1996; Jenkins, 1994; Lenschow et al., 1996). Similarly, blockade of the CD4L/CD40 pathway can lead to tolerance induction (Foy et al., 1996), whereas blockade of CTLA-4 enhances the immune response (Chambers et al., 2001). Another key consequence of antigen triggering of CD4+ T cells is the further differentiation of the CD4+ T cells into the functional distinct peripheral Th1 and Th2 subsets (Coffman and Mosmann, 1991; Mossman et al., 1986, 1989). The elaboration of distinct cytokines by the Th subsets provides a second level of control intrinsic to the outgrowth and function of CD4+ T cells (Fitch et al., 1993). In this regard, interferon-γ (IFN) secreted by Th1 cells is known to downregulate the differentiation and function of Th2 cells and, conversely, IL-4. Transforming growth factor-β (TGFβ) and interleukin-10 (IL-10) inhibits Th1 cell differentiation (Fitch et al, 1993; Mosmann et al., 1989; Seder and Paul, 1994). In addition, other cytokine secreting Th subsets capable of secreting the immunosuppressive cytokines IL-10 and/or TGFβ but no IL-4 (Tr1 or Tr3 cells) (Groux et al., 1997; Levings and Roncarolo, 2000; Roncarolo and Levings, 2000) have been observed. A widely prevalent view is that the balance between the emergence of Th1 and Th2 as well as other Th cytokine secreting regulatory Tr1 and Tr3 cells following antigen activation plays a major role in controlling the outgrowth and functions of both self-reactive and foreign reactive T cells (Charlton and Lafferty, 1995; Del Prete, 1998; O'Garra and Vieira, 2004). These Th1, Th2, Tr1, and Tr3 cells can be collectively grouped into one set of regulatory T cells (Tregs) that are specifically induced by antigen but mediate suppression nonspecifically by releasing cytokines. Taken together, controlling the magnitude and class of immune responses ensures optimal function of an immune response by (1) preventing the collateral damage caused by too much reactivity or (2) avoiding inadequate immunity due to too little reactivity. In addition, regulating the class of immune responses by skewing the immune response toward secretion of Th1 or Th2 cytokines also contributes to controlling the outgrowth of pathogenic self-reactive Th1 or Th2 cells.

## 2.2 Regulation by Discriminating Self from Non-self

In contrast to the mechanisms controlling the magnitude and class of immune responses mediated by intrinsic mechanisms including either antigen-induced T cell death or secretion of cytokines, the control mechanisms involved in discriminating self from non-self is largely mediated by suppressor cells via cell surface receptors with the cognitive capacity to perceive molecular distinctions between self and non-self. It is of interest that cells involved in innate immunity, including natural killer (NK) cells and various antigen-presenting dendritic cells and macrophages, employ receptors highly homologous to the more ancient mannose, lectin, and Toll-like receptors (TLRs), which recognize molecular patterns expressed by pathogens present in the environment. The TLRs and their signaling pathways are represented in such diverse creatures as in mammals, fruit flies, and plants. Moreover, 10 members of the TLR family have been identified in humans, and several of them appear to recognize specific microbial products, including lipopolysaccharide, bacterial lipoproteins, peptidoglycan, and bacterial DNA (Janeway and Medzhitov, 2002). These receptors thus recognize distinct sets of molecular patterns and repeated motifs expressed in a broad variety of foreign pathogens but not found in vertebrates (Janeway and Medzhitov, 2002; Medzhitov and Janeway, 2000, 2002). Because these receptors are expressed on macrophages and dendritic cells, in addition to their role in innate immune responses they may also trigger adaptive immune responses via functioning as APCs (Medzhitov and Janeway, 2002; Pesare and Medzhitov, 2003). In contrast to the innate systems, the receptors employed by the adaptive immune system are the clonally distributed antigen receptors expressed on T cells (TCRs) or B cells (immunoglobulin, or Ig).

At the level of T cell regulation, specific suppression of self-reactive cells could result from recognition by the TCRs on suppressor cells of unique sequences derived from the hypervariable regions (idiotypes) of TCRs or Igs expressed by activated self-reactive T or B cells (Batchelor et al., 1989; Gammon and Sercarz, 1990). The idiotypic model predicts that the size of the repertoire for suppressor cells would be comparable to the repertoire for the antigen-specific effector T cells, which by itself is larger than $10^9$ of the clonal TCR diversities. The idiotypic model of regulation was postulated by Jerne (1974) and has been investigated during the past several decades (Dorf and Benacerraf, 1984; Gammon et al., 1990), but little evidence has emerged in support of the idiotypic model (Bloom et al., 1992; Janeway, 1988; Moller, 1988). More recently, an alternative model has been proposed based on the evidence that regulatory T cells can achieve self non-self discrimination by recognizing a set of surrogate molecules which are expressed on the target T cells as the consequence of intermediate affinity/avidity interactions between TCRs on activated T cells and MHC/antigen-peptides presented by APCs (Jiang et al., 2005; Lohse et al., 1989). Because the pathogenic self-reactive TCR repertoire is enriched in intermediate affinity/avidity T cells but devoid in high affinity/avidity T cells, whereas TCR repertoire to foreign antigens is enriched in high affinity/avidity T cells, the selective downregulation of intermediate affinity/avidity T cells to both self and foreign antigens suppresses the autoimmune TCR repertoire yet results in preservation and/or enrichment in foreign reactive T cells. Conceptually, the "affinity/avidity model" contains some elements of "ergotypic regulation" in that both types of regulation recognize the consequence of T cell activation. It differs from "ergotypic regulation," however, because the "ergotypic regulation" does not consider the affinity/avidity interactions of T cell activation (Cohen et al., 2004; Lohse et al., 1989). The "affinity/avidity model" thus represents a simple yet efficient means to discriminate self from non-self and is discussed in more detail below (Jiang and Chess, 2000; Jiang et al., 2005).

We emphasize that the biological function of peripheral regulatory mechanisms involved in either control of the magnitude and class of immune responses or in self/non-self discrimination has direct clinical relevance to our current understanding of a wide spectrum of disorders including those resulting in autoimmune disease, transplantation rejection, inappropriate responses to infectious pathogens, and failure to induce effective antitumor immunity. It is likely that precise understanding of the molecular and cellular mechanisms involved in immunoregulation will permit the generation of novel strategies to prevent and/or treat immune-mediated diseases. We now focus on the major suppressor T cell subsets, which function in concert with the intrinsic immunoregulatory mechanisms to regulate immune responses.

# 3    Suppressor Cell Subsets

There is increasing evidence that superimposed on the intrinsic mechanisms of homeostatic regulation are extrinsic regulatory mechanisms mediated by distinct subsets of regulatory NKT, CD4+, and CD8+ T suppressor cells. Each of the suppressor T cell subsets is characterized by distinct receptors, employs different effector mechanisms, and functions at different stages during the peripheral immune responses (Jiang and Chess, 2004b). Moreover, these subsets either predominantly control the magnitude and class of immune responses (Th1 or Th2 and humoral or cellular), or they are primarily involved in self/non-self discrimination by selectively downregulating potentially pathogenic self-reactive T cells. For example, the NKT cells and CD4+ CD25+ regulatory cells (Tregs) exist as "natural suppressor cells" prior to antigen activation and primarily function during the early "innate" and/or primary immune responses. These cells predominantly effect suppression by regulating the magnitude or class of immune responses. In contrast, the regulatory function of the Qa-1-dependent CD8+ T cells is not observed in naive animals prior to antigen encounter. However, like classical immunocompetent CD8+ T cells reactive to viruses or tumor cells, the CD8+ Tregs are induced to differentiate into effector cells during the primary immune response and function as suppressor cells during the secondary and memory phases of immunity (Hu et al., 2004; Jiang et al., 2003, 2005; Panoutsakopoulou et al., 2004). These CD8+ T cells are primarily involved in self/non-self discrimination.

## 3.1    NK T Cells

The NKT cells are endowed with properties of natural killer (NK) cells as well as expressing αβ TCRs that are comprised of an invariant TCRα chain paired preferentially to various Vβ chains (Bendelac, 1995). The human NKT cells specifically recognize glycolipid molecules structurally related to the glycolipid α-galactosylceramide (α-GalCer), often expressed by various pathogenic microorganisms as well as autologous cells. α-GalCer binds to the MHC molecule CD1d, a molecule expressed by APCs, and trigger the NKT cells to kill and secrete cytokines (Bendelac et al., 1995, 1997; Mattner et al., 2005). Although NKT cells were originally functionally defined by their capacity to lyse tumor cells (Cui et al., 1997), they were later found also to be involved in the regulation of autoimmune diseases (Godfrey et al., 2000; Gombert et al., 1996). These in vivo roles in immune responses are linked to the observations that following TCR antigen triggering in vivo NKT cells develop augmented killer cell activity and secrete large amounts of cytokines, including IL-4 and IFNγ, as well as TGFβ and IL-10 (Bendelac et al., 1995, 1997; D'Orazio and Niederkorn, 1998; Sharif et al., 2002), which are known to be important in mediating and controlling inflammation during innate and adaptive immune responses.

Prominent among the murine disease models affected by NKT cells are those primarily induced by Th1 cells, including diabetes and multiple sclerosis models (Furlan et al., 2003; Godfrey et al., 2000; Singh et al., 2001). In these diseases the evidence strongly suggests that the Th2-favoring cytokines IL-4 and IL-10, secreted by NKT cells. play an important role (Baxter et al., 1997; Hammond et al., 1998; Sharif et al., 2002). Adoptive transfer of cell populations enriched for NKT cells prevents type 1 diabetes (T1D) in NOD mouse recipients (Baxter et al., 1997; Falcone et al., 1999; Sharif et al., 2001). Moreover, depletion of NKT cells early in the evolution of T1D in the NOD mice accelerates the onset of T1D (Frey and Rao, 1999). Similarly, in models of colitis or multiple sclerosis (experimental autoimmune encephalitis, or EAE), depletion of NKT cells accelerates the onset of disease, whereas in vivo activation of NKT cells by treatment with the glycolipid ligand induces significant improvement or prevents disease.

Moreover, reduced numbers or altered functions of NKT cells have been correlated with autoimmune disease in humans. For example, diabetic individuals have a lower frequency of the NKT cell-specific $V\alpha24$–JaQ TCRs in comparison with their nondiabetic monozygotic twins (Wilson et al., 1998). These studies suggest a role of NKT cells in natural protection against T1D. On the other hand, NKT cell frequency and IL-4 production are conserved during the course of human T1D. These results do not necessarily refute the hypothesis that NKT cell defects are involved in T1D but may indicate that immunoregulation in T1D and other autoimmune diseases is mediated by interacting subsets of immunoregulatory cells functioning in concert (Lee et al., 2002).

## 3.2    CD4+CD25+ T Regulators

The CD4+CD25+ T regs were originally identified during the course of attempts to define the pathogenic autoimmunity that occurs in lymphopenic thymectomized mice and genetically immunodeficient mice (Sakaguchi et al., 1985, 1995). These experiments, pioneered by Sakaguchi and Shevach, led to the idea that CD4+ suppressor cells were a key cellular element in immunoregulation (Lafaille et al., 1994; Sakaguchi et al, 1985, 1995; Shevach, 2001). The early experimental protocols initially in vivo took advantage of the observations that mice deficient in T cells either following neonatal thymectomy (2–4 days of age) or by adoptive transfer of CD25+ cell-depleted/CD4+ T cells or thymocytes into nu/nu mice developed various autoimmune-like diseases accompanied by a systemic wasting disease. Importantly, the autoimmune disease in these immunodeficient mice was abrogated by adoptive transfer of normal splenic CD4+ T cells. This suppression of diseases was not observed if the CD4+ T cells were depleted of CD25+ T cells, and direct reconstitution with the CD4+CD25+ population inhibited autoimmune development (Sakaguchi, 2000l; Shevach, 2000). It is of interest that CD4+CD25+ T regs emerge from the thymus and populate peripheral lymphoid organs within a few days of life and are thought to persist into adulthood as naturally occurring Tregs; and although they are present in small numbers, they are perhaps responsible for preventing autoimmunity in immunocompetent adults.

In this regard, seminal studies by Lafaille and colleagues stressed the potency of CD4+ suppressor cells. These studies evaluated mice that were transgenic (TG) for a pathogenic myelin-specific TCR known to induce EAE, a disease model of human multiple sclerosis. Despite the fact that > 95% of the T cells in these mice express pathogenic TCRs, the mice do not spontaneously develop EAE. However, if the TCR TG mice are mated with rag–/– mice (which are devoid of any nontransgenic T cells or B cells), the resulting animals develop florid EAE. These studies suggest that a small number of nontransgenic lymphocytes, present in TCR TG mice, suppress the induction of EAE (Lafaille et al., 1994). Subsequently, Lafaille found that both CD4+CD25+ and CD4+CD25– cells could mediate this suppression (Curotto et al., 2002; Furtado et al., 2001).

The naturally occurring CD4+CD25+ T regs have been hypothesized to represent a lineage-specific suppressor population arising directly from the thymus (Sakaguchi, 2000). In this regard, it was observed that a cloned transcription factor, Foxp3, a member of the forkhead family of DNA binding transcription factors (Schubert et al., 2001), is highly expressed in the naturally occurring CD4+CD25+ T regs (Fontenot et al., 2003; Khattri et al., 2003). Moreover, mutational defects in the Foxp3 gene result in the fatal autoimmune and inflammatory disorder of the "scurfy mouse" and in the clinical and molecular features of the immunodysregulation/polyendocrinopathy/enteropathy/X-linked (IPEX) syndrome in humans. Both scurfy mice and IPEX patients have defects in T cell activation and reduced numbers, as well as defective in vitro suppressor functions mediated by the CD4+CD25+ T cells (Bennett et al., 2001; Brunkow et al., 2001; Wildin et al., 2001). In Foxp3 overexpressing mice, both CD4+CD25– and CD4-CD8+ T cells show suppressive activity, suggesting that expression of Foxp3 is linked to suppressor functions (Fontenot et al., 2003). Taken together, these data support the idea that Foxp3 may define the subset of naturally occurring CD4+ suppressor T cells. However, the findings that Foxp3 can be expressed in CD4+CD25– cells following activation and are also expressed in activated CD8+ T cells suggest that Foxp3 may be functionally linked to suppression (Chen et al., 2003; Cosmi et al., 2003; Manavalan et al., 2004; Walker et al., 2003).

The relation of Foxp3 and suppression is supported by recent evidence that expression of Foxp3 is a subset of $\alpha\beta$ of T cells, irrespective to CD25 expression (Fontenot et al., 2005; Rudensky, 2005). Thus, in experiments in which the Foxp3 gene was "marked" by in-frame insertion of the green fluorescent protein (GFP) and used to isolate CD4+CD25hi Foxp3 gfp+ and CD4+CD25lo Foxp3 gfp+ populations as well as CD4+ Foxp3 gfp– cells, it was shown that Foxp3 expression correlated with suppressor activity of CD4+ Tregs, regardless of whether the Tregs expressed CD25. These observations are consistent with the idea that Foxp3 is linked to the function of suppression (Fontenot et al., 2005). Furthermore, these studies unequivocally demonstrated that CD25 could not be used as a sole marker to define "specialized" T regs. The observations that CD25– T cells gain suppressive activity when transfected with Foxp3 gene (Khattri et al., 2003; Zelenay et al., 2005) further support this notion. In this regard, conclusions derived from the prior studies of CD4+ Tregs that rely solely on the assumption that CD25 is the marker to define the CD4+ Treg phenotype may have to be reexamined and/or reinterpreted. Importantly, the relation between the CD4+Foxp3 Tregs and other regulatory subsets of CD4+ T cells, including the CD25+ Th1 versus Th2 cells or Tr1 versus Tr3 cells, can now be approached experimentally in a variety of experimental systems and animal models. For example, it should be experimentally possible to sort out, in a population of CD4+CD25+ T regs, what suppressive functions are associate with CD25+ Th1 versus Th2 cells or Tr1 versus Tr3 cells and which suppressive functions are actually mediated by the "specialized" or "certified" CD4+Foxp3+ T regs. These studies more readily permit interpretation of the mechanisms of either positive or negative effects of CD4+CD25+ Tregs in adoptive transfer experiments (Horwitz et al., 2004). This is of clinical importance because clinical interventions employing adoptive transfer of CD4+CD25+ T regs, already effective in animal models, is being actively envisioned as immunosuppressive therapy in humans (Chai et al., 2005; Horwitz et al., 2004; Kelsen et al., 2005). In addition, it is important to point out that many other cell surface molecules that have been proposed to distinguish CD4+CD25+ Tregs, including CTLA-4, the glucocorticoid-induced tumor necrosis factor (TNF) receptor family-related gene (GITR), and CD45RO (Lafaille et al, 2002; Piccinillo and Thornton, 2004; Sakaguchi, 2000; Shevach, 2002), may now be correlated with Foxp3 expression.

Over the past several years numerous studies have sought to define the specificity and mechanism of suppression mediated by the CD4+CD25+ Tregs (Bach, 2003; Sakaguchi, 2000; Shevach et al., 2001; von Herrath and Harrison, 2003). For example, CD4+CD25+ Tregs are specifically induced by a wide variety of conventional self and foreign antigens to mediate suppression nonspecifically. Thus, CD4+CD25+ Tregs induced by antigen X suppress immune responses induced not only by antigen X but also those by antigens Y or Z. Similarly, CD4+CD25+ Tregs induced by antigen Y or Z nonspecifically suppress immune responses induced by antigen X or Y or Z (Shevach et al., 2001; Thornton and Shevach, 2000). Thus, although CD4+CD25+ T cells require antigen activation via their TCRs to become suppressive, once activated their suppressor effector function is completely nonspecific (Thornton and Shevach, 2000). Although some studies indicate that regulation of the "specialized CD4+ Tregs" is "antigen-specific" (Sakaguchi, 2004), it does not address "if and how" the regulation involves self/non-self discrimination (Jiang and Chess, 2006) or if the specificity is achieved by local release of cytokines at the site of inflammation.

The mechanism of suppression mediated by the CD4+CD25+ Tregs is unknown although both contact-dependent and lymphokine-dependent mechanisms have been proposed to be involved (Sakaguchi, 2004, Shevach, 2000). Interestingly, the cell surface molecules thought to be involved in the contact-dependent mechanism include CD80 (B7) on APCs and CTLA-4 molecules on Tregs, which are also known to be important in the intrinsic homeostatic regulatory mechanisms. Because of lack of the specificity in regulating the immune response, it has been surmised that the CD4+CD25+ Tregs, like the Th1, Th2, Tr1, and Tr3 cells, function predominantly to control the magnitude and class of the immune response (Cohn, 2004; von Herrath and Harrison, 2003).

In summary, although the "CD4+CD25+ Tregs" are specifically induced by a variety of self or foreign antigens during the afferent phase of the immune response, there is no evidence that the suppressor effector function mediated by these cells is antigen-specific. It remains intriguing that the suppression mediated by the CD4+ Tregs is, at least, in part, cell contact-dependent and presumably involves as yet unidentified cell surface receptors (Shevach, 2002). Clearly, identification of these contact-dependent molecules will greatly enhance our basic understanding of the precise mechanisms employed by the CD4+ Tregs to regulate immune responses and will give us greater insight into whether the CD4+ Tregs are involved in self/non-self discrimination. Taken together, although CD4+ Tregs are likely to be critically important in controlling the magnitude and class of immune responses (Cohn, 2004), the evidence that these cells are directly involved in self/non-self discrimination is uncertain.

## 3.3   CD8+ Suppressor T Cells Defined by MHC Class 1b (Qa-1 or HLA-E) Restricted TCRs

At some level immunoregulation by suppressor cells must ultimately deal with the fundamental issue of self/non-self discrimination and play a role in selectively inhibiting the outgrowth of potentially pathogenic self-reactive T cells while simultaneously preserving and/or facilitating the immune response to foreign antigens. The search for suppressor cell pathways that may serve this function has been a long sought goal. In this regard, during the past few years a pathway of immunoregulation mediated by the murine Qa-1 restricted CD8+ T cells has been delineated (Jiang and Chess, 2000, 2004a; Jiang et al., 1995, 1998, 2001, 2003, 2005). Emerging evidence indicates that the strategy employed by the CD8+ suppressor T cells to effect this goal involves "negative selection" of some but not all T cell clones in the periphery based on the recognition of surface surrogate structures expressed on certain T cells as a function of affinity/avidity of antigen activation (Jiang et al., 2005). In contrast to the CD4+CD25+ T cells, most of the CD8+ regulatory T cells utilize their MHC restricted αβ TCRs to recognize target cells during the effector phase of suppression. Interestingly, the MHC restriction element employed by the regulatory CD8+ T cells is the nonclassical MHC class Ib molecule: Qa-1 in mice and HLA-E in humans (Jiang and Chess, 2000; Jiang et al., 1995; Li et al., 2001; Ware et al., 1995).

The discovery of this pathway was initiated by studies showing that CD8+ T cells participate in vivo in the resistance to disease induced during the natural history of EAE, a well studied murine model of the human disease multiple sclerosis (Jiang et al., 1992; Koh et al., 1992). For example, EAE is induced by myelin proteins such as myelin basic protein (MBP), which activates encephalitogenic CD4+ Th1 cells. In the B10PL strain, the mice completely recover from the first episode of EAE and these recovered mice are highly resistant to reinitiation of EAE by secondary immunization. If these protected mice are then depleted of CD8+ T cells using monoclonal anti-CD8 antibodies, the protection is reversed and the mice develop clinical EAE upon reimmunization with MBP (Jiang et al., 1992). Furthermore, mice depleted of CD8+ T cells during the initial induction of EAE and allowed to recover normal levels of CD8+ T cells are not resistant and develop EAE again upon rechallenge with MBP. Thus, CD8+ T cells require priming during the first episode of EAE to regulate CD4+ T cells triggered by secondary MBP stimulation in vivo. Moreover, when CD8−/− "knockout" mice are bred with the EAE-susceptible PL/J strain, the CD8−/− mice develop more chronic EAE than the wild-type PL/J mice, reflected by a higher frequency of relapses (Koh et al., 1992). The experiments provide evidence that CD8+ T cells play a key role in both inducing resistance to autoimmune EAE and abrogating or suppressing recurrent relapsing episodes of pathogenic autoimmunity in vivo.

Importantly, the Qa-1 restricted CD8+ suppressor cells suppress disease activity by preferentially downregulating the pathogenic autoreactive T cell clones that induce disease. Moreover, adoptive transfer of suppressor cells or induction of the suppressor cells by T cell vaccination prevents disease. For example, Qa-1 expressing MBP-specific CD4+ T cells can serve as vaccine T cells and be used to induce regulatory CD8+ T cells in vivo. This T cell vaccination procedure protects animals from developing EAE, and the protection is abrogated by depletion of the Qa-1-restricted CD8+ T cells in vivo (Jiang et al., 2001; Panoutsakopoulou et al., 2004). These studies suggest that during the natural history of EAE in mice or during the remission of multiple sclerosis in humans CD8+ suppressor cells emerge that fine-tune the self-reactive TCR repertoire that is preferentially activated by MBP in vivo by selective downregulation of the potentially pathogenic self-reactive T cells (Jiang et al., 2003). The finding that CD8+ suppressor cells were restricted by the MHC Ib molecule, Qa-1 or HLA-E, was intriguing and suggested that these nonclassical MHC Ib molecules play a role in vivo in restricting the suppression mediated by CD8+ T cells. Indeed subsequent experiments in vivo showed that the suppression mediated by CD8+ T suppressor cells is blocked by antibodies to the MHC Ib Qa-1 molecule. These biological functions of the Qa-1-restricted CD8+ T cell pathway has been confirmed in recent experiments showing that molecularly engineered Qa-1-deficient (Qa-1−/−) mice develop severe neurological symptoms of EAE when exposed to the myelin-associated peptides and fail to develop resistance to EAE that normally develops in wild-type mice after immunization with myelin-associated antigens (Hu, 2004). Moreover, this failure of resistance to EAE in Qa-1−/− mice is associated with the escape of Qa-1-deficient CD4 cells from CD8+ T cell suppression, and resistance to disease is restored by reexpression of Qa-1 (Hu et al., 2004).

Recently, the Qa-1-dependent regulatory CD8 pathway has begun to be translated from mice to humans with the in vitro findings that human CD8+ T cells can be induced to differentiate into regulatory cells (Ware et al., 1995) whose function is dependent on HLA-E, the human homologue of Qa-1 (Jiang and Chess, 2000; Li et al., 2001). Moreover, HLA-E expressed on the surface of transfectants have been observed to bind a variety of endogenous self-peptides, including MHC Ia peptides, cytokine peptides, and heat shock peptides; and certain HLA-E/peptide complexes can induce HLA-E-restricted human CD8+ cytotoxic T lymphocytes

(CTLs). Taken together, the data provide evidence that an immunoregulatory pathway exists in humans that is analogous to the murine immunoregulatory pathway mediated by Qa-1-dependent CD8+ T cells.

To understand further the biology of suppression mediated by CD8+ T cells, the analysis of the regulation of the immune response to self-antigens such as MBP was extended to the analysis of the regulation of responses to conventional antigens (Jiang et al., 2005). Particularly revealing were studies of the immune response to the antigen Hen Egg Lysozyme (HEL) in both wild type (WT) and HEL transgenic (TG) animals (Jiang et al., 2005). Because HEL represents a foreign antigen in WT mice but a self-antigen in HEL TG mice, direct comparison of the immune response to HEL in these two types of mouse permits studies of the cellular and molecular mechanisms by which the CD8+ T cells regulate the immune response to both self and foreign antigen at a biological system level. First, it was found that the tolerance or unresponsiveness to HEL normally observed in TG mice is abrogated, in part, by in vivo treatment with monoclonal antibodies (mAbs) to either CD8 or Qa-1. These findings showed that the normal tolerance to self-antigens is not entirely a consequence of central negative selection in the thymus but is also dependent on regulatory CD8+ T cells in the periphery.

The studies of regulatory CD8+ T cells in WT mice, where HEL functions as a foreign antigen and elicits a vigorous secondary response to HEL characterized by increased overall affinity/avidity (affinity maturation), were quite surprising. Instead of suppressing the overall response to HEL, the CD8+ T cells augmented the affinity maturation of the T cell response. In fact, treatment of WT animals with mAbs to CD8 or Qa-1 abrogates the affinity maturation in vivo. On the surface, the findings that the regulatory Qa-1-dependent CD8+ T cells are involved in both the development of peripheral tolerance to self-antigen and the affinity maturation of T cells to foreign antigens seems paradoxical. However, the paradox is illuminated by the evidence that the strategy employed by the Qa-1-dependent CD8+ T cells to accomplish these tasks in vivo is to downregulate selectively the T cell clones that are of intermediate affinity/avidity to both self and foreign antigens. Thus, after testing a panel of HEL T cell clones for their susceptibility to downregulation by CD8+ T cells in vitro, it was observed that CD8+ T cells preferentially downregulate target clones of intermediate affinity/avidity and tend to exclude clones of high and low affinity/avidity from the downregulation.

# 4    Affinity/Avidity Model of Peripheral T Cell Regulation

These in vitro and in vivo studies have thus led to the formulation of an "affinity/avidity model of peripheral T cell regulation" in which the susceptibility of activated T cells to the downregulation by the CD8+ T cells is determined by the affinity/avidity interactions during the initial T cell activation (Jiang and Chess, 2000l Jiang et al., 2005). The model envisions that the immune system achieves the goal of self/non-self discrimination by selective downregulation of the intermediate affinity T cell clones specific to both self and foreign antigens via perceiving the affinity/avidity interactions between the TCRs on T cells, activated by any antigens, and MHC/MHC peptide complexes presented by APCs. The model is based on the notion that because the compositions of the naive peripheral TCR repertoires to self and foreign antigens are different owing to thymic negative selection, the biological consequences of selective downregulation of the intermediate affinity/avidity T cells to self and foreign antigens are also different. Intrathymic deletion of high affinity/avidity self-reactive T cell clones generates a truncated peripheral self-reactive repertoire that does not possess high affinity/avidity clones compared with the peripheral foreign-reactive repertoire. Therefore, by selectively downregulating T cells with intermediate affinity/avidity to self, which include the potentially pathogenic self-reactive T cells (Han et al., 2005; Jiang et al., 2003), the immune system controls peripheral autoimmune disease. On the other hand, the selective downregulation of T cells with intermediate affinity/avidity to foreign indirectly promotes the growth of T cells with high affinity/avidity to foreign antigens, which are essential for effective immunity to infectious pathogens (Jiang et al., 2005). Thus, by selectively downregulating intermediate affinity/avidity T cells to both self and foreign antigens, the immune system ensures peripheral self-tolerance and optimizes the capacity to mediate effective anti-infection immunity (Jiang et al., 2005). This forms the conceptual framework for a new paradigm to explain, at a biological system level, how the immune system achieves the goal of self non-self discrimination during the adaptive immune response without the need to distinguish self from non-self in the periphery at the level of T cell regulation.

This regulatory pathway on which the affinity/avidity model is based is composed of a series of sequential cellular events. It is initiated by activation of naive T cells during the primary immune response in which the TCR on T cells interacts with MHC/antigen peptide complexes presented by conventional APCs. One of the consequences of the initial T cell activation is the differential expression of a specific "target antigen," which, at least, includes a "Qa-1/self-peptide complex" on the surface of target T cells. Importantly, expression of the "target antigen," which is recognized by the TCRs on regulatory CD8+ T cells, is determined by the affinity/avidity during T cell activation, regardless of which antigen has triggered the target T cells. In this regard, because T cells are not professional APCs, the professional APCs (e.g., dendritic cells) may be recruited and function to provide co-stimulatory molecules during the induction phase of the regulatory CD8+ T cells. The "target antigen" expressed on certain activated T cells triggers the regulatory CD8+ T cells to differentiate into effector cells, which in turn downregulate any activated T cells expressing the same target antigen. At the present time, the only distinguishable phenotype between the Qa-1-dependent CD8+ T cells and the conventional CD8+ T cells is the usage of a distinct set of $\alpha\beta$ TCRs, which specifically recognize the Qa-1/self-peptide complex differentially expressed on susceptible target T cells. It is unlikely that Qa-1-dependent CD8+ T cells represent a functionally distinct lineage of CD8+ T cells. In this regard, like the conventional CD8+ T cells, the Qa-1-dependent regulatory CD8+

T cells function to mediate suppression by cell contact-dependent cytolysis and/or local cytokine release. Unlike the conventional CD8+ T cells that recognize viral or tumor antigens presented by MHC Ia molecules, the regulatory CD8+ T cells recognize self-peptide(s) presented by MHC class Ib molecules preferentially expressed on intermediate affinity/avidity T cells as a consequence of T cell activation. Finally, the Qa-1-dependent CD8+ T cell pathway may represent one example of how the immune system utilizes a unified mechanism of recognizing the consequence of immune responses—based on the affinity/avidity interactions of T cell activation—to regulate the adaptive immunity to both self and foreign antigens. The conceptual framework of the "affinity/avidity model" may also well be suited for other peripheral regulatory pathways. For example, at the present time the specificity of the regulation mediated by CD4+ Tregs is unclear. As a consequence, there is no conceptual framework in which to explain uniformly their function in vivo, which must involve selective suppression of some but not all immune responses (Bach, 2003; Cohn, 2004; Sakaguchi, 2000; Shevach, 2001; von Herrath and Harrison, 2003). Because the effector phase of suppression is mediated in part by contact-dependent cell surface interactions, it is conceivable that subsets of CD4+ Tregs may, like the Qa-1-dependent CD8+ T cells, preferentially downregulate intermediate affinity/avidity T cells but employ different recognition and effector mechanisms.

## *References*

Bach, J. F. (2003) Regulatory T cells under scrutiny. *Nat. Rev. Immunol.* 3(3):189-198.

Batchelor, J. R., Lombardi, G., Lechler, R. I. (1989) Speculations on the specificity of suppression. *Immunol. Today* 10(2):37-40.

Baxter, A. G., Kinder, S. J., Hammond, K. J., et al. (1997) Association between alphabetaTCR+CD4-CD8-T-cell deficiency and IDDM in NOD/Lt mice. *Diabetes* 46(4):572-582.

Bendelac, A. (1995) Mouse NK1+ T cells. *Curr. Opin. Immunol.* 7(3):367-374.

Bendelac, A., Lantz, O., Quimby, M. E., et al. (1995) CD1 recognition by mouse NK1+ T lymphocytes. *Science* 268(5212):863-865.

Bendelac, A., Rivera, M. N., Park, S. H., Roark, J. H. (1997) Mouse CD1-specific NK1 T cells: development, specificity, and function. *Annu. Rev. Immunol.* 15:535-562.

Bennett, C. L., Christie, J., Ramsdell, F., et al. (2001) The immune dysregulation, polyendocrinopathy, enteropathy, X-linked syndrome (IPEX) is caused by mutations of FOXP3. *Nat. Genet.* 27(1):20-21.

Bevan, M. J. (1977) In a radiation chimaera, host H-2 antigens determine immune responsiveness of donor cytotoxic cells. *Nature* 269(5627):417-418.

Bevan, M. J., Fink, P. J. (1978) The influence of thymus H-2 antigens on the specificity of maturing killer and helper cells. *Immunol. Rev.* 1978;42:3-19.

Bloom, B. R., Salgame, P., Diamond, B. (1992) Revisiting and revising suppressor T cells. *Immunol. Today* 13(4):131-136.

Bouneaud, C., Kourilsky, P., Bousso, P. (2000) Impact of negative selection on the T cell repertoire reactive to a self-peptide: a large fraction of T cell clones escapes clonal deletion. *Immunity* 13(6):829-840.

Boussiotis, V. A., Freeman, G. J., Gribben, J. G., Nadler, L. M. (1996) The role of B7-1/B7-2:CD28/CLTA-4 pathways in the prevention of anergy, induction of productive immunity and down-regulation of the immune response. *Immunol. Rev.* 153:5-26.

Brunkow, M. E., Jeffery, E. W., Hjerrild, K. A., et al. (2001) Disruption of a new fork-head/winged-helix protein, scurfin, results in the fatal lymphoproliferative disorder of the scurfy mouse. *Nat. Genet.* 27(1):68-73.

Caux, C., Massacrier, C., Vanbervliet, B., et al. (1994) Activation of human dendritic cells through CD40 cross-linking. *J. Exp. Med.* 180(4):1263-1272.

Chai, J. G., Xue, S. A., Coe, D., et al. (2005) Regulatory T cells, derived from naive CD4+CD25− T cells by in vitro Foxp3 gene transfer, can induce transplantation toler-ance. *Transplantation* 79(10):1310-1316.

Chambers, C. A., Kuhns, M. S., Egen, J. G., Allison, J. P. (2001) CTLA-4-mediated inhi-bition in regulation of T cell responses: mechanisms and manipulation in tumor immunotherapy. *Annu. Rev. Immunol* 19:565-594.

Charlton, B., Lafferty, K. J. (1995) The Th1/Th2 balance in autoimmunity. *Curr. Opin. Immunol.* 7(6):793-798.

Chen, W., Jin, W., Hardegen, N., et al. (2003) Conversion of peripheral CD4+CD25− naive T cells to CD4+CD25+ regulatory T cells by TGF-beta induction of transcription factor Foxp3. *J. Exp. Med.* 198(12):1875-1886.

Coffman, R. L., Mosmann, T. R. (1991) CD4+ T-cell subsets: regulation of differentiation and function. *Res. Immunol.* 142(1):7-9.

Cohen, I. R., Quintana, F. J., Mimran, A. (2004) Tregs in T cell vaccination: exploring the regulation of regulation. *J. Clin. Invest.* 114(9):1227-1232.

Cohn, M. (2004) Whither T-suppressors: if they didn't exist would we have to invent them? *Cell. Immunol.* 227(2):81-92.

Cosmi, L., Liotta, F., Lazzeri, E., et al. (2003) Human CD8+CD25+ thymocytes share phenotypic and functional features with CD4+CD25+ regulatory thymocytes. *Blood* 102(12):4107-4114.

Cui, J., Shin, T., Kawano, T., et al. (1997) Requirement for Valpha14 NKT cells in IL-12-mediated rejection of tumors. *Science* 278(5343):1623-1626.

Curotto de Lafaille, M. A., Lafaille, J. J. (2002) CD4(+) regulatory T cells in autoimmu-nity and allergy. *Curr. Opin. Immunol.* 14(6):771-778.

Davis, M. M., Boniface, J. J., Reich, Z., et al. (1998) Ligand recognition by alpha beta T cell receptors. *Annu. Rev. Immunol.* 16:523-544.

Del Prete, G. (1998) The concept of type-1 and type-2 helper T cells and their cytokines in humans. *Int. Rev. Immunol.* 16(3-4):427-455.

D'Orazio, T. J., Niederkorn, J. Y. (1998) A novel role for TGF-beta and IL-10 in the induction of immune privilege. *J. Immunol.* 160(5):2089-2098.

Dorf, M. E., Benacerraf, B. (1984) Suppressor cells and immunoregulation. *Annu. Rev. Immunol.* 2:127-158.

Durie, F. H., Foy, T. M., Masters, S. R., et al. (1994) The role of CD40 in the regulation of humoral and cell-mediated immunity. *Immunol. Today* 15(9):406-411.

Falcone, M., Yeung, B., Tucker, L., et al. (1999) A defect in interleukin 12-induced acti-vation and interferon gamma secretion of peripheral natural killer T cells in nonobese diabetic mice suggests new pathogenic mechanisms for insulin-dependent diabetes mellitus. *J. Exp. Med.* 190(7):963-972.

Fitch, F. W., McKisic, M. D., Lancki, D. W., Gajewski, T. F. (1993) Differential regulation of murine T lymphocyte subsets. *Annu. Rev. Immunol.* 11:29-48.

Fontenot, J. D., Gavin, M. A., Rudensky, A. Y. (2003) Foxp3 programs the development and function of CD4+CD25+ regulatory T cells. *Nat. Immunol.* 4(4):330-336.

Fontenot, J. D., Rasmussen, J. P., Williams, L. M., et al. (2005) Regulatory T cell lineage specification by the forkhead transcription factor foxp3. *Immunity* 22(3):329-341.

Foy, T. M., Aruffo, A., Bajorath, J., et al. (1996) Immune regulation by CD40 and its ligand GP39. *Annu. Rev. Immunol.* 14:591-617.

Frey, A. B., Rao, T. D. (1999) NKT cell cytokine imbalance in murine diabetes mellitus. *Autoimmunity* 29(3):201-214.

Furlan, R., Bergami, A., Cantarella, D., et al. (2003) Activation of invariant NKT cells by alphaGalCer administration protects mice from MOG35-55-induced EAE: critical roles for administration route and IFN-gamma. *Eur. J. Immunol.* 33(7):1830-1838.

Furtado, G. C., Olivares-Villagomez, D., Curotto de Lafaille, M. A., et al. (2001) Regulatory T cells in spontaneous autoimmune encephalomyelitis. *Immunol. Rev.* 182:122-134.

Gammon, G., Sercarz, E. (1990) Does the presence of self-reactive T cells indicate the breakdown of tolerance? *Clin. Immunol. Immunopathol.* 56(3):287-297.

Garcia, K. C., Degano, M., Pease, L. R., et al. (1998) Structural basis of plasticity in T cell receptor recognition of a self peptide-MHC antigen. *Science* 279(5354):1166-1172.

Godfrey, D. I., Hammond, K. J., Poulton, L. D., et al. (2000) NKT cells: facts, functions and fallacies. *Immunol. Today* 21(11):573-583.

Gombert, J. M., Herbelin, A., Tancrede-Bohin, E., et al. (1996) Early quantitative and functional deficiency of NK1+-like thymocytes in the NOD mouse. *Eur. J. Immunol.* 26(12):2989-2998.

Green, D. R., Flood, P. M., Gershon, R. K. (1983) Immunoregulatory T-cell pathways. *Annu. Rev. Immunol.* 1:439-463.

Groux, H., O'Garra. A., Bigler, M., et al. (1997) A CD4+ T-cell subset inhibits antigen-specific T-cell responses and prevents colitis. *Nature* 389(6652):737-742.

Hammond, K. J., Poulton, L. D., Palmisano, L. J., et al. (1998) Alpha/beta-T cell receptor (TCR)+CD4-CD8-(NKT) thymocytes prevent insulin-dependent diabetes mellitus in nonobese diabetic (NOD)/Lt mice by the influence of interleukin (IL)-4 and/or IL-10. *J. Exp. Med.* 187(7):1047-1056.

Han, B., Serra, P., Yamanouchi, J., et al. (2005) Developmental control of CD8 T cell-avidity maturation in autoimmune diabetes. *J. Clin. Invest.* 115(7):1879-1887.

Hengartner, H., Odermatt, B., Schneider, R., et al. (1988) Deletion of self-reactive T cells before entry into the thymus medulla. *Nature* 336(6197):388-390.

Horwitz, D. A., Zheng, S. G., Gray, J. D., et al. (2004) Regulatory T cells generated ex vivo as an approach for the therapy of autoimmune disease. *Semin. Immunol.* 16(2):135-143.

Hu, D., Ikizawa, K., Lu, L., et al. (2004) Analysis of regulatory CD8 T cells in Qa-1-deficient mice. *Nat. Immunol.* 5(5):516-523.

Janeway, C. A., Jr. (1988) Do suppressor T cells exist? A reply [editorial]. *Scand. J. Immunol.* 27(6):621-623.

Janeway, C. A., Jr., Medzhitov, R. (2002) Innate immune recognition. *Annu. Rev. Immunol.* 20:197-216.

Jenkins, M. K. (1994) The ups and downs of T cell costimulation. *Immunity* 1(6):443-446.

Jerne, NK. (1974) The immune system: a web of V-domains. *Harvey Lect.* 70(Series):93-110.

Jiang, H., Chess, L. (2000) The Specific Regulation Of Immune Responses by CD8+ T Cells Restricted by the MHC Class IB Molecule, QA-1. *Annu. Rev. Immunol.* 18:185-216.

Jiang, H., Chess, L. (2004a) An integrated model of immunoregulation mediated by regulatory T cell subsets. *Adv. Immunol.* 83:253-288.

Jiang, H., Chess, L. (2004b) An integrated view of suppressor T cell subsets in immunoregulation. *J. Clin. Invest.* 114(9):1198-1208.

Jiang, H., Chess, L. (2006) Regulation of immune responses by T cells. *N. Engl. J. Med.* 354(11):1166-1176.

Jiang, H., Zhang, S. I., Pernis, B. (1992) Role of CD8+ T cells in murine experimental allergic encephalomyelitis. *Science* 256(5060):1213-1215.

Jiang, H., Ware, R., Stall, A., et al. (1995) Murine CD8+ T cells that specifically delete autologous CD4+ T cells expressing V beta 8 TCR: a role of the Qa-1 molecule. *Immunity* 2(2):185-194.

Jiang, H., Kashleva, H., Xu, L. X., et al. (1998) T cell vaccination induces T cell receptor Vbeta-specific Qa-1-restricted regulatory CD8(+) T cells. *Proc. Natl. Acad. Sci. U S A* 95(8):4533-4537.

Jiang, H., Braunstein, N. S., Yu, B., et al. (2001) CD8+ T cells control the TH phenotype of MBP-reactive CD4+ T cells in EAE mice. *Proc. Natl. Acad. Sci. U S A* 98(11): 6301-6306.

Jiang, H., Curran, S., Ruiz-Vazquez, E., et al. (2003) Regulatory CD8+ T cells fine-tune the myelin basic protein-reactive T cell receptor V beta repertoire during experimental autoimmune encephalomyelitis. *Proc. Natl. Acad. Sci. U S A* 100(14):8378-8383.

Jiang, H., Wu, Y., Liang, B., et al. (2005) An affinity/avidity model of peripheral T cell regulation. *J. Clin. Invest.* 115(2):302-312.

Kappler, J. W., Roehm, N., Marrack, P. (1987) T cell tolerance by clonal elimination in the thymus. *Cell* 49(2):273-280.

Kelsen, J., Agnholt, J., Hoffmann, H. J., et al. (2005) FoxP3(+)CD4(+)CD25(+) T cells with regulatory properties can be cultured from colonic mucosa of patients with Crohn's disease. *Clin. Exp. Immunol.* 141(3):549-557.

Khattri, R., Cox, T., Yasayko, S. A., Ramsdell, F. (2003) An essential role for Scurfin in CD4+CD25+ T regulatory cells. *Nat. Immunol.* 4(4):337-342.

Klaus, S. J., Pinchuk, L. M., Ochs, H. D., et al. (1994) Costimulation through CD28 enhances T cell-dependent B cell activation via CD40-CD40L interaction. *J. Immunol.* 152(12):5643-5652.

Koh, D-R., Fung-Leung, W-P., Ho, A., et al. (1992) Less mortality but more relapses in experimental allergic encephalomyelitis in CD8−/− mice. *Science* 256:1210-1213.

Koulova, L., Clark, E. A., Shu, G., Dupont, B. (1991) The CD28 ligand B7/BB1 provides costimulatory signal for alloactivation of CD4+ T cells. *J. Exp. Med.* 173(3):759-762.

Kuchroo, V. K., Anderson, A. C., Waldner, H., et al. (2002) T cell response in experimental autoimmune encephalomyelitis (EAE): role of self and cross-reactive antigens in shaping, tuning, and regulating the autopathogenic T cell repertoire. *Annu. Rev. Immunol.* 20:101-123.

Lafaille, J. J., Nagashima, K., Katsuki, M., Tonegawa, S. (1994) High incidence of spontaneous autoimmune encephalomyelitis in immunodeficient anti-myelin basic protein T cell receptor transgenic mice. *Cell* 78(3):399-408.

Lederman, S., Yellin, M. J., Krichevsky, A., et al. (1992) Identification of a novel surface protein on activated CD4+ T cells that induces contact-dependent B cell differentiation (help). *J. Exp. Med.* 175(4):1091-1101.

Lee, P. T., Putnam, A., Benlagha, K., et al. (2002) Testing the NKT cell hypothesis of human IDDM pathogenesis. *J. Clin. Invest.* 110(6):793-800.

Lenardo, M., Chan, K. M., Hornung, F., et al. (1999) Mature T lymphocyte apoptosis— immune regulation in a dynamic and unpredictable antigenic environment. *Annu. Rev. Immunol.* 17:221-253.

Lenschow, D. J., Walunas, T. L., Bluestone, J. A. (1996) CD28/B7 system of T cell costimulation. *Annu. Rev. Immunol.* 14:233-258.

Levings, M. K., Roncarolo, M. G. (2000) T-regulatory 1 cells: a novel subset of CD4 T cells with immunoregulatory properties. *J. Allergy Clin. Immunol.* 106(1 Pt 2): S109-S112.

Li, J., Goldstein, I., Glickman-Nir, E., et al. (2001) Induction of TCR Vbeta-specific CD8+ CTLs by TCR Vbeta-derived peptides bound to HLA-E. *J. Immunol.* 167(7):3800-3808.

Lohse, A. W., Mor, F., Karin, N., Cohen, I. R. (1989) Control of experimental autoimmune encephalomyelitis by T cells responding to activated T cells. *Science* 244(4906):820-822.

Manavalan, J. S., Kim-Schulze, S., Scotto, L., et al. (2004) Alloantigen specific CD8+CD28– FOXP3+ T suppressor cells induce ILT3+ ILT4+ tolerogenic endothelial cells, inhibiting alloreactivity. *Int. Immunol.* 16(8):1055-1068.

Mason, D. (1998) A very high level of crossreactivity is an essential feature of the T-cell receptor. *Immunol. Today* 19:395–404.

Mattner, J., Debord, K. L., Ismail, N., et al. (2005) Exogenous and endogenous glycolipid antigens activate NKT cells during microbial infections. *Nature* 434(7032):525-529.

Medzhitov, R., Janeway, C., Jr. (2000) Innate immunity. *N. Engl. J. Med.* 343(5):338.

Medzhitov, R., Janeway, C. A., Jr. (2002) Decoding the patterns of self and nonself by the innate immune system. *Science* 296(5566):298-300.

Moller, G. (1988) Do suppressor T cells exist? *Scand. J. Immunol.* 27(3):247-250.

Mosmann, T. R., Cherwinski, H., Bond, M. W., et al. (1986) Two types of murine helper T cell clone. I. Definition according to profiles of lymphokine activities and secreted proteins. *J. Immunol.* 136(7):2348-2357.

Mosmann, T. R., Coffman, R. L. (1989) TH1 and TH2 cells: different patterns of lymphokine secretion lead to different functional properties. *Annu. Rev. Immunol.* 7:145-173.

Nossal, G. J. (1994) Negative selection of lymphocytes. *Cell* 76(2):229-239.

O'Garra, A., Vieira, P. (2004) Regulatory T cells and mechanisms of immune system control. *Nat. Med.* 10(8):801-805.

Panoutsakopoulou, V., Huster, K. M., McCarty, N., et al. (2004) Suppression of autoimmune disease after vaccination with autoreactive T cells that express Qa-1 peptide complexes. *J. Clin. Invest.* 113:1218-1224.

Pasare, C., Medzhitov, R. (2003) Toll pathway-dependent blockade of CD4+CD25+ T cell-mediated suppression by dendritic cells. *Science* 299(5609):1033-1036.

Piccirillo, C. A., Thornton, A. M. (2004) Cornerstone of peripheral tolerance: naturally occurring CD4+CD25+ regulatory T cells. *Trends Immunol.* 25(7):374-380.

Roncarolo, M. G., Levings, M. K. (2000) The role of different subsets of T regulatory cells in controlling autoimmunity. *Curr. Opin. Immunol.* 12(6):676-683.

Rudensky, A. (2005) Foxp3 and dominant tolerance. *Philos. Trans. R. Soc. Lond. B Biol. Sci.* 360(1461):1645-1646.

Sakaguchi, S. (2000) Regulatory T cells: key controllers of immunologic self-tolerance. *Cell* 101(5):455 458.

Sakaguchi, S. (2004) Naturally arising CD4+ regulatory t cells for immunologic self-tolerance and negative control of immune responses. *Annu. Rev. Immunol.* 22:531-562

Sakaguchi, S., Fukuma, K., Kuribayashi, K., Masuda, T. (1985) Organ-specific autoimmune diseases induced in mice by elimination of T cell subset. I. Evidence for the active participation of T cells in natural self-tolerance; deficit of a T cell subset as a possible cause of autoimmune disease. *J. Exp. Med.* 161(1):72-87.

Sakaguchi, S., Sakaguchi, N., Asano, M., et al. (1995) Immunologic self-tolerance maintained by activated T cells expressing IL-2 receptor alpha-chains (CD25): breakdown of a single mechanism of self-tolerance causes various autoimmune diseases. *J. Immunol.* 155(3):1151-1164.

Sandberg, J. K., Franksson, L., Sundback, J., et al. (2000) T cell tolerance based on avidity thresholds rather than complete deletion allows maintenance of maximal repertoire diversity. *J. Immunol.* 165(1):25-33.

Savage, P. A., Boniface, J. J., Davis, M. M. (1999) A kinetic basis for T cell receptor repertoire selection during an immune response. *Immunity* 10(4):485-492.

Schubert, L. A., Jeffery, E., Zhang, Y., et al. (2001) Scurfin (FOXP3) acts as a repressor of transcription and regulates T cell activation. *J. Biol. Chem.* 276(40):37672-37679.

Seder, R. A., Paul, W. E. (1994) Acquisition of lymphokine-producing phenotype by CD4+ T cells. *Annu. Rev. Immunol.* 12:635-673.

Sharif, S., Arreaza, G. A., Zucker, P., et al. (2001) Activation of natural killer T cells by alpha-galactosylceramide treatment prevents the onset and recurrence of autoimmune type 1 diabetes. *Nat. Med.* 7(9):1057-1062.

Sharif, S., Arreaza, G. A., Zucker, P., Delovitch, T. L. (2002) Regulatory natural killer T cells protect against spontaneous and recurrent type 1 diabetes. *Ann. N.Y. Acad. Sci.* 958:77-88.

Shevach, E. M. (2000) Regulatory T cells in autoimmmunity. *Annu. Rev. Immunol.* 18: 423-449.

Shevach, E. M. (2001) Certified professionals: CD4(+)CD25(+) suppressor T cells. *J. Exp. Med.* 193(11):F41-F46.

Shevach, E. M. (2002) CD4+ CD25+ suppressor T cells: more questions than answers. *Nat. Rev. Immunol.* 2(6):389-400.

Shevach, E. M., McHugh, R. S., Piccirillo, C. A., Thornton, A. M. (2001) Control of T-cell activation by CD4+ CD25+ suppressor T cells. *Immunol. Rev.* 182:58-67.

Shortman, K., Egerton, M., Spangrude, G. J., Scollay, R. (1990) The generation and fate of thymocytes. *Semin. Immunol.* 2(1):3-12.

Singh, A. K., Wilson, M. T., Hong, S., et al. (2001) Natural killer T cell activation protects mice against experimental autoimmune encephalomyelitis. *J. Exp. Med.* 194(12):1801-1811.

Sprent, J., Webb, S. R. (1995) Intrathymic and extrathymic clonal deletion of T cells. *Curr. Opin. Immunol.* 7(2):196-205.

Thornton, A. M., Shevach, E. M. (2000) Suppressor effector function of CD4+CD25+ immunoregulatory T cells is antigen nonspecific. *J. Immunol.* 2000;164(1):183-90.

Von Boehmer, H., Kisielow, P. (1990) Self-nonself discrimination by T cells. *Science* 248(4961):1369-1373.

Von Herrath, M. G., Harrison, L. C. (2003) Antigen-induced regulatory T cells in autoimmunity. *Nat. Rev. Immunol.* 3(3):223-232.

Waldmann, H. (1978) The influence of the major histocompatibility complex on the function of T-helper cells in antibody formation. *Immunol. Rev.* 42:202-223.

Walker, M. R., Kasprowicz, D.J., Gersuk, V.H., et al. (2003) Induction of FoxP3 and acquisition of T regulatory activity by stimulated human CD4+CD25− T cells. *J. Clin. Invest.* 112(9):1437-1443.

Ware, R., Jiang, H., Braunstein, N., et al. (1995) Human CD8+ T lymphocyte clones specific for T cell receptor V beta families expressed on autologous CD4+ T cells. *Immunity* 2(2):177-184.

Wildin, R. S., Ramsdell, F., Peake, J., et al. (2001) X-linked neonatal diabetes mellitus, enteropathy and endocrinopathy syndrome is the human equivalent of mouse scurfy. *Nat. Genet.* 27(1):18-20.

Wilson, S. B., Kent, S. C., Patton, K. T., et al. (1998) Extreme Th1 bias of invariant Valpha24JalphaQ T cells in type 1 diabetes. *Nature* 391(6663):177-181.

Zelenay, S., Lopes-Carvalho, T., Caramalho, I., et al. (2005) Foxp3+ CD25– CD4 T cells constitute a reservoir of committed regulatory cells that regain CD25 expression upon homeostatic expansion. *Proc. Natl. Acad. Sci. U S A* 102(11):4091-4096.

Zinkernagel, R. M. (1978) Thymus and lymphohemopoietic cells: their role in T cell maturation in selection of T cells' H-2-restriction-specificity and in H-2 linked Ir gene control. *Immunol. Rev.* 42:224-270.

# 4
# Anti-Ergotypic Regulation of the Immune Response

Francisco J. Quintana and Irun R. Cohen

## 1   Introduction

The mammalian immune system is characterized by the generation of a large repertoire of B and T cell receptors (BCRs and TCRs, respectively) through a complex combinatorial process (Jung and Alt, 2004). This combinatorial process, which is responsible for the generation of BCR and TCR diversity, also leads to the generation of self-reactive clones that could potentially trigger autoimmune disorders. Many of these autoreactive clones are removed from the mature immune repertoire by negative selection (Kyewski and Klein, 2006), but many self-reactive clones are positively selected and populate the healthy repertoire (Quintana and Weiner, 2006). Indeed, it has been proposed that natural autoimmunity to certain self-antigens has a physiological function (Cohen, 1992; Cohen and Young, 1991). In any case, autoimmunity, whether generated by design or accident, must still be kept under the control of the regulatory activities of the various cell types. Moreover, it is reasonable to suppose that even immunity to foreign antigens must be controlled.

Based on their cell surface markers and their profile of cytokine secretion, several cell populations with regulatory activities have been identified: CD4+CD25+ T cells (Sakaguchi, 2004), Th3 cells (Chen et al., 1994; Weiner et al., 1994), Tr1 cells (Groux et al., 1997), Qa-1-restricted CD8+ T cells (Sarantopoulos et al., 2004), and natural killer (NK) cells (Kronenberg, 2005) among others. In addition, the study of T cell vaccination (TCV) as a method to treat autoimmune disorders (Ben-Nun et al., 1981) has led to the identification of two regulatory populations (Cohen, 2001): anti-idiotypic (Lider et al., 1988) and anti-ergotypic T cells (Lohse et al., 1989). Anti-idiotypic responses are directed against antigen determinants that are clone-specific, exemplified by the CDR3 region of the TCR. Hence, the anti-idiotypic response directed against a myelin basic protein (MBP)-specific clone does not cross-react with a clone that carries a different TCR and therefore shows different antigen specificity (Lider et al., 1988). The anti-ergotypic response, in contrast, recognizes antigenic determinants derived from activation markers (ergotopes), such as the CD25 molecule, which is upregulated by activated T cells (*ergon* means work or activity in Greek). Thus, the anti-ergotypic response targets syngeneic activated, but not resting, T cells regardless of their specificity. In this chapter we analyze the characteristics of anti-ergotypic regulation (Lohse et al., 1989).

## 2  Anti-Ergotypic Response

The anti-ergotypic response can be analyzed based on its three basic components: the target T cell, the ergotope, and the anti-ergotypic T cell.

### 2.1  Target T Cells

To participate in regulatory anti-ergotypic (or anti-idiotypic) responses, a T cell must be capable of processing and presenting to anti-ergotypic regulators its ergotopes, usually complexed to major histocompatibility complex class I (MHC I) and/or MHC II molecules. Human and rat T cells express both MHC I and MHC II molecules (Indiveri et al., 1980) and the co-stimulatory molecules CD80 and CD86 (Azuma et al., 1993). Therefore they can activate MHC-restricted anti-ergotypic (and anti-idiotypic) CD8+ or CD4+ T cell responses. Mouse T cells, however, do not express MHC II, but they do express the nonclassical MHC I-Qa molecule. MHC I-Qa is known to mediate T–T regulatory interactions (Jiang and Chess, 2000); therefore MHC I-Qa might mediate anti-ergotypic regulation in the mouse. Notably, surface expression of MHC II, CD80, and CD86 is upregulated upon TCR-triggered activation (Azuma et al., 1993; Indiveri et al., 1980); activated T cells are better antigen-presenting cells (APCs) and make better T cell vaccines (Ben-Nun et al., 1981). The increased APC function of activated T cells suggests that the participation in anti-ergotypic regulatory networks is part of the T cell activation program, probably to guarantee tight control of the T cell response induced to all antigens—foreign and self.

Thus, the ability of the target T cell to process self-peptides and efficiently present them in MHC I and MHC II molecules conditions its control by anti-ergotypic regulators and shapes the nature of the anti-ergotypic response, irrespective of whether it is mediated by CD4+ or CD8+ cells. Of note, although many anti-ergotypic regulators characterized so far are MHC I- or MHC II-restricted, non-MHC-restricted anti-ergotypic T cells have also been described, most of them displaying a TCRγδ+ phenotype (see below).

## 2.2   Ergotope

By definition, an ergotope is an activation marker, a molecule whose level of expression—and ultimately its presentation to anti-ergotypic regulators—is upregulated during the course of T cell activation. Two types of ergotope have been identified so far: T cell-restricted ergotopes and T cell-shared ergotopes.

### 2.2.1   T Cell-Restricted Ergotopes

The $\alpha$-chain of the interleukin-2 (IL-2) receptor (CD25) is an example of a T cell-restricted ergotope because CD25 is expressed only by T cells. Anti-ergotypic T cell responses to CD25 are well documented. Mor et al. showed that vaccination of rats with immunogenic peptides derived from the CD25 sequence inhibits the subsequent induction of experimental autoimmune encephalomyelitis (EAE) triggered with gpMBP (Mor et al., 1996). Moreover, a DNA vaccine coding for the full-length CD25 could inhibit the development of adjuvant arthritis (AA) (Mimran et al., 2004). Control peptides or DNA vaccines derived from the CD132 molecule (the $\gamma$–subunit of the IL-2 receptor whose expression levels are not affected by the state of activation of the T cell) did not have a significant effect on EAE or AA progression (Mimran et al., 2004; Mor et al., 1996). The importance of CD25 as an ergotope has been recently highlighted by the isolation of CD25-specific anti-ergotypic regulatory T cells from multiple sclerosis (MS) patients treated by TCV (Hong et al., 2006).

### 2.2.2   Shared Ergotopes

Data suggest that the 60-kDa heat shock protein (HSP60) is an ergotope targeted by regulatory T–T interactions (Cohen et al., 2004). HSP60 is an endogenous immunomodulatory molecule that exerts a diverse array of effects on the innate and adaptive arms of the immune system (Quintana and Cohen, 2005; van Eden et al., 2005). The detection of HSP60-reactive T cells and antibodies in healthy individuals and the demonstration that immunity to HSP60 might participate in the pathogenesis and control of arthritis, diabetes, and Behçet's disease, among other diseases, highlights its importance for immune homeostasis (Quintana and Cohen, 2005; van Eden et al., 2005). However, more relevant for the present discussion is the finding that HSP60 is upregulated by activated T cells (Cohen et al., unpublished data). HSP60-derived epitopes are processed and presented by activated T cells, which can therefore stimulate syngenic HSP60-specific T cells (unpublished data). However, HSP60 expression is not restricted to T cells; HSP60 is expressed by many cell types under stress (Macario and Conway de Macario, 2005). Thus, HSP60 constitutes an example of a molecule that is widely expressed during the stress response but can participate in anti-ergotypic regulation only when expressed by activated T cells, whose augmented APC function allows it to trigger anti-ergotypic regulators. Note that under certain conditions HSP60 might localize to the cell membrane (Pfister et al., 2005). Taken together with the non-MHC-restricted reactivity of TCRγδ+ cells to HSP60, surface

HSP60 might also account for the activation of at least some TCRγδ+ anti-ergotypic regulators.

Shared ergotopes also provide the vehicle for the extension of anti-ergotypic regulation to non-T cell targets. B cells, for example, upregulate their HSP60 levels upon activation. Thus, HSP60 anti-ergotypic regulators could target activated B cells to control their expansion or function.

All in all, and regardless of whether a particular molecule is expressed only by T cells or enjoys a more promiscuous pattern of expression, any self-antigen can participate in anti-ergotypic regulatory interactions so long as it fulfills the three following criteria: (1) The expression levels of the candidate ergotope are upregulated in activated T cells. (2) The candidate ergotope is presented on the surface of the target T cells, either intact or processed and complexed to MHC molecules. (3) Regulatory T cells recognize and can be activated by the putative ergotope.

## 2.3   Anti-Ergotypic T Cells

The anti-ergotypic response can be mediated by a heterogeneous group of cells that can change their composition according to the immune state of the individual.

### 2.3.1   Anti-Ergotypic T Cell Responses in Nonimmunized Individuals

Anti-ergotypic responses are detectable in the thymus of 1-day-old rats, suggesting that their generation is independent of antigen priming (Mimran et al., 2005). The anti-ergotypic T cells detectable in newborn rats bear a CD8+ phenotype and include both TCRαβ+ and TCRγδ+ T cells. These two populations differ in terms of their profile of cytokine secretion and MHC and their co-stimulation requirements for activation: TCRαβ+ anti-ergotypic cells proliferated but did not secret detectable cytokines, whereas TCRγδ+ anti-ergotypic cells secreted interferon-γ (IFNγ) and tumor necrosis factor-α (TNFα) in response to activated T cells (Mimran et al., 2005). The response of the TCRαβ+ CD8+ anti-ergotypic T cells was MHC I-restricted and B7-CD28-dependent; the response of the TCRγδ+ anti-ergotypic T cells was B7-CD28-dependent but was not inhibited by antibodies to classical MHC I or MHC II molecules (Mimran et al., 2005).

Anti-ergotypic CD8+ T cells can also be isolated from healthy humans. Of note, these CD8+ anti-ergotypic T cells could recognize antigen-activated but not phytohemagglutinin-activated autologous CD4+ T cells (Correale et al., 1997). CD8+ TCRαβ+ and CD8+ TCRγδ+ T cells differed in their MHC requirements and pattern of cytokine secretion: CD8+ TCRαβ+ cells were MHC I-restricted and secreted IFNγ, TNFα/β, and TGFβ, whereas CD8+ TCRγδ+ cells were not MHC I-restricted and secreted IFNγ and TNFα/β but not TGFβ (Correale et al., 1997). Thus, the anti-ergotypic response in nonimmunized individuals is driven by CD8+ TCRαβ+ and TCRγδ+ CD8+ or CD4+ T cells; the TCRγδ+ cell compartment recognizes its target T cells by a non-MHC-mediated mechanism. The presence of anti-ergotypic T cell reactivity in nonimmunized subjects suggests that this type of regulation is of importance for controlling the healthy immune system under physiological conditions.

However, one must keep in mind that nonimmunized healthy rats and humans are not immunologically naive as they are under the constant stimulation of environmental antigens and commensal microbes.

### 2.3.2  Anti-Ergotypic Responses in Immunized Individuals

The vaccination of rats with complete Freund's adjuvant (CFA) triggers strong anti-ergotyic reactivity (Lohse et al., 1993). This immunization-induced anti-ergotypic response probably results from a process of self-vaccination with the T cell clones expanded by the CFA. As we already mentioned, T cell activation leads to the acquisition of APC function by the activated T cells (Cohen et al., 2004). Thus the documentation of anti-ergotypic responses following vaccination with a strong immunogen suggests that anti-ergotypic regulators participate in the resolution phase of an immune response, thereby limiting hyperimmune pathology. Indeed, the anti-ergotypic T cells of naive rats secrete IFNγ, a cytokine that has been reported to play two complementary roles in T cell immunity: It initially boosts induction of the T cell response and later triggers its resolution (Pfister et al., 2005).

### 2.3.3  Anti-Ergotypic Responses Triggered by Therapeutic Vaccination

Several studies suggest that that the anti-ergotypic response is depressed in autoimmune disorders. Rats at the peak of the experimental autoimmune disease AA show decreased anti-ergotypic responses (Mimran et al., 2004). Similarly, patients suffering from MS showed decreased proliferation to autologous activated T cells (Hafler et al., 1985). Immunization regimens aimed at strengthening natural anti-ergotypic regulatory networks are therefore expected to be beneficial in the control of autoimmune disorders. Indeed, vaccination with activated T cells of a nonrelevant specificity (Lohse et al., 1989, 1993) or with defined ergotopes (Mimran et al., 2004; Mor et al., 1996) have been shown to control the experimental autoimmune diseases AA and EAE. DNA vaccination with CD25 led to the induction of CD4+ and CD8+ TCRαβ+ and TCRγδ+ anti-ergotypic T cells that secreted IL-10 but not IFNγ; in contrast, the anti-ergotypic regulators detected in nonimmunized rats secrete IFNγ and TNFα but not IL-10 (Mimran et al., 2004). The TCRαβ+ anti-ergotypic regulators were MHC I- and MHC II-restricted, but the anti-ergotypic interactions mediated by TCRγδ+ cells were MHC-independent (Mimran et al., 2004).

Most of the human anti-ergotypic responses characterized so far were part of studies designed to study the effectiveness of TCV for the treatment of MS. TCV seems to have significant effects on the TCRαβ+ and TCRγδ+ cell compartments. Stinissen and coworkers described the upregulation of TCRγδ+-mediated anti-ergotypic responses following TCV (Stinissen et al., 1998). This stimulation of TCRγδ+ anti-ergotypic activity was accompanied by a shift in the TCRγδ+ repertoire from Vγ2+/Vδ2+ to Vγ1+/V δ1+ and by the production of high levels of IL-2, TNFα, and IL-10 by following stimulation with activated autologous T cells (Stinissen et al., 1998).

In addition, Zhang and coworkers studied anti-ergotypic and anti-idiotypic responses following TCV in MS patients by generating vaccine-reactive CD4+

T cell lines (Hong et al., 2006). The authors noted that all the antivaccine reactivity mediated by CD4+ TCRαβ+ cells was indeed anti-ergotypic and directed against epitopes derived from the CD25 molecule (Hong et al., 2006). Under their experimental conditions, none of these lines could be raised from healthy controls (Hong et al., 2006). These CD4+ anti-ergotypic regulators were MHC II-restricted and produced IL-4 and IL-10 upon stimulation with activated autologous T cells (Hong et al., 2006).

In conclusion, the above results demonstrate that anti-ergotypic T cells are part of the repertoire of naive individuals and comprise CD4+ and CD8+ TCRγδ+ and TCRαβ+ cells. TCRγδ+ regulators seem to be MHC-independent. In a healthy immune system, these regulatory populations are activated following the induction of T cell immunity and probably contribute to the contraction phase or resolution of the immune response. Anti-ergotypic regulation is depressed in autoimmune disorders, but regulatory TCRγδ+ or TCRαβ+ populations can be boosted by vaccination with activated T cells or defined ergotopes. However, vaccination not only expands preexisting anti-ergotypic regulators, it also changes their phenotypic characteristics.

# 3    Overlap and Interactions With Other Regulatory Populations

The definition of anti-ergotypic T cells is based on their reactivity against syngeneic activated but not resting T cells (Cohen et al., 2004). This definition puts under the category of anti-ergotypic regulators several groups of cells that differ in their surface markers and cytokine secretion profiles. Moreover, this definition does not exclude an overlap between anti-ergotypic regulators and other classes of regulatory T cells. Indeed, an overlap between human CD4+CD25+ regulatory T cells and anti-ergotypic regulators has been suggested.

Sakaguchi and coworkers identified a population of regulatory CD4+ T cells characterized by the expression of high levels of surface CD25+ (Sakaguchi, 2004). These CD4+CD25+ T cells showed low proliferative responses upon TCR triggering in vitro and can inhibit the proliferation of other T cells by a contact-dependent mechanism (Sakaguchi, 2004). In vivo, CD4+CD25+ T cells regulate a diverse array of T and B cell responses (Sakaguchi, 2004). Regulatory CD4+CD25+ T cells display a highly self-reactive T cell repertoire that partially overlaps that of the pathogenic T cells they control (Hsieh and Rudensky, 2005; Hsieh et al., 2006). At the molecular level, the regulatory CD4+CD25+ T cells were found to express the forkhead transcription factor FoxP3 (Hori et al., 2003); however, not all the CD4+CD25+ T cells have a regulatory phenotype and not all the FoxP3+ T cells are CD4+CD25+ cells (Fontenot et al., 2005).

The first study suggesting an overlap between the anti-ergotypic and the CD4+CD25+ regulatory T cell populations was carried out by Vandenbark and coworkers, who studied the specificity of the CD4+CD25+ regulatory T cell compartment in healthy controls and MS patients (Buenafe et al., 2004). They reported that the CD4+CD25+ T cell compartment includes anti-idiotypic clones and clones reactive with TCR CDR2 determinants from the germline V gene repertoire. Although anti-ergotypic reactivity was not directly studied, the isolation of CDR2-specific CD4+CD25+ T cell clones suggested that some anti-ergotypic regulators might belong to the CD4+CD25+ T cell class. In addition, this work was the first to demonstrate that the human CD4+CD25+ T cell compartment includes T cell regulators involved in T–T interactions.

The group of Jingwu Zhang has analyzed the contribution of anti-ergotypic regulators to the CD4+CD25+ regulator T cells induced in MS patients by TCV (Hong et al., 2006). They reported that almost all of the CD4+ T cell reactivity directed against the vaccinating T cell clone is anti-ergotypic. Moreover, the anti-ergotypic regulators were CD4+CD25+ T cells that could be classified according to their expression of FoxP3 (Hong et al., 2006). CD4+CD25+ FoxP3+T cells exerted their regulatory activity by secreting IL-10; IL-10–blocking antibodies could inhibit the regulatory effects (Hong et al., 2006). CD4+CD25+ FoxP3– T cells, however, secreted both IFNγ and IL-10, but anti-IL-10 neutralizing antibodies had no effect on the inhibitory activities. Taken together, these results suggest that a fraction of anti-ergotypic regulators are CD4+CD25+ T cells. Note that such CD4+CD25+ anti-ergotypic regulatory T cells were not detected in samples obtained from the same patients before TCV (Hong et al., 2006). Moreover, the anti-ergotypic response of naive rats was not affected by the depletion of CD4+CD25+ T cells (Mimran et al., 2005), suggesting that CD4+CD25+ anti-ergotypic T cells might be expanded by TCV but might not play a leading role in the regulation of the immune response of naive individuals.

An alternative, nonexclusive look at the relation between CD4+CD25+ regulatory T cells and the anti-ergotypic regulators results from the observation made by Baecher-Allan and colleagues, who demonstrated that when compared to CD4+CD25– T cells the regulatory CD4+CD25+ T cells express higher levels of the HLA-DR MHC II molecule (Baecher-Allan et al., 2001). In other words, regulatory CD4+CD25+ T cells, by upregulating MHC II-associated ergotopes, might serve as targets for anti-ergotypic T cells. It is tempting to speculate that the formation of T–T interactions between anti-ergotypic regulators and CD4+CD25+ T cell regulators might lead to the cross-regulation of these two cell populations (Fig. 4.1). This interaction might therefore contribute to the fine-tuning and coordination of various immunoregulatory mechanisms. In other words, a direct interaction between anti-ergotypic and CD4+CD25+ regulatory T cells might lead to the mutual regulation of the regulators.

# 4    Conclusions

We have summarized and discussed the basic principles of anti-ergotypic immunoregulation. Anti-ergotypic regulators are generated in the thymus and do not require activation by external antigens. Anti-ergotypic regulators are naturally expanded during antigen-specific immune responses but can also be boosted by

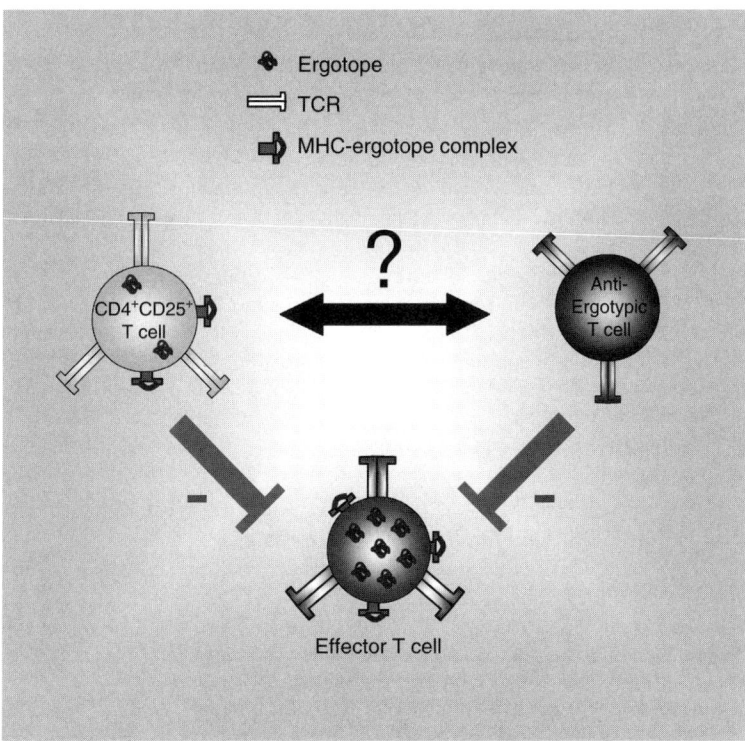

FIGURE 4.1. Cross-talk between CD4+ CD25+ T cells and anti-ergotypic T cells. CD4+ CD25+ T cells and anti-ergotypic T cells are two types of T cell that have been shown to downregulate the activities of effector T cells. However, it is conceivable that the two types of regulatory T cell might also mutually regulate each other—regulators regulating regulators. TCR, T cell receptor; MHC, major histocompatibility complex.

TCV or immunization with defined ergotopes administered as peptides or DNA vaccines. The anti-ergotypic regulatory response is mediated by several cell populations, among which are the CD4+CD25+ regulatory T cells. Indeed, CD4+CD25+ T cells can themselves be targeted by anti-ergotypic regulators as a consequence of their expression of relatively high levels of the MHC II molecule HLA-DR. Thus, we propose that the cross-talk between the anti-ergotypic and the CD4+CD25+ regulatory T cell populations fine-tunes the immunoregulatory mechanisms in an individual and so regulates the regulators (Fig. 4.1).

Many basic questions related to the biology of anti-ergotypic T cells are still unanswered. Probably the most compelling addresses the fate of the targeted T cells. Answering these and other questions will guide our steps toward a better understanding of the biology of anti-ergotypic regulatory networks and might eventually lead to its exploitation in the design of new therapies for autoimmune disorders based on the strengthening of preexisting immunoregulatory networks.

# References

Azuma, M., Yssel, H., Phillips, J. H., et al. (1993) Functional expression of B7/BB1 on activated T lymphocytes. *J. Exp. Med.* 177:845-850.

Baecher-Allan, C., Brown, J. A., Freeman, G. J., Hafler, D. A. (2001) CD4+CD25 high regulatory cells in human peripheral blood. *J. Immunol.* 167:1245-1253.

Ben-Nun, A., Wekerle, H., Cohen, I. R. (1981) Vaccination against autoimmune encephalomyelitis with T-lymphocyte line cells reactive against myelin basic protein. *Nature* 292:60-61.

Buenafe, A. C., Tsaknaridis, L., Spencer, L., et al. (2004) Specificity of regulatory CD4+CD25+ T cells for self-T cell receptor determinants. *J. Neurosci. Res.* 76:129-140.

Chen, Y., Kuchroo, V. K., Inobe, J., et al. (1994) Regulatory T cell clones induced by oral tolerance: suppression of autoimmune encephalomyelitis. *Science* 265:1237-1240.

Cohen, I. R. (1992) The cognitive paradigm and the immunological homunculus. *Immunol. Today* 13;490-494.

Cohen, I. R. (2001) T-cell vaccination for autoimmune disease: a panorama. *Vaccine* 20:706-710.

Cohen, I. R., Quintana, F. J., Mimran, A. (2004) Tregs in T cell vaccination: exploring the regulation of regulation. *J. Clin. Invest.* 114:1227-1232.

Cohen, I. R., Young, D. B. (1991) Autoimmunity, microbial immunity and the immunological homunculus. *Immunol. Today* 12:105-110.

Correale, J., Rojany, M., Weiner, L. P. (1997) Human CD8+ TCR-alpha beta(+) and TCR-gamma delta(+) cells modulate autologous autoreactive neuroantigen-specific CD4+ T-cells by different mechanisms. *J. Neuroimmunol.* 80:47-64.

Fontenot, J. D., Rasmussen, J. P., Williams, L. M., et al. (2005) Regulatory T cell lineage specification by the forkhead transcription factor foxp3. *Immunity* 22:329-341.

Groux, H., O'Garra, A., Bigler, M., et al. (1997) A CD4+ T-cell subset inhibits antigen-specific T-cell responses and prevents colitis. *Nature* 389:737-742.

Hafler, D. A., Buchsbaum, M., Weiner, H. L. (1985) Decreased autologous mixed lymphocyte reaction in multiple sclerosis. *J. Neuroimmunol.* 9:339-347.

Hong, J., Zang, Y. C., Nie, H., Zhang, J. Z. (2006). CD4+ regulatory T cell responses induced by T cell vaccination in patients with multiple sclerosis. *Proc. Natl. Acad. Sci. U S A* 103:5024-5029.

Hori, S., Nomura, T., Sakaguchi, S. (2003) Control of regulatory T cell development by the transcription factor Foxp3. *Science* 299:1057-1061.

Hsieh, C. S., Rudensky, A. Y. (2005) The role of TCR specificity in naturally arising CD25+ CD4+ regulatory T cell biology. *Curr. Top. Microbiol. Immunol.* 293:25-42.

Hsieh, C. S., Zheng, Y., Liang, Y., et al. (2006) An intersection between the self-reactive regulatory and nonregulatory T cell receptor repertoires. *Nat. Immunol.* 7:401-410.

Indiveri, F., Wilson, B. S., Russo, C., et al. (1980) Ia-like antigens on human T lymphocytes: relationship to other surface markers, role in mixed lymphocyte reactions, and structural profile. *J. Immunol.* 125:2673-2678.

Jiang, H., Chess, L. (2000) The specific regulation of immune responses by CD8+ T cells restricted by the MHC class Ib molecule, Qa-1. *Annu. Rev. Immunol.* 18:185-216.

Jung, D., Alt, F. W. (2004) Unraveling V(D)J recombination; insights into gene regulation. *Cell* 116:299-311.

Kronenberg, M. (2005) Toward an understanding of NKT cell biology: progress and paradoxes. *Annu. Rev. Immunol.* 23:877-900.

Kyewski, B., Klein, L. (2006) A central role for central tolerance. *Annu. Rev. Immunol.* 24:571-606.

Lider, O., Reshef, T., Beraud, E., et al. (1988) Anti-idiotypic network induced by T cell vaccination against experimental autoimmune encephalomyelitis. Science 239:181-183.

Lohse, A. W., Mor, F., Karin, N., Cohen, I. R. (1989) Control of experimental autoimmune encephalomyelitis by T cells responding to activated T cells. *Science* 244:820-822.

Lohse, A. W., Spahn, T. W., Wolfel, T., et al. (1993) Induction of the anti-ergotypic response. *Int. Immunol.* 5:533-539.

Macario, A. J., Conway de Macario, E. (2005) Sick chaperones, cellular stress, and disease. *N Engl J Med.* 353, 1489-1501.

Mimran, A., Mor, F., Carmi, P., et al. (2004) DNA vaccination with CD25 protects rats from adjuvant arthritis and induces an antiergotypic response. *J. Clin. Invest.* 113: 924-932.

Mimran, A., Mor, F., Quintana, F. J., Cohen I. R. (2005) Anti-ergotypic T cells in naïve rats. *J. Autoimmun.* 24:191-201.

Mor, F., Reizis, B., Cohen, I. R., Steinman L. (1996) IL-2 and TNF receptors as targets of regulatory T-T interactions: isolation and characterization of cytokine receptor-reactive T cell lines in the Lewis rat. *J. Immunol.* 157:4855-4861.

Pfister, G., Stroh, C. M., Perschinka, H., et al. (2005) Detection of HSP60 on the membrane surface of stressed human endothelial cells by atomic force and confocal microscopy. *J. Cell Sci.* 118:1587-1594.

Quintana, F. J., Cohen, I. R. (2005) Heat shock proteins as endogenous adjuvants in sterile and septic inflammation. *J. Immunol.* 175:2777-2782.

Quintana, F. J., Weiner, H. L. (2006) Understanding natural and pathological autoimmunity. *J. Neuroimmunol.* 10:10.

Sakaguchi, S. (2004) Naturally arising CD4+ regulatory T cells for immunologic self-tolerance and negative control of immune responses. *Annu. Rev. Immunol.* 22:531-562.

Sarantopoulos, S., Lu, L., Cantor, H. (2004) Qa-1 restriction of CD8+ suppressor T cells. *J. Clin. Invest.* 114:1218-1221.

Stinissen, P., Zhang, J., Vandevyver, C., et al. (1998) Gammadelta T cell responses to activated T cells in multiple sclerosis patients induced by T cell vaccination. *J. Neuroimmunol.* 87:94-104.

Van Eden, W., van der Zee, R., Prakken, B. (2005) Heat-shock proteins induce T-cell regulation of chronic inflammation. *Nat. Rev. Immunol.* 5:318-330.

Weiner, H. L., Friedman, A., Miller, A., et al. (1994). Oral tolerance: immunologic mechanisms and treatment of animal and human organ-specific autoimmune diseases by oral administration of autoantigens. *Annu. Rev. Immunol.* 12:809-837.

# 5
# How B Cells Contribute to Multiple Sclerosis Pathology

Kevin C. O'Connor, Sunil V. Cherry, and David A. Hafler

## 1   Introduction

Multiple sclerosis (MS) is an inflammatory disease of the central nervous system (CNS) white matter likely to be of autoimmune origin. The defining pathological feature of MS is the demyelinated lesion, which is found throughout the brain and spinal cord. Autoimmune dysregulation is viewed as the major contributor to tissue damage. A widely held model of MS immunopathology (Noseworthy et al. 2000) suggests that activated autoreactive T cells recognizing myelin antigens in the immune periphery cross the CNS endothelium, triggering a cascade of events that result in white matter inflammation and tissue destruction. Inside the CNS, release of local cytokines, chemokines, and matrix metalloproteins support the recruitment of subsequent waves of infiltrating effector cells, including monocytes, and B cells. Mechanisms of myelin destruction and axonal damage are likely to be multiple and include direct effects of pro-inflammatory cytokines, complement-fixing antibodies, antigen-specific and antigen-nonspecific cytotoxicity, and apoptosis.

We describe the role of B cells and their products, antibodies, with particular emphasis on the pathology of MS and other related neuroinflammatory diseases. We begin with an overview of B cell development and a presentation of B cell activation, both providing a foundation for a discussion of B cell proliferation in the CNS. We then discuss B cell-mediated autoimmune disease, followed by a detailed presentation on the respective roles of B cells and antibodies in MS. We conclude with the most recent information on B cell-targeted MS therapy.

## 2   B Cell Development and Activation

Development of B lymphocytes occurs in several stages marked by gene rearrangement and the surface expression of specific gene products. The first phase, which occurs in the bone marrow, has been considered antigen-independent, although current evidence supports the view that self-antigen presented in the bone marrow provides a positive signal to developing B cells (Pillai 1999). During maturation, pluripotent stem cells give rise to early pro-B cells concomitant with the rearrangement of $D_H$ and $J_H$ gene segments. Early pro-B cells proceed to late pro-B cells marked by rearrangement of the $V_H$ gene segment to $D_H J_H$. Then late pro-B cells give rise to pre-B cells, which express the rearranged heavy chain paired with the surrogate light chain (SLC). Surface expression of a heavy chain signals the cell to begin rearrangement of the light chain marked by $V_L$ and $J_L$ joining. Finally, the rearranged light chain displaces the SLC by pairing with the heavy chain. The immature B cell, which now has surface expression of immunoglobulin M (IgM) and IgD, migrates from the bone marrow to the periphery. This process, in some cases, results in the production of self-reactive B cells. The elimination or inactivation of B cells producing self-reactive antibodies is required to prevent autoimmune disease. Several mechanisms for balancing tolerance to self and self-reactivity have been elucidated, including deletion, anergy, and receptor editing. During the receptor editing process, secondary light chain gene rearrangements replace existing light chains, generating receptors with altered specificities. B cells develop to maturity if the replacement generates a nonself-reactive receptor.

Once on patrol in the periphery, the naive B cell awaits an encounter with antigen. Naive B cells may be identified by surface expression of CD19, CD20, and IgM and their nonmutated variable region gene segments. In a T cell-dependent immune response, antigen-activated B cells migrate into the primary follicles of the peripheral lymphatic organs. In this network of follicular dendritic cells (FDCs), B cells are induced to expand clonally, leading to the formation of a structure called the germinal center (GC), which is a specialized site of the secondary lymphoid follicles where B cells undergo somatic mutation and differentiate. The formation of GCs is initiated when antigen-specific T and B cells make physical contact with T cell-rich zones of secondary lymphoid organs and the activated B cells enter the primary follicles. These B cells, called centroblasts, divide and expand vigorously, forming the dark zone of the GC. The proliferation of centroblasts increases the number of antigen-specific B cells and compresses the surrounding uninvolved follicular cells to form a mantle zone around the new GC.

During proliferation, both somatic hypermutation of rearranged Ig genes and Ig class switching are also activated in centroblasts. Antigen presented on FDCs ensures that only those B cells with high-affinity receptors are selected to differentiate into plasma cells and memory cells. Variable (V) regions of antibodies selected for increased affinity show a characteristic pattern of somatic mutations. Replacement mutations are mainly seen in the complementarity-determining regions (CDRs), whereas silent mutations accumulate in the framework residues (FRs). Thus, sequence analysis of V regions can be used as an indicator of whether B cells have gone through an antigen-driven process of affinity maturation. The degree to which mutation has occurred can be measured by comparing a V region to its germline gene. Homologies can be as low as 85%, which translates to approximately 40 base changes.

As centroblasts further mature, they cease dividing and become small centrocytes. Centrocytes move out of the dark zone into an area of the GC called the light zone, which also contains a rich network of FDCs and CD4+ T cells. FDCs retain native unprocessed antigen as immune complexes on their surface and trap antigen-specific B cells. Centrocytes specific to the original antigen take it up from FDCs, process it, and present peptides to T cells. T cells that recognize the antigenic peptides deliver two types of stimuli that result in the proliferation and differentiation of centrocytes. The contact-initiated stimulus is mainly mediated through the CD40 ligand, CD28, and CTLA4 expressed by activated T cells and CD40, CD80 (B7.1), and CD86 (B7.2) expressed on B cells. The soluble signal is delivered by cytokines. The positively selected centrocytes finally leave the GC to become either memory B cells or plasma cells. Plasma cells are terminally differentiated and nondividing, and they synthesize thousands of specific antibodies each second (Shapiro-Shelef and Calame, 2005). Centrocytes that have become autoreactive or have lost their antigen-binding specificity because of somatic hypermutations are negatively selected by the lack of proper T cell help and die via apoptosis.

## 3    B Cells in Autoimmunity

By definition, immunological mechanisms play a role in the patholology of autoimmune diseases. Yet, there are few situations in human immunology where the detailed pathogenesis attributed to autoreactive B cells is understood. In fact, the pathogenic relations among B cells, autoantibodies, and clinical manifestations of disease have been revealed for only a few of the many autoimmune diseases. Even in the absence of a direct link to pathology, autoantibodies serve as useful diagnostic and prognostic indicators. Such is the case of autoimmune diabetes, in which autoantibodies recognizing insulin, GAD, and IA2 aid disease prediction, diagnosis, and severity. However, thus far, none of these autoantibodies is directly implicated in the pathology. The same is true for systemic lupus erythematosus (SLE), considered a classic B cell-mediated autoimmune disease in which autoantibodies to nuclear components are diagnostic; but again, their firm link to pathology remains to be established.

In certain situations autoantibodies are soundly linked to disease pathology. A well characterized example is myasthenia gravis (MG), in which autoantibodies, in most cases, are directed toward the acetylcholine receptor (AChR). These pathogenic autoantibodies alter AChR organization associated with the neural synapse and functionally block neural transmission, resulting in paralysis of the muscle (Vincent, 2002). Other examples are the autoimmune blistering diseases pemphigus and bullous pemphigoid, which are characterized by the presence of autoantibodies against specific adhesion molecules of the skin and mucous membranes (Nousari and Anhalt, 1999). The skin blisters on patients affected by bullous pemphigoid are directly related to complement-fixing autoantibodies that recognize critical components of dermoepidermal adhesion (Liu et al., 1995). Passive transfer of human IgG autoantibodies from patients with pemphigus into mice reproduces the cutaneous disease (Anhalt et al., 1982). This finding is notable, as in few human autoimmune diseases can passive transfer of autoantibodies alone reproduce tissue injury in an animal. Anti-GBM in Goodpasture's syndrome, anti-TSH in Grave's disease, and anti-RBC in autoimmune hemolytic anemia are also among the antibodies directly involved in autoimmune disease pathogenesis. The mechanisms of pathology in these well characterized B cell-mediated diseases provides a rough outline of mechanisms that may explain MS.

## 4    B Cells in Multiple Sclerosis

In human immunology, establishing large collections of specimens representing a heterogeneous disease such as MS is challenging. MS is a heterogeneous disease, with significant variation of symptoms, neuropathology, and immune system involvement within and among patients. Studies in this field usually include subsets of MS patients. Thus, the conclusions may not be universal for all types of MS, and the reader should be aware that the immunological characteristics discussed do not apply to all patients with MS.

### 4.1    Teachings of the Animal Models

Experimental autoimmune encephalomyelitis (EAE) is a disease characterized by multiple sclerotic plaques and perivascular inflammatory infiltrates in the CNS white matter of primarily rodents, resulting in paralytic neurological disability. The demyelinating condition is induced by the immunization of susceptible animals with various myelin antigens such as myelin basic protein (MBP), myelin oligodendrocyte glycoprotein (MOG), and proteolipid protein (PLP) with Freund's adjuvant (Lannes-Vieira et al., 1994). The presentation of peptides derived from these antigens to CD4+ T cells is a minimal requirement for the development of inflammatory foci in the CNS (Kuchroo et al., 2002).

The role of B cells in EAE is incompletely understood. Early studies suggest that B cells are not necessary for disease expression. Indeed, passive transfer of B cells and antibodies by themselves from EAE animals to naive animals does not induce disease. Furthermore, EAE can be induced in B cell-deficient mice (Hjelmstrom et al., 1998; Wolf et al., 1996). Spontaneous lesions develop in MBP T cell receptor (TCR) transgenic RAG(–/–) mice, which lack B cells (Lafaille et al., 1994).

However, autoantibodies to myelin antigens may nevertheless be highly relevant in CNS demyelinating diseases in the more complex setting of an intact immune system. Indeed, the severity of EAE is exacerbated by administration of myelin antibodies after disease is induced (Linington et al., 1988) with significance attributed to complement-fixing antibodies and B cells (Piddlesden et al., 1993). The potential of MOG autoantibodies to induce severe demyelination and oligodendrocyte loss was convincingly demonstrated by passive transfer of anti-MOG monoclonal antibody (mAb) (8-18C5) into mice or rats with mild clinical signs of EAE (Piddlesden et al., 1993; Schluesener et al., 1987), and demyelination was shown to be dependent on the coordinate action of myelin-specific T cells and autoantibodies. Antibodies to MOG also appear to play an important role in the chronic, relapsing–remitting disease process in the marmoset model of EAE induced by immunization with the extracellular (EC) domain of MOG (Genain et al., 1995; Raine et al., 1999). Large demyelinated lesions resembling MS plaques were found in this model, characterized by deposition of complement components and immunoglobulin (including those directed to MOG) and the uptake of myelin debris by macrophages. MOG-specific antibodies participate in attacking the myelin membrane by triggering complement- and antibody-dependent cell-mediated cytotoxicity (ADCC)-dependent effector mechanisms (Genain et al., 1999; Storch et al., 1998).

Another role for B cells in EAE is that of recovery. Several studies indicate the existence of distinct B cell subsets that suppress the progression of and/or enhance the recovery from acquired immune-mediated inflammation. Indeed, by examining the effect of B cells on a murine model of EAE, Janeway and colleagues (Wolf et al., 1996) realized that an immunoregulatory B cell subset played a role in recovery. Among several roles, interleukin-10 (IL-10) is an anti-inflammatory cytokine that has a strong suppressive effect on Th1 lymphocytes; B cells producing IL-10 can modulate T cell responses and have thus been investigated in this light in EAE models. By producing IL-10, autoreactive B cells mediate the recovery phase of MOG-induced murine EAE (Fillatreau et al., 2002). There has been speculation that B cells in MS mediate repair rather than damage, which lends support to this view. Moreover, work examining EAE resident B cells in detail is warranted, as it is now insufficient.

## 4.2    Autoantibodies in MS

It is likely that both neurodegeneration and inflammation contribute to the pathology of MS. In seminal work underscoring the significance of both inflammatory and noninflammatory mechanisms in MS pathology, Lassman and colleagues

(Lucchinetti et al., 2000) catalogued lesions and demonstrated the existence of four fundamentally different patterns of demyelination in humans. Although two patterns (I and II) were similar to T cell-mediated or T cell plus antibody-mediated autoimmune encephalomyelitis, the other patterns (III and IV) were highly suggestive of a primary oligodendrocyte dystrophy, reminiscent of demyelination induced by a virus or toxin rather than one involving the immune system. Pattern II, which was partially characterized by the presence of antibodies and complement, was observed in approximately 50% of the active lesions studied.

The presence of oligoclonal immunoglobulin in the CNS of patients with MS remains the hallmark of the B cell's putative participation in MS. Although often absent during the early stages of the disease, oligoclonal bands (OCBs) can eventually be detected in more than 80% of patients with clinically definite MS. Their presence indicates an intrathecal clonal expansion of B cells seemingly driven by antigen. Although described more than a half century ago (Kabat et al., 1950), the specificity of the intrathecal immunoglobulin is unclear. OCBs can also be identified in the cerebrospinal fluid (CSF) of patients with other inflammatory diseases of infectious origin, such as subacute sclerosing panencephalitis (SSPE) (Mattson et al., 1980) and varicella zoster (Vartdal et al., 1982). Unlike the situation in MS, the disease-related antigen drives the production of the OCBs, and these immunoglobulins react with the same antigen.

As autoantibodies have been shown to be useful markers in some autoimmune diseases, these molecules have also been examined for their suitability as biomarkers in MS. Many investigations have focused on popular candidate antigens. Consequently, there is an abundance of reports concerning autoantibodies in MS (reviewed in O'Connor et al., 2001; Ziemssen and Ziemssen, 2005). Although much of this work has focused on the role of autoantibodies in the immunopathogenesis of MS, most are still only putative biomarkers because a direct role in the pathology has not been reported. That said, no study has clearly demonstrated that autoantibodies in MS are either reliable biomarkers or contributors to the disease pathology. It is likely that no single biomarker will be conclusive; rather, an arrangement of at least several biomarkers in the form of autoantibodies or other molecules, forming a profile or fingerprint, will provide valuable information.

The possibility that CSF iommunoglobulin in patients with MS is generated as a response to myelin self-antigens has been investigated in detail. Antibodies specific for MBP have been identified in the CSF (Panitch et al., 1980; Warren and Catz, 1987, 1999). Autoantibodies reactive with myelin-associated glycoprotein (MAG) (Moller et al., 1989), the enzyme transaldoase (TAL) (Banki et al., 1994), MOG (Xiao et al., 1991), and oligodendrocyte-specific protein (OSP) (Bronstein et al., 1999) have also been detected in the CSF of a subset of patients with MS. One series of reports (Villar et al., 2003) demonstrated that individuals with MS who have intrathecal synthesis of IgM were likely to progress from relapsing-remitting MS (RRMS) to a more severe progressive form. The presence of intrathecal IgM correlates with progression from initial stages to clinically definite MS (Villar et al., 2002b); higher EDSS (expanded disability status scale) scores (Villar et al., 2002a) and a fraction of this IgM reactive with myelin lipids suggest a more aggressive course (Villar et al., 2005).

Serum anti-MBP antibodies have been reported (Reindl et al., 1999; Terryberry et al., 1998), though not in all studies (Brokstad et al., 1994; Colombo et al., 1997; Olsson et al., 1990). There are reports detecting serum autoantibodies against recombinant TAL (Banki et al., 1994), and B cells secreting autoantibodies directed against PLP have been identified in the peripheral blood of patients with MS (Sun et al., 1991). Serum autoantibodies to other putative MS antigens, such as MOBP, CNPase, αß-crystalline, and S-100ß, are under investigation (Schmidt, 1999). Autoantibodies directed against nonmyelin antigens have also been detected in the sera of some patients with MS. Anti-nuclear antibodies (ANA) (Barned et al., 1995), anti-cardiolipin antibodies (Colaco et al., 1987), and antibodies against beta-2-glycoprotein I (Roussel et al., 2000) are detected in MS sera more frequently than in normal sera. Elevated levels of antibodies to a large panel of organ- and non-organ-specific antigens were reported in patients with MS relative to controls (Spadaro et al., 1999).

The search for autoantibodies to MOG and MBP has received increased attention in recent years. Our own studies indicate that the antibodies to these chief candidate myelin antigens are not readily detectable in the serum and CSF of patients with MS (O'Connor et al., 2003, 2005). Rather, we have shown that patients with acute disseminated encephalomyelitis (ADEM) and induced encephalomyelitis have these autoantibodies (O'Connor et al., 2003; unpublished results). The difficulty in consistently detecting autoantibodies in the periphery of patients with MS is that they may be produced locally in the CNS at the site of injury. Indeed, autoantibodies to MOG were found bound to disintegrating myelin segments in MS lesions (Genain et al., 1999) and in IgG eluates from MS lesions (O'Connor et al., 2005). IgG isolated from CNS tissue binds to MBP in solid-phase assays (Warren and Catz, 1993). Collectively, these results support the hypothesis that pathologically significant antibodies may be produced in the CNS.

Berger and colleagues have reported a series of studies in which they examined serum and CSF samples using Western blot analysis of the recombinant MOG extracellular domain. Antibodies to MOG were detected in 38% of MS patients, 53% of patients with other inflammatory CNS diseases, and 3% of patients with noninflammatory CNS diseases. Detection of antibodies to MBP was also not disease-specific, as they were detected in 28% of MS patients, 47% of patients with other inflammatory CNS diseases, and 60% of patients with rheumatoid arthritis. The major limitations of this method are that large amounts of recombinant protein were loaded per lane and that no control proteins were included on these blots to assess the level of background binding, an issue made relevant by the subjective nature of colorimetric detection. A study by Berger et al. reported that the development of clinically definite MS can be predicted by the presence of serum IgM antibodies to MOG in patients with a clinically isolated syndrome (Berger et al., 2003). However, a study by another group failed to confirm these findings (Lampasona et al., 2004).

Differences in patient populations are often cited as the cause for such discrepancies, but the basis for incongruity in these results is likely to lie in the antigen preparation and/or the assay conditions. Extensive studies in EAE models have demonstrated that only antibodies that recognize folded MOG protein are pathogenic, and antibodies that solely bind to denatured protein or short synthetic peptides fail to induce demyelination (Brehm et al., 1999; von Budingen et al., 2004). The co-crystal structure of MOG and the pathogenic murine 8-18C5 antibody also revealed that the antibody binds to a surface created by three discontinuous loops and the N-terminus (Breithaupt et al., 2003). Current techniques permit detection of autoantibodies directed against linear epitopes only on denatured proteins (Western blot) or do not adequately discriminate between denatured and folded proteins (enzyme-linked immunosorbent assay, or ELISA) and are thus not optimized for the detection of conformation-dependent autoantibodies (Mathey et al., 2004). To circumvent these constraints, we developed a novel approach for the identification of autoantibodies using specific and sensitive assays employing folded and glycosylated protein such that conformation-dependent epitopes are present (O'Connor et al., 2005).

Arguably, the most significant information to date regarding autoantibodies in the presence of neuroinflammation comes from optic neuritis (ON) and its models. There is a strong relationship between ON, formerly known as Devic's disease, and MS. In approximately 25% of patients, the first clinical presentation of MS is in the form of ON. Moreover, nearly two-thirds of patients with MS develop some form of ON (Sorensen et al., 1999). ON differs from MS in that its immunopathology is restricted to the spinal cord and optic nerves. Serum IgG in approximately 75% of ON patients bind to the abluminal face of microvessels on mouse CNS tissue; conversely, the sera of patients with MS do not (Lennon et al., 2004). This binding is specifically directed toward the aquaporin-4 (AQP4) water channel (Lennon et al., 2005), demonstrating that a particular form of demyelinating disease can be associated with autoantibodies against a defined structure.

# 5   B Cell Antigen Target Discovery

An alternative strategy for identifying antigens in MS is to use a broad approach rather than simply test popular candidate antigens. A study of CSF-derived IgG using phage-displayed random peptides identified reacting antigens unrelated to any known protein. IgG from different MS patients bound different peptides, leading the authors to speculate that the CSF antibody repertoire is individual-specific (Cortese et al., 1996). Another study examined the specificity of two recombinant antibodies derived from the CNS lesions of patients with MS. The specificity of these antibodies was determined to be directed toward double-stranded DNA (Williamson et al., 2001). Though the finding is significant, the antibodies in this study were derived from the total RNA isolated from an entire lesion area. Thus, the antibodies were constructed without knowledge of the natural $V_H$ and $V_L$ pairing, and no information was provided on the oligoclonal nature of individual B cells in the lesions. Interestingly, a cDNA expression library derived from oligodendrocytes containing more than 2 million clones did not reveal any specific binding to these antibodies.

Another systematic approach to identifying novel antigen targets used serum immunoglobulin to bind the surface of CNS-related cell lines. MS was distinguishable from controls in many cases, but the cell surface molecules have not yet been identified (Lily et al., 2004). High-throughput analysis is likely to become a preferred autoantibody search strategy; antigen microarrays are becoming available and afford speed and multiplexing (Robinson et al., 2002). Application of this technology to antigen discovery in MS has revealed that autoimmune responses to several myelin lipids are a feature in some patients with MS (Kanter et al., 2006).

# 6    B Cells at the Site of Myelin Destruction: CSF and the CNS Tissue

Further evidence for the role of antibodies in MS comes from their presence in the compartment most closely associated with MS, the CNS. An abnormal humoral immune response in the CNS of patients with MS has been well described (reviewed by Cross et al., 2001; O'Connor et al., 2001). The evidence includes intrathecal immunoglobulin synthesis (Kabat et al., 1950) and the presence of IgG oligoclonal bands (Lowenthal et al., 1960), which are used as a diagnostic test (Johnson et al., 1977). B cells and plasma cells have been found in active and late MS lesions (Esiri, 1977; Prineas and Connell, 1978; Prineas and Wright, 1978), and antibodies are found in active MS lesions (Mehta et al., 1981; Storch et al., 1998), as are immunoglobulin transcripts (Lock et al., 2002) and evidence of B cell differentiation (Corcione et al., 2004). An example of an MS lesion showing characteristic demyelination and a perivascular immune cell infiltrate comprising B cells is shown in Figures 5.1A and 5.1B, respectively.

Investigations of B cells in both CSF and CNS tissue lesions of MS patients suggest a T cell-mediated, antigen-driven clonal expansion of B cells. Molecular studies designed to characterize the variable regions ($V_H$) of IgG expressed in MS plaques (Baranzini et al., 1999; Owens et al., 1998; Smith-Jensen et al., 2000; Williamson et al., 2001) and CSF (Colombo et al., 2000; Owens et al., 2003; Qin et al., 1998) revealed that the variable regions are biased in family representation, are extensively mutated, and are oligoclonal. Induction of immune effector mechanisms by these antibodies is evidenced by: (1) capping of surface IgG on macrophages involved in myelin breakdown (Prineas and Graham, 1981); (2) co-deposition of IgG and complement, particularly the activated terminal lytic complex (Gay et al., 1997; Lucchinetti et al., 2000; Storch et al., 1998) at plaque borders; and (3) the presence in CSF of membrane attack complex-enriched membrane vesicles, indicating a role for complement-mediated injury in MS (Scolding et al., 1989).

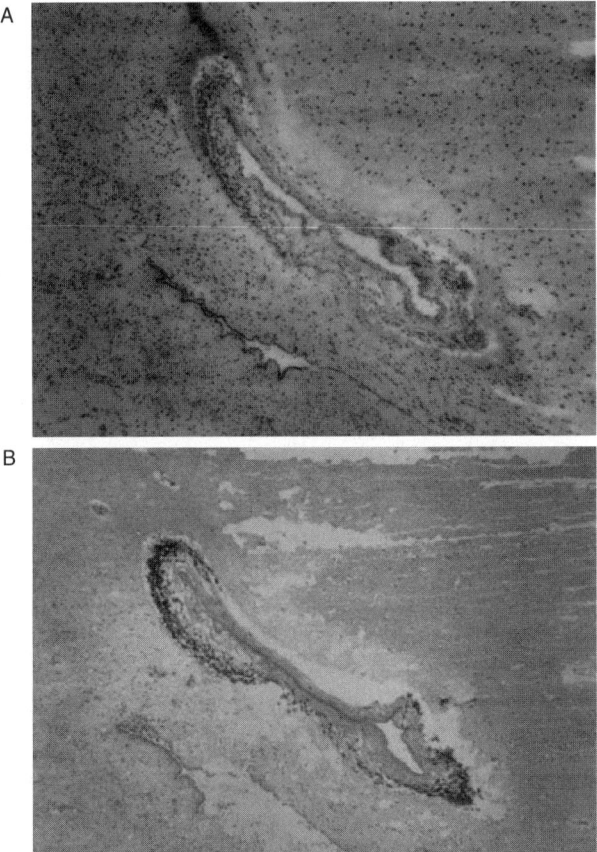

FIGURE 5.1. The perivascular infiltrate proximal to demyelinated CNS tissue includes CD19+ B cells. Perivascular infiltration of CD-19-positive B cells in a demyelinated area of an MS brain. A. An MS lesion stained with luxol fast blue indicates a large area of demyelination proximal to the vasculature. B. Anti-CD19 staining indicates the presence of B cells in the infiltrate. (Histopathology provided by Gerald P. Bailey, MD, PhD and Kevin C. O'Connor, PhD, Brigham and Women's Hospital and Harvard Medical School, Boston, MA.) (*See Color Plates between pages 256-257.*)

## 6.1   B Cells in the CSF of Patients with MS

The attention given to the investigation of B cells in the CSF is warranted, as it has been known for decades that intrathecally produced oligoclonal immunoglobulin bands (OGBs) are present in the CSF of many patients with MS. Strategies set to examine CSF immunoglobulins have demonstrated that a subset of the B cells are clonally expanded, have class-switched, and contain somatic mutations—all of which are characteristic of an antigen-driven response. To date, a direct link between these cell populations and the OGBs has not been found.

Moreover, the antigen to which they are directed is unknown. The reverse is true in other neurological diseases where OGBs were shown to be directed against the agent that causes disease. Examples include the CSF OGBs present during varicella-zoster virus (VZV) vasculopathy, which are directed against VZV (Burgoon et al., 2003), and measles virus-specific OGBs from the CSF of patients with SSPE (Vandvik et al., 1976). Furthermore, IgG derived from expanded cells in SSPE CNS tissue recognize measles virus (Burgoon et al., 2005).

The collective work examining B cells in MS CSF has confirmed that these cells are experienced. Among the seven families in the human VH repertoire, VH3 and VH4 are preferentially expressed in MS CSF (Colombo et al., 2000; Qin et al., 1998; Ritchie et al., 2004). Conversely, in peripheral blood from normal healthy individuals, each family is approximately represented in equivalence with their composition in the germline: There are seven families in the human $V_H$ repertoire. In the periphery of a normal individual, each is expressed in a distribution ordered in approximate accordance with its germline complexity (Brezinschek et al., 1995). Families VH1, VH3, and VH4 make up the majority, comprising 21.5%, 43.1%, and 21.5%, respectively (Cook and Tomlinson, 1995). When the repertoire of an individual is examined longitudinally, changes are not observed (Baranzini et al., 1999). Expansion of B cell clones is another feature of the B cell population in MS CSF. Moreover, among these expanded CSF B cells are clonal variants of the $V_H$ genes that differ by several mutations (Colombo et al. 2000; Ritchie et al. 2004). Receptor editing has also been observed among CSF B cells from patients with MS (Monson et al., 2005). The degree of somatic mutation is often extensive; the homology to the germline can be as low as 85% (Ritchie et al., 2004). Collectively, these characteristics observed in the CSF B cell population indicates an ongoing process of intraclonal diversification due to affinity maturation.

The clonal populations in the CSF include, both CD19+ B cells and CD138+ plasma cells. In a few cases, clones identified in the CSF have been found among the peripheral blood monocytic cells (PBMCs) (Colombo et al., 2000). On the whole, plasma cells comprise most of the clonal cells in the CSF (Ritchie et al., 2004). Identical clones may be found to exist in both the CD19+ and CD138+ populations, but this has been reported in limited cases. Moreover, this phenomenon must be viewed carefully, as CD138+ cells still express CD19, so distinction of the populations is subjective.

Similar to the circumstances in MS, B cells present in the CSF of patients with ON maintain the characteristics of an antigen-driven response (Haubold et al., 2004). They display considerable clonal expansion, and the CD19 and CD138 populations share clones. Variable region usage is skewed, and a significant degree of somatic mutation is observed.

The distribution of naive, memory, and activated (plasma or plasma blasts) B cells in the periphery of patients with MS, patients with other inflammatory neurological diseases (OINDs), and normals are comparable. However, B cell subset distribution is distorted in the CSF of patients with MS. CD138+ plasma cells are enriched in a subset of patients with MS. Double-positive, IgD+, CD27+, naive B cells are more prevalent in the periphery than in CSF. Activated CD38+ B cells are equally distributed between these two compartments (Corcione et al., 2004). The IgD+ CD27+ subset of memory B cells is enriched in the CSF relative to the periphery, as are the CD27 CD38 high set representing cells in the process of differentiating toward memory cells (Corcione et al., 2004). These distributions are not exclusive to MS, as they are also found OINDs. Centroblasts are often identified through their expression of CD77 and high levels of CD38; as centroblasts mature toward centrocytes, they lose expression of CD77. Centroblasts have been found in the CSF of patients with MS but are not present in matched peripheral blood, whereas centrocytes are commonly found in these patients, controls, and normals. Centroblasts in the CSF of patients with MS also express the proliferation marker Ki67.

## 6.2  B Cells and Antibodies in MS CNS Lesions

Under normal circumstances the brain is an immune privileged site, meaning the normal immune function does not occur in this tissue. MS is unlike many other autoimmune diseases in that the site of damage, the CNS tissue, is immune privileged. It is understood that during the disease process the blood-brain barrier is compromised and immune cells breach the barrier. It is unclear whether the antibodies in MS lesions are produced in the periphery and then deposited in the CNS, which may act as a "sink," or the B cell response in MS is partly compartmentalized, and the antibodies are produced locally.

Little is known regarding the specificity of anti-myelin antibodies derived from the CNS plaque tissue of patients with MS. The humoral response in the CNS toward the immunodominant antigens MBP and MOG has been examined. Antibodies to MOG were found bound to disintegrating myelin segments in MS lesions (Genain et al. 1999), providing evidence that myelin autoantibodies may, in part, mediate CNS tissue damage. IgG isolated from CNS tissue binds to MBP, and an MBP(83-97) peptide was capable of significantly inhibiting this binding (Warren et al. 1993). IgG isolated directly from CNS lesions binds to glycosylated MOG (O'Connor et al. 2005). These advances are important and have added to our knowledge of the humoral response in MS. However, these and several other antigens identified as associated with MS: (1) do not adsorb oligoclonal bands; (2) have not been demonstrated to be produced in lesion sites; and (3) have an as yet unknown role in the pathology and etiology of the disease.

Analysis similar to that of immunoglobulin genes in the CSF has been performed with CNS tissue as the source of the genes. As with the CSF, analyses of immunoglobulin libraries revealed a restricted use of the VH3 and VH4 families in MS plaque tissue (Owens et al., 1998; Smith-Jensen et al., 2000). Bias toward usage of the VH1 and VH4 families also can occur (Baranzini et al. 1999). Such transcripts are not found in normal CNS tissue. Expanded clones and clonal variants are often present in lesions that are separated in the tissue (Owens et al., 1998). Extensive somatic mutation accumulating in the CDRs is another feature of these immunoglobulin genes. Clonally expanded populations found in lesions appear to be absent in the peripheral blood (Owens et al., 2001). Because this study used a single subject, additional cases are needed to confirm the finding, it is also possible that the clones present in the CNS are present at extremely low levels in the periphery, but the lesion immunoglobulin repertoire was, in this case, not reflected in the peripheral blood, suggesting that the response in the CNS is compartmentalized. The isotypes produced by B cells in MS CNS lesions include IgG and in one report IgA (Zhang et al., 2005); both seem to share the characteristics of an antigen-driven response. Studies to evaluate the presence of IgM and its relation to the IgG and IgA present in the lesions have not been reported.

The cells present in these lesions are likely to be the source of the resident immunoglobulin deposits (O'Connor et al., 2005), although direct proof of this assumption has yet to be reported. It follows that clonal B cells may produce IgG that forms OCBs. IgG isolated from MS lesions and CNS lesions caused by infection, as is the case in SSPE, both display OCBs (Owens et al., 1997). Thus, in SSPE, abnormal expression of germline V segments, overrepresentation of particular sequences that correspond to the OCBs, and substantial somatic mutation of most clones from the germline, which, taken together, constitute features of antigen-driven selection in the IgG response (Burgoon et al., 1999). Thus, the OCBs in the CSF, the immunoglobulin deposits, and the B cells producing each are primarily directed toward the cause of SSPE, the measles virus and its nucleocapsid protein (Burgoon et al., 2001, 2005).

## 6.3    CNS Support of Lymphoid-like Structures

In several autoimmune diseases, such as rheumatoid arthritis (Young et al., 1984), massive infiltrations of mononuclear cells are seen in the affected tissue. Often these infiltrates, not associated with lymphoid tissue, form GC-like structures (Schroder et al., 1996) in which FDCs as well as T and B cells have been demonstrated. Such ectopic germinal centers or ectopic lymphoid aggregates are often present in the tissue of patients with RA, Sjogren's syndrome, Crohn's disease, Hasimoto's thyroiditis, and others (reviewed in Hjelmstrom, 2001; Weyland et al., 2001). These structures are believed to be a source of the autoreactive B cells and antibodies and to participate in maintenance of the autoimmune response.

The question arises whether in nonlymphoid tissue a microenvironment can be built up that allows antigen-dependent development and differentiation of B cells. Thus, in MS the question remains whether peripherally activated B cells are selectively recruited to the CNS tissue or B cells are recruited, mature locally, and differentiate into plasma cells emulating a germinal center.

A compartmentalized B cell response appears to occur in the CNS. All stages of B cell differentiation, which are usually observed only in secondary lymphoid organs, appear to occur in a favorable microenvironment arranged in the CNS. In the germinal centers of lymphoid follicles, antigen-activated B cells undergo clonal expansion and differentiation into memory B cells and plasma cells. The support structure required for this process to occur includes T cells and FDCs. FDCs present antigen and provide survival and proliferation signals to B cells.

These cells also produce a B cell chemoattractant, CXCL13, which regulates migration of B cells. FDCs can be recognized by their expression of CD35. CXCL13-expressing, CD35-positive FDCs associated with B cells have been observed in the meninges of a subset of patients with MS (Serafini et al., 2004). These ectopic lymphoid follicles do not appear to contain high endothelial vessels and T cell zones. However, the B cells present express Ki67, a lymphocyte proliferation marker. These findings indicate that germinal center-like structures in lymphoid follicles can develop in the CNS.

## 6.4   Strategies to Elucidate the Specificity of B Cells in the CNS

Thestudies concerning B cell lineages in MS are somewhat limited. Although many have focused on molecular characterization of the immunoglobulin genes, few of them have focused on the periphery; most have been concerned with the CSF and the CNS tissue. Before the appearance of the single-cell polymerase chain reaction (PCR), such analyses were typically performed by constructing a library of immunoglobulin variable region genes, followed by sequence analysis of the library elements. Both mRNA and DNA can be used for the PCR template. The advantage of mRNA is that it is abundant in cells, but with the caveat that activated B cells and plasma cells contain disproportionately more mRNA; thus, significant bias in the library is inevitable. The use of DNA circumvents this constraint, but only one copy from each cell is present, making PCR less reliable and the resulting library possibly nonrepresentative. Moreover, the primers do not include the constant region but rather the J, so the isotype information is lost, and occasionally somatic mutations in the J are lost owing to the primer. Preparations of multiple cDNA libraries from adjacent sections of tissue or aliquots of CSF cells or multiple PCR amplifications allow conclusions to be drawn regarding clonal expansion (Qin et al. 1998, 2003). Absence of such a strategy does not allow clonal expansion to be differentiated from PCR amplification of the same template, typically provided by mRNA-rich plasma cells. Isolation of single cells by either the fluorescence-activated cell sorting (FACS) or microdissection, followed by single-cell PCR permits the most comprehensive analysis of B cells from particular compartments. The restrictions and limitations imposed by libraries are eradicated.

## 7   B Cell-Directed Therapy for MS

The role of B cells in MS pathophysiology has led clinician-scientists to use them as targets in the management and treatment of MS. One novel approach along this line has been use of the chimeric monoclonal antibody rituximab. The antigen to which rituximab binds, CD 20, is expressed by most human B lymphocytes but not plasma cells. Rituximab (Rituxan; Genentech, South San Francisco, CA, USA and Biogen Idec, Cambridge, UK) is a human-mouse chimeric monoclonal antibody that targets the B cell CD20 antigen and causes rapid and specific B-cell depletion. Rituximab was approved in the United States in 1997 to treat low-grade or follicular, relapsed or refractory, CD20-positive B cell non-Hodgkin's lymphoma. Since then, the use of rituximab against autoimmune disorders, such as rheumatoid arthritis, immune thrombocytopenic purpura, autoimmune hemolytic anemia, systemic lupus erythematosus, and multiple sclerosis, is also under investigation. Although the exact in vivo mechanisms of action have not been clearly elucidated, in vitro data indicate that the drug induces apoptosis, complement-dependent cytotoxicity, and antibody-dependent cellular toxicity. Currently, rituximab is indicated in both follicular and aggressive B cell non-Hodgkin's lymphoma (Cartron et al., 2004). Early studies indicate effectiveness in MS patients. Beneficial clinical effects were demonstrated in two case studies of active RRMS patients given rituximab in which EDSS scores improved, T1-weighted magnetic resonance imaging lesion progression was hindered, and peripheral blood and CSF B cells were depleted (Petereit and Rubbert, 2005; Stuve et al., 2005). However, a larger study conducted by Monson et al. with PPMS patients showed that rituximab does not as efficiently deplete CSF B cells as their peripheral blood counterparts, probably because CSF B cells are highly activated advanced-memory or plasma B cells that express little or no CD20 (Monson et al., 2005). Larger studies are needed to determine the exact role of rituximab in the treatment of patients with MS.

## 8    Conclusion

The belief that MS is only a T cell-mediated autoimmune disease must be revised. The most common pattern of MS lesion pathology is characterized by the significant presence of components of the humoral immune system. The presence of oligoclonal bands and of B cell clonal expansion in CSF and the CNS tissue of MS patients supports the view that an ongoing antigen-driven humoral immune response is occurring. Although an unambiguous role for B cells and antibodies in the pathology of MS has not been defined, our current evidence for their relevance to the disease continues to motivate investigation that will surely provide empirical data for their role in the disease process.

*Acknowledgments.* The authors thank Jaylyn Olivo of the Brigham and Women's Hospital Editorial Service and Ms. Shannon McArdel for their editorial review.

## *References*

Anhalt, G. J., Labib R. S., Voorhees, J. J., et al. (1982) Induction of pemphigus in neonatal mice by passive transfer of IgG from patients with the disease. *N. Engl. J. Med.* 306:1189-1196.

Banki, K., Colombo, E., Sia, F., et al. (1994) Oligodendrocyte-specific expression and autoantigenicity of transaldolase in multiple sclerosis. *J. Exp. Med.* 180:1649-1663.

Baranzini, S. E., Jeong M. C., Butunoi, C., et al. (1999) B cell repertoire diversity and clonal expansion in multiple sclerosis brain lesions. *J. Immunol.* 163:5133-5144.

Barned, S., Goodman, A. D., Mattson, D. H. (1995) Frequency of anti-nuclear antibodies in multiple sclerosis. *Neurology* 45:384-385.

Berger, T., Rubner, P., Schautzer, F., et al. (2003) Antimyelin antibodies as a predictor of clinically definite multiple sclerosis after a first demyelinating event. *N. Engl. J. Med.* 349:139-145.

Brehm, U., Piddlesden, S. J., Gardinier, M. V., Linington, C. (1999) Epitope specificity of demyelinating monoclonal autoantibodies directed against the human myelin oligodendrocyte glycoprotein (MOG). *J. Neuroimmunol.* 97:9-15.

Breithaupt, C., Schubart, A., Zander, H., et al. (2003) Structural insights into the antigenicity of myelin oligodendrocyte glycoprotein. *Proc. Natl. Acad. Sci. U S A* 100: 9446-9451.

Brezinschek, H. P., Brezinschek, R. I., Lipsky, P. E. (1995) Analysis of the heavy chain repertoire of human peripheral B cells using single-cell polymerase chain reaction. *J. Immunol.* 155:190-202.

Brokstad, K. A., Page, M., Nyland, H., Haaheim, L. R. (1994) Autoantibodies to myelin basic protein are not present in the serum and CSF of MS patients. *Acta Neurol. Scand.* 89:407-411.

Bronstein, J. M., Lallone, R. L., Seitz, R. S., et al. (1999) A humoral response to oligodendrocyte-specific protein in MS: a potential molecular mimic. *Neurology* 53:154-161.

Burgoon, M. P., Hammack, B. N., Owens, G. P., et al. (2003) Oligoclonal immunoglobulins in cerebrospinal fluid during varicella zoster virus (VZV) vasculopathy are directed against VZV. *Ann Neurol* 54:459-463.

Burgoon, M. P., Keays, K. M., Owens, G. P., et al. (2005) Laser-capture microdissection of plasma cells from subacute sclerosing panencephalitis brain reveals intrathecal disease-relevant antibodies. *Proc. Natl. Acad. Sci. U S A* 102:7245-7250.

Burgoon, M. P., Owens, G. P., Carlson, S., et al. (2001) Antigen discovery in chronic human inflammatory central nervous system disease: panning phage-displayed antigen libraries identifies the targets of central nervous system-derived IgG in subacute sclerosing panencephalitis. *J. Immunol.* 167:6009-6014.

Burgoon, M. P., Owens, G. P., Smith-Jensen, T., et al. (1999) Cloning the antibody response in humans with inflammatory central nervous system disease: analysis of the expressed IgG repertoire in subacute sclerosing panencephalitis brain reveals disease-relevant antibodies that recognize specific measles virus antigens. *J. Immunol.* 163: 3496-3502.

Cartron, G., Watier, H., Golay, J., Solal-Celigny, P. (2004) From the bench to the bedside: ways to improve rituximab efficacy. *Blood* 104:2635-2642.

Colaco, C. B., Scadding, G. K., Lockhart, S. (1987) Anti-cardiolipin antibodies in neurological disorders: cross-reaction with anti-single standard DNA activity. *Clin. Exp. Immunol.* 68:313-319.

Colombo, E., Banki, K., Tatum, A. H., (1997) Comparative analysis of antibody and cell-mediated autoimmunity to transaldolase and myelin basic protein in patients with multiple sclerosis. *J. Clin. Invest.* 99:1238-1250.

Colombo, M., Dono, M., Gazzola, P., et al. (2000) Accumulation of clonally related B lymphocytes in the cerebrospinal fluid of multiple sclerosis patients. *J. Immunol.* 164:2782-2789.

Cook, G. P., Tomlinson, I. M. (1995) The human immunoglobulin VH repertoire. *Immunol. Today* 16:237-242.

Corcione, A., Casazza, S., Ferretti, E., et al. (2004) Recapitulation of B cell differentiation in the central nervous system of patients with multiple sclerosis. *Proc. Natl. Acad. Sci. U S A* 101:11064-11069.

Cortese, I., Tafi, R., Grimaldi, L. M., et al. (1996) Identification of peptides specific for cerebrospinal fluid antibodies in multiple sclerosis by using phage libraries. *Proc. Natl. Acad. Sci. U S A* 93:11063-11067.

Cross, A. H., Trotter, J. L., Lyons, J. (2001) B cells and antibodies in CNS demyelinating disease. *J. Neuroimmunol.* 112:1-14.

Esiri, M. M. (1977) Immunoglobulin-containing cells in multiple-sclerosis plaques. *Lancet* 2:478.

Fillatreau, S., Sweenie, C. H., McGeachy, M. J., et al. (2002) B cells regulate autoimmunity by provision of IL-10. *Nat. Immunol.* 3:944-950.

Gay, F. W., Drye, T. J., Dick, G. W., Esiri, M. M. (1997) The application of multifactorial cluster analysis in the staging of plaques in early multiple sclerosis: identification and characterization of the primary demyelinating lesion. *Brain* 120(Pt 8):1461-1483.

Genain, C. P., Cannella, B., Hauser, S. L., Raine, C. S. (1999) Identification of autoantibodies associated with myelin damage in multiple sclerosis. *Nat. Med.* 5:170-175.

Genain, C. P., Nguyen, M. H., Letvin, N. L., et al. (1995) Antibody facilitation of multiple sclerosis-like lesions in a nonhuman primate. *J. Clin. Invest.* 96:2966-2974.

Haubold, K., Owens, G. P., Kaur, P., et al. (2004) B-lymphocyte and plasma cell clonal expansion in monosymptomatic optic neuritis cerebrospinal fluid. *Ann. Neurol.* 56: 97-107.

Hjelmstrom, P. (2001) Lymphoid neogenesis: de novo formation of lymphoid tissue in chronic inflammation through expression of homing chemokines. *J. Leukoc. Biol.* 69:331-339.

Hjelmstrom, P., Juedes, A. E., Fjell, J., Ruddle, N. H. (1998) B-cell-deficient mice develop experimental allergic encephalomyelitis with demyelination after myelin oligodendrocyte glycoprotein sensitization. *J. Immunol.* 161:4480-4483.

Johnson, K. P., Arrigo, S. C., Nelson, B. J., Ginsberg, A. (1977) Agarose electrophoresis of cerebrospinal fluid in multiple sclerosis: a simplified method for demonstrating cerebrospinal fluid oligoclonal immunoglobulin bands. *Neurology* 27:273-277.

Kabat, E. A., Freedman, D. A., Murray, J. P., Knaub, V. (1950) A study of the crystalline albumin, gamma globulin and the total protein in the cerebrospinal fluid of one hundred cases of multiple sclerosis and other diseases. *Am. J. Med. Sci.* 219:55-64.

Kanter, J. L., Narayana, S., Ho, P. P., et al. (2006) Lipid microarrays identify key mediators of autoimmune brain inflammation. *Nat. Med.* 12:138-143.

Kuchroo, V. K., Anderson, A. C., Waldner, H., et al. (2002) T cell response in experimental autoimmune encephalomyelitis (EAE): role of self and cross-reactive antigens in shaping, tuning, and regulating the autopathogenic T cell repertoire. *Annu. Rev. Immunol.* 20:101-123.

Lafaille, J. J., Nagashima, K., Katsuki, M., Tonegawa, S. (1994) High incidence of spontaneous autoimmune encephalomyelitis in immunodeficient anti-myelin basic protein T cell receptor transgenic mice. *Cell* 78:399-408.

Lampasona, V., Franciotta, D., Furlan, R., et al. (2004) Similar low frequency of anti-MOG IgG and IgM in MS patients and healthy subjects. *Neurology* 62:2092-2094.

Lannes-Vieira, J., Gehrmann, J., Kreutzberg, G. W., Wekerle, H. (1994) The inflammatory lesion of T cell line transferred experimental autoimmune encephalomyelitis of the Lewis rat: distinct nature of parenchymal and perivascular infiltrates. *Acta Neuropathol. (Berl.)* 87:435-442.

Lennon, V. A., Kryzer, T. J., Pittock, S. J., et al. (2005) IgG marker of optic-spinal multiple sclerosis binds to the aquaporin-4 water channel. *J. Exp. Med.* 202:473-477.

Lennon, V. A., Wingerchuk, D. M., Kryzer, T. J., et al. (2004) A serum autoantibody marker of neuromyelitis optica: distinction from multiple sclerosis. *Lancet* 364:2106-2112.

Lily, O., Palace, J., Vincent, A. (2004) Serum autoantibodies to cell surface determinants in multiple sclerosis: a flow cytometric study. *Brain* 127(Pt 2):269-279.

Linington, C., Bradl, M., Lassmann, H., et al. (1988) Augmentation of demyelination in rat acute allergic encephalomyelitis by circulating mouse monoclonal antibodies directed against a myelin/oligodendrocyte glycoprotein. *Am. J. Pathol.* 130:443-454.

Liu, Z., Giudice, G. J., Swartz, S. J., et al. (1995) The role of complement in experimental bullous pemphigoid. *J. Clin. Invest.* 95:1539-1544.

Lock, C., Hermans, G., Pedotti, R., et al. (2002) Gene-microarray analysis of multiple sclerosis lesions yields new targets validated in autoimmune encephalomyelitis. *Nat. Med.* 8:500-508.

Lowenthal, A., Van Sande, M., Karcher, D. (1960) The differential diagnosis of neurological diseases by fractionating electrophoretically the CSF G-globulins. *J. Neurochem.* 235:229-233.

Lucchinetti, C., Bruck, W., Parisi, J., et al. (2000) Heterogeneity of multiple sclerosis lesions: implications for the pathogenesis of demyelination. *Ann. Neurol.* 47:707-717.

Mathey, E., Breithaupt, C., Schubart, A., Linington, C. (2004) Sorting the wheat from the chaff: identifying demyelinating components of the myelin oligodendrocyte glycoprotein (MOG)-specific autoantibody repertoire. *Eur. J. Immunol.* 34:2065-2071.

Mattson, D. H., Roos, R. P., Arnason, B. G. (1980) Isoelectric focusing of IgG eluted from multiple sclerosis and subacute sclerosing panencephalitis brains. *Nature* 287:335-337.

Mehta, P. D., Frisch, S., Thormar, H., et al. (1981) Bound antibody in multiple sclerosis brains. *J. Neurol. Sci.* 49:91-98.

Moller, J. R., Johnson, D., Brady, R. O., et al. (1989) Antibodies to myelin-associated glycoprotein (MAG) in the cerebrospinal fluid of multiple sclerosis patients. *J. Neuroimmunol.* 22:55-61.

Monson, N. L., Brezinschek, H. P., Brezinschek, R. I., et al. (2005) Receptor revision and atypical mutational characteristics in clonally expanded B cells from the cerebrospinal fluid of recently diagnosed multiple sclerosis patients. *J. Neuroimmunol.* 158:170-181.

Monson, N. L., Cravens, P. D., Frohman, E. M., et al. (2005) Effect of rituximab on the peripheral blood and cerebrospinal fluid B cells in patients with primary progressive multiple sclerosis. *Arch. Neurol.* 62:258-264.

Noseworthy, J. H., Lucchinetti, C., Rodriguez, M., Weinshenker, B. G. (2000) Multiple sclerosis. *N. Engl. J. Med.* 343:938-952.

Nousari, H. C., Anhalt, G. J. (1999) Pemphigus and bullous pemphigoid. *Lancet* 354:667-672.

O'Connor, K. C., Appel, H., Bregoli, L., et al. (2005) Antibodies from inflamed central nervous system tissue recognize myelin oligodendrocyte glycoprotein. *J. Immunol.* 175:1974-1982.

O'Connor, K. C., Bar-Or, A., Hafler, D. A. (2001) The neuroimmunology of multiple sclerosis: possible roles of T and B lymphocytes in immunopathogenesis. *J. Clin. Immunol.* 21:81-92.

O'Connor, K. C., Chitnis, T., Griffin, D. E., et al. (2003) Myelin basic protein-reactive autoantibodies in the serum and cerebrospinal fluid of multiple sclerosis patients are characterized by low-affinity interactions. *J. Neuroimmunol.* 136:140-148.

Olsson, T., Baig, S., Hojeberg, B., Link, H. (1990) Antimyelin basic protein and antimyelin antibody-producing cells in multiple sclerosis. *Ann. Neurol.* 27:132-136.

Owens, G. P., Burgoon, M. P., Anthony, J., et al. (2001) The immunoglobulin G heavy chain repertoire in multiple sclerosis plaques is distinct from the heavy chain repertoire in multiple sclerosis plaques is distinct from the heavy chain repertoire in peripheral blood lymphocytes. *Clin. Immunol.* 98:258-263.

Owens, G. P., Burgoon, M. P., Devlin, M. E., Gilden, D. H. (1997) Extraction and purification of active IgG from SSPE and MS brain. *J. Virol. Methods* 68:119-125.

Owens, G. P., Kraus, H., Burgoon, M. P., et al. (1998) Restricted use of VH4 germline segments in an acute multiple sclerosis brain. *Ann. Neurol.* 43:236-243.

Owens, G. P., Ritchie, A. M., Burgoon, M. P., et al. (2003) Single-cell repertoire analysis demonstrates that clonal expansion is a prominent feature of the B cell response in multiple sclerosis cerebrospinal fluid. *J. Immunol.* 171:2725-2733.

Panitch, H. S., Hafler, D. A., Johnson, K. P. (1980) Antibodies to myelin basic protein in cerebrospinal fluid of patients with multiple sclerosis. In: Baur, H. J., Poser, S., Ritter, G. (eds) *Progress in Multiple Sclerosis Research*. Berlin, Springer-Verlag, pp. 98-105.

Petereit, H. F., Rubbert, A. (2005) Effective suppression of cerebrospinal fluid B cells by rituximab and cyclophosphamide in progressive multiple sclerosis. *Arch. Neurol.* 62:1641-1642.

Piddlesden, S. J., Lassmann, H., Zimprich, F., et al. (1993) The demyelinating potential of antibodies to myelin oligodendrocyte glycoprotein is related to their ability to fix complement. *Am. J. Pathol.* 143:555-564.

Pillai, S. (1999) The chosen few? Positive selection and the generation of naive B lymphocytes. *Immunity* 10:493-502.

Prineas, J. W., Connell, F. (1978) The fine structure of chronically active multiple sclerosis plaques. *Neurology* 28(Pt 2):68-75.

Prineas, J. W., Graham, J. S. (1981) Multiple sclerosis: capping of surface immunoglobulin G on macrophages engaged in myelin breakdown. *Ann. Neurol.* 10:149-158.

Prineas, J. W., Wright, R. G. (1978) Macrophages, lymphocytes, and plasma cells in the perivascular compartment in chronic multiple sclerosis. *Lab. Invest.* 38:409-421.

Qin, Y., Duquette, P., Zhang, Y., et al. (2003) Intrathecal B-cell clonal expansion, an early sign of humoral immunity, in the cerebrospinal fluid of patients with clinically isolated syndrome suggestive of multiple sclerosis. *Lab. Invest.* 83:1081-1088.

Qin, Y., Duquette, P., Zhang, Y., et al. (1998) Clonal expansion and somatic hypermutation of V(H) genes of B cells from cerebrospinal fluid in multiple sclerosis. *J. Clin. Invest.* 102:1045-1050.

Raine, C. S., Cannella, B., Hauser, S. L., Genain, C. P. (1999) Demyelination in primate autoimmune encephalomyelitis and acute multiple sclerosis lesions: a case for antigen-specific antibody mediation. *Ann. Neurol.* 46:144-160.

Reindl, M., Linington, C., Brehm, U., et al. (1999) Antibodies against the myelin oligodendrocyte glycoprotein and the myelin basic protein in multiple sclerosis and other neurological diseases: a comparative study. *Brain* 122(Pt 11):2047-2056.

Ritchie, A. M., Gilden, D. H., Williamson, R. A., et al. (2004) Comparative analysis of the CD19+ and CD138+ cell antibody repertoires in the cerebrospinal fluid of patients with multiple sclerosis. *J. Immunol.* 173:649-656.

Robinson, W. H., DiGennaro, C., Hueber, W., et al. (2002) Autoantigen microarrays for multiplex characterization of autoantibody responses. *Nat. Med.* 8:295-301.

Roussel, V., Yi, F., Jauberteau, M. O., et al. (2000) Prevalence and clinical significance of anti-phospholipid antibodies in multiple sclerosis: a study of 89 patients. *J. Autoimmun.* 14:259-265.

Schluesener, H. J., Sobel, R. A., Linington, C., Weiner, H. L. (1987) A monoclonal anti-body against a myelin oligodendrocyte glycoprotein induces relapses and demyelination in central nervous system autoimmune disease. *J. Immunol.* 139:4016-4021.

Schmidt, S. (1999) Candidate autoantigens in multiple sclerosis. *Mult. Scler.* 5:147-160.

Schroder, A. E., Greiner, A., Seyfert, C., Berek, C. (1996) Differentiation of B cells in the nonlymphoid tissue of the synovial membrane of patients with rheumatoid arthritis. *Proc. Natl. Acad. Sci. U S A* 93:221-225.

Scolding, N. J., Morgan, B. P., Houston, W. A., et al. (1989) Vesicular removal by oligo-dendrocytes of membrane attack complexes formed by activated complement. *Nature* 339:620-622.

Serafini, B., Rosicarelli, B., Magliozzi, R., et al. (2004) Detection of ectopic B-cell folli-cles with germinal centers in the meninges of patients with secondary progressive mul-tiple sclerosis. *Brain Pathol.* 14:164-174.

Shapiro-Shelef, M., Calame, K. (2005) Regulation of plasma-cell development. *Nat. Rev. Immunol.* 5:230-242.

Smith-Jensen, T., Burgoon, M. P., Anthony, J., et al. (2000) Comparison of immunoglob-ulin G heavy-chain sequences in MS and SSPE brains reveals an antigen-driven response. *Neurology* 54:1227-1232.

Sorensen, T. L., Frederiksen, J. L., Bronnum-Hansen, H., Petersen, H. C. (1999) Optic neuritis as onset manifestation of multiple sclerosis: a nationwide, long-term survey. *Neurology* 53:473-478.

Spadaro, M., Amendolea, M. A., Mazzucconi, M. G., et al. (1999) Autoimmunity in mul-tiple sclerosis: study of a wide spectrum of autoantibodies. *Mult. Scler.* 5:121-125.

Storch, M. K., Piddlesden, S., Haltia, M., et al. (1998) Multiple sclerosis: in situ evidence for antibody- and complement-mediated demyelination. *Ann. Neurol.* 43:465-471.

Stuve, O., Cepok, S., Elias, B., et al. (2005) Clinical stabilization and effective B-lympho-cyte depletion in the cerebrospinal fluid and peripheral blood of a patient with fulminant relapsing-remitting multiple sclerosis. *Arch. Neurol.* 62:1620-1623.

Sun, J. B., Olsson, T., Wang, W. Z., et al. (1991) Autoreactive T and B cells responding to myelin proteolipid protein in multiple sclerosis and controls. *Eur. J. Immunol.* 21:1461-1468.

Terryberry, J. W., Thor, G., Peter, J. B. (1998) Autoantibodies in neurodegenerative diseases: antigen-specific frequencies and intrathecal analysis. *Neurobiol. Aging* 19:205-216.

Vandvik, B., Norrby, E., Nordal, H. J., Degre, M. (1976) Oligoclonal measles virus-spe-cific IgG antibodies isolated from cerebrospinal fluids, brain extracts, and sera from patients with subacute sclerosing panencephalitis and multiple sclerosis. *Scand. J. Immunol.* 5:979-992.

Vartdal, F., Vandvik, B., Norrby, E. (1982) Intrathecal synthesis of virus-specific oligo-clonal IgG, IgA and IgM antibodies in a case of varicella-zoster meningoencephalitis. *J. Neurol. Sci.* 57:121-132.

Villar, L. M., Masjuan, J., Gonzalez-Porque, P., et al. (2002) Intrathecal IgM synthesis in neurologic diseases: relationship with disability in MS. *Neurology* 58:824-826.

Villar, L. M., Masjuan, J., Gonzalez-Porque, P., et al. (2002) Intrathecal IgM synthesis pre-dicts the onset of new relapses and a worse disease course in MS. *Neurology* 59:555-559.

Villar, L. M., Masjuan, J., Gonzalez-Porque, P., et al. (2003) Intrathecal IgM synthesis is a prognostic factor in multiple sclerosis. *Ann. Neurol.* 53:222-226.

Villar, L. M., Sadaba, M. C., Roldan, E., et al. (2005) Intrathecal synthesis of oligoclonal IgM against myelin lipids predicts an aggressive disease course in MS. *J. Clin. Invest.* 115:187-194.

Vincent, A. (2002) Unravelling the pathogenesis of myasthenia gravis. *Nat. Rev. Immunol.* 2:797-804.

Von Budingen, H. C., Hauser, S. L., Ouallet, J. C., et al. (2004) Epitope recognition in the myelin/oligodendrocyte glycoprotein differentially influences disease phenotype and antibody effector functions in autoimmune demyelination. *Eur. J. Immunol.* 34: 2072-2083.

Warren, K. G., Catz, I. (1987) A correlation between cerebrospinal fluid myelin basic protein and anti-myelin basic protein in multiple sclerosis patients. *Ann. Neurol.* 21:183-189.

Warren, K. G., Catz, I. (1993) Autoantibodies to myelin basic protein within multiple sclerosis central nervous system tissue. *J. Neurol. Sci.* 115:169-176.

Warren, K. G., Catz, I. (1999) An extensive search for autoantibodies to myelin basic protein in cerebrospinal fluid of non-multiple-sclerosis patients: implications for the pathogenesis of multiple sclerosis. *Eur. Neurol.* 42:95-104.

Weyland, C. M., Kurtin, P. J., Goronzy, J. J. (2001) Ectopic lymphoid organogenesis. *Am. J. Pathol.* 159:787-792.

Williamson, R. A., Burgoon, M. P., Owens, G. P., et al. (2001) Anti-DNA antibodies are a major component of the intrathecal B cell response in multiple sclerosis. *Proc. Natl. Acad. Sci. U S A* 98:1793-1798.

Wolf, S. D., Dittel, B. N., Hardardottir, F., Janeway, C. A., Jr. (1996) Experimental autoimmune encephalomyelitis induction in genetically B cell-deficient mice. *J. Exp. Med.* 184:2271-2278.

Xiao, B. G., Linington, C., Link, H. (1991) Antibodies to myelin-oligodendrocyte glycoprotein in cerebrospinal fluid from patients with multiple sclerosis and controls. *J. Neuroimmunol.* 31:91-96.

Young, C. L., Adamson, T. C., 3rd, Vaughan, J. H., Fox, R. I. (1984) Immunohistologic characterization of synovial membrane lymphocytes in rheumatoid arthritis. *Arthritis Rheum.* 27:32-39.

Zhang, Y., Da, R. R., Hilgenberg, L. G., et al. (2005) Clonal expansion of IgA-positive plasma cells and axon-reactive antibodies in MS lesions. *J. Neuroimmunol.* 167:120-130.

Ziemssen, T., Ziemssen, F. (2005) The role of the humoral immune system in multiple sclerosis (MS) and its animal model experimental autoimmune encephalomyelitis (EAE). *Autoimmun. Rev.* 4:460-467.

# 6
# Molecular Basis for Induction of Tolerance in Type 1 Diabetes

Sophie Candon, Lucienne Chatenoud, and Jean-François Bach

## 1   Introduction

Type 1 diabetes (T1D) is caused by the progressive and selective destruction of insulin-producing pancreatic β cells by autoreactive T lymphocytes producing Th1-type cytokines such as interferon-γ (IFNγ), interleukin-2 (IL-2), and tumor necrosis factor-α (TNFα) (Bach, 1994; Delovitch and Singh, 1997). Although the pathogenesis of autoimmune T1D has been extensively studied, the precise cellular and molecular mechanisms involved in the initiation and progression of β cell destruction remain unclear. Animal models of spontaneous autoimmune diabetes, such as the nonobese diabetic (NOD) mouse and the BioBreeding (BB) rat have greatly enhanced our understanding of the pathogenic mechanisms involved in this disease (Lang and Bellgrau, 2004; Solomon and Sarvetnick, 2004). The disease results from the activation of both β cell-specific CD4+ and CD8+ T lymphocytes, as demonstrated by their ability to induce disease upon adoptive transfer into syngeneic immunodeficient hosts (i.e., NOD neonates, adult irradiated NOD mice, and NOD SCID mice) (Haskins and Wegmann, 1996; Wong, 1996). The development of the disease is also B cell-dependent (Akashi et al., 1997; Noorchashm et al., 1997; Serreze, 1996), although anti-islet autoantibodies found both in NOD mice and patients with T1D are not pathogenic (Myers et al., 1998; Yu et al., 2000). Early in the disease process, a small number of β cell autoantigens are the target of the autoimmune process. In young NOD mice, T cells reactive to the insulin B chain (Daniel and Wegmann, 1996; Daniel et al., 1995; Wegmann, 1994a,b), glutamic acid decarboxylase 65 (GAD65) (Baekkeskov et al., 1990; Honeyman et al., 1993; Panina-Bordignon et al., 1995; Tisch et al., 1993), and glucose-6 phosphatase catalytic subunit related protein (IGRP) (Lieberman et al., 2003) are detected in lymphoid organs and/or islet-infiltrating cells. T cells reactive to other antigens such as insulinoma-associated protein 2 (IA-2) (Dotta et al., 1999; Trembleau et al., 2000) and heat shock protein 60 (HSP60) (Elias et al., 1994) are recruited over time through epitope spreading, thus amplifying the overall autoimmune response toward β islets. T cell clones specific to these various autoantigens have been derived that transfer diabetes as efficiently as polyclonal diabetogenic CD4+ and CD8+ cells. Islet-reactive T cells are present in diabetic patients as well as in the blood of healthy individuals (Danke et al., 2004; Lohmann et al., 1996). The peaceful coexistence of such "physiological" autoreactive T cells with the organ expressing the target autoantigens reflects the concept of tolerance that, under peculiar incompletely understood circumstances, can be broken, leading to the development of diabetes.

## 2    Maintenance and Loss of Tolerance to β Cell Autoantigens

### 2.1    Central Tolerance

Central tolerance applies to immature T cells (Ohashi, 2003; Venanzi et al., 2004) during their stochastic development in the thymus and removes self-reactive lymphocytes from the repertoire. It is believed that T cells showing high affinity for peripheral autoantigens expressed in the thymus, including β cell autoantigens such as insulin (Vafiadis et al., 1997), I-A2, and GAD65 (Hanahan, 1998), are the ones deleted through negative selection (Mathis and Benoist, 2004). The protection from diabetes in both NOD mice and BB rats afforded by intrathymic grafting of syngeneic or allogenic β islets is in favor of such a deletional mechanism (Baumann et al., 1995; Charlton et al., 1994; Gerling et al., 1992; Herold et al., 1992; Koevary and Blomberg, 1992; Nomura et al., 1993; Posselt et al., 1992). Similarly, transgenic expression of a viral antigen (which in this setting may be considered an autoantigen) in the thymus and in the β cells, is associated with a delayed appearance of diabetes after infection with the virus, in contrast to the early-onset disease observed when the viral antigen is exclusively expressed in β cells (von Herrath et al., 1994).

Work on the AIRE gene has highlighted the role of ectopic expression of autoantigens in thymic selection. The primary mediators of negative selection are thymic medullary epithelial cells (MECs), a heterogeneous radio-resistant stromal cell fraction (Kishimoto and Sprent, 1999; Klein et al., 1998; Oukka et al., 1996a,b). Promiscuous expression of transcripts encoding tissue-specific proteins, including insulin, in MECs is controlled by AIRE, a transcription factor with transactivation potential (Anderson et al., 2002, 2005). Loss of function of the AIRE gene in humans is responsible for autoimmune polyendocrinopathy-candidiasis-ectodermal dystrophy (APECED) (Vogel et al., 2002). Using double antigen/T cell receptor (Ag/TCR) transgenic systems (Anderson et al., 2005; Liston et al., 2003, 2004a), it was recently shown that AIRE essentially controls clonal deletion of self-reactive thymocytes. Finally, it has been proposed that the higher levels of thymic proinsulin expression observed in individuals carrying the protective class III variable number tandem repeat (VNTR) located at the 5' end of the insulin gene are responsible for the dominant protection linked to the human insulin-dependent diabetes mellitus locus 2 (IDDM2) through more efficient deletion of proinsulin-reactive T cells (Pugliese et al., 1997).

The critical events involved in the loss of T cell tolerance to β cell proteins remain poorly defined. In the NOD model, it is believed that inefficient negative selection in the thymus increases the frequency of β cell-specific T cells in the periphery (Benoist and Mathis, 1997). In fact, the general propensity toward autoimmunity of the NOD mouse strain, as shown by the frequent association to T1D of spontaneous autoimmune sialitis, dacryoadenitis, and thyroiditis and to an increased susceptibility to other experimentally induced autoimmune diseases (Baker et al., 1995; Baxter et al., 1994; Boulard et al., 2002; Oldenborg et al., 2002; Rivero et al., 1998; Wicker et al., 1995; Zaccone et al., 2002) is thought to be linked to an intrinsic resistance of NOD thymocytes to clonal deletion (Choisy-Rossi et al., 2004; Kishimoto and Sprent, 2001; Lesage et al., 2002; Liston et al., 2004b; Zipris et al., 1991). Mature NOD thymocytes have altered sensitivity to deletion mediated by either anti-CD3 antibodies or superantigens in vitro and in vivo (Kishimoto and Sprent, 2001), and this effect appears to involve both Fas-dependent and Fas-independent pathways. The expression of the pro-apoptotic Bcl2-family inhibitor Bim, a mediator of thymic clonal deletion (Bouillet et al., 2002) is poorly induced in NOD thymocytes as compared to their B10 counterparts, suggesting a defect in the Bim pathway (Liston and Goodnow, 2005; Liston et al., 2004b). Several loci conferring susceptibility to defective thymic deletion have been identified (Liston et al., 2004b; Zucchelli et al., 2005), some of which co-localize with regions previously identified as diabetes susceptibility loci (Ghosh et al., 1993).

As mentioned before, the correlation between the number of copies of a particular VNTR located at the 5' end of the insulin gene (IDDM2) levels of insulin expression in the thymus and incidence of diabetes (Pugliese et al., 1997; Vafiadis et al., 1997) suggests that inefficient clonal deletion of insulin-reactive T cells when thymic levels of insulin expression are low might allow more of these insulin-reactive cells to exit the thymus. The establishment of a mouse model with graded thymic insulin expression (Chentoufi and Polychronakos, 2002) and the recent finding that IDDM2-deficient mice develop more aggressive diabetes than wild-type mice argues for the biological relevance of this correlation (Dubois-Lafforgue et al., 2002; Moriyama et al., 2003; Nakayama et al., 2005).

Although high frequencies of β cell (GAD, insulin, IGRP, HSP60)-reactive T cells have been identified in the NOD mouse using specific I-A$^{g7}$ tetramers

(Lieberman et al., 2004; Mallone et al., 2004; Reijonen et al., 2002, 2003; You et al., 2003), the recent development of HLA class I and II tetramers (Danke et al., 2005; Mallone et al., 2004; Oling et al., 2005; Reijonen et al., 2004; Yang et al., 2006) has not yet clearly demonstrated higher frequencies of β cell-reactive T cells in diabetic patients compared to normal individuals, which would support a role for defective thymic deletion in the pathogenesis of human T1D.

## 2.2    Peripheral Tolerance

Negative selection is only partial because autoantigen-reactive T cells are found in the peripheral blood of individuals without autoimmune disease (Lohmann et al., 1996; Semana et al., 1999). Self-reactive T cells, after their emigration from the thymus, are kept under control by mechanisms of peripheral tolerance that include indifference, anergy, clonal deletion, and suppression by specialized subsets of T cells, namely regulatory T cells (Walker and Abbas, 2002; Bach and Chatenoud, 2001). The first three mechanisms (referred to as passive mechanisms of tolerance) have been investigated essentially in TCR transgenic models because they are difficult, if not impossible, to address in experimental systems with polyclonal T cell repertoires (Lohr et al., 2005).

### 2.2.1    Passive Peripheral Tolerance

Clear evidence for indifference or ignorance came from double transgenic mice expressing an antigen under the control of the rat insulin promoter (RIP) and a major histocompatibility complex I (MHC I)-restricted relevant TCR. In this type of experimental setting, the transgenic T cells ignored their target antigen in the β cells while remaining fully reactive toward the exogenously introduced antigen (Ohashi et al., 1991). In another, similar experimental setting, tolerance through indifference was reversed by the transgenic expression of IL-2 in islet β cell (Heath et al., 1992). Such ignorance was thought to be due to the fact that islet β cells are poor antigen-presenting cells (APCs), unable to provide necessary co-stimulatory signals necessary for the activation of naive CD8+ T cells. Such unresponsiveness to β cell-expressed antigen can indeed be overcome by the co-expression of B7.1 by the β cell (von Herrath et al., 1995; Wong et al., 1995). Alternatively, uptake of the antigen and presentation by professional APCs may be too quantitatively insufficient to be operational. Indifference is not, however, a general feature of CD8 transgenic T cells, as diabetes occurred spontaneously and rapidly in a similar double transgenic model using a different β cell-expressed antigen, the influenza hemagglutinin (HA) (Morgan et al., 1996). It was proposed that differences affecting T cell triggering, such as the level of antigen presentation, the affinity of the specific TCR for its cognate MHC/peptide complex, or co-stimulation requirements might account for the lack of tolerance in this model.

Apoptosis not only mediates negative selection of autoreactive T cells during thymic ontogeny, it also regulates immune homeostasis in the periphery. Upon activation by antigens, mature T cells may undergo clonal deletion by a mechanism of activation-induced cell death (AICD) involving CD95/CD95L interactions and TNFα receptor (TNFR) signaling. Integrated signaling through the TCR and co-stimulatory pathways must exceed a threshold level to trigger an AICD response, and this threshold is higher than the stimulatory levels required to activate imunological effector functions (Girao et al., 1997; Ucker et al., 1992). Clonal deletion of autoreactive T cells in the periphery was initially demonstrated in RIP-OVA transgenic mice expressing large amounts of ovalbumin (OVA) in pancreatic β cells and transferred with transgenic OT-I T cells (Kurts et al., 1999). However, other groups analyzed double antigen/TCR transgenic mice expressing HA in β cells and various transgenic MHC II-restricted HA-specific TCRs. In none of these models was there any evidence for efficient peripheral T cell tolerance. In all cases, spontaneous activation of transgenic T cells was observed that led to variable disease patterns ranging from benign insulitis with no disease (Scott et al., 1994) to moderate diabetes (Lanoue et al., 1997; Scott et al., 1994) or fulminant diabetes (Radu et al., 1999). These differences were ascribed to either the non-MHC genetic background in which the transgenes were expressed or the differential affinities of the TCRs for their cognate MHC/peptide complexes.

Anergy is a state of long-lived functional T cell unresponsiveness with regard to proliferation and IL-2 production that was first described with T cell lines stimulated in vitro with strong TCR-mediated stimuli (anti-CD3 antibody or cognate antigen) in the absence of co-stimulatory molecules (Jenkins et al., 1990; Macian et al., 2002; Schwartz, 1990). In vivo, anergy appears not simply as the result of antigen recognition without co-stimulation but an active process induced by particular conditions of self-antigen presentation (Schwartz, 2003). Although anergy has clearly been identified as a mechanism by which immunotherapy can restore tolerance to β cell antigens in NOD mice (as discussed below), little evidence in favor of a role for anergy as a mechanism of tolerance to pancreatic autoantigens has been reported. Notably, in a TCR transgenic model, it was shown that one mechanism of MHC I-mediated T1D resistance is the ability to anergize β cell-autoreactive CD8 T cells (Choisy-Rossi et al., 2004). TCR transgenic AI4 CD8+ T cells that are $K^d$-restricted and pathogenic on the NOD $H2^{g7}$ background were efficiently selected on an $H2^b$ background and, interestingly, anergized through TCR downmodulation, thus leading to complete resistance of (NOD.$H2^b$ X NOD.AI4αβ Tg) F1 mice to diabetes.

Thus, taken together, these various transgenic models show that peripheral self-tolerance through indifference, deletion, or anergy is not a common phenomenon; and when it occurs, indifference and peripheral deletion appear to be the main mechanisms.

Extensive indirect evidence for ineffective "passive" mechanisms of peripheral tolerance in NOD mice has been provided (Arreaza et al., 2003; Dahlen et al., 2000; Feili-Hariri and Morel, 2001; Grohmann et al., 2003; Kreuwel et al., 2001; Prasad and Goodnow, 2002; Rosmalen et al., 2002; Serreze and Chen, 2005; Serreze et al., 1993; Strid, 2001). On one hand, several studies have suggested that APCs might be unable to fulfill the requirements for optimal T cell stimulation leading to AICD (defective myelopoieisis leading to impaired maturation of macrophages and dendritic cells and impaired co-stimulation, enhanced expression of prostaglandins by APCs suppressing T cell IL-2-mediated signal transduction). On the other hand, abnormalities in signal transduction pathways in the T cells themselves have been reported. Indeed, NOD lymphocytes are relatively resistant to the induction of apoptosis by several stimuli. Direct evidence for defective AICD comes from the observation of impaired peripheral deletion of CD8+ T cells in transgenic NOD mice expressing HA in their β cells (Kreuwel et al., 2001).

### 2.2.2    Immunoregulation

In the NOD model, there is a significant delay (4–8 weeks) between the first signs of autoreactive T cell activation at 3 to 4 weeks of age when intra-islet inflammatory infiltration begins and the onset of significant islet cell destruction resulting in overt diabetes at 3 to 5 months of age. Data have accumulated to suggest that this "prediabetic" phase is due to T cell-mediated regulatory mechanisms countering the anti-islet immune response (Bach, 1994; Delovitch and Singh, 1997). Initial evidence for the existence of specialized T cell suppressor subsets in the NOD model stems from co-transfer experiments showing that young prediabetic mice harbor a CD4+ T population that fully prevents disease when co-transferred with syngeneic diabetogenic cells into immunodeficient recipients (Boitard et al., 1989; Hutchings and Cooke, 1990). Later, it was shown that the protective or suppressor capacity is concentrated in the CD4 + CD62L+ compartment (Herbelin et al., 1998; Lepault and Gagnerault, 2000; Lepault et al., 1995). Similar observations were made using CD25 as a marker of regulatory T cells. The highest regulatory activity was found in the CD4 + CD25 + CD62L+ subset (Salomon et al., 2000). In fact, regulatory activity is found in both the CD62L+ and CD25+ subsets of CD4+ T lymphocytes, and these two subsets of regulatory T cells only partially overlap in young NOD mice (Alyanakian et al., 2003). CD4+CD25+CD62L+ regulatory T cells originate in the thymus (Boitard et al., 1989; Dardenne et al., 1989) but can also be generated in the periphery through adaptative mechanisms (Akbar et al., 2003; Bach and Chatenoud, 2001; Bluestone and Abbas, 2003; Walker et al., 2003). Support for their thymic origin comes from studies showing that thymectomy of NOD females at weaning (3 weeks of age) accelerates disease onset (Dardenne et al., 1989) and that mature HSA+αβTCR+CD4+CD62L+ thymocytes from prediabetic NOD mice fully prevent diabetes transfer (Herbelin et al., 1998). The functional absence of B7.1, B7.2, CD40, CD40L, or CD28 molecules results in a significant decrease in the frequency/function of CD4+CD25+ regulatory T cells in the periphery and the induction of autoimmunity (Boden et al., 2003; Salomon et al., 2000), which indicates a role for co-stimulation in the development of CD4+CD25+ regulatory T cells. NOD-CD28–/– and NOD-B7.1/B7.2–/– mice develop diabetes more rapidly than their NOD control littermates, whereas infusion of wild-type syngeneic CD25+ T cells back into CD28–/– NOD recipients prevents disease, suggesting that CD4+ CD25+ regulatory T cells control β islet cell autoimmunity (Salomon et al., 2000).

The naturally occurring CD4+ CD25+ T cells present in NOD mice resemble those described in other models, such as the polyautoimmune syndrome induced by day 3 thymectomy in BALB/c mice (Asano et al., 1996; Sakaguchi et al., 1995; Suri-Payer et al., 1996, 1998) or the diabetes observed after thymectomy and sublethal irradiation of PVG rats (Saoudi et al., 1996; Seddon and Mason, 1999; Stephens and Mason, 2000). In particular, they express high levels of the transcription factor foxp3 (Herman et al., 2004; Pop et al., 2005; Chatenoud, unpublished observation), at present considered to be the most specific "candidate" marker of regulatory T cells (Fontenot and Rudensky, 2005; Gavin and Rudensky, 2003; Ramsdell, 2003), as well as other cell surface markers such as CTLA4 (McHugh et al., 2002), GITR (McHugh et al., 2002; Tone et al., 2003), membrane transforming growth factor-$\beta$ (TGF$\beta$) (Nakamura et al., 2001) and CD103 (Lehmann et al., 2002; Zelenika et al., 2002). In vitro, they effectively suppress the anti-CD3-induced proliferation and cytokine production of autologous CD4+CD25-T cells (Alyanakian et al., 2003; You et al., 2005, 2006).

In the NOD model, the in vivo suppressive ability of naturally occurring CD4+CD25+ regulatory T cells (isolated from the spleen or the thymus of wild-type NOD mice) upon co-transfer into immunocompromised hosts is rather inefficient compared to other models, requiring high numbers of regulatory T cells (Herbelin et al., 1998; Lepault and Gagnerault, 2000; Salomon et al., 2000). A low precursor frequency of regulatory T cells specific to pancreatic self-antigens is a likely explanation, as regulatory T cells specific to a pancreatic self-antigen isolated from NOD-BDC2.5 TCR transgenic mice more efficiently inhibit diabetes after activation and co-transfer with diabetogenic T cells in immunodeficient NOD mice than activated polyclonal regulatory T cells from NOD mice (Kanagawa et al., 2002; Tang et al., 2004). This is in keeping with the observation that adoptive transfer of freshly isolated or activated NOD CD4+CD25+ T cells into immunocompetent hosts, such as CD28–/– NOD mice, can delay diabetes only when high numbers of cells are infused (> $5 \times 10^6$ cells) (Salomon et al., 2000), whereas low numbers ($5 \times 10^5$) of activated BDC2.5 CD25+ regulatory T cells completely prevent diabetes (Tang et al., 2004). These results suggest that in the NOD model regulatory T cells need to be activated antigen-specifically at the site of the autoimmune process to exert their suppressive function. Indeed, data suggest that regulatory T cells act directly in the pancreatic islet, rather than in the pancreatic lymph node, thought to be the site of the initial priming of autoreactive T cells (Herman et al., 2004). Furthermore, CD4+CD25+ regulatory T cells responsive to a single autoantigen, such as those of BDC2.5 mice, can inhibit diabetes mediated by pathogenic T cell responses to multiple autoantigens.

The protective effect of CD4+CD62L+ T cells is not abrogated by treatment with antibodies to IL-4 and IL-10 (Lepault and Gagnerault, 2000). Furthermore, IL-4–/– mice do not show accelerated diabetes onset (Wang et al., 1998). However, anti-TGF$\beta$ antibodies inhibit the protective effect of CD4+CD25+ CD62L+ regulatory T cells from prediabetic NOD mice cells in co-transfer experiments (You et al., 2006). In vitro, we have found that the inhibition of CD4+CD25− T cells mediated by CD4+CD25+ regulatory T cells isolated from young NOD mice was abrogated by an anti-TGFβ antibody in a dose-dependent fashion (You, 2005, 2006). Overall, our results indicate that in the NOD model naturally occuring regulatory T cells mediate in vitro and in vivo suppression through mechanisms involving TGFβ.

Despite the presence of CD4+CD25+ regulatory T cells during the prediabetic phase, NOD mice go on to develop T1D. Defects in naturally occurring regulatory T cells may play a role in this process. A general decrease in the numbers of CD4+CD25+ natural regulatory T cells in NOD mice compared to other strains has been reported by some groups (Salomon et al., 2000; Wu et al., 2001). Another study, however, found normal numbers of these cells in the thymus and lymphoid organs, including pancreatic lymph nodes, of prediabetic 6-week-old NOD females compared to age-matched females from three nonautoimmune strains (CBA, BALB/c, B6) (Berzins et al., 2003). Therefore, it is still not clear whether quantitative deficiencies in the CD4+CD25+ compartment may partly explain the loss of tolerance in the NOD mouse. However, data from Adorini's group and our own suggest that a decline in regulatory T cell function during the progression from insulitis to diabetes may contribute to disease establishment (Gregori et al., 2003; You et al., 2005). In fact, CD4+CD25+ regulatory T cells from overtly diabetic NOD mice (16 to 20 weeks old) seem less potent suppressors than those from prediabetic 6- to 8-week-old mice. The progressive decline of the suppressive functions of the CD4+CD25+CD62+ regulatory T cells is associated with a decrease in the proportion of cells expressing foxp3 and TGFβ in this subset (Pop et al., 2005). In the absence of more in-depth knowledge of the molecular mechanisms involved in the suppression mediated by regulatory T cells, it is difficult to address why and how CD4+CD25+ T cells lose their suppressive ability with aging. Interestingly, there is now evidence that reduced sensitivity of diabetogenic T cells to immunoregulation might participate in the progression from nondestructive insulitis to overt diabetes (Gregori et al., 2003; You et al., 2005).

CD4+CD25+ regulatory T cells are not the only regulatory T cell population in NOD mice. We identified a CD4+CD25−CD62L+ subset capable of inhibiting disease transfer when co-injected into NOD-SCID mice with diabetogenic T cells (You, 2004). CD4+CD25−CD62L+ regulatory T cells have also been described in a model of diabetes induced by thymectomy and sublethal irradiation of PVG rats (Saoudi et al., 1996; Seddon and Mason, 1999; Stephens and Mason, 2000). Moreover, in the NOD-BDC2.5 TCR transgenic model, T cells expressing low levels of BDC2.5 TCR and displaying functional features of regulatory T cells were found in both the CD25+ and CD25− subsets (Kanagawa et al., 2002). We have demonstrated that diabetes onset in BDC2.5-NOD-*rag*−/− mice devoid of regulatory T cells is completely prevented by infusion of polyclonal NOD CD4+CD62L+ T cells but only marginally so by infusion of NOD CD4+CD25+ T cells (You et al., 2004). At variance with CD4+CD25+ T cells, CD4+CD25−CD62L+ T cells found in NOD mice are not anergic, not suppressive in vitro

(Alyanakian et al., 2003), and they do not express GITR, CD103, or foxp3 (Chatenoud, unpublished observations). However, they were shown to produce TGFβ but no IL-4 or IL-10 (Alyanakian et al., 2003). Overall, it clearly appears that in the NOD model CD4+CD25+ and CD25−CD62L+ T cells can afford regulation separately but also probably synergistically.

The presence in humans of CD4+CD25+ T cells exhibiting features similar to those of their mouse counterparts has recently been demonstrated (Baecher-Allan et al., 2001; Dieckmann et al., 2001; Stephens et al., 2001). Their suppressive function, tested in proliferation assays, seems to segregate a minor subset of cells exhibiting the CD25$^{high}$ phenotype (Baecher-Allan et al., 2004). Regulatory T cells likely play an important role in keeping autoreactive T cells in check, as identification of certain CD4+ self-reactive T cells, including GAD65-specific T cells, in the peripheral blood of healthy individuals using HLA class II tetramers, which requires an ex vivo antigen-specific expansion step, is possible only after depletion of the CD4+CD25+ T cell subset (Danke et al., 2004). Conflicting data have been reported concerning the frequencies and functional properties of these regulatory T cells in patients with T1D. One study reported a quantitative defect of peripheral blood CD4+CD25+ T cells in patients with newly diagnosed T1D, but the in vitro suppressive ability of this particular T cell subset was not tested (Kukreja et al., 2002). Another report showed that although proportions of CD4+CD25+ T cells were normal in patients with recent-onset adult T1D the in vitro suppressive ability of the regulatory T cells in this population was markedly reduced compared with control subjects (Lindley et al., 2005). In contrast, another group showed no significant differences in the number or in vitro regulatory function of CD4+CD25$^{high}$ cells in chronic T1D subjects (Putnam et al., 2005). Differences in the selected groups of patients (recent-onset versus chronic T1D) might account for these contrasting results. Evidence in humans indicates that CD4+CD25+ regulatory T cells may be generated in the periphery after activation of CD4+CD25− T cells, at least in vitro (Akbar et al., 2003; Walker et al., 2003). This is particularly interesting in light of recent work that we discuss below showing that the effect of immunointervention therapies tested in the NOD model and in humans correlate with an increase in CD4+CD25+ T cells (Belghith et al., 2003; Feili-Hariri et al., 2002; Goudy et al., 2003; Gregori et al., 2002; Mukherjee et al., 2003).

In summary, the NOD mouse has several defects in thymic selection processes, but it remains to be demonstrated that these defects alter the repertoire of mature T cells in a manner that directly promotes T1D. Extensive evidence of ineffective peripheral tolerance induction/maintenance has also accumulated. In particular, the progressive failure of regulatory T cells in keeping in check autoreactive T cells in the periphery appears as an important factor participating to T1D progression. Better comprehension of the phenotype and mode of action of regulatory T cells will allow the development of therapeutic strategies aimed at reestablishing an adequate balance between regulatory T cells and pathogenic T cells.

# 3    Approaches to Restore Self-Tolerance to β Cell Antigens

Because of the many similarities of autoimmune diabetes in NOD mice with human T1D (regarding genetic susceptibility, disease pathogenesis, and candidate autoantigens), the NOD mouse has been used extensively as a preclinical tool for the development of immunotherapy. Two major checkpoints have been defined in the pathogenesis of diabetes in NOD mice. The first checkpoint, between 3 and 4 weeks of age (concomitant with the appearance of measurable reactivity to pancreatic islet antigens) regulates the onset of insulitis and is controlled by APC subsets and the expression of adhesion molecules and integrins. The second checkpoint, active between 8 and 12 weeks, regulates the progression from "controlled," benign insulitis to destructive, "aggressive" insulitis and the development of glucose dysregulation and clinical overt diabetes (Andre et al., 1996). Strategies aiming at preventing the loss of tolerance before the appearance of any measurable autoreactivity (before checkpoint 1) would have to be carried out in subjects with an increased genetic risk of developing T1D and without islet autoantibodies. Considering that only a subset of these individuals develop diabetes, such strategies should have minimal risks and side effects (Gianani and Eisenbarth, 2005). A more realistic timing of immunotherapy in T1D is to act after tolerance to β cell antigens has been broken, either before checkpoint 2 at an early prediabetic phase (in patients with islet autoantibodies and intact insulin secretion) or later when the destruction of β cells has been initiated (islet autoantibody-positive prediabetic patients with loss of first phase insulin secretion or recently diagnosed patients). Many immune intervention approaches have proven effective in NOD mice during the prediabetic phase, but few strategies are able to reverse the disease once it is clinically overt (Shoda et al., 2005). Initial clinical therapies targeting pathogenic T cell effectors used immunosuppressive agents such as cyclosporin A (Feutren et al., 1986), prednisone (Elliott et al., 1981), azathioprine (Silverstein et al., 1988), and anti-thymocyte globulins (Eisenbarth et al., 1985). These treatments afforded a modest delay in diabetes onset or even (in the case of cyclosporin) reversed diabetes, but they were associated with significant side effects, thus precluding prolonged therapy. One major focus of research has been to develop approaches that restore tolerance to self-antigens in patients with autoimmune diseases without nonspecific immunosuppression. The demonstration of a role for regulatory T cells in the pathogenesis of diabetes in NOD mice prompted strategies aimed at reestablishing the functional balance between pathogenic effector T cells and regulatory T cells through enhancement/induction of the immunoregulatory capacities of the immune system. As we shall see below, various types of adaptative regulatory T cells can be induced in vivo (Filippi et al., 2005) (Table 6.1): CD4+CD25+ regulatory T cells resembling their thymic naturally occurring counterparts, IL-10-secreting CD4+ T cells (Tr1 cells) (Roncarolo et al., 2001), TGFβ-secreting CD4+ T cells (Th3) (Weiner, 2001a), IL-4-secreting Th2-like cells (Mosmann and Coffman, 1989), and CD8+ γδ cells (Harrison et al., 1996). Alternatively, protocols to expand certain subsets of regulatory T cells ex vivo have recently been developed, and the in vivo therapeutic proprieties of these expanded regulatory T cells are currently being analyzed.

TABLE 6.1. Diversity of Regulatory T Cells.

| Cell phenotype | Independent lineage | Site of generation | Mode of induction | Mechanism of suppression | Expression of foxp3 |
|---|---|---|---|---|---|
| Natural CD4+ CD25+ regulatory T cell | Yes | Thymus | Naturally occuring | Cell contact (T-T, T-APC) Cytokine-independent in vitro | Yes |
| Adaptative CD4+CD25+ regulatory T cell | No | Periphery | CD3 antibody | Cell contact? Cytokine-dependent TGF-β | Yes |
| CD4+ Th2 | No | Periphery | Various routes and forms of autoantigen delivery (DNA vaccine, IV soluble antigen) | Cytokine-dependent IL-4, IL-10 | No |
| CD4+ Th3 | No | Periphery | Mucosal, nasal, oral immu-nization | Cytokine-dependent TGF-β | No |
| CD4+ Tr1 | No | Periphery | Mucosal, nasal, oral immu-nization | Cell contact? Cytokine-dependent IL-10 | No |

## 3.1    Vaccination to β Cell Autoantigens

A direct approach to immune modulation is the induction of hyporesponsiveness to the self-antigen that causes the disease through introduction of this self-antigen in a tolerizing vaccination regimen, such as intravenous or intranasal injection, subcutaneous immunization in incomplete Freund's adjuvant (IFA), oral feeding, or DNA vaccination. Such immunization with self-antigens has been shown to pre-vent T1D development through the induction of autoreactive regulatory T cells leading to deviation of the autoimmune response from an aggressive to an innocu-ous one. In many of the reported studies, antigen-induced protection is, in fact, transferable to naive recipients by αβ CD4+ T cells, supporting the notion of an active tolerance phenomenon (Elias and Cohen, 1995; Muir, 1995; Tian et al., 1996a; Zhang et al., 1991). The phenotype of the induced regulatory T cells depends on the strategy employed to deliver the antigen. Other mechanisms involved in the induction of tolerance include anergy through downregulation of specific TCRs, and peripheral clonal deletion. Most of the strategies tested so far in the NOD model are effective during the prediabetic phase (Shoda et al., 2005).

### 3.1.1    Various Tolerogens and Protocols

The first vaccinal approach tested in NOD mice used GAD, against which T cell reactivity can be detected as early as 4 weeks of age, as well as autoantibodies soon after initiation of the autoimmune insulitis (Tisch et al., 1993). GAD 65 was initially injected in young NOD mice intravenously (Kaufman, 1993) and intrathymically (Tisch et al., 1993), resulting in prevention of diabetes onset. Similar results in various mouse models of autoimmune diabetes were subsequently reported using GAD65 protein or peptides in IFA (Tisch et al., 1998, 1999), GAD67 (isoform) (Elliott et al., 1994), GAD-derived peptides (Sai et al., 1996) administered nasally (Tian et al., 1996a) or plasmidic GAD DNA (Balasa et al., 2001; Li and Escher, 2003; Tisch et al., 2001), anti-GAD monoclonal antibodies (Menard et al., 1999), GAD antisense DNA (Yoon et al., 1999), or vaccinia virus-expressing GAD (Jun et al., 2002). In the case of vaccination with GAD-derived peptides, some studies reported an absence of effect on disease incidence after immunization or even exacerbation of the disease (Zechel, 1998a,b). A recombinant form of GAD65 formulated in alum was successfully tested in NOD mice (Ramiya, 1996), which led to a safety and efficacy Phase II study in patients with latent autoimmune diabetes in adults (LADA) (Agardh et al., 2005). No definitive conclusion concerning the effect of this therapy on β cell function can be drawn from this preliminary study.

HSP60, particularly its p277 epitope, protects mice exposed to multiple low doses of streptozocin and NOD mice from insulitis and diabetes (Elias and Cohen, 1995, 1996; Elias et al., 1991, 1994, 1997), with a corresponding shift of splenic Th1 cells toward a tolerogenic Th2 phenotype. Further support for the protective effect of HSP60 comes from the finding that overexpression of p277 in transgenic mice renders the animals resistant to diabetes (Birk et al., 1996). Remarkably, p277-based therapy is able to prevent diabetes when administered at a late prediabetic stage (up to 17 weeks) (Elias et al., 1994) and to arrest streptozocin-induced diabetes (Elias and Cohen, 1996). Based on these studies and the report of increased T cell responses to HSP60 and p277 in T1D patients (Abulafia-Lapid et al., 1999), a human vaccine, DiaPep277, was developed and tested in a placebo-controlled Phase II trial with recently diagnosed T1D patients (Raz et al., 2001). It was shown that at 10 months of follow-up, β cell function was significantly preserved in the DiaPep277 group, and T cells exposed to p277 had shifted at the end of the trial from a Th1 to a Th2 response. It was recently suggested that the preservation of β cell function afforded by the DiaPep277 vaccine might be related to uneven stimulation of different Toll-like receptors (TLRs) (Zanin-Zhorov et al., 2005).

The best known and most studied β cell-specific antigen is insulin. Anti-insulin antibodies can be found early during the prediabetic phase of human patients. In NOD mice, around 50% of T cells infiltrating the pancreatic islets react with insulin (Wegmann et al., 1994a). Protection was reported with intravenous, subcutaneous, oral, and nasal administration of insulin (Atkinson et al., 1990; Bergerot et al., 1997; Bertrand et al., 1992; Daniel and Wegmann, 1996; Gotfredsen, 1985; Hanninen, 2000; Harrison et al., 1996; Karounos, 1997; Muir, 1995; Zhang et al., 1991). The initial use of the metabolically active hormone (Atkinson et al., 1990; Bertrand et al., 1992; Gotfredsen, 1985) precluded clear interpretation of the data owing to the "β cell rest" afforded by insulin that has been associated with decreased expression of some islet autoantigens (Anastasi et al., 1999). The fact that the same protective effect was later observed upon administration of the nonmetabolically active insulin B chain strongly argues for the tolerogenic potential of insulin as a vaccine (Karounos, 1997; Muir, 1995; Ploix et al., 1998; Polanski et al., 1997). Studies performed in mice vaccinated with oral insulin have shown that insulin B chain-reactive regulatory T cells secreting IL-4 and acting as bystander suppressors mediated diabetes prevention. These results led to the Diabetes Prevention Trial-1 (DPT1) in which prediabetic relatives of diabetic patients were treated with oral or parenteral insulin. This trial failed to prevent progression to overt diabetes and had no effect on the autoimmune process (DPT1 DiabetesStudyGroup, 2002; Pozzilli, 2002; Pozzilli et al., 2000). Another trial conducted in prediabetic patients treated with oral insulin reported similar results (Chaillous et al., 2000). Although the reasons for these failures are unclear, an insufficient dose and the bioavailability of insulin are thought to be likely factors (Shoda et al., 2005), as well as insufficient numbers of induced regulatory T cells, concomitant activation of pathogenic T cells, and inefficient homing of the regulatory T cells to the target organ (although it is not clear whether regulatory T cells were induced at all). It is, however, noteworthy that a subgroup of patients with the highest titers of insulin autoantibodies did in fact respond to treatment with oral insulin (Skyler et al., 2005). The demonstration that aerosol insulin administered to NOD mice at a late prediabetic stage reduced the frequency of diabetes onset through induction of protective CD8 $\gamma\delta$ T cells (Harrison et al., 1996) led to a pilot study testing the safety and efficacy of intranasal administration of insulin in prediabetic individuals (Harrison et al., 2004). Although no effect on diabetes incidence could be formally demonstrated, it was shown that T cell-mediated responses to insulin were diminished in insulin-treated patients compared to placebo-treated subjects, and the titers of anti-insulin autoantibodies were increased, suggesting a shift from a Th1- to a Th2-type response. Epitope mapping of insulin chains has revealed that the B chain contains an immunodominant peptide $InsB_{9-23}$. In the NOD mouse, among the T cells infiltrating the pancreatic islets and reactive to insulin, 97% respond to $InsB_{9-23}$. This epitope seems to be particularly efficient in T1D prevention in mouse models (Daniel and Wegmann, 1996; Heath et al., 1999; Ramiya, 1996; Urbanek-Ruiz et al., 2001). A Phase II trial is currently being held using a variant of $InsB_{9-23}$ as a vaccine in early diabetes (Alleva et al., 2006).

IGRP has recently been identified in the NOD mouse as the target of CD8 pathogenic T cells detected in early islet infiltrates of 9-week-old NOD mice (Lieberman et al., 2003). Many T IGRP epitopes are targeted, but CD8 T cells reactive against the $IGRP_{206-215}$ epitope appear prevalent (Han et al., 2005). An increase in the avidity of these IGRP-reactive T cells is associated with progression of the disease (Amrani et al., 2000). Treatment of prediabetic or diabetic NOD mice with high doses of $IGRP_{206-215}$, despite inducing nearly complete deletion of intra-islet $IGRP_{206-215}$-reactive CD8+ T cells, was ineffective at preventing or inhibiting the disease (Han et al., 2005). The expansion of CD8+ T cells reactive to subdominant IGRP epitopes and involved in disease progression was proposed as a mechanism to explain this paradoxical inefficiency. By contrast, several altered variants of $IGRP_{206-215}$ afforded protection from diabetes through deletion of high-avidity $IGRP_{206-215}$-reactive CD8 T cells and expansion of their low-avidity, nonpathogenic counterparts. Interestingly, the human IGRP gene (Martin et al., 2001) maps to a diabetes susceptibility locus on chromosome 2, IDDM7 (Pociot and McDermott, 2002), suggesting that IGRP is also involved in human T1D. The recent identification of IGRP epitopes restricted by HLA-$A2^*0201$ (Takaki et al., 2006), HLA-DR401, or HLA-DR301 (Yang et al., 2006), and the development of HLA class I and class II tetramers will certainly allow evaluation of the role of IGRP in the pathogenesis of human T1D. A first set of data suggests that IGRP-reactive CD4+ T cells can be found in both diabetic patients and healthy individuals (Yang et al., 2006).

### 3.1.2    Mechanisms of Tolerance Induction

Most of these antigen-based therapies prevent or delay the onset of the disease through induction of regulatory Th2, Th3, or Tr1 cells and immune deviation. This is well illustrated by the shift of autoantibody responses to GAD (Tian et al., 1996a,b) and HSP60 (Elias et al., 1997) toward the Th2-dependent IgG1 isotype, and the in vitro T cell response to the tolerogen that results in preferential IL-4, IL-5, and IL-10 production (Elias et al., 1997; Homann et al., 1999; Tian et al., 1997). The role of Th2 cytokines is also supported by the fact that IL-4−/− NOD mice resist GAD-induced protection even though they do not exhibit any acceleration of the spontaneous disease (Tisch et al., 1999). Data obtained with altered peptide ligand (APL) variants of insulin-derived peptides suggest that APLs may also act through immune deviation by favoring the induction of adaptative Th2-like regulatory T cells. In fact, the protection from diabetes afforded by NBI-6024, an APL of InsB9-23, is associated with the induction of a T cell response cross-reactive with the native InsB9-23 peptide and the production of IL-4 and IL-10 (Alleva et al., 2002). Mucosal (oral or intranasal) administration of autoantigens favors antigenic presentation through mucosal dendritic cells (DCs), which seem to induce regulatory T cells resembling Tr1 and Th3 cells preferentially (Weiner, 2001b). Following nasal administration of aerosolized insulin, diabetes protection is associated with production of IL-10 in pancreatic lymph nodes by protective γδ CD8 T cells (Hanninen, 2000; Harrison et al., 1996). Other data obtained with oral insulin suggested, by contrast, that diabetes protection may be related to the induction of Tr1 CD4 regulatory T cells (Bergerot et al., 1997; Zhang et al., 1991), a difference perhaps due to the degradation of insulin in the digestive tract (Hanninen, 2000). Finally, some strategies seem to induce TGFβ-producing Th3 regulatory T cells, as shown for insulin administered orally (Hancock, 1995). These peculiarities might be linked to the nature of the tolerogen or to its route of administration.

   Protection from diabetes is associated with bystander suppression: The induced regulatory T cells are activated in the tissue or draining lymph node where the tolerizing antigen is expressed and suppress autoreactive T cells independently of their antigenic specificity. Bystander suppression is presumably mediated by the production of IL-4, IL-10, and TGFβ (Bach, 2001; Chatenoud, 2001; Homann et al., 1999; Maron et al., 1999; Polanski et al., 1997; Tian et al., 1997, 2002; von Herrath and Harrison, 2003; Weiner, 2001a). Thus, NOD mice injected with GAD at an early age show decreased proliferation to this autoantigen as a reflection of the restoration of self-tolerance. This unresponsiveness is not restricted to the tolerogen but spreads to other β cell autoantigens, notably insulin and HSP60 (Kaufman, 1993; Tisch et al., 1993).

Mechanisms other than the induction of adaptative regulatory T cells, such as anergy or deletion, may participate in β cell antigen-induced tolerance. Parenterally administered soluble peptides are usually cleared within 2 days (Metzler et al., 2000). They induce a weak and short-lived activation state that, in the absence of co-stimulatory signals, leads to anergy and deletion (Aichele, 1994; Toes and Offringa, 1996). In double transgenic mice, it was thus shown that intravenous injection of a tolerogenic HA peptide could induce tolerance so long as injections were regularly repeated (Bercovici et al., 1999, 2000). Interestingly, tolerance was associated with massive peptide-induced apoptosis of transgenic T cells in the thymus and the periphery, as demonstrated in thymectomized mice. Remnant transgenic T cells were shown to be unable to proliferate in response to the antigen and thus were "anergic" (Bercovici et al., 1999, 2000). This unresponsiveness was not reversible in the presence of exogenous IL-2 (Bercovici et al., 1999). Anergy was also incriminated in a similar model using HA peptide plus an anti-CD4 antibody; but here again there was no reversal after IL-2 addition, and "anergic" cells were shown to produce IL-10 (Lanoue et al., 1997). Interestingly, it was recently shown, as mentioned above, that in nontransgenic polyclonal conditions APLs of $IGRP_{206-215}$ can prevent the onset of diabetes through deletion of high-avidity $IGRP_{206-215}$-reactive CD8 T cells (Han et al., 2005).

All the evidence presented above tends to indicate that tolerization by β cell autoantigens essentially acts through induction of regulatory T cells acting as bystander suppressors, leading to immune deviation and eventually diabetes protection. Anergy or deletion may occur in some models, but this does not exclude the role of immunoregulation, which could be more prevalent in nontransgenic mice with a full T cell repertoire. Although many of the antigen-based strategies tested in the NOD mouse were proven to be effective at preventing/delaying the onset of diabetes, there are also numerous reports of inefficient "vaccination" protocols, including insulin-based therapy (Shoda et al., 2005). On the other hand, a few studies have reported exacerbation of the disease (Cetkovic-Cvrlje et al., 1997; Trembleau et al., 2000; Weaver et al., 2001; Wilson et al., 2001; Zechel, 1998a,b), which raises important issues regarding translation of such strategies to the clinic. The risk of accelerating diabetes onset due to induction of undesirable idiosyncratic sensitization, rather than tolerance, is illustrated by the induction of diabetes observed in double transgenic mice expressing OVA in their β cells and an anti-OVA TCR after oral administration of large doses of the antigen (Blanas et al., 1996). Another important issue illustrated by the fatal anaphylaxis reported in NOD mice after subcutaneous injections of insulin (Liu et al., 2002) is the induction of hypersensitivity associated with the immune deviation of the autoimmune process toward a Th2 response. In any case, these safety issues and the disappointing results of the human trials using insulin as a tolerogen highlight the need for a better understanding of the cellular and molecular mechanisms involved in tolerance induction and improvement of the strategies.

### 3.1.3 Improvement of Antigen-Based Therapy

Many recently reviewed parameters may influence the efficacy of antigen-dependent immunotherapy (Goudy and Tisch, 2005). First, the number of induced regulatory T cells is critical for mediating protection, particularly at late stages of the disease when the frequency of diabetogenic T cells is high. Induction of sufficient numbers of regulatory T cells, which presumably depends on the size of the pool of naive T cell precursors for a given antigen, might require targeting β cell antigens that are subdominant or even ignored during the disease process, for which sufficient numbers of naive precursors can be recruited. An in-depth analysis of the naturally processed and presented β cell peptides (Arif et al., 2004; Hassainya et al., 2005), as well as a better understanding of β cell antigen processing, might allow us to identify such "cryptic epitopes."

Adjuvants can modulate the nature of an immune response toward a vaccinating antigen. For example, cholera toxin B (CTB) enhances the protective effect of oral insulin in prediabetic NOD mice by favoring the induction of insulin-specific Tr1 regulatory T cells (Ploix, et al., 1999). Coupling of CTB to GAD also reduces the antigen dose up to 100-fold (Petersen et al., 2003; Ploix et al., 1999). Similarly, as mentioned earlier, alum, typically used to favor Th2 responses, has been successfully tested with GAD65 in NOD mice (Ramiya, 1996). Promising results have been obtained in humans as well (Agardh et al., 2005). "Tolerogenic" DCs obtained, for example, through culture with an analogue of 1,25-dihydroxyvitamin $D_3$ [$1,25(OH)_2D_3$] or dexamethasone, favor IL-10 production after activation, thus promoting the induction of Tr1 or foxp3+CD4+CD25+ regulatory T cells (Adorini et al., 2003, 2004; Jonuleit et al., 2000; Pedersen et al., 2004; Penna et al., 2005; Steinman et al., 2003). Treatment of NOD mice with an analogue of $1,25(OH)_2D_3$ arrests the progression of insulitis and prevents diabetes development (Gregori et al., 2002). This effect is associated with an enhanced frequency of CD4+CD25+ regulatory T cells in the pancreatic lymph nodes that are able to inhibit the T cell response to the pancreatic autoantigen IA-2 and to delay significantly diabetes induced by the transfer of pathogenic CD4+CD25− cells. Other agents acting on DC differentiation and maturation, such as the Flt3 ligand, may also prove to be good adjuvants (Chilton et al., 2004).

As previously discussed, regulatory T cells can exert their suppressive functions through various cytokines in vivo. Studies have suggested that IL-4 and IL-10 may promote their differentiation and/or increase their functional abilities (Goudy et al., 2003; Yamamoto et al., 2001). We have observed that retroviral transduction of CD4+CD62L+ splenocytes from overtly diabetic or prediabetic NOD mice with murine IL-4 potentiate their capacity to inhibit disease transfer

(Yamamoto et al., 2001). DNA vaccine studies have further shown that the addition of plasmids encoding IL-4 to a GAD DNA vaccine enhances protection against diabetes (Tisch et al., 2001; Weaver et al., 2001). Similarly, it has been shown that systemic treatment with a nonpathogenic recombinant adeno-associated virus expressing IL-10 prevents diabetes in NOD mice (Goudy, 2001) and that this effect is associated with an increase of the percentage of CD4+CD25+ T cells in the spleen (Goudy et al., 2003). However, because of paradoxical effects of the administration of IL-10 to NOD mice on diabetes progression (Moritani et al., 1994; Pennline et al., 1994; Wogensen et al, 1994), the use of this cytokine as an adjuvant presents potential risks in T1D therapy. A role for TGFβ in the development of regulatory T cells in vivo has been suggested by the observation that transgenic expression of TGFβ restricted to pancreatic islets of NOD mice inhibits the development of diabetes and promotes the expansion of CD4+CD25+ T cells (Peng et al., 2004). Again, given the pleiotropic nature of this cytokine and the fact that it could actually induce spontaneous diabetes when expressed with TNFα in the pancreas (Filippi et al., 2005), using TGFβ as an adjuvant might not be beneficial.

Another strategy to enhance the efficiency of antigen-based therapy has recently been proposed, using peptide-immunoglobulin (Ig) or MHC-peptide-Ig chimeric molecules. Peptide-Ig molecules such as $InsB_{9-23}$-Ig, increases the in vivo half-life of the peptide, thus improving its APC uptake and presentation to T cells (Gregg et al., 2005). Treatment with $InsB_{9-23}$-Ig protected young NOD mice by triggering IL-10 production by APCs and expanding IL-10-producing T regulatory cells. Importantly, $InsB_{9-23}$-Ig was totally ineffective in older prediabetic mice (10 weeks), thus calling into question the relevance of this approach in human patients. More interestingly, Ig fusion molecules of MHC II/peptide complexes have been designed and used in a double antigen-TCR transgenic system (Casares et al., 2002). The MHC/peptide dimers were able not only to prevent the onset of diabetes but also reverse the disease in diabetic animals through induction of anergy in the autoreactive effectors and stimulation of Tr1 regulatory T cells. Similarly, specific in vivo treatment with BDC2.5 peptide p31-I-A$^{g7}$ dimers protected mice from diabetes mediated by the adoptive transfer of transgenic diabetogenic BDC2.5 CD4+ T cells (Masteller et al., 2003). Protection appeared to be associated with increased cell death of pathogenic effectors and induction of hypoproliferative Tr1 cells. Importantly, the p31-I-A$^{g7}$ therapy was unable to prevent or delay diabetes caused by diabetogenic spleen cells from NOD mice or reverse diabetes in newly diabetic NOD mice, possibly because of a too-low frequency of p31 I-A$^{g7}$-reactive T cells. These results underscore the fact that the nature of the epitopic determinant used for antigen-based therapy is critical and that the identification of naturally processed epitopes is needed.

## 3.2   CD3 Antibody Treatment

Anti-T cell monoclonal antibodies have a dual capacity to promote both immunosuppression and immune tolerance. Although their immunosuppressive effect is transient, these antibodies can reverse established autoimmunity or prevent an autoimmune relapse by establishing long-term antigen-specific unresponsiveness that persists in the absence of immunosuppression (Chatenoud, 2002). Antibodies to CD3 have unique tolerogenic properties in NOD mice (Chatenoud et al., 1994, 1997). Treating NOD mice that have full-blown diabetes with low doses of the CD3 antibody 145-2C11 induces diabetes remission in the absence of exogenous insulin supply in 60% to 80% of mice (Chatenoud et al., 1994, 1997). The remission is complete and definitive (follow-up of > 6 months) and relies on an antigen-specific effect because CD3 antibody-treated NOD mice do not destroy syngeneic islet grafts and are fully responsive to other unrelated tissue antigens (e.g., alloantigens) (Chatenoud et al., 1994). CD3 antibody-induced tolerance is most effective in mice with recent-onset disease (14–20 weeks old) during the 1 to 2 weeks following diabetes onset before the β cells are completely destroyed. No diabetes retardation is seen if the CD3 antibody is administered earlier, a finding at variance with other evidence found in immune intervention procedures, with the exception of neonatal injection of the CD3 antibody itself, which provides long-term and complete protection (the reasons for which are still unclear) (Hayward, 1989). The only other strategy as effective as anti-CD3 antibodies for overt disease is the administration of anti-lymphocyte serum (Maki et al., 1992). The tolerogenic effect of anti-CD3 therapy is observed with intact mitogenic antibodies (Chatenoud, 1994) and with nonmitogenic and well tolerated F(ab′)2 fragments (Chatenoud et al., 1997).

In CD3 antibody-treated mice, insulitis is cleared rapidly (within 2–3 days) and glycemia normales within 2 to 3 weeks following a 5-day treatment. The mechanism of protection is not the result of massive deletion of autoreactive cells in protected animals. The CD3 antibody, and even F(ab′)2 fragments (although to a lesser degree) do induce some T cell depletion, but such depletion is incomplete and transient. Antigenic modulation (internalization) of the TCR CD3 complex is another possible mode of action, but it is also transient and vanishes as soon as the antibody disappears from the serum (Chatenoud, 1982). It seems that immunoregulatory mechanisms resembling those present in young prediabetic NOD mice are induced and/or restored in anti-CD3-treated animals (Bach and Chatenoud, 2001; Belghith et al., 2003; Chatenoud et al., 2001). A single injection of cyclophosphamide rapidly and reproducibly reverses the anti-CD3-induced tolerance, indicating the active nature of the tolerance restored in the animals. Co-transfer experiments have demonstrated the presence of CD4+CD25+ cells in the spleen of CD3-treated mice (Belghith et al., 2003) that effectively inhibit the transfer of disease by diabetogenic T lymphocytes to NOD-SCID mice. These anti-CD3-induced protective CD4+CD25+ cells are also found in the lymph nodes and especially in the draining lymph nodes of the target organ. Their suppressive potential is also functional in vitro when tested in classical co-cultures with splenic diabetogenic T cells from overtly diabetic NOD mice. Suppression is TGFβ-dependent both in vitro and in vivo (Belghith et al., 2003). Remarkably, anti-CD3 therapy also reverses diabetes in CD28-deficient NOD mice lacking naturally occuring CD25+ regulatory T cells. Overall, these results indicate that upon treatment with an anti-CD3 antibody a subset of regulatory CD4+CD25+ T cells are induced that are capable of restoring long-lasting tolerance to β islet antigens in recent-onset diabetic mice. However, the precise mechanisms by which engagement of T cells with FcR nonbinding anti-CD3 antibodies leads to regulatory T cell development and expansion are still unknown.

These results in the NOD model have led to two clinical trials of anti-CD3 treatment in patients with recent-onset diabetes (Herold et al., 2002; Keymeulen et al., 2005). One Phase I/II trial tested the effect of a single 14-day course of the non-FcR-binding anti-CD3 monoclonal antibody hOKT3γ1 (Ala-Ala) in newly diagnosed T1D patients (Herold et al., 2002). A decrease in the rate of loss of insulin production associated with a significant increase in serum IL-10 and IL-5 was observed over a 2-year period. IL-10-producing CD4+ T cells were found after treatment, a subset of which also secreted TGFβ (Herold et al., 2003). A second Phase II placebo-controlled trial used another non-FcR-binding humanized antibody, ChAGly-CD3, administered over the course of 6 days (Keymeulen et al., 2005). The treatment resulted in better maintenance of β cell function in ChAGly-CD3-treated patients. The effect was more pronounced in patients with better initial residual β cell function, with lower needs for insulin at the end of the trial to obtain adequate glycemic control. In both studies, CD3 antibody treatment was well tolerated, although some minor side effects linked to T cell activation and short-term immunosuppression were observed.

## 3.3  Ex Vivo Expansion of Regulatory T Cells

Although CD4+CD25+ regulatory T cells have been shown to be anergic in vitro following TCR stimulation, multiple methods of manipulating both their numbers and their activation status are now available (Horwitz et al., 2004). Some of them have been applied to the therapy of spontaneous T1D in the NOD model. CD4+CD25+ regulatory T cells isolated from NOD or NOD-BDC2.5 TCR transgenic mice and expanded in vitro with a combination of anti-CD3 and anti-CD28 antibodies with high doses of IL-2 (Tang et al., 2004) were shown to maintain the phenotypic features of conventional regulatory T cells (CD25+, CD62L+, CTLA-4+, GITR+, foxp3+), produce IL-10 and TGFβ, and display suppressive ability in vitro and in vivo. Importantly, regulatory T cells expanded from transgenic BDC2.5 islet-specific CD4+ lymphocytes that could prevent diabetes when transferred into CD28−/− NOD mice and reverse diabetes in new-onset diabetic NOD mice, whereas expanded polyclonal NOD regulatory T cells in similar numbers could not. This strongly suggests that to exert their suppressive function expanded regulatory T cells, despite TCR stimulation during their expansion, need to be reactivated in an antigen-specific manner at the inflammatory site. Antigen-specific regulatory T cells have also been successfully expanded from NOD mice using IL-2, anti-CD28 antibodies, and beads coated with recombinant MHC II/peptide complexes. Here again the antigen-specific regulatory T cells were more effective for suppressing diabetes than polyclonal regulatory T cells (Masteller et al., 2005). Functional autoantigen-specific regulatory T cells have also been produced after culture of NOD-BDC2.5 CD4+CD25+ T cells with mature bone marrow NOD DCs pulsed with p31 peptide (which stimulates BDC2.5 T cells) in the presence of IL-2 (Tarbell et al., 2004). As few as $5 \times 10^3$ of the expanded transgenic regulatory T cells could block autoimmunity caused by diabetogenic T cells in NOD mice, whereas $10^5$ polyclonal NOD CD4+CD25+ T cells could not. Overall, these results highlight the fact that, like conventional CD4+CD25+ regulatory T cells, ex vivo expanded regulatory T cells responsive to a single autoantigen can inhibit autoreactivity to multiple β cell antigens so long as their suppressive ability has been locally activated in an antigen-specific manner.

Expansion of human polyclonal CD4+CD25+ T cells using anti-CD3- and anti-CD28-coated beads with IL-2 has also been reported. These expanded regu-latory T cells maintained their phenotype and were shown to be suppressive in vitro in allogenic mixed lymphocyte reactions (Godfrey et al., 2004; Hoffmann et al., 2004). The efficiency of large-scale expansion of human regulatory T cells will certainly allow more detailed biological studies of human regulatory T cells and facilitate their evaluation as potential therapeutic agents.

# 4    Conclusion

It appears that a number of therapeutic procedures can be proposed to restore self-tolerance to β cells in T1D. It is remarkable that such restoration of tolerance may be considered for a disease in which many β cell autoantigens are known to be involved. In the case of β cell antigen-based therapy, tolerance is mediated by regulatory T cells induced by the tolerogen, whatever its role in the initiation, and act through bystander suppression. In the case of CD3 antibodies, which are by themselves non-antigen-specific, tolerance specificity seems afforded by the autoantigens expressed by the remnant β cells. The data and concepts discussed here for T1D may also apply to other autoimmune diseases, particularly those that are Th1-mediated. In terms of basic understanding, they provide new insights into the cellular and molecular mechanisms of tolerance to self-antigens. From the therapeutic point of view, they open new perspectives that are now being investigated in the clinic with the hope of proposing specific, safe therapeutic approaches.

## References

Abulafia-Lapid, R., Elias, D., et al. (1999) T cell proliferative responses of type 1 diabetes patients and healthy individuals to human hsp60 and its peptides. *J. Autoimmun.* 12:121-129.

Adorini, L., Giarratana, N., Penna, G. (2004) Pharmacological induction of tolerogenic dendritic cells and regulatory T cells. *Semin. Immunol.* 16:127-134.

Adorini, L., Penna, G., Giarratana, N., Uskokovic, M. (2003) Tolerogenic dendritic cells induced by vitamin D receptor ligands enhance regulatory T cells inhibiting allograft rejection and autoimmune diseases. *J. Cell. Biochem.* 88:227-233.

Agardh, C. D., Cilio, C. M., Lethagen, A., et al. (2005) Clinical evidence for the safety of GAD65 immunomodulation in adult-onset autoimmune diabetes. *J. Diabetes Complications* 19:238-246.

Aichele, P. (1994) Peptide-induced T-cell tolerance. *Proc. Natl. Acad. Sci. U S A* 91: 444-448.

Akashi, T., Nagafuchi, S., Anzai, K., et al. (1997) Direct evidence for the contribution of B cells to the progression of insulitis and the development of diabetes in non-obese diabetic mice. *Int. Immunol.* 9:1159-1164.

Akbar, A. N., Taams, L. S., Salmon, M., Vukmanovic-Stejic, M. (2003) The peripheral generation of CD4+ CD25+ regulatory T cells. *Immunology* 109: 319-325.

Alleva, D. G., Maki, R. A., Putnam, A. L., et al. (2006) Immunomodulation in type 1 diabetes by NBI-6024, an altered peptide ligand of the insulin B epitope. *Scand. J. Immunol.* 63:59-69.

Alleva, D. G., Gaur, A., Jin, L., Wegmann, D., et al. (2002) Immunological characterization and therapeutic activity of an altered-peptide ligand, NBI-6024, based on the immunodominant type 1 diabetes autoantigen insulin B-chain (9-23) peptide. *Diabetes* 51:2126-2134.

Alyanakian, M. A., You, S., Damotte, D., et al. (2003) Diversity of regulatory CD4+ T cells controlling distinct organ-specific autoimmune diseases. *Proc. Natl. Acad. Sci. U S A* 100:15806-15811.

Amrani, A., Verdaguer, J., Serra, P., et al. (2000) Progression of autoimmune diabetes driven by avidity maturation of a T-cell population. *Nature* 406:739-742.

Anastasi, E., Dotta, F., Tiberti, C., et al. (1999) Insulin prophylaxis down-regulates islet antigen expression and islet autoimmunity in the low-dose Stz mouse model of diabetes. *Autoimmunity* 29:249-256.

Anderson, M. S., Venanzi, E. S., Chen, Z., et al. (2005) The cellular mechanism of Aire control of T cell tolerance. *Immunity* 23:227-239.

Anderson, M. S., Venanzi, E. S., Klein, L., et al. (2002) Projection of an immunological self shadow within the thymus by the aire protein. *Science* 298:1395-1401.

Andre, I., Gonzalez, A., Wang, B., et al. (1996) Checkpoints in the progression of autoimmune disease: lessons from diabetes models. *Proc. Natl. Acad. Sci. U S A* 93:2260-2263.

Arif, S., Tree, T. I., Astill, T. P., et al. (2004) Autoreactive T cell responses show proinflammatory polarization in diabetes but a regulatory phenotype in health. *J. Clin. Invest.* 113:451-463.

Arreaza, G., Salojin, K., Yang, W., et al. (2003) Deficient activation and resistance to activation-induced apoptosis of CD8+ T cells is associated with defective peripheral tolerance in nonobese diabetic mice. *Clin. Immunol.* 107:103-115.

Asano, M., Toda, M., Sakaguchi, N., Sakaguchi, S. (1996) Autoimmune disease as a consequence of developmental abnormality of a T cell subpopulation. *J. Exp. Med.* 184:387-396.

Atkinson, M. A., Maclaren, N. K., Luchetta, R. (1990) Insulitis and diabetes in NOD mice reduced by prophylactic insulin therapy. *Diabetes* 39:933-937.

Bach, J. F. (1994) Insulin-dependent diabetes mellitus as an autoimmune disease. *Endocr. Rev.* 15:516-542.

Bach, J. F. (2001) Immunotherapy of insulin-dependent diabetes mellitus. *Curr. Opin. Immunol.* 13:601-605.

Bach, J. F., Chatenoud, L. (2001) Tolerance to islet autoantigens in type 1 diabetes. *Annu. Rev. Immunol.* 19:131-161.

Baecher-Allan, C., Viglietta, V., Hafler, D. A. (2004) Human CD4+CD25+ regulatory T cells. *Semin. Immunol.* 16:89-98.

Baecher-Allan, C., Brown, J. A., Freeman, G. J., Hafler, D. A. (2001) CD4+CD25[high] regulatory cells in human peripheral blood. *J. Immunol.* 167:1245-1253.

Baekkeskov, S., Aanstoot, H. J., Christgau, S., et al. (1990) Identification of the 64K autoantigen in insulin-dependent diabetes as the GABA-synthesizing enzyme glutamic acid decarboxylase. *Nature* 347:151-156.

Baker, D., Rosenwasser, O. A., O'Neill, J. K., Turk, J. L. (1995) Genetic analysis of experimental allergic encephalomyelitis in mice. *J. Immunol.* 155:4046-4051.

Balasa, B., Boehm, B. O., Fortnagel, A., et al. (2001) Vaccination with glutamic acid decarboxylase plasmid DNA protects mice from spontaneous autoimmune diabetes and B7/CD28 costimulation circumvents that protection. *Clin. Immunol.* 99:241-252.

Baumann, E. E., Buckingham, F., Herold, K. C. (1995) Intrathymic transplantation of islet antigen affects CD8+ diabetogenic T-cells resulting in tolerance to autoimmune IDDM. *Diabetes* 44:871-877.

Baxter, A. G., Horsfall, A. C., Healey, D., et al. (1994) Mycobacteria precipitate an SLE-like syndrome in diabetes-prone NOD mice. *Immunology* 83:227-231.

Belghith, M., Bluestone, J. A., Barriot, S., et al. (2003) TGF-beta-dependent mechanisms mediate restoration of self-tolerance induced by antibodies to CD3 in overt autoimmune diabetes. *Nat. Med.* 9:1202-1208.

Benoist, C., Mathis, D. (1997) Selection for survival? *Science* 276:2000-2001.

Bercovici, N., Delon, J., Cambouris, C., et al. (1999) Chronic intravenous injections of antigen induce and maintain tolerance in T cell receptor-transgenic mice. *Eur. J. Immunol.* 29:345-354.

Bercovici, N., Heurtier, A., Vizler, C., et al. (2000) Systemic administration of agonist peptide blocks the progression of spontaneous CD8-mediated autoimmune diabetes in transgenic mice without bystander damage. *J. Immunol.* 165:202-210.

Bergerot, I., Ploix, C., Petersen, J., et al. (1997) A cholera toxoid-insulin conjugate as an oral vaccine against spontaneous autoimmune diabetes. *Proc. Natl. Acad. Sci. U S A* 94:4610-4614.

Bertrand, S., De Paepe, M., Vigeant, C., Yale, J. F. (1992) Prevention of adoptive transfer in BB rats by prophylactic insulin treatment. *Diabetes* 41:1273-1277.

Berzins, S. P., Venanzi, E. S., Benoist, C., Mathis, D. (2003) T-cell compartments of prediabetic NOD mice. *Diabetes* 52:327-334.

Birk, O. S., Douek, D. C., Elias, D., et al. (1996). A role of Hsp60 in autoimmune diabetes: analysis in a transgenic model. *Proc. Natl. Acad. Sci. U S A* 93:1032-1037.

Blanas, E., Carbone, F. R., Allison, J., et al. (1996) Induction of autoimmune diabetes by oral administration of autoantigen. *Science* 274:1707-1709.

Bluestone, J. A., Abbas, A. K. (2003) Natural versus adaptive regulatory T cells. *Nat. Rev. Immunol.* 3:253-257.

Boden, E., Tang, Q., Bour-Jordan, H., Bluestone, J. A. (2003) The role of CD28 and CTLA4 in the function and homeostasis of CD4+CD25+ regulatory T cells. *Novartis Found. Symp.* 252:55-63; discussion 63-56, 106-114.

Boitard, C., Yasunami, R., Dardenne, M., Bach, J. F. (1989) T cell-mediated inhibition of the transfer of autoimmune diabetes in NOD mice. *J. Exp. Med.* 169:1669-1680.

Charlton, B., Taylor-Edwards, C., Tisch, R., Fathman, C. G. (1994) Prevention of diabetes and insulitis by neonatal intrathymic islet administration in NOD mice. *J. Autoimmun.* 7:549-560.

Chatenoud, L. (1982) Human in vivo antigenic modulation induced by the anti-T cell OKT3 monoclonal antibody. *Eur. J. Immunol.* 12:979-982.

Chatenoud, L. (2001) Restoration of self-tolerance is a feasible approach to control ongoing beta-cell specific autoreactivity: its relevance for treatment in established diabetes and islet transplantation. *Diabetologia* 44:521-536.

Chatenoud, L. (2002) The use of monoclonal antibodies to restore self-tolerance in established autoimmunity. *Endocrinol. Metab. Clin. North Am.* 31:457-475, ix.

Chatenoud, L., Primo, J., Bach, J. F. (1997) CD3 antibody-induced dominant self tolerance in overtly diabetic NOD mice. *J. Immunol.* 158:2947-2954.

Chatenoud, L., Salomon, B., Bluestone, J. A. (2001) Suppressor T cells—they're back and critical for regulation of autoimmunity! *Immunol. Rev.* 182:149-163.

Chatenoud, L., Thervet, E., Primo, J., Bach, J. F. (1994) Anti-CD3 antibody induces long-term remission of overt autoimmunity in nonobese diabetic mice. *Proc. Natl. Acad. Sci. U S A* 91:123-127.

Chentoufi, A. A., Polychronakos, C. (2002) Insulin expression levels in the thymus modulate insulin-specific autoreactive T-cell tolerance: the mechanism by which the IDDM2 locus may predispose to diabetes. *Diabetes* 51:1383-1390.

Chilton, P. M., Rezzoug, F., Fugier-Vivier, I., et al. (2004) Flt3-ligand treatment prevents diabetes in NOD mice. *Diabetes* 53:1995-2002.

Choisy-Rossi, C. M., Holl, T. M., Pierce, M. A., et al. (2004) Enhanced pathogenicity of diabetogenic T cells escaping a non-MHC gene-controlled near death experience. *J. Immunol.* 173:3791-3800.

Dahlen, E., Hedlund, G., Dawe, K. (2000) Low CD86 expression in the nonobese diabetic mouse results in the impairment of both T cell activation and CTLA-4 up-regulation. *J. Immunol.* 164:2444-2456.

Daniel, D., Wegmann, D. R. (1996) Protection of nonobese diabetic mice from diabetes by intranasal or subcutaneous administration of insulin peptide B-(9-23). *Proc. Natl. Acad. Sci. U S A* 93:956-960.

Daniel, D., Gill, R. G., Schloot, N., Wegmann, D. (1995) Epitope specificity, cytokine production profile and diabetogenic activity of insulin-specific T cell clones isolated from NOD mice. *Eur. J. Immunol.* 25:1056-1062.

Danke, N. A., Yang, J., Greenbaum, C., Kwok, W. W. (2005) Comparative study of GAD65-specific CD4+ T cells in healthy and type 1 diabetic subjects. *J. Autoimmun.* 25:303-311.

Danke, N. A., Koelle, D. M., Yee, C., et al. (2004) Autoreactive T cells in healthy individuals. *J. Immunol.* 172:5967-5972.

Dardenne, M., Lepault, F., Bendelac, A., Bach, J. F. (1989) Acceleration of the onset of diabetes in NOD mice by thymectomy at weaning. *Eur. J. Immunol.* 19: 889-895.

Delovitch, T. L., Singh, B. (1997) The nonobese diabetic mouse as a model of autoimmune diabetes: immune dysregulation gets the NOD. *Immunity* 7:727-738.

Dieckmann, D., Plottner, H., Berchtold, S., et al. (2001) Ex vivo isolation and characterization of CD4(+)CD25(+) T cells with regulatory properties from human blood. *J. Exp. Med.* 193:1303-1310.

Dotta, F., Dionisi, S., Viglietta, V., et al. (1999) T-cell mediated autoimmunity to the insulinoma-associated protein 2 islet tyrosine phosphatase in type 1 diabetes mellitus. *Eur. J. Endocrinol.* 141:272-278.

DPT1 Diabetes Study Group (2002) Effects of insulin in relatives of patients with type 1 diabetes mellitus. *N. Engl. J. Med.* 346:1685-1691.

Dubois-Lafforgue, D., Mogenet, L., Thebault, K., et al. (2002) Proinsulin 2 knockout NOD mice: a model for genetic variation of insulin gene expression in type 1 diabetes. *Diabetes* 51(Suppl 3):S489-S493.

Eisenbarth, G. S., Srikanta, S., Jackson, R., et al. (1985) Anti-thymocyte globulin and prednisone immunotherapy of recent onset type 1 diabetes mellitus. *Diabetes Res.* 2:271-276.

Elias, D., Cohen, I. R. (1995) Treatment of autoimmune diabetes and insulitis in NOD mice with heat shock protein 60 peptide p277. *Diabetes* 44:1132-1138.

Elias, D., Cohen, I. R. (1996) The hsp60 peptide p277 arrests the autoimmune diabetes induced by the toxin streptozotocin. *Diabetes* 45:1168-1172.

Elias, D., Reshef, T., Birk, O. S., et al. (1991) Vaccination against autoimmune mouse diabetes with a T-cell epitope of the human 65-kDa heat shock protein. *Proc. Natl. Acad. Sci. U S A* 88:3088-3091.

Elias, D., Prigozin, H., Polak, N., et al. (1994) Autoimmune diabetes induced by the beta-cell toxin STZ:immunity to the 60-kDa heat shock protein and to insulin. *Diabetes* 43:992-998.

Elias, D., Meilin, A., Ablamunits, V., et al. (1997) Hsp60 peptide therapy of NOD mouse diabetes induces a Th2 cytokine burst and downregulates autoimmunity to various beta-cell antigens. *Diabetes* 46:758-764.

Elliott, J. (1994) Immunization with the larger isoform of mouse glutamic acid decarboxylase (GAD67) prevents autoimmune diabetes in NOD mice. *Diabetes* 43: 1494-1499.

Elliott, R. B., Crossley, J. R., Berryman, C. C., James, A. G. (1981) Partial preservation of pancreatic beta-cell function in children with diabetes. *Lancet* 2:1-4.

Feili-Hariri, M., Morel, P. A. (2001) Phenotypic and functional characteristics of BM-derived DC from NOD and non-diabetes-prone strains. *Clin. Immunol.* 98:133-142.

Feili-Hariri, M., Falkner, D. H., Morel, P. A. (2002) Regulatory Th2 response induced following adoptive transfer of dendritic cells in prediabetic NOD mice. *Eur. J. Immunol.* 32:2021-2030.

Feutren, G., Papoz, L., Assan, R., Vialettes, B., et al. (1986) Cyclosporin increases the rate and length of remissions in insulin-dependent diabetes of recent onset. Results of a multicentre double-blind trial. *Lancet* 19:119-124.

Filippi, C., Bresson, D., von Herrath, M. (2005) Antigen-specific induction of regulatory T cells for type 1 diabetes therapy. *Int. Rev. Immunol.* 24:341-360.

Fontenot, J. D., Rudensky, A. Y. (2005) A well adapted regulatory contrivance: regulatory T cell development and the forkhead family transcription factor Foxp3. *Nat. Immunol.* 6:331-337.

Gavin, M., Rudensky, A. (2003) Control of immune homeostasis by naturally arising regulatory CD4+ T cells. *Curr. Opin. Immunol.* 15:690-696.

Gerling, I. C., Serreze, D. V., Christianson, S. W., Leiter, E. H. (1992) Intrathymic islet cell transplantation reduces beta-cell autoimmunity and prevents diabetes in NOD/Lt mice. *Diabetes* 41:1672-1676.

Ghosh, S., Palmer, S. M., Rodrigues, N. R., et al. (1993) Polygenic control of autoimmune diabetes in nonobese diabetic mice. *Nat. Genet.* 4:404-409.

Gianani, R., Eisenbarth, G. S. (2005) The stages of type 1A diabetes: 2005. *Immunol. Rev.* 204:232-249.

Girao, C., Hu, Q., Sun, J., Ashton-Rickardt, P. G. (1997) Limits to the differential avidity model of T cell selection in the thymus. *J. Immunol.* 159:4205-4211.

Godfrey, W. R., Ge, Y. G., Spoden, D. J., et al. (2004) In vitro-expanded human CD4(+) CD25(+) T-regulatory cells can markedly inhibit allogeneic dendritic cell-stimulated MLR cultures. *Blood* 104:453-461.

Gotfredsen, C. F. (1985) Reduction of diabetes incidence of BB Wistar rats by early prophylactic insulin treatment of diabetes-prone animals. *Diabetologia* 28:933-935.

Goudy, K., Song, S., Wasserfall, C., et al. (2001) Adeno-associated virus vector-mediated IL-10 gene delivery prevents type 1 diabetes in NOD mice. *Proc. Natl. Acad. Sci. U S A* 98:13913-13918.

Goudy, K. S., Tisch, R. (2005) Immunotherapy for the prevention and treatment of type 1 diabetes. *Int. Rev. Immunol.* 24:307-326.

Goudy, K. S., Burkhardt, B. R., Wasserfall, C., et al. (2003) Systemic overexpression of IL-10 induces CD4+CD25+ cell populations in vivo and ameliorates type 1 diabetes in nonobese diabetic mice in a dose-dependent fashion. *J. Immunol.* 171:2270-2278.

Gregg, R. K., Bell, J. J., Lee, H. H., et al. (2005) IL-10 diminishes CTLA-4 expression on islet-resident T cells and sustains their activation rather than tolerance. *J. Immunol.* 174:662-670.

Gregori, S., Giarratana, N., Smiroldo, S., Adorini, L. (2003) Dynamics of pathogenic and suppressor T cells in autoimmune diabetes development. *J. Immunol.* 171: 4040-4047.

Gregori, S., Giarratana, N., Smiroldo, S., et al. (2002) A 1alpha,25-dihydroxyvitamin D(3) analog enhances regulatory T-cells and arrests autoimmune diabetes in NOD mice. *Diabetes* 51:1367-1374.

Grohmann, U., Fallarino, F., Bianchi, R., et al. (2003) A defect in tryptophan catabolism impairs tolerance in nonobese diabetic mice. *J. Exp. Med.* 198:153-160.

Han, B., Serra, P., Amrani, A., et al. (2005) Prevention of diabetes by manipulation of anti-IGRP autoimmunity: high efficiency of a low-affinity peptide. *Nat. Med.* 11: 645-652.

Hanahan, D. (1998) Peripheral-antigen-expressing cells in thymic medulla: factors in self-tolerance and autoimmunity. *Curr. Opin. Immunol.* 10:656-662.

Hancock, W. W. (1995) Suppression of insulitis in non-obese diabetic (NOD) mice by oral insulin administration is associated with selective expression of interleukin-4 and −10, transforming growth factor-beta, and prostaglandin-E. *Am. J. Pathol.* 147: 1193-1199.

Hanninen, A. (2000) Gamma delta T cells as mediators of mucosal tolerance: the autoimmune diabetes model. *Immunol. Rev.* 173:109-119.

Harrison, L. C., Dempsey-Collier, M., Kramer, D. R., Takahashi, K. (1996) Aerosol insulin induces regulatory CD8 gamma delta T cells that prevent murine insulin-dependent diabetes. *J. Exp. Med.* 184:2167-2174.

Harrison, L. C., Honeyman, M. C., Steele, C. E., et al. (2004) Pancreatic beta-cell function and immune responses to insulin after administration of intranasal insulin to humans at risk for type 1 diabetes. *Diabetes Care* 27:2348-2355.

Haskins, K., Wegmann, D. (1996) Diabetogenic T-cell clones. *Diabetes* 45:1299-1305.

Hassainya, Y., Garcia-Pons, F., Kratzer, R., et al. (2005) Identification of naturally processed HLA-A2-restricted proinsulin epitopes by reverse immunology. *Diabetes* 54:2053-2059.

Hayward, A. R. (1989) Neonatal injection of CD3 antibody into nonobese diabetic reduces the incidence of insulitis and diabetes. *J Immunol.* 143:1555-1559.

Heath, V. L., Hutchings, P., Fowell, D. J., et al. (1999) Peptides derived from murine insulin are diabetogenic in both rats and mice, but the disease-inducing epitopes are different: evidence against a common environmental cross-reactivity in the pathogenicity of type 1 diabetes. *Diabetes* 48:2157-2165.

Heath, W. R., Allison, J., Hoffmann, M. W., et al. (1992) Autoimmune diabetes as a consequence of locally produced interleukin-2. *Nature* 359:547-549.

Herbelin, A., Gombert, J. M., Lepault, F., et al. (1998) Mature mainstream TCR alpha beta+CD4+ thymocytes expressing L-selectin mediate "active tolerance" in the nonobese diabetic mouse. *J. Immunol.* 161:2620-2628.

Herman, A. E., Freeman, G. J., Mathis, D., Benoist, C. (2004) CD4+CD25+ T regulatory cells dependent on ICOS promote regulation of effector cells in the prediabetic lesion. *J. Exp. Med.* 199:1479-1489.

Herold, K. C., Montag, A. G., Buckingham, F. (1992) Induction of tolerance to autoimmune diabetes with islet antigens. *J. Exp. Med.* 176:1107-1114.

Herold, K. C., Burton, J. B., Francois, F., et al. (2003) Activation of human T cells by FcR nonbinding anti-CD3 mAb, hOKT3gamma1(Ala-Ala). *J. Clin. Invest.* 111:409-418.

Herold, K. C., Hagopian, W., Auger, J. A., et al. (2002) Anti-CD3 monoclonal antibody in new-onset type 1 diabetes mellitus. *N. Engl. J. Med.* 346:1692-1698.

Hoffmann, P., Eder, R., Kunz-Schughart, L. A., et al. (2004) Large-scale in vitro expansion of polyclonal human CD4(+)CD25high regulatory T cells. *Blood* 104:895-903.

Homann, D., Holz, A., Bot, A., et al. (1999) Autoreactive CD4+ T cells protect from autoimmune diabetes via bystander suppression using the IL-4/Stat6 pathway. *Immunity* 11:463-472.

Honeyman, M. C., Cram, D. S., Harrison, L. C. (1993). Glutamic acid decarboxylase 67-reactive T cells: a marker of insulin-dependent diabetes. *J. Exp. Med.* 177:535-540.

Horwitz, D. A., Zheng, S. G., Gray, J. D., et al. (2004) Regulatory T cells generated ex vivo as an approach for the therapy of autoimmune disease. *Semin. Immunol.* 16:135-143.

Hutchings, P. R., Cooke, A. (1990) The transfer of autoimmune diabetes in NOD mice can be inhibited or accelerated by distinct cell populations present in normal splenocytes taken from young males. *J. Autoimmun.* 3:175-185.

Jenkins, M. K., Chen, C. A., Jung, G., et al. (1990) Inhibition of antigen-specific proliferation of type 1 murine T cell clones after stimulation with immobilized anti-CD3 monoclonal antibody. *J. Immunol.* 144:16-22.

Jonuleit, H., Schmitt, E., Schuler, G., et al. (2000) Induction of interleukin 10-producing, nonproliferating CD4(+) T cells with regulatory properties by repetitive stimulation with allogeneic immature human dendritic cells. *J. Exp. Med.* 192:1213-1222.

Jun, H. S., Chung, Y. H., Han, J., et al. (2002) Prevention of autoimmune diabetes by immunogene therapy using recombinant vaccinia virus expressing glutamic acid decarboxylase. *Diabetologia* 45:668-676.

Kanagawa, O., Militech, A., Vaupel, B. A. (2002) Regulation of diabetes development by regulatory T cells in pancreatic islet antigen-specific TCR transgenic nonobese diabetic mice. *J. Immunol.* 168:6159-6164.

Karounos, D. G. (1997) Metabolically inactive insulin analog prevents type I diabetes in prediabetic NOD mice. *J. Clin. Invest.* 100:1344-1348.

Kaufman, D. L. (1993) Spontaneous loss of T-cell tolerance to glutamic acid decarboxylase in murine insulin-dependent diabetes. *Nature* 366:69-72.

Keymeulen, B., Vandemeulebroucke, E., Ziegler, A. G., et al. (2005) Insulin needs after CD3-antibody therapy in new-onset type 1 diabetes. *N. Engl. J. Med.* 352:2598-2608.

Kishimoto, H., Sprent, J. (1999) Several different cell surface molecules control negative selection of medullary thymocytes. *J. Exp. Med.* 190:65-73.

Kishimoto, H., Sprent, J. (2001) A defect in central tolerance in NOD mice. *Nat. Immunol.* 2:1025-1031.

Klein, L., Klein, T., Ruther, U., Kyewski, B. (1998) CD4 T cell tolerance to human C-reactive protein, an inducible serum protein, is mediated by medullary thymic epithelium. *J. Exp. Med.* 188:5-16.

Koevary, S. B., Blomberg, M. (1992) Prevention of diabetes in BB/Wor rats by intrathymic islet injection. *J. Clin. Invest.* 89:512-516.

Kreuwel, H. T., Biggs, J. A., Pilip, I. M., et al. (2001) Defective CD8+ T cell peripheral tolerance in nonobese diabetic mice. *J. Immunol.* 167:1112-1117.

Kukreja, A., Cost, G., Marker, J., et al. (2002) Multiple immuno-regulatory defects in type-1 diabetes. *J. Clin. Invest.* 109:131-140.

Kurts, C., Sutherland, R. M., Davey, G., et al. (1999) CD8 T cell ignorance or tolerance to islet antigens depends on antigen dose. *Proc. Natl. Acad. Sci. U S A* 96:12703-12707.

Lang, J., Bellgrau, D. (2004) Animal models of type 1 diabetes: genetics and immunological function. *Adv. Exp. Med. Biol.* 552:91-116.

Lanoue, A., Bona, C., von Boehmer, H., Sarukhan, A. (1997) Conditions that induce tolerance in mature CD4+ T cells. *J. Exp. Med.* 185:405-414.

Lehmann, J., Huehn, J., de la Rosa, M., et al. (2002) Expression of the integrin alpha Ebeta 7 identifies unique subsets of CD25+ as well as CD25– regulatory T cells. *Proc. Natl. Acad. Sci. U S A* 99:13031-13036.

Lepault, F., Gagnerault, M. C. (2000) Characterization of peripheral regulatory CD4+ T cells that prevent diabetes onset in nonobese diabetic mice. *J. Immunol.* 164:240-247.

Lepault, F., Gagnerault, M. C., Faveeuw, C., et al. (1995) Lack of L-selection expression by cells transferring diabetes in NOD mice: insights into the mechanisms involved in diabetes prevention by Mel-14 antibody treatment. *Eur. J. Immunol.* 25:1502-1507.

Lesage, S., Hartley, S. B., Akkaraju, S., et al. (2002) Failure to censor forbidden clones of CD4 T cells in autoimmune diabetes. *J. Exp. Med.* 196:1175-1188.

Li, A. F., Escher, A. (2003) Intradermal or oral delivery of GAD-encoding genetic vaccines suppresses type 1 diabetes. *DNA Cell Biol.* 22:227-232.

Lieberman, S. M., Takaki, T., Han, B., et al. (2004) Individual nonobese diabetic mice exhibit unique patterns of CD8+ T cell reactivity to three islet antigens, including the newly identified widely expressed dystrophia myotonica kinase. *J. Immunol.* 173:6727-6734.

Lieberman, S. M., Evans, A. M., Han, B., et al. (2003) Identification of the beta cell antigen targeted by a prevalent population of pathogenic CD8+ T cells in autoimmune diabetes. *Proc. Natl. Acad. Sci. U S A* 100:8384-8388.

Lindley, S., Dayan, C. M., Bishop, A., et al. (2005) Defective suppressor function in CD4(+)CD25(+) T-cells from patients with type 1 diabetes. *Diabetes* 54:92-99.

Liston, A., Goodnow, C. C. (2005) Genetic lesions in thymic T cell clonal deletion and thresholds for autoimmunity. *Novartis Found. Symp.* 267:180-192; discussion 192-189.

Liston, A., Lesage, S., Wilson, J., et al. (2003) Aire regulates negative selection of organ-specific T cells. *Nat. Immunol.* 4:350-354.

Liston, A., Gray, D. H., Lesage, S., et al. (2004a) Gene dosage-limiting role of Aire in thymic expression, clonal deletion, and organ-specific autoimmunity. *J. Exp. Med.* 200:1015-1026.

Liston, A., Lesage, S., Gray, D. H., et al. (2004b) Generalized resistance to thymic deletion in the NOD mouse; a polygenic trait characterized by defective induction of Bim. *Immunity* 21:817-830.

Liu, E., Moriyama, H., Abiru, N., et al. (2002) Anti-peptide autoantibodies and fatal anaphylaxis in NOD mice in response to insulin self-peptides B:9-23 and B:13-23. *J. Clin. Invest.* 110:1021-1027.

Lohmann, T., Leslie, R. D., Londei, M. (1996) T cell clones to epitopes of glutamic acid decarboxylase 65 raised from normal subjects and patients with insulin-dependent diabetes. *J. Autoimmun.* 9:385-389.

Lohr, J., Knoechel, B., Nagabhushanam, V., Abbas, A. K. (2005) T-cell tolerance and autoimmunity to systemic and tissue-restricted self-antigens. *Immunol. Rev.* 204:116-127.

Macian, F., Garcia-Cozar, F., Im, S. H., et al. (2002) Transcriptional mechanisms underlying lymphocyte tolerance. *Cell* 109:719-731.

Maki, T., Ichikawa, T., Blanco, R., Porter, J. (1992) Long-term abrogation of autoimmune diabetes in nonobese diabetic mice by immunotherapy with anti-lymphocyte serum. *Proc. Natl. Acad. Sci. U S A* 89:3434-3438.

Mallone, R., Kochik, S. A., Laughlin, E. M., et al. (2004) Differential recognition and activation thresholds in human autoreactive GAD-specific T-cells. *Diabetes* 53: 971-977.

Maron, R., Melican, N. S., Weiner, H. L. (1999) Regulatory Th2-type T cell lines against insulin and GAD peptides derived from orally- and nasally-treated NOD mice suppress diabetes. *J. Autoimmun.* 12:251-258.

Martin, C. C., Bischof, L. J., Bergman, B., et al. (2001) Cloning and characterization of the human and rat islet-specific glucose-6-phosphatase catalytic subunit-related protein (IGRP) genes. *J. Biol. Chem.* 276:25197-25207.

Masteller, E. L., Warner, M. R., Tang, Q., et al. (2005) Expansion of functional endogenous antigen-specific CD4+CD25+ regulatory T cells from nonobese diabetic mice. *J. Immunol.* 175:3053-3059.

Masteller, E. L., Warner, M. R., Ferlin, W., et al. (2003) Peptide-MHC class II dimers as therapeutics to modulate antigen-specific T cell responses in autoimmune diabetes. *J. Immunol.* 171:5587-5595.

Mathis, D., Benoist, C. (2004) Back to central tolerance. *Immunity* 20:509-516.

McHugh, R. S., Whitters, M. J., Piccirillo, C. A., et al. (2002) CD4(+)CD25(+) immunoregulatory T cells: gene expression analysis reveals a functional role for the glucocorticoid-induced TNF receptor. *Immunity* 16:311-323.

Menard, V., Jacobs, H., Jun, H. S., et al. (1999) Anti-GAD monoclonal antibody delays the onset of diabetes mellitus in NOD mice. *Pharm. Res.* 16:1059-1066.

Metzler, B., Anderton, S. M., Manickasingham, S. P., Wraith, D. C. (2000) Kinetics of peptide uptake and tissue distribution following a single intranasal dose of peptide. *Immunol. Invest.* 29:61-70.

Morgan, D. J., Liblau, R., Scott, B., et al. (1996) CD8(+) T cell-mediated spontaneous diabetes in neonatal mice. *J. Immunol.* 157:978-983.

Moritani, M., Yoshimoto, K., Tashiro, F., et al. (1994) Transgenic expression of IL-10 in pancreatic islet A cells accelerates autoimmune insulitis and diabetes in non-obese diabetic mice. *Int. Immunol.* 6:1927-1936.

Moriyama, H., Abiru, N., Paronen, J., et al. (2003) Evidence for a primary islet autoantigen (preproinsulin 1) for insulitis and diabetes in the nonobese diabetic mouse. *Proc. Natl. Acad. Sci. U S A* 100:10376-10381.

Mosmann, T. R., Coffman, R. L. (1989) TH1 and TH2 cells: different patterns of lymphokine secretion lead to different functional properties. *Annu. Rev. Immunol.* 7:145-173.

Muir, A. (1995) Insulin immunization of nonobese diabetic mice induces a protective insulitis characterized by diminished intraislet interferon-gamma transcription. *J. Clin. Invest.* 95:628-634.

Mukherjee, R., Chaturvedi, P., Qin, H. Y., Singh, B. (2003) CD4+CD25+ regulatory T cells generated in response to insulin B:9-23 peptide prevent adoptive transfer of diabetes by diabetogenic T cells. *J. Autoimmun.* 21:221-237.

Myers, M. A., Laks, M. R., Feeney, S. J., et al. (1998) Antibodies to ICA512/IA-2 in rodent models of IDDM. *J. Autoimmun.* 11:265-272.

Nakamura, K., Kitani, A., Strober, W. (2001) Cell contact-dependent immunosuppression by CD4(+)CD25(+) regulatory T cells is mediated by cell surface-bound transforming growth factor beta. *J. Exp. Med.* 194:629-644.

Nakayama, M., Abiru, N., Moriyama, H., et al. (2005) Prime role for an insulin epitope in the development of type 1 diabetes in NOD mice. *Nature* 435:220-223.

Nomura, Y., Stein, E., Mullen, Y. (1993) Prevention of overt diabetes and insulitis by intrathymic injection of syngeneic islets in newborn nonobese diabetic (NOD) mice. *Transplantation* 56:638-642.

Noorchashm, H., Noorchashm, N., Kern, J., et al. (1997) B-cells are required for the initiation of insulitis and sialitis in nonobese diabetic mice. *Diabetes* 46:941-946.

Ohashi, P. S. (2003) Negative selection and autoimmunity. *Curr. Opin. Immunol.* 15:668-676.

Ohashi, P. S., Oehen, S., Buerki, K., et al. (1991) Ablation of "tolerance" and induction of diabetes by virus infection in viral antigen transgenic mice. *Cell* 65:305-317.

Oldenborg, P. A., Gresham, H. D., Chen, Y., et al. (2002) Lethal autoimmune hemolytic anemia in CD47-deficient nonobese diabetic (NOD) mice. *Blood* 99:3500-3504.

Oling, V., Marttila, J., Ilonen, J., et al. (2005) GAD65- and proinsulin-specific CD4+ T-cells detected by MHC class II tetramers in peripheral blood of type 1 diabetes patients and at-risk subjects. *J. Autoimmun.* 25:235-243.

Oukka, M., Cohen-Tannoudji, M., Tanaka, Y., et al. (1996a) Medullary thymic epithelial cells induce tolerance to intracellular proteins. *J. Immunol.* 156:968-975.

Oukka, M., Colucci-Guyon, E., Tran, P. L., et al. (1996b) CD4 T cell tolerance to nuclear proteins induced by medullary thymic epithelium. *Immunity* 4:545-553.

Panina-Bordignon, P., Lang, R., van Endert, P. M., et al. (1995) Cytotoxic T cells specific for glutamic acid decarboxylase in autoimmune diabetes. *J. Exp. Med.* 181:1923-1927.

Pedersen, A. E., Gad, M., Walter, M. R., Claesson, M. H. (2004) Induction of regulatory dendritic cells by dexamethasone and 1 alpha,25-dihydroxyvitamin D(3). *Immunol. Lett.* 91:63-69.

Peng, Y., Laouar, Y., Li, M. O., et al. (2004) TGF-beta regulates in vivo expansion of Foxp3-expressing CD4+CD25+ regulatory T cells responsible for protection against diabetes. *Proc. Natl. Acad. Sci. U S A* 101:4572-4577.

Penna, G., Roncari, A., Amuchastegui, S., et al. (2005) Expression of the inhibitory receptor ILT3 on dendritic cells is dispensable for induction of CD4+ Foxp3+ regulatory T cells by 1,25-dihydroxyvitamin D₃. *Blood* 106:3490-3497.

Pennline, K. J., Roque-Gaffney, E., Monahan, M. (1994) Recombinant human IL 10 prevents the onset of diabetes in the nonobese diabetic mouse. *Clin. Immunol. Immunopathol.* 71:169-175.

Petersen, J. S., Bregenholt, S., Apostolopolous, V., et al. (2003) Coupling of oral human or porcine insulin to the B subunit of cholera toxin (CTB) overcomes critical antigenic differences for prevention of type I diabetes. *Clin. Exp. Immunol.* 134:38-45.

Ploix, C., Bergerot, I., Fabien, N., et al. (1998) Protection against autoimmune diabetes with oral insulin is associated with the presence of IL-4 type 2 T-cells in the pancreas and pancreatic lymph nodes. *Diabetes* 47:39-44.

Ploix, C., Bergerot, I., Durand, A., et al. (1999) Oral administration of cholera toxin B-insulin conjugates protects NOD mice from autoimmune diabetes by inducing CD4+ regulatory T-cells. *Diabetes* 48:2150-2156.

Pociot, F., McDermott, M. F. (2002) Genetics of type 1 diabetes mellitus. *Genes Immun.* 3:235-249.

Polanski, M., Melican, N. S., Zhang, J., Weiner, H. L. (1997) Oral administration of the immunodominant B-chain of insulin reduces diabetes in a co-transfer model of diabetes in the NOD mouse and is associated with a switch from Th1 to Th2 cytokines. *J. Autoimmun.* 10:339-346.

Pop, S. M., Wong, C. P., Culton, D. A., et al. (2005) Single cell analysis shows decreasing FoxP3 and TGFbeta1 coexpressing CD4+CD25+ regulatory T cells during autoimmune diabetes. *J. Exp. Med.* 201:1333-1346.

Posselt, A. M., Barker, C. F., Friedman, A. L., Naji, A. (1992) Prevention of autoimmune diabetes in the BB rat by intrathymic islet transplantation at birth. *Science* 256:1321-1324.

Pozzilli, P. (2002) The DPT-1 trial: a negative result with lessons for future type 1 diabetes prevention. *Diabetes Metab. Res. Rev.* 18:257-259.

Pozzilli, P., Pitocco, D., Visalli, N., et al. (2000) No effect of oral insulin on residual beta-cell function in recent-onset type I diabetes (the IMDIAB VII). IMDIAB Group. *Diabetologia* 43:1000-1004.

Prasad, S. J., Goodnow, C. C. (2002) Intrinsic in vitro abnormalities in dendritic cell generation caused by non-MHC non-obese diabetic genes. *Immunol. Cell Biol.* 80:198-206.

Pugliese, A., Zeller, M., Fernandez, A., Jr., et al. (1997) The insulin gene is transcribed in the human thymus and transcription levels correlated with allelic variation at the INS VNTR-IDDM2 susceptibility locus for type 1 diabetes. *Nat. Genet.* 15:293-297.

Putnam, A. L., Vendrame, F., Dotta, F., Gottlieb, P. A. (2005) CD4+CD25[high] regulatory T cells in human autoimmune diabetes. *J. Autoimmun.* 24:55-62.

Radu, D. L., Brumeanu, T. D., McEvoy, R. C., et al. (1999) Escape from self-tolerance leads to neonatal insulin-dependent diabetes mellitus. *Autoimmunity* 30:199-207.

Ramiya, V. (1996) Antigen based therapies to prevent diabetes in NOD mice. *J. Autoimmun.* 9:349-356.

Ramsdell, F. (2003) Foxp3 and natural regulatory T cells: key to a cell lineage? *Immunity* 19:165-168.

Raz, I., Elias, D., Avron, A., et al. (2001) Beta-cell function in new-onset type 1 diabetes and immunomodulation with a heat-shock protein peptide (DiaPep277): a randomised, double-blind, phase II trial. *Lancet* 358:1749-1753.

Reijonen, H., Kwok, W. W., Nepom, G. T. (2003) Detection of CD4+ autoreactive T cells in T1D using HLA class II tetramers. *Ann. N. Y. Acad. Sci.* 1005:82-87.

Reijonen, H., Novak, E. J., Kochik, S., et al. (2002) Detection of GAD65-specific T-cells by major histocompatibility complex class II tetramers in type 1 diabetic patients and at-risk subjects. *Diabetes* 51:1375-1382.

Reijonen, H., Mallone, R., Heninger, A. K., et al. (2004) GAD65-specific CD4+ T-cells with high antigen avidity are prevalent in peripheral blood of patients with type 1 diabetes. *Diabetes* 53:1987-1994.

Rivero, V. E., Cailleau, C., Depiante-Depaoli, M., et al. (1998) Non-obese diabetic (NOD) mice are genetically susceptible to experimental autoimmune prostatitis (EAP). *J. Autoimmun.* 11:603-610.

Roncarolo, M. G., Bacchetta, R., Bordignon, C., et al. (2001) Type 1 T regulatory cells. *Immunol. Rev.* 182:68-79.

Rosmalen, J. G., van Ewijk, W., Leenen, P. J. (2002) T-cell education in autoimmune diabetes: teachers and students. *Trends Immunol.* 23:40-46.

Sai, P., Rivereau, A. S., Granier, C., et al. (1996) Immunization of non-obese diabetic (NOD) mice with glutamic acid decarboxylase-derived peptide 524-543 reduces cyclophosphamide-accelerated diabetes. *Clin. Exp. Immunol.* 105:330-337.

Sakaguchi, S. (2000) Regulatory T cells: key controllers of immunologic self-tolerance. *Cell* 101:455-458.

Sakaguchi, S. (2004) Naturally arising CD4+ regulatory T cells for immunologic self-tolerance and negative control of immune responses. *Annu. Rev. Immunol.* 22:531-562.

Sakaguchi, S., Sakaguchi, N., Asano, M., et al. (1995) Immunologic self-tolerance maintained by activated T cells expressing IL-2 receptor alpha-chains (CD25): breakdown of a single mechanism of self-tolerance causes various autoimmune diseases. *J. Immunol.* 155:1151-1164.

Salomon, B., Lenschow, D. J., Rhee, L., et al. (2000) B7/CD28 costimulation is essential for the homeostasis of the CD4+CD25+ immunoregulatory T cells that control autoimmune diabetes. *Immunity* 12:431-440.

Saoudi, A., Seddon, B., Fowell, D., Mason, D. (1996) The thymus contains a high frequency of cells that prevent autoimmune diabetes on transfer into prediabetic recipients. *J. Exp. Med.* 184:2393-2398.

Schwartz, R. H. (1990) A cell culture model for T lymphocyte clonal anergy. *Science* 248:1349-1356.

Schwartz, R. H. (2003) T cell anergy. *Annu. Rev. Immunol.* 21:305-334.

Scott, B., Liblau, R., Degermann, S., et al. (1994) A role for non-MHC genetic polymorphism in susceptibility to spontaneous autoimmunity. *Immunity* 1:73-83.

Seddon, B., Mason, D. (1999) Regulatory T cells in the control of autoimmunity: the essential role of transforming growth factor beta and interleukin 4 in the prevention of autoimmune thyroiditis in rats by peripheral CD4(+)CD45RC-cells and CD4(+)CD8(−) thymocytes. *J. Exp. Med.* 189:279-288.

Semana, G., Gausling, R., Jackson, R. A., Hafler, D. A. (1999) T cell autoreactivity to proinsulin epitopes in diabetic patients and healthy subjects. *J. Autoimmun.* 12:259-267.

Serreze, D. V., Chen, Y. G. (2005) Of mice and men: use of animal models to identify possible interventions for the prevention of autoimmune type 1 diabetes in humans. *Trends Immunol.* 26:603-607.

Serreze, D. V., Gaedeke, J. W., Leiter, E. H. (1993) Hematopoietic stem-cell defects underlying abnormal macrophage development and maturation in NOD/Lt mice: defective regulation of cytokine receptors and protein kinase C. *Proc. Natl. Acad. Sci. U S A* 90:9625-9629.

Serreze, D. V., Chapman, H. D., Varnum, D. S., et al. (1996) B lymphocytes are essential for the initiation of T cell-mediated autoimmune diabetes: analysis of a new "speed congenic" stock of NOD.Ig mu null mice. *J. Exp. Med.* 184:2049-2053.

Shoda, L. K., Young, D. L., Ramanujan, S., et al. (2005) A comprehensive review of interventions in the NOD mouse and implications for translation. *Immunity* 23:115-126.

Silverstein, J., Maclaren, N., Riley, W., et al. (1988) Immunosuppression with azathioprine and prednisone in recent-onset insulin-dependent diabetes mellitus. *N. Engl. J. Med.* 319:599-604.

Skyler, J. S., Krischer, J. P., Wolfsdorf, J., et al. (2005) Effects of oral insulin in relatives of patients with type 1 diabetes: the Diabetes Prevention Trial—Type 1. *Diabetes Care* 28:1068-1076.

Solomon, M., Sarvetnick, N. (2004) The pathogenesis of diabetes in the NOD mouse. *Adv. Immunol.* 84:239-264.

Steinman, R. M., Hawiger, D., Nussenzweig, M. C. (2003) Tolerogenic dendritic cells. *Annu. Rev. Immunol.* 21:685-711.

Stephens, L. A., Mason, D. (2000) CD25 is a marker for CD4+ thymocytes that prevent autoimmune diabetes in rats, but peripheral T cells with this function are found in both CD25+ and CD25- subpopulations. *J. Immunol.* 165:3105-3110.

Stephens, L. A., Mottet, C., Mason, D., Powrie, F. (2001) Human CD4(+)CD25(+) thymocytes and peripheral T cells have immune suppressive activity in vitro. *Eur. J. Immunol.* 31:1247-1254.

Strid, J., Lopes, L., Marcinkiewicz, J., et al. (2001) A defect in bone marrow derived dendritic cell maturation in the nonobesediabetic mouse. *Clin. Exp. Immunol.* 123:375-381.

Suri-Payer, E., Kehn, P. J., Cheever, A. W., Shevach, E. M. (1996) Pathogenesis of post-thymectomy autoimmune gastritis: identification of anti-H/K adenosine triphosphatase-reactive T cells. *J. Immunol.* 157:1799-1805.

Suri-Payer, E., Amar, A. Z., Thornton, A. M., Shevach, E. M. (1998) CD4+CD25+ T cells inhibit both the induction and effector function of autoreactive T cells and represent a unique lineage of immunoregulatory cells. *J. Immunol.* 160:1212-1218.

Takaki, T., Marron, M. P., Mathews, C. E., et al. (2006) HLA-A*0201-restricted T cells from humanized NOD mice recognize autoantigens of potential clinical relevance to type 1 diabetes. *J. Immunol.* 176:3257-3265.

Tang, Q., Henriksen, K. J., Bi, M., et al. (2004). In vitro-expanded antigen-specific regulatory T cells suppress autoimmune diabetes. *J. Exp. Med.* 199:1455-1465.

Tarbell, K. V., Yamazaki, S., Olson, K., et al. (2004). CD25+ CD4+ T cells, expanded with dendritic cells presenting a single autoantigenic peptide, suppress autoimmune diabetes. *J. Exp. Med.* 199:1467-1477.

Tian, J., Lehmann, P. V., Kaufman, D. L. (1997) Determinant spreading of T helper cell 2 (Th2) responses to pancreatic islet autoantigens. *J. Exp. Med.* 186:2039-2043.

Tian, J., Olcott, A. P., Kaufman, D. L. (2002) Antigen-based immunotherapy drives the precocious development of autoimmunity. *J. Immunol.* 169:6564-6569.

Tian, J., Atkinson, M. A., Clare-Salzler, M., et al. (1996a) Nasal administration of glutamate decarboxylase (GAD65) peptides induces Th2 responses and prevents murine insulin-dependent diabetes. *J. Exp. Med.* 183:1561-1567.

Tian, J., Clare-Salzler, M., Herschenfeld, A., et al. (1996b) Modulating autoimmune responses to GAD inhibits disease progression and prolongs islet graft survival in diabetes-prone mice. *Nat. Med.* 2:1348-1353.

Tisch, R., Wang, B., Serreze, D. V. (1999) Induction of glutamic acid decarboxylase 65-specific Th2 cells and suppression of autoimmune diabetes at late stages of disease is epitope dependent. *J. Immunol.* 163:1178-1187.

Tisch, R., Liblau, R. S., Yang, X. D., et al. (1998) Induction of GAD65-specific regulatory T-cells inhibits ongoing autoimmune diabetes in nonobese diabetic mice. *Diabetes* 47:894-899.

Tisch, R., Yang, X. D., Singer, S. M., et al. (1993) Immune response to glutamic acid decarboxylase correlates with insulitis in non-obese diabetic mice. *Nature* 366:72-75.

Tisch, R., Wang, B., Weaver, D. J., et al. (2001) Antigen-specific mediated suppression of beta cell autoimmunity by plasmid DNA vaccination. *J. Immunol.* 166:2122-2132.

Toes, R., Offringa, R. (1996) Peptide vaccination can lead to. *Proc. Natl. Acad. Sci. USA* 93:7855-7860.

Tone, M., Tone, Y., Adams, E., et al. (2003) Mouse glucocorticoid-induced tumor necrosis factor receptor ligand is costimulatory for T cells. *Proc. Natl. Acad. Sci. USA* 100:15059-15064.

Trembleau, S., Penna, G., Gregori, S., et al. (2000) Early Th1 response in unprimed nonobese diabetic mice to the tyrosine phosphatase-like insulinoma-associated protein 2, an autoantigen in type 1 diabetes. *J. Immunol.* 165:6748-6755.

Ucker, D. S., Meyers, J., Obermiller, P. S. (1992) Activation-driven T cell death. II. Quantitative differences alone distinguish stimuli triggering nontransformed T cell proliferation or death. *J. Immunol.* 149:1583-1592.

Urbanek-Ruiz, I., Ruiz, P. J., Paragas, V., et al. (2001) Immunization with DNA encoding an immunodominant peptide of insulin prevents diabetes in NOD mice. *Clin. Immunol.* 100:164-171.

Vafiadis, P., Bennett, S. T., Todd, J. A., et al. (1997) Insulin expression in human thymus is modulated by INS VNTR alleles at the IDDM2 locus. *Nat. Genet.* 15:289-292.

Venanzi, E. S., Benoist, C., Mathis, D. (2004) Good riddance: thymocyte clonal deletion prevents autoimmunity. *Curr. Opin. Immunol.* 16:197-202.

Vogel, A., Strassburg, C. P., Obermayer-Straub, P., et al. (2002) The genetic background of autoimmune polyendocrinopathy-candidiasis-ectodermal dystrophy and its autoimmune disease components. *J. Mol. Med.* 80:201-211.

Von Herrath, M. G., Harrison, L. C. (2003) Antigen-induced regulatory T cells in autoimmunity. *Nat. Rev. Immunol.* 3:223-232.

Von Herrath, M. G., Dockter, J., Oldstone, M. B. (1994) How virus induces a rapid or slow onset insulin-dependent diabetes mellitus in a transgenic model. *Immunity* 1:231-242.

Von Herrath, M. G., Guerder, S., Lewicki, H., et al. (1995) Coexpression of B7-1 and viral ("self") transgenes in pancreatic beta cells can break peripheral ignorance and lead to spontaneous autoimmune diabetes. *Immunity* 3:727-738.

Walker, L. S., Abbas, A. K. (2002) The enemy within: keeping self-reactive T cells at bay in the periphery. *Nat. Rev. Immunol.* 2:11-19.

Walker, M. R., Kasprowicz, D. J., Gersuk, V. H., et al. (2003) Induction of FoxP3 and acquisition of T regulatory activity by stimulated human CD4+CD25- T cells. *J. Clin. Invest.* 112:1437-1443.

Wang, B., Gonzalez, A., Hoglund, P., et al. (1998) Interleukin-4 deficiency does not exacerbate disease in NOD mice. *Diabetes* 47:1207-1211.

Weaver, D. J., Jr., Liu, B., Tisch, R. (2001) Plasmid DNAs encoding insulin and glutamic acid decarboxylase 65 have distinct effects on the progression of autoimmune diabetes in nonobese diabetic mice. *J. Immunol.* 167:586-592.

Wegmann, D. R., Norbury-Glaser, M., Daniel, D. (1994a) Insulin-specific T cells are a predominant component of islet infiltrates in pre-diabetic NOD mice. *Eur. J. Immunol.* 24:1853-1857.

Wegmann, D. R., Gill, R. G., Norbury-Glaser, M., et al. (1994b) Analysis of the spontaneous T cell response to insulin in NOD mice. *J. Autoimmun.* 7:833-843.

Weiner, H. L. (2001a) Oral tolerance: immune mechanisms and the generation of Th3-type TGF-beta-secreting regulatory cells. *Microbes Infect.* 3:947-954.

Weiner, H. L. (2001b) The mucosal milieu creates tolerogenic dendritic cells and T(R)1 and T(H)3 regulatory cells. *Nat. Immunol.* 2:671-672.

Wicker, L. S., Todd, J. A., Peterson, L. B. (1995) Genetic control of autoimmune diabetes in the NOD mouse. *Annu. Rev. Immunol.* 13:179-200.

Wilson, S. S., White, T. C., DeLuca, D. (2001) Therapeutic alteration of insulin-dependent diabetes mellitus progression by T cell tolerance to glutamic acid decarboxylase 65 peptides in vitro and in vivo. *J. Immunol.* 167:569-577.

Wogensen, L., Lee, M. S., Sarvetnick, N. (1994) Production of interleukin 10 by islet cells accelerates immune-mediated destruction of beta cells in nonobese diabetic mice. *J. Exp. Med.* 179:1379-1384.

Wong, F. S., Visintin, I., Wen, L., et al. (1996) CD8 T cell clones from young nonobese diabetic (NOD) islets can transfer rapid onset of diabetes in NOD mice in the absence of CD4 cells. *J. Exp. Med.* 183:67-76.

Wong, S., Guerder, S., Visintin, I., et al. (1995) Expression of the co-stimulator molecule B7-1 in pancreatic beta-cells accelerates diabetes in the NOD mouse. *Diabetes* 44:326-329.

Wu, Q., Salomon, B., Chen, M., et al. (2001) Reversal of spontaneous autoimmune insulitis in nonobese diabetic mice by soluble lymphotoxin receptor. *J. Exp. Med.* 193:1327-1332.

Yamamoto, A. M., Chernajovsky, Y., Lepault, F., et al. (2001) The activity of immunoregulatory T cells mediating active tolerance is potentiated in nonobese diabetic mice by an IL-4-based retroviral gene therapy. *J. Immunol.* 166:4973-4980.

Yang, J., Danke, N. A., Berger, D., et al. (2006) Islet-specific glucose-6-phosphatase catalytic subunit-related protein-reactive CD4+ T cells in human subjects. *J. Immunol.* 176:2781-2789.

Yoon, J. W., Yoon, C. S., Lim, H. W., et al. (1999) Control of autoimmune diabetes in NOD mice by GAD expression or suppression in beta cells. *Science* 284:1183-1187.

You, S., Slehoffer, G., Barriot, S., et al. (2004) Unique role of CD4+CD62L+ regulatory T cells in the control of autoimmune diabetes in T cell receptor transgenic mice. *Proc. Natl. Acad. Sci. USA* 101(Suppl 2):14580-14585.

You, S., Thieblemont, N., Alyanakian, M. A., et al. (2006) TGF-beta and T-cell mediated immunoregulation in the control of autoimmune diabetes. *Annu Rev Immunol*, 212: 185-202.

You, S., Chen, C., Lee, W. H., et al. (2003) Detection and characterization of T cells specific for BDC2.5 T cell-stimulating peptides. *J. Immunol.* 170:4011-4020.

You, S., Belghith, M., Cobbold, S., et al. (2005) Autoimmune diabetes onset results from qualitative rather than quantitative age-dependent changes in pathogenic T-cells. *Diabetes* 54:1415-1422.

Yu, J., Yu, L., Bugawan, T. L., et al. (2000) Transient antiislet autoantibodies: infrequent occurrence and lack of association with "genetic" risk factors. *J. Clin. Endocrinol. Metab.* 85:2421-2428.

Zaccone, P., Fehervari, Z., Blanchard, L., et al. (2002) Autoimmune thyroid disease induced by thyroglobulin and lipopolysaccharide is inhibited by soluble TNF receptor type I. *Eur. J. Immunol.* 32:1021-1028.

Zanin-Zhorov, A., Bruck, R., Tal, G., et al. (2005) Heat shock protein 60 inhibits Th1-mediated hepatitis model via innate regulation of Th1/Th2 transcription factors and cytokines. *J. Immunol.* 174:3227-3236.

Zechel, M. (1998a) Characterization of novel T-cell epitopes on 65 kDa and 67 kDa glutamic acid decarboxylase relevant in autoimmune responses in NOD mice. *J. Autoimmun.* 11:83-95.

Zechel, M. (1998b) Epitope dominance: evidence for reciprocal determinant spreading to glutamic acid decarboxylase in non-obese diabetic mice. *Immunol. Rev.* 164:111-118.

Zelenika, D., Adams, E., Humm, S., et al. (2002) Regulatory T cells overexpress a subset of Th2 gene transcripts. *J. Immunol.* 168:1069-1079.

Zhang, Z. J., Davidson, L., Eisenbarth, G., Weiner, H. L. (1991) Suppression of diabetes in nonobese diabetic mice by oral administration of porcine insulin. *Proc. Natl. Acad. Sci. USA* 88:10252-10256.

Zipris, D., Lazarus, A. H., Crow, A. R., et al. (1991) Defective thymic T cell activation by concanavalin A and anti-CD3 in autoimmune nonobese diabetic mice: evidence for thymic T cell anergy that correlates with the onset of insulitis. *J. Immunol.* 146: 3763-3771.

Zucchelli, S., Holler, P., Yamagata, T., et al. (2005) Defective central tolerance induction in NOD mice: genomics and genetics. *Immunity* 22:385-396.

# 7
# Co-stimulation Regulation of Immune Tolerance and Autoimmunity

Chen Dong, Roza Nurieva, and Natalia Martin-Orozco

## 1   Introduction

CD4+ helper T (Th) cells play crucial roles in immunity against microbial pathogens and are also key mediators in autoimmune diseases. When they encounter appropriate antigen-presenting cells (APCs), Th cells are activated, undergo clonal expansion, and subsequently differentiate into effector cells. Effector Th cells were traditionally classified as Th1 cells, which produce interferon-$\gamma$ (IFN$\gamma$) and tumor necrosis factor-$\alpha$ (TNF$\alpha$), and Th2 cells, which produce interleukin-4 (IL-4), IL-5, and IL-13 cytokines—which regulate cell-mediated or humoral immunity, respectively (Dong and Flavell, 2000a,b). Recently, a novel lineage of Th cells that express

121

IL-17 was found to play important roles in inflammatory autoimmune diseases in the mouse (Langrish et al., 2005; Park et al., 2005).

Th cell activation and function are tightly regulated to ensure effective elimination of pathogens while maintaining tolerance to self-tissues. In contrast to the robust responses by Th cells to pathogen-associated antigens, T cell tolerance to self-antigens must be maintained to prevent autoimmune diseases. This is initially mediated by intrathymic deletion of self-reactive T cells, a process called negative selection. In addition, peripheral tolerance mechanisms exist. Previous studies indicate that tissue antigens presented by immature or tolerogenic dendritic cells in the resting states lead to antigen-specific T cell tolerance in the form of clonal deletion or anergy (Heath and Carbone, 2001; Steinman et al., 2003). Furthermore, a specialized regulatory T (Treg) cell population that expresses FoxP3 transcription factor critically inhibits T cell activation and prevents autoimmune reactions (Fontenot and Rudensky, 2005).

In addition to the major histocompatibility complex (MHC)-peptide complexes, more co-stimulatory signals are provided by APCs to regulate peripheral T cell activation (Schwartz, 2003). B7.1 and B7.2, which engage CD28 receptor on naive T cells, are by far the most important in T cell activation. They are highly upregulated on APCs by infectious agents (Janeway and Medzhitov, 2002). Stimulation of Th1 clones with a T cell receptor (TCR) agonist in the absence of CD28 co-stimulation resulted in T cell anergy (Mueller et al., 1989; Schwartz, 2003). Mice deficient in CD28 or both B7.1 and B7.2 were found to be highly resistant to autoimmune disease models (Greenwaldet et al., 2005). A second receptor for B7, CTLA4, is induced on activated T cells and serves as a negative regulator of T cell activation and proliferation (Chambers and Allison, 1999). Mice deficient in CTLA4 die at the neonatal stage owing to massive T cell activation and infiltration into tissues.

In the past few years, the B7 and CD28 families have been rapidly expanded to include novel co-stimulatory molecules (Table 7.1). Here we summarize our work and that of others in the field on the roles of these molecules in the regulation of immune tolerance and autoimmunity.

TABLE 7.1. Brief Summary of B7 and CD28 Family Members.

| Ligand | Expression | Receptor | Function |
|---|---|---|---|
| B7.1/B7.2 (CD80/CD86) | Activated APC | CD28/CTLA4 | T cell activation and tolerance |
| B7h (B7RP-1) | B cells, macrophages, and nonlymphoid tissues | ICOS | T cell activation and function |
| B7-H1 (PD-L1)/ B7DC (PD-L2) | APCs and nonlymphoid tissues | PD-1 | Inhibition of T cell proliferation, and immune tolerance |
| B7-H3 | Lymphoid and nonlymphoid tissues | Unknown | Inhibition of T cell activation |
| B7S1 | Lymphoid and nonlymphoid tissues | Unknown | Inhibition of T cell activation |
| HVEM | Lymphoid and nonlymphoid tissues | BTLA | Inhibition of T cells and B cells |

APCs, antigen-presenting cells.

# 2   Regulation of T Cell Tolerance and Autoimmunity by ICOS and B7h

Inducible co-stimulator (ICOS), the third member of the CD28 family, is induced after T cell activation (Hutloff et al., 1999). The ligand for ICOS, B7h (also named B7RP-1) is constitutively expressed on certain APCs, such as B cells and macrophages, and can be induced in nonlymphoid tissues and cells by inflammatory stimuli (Swallow et al., 1999; Yoshinaga et al., 1999). In contrast to some other members of the CD28 family (B7.1/B7.2-CD28/CTLA4 and PDL1/PDL2-PD-1), ICOS and B7h are the only receptor and ligand for the other, respectively (Nurieva et al., 2003b).

## 2.1   ICOS-B7h Interaction During Th Cell Activation and Differentiation

In recent years, analysis of mice deficient in ICOS or B7h has revealed the role of this co-stimulatory pathway in T cell activation and function (Dong and Nurieva, 2003). Experiments using either ICOS-deficient T cells or B7h-deficient APCs have shown a role of B7h-ICOS interaction during T cell activation, where reduced IL-2 production and proliferation was found in the absence of this pathway (Dong et al., 2001a; Nurieva et al., 2003b). In addition, disruption of ICOS-B7h interaction selectively impaired IL-4 production in activated T cells during in vitro differentiation or after in vivo priming with protein antigen in complete Freund's adjuvant (CFA) or alum (Dong et al., 2001a; Nurieva et al., 2003b). The transcriptional mechanisms whereby ICOS-B7h interaction regulates Th2 differentiation and IL-4 expression in effector T cells were recently studied (Nurieva et al., 2003a; Nurieva, 2003b). Effector T cells generated in the absence of ICOS-B7h interaction exhibited selective deficiency of c-Maf transcription factor that regulates IL-4 gene expression (Ho et al., 1998; Kim, 1999 #16). Moreover, c-Maf overexpression restored the IL-4 defect in T cells activated in the absence of ICOS-B7h interaction (Nurieva al., 2003a: Nurieva, 2003 b). c-Maf expression in effector cells was regulated not directly by ICOS but by IL-4 during Th differentiation (Nurieva et al., 2003a). ICOS co-stimulation potentiated TcR- and CD28-mediated initial IL-4 production, possibly through enhancement of NFATc1 expression (Nurieva et al., 2003a).

Regulation of Th2 cytokine expression by ICOS appears to be a crucial protective mechanism in experimental autoimmune encephalomyelitis (EAE) (Table 7.2). In this model, ICOS−/− mice and those treated with an ICOS blocking antibody during the T cell priming phase developed greatly enhanced disease (Dong et al., 2001a; Rottman et al., 2001). Furthermore, when subject to EAE, knockout mice had massive infiltrates; and flow cytometric analysis revealed more CD4+ T cells in the central nervous system (CNS) (Dong et al., 2001a). These results are consistent to the roles of Th2 cytokines in negative regulation of Th1 differentiation and Th differentiation into IL-17-expressing cells (Park et al., 2005).

TABLE 7.2. Roles of ICOS and B7h in Immune Tolerance and Autoimmunity.

| Disease | Manipulation | Results | References |
|---|---|---|---|
| EAE | ICOS KO, ICOS-Ig | ICOS KO or blockade during priming exacerbates disease. | Dong et al., 2001; Rottman et al., 2001 |
| Lupus | Anti-B7h mAbs | B7h blockade before the onset of renal disease significantly delayed the onset of proteinuria and prolonged survival and significantly decreased IgG autoantibody production. B7h blockade after onset of proteinuria prevented disease progression. | Iwai et al., 2003 |
| CIA | ICOS KO, anti-B7h mAbs | Blockade with anti-B7h mAbs significantly ameliorated the disease. ICOS KO mice resistant to CIA exhibit reduced IgM and IgG2a autoantibodies production and decreased IL-17 production by T cells. | Iwai et al., 2002; Nurieva, 2005 |
| Diabetes | Anti-ICOS | Blockade of ICOS altered the balance between T effector and Treg cells, resulting in progression from prediabetic insulitis to diabetes. | Herman et al., 2004 |

EAE, experimental autoimmune encephalitis; CIA, collagen-induced arthritis; mAbs, monoclonal antibodies; ICOS, inducible co-stimulator.

## 2.2   Regulation of Humoral Immunity by ICOS

Analysis of ICOS- and B7h-deficient mice also reveals the critical function of ICOS and B7h in humoral immunity (Dong et al., 2001a; McAdam et al., 2001; Nurieva et al., 2003b; Tafuri et al., 2001; Wong et al., 2003). First, mice deficient in ICOS or B7h, upon immunization with protein antigens, exhibited reduced antigen-specific immunoglobulin G1 (IgG1), IgG2a, and IgE production (Dong et al., 2001a; McAdam et al., 2001; Nurieva et al., 2003b; Tafuri et al., 2001; Wong et al., 2003). These results suggest a role for ICOS and B7h interaction in immunoglobulin class-switching by B cells. It was reported in one study that ICOS regulated CD40L upregulation on T cells (McAdam et al., 2001); this result, however, could not be confirmed by others (Dong et al., 2001b; Mak et al., 2003). Moreover, ICOS-B7h appears to be important for optimal IgM production in both mouse and human (Grimbacher et al., 2003; Nurieva et al., 2003b), suggesting that ICOS may function beyond class switching. IL-4 is an important Th cytokine regulating B cell proliferation and class switching to IgG1 and IgE. Although ICOS may regulate IgE production in an IL-4-dependent manner, IL-4 does not appear to regulate IgM production (our unpublished data); there may exist additional mechanism(s) through which ICOS regulates humoral immunity.

A hallmark of T-dependent humoral responses is the formation of germinal centers. ICOS is highly expressed on germinal center (GC) T cells and its ligand is constitutively expressed on B cells (Hutloff et al., 1999; Swallow et al., 1999). Interestingly ICOS-deficient mice were found defective in primary GC and were unable to form secondary GC (Dong et al., 2001b; Tafuri et al., 2001; Wong et al., 2003). Recently, a population of T-helper cells called follicular B cell helper T cells ($T_{FH}$) has emerged as a Th subset regulating GC reactions and humoral immunity. These cells express CXCR5 and high levels of ICOS et al., 2005b). Moreover, ICOS-B7h interaction is required for the generation of $T_{FH}$ cells; mice deficient in ICOS or treated with anti-B7h monoclonal antibodies (mAbs) showed impaired development of CXCR5+ $T_{FH}$ cells as well as PNA+B220+ GC B cells in the spleen in response to primary or secondary immunizations (Akiba et al., 2005). CXCR5+ cells do not express Th2 cytokines (IL-4, IL-5, IL-10) (Akiba et al., 2005); instead they produce IL-21, a cytokine important for B cell maturation (Leonard and Spolski, 2005; Ozaki, 2004 #27). ICOS-deficient mice also exhibited defective IL-21 expression (our unpublished data). In addition, Vanuesa et al. found that roquin, a RNA-binding E3 ubiquitin ligase, negatively regulates ICOS expression and the development of $T_{FH}$ cells (Vinuesa et al., 2005). Mutation of this gene in Sanroque mice resulted in increased expression of ICOS, enhanced development of $T_{FH}$ cells, and overproduction of IL-21 (Vinuesa et al., 2005a). These results indicate that ICOS and B7h play a critical role in the development of CXCR5+ $T_{FH}$, which are involved in GC reactions and in enhancement of antibody responses. ICOS may thus regulate humoral immunity through both Th2 and $T_{FH}$ cells.

Interestingly, Sanroque mice spontaneously develop high titers of autoantibodies and a pattern of pathology consistent with lupus (Vinuesa et al., 2005a), suggesting that ICOS may play a pathogenic function in lupus. Indeed, this was the case when Iwai et al. showed that administration of anti-B7h mAb before the initiation of renal disease significantly delayed the onset of proteinuria, prolonged survival, and effectively inhibited all subclasses of IgG autoantibody production (Iwai et al., 2003).

## 2.3   ICOS Regulates IL-17 Expression in Arthritis Disease

Collagen-induced arthritis (CIA), the most commonly used animal model for human rheumatoid arthritis (RA) disease, is characterized by both autoantibody production and inflammation in the joint tissue. In striking contrast to EAE, in which absence of ICOS enhances autoimmune disease, ICOS is essential in the pathogenesis of the CIA model. ICOS-deficient mice as well as those receiving an anti-B7h antibody exhibited resistance to CIA (Dong and Nurieva, 2003; Iwai et al., 2002). Compared to ICOS+/+ mice, ICOS−/− mice exhibited greatly reduced anti-collagen IgM and IgG2a antibody titers, indicating that autoantibody production was compromised (Dong and Nurieva, 2003). In T cells, ICOS deficiency did not significantly reduce activation or affect the expression of IFNγ and TNFα. Instead, ICOS is required for production of IL-17 (Dong and Nurieva, 2003).

Interleuin-17 is a pro-inflammatory cytokine secreted by activated CD4+ T cells and is frequently found in RA synovium (Chabaud, 1999 #23, Fossiez, 1996 #22, Yao et al., 1995). IL-17 produced by CD4+ cells in the joint tissue may contribute to the inflammatory process in RA. To support this idea, overexpression of IL-17 in the joints resulted in inflammatory infiltration of the synovium and aggressive cartilage degradation (Lubberts et al., 2002). Blockade of IL-17 action with a soluble receptor significantly reduced the CIA, including clear suppression of joint damage (Lubberts et al., 2001). Thus, defective IL-17 production in the ICOS deficient mice may contribute to lack of joint inflammation in these mice. Recently, Th cells expressing IL-17 were found to be a novel lineage of Th cells, distinct from the traditional Th1 and Th2 cells (Harrington et al., 2005; Park et al., 2005). Park et al. indicated that generation of these cells required CD28 and ICOS co-stimulation and IL-23 cytokines but was independent of the cytokine and transcription programs normally associated with Th1 and Th2 differentiation (Park et al., 2005). Furthermore, ICOS not only regulates the priming of Th cells to differentiate into IL-17-producing cells, it also is important for IL-17 expression during the effector phase (our unpublished data).

## 2.4    Function of ICOS in Regulatory T Cells

Immune tolerance is regulated by antiinflammatory regulatory T cells. Early studies found that anti-ICOS co-stimulated human T cells in production of IL-10 (Hutloff et al., 1999), a key antiinflammatory cytokine produced by regulatory T cells (Tregs). Lohning et al. characterized ICOS-expressing cells obtained from human secondary lymphoid tissues and found that whereas T cells expressing medium levels of ICOS secreted Th2 cytokines IL-4, IL-5, and IL-13, those expressing high levels of ICOS predominantly produced IL-10 (Lohning et al., 2003). The role of ICOS in stimulating IL-10 production may also be important in the generation of Tregs. Development of IL-10-producing Treg cells in a mouse asthma model was dependent on the ICOS-ICOS-ligand pathway (Akbari et al., 2002). Previous studies have shown that IL-10 is a critical cytokine for induction of oral and nasal tolerance (Faria, 2003 #34; Massey et al., 2002). Miyamoto et al. found that ICOS plays an essential role in the regulation of mucosal tolerance (Miyamoto et al., 2005). Stimulation of ICOS promoted the function of IL-10-producing Treg cells and the induction of oral and nasal tolerance.

Herman et al. reported that Treg cells play a critical role in preventing the progression of type 1 diabetes (Herman et al., 2004). Treg cells in pancreatic islets but not in draining lymph nodes highly expressed ICOS and IL-10 transcripts and coexisted in balance with aggressive effector T cells in pancreatic lesions. Blockade of ICOS in this model resulted in loss of this balance and rapid onset of diabetes. These findings suggest that Treg cells may suppress autoimmune development in an ICOS-dependent manner.

## 3    Regulation of T Cell Tolerance by PD-1 and Its Ligands

Programmed death-1 (PD-1), a member of the CD28 superfamily (Ishida et al., 1992), is expressed on activated T cells, B cells, and monocytes (Agata et al., 1996; Yamazaki et al., 2002). The cytoplasmic domain of PD-1 contains an immunoreceptor tyrosine-based inhibitory motif (ITIM) and an immunoreceptor tyrosine-based switch motif (ITSM) (Finger et al., 1997; Shinohara et al., 1994), which appear to be responsible for the negative signaling on T and B cells by recruiting the phosphatase SHP-2 [SRC homology 2(SH2)-domain-containing protein tyrosine phosphatase 2] (Latchman et al., 2001; Okazaki et al., 2001). Recent genetic studies have supported an inhibitory function for PD-1 in providing a negative co-stimulatory signal to T and B cells. PD-1 knockout mouse develop autoimmune sequelae. Interestingly, the genetic background influences the autoimmune phenotype. For example, knockout of *Pdcd-1* on the C57BL/6 background leads to arthritis and lupus-like glomerulonephritis (Nishimura et al., 1999), whereas in Balb/c mice knockout of *Pdcd-1* yields dilated cardiomyopathy with the presence of elevated titers of anti-cardiac troponin I autoantibodies (Nishimura et al., 2001; Okazaki et al., 2003). These results indicate that PD-1, like CTLA4, plays a crucial role in maintaining immune tolerance.

PD-1 has two ligands belonging to the B7 superfamily: PD-L1 (B7-H1) and PD-L2 (B7-DC) (Dong et al., 1999; Freeman et al., 2000; Latchman et al., 2001; Tseng et al., 2001). PD-L1 mRNA is broadly expressed in various human and mouse tissues, such as heart, placenta, muscle, fetal liver, splenic lymph nodes, and thymus for both species. In humans, PDL-1 protein expression has been found in human endothelial cells (Chen et al., 2005; de Haij et al., 2005; Mazanet and Hughes, 2002), myocardium (Brown et al., 2003), syncytiotrophoblasts (Brown et al., 2003; Petroff et al., 2002), resident macrophages of some tissues or in macrophages that have been activated with IFNγ or TNFα (Latchman et al., 2001), and in various tumors (Dong and Chen, 2003; Dong et al., 1999). In the mouse, PD-L1 protein expression is found in heart endothelium, islet cells of the pancreas, small intestines, and placenta (Liang et al., 2003). In mouse hematopoietic cells, PD-L1 is constitutively expressed on T cells, B cells, macrophages, and dendritic cells (DCs) and can be upregulated upon activation (Yamazaki et al., 2002). PDL-2 mRNA is found in heart, placenta, lung, liver, muscle, pancreas, splenic lymph nodes, and thymus for both species and in brain and kidney only in the mouse (Latchman et al., 2001). On the other hand, PDL-2 protein expression is only found in macrophages and DCs and can be upregulated upon activation with IFNγ, granulocyte/macrophage colony-stimulating factor (GM-CSF), and IL-4 (Yamazaki et al., 2002). Macrophages represent an interesting case where Th1 cytokines regulate the expression of PD-L1 expression, and Th2 cytokines regulate PD-L2 expression (Loke and Allison, 2003). Although both PD-L1 and PD-L2 bind to PD-1, results from various groups exist to suggest a secondary positive receptor for either molecule (Dong et al., 1999; Tseng et al., 2001).

## 3.1   PD-1 in Type 1 Diabetes

In the NOD model of type 1 diabetes (T1D), the administration of anti-PD1 or anti-PD-L1 antibody to neonatal and adult prediabetic NOD male or female mice accelerated the onset of diabetes (Ansari and Abdi, 2003), whereas anti-PD-L2 did not modify the progression of T1D, indicating that PD-1/PD-L1 interaction negatively regulates the initiation and progression of the disease. In accordance with this observation, the *Pdcd-1–/–* mice in the NOD background exhibited early diabetes onset (5–10 weeks of age) with 100% penetrance in males and females (Wang et al., 2005). Interestingly, the insulitic lesions of *Pdcd-1–/–* NOD mice showed more CD8 T cells inside the islets when compared to NOD mice, indicating that PD-1 deficiency may result in enhanced autoreactive CD8 T cell priming in vivo (Wang et al., 2005) (Table 7.3).

Although the above results indicate PD-1/PD-L1 interaction in regulation of immune tolerance against T1D, PD-L1 constitutively expressed in pancreatic islets appears paradoxically to support autoimmune destruction. Transgenic expression of PD-L1 in islet cells in C57BL/6 mice peculiarly provokes spontaneous diabetes in 7% to 14% of the mice 3 to 6 weeks of age (Subudhi et al., 2004) (Table 7.3). Moreover, the double transgenic *RIP.B7-H1/mOVA* causes spontaneous diabetes in 30% of the animals with a similar onset. The transfer of low doses of CD8 T cells (*OT-1*) specific against OVA peptide (SIINFEKL) into the *RIP.B7-H1/mOVA* animals causes diabetes with 100% incidence between days 10 and 20 after transfer. Moreover, the proliferation of OT-1 cells due to primming in pancreatic lymph nodes was accelerated in *RIP.B7-H1/mOVA* mice when compared to the RIP.OVA control mice, and treatment with anti-PD-1 blocking antibody did not reduce this accelerated proliferation. These results indicate that PD-L1 activation of autoreactive CD8 T cells occurs independent of PD-1.

## 3.2    PD-1 in Experimental Autoimmune Encephalomyelitis

In the EAE disease model, PD-1, PD-L1, and PD-L2 expression was found in infiltrating mononuclear cells in the meninges of mice (Liang et al., 2003; Salama et al., 2003). PD-L1 expression was upregulated on the endothelium surrounding the cell infiltrates on astrocytes and microglia, and PDL2 was found on macrophages at these sites. Antibody blockade of PD-1 at the time of immunization accelerated the onset of disease and increased the severity of symptoms (Salama et al., 2003). However, PD-1 blockade 10 days after immunization did not have any effect on the progression of EAE, indicating that PD-1 dampens the response of autoreactive T cells during priming and initial expansion phase. Interestingly, blockade of PD-L1 during this same time period did not alter the progression of EAE, but the blockade of PD-L2 provoked more severe disease with the same incidence and time of onset (Salama et al., 2003). These results indicate a crucial role of PD-1/PD-L2 interaction in negative regulation of EAE disease. However, analysis of the PDL-1-deficient mice revealed that these mice developed more severe EAE on an 129Sv background, which is normally resistant to EAE development (Latchman et al., 2004). These mice exhibited an increase in antigen-specific CD4 T cells in the CNS, indicating that PDL-1 engagement to PD-1 may control the expansion of autoreactive T cells at the tissue sites. Thus, in the EAE model, PDL-2 is involved in controlling the priming and early expansion of autoreactive T cells at the lymph node sites, and PDL-1 may play a role at the effector phase, regulating the chronic inflammation and destruction of the target tissue (Table 7.3).

## 3.3   PD-1 in Inflammatory Bowel Disease

Inflammatory bowel disease (IBD) can be induced in lymphopenic mice by the transfer of naive T cells. In this model as well as in human patients with IBD, the expression of PD-1 and its ligands was found to be upregulated in mononuclear cells of lamina propia from inflamed gut mucosa (Kanai et al., 2003) (Table 7.3). The antibody blockade of PD-L1 after the transfer of CD45RBhi cells, but not PD-L2, reduced the wasting disease with colitis, abrogated leukocyte infiltration, and reduced the production of IFN$\gamma$, IL-2, and TNF$\alpha$ but not IL-4 or IL-10, by lamina propia CD4+ T cells (Kanai et al., 2003). Therefore PDL-1 co-stimulation may promote mucosal inflammation in the gut. A recent study suggested the regulation of colitis by PD-1 through a population of regulatory T cells (CD4+CD25−PD-1+), where the co-transfer of CD45RBhi T cells with CD4+CD25−PD-1+ cells into SCID mice reduced the severity of colitis (Totsuka et al., 2005). Interestingly, this population of regulatory T cells produced large amounts of IL-4 and IL-10 when restimulated in vitro with anti-CD3; however, blockade of either of these cytokines did not prevent the suppression activity of these cells, leaving the question open as how PD-1 in this cells regulate the autoimmune process (Totsuka et al., 2005).

## 3.4   PD-1 in Autoimmune Hepatitis

Concanavalin A (ConA)-induced hepatitis is an experimental murine model of autoimmune hepatitis, where the hepatocyte injury is caused by infiltrating lymphocytes including CD4, CD8, and NKT cells (Tiegs et al., 1992). Mice deficient in PD-L1 exhibited accelerated hepatitis upon ConA treatment compared to wild-type mice (Dong et al., 2004). In addition, the hepatic lesion was more extensive in PD-L1−/− mice and accelerated localization of CD8 T cells to the site was observed on day 1 after ConA treatment (Dong et al., 2004). In the meantime, decreased apoptosis of CD8 T cells in the liver was observed in PD-L1-deficient mice (Dong et al., 2004). Thus, PD-L1 is controlling the effector function of CD8 T cells and its longevity in the liver. It is unclear at this stage whether PD-L1 exerts its function through engaging PD-1 (Table 7.3).

## 3.5   PD-1 in Systemic Lupus Erythematosus

Pdcd-1 knockout mice developed arthritis and lupus-like glomerulonephritis on the C57BL/6 background (Nishimura et al., 1999). Incorporation of the PD-1 deficiency with lpr mutation led to accelerated disease onset and increased the severity of symptoms (Nishimura et al., 1999). Tasuju Honjo's group developed a mouse doubly deficient for the low affinity receptor for IgG (Fc$\gamma$RIIB) and PD-1 on the Balb/c background. This mouse develops hydronephrosis with autoantibodies (autoAbs) against urothelial antigens, including uroplakin IIIa, deposited on the urothelial cells of the urinary bladder, which is a complication seen in human patients with systemic lupus erythematosus (SLE) and Sjögren's

TABLE 7.3. Role of PD-1 in Autoimmune Diseases.

| Disease | Intervention | Result | Mechanism | References |
|---|---|---|---|---|
| Type 1 diabetes (T1D) | a. Anti-PD-1, anti-PDL1, *Pdcd1* null mutation in NOD mice | Accelerated diabetes. Increased IFNγ producing T cells. Increased CD8 cells in lesion. | PD-1 and PDL1 inhibits autoreactive T cell priming. | Ansari & Abdi, 2003, Wang et al., 2005 |
| | b. Transgenic expression of PDL1 in islets of C57BL/6 mice | Spontaneous diabetes. | T cell-dependent. PDL1 promotes priming and effector function of autoreactive CD8 T cells. | Subudhi et al., 2004 |
| EAE | a. Anti-PD-1 and anti-PDL-2 at disease induction | Accelerated onset of disease and increased severity of symptoms. | PD-1 and PDL2 dampens the response of autoreactive T cells during priming and initial expansion. | Salama et al., 2003 |
| | PDL1 null mutation | Susceptible to EAE on 129Sv. Increased MOG-specific CD4 T cells. | PDL1 suppresses autoreactive T cells in the target tissue. | Latchman et al., 2004 |
| IBD | Anti-PDL-1 at the time of the cell transfer | Reduced colitis, abrogated leukocyte infiltration and T cell function | PDL-1 promotes mucosal inflammation in the gut. | Kanai et al., 2003 |
| ConA-induced hepatitis | PDL-1 null mutation | Accelerated hepatitis and localization of CD8 T cells to the liver. Decreased CD8 T cell apoptosis. | PDL-1 regulates CD8 T cell effector function and survival. | Dong et al., 2004 |
| Lupus | a. PD-1 null mutation | Glomerulon-ephritis. | Breakdown of immune tolerance. | Nishimura et al., 1999 |
| | b. PD-1 and FcR2b double mutation | Hydronephrosis with anti-uroplakin IIIa and antinuclear autoAbs. | Co-inhibition by PD-1 and FcγRIIB on autoreactive B cells. | Okazaki et al., 2005 |

syndrome (Okazaki et al., 2005). In addition, ~15% of the double knockout mice produced antinuclear autoAbs (Okazaki et al., 2005). These results indicate that the negative signals provided by FcγRIIB and PD-1 regulate tolerance to SLE. Future experiments are needed to clarify the regulation by these two molecules on B versus T cells in this crucial regulation (Table 7.3).

The human *PDCD1* gene is located on chromosome 2q37. In a search for a susceptibility locus for SLE, a single nucleotide polymorphism in 12% of Europeans and 7% of Mexicans patients with SLE, called PD1.3, was found at nucleotide 7146 in an enhancer-like structure in the fourth intron of *PDCD1* (Prokunina et al., 2002). The mutation disrupts the predicted DNA-binding site for the runt-related transcription factor RUNX-1 or AML-1. Interestingly, this same polymorphism was also associated with risk of T1D (Nielsen et al., 2003), with RA (Prokunina et al., 2004), and with multiple sclerosis (Kroner et al., 2005).

# 4   B7-H3, B7S1, and BTLA

## 4.1   B7-H3

Several novel B7 and CD28 superfamily members were discovered to regulate immune tolerance and autoimmunity. B7-H3 was first identified in human dendritic cells activated by inflammatory cytokines (Chapoval et al., 2001). Mouse B7-H3 appears to be more broadly expressed in lymphoid and nonlymphoid tissues, and its expression on dendritic cells was found further upregulated by lipopolysaccharide (LPS) (Prasad et al., 2004; Sun et al., 2002). Both human and mouse B7-H3 recombinant proteins were reported to bind to an unidentified receptor expressed on activated but not naive T cells (Chapoval et al., 2001; Sun et al., 2002).

Human B7-H3 was initially reported to co-stimulate proliferation of both CD4+ and CD8+ T cells and selectively enhance IFNγ production in the presence of T cell receptor signaling (Chapoval et al., 2001). However, this function was not reproducible in murine systems. Mouse B7-H3 moderately reduced proliferation of T cells and their IL-2 production (Prasad et al., 2004). In vivo, mice deficient in B7-H3 or treated with an antagonistic antibody exhibited greatly enhanced EAE disease characterized by excessive inflammatory infiltrates in the CNS (Prasad et al., 2004; Suh et al., 2003). Interestingly, deficiency in B7-H3 did not affect Th2 responses or eosinophilia in an asthma model (Suh et al., 2003). The late-onset increase of autoantibodies in B7-H3-deficient animals (Suh et al., 2003) suggests a modest role of this pathway in maintaining immune tolerance to self-antigens.

## 4.2   B7S1/B7-H4/B7x

B7S1, also called B7-H4 and B7x, was simultaneously discovered by three groups (Prasad et al., 2003; Sica et al., 2003; Zang et al., 2003). Mouse B7S1, most similar to B7-H3 in sequence, is also broadly expressed in tissues and appears to be upregulated in certain tumors (Prasad et al., 2003; Sica et al., 2003; Zang et al., 2003). B7S1 engages to an unidentified receptor on activated but not naive T cells, which is distinct from CD28, CTLA4, PD-1, and BTLA (Prasad et al., 2003; Sedy et al., 2005; Sica et al., 2003).

In vitro studies consistently indicate that B7S1 stimulation inhibits T cell proliferation and IL-2 production (Prasad et al., 2003; Sica et al., 2003; Zang et al.,
2003). A blocking antibody to B7S1 enhanced T cell activation in vitro (Prasad
et al., 2003). Limited literature exists regarding B7S1 function in vivo. Anti-B7S1
moderately enhanced T cell immunity in an immunization experiment but greatly
exacerbated EAE (Prasad et al., 2003), suggesting that B7S1 is a negative regulator in autoimmune disease. However, the role of B7S1 in immune tolerance has
not been documented.

## 4.3   BTLA

BTLA was first identified as a member of the CD28 family whose mRNA is selectively expressed on Th1 cells (Watanabe et al., 2003). Further analysis revealed
that BTLA protein is constitutively expressed by B cells and dendritic cells
(Hurchla et al., 2005). Initially reported to bind to B7S1, BTLA was found by two
groups to bind to a member of the TNF family: HVEM (Gonzalez et al., 2005;
Sedy et al., 2005).

BTLA has similarity in its cytoplasmic region with CTLA4 and PD-1 and contains two copies of the ITIM motif that associate with SHP-1 and SHP-2 (Watanabe
et al., 2003). Engagement of BTLA on CD4+ and CD8+ T cells resulted in attenuation of antigen-driven proliferation (Krieg et al., 2005), supporting the idea that it
serves as an inhibitory receptor on T cells. BTLA deficiency resulted in augmented
antibody responses in vivo and moderately enhanced EAE disease (Watanabe et al.,
2003). Future studies are needed to reveal the function of this molecule in the
various immune tolerance and autoimmunity models.

## 5   Conclusion

Studies from many investigators have revealed "rich" co-stimulatory signals in
T cell regulation. ICOS joins CD28 to regulate T cell activation and effector function positively. ICOS-B7h interaction plays a crucial role in regulation in autoimmune diseases. On the other hand, negative co-stimulatory pathways have rapidly
expanded. PD-1 has emerged as a crucial regulator in T cell tolerance. B7-H3,
B7S1, and BTLA appear essential not for controlling immune tolerance but important to fine-tune T cell activity in autoimmune disease models. Based on the
advances made by these studies, a major concept has become obvious that T cell
tolerance or activation is determined by a combination of co-stimulation signals
(Fig. 7.1). In the resting state, negative co-stimulatory molecules may selectively
or collectively maintain T cell activation thresholds to prevent activation of autoreactive T cells. During infection, ligands for CD28 and ICOS are greatly upregulated on APCs, which overcomes the constitutive tolerance mechanisms mediated
by negative co-stimulation and results in activation of pathogen-specific T cells.

Modulation of these co-stimulatory pathways has remained one of the
most attractive approaches to inducing T cell tolerance against self-antigens.

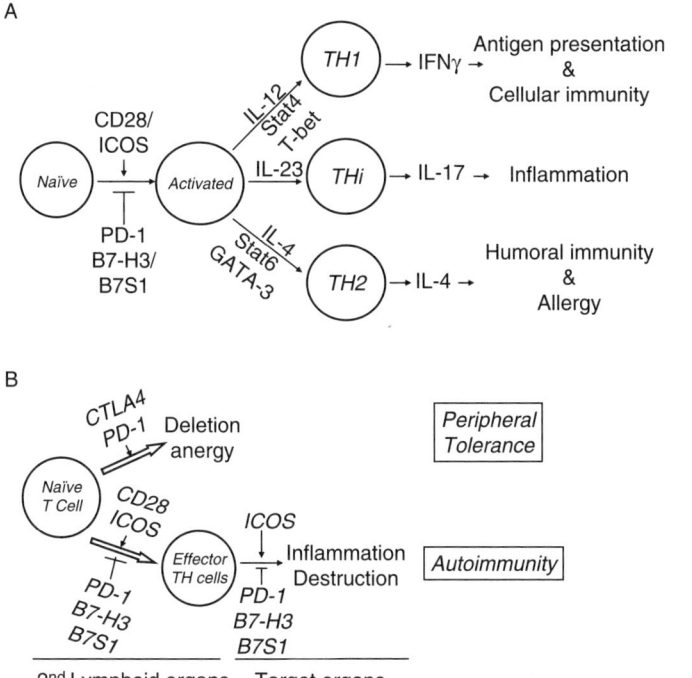

FIGURE 7.1. Regulation of T cell activation and tolerance by co-stimulatory molecules.

However, more studies are needed in at least the following scenarios to make this goal feasible.

1. Elucidate the context in which each co-stimulatory pathway functions. Is there specific regulation of the expression of each B7 family member? Are they uniquely distributed in a distinct dendritic cell subset or in various tissues, or are they regulated by exogenous and endogenous inflammatory and antiinflammatory signals?

2. Examine the redundancy of various co-stimulatory pathways. CD28 and ICOS have been shown to have overlapping function in humoral immunity (Suh et al., 2004). Is there any redundancy in the growing negative co-stimulatory pathways?

3. Illustrate T cell regulation by co-stimulatory molecules. It is still unclear how T cells respond to costimulation. What are the signal transduction and transcriptional regulation that account for the biological effect of co-stimulation regulation?

4. Compare co-stimulation regulation across species. Although there are considerable advances in our understanding in the mouse, the roles of co-stimulation in human are needed before applying them to clinical practice.

*Acknowledgments.* We thank many colleagues for scientific contribution and supports from the NIH (C.D.). C.D. is an Arthritis Investigator of the Arthritis Foundation, a Cancer Research Institute Investigator, and an MD Anderson Cancer Center Trust Fellow. R.N. received a Postdoctoral Fellowship from the Arthritis Foundation and a Scientist Development Grant from the American Heart Association.

## References

Agata, Y., Kawasaki, A., Nishimura, H., et al. (1996) Expression of the PD-1 antigen on the surface of stimulated mouse T and B lymphocytes. *Int. Immunol.* 8(5):765-772.

Akbari, O., Freeman, G. J., Meyer, et al. (2002) Antigen-specific regulatory T cells develop via the ICOS-ICOS-ligand pathway and inhibit allergen-induced airway hyperreactivity. *Nat. Med.* 8(9):1024-1032.

Akiba, H., Takeda, K., Kojima, Y., et al. (2005) The role of ICOS in the CXCR5+ follicular B helper T cell maintenance in vivo. *J. Immunol.* 175(4):2340-2348.

Ansari, M. J., Abdi, R. (2003) Emerging immunomodulatory therapies targeting the costimulatory pathways for the prevention of transplant rejection. *Drugs* 6(10):964-969.

Brown, J. A., Dorfman, D. M., Ma, F. R., et al. (2003) Blockade of programmed death-1 ligands on dendritic cells enhances T cell activation and cytokine production. *J. Immunol.* 170(3):1257-1266.

Chabaud, M., Durand, J. M., Buchs, N., et al. (1999) Human interleukin-17: A T cell-derived proinflammatory cytokine produced by the rheumatoid synovium. *Arthritis Rheum.* 42:963-970.

Chambers, C. A., Allison, J. P. (1999) Costimulatory regulation of T cell function. *Curr. Opin. Cell Biol.* 11(2):203-210.

Chapoval, A. I., Ni, J., Lau, J. S., et al. (2001) B7-H3: a costimulatory molecule for T cell activation and IFN-gamma production. *Nat. Immunol.* 2(3):269-274.

Chen, Y., Li, J., Zhang, J., et al. (2005) Sinomenine inhibits B7-H1 and B7-DC expression on human renal tubular epithelial cells. *Int. Immunopharmaco.*, 5(9):1446-1457.

De Haij, S., Woltman, A. M., Trouw, L. A., et al. (2005) Renal tubular epithelial cells modulate T-cell responses via ICOS-L and B7-H1. *Kidney Int.* 68(5):2091-2102.

Dong, C., Flavell, R. A. (2000a) Cell fate decision: T-helper 1 and 2 subsets in immune responses. *Arthritis Res.* 2(3):179-188.

Dong, C., Flavell, R. A. (2000b) Control of T helper cell differentiation—in search of master genes. *Sci STKE* 2000(49):E1.

Dong, C., Juedes, A. E., Temann, U. A., et al. (2001a). ICOS co-stimulatory receptor is essential for T-cell activation and function. *Nature* 409(6816):97-101.

Dong, C., Nurieva, R. I. (2003). Regulation of immune and autoimmune responses by ICOS. *J/ Autoimmun.* 21(3):255-260.

Dong, C., Temann, U. A., Flavell, R. A. (2001b) Cutting edge: critical role of inducible costimulator in germinal center reactions. *J. Immunol.* 166(6):3659-3662.

Dong, H., Chen, L. (2003) B7-H1 pathway and its role in the evasion of tumor immunity. *J. Mol. Med.* 81(5):281-287.

Dong, H., Zhu, G., Tamada, K., Chen, L. (1999) B7-H1, a third member of the B7 family, co-stimulates T-cell proliferation and interleukin-10 secretion. *Nat. Med.* 5(12):1365-1369.

Dong, H., Zhu, G., Tamada, K., et al. (2004) B7-H1 determines accumulation and deletion of intrahepatic CD8(+) T lymphocytes. *Immunity* 20(3):327-336.

Faria, A. M., Maron, R., Ficker. S. M., et al. (2003) Oral tolerance induced by continuous feeding: enhanced up-regulation of transforming growth factor-beta/interleukin-10 and suppression of experimental autoimmune encephalomyelitis. *J. Autoimmun.* 20:135-145.

Finger, L. R., Pu, J., Wasserman, R., et al. (1997). The human PD-1 gene: complete cDNA, genomic organization, and developmentally regulated expression in B cell progenitors. *Gene* 197(1-2):177-187.

Fossiez, F., Djossou, O., Chomarat, P., et al. (1996) T cell interleukin-17 induces stromal cells to produce proinflammatory and hematopoietic cytokines. *J. Exp. Med.* 183:2593-2603.

Fontenot, J. D., Rudensky, A. Y. (2005) A well adapted regulatory contrivance: regulatory T cell development and the forkhead family transcription factor Foxp3. *Nat. Immunol.* 6(4):331-337.

Freeman, G. J., Long, A. J., Iwai, Y., et al. (2000) Engagement of the PD-1 immunoinhibitory receptor by a novel B7 family member leads to negative regulation of lymphocyte activation. *J. Exp. Med.* 192(7):1027-1034.

Gonzalez, L. C., Loyet, K. M., Calemine-Fenaux, J., et al. (2005) A coreceptor interaction between the CD28 and TNF receptor family members B and T lymphocyte attenuator and herpesvirus entry mediator 10.1073/pnas.0409071102. *Proc. Natl. Acad. Sci. U S A* 102(4):1116-1121.

Greenwald, R. J., Freeman, G. J., Sharpe, A. H. (2005) The B7 family revisited. *Annu. Rev. Immunol.* 23(1);515-548.

Grimbacher, B., Hutloff, A., Schlesier, M., et al. (2003) Homozygous loss of ICOS is associated with adult-onset common variable immunodeficiency. *Nat. Immunol.* 4(3):261-268.

Harrington, L. E., Hatton, R. D., Mangan, P. R., et al. (2005) Interleukin 17-producing CD4+ effector T cells develop via a lineage distinct from the T helper type 1 and 2 lineages. *Nat. Immunol.* 6(11):1123-1132.

Heath, W. R., Carbone, F. R. (2001) Cross-presentation, dendritic cells, tolerance and immunity. *Annu. Rev. Immunol.* 19:47-64.

Herman, A. E., Freeman, G. J., Mathis, D., Benoist, C. (2004) CD4+CD25+ T regulatory cells dependent on ICOS promote regulation of effector cells in the prediabetic lesion. *J. Exp. Med.* 199(11):1479-1489.

Ho, I. C., Lo, D., Glimcher, L. H. (1998). c-maf promotes T helper cell type 2 (Th2) and attenuates Th1 differentiation by both interleukin 4-dependent and -independent mechanisms. *J. Exp. Med.* 188(10):1859-1866.

Hurchla, M. A., Sedy, J. R., Gavrielli, M., et al. (2005) B and T lymphocyte attenuator exhibits structural and expression polymorphisms and is highly induced in anergic CD4+ T cells *J. Immunol.* 174(6):3377-3385.

Hutloff, A., Dittrich, A. M., Beier, K. C., et al. (1999) ICOS is an inducible T-cell costimulator structurally and functionally related to CD28. *Nature* 397:263-266.

Ishida, Y., Agata, Y., Shibahara, K., Honjo, T. (1992) Induced expression of PD-1, a novel member of the immunoglobulin gene superfamily, upon programmed cell death. *EMBO J.* 11(11):3887-3895.

Iwai, H., Abe, M., Hirose, S., et al. (2003) Involvement of inducible costimulator-B7 homologous protein costimulatory pathway in murine lupus nephritis. *J. Immunol.* 171(6):2848-2854.

Iwai, H., Kozono, Y., Hirose, S., et al. (2002) Amelioration of collagen-induced arthritis by blockade of inducible costimulator-B7 homologous protein costimulation. *J. Immunol.* 169(8):4332-4339.

Janeway, C. A., Jr., Medzhitov, R. (2002) Innate immune recognition. *Annu. Rev. Immunol.* 20:197-216.

Kanai, T., Totsuka, T., Uraushihara, K., et al. (2003) Blockade of B7-H1 suppresses the development of chronic intestinal inflammation. *J. Immunol.* 171(8):4156-4163.

Kim, J. I., Ho, I. C., Grusby, M. J., Glimcher, L. H. (1999) The transcription factor c-Maf controls the production of interleukin-4 but not other Th2 cytokines. *Immunity* 10: 745-751.

Krieg, C., Han, P., Stone, R., et al. (2005) Functional analysis of B and T lympho-cyte attenuator engagement on CD4+ and CD8+ T cells *J. Immunol,* 175(10):6420-6427.

Kroner, A., Mehling, M., Hemmer, B., et al. (2005) A PD-1 polymorphism is associated with disease progression in multiple sclerosis. *Ann. Neurol.* 58(1):50-57.

Langrish, C. L., Chen, Y., Blumenschein, W. M., et al. (2005) IL-23 drives a pathogenic T cell population that induces autoimmune inflammation. *J. Exp. Med.* 201(2):233-240.

Latchman, Y., Wood, C. R., Chernova, T., et al. (2001) PD-L2 is a second ligand for PD-1 and inhibits T cell activation. *Nat. Immunol.* 2(3):261-268.

Latchman, Y. E., Liang, S. C., Wu, Y., et al. (2004) PD-L1-deficient mice show that PD-L1 on T cells, antigen-presenting cells, and host tissues negatively regulates T cells 10.1073/pnas.0307252101. *Proc. Natl. Acad. Sci. U S A* 101(29):10691-10696.

Leonard, W. J., Spolski, R. (2005) Interleukin-21: a modulator of lymphoid proliferation, apoptosis and differentiation. *Nat. Rev. Immunol.* 5(9):688-698.

Liang, S. C., Latchman, Y. E., Buhlmann, J. E., et al. (2003) Regulation of PD-1, PD-L1, and PD-L2 expression during normal and autoimmune responses. *Eur. J. Immunol.* 33(10):2706-2716.

Lohning, M., Hutloff, A., Kallinich, T., et al. (2003) Expression of ICOS in vivo defines CD4+ effector T cells with high inflammatory potential and a strong bias for secretion of interleukin 10. *J. Exp. Med.* 197(2):181-193.

Loke, P., Allison, J. P. (2003) PD-L1 and PD-L2 are differentially regulated by Th1 and Th2 cells. *Proc. Natl. Acad. Sci. U S A* 100(9):5336-5341.

Lubberts, E., Joosten, L. A., Oppers, B., et al. (2001) IL-1-independent role of IL-17 in synovial inflammation and joint destruction during collagen-induced arthritis. *J. Immunol.* 167(2):1004-1013.

Lubberts, E., Joosten, L. A., van de Loo, F. A., et al. (2002) Overexpression of IL-17 in the knee joint of collagen type II immunized mice promotes collagen arthritis and aggravates joint destruction. *Inflamm. Res.* 51(2):102-104.

Mak, T. W., Shahinian, A., Yoshinaga, S. K., et al. (2003) Costimulation through the inducible costimulator ligand is essential for both T helper and B cell functions in T cell-dependent B cell responses. *Nat. Immunol.* 4(8):765-772.

Massey, E. J., Sundstedt, A., Day, M. J., et al. (2002) Intranasal peptide-induced peripheral tolerance: the role of IL-10 in regulatory T cell function within the context of experimental autoimmune encephalomyelitis. *Vet. Immunol. Immunopathol.* 87(3-4): 357-372.

Mazanet, M. M., Hughes, C. C. (2002) B7-H1 is expressed by human endothelial cells and suppresses T cell cytokine synthesis. *J. Immunol.* 169(7):3581-3588.

McAdam, A. J., Greenwald, R. J., Levin, M. A., et al. (2001) ICOS is critical for CD40-mediated antibody class switching. *Nature* 409:102-105.

Miyamoto, K., Kingsley, C. I., Zhang, X., et al. (2005) The ICOS molecule plays a crucial role in the development of mucosal tolerance. *J. Immunol.* 175(11):7341-7347.

Mueller, D. L., Jenkins, M. K., Schwartz, R. H. (1989) Clonal expansion versus functional clonal inactivation: a costimulatory signalling pathway determines the outcome of T cell antigen receptor occupancy. *Annu. Rev. Immunol.* 7:445-480.

Nielsen, C., Hansen, D., Husby, S., et al. (2003) Association of a putative regulatory poly-morphism in the PD-1 gene with susceptibility to type 1 diabetes. *Tissue Antigens* 62(6):492-497.

Nishimura, H., Nose, M., Hiai, H., et al. (1999) Development of lupus-like autoimmune diseases by disruption of the PD-1 gene encoding an ITIM motif-carrying immunore-ceptor. *Immunity* 11(2):141-151.

Nishimura, H., Okazaki, T., Tanaka, Y., et al. (2001) Autoimmune dilated cardiomyopathy in PD-1 receptor-deficient mice. *Science* 291(5502):319-322.

Nurieva, R. I. (2005) Regulation of immune and autoimmune responses by ICOS-B7h interaction. *Clin. Immunol.* 115(1):19-25.

Nurieva, R. I., Duong, J., Kishikawa, H., et al. (2003a) Transcriptional regulation of th2 differentiation by inducible costimulator. *Immunity* 18(6):801-811.

Nurieva, R. I., Mai, X. M., Forbush, K., et al. (2003b) B7h is required for T cell activation, differentiation, and effector function. *Proc. Natl. Acad. Sci. U S A* 100(24): 14163-14168.

Okazaki, T., Maeda, A., Nishimura, H., et al. (2001) PD-1 immunoreceptor inhibits B cell receptor-mediated signaling by recruiting src homology 2-domain-containing tyrosine phosphatase 2 to phosphotyrosine. *Proc. Natl. Acad. Sci. U S A* 98(24):13866-13871.

Okazaki, T., Otaka, Y., Wang, J., et al. (2005) Hydronephrosis associated with antiurothe-lial and antinuclear autoantibodies in BALB/c-Fcgr2b-/-Pdcd1-/- mice. *J Exp Med* 202(12):1643-1648.

Okazaki, T., Tanaka, Y., Nishio, R., et al. (2003) Autoantibodies against cardiac troponin I are responsible for dilated cardiomyopathy in PD-1-deficient mice. *Nat. Med.* 9(12):1477-1483.

Ozaki, K., Spolski, R., Ettinger, R., et al. (2004) Regulation of B cell differentiation and plasma cell generation by IL-21, a novel inducer of Blimp-1 and Bcl-6. *J. Immunol.* 173:5361-5371.

Park, H., Li, Z., Yang, X. O., et al. (2005) A distinct lineage of CD4 T cells regulates tis-sue inflammation by producing interleukin 17. *Nat. Immunol.* 6(11):1133-1141.

Petroff, M. G., Chen, L., Phillips, T. A., Hunt, J. S. (2002) B7 family molecules: novel immunomodulators at the maternal-fetal interface. *Placenta* 23(Suppl A):S95-S101.

Prasad, D. V., Richards, S., Mai, X. M., Dong, C. (2003) B7S1, a novel B7 family mem-ber that negatively regulates T cell activation. *Immunity* 18(6):863-873.

Prasad, D. V. R., Nguyen, T., Li, Z., et al. (2004) Mouse B7-H3 is a negative regulator of T cells. *J. Immunol.* 173:2500-2506.

Prokunina, L., Castillejo-Lopez, C., Oberg, F., et al. (2002) A regulatory polymorphism in PDCD1 is associated with susceptibility to systemic lupus erythematosus in humans. *Nat. Genet.* 32(4):666-669.

Prokunina, L., Padyukov, L., Bennet, A., et al. (2004) Association of the PD-1.3A allele of the PDCD1 gene in patients with rheumatoid arthritis negative for rheumatoid factor and the shared epitope. *Arthritis Rheum.* 50(6):1770-1773.

Rottman, J. B., Smith, T., Tonra, J. R., et al. (2001) The costimulatory molecule ICOS plays an important role in the immunopathogenesis of EAE. *Nat. Immunol.* 2(7), 605-611.

Salama, A. D., Chitnis, T., Imitola, J., et al. (2003) Critical role of the programmed death-1 (PD 1) pathway in regulation of experimental autoimmune encephalomyelitis. *J. Exp. Med.* 198(1):71-78.

Schwartz, R. H. (2003) T cell anergy. *Annu. Rev. Immunol.* 21:305-334.

Sedy, J. R., Gavrieli, M., Potter, K. G., et al. (2005) B and T lymphocyte attenuator regulates T cell activation through interaction with herpesvirus entry mediator. *Nat. Immunol.* 6(1):90-98.

Shinohara, T., Taniwaki, M., Ishida, Y., et al. (1994) Structure and chromosomal localization of the human PD-1 gene (PDCD1). *Genomics* 23(3):704-706.

Sica, G. L., Choi, I. H., Zhu, G., et al. (2003) B7-H4, a molecule of the B7 family, negatively regulates T cell immunity. *Immunity* 18(6):849-861.

Steinman, R. M., Hawiger, D., Nussenzweig, M. C. (2003) Tolerogenic dendritic cells. *Annu. Rev. Immunol.* 21(1):685-711.

Subudhi, S. K., Zhou, P., Yerian, L. M., et al. (2004) Local expression of B7-H1 promotes organ-specific autoimmunity and transplant rejection. *J. Clin. Invest.* 113(5): 694-700.

Suh, W. K., Gajewska, B. U., Okada, H., et al. (2003) The B7 family member B7-H3 preferentially down-regulates T helper type 1-mediated immune responses. *Nat. Immunol.* 4(9):899-906.

Suh, W-K., Tafuri, A., Berg-Brown, N. N., et al. (2004) The inducible constimulator plays the major costimulatory role in humoral immune responses in the absence of CD28. *J. Immunol.* 172(10): 5917-5923.

Sun, M., Richards, S., Prasad, D. V., et al. (2002) Characterization of mouse and human B7-H3 genes. *J. Immunol.* 168(12):6294-6297.

Swallow, M. M., Wallin, J. J., Sha, W. C. (1999) B7h, a novel costimulatory homolog of B7.1 and B7.2, is induced by TNFa. *Immunity* 11:423-432.

Tafuri, A., Shahinian, A., Bladt, F., et al. (2001) ICOS is essential for effective T-helper-cell responses. *Nature* 2001(409):105-109.

Tiegs, G., Hentschel, J., Wendel, A. (1992) A T cell-dependent experimental liver injury in mice inducible by concanavalin A. *J. Clin. Invest.* 90(1):196-203.

Totsuka, T., Kanai, T., Makita, S., et al. (2005) Regulation of murine chronic colitis by CD4+CD25-programmed death-1+T cells. *Eur. J. Immunol.* 35(6):1773-1785.

Tseng, S. Y., Otsuji, M., Gorski, K., et al. (2001) B7-DC, a new dendritic cell molecule with potent costimulatory properties for T cells. *J. Exp. Med.* 193(7):839-846.

Vinuesa, C. G., Cook, M. C., Angelucci, C., et al. (2005a) A RING-type ubiquitin ligase family member required to repress follicular helper T cells and autoimmunity. *Nature* 435(7041):452-458.

Vinuesa, C. G., Tangye, S. G., Moser, B., Mackay, C. R. (2005b) Follicular B helper T cells in antibody responses and autoimmunity. *Nat. Rev. Immunol.* 5(11):853-865.

Wang, J., Yoshida, T., Nakaki, F., et al. (2005) Establishment of NOD-Pdcd1-/- mice as an efficient animal model of type I diabetes. *Proc. Natl. Acad. in U S A*, 102(33): 11823-11828.

Watanabe, N., Gavrieli, M., Sedy, J. R., et al. (2003) BTLA is a lymphocyte inhibitory receptor with similarities to CTLA-4 and PD-1. *Nat. Immunol.* 4(7): 670-679.

Wong, S. C., Oh, E., Ng, C. H., Lam, K. P. (2003) Impaired germinal center formation and recall T-cell-dependent immune responses in mice lacking the costimulatory ligand B7-H2. *Blood* 102(4):1381-1388.

Yamazaki, T., Akiba, H., Iwai, H., et al. (2002) Expression of programmed death 1 ligands by murine T cells and APC. *J. Immunol.* 169(10):5538-5545.

Yao, Z., Painter, S. L., Fanslow, W. C., et al. (1995) Human IL-17: a novel cytokine derived from T cells. *J. Immunol.* 155(12):5483-5486.

Yoshinaga, S. K., Whoriskey, J. S., Khare, S. D., et al. (1999) T-cell co-stimulation through B7RP-1 and ICOS. *Nature* 402(6763):827-832.

Zang, X., Loke, P., Kim, J., et al. (2003) B7x: a widely expressed B7 family member that inhibits T cell activation. *Proc. Natl. Acad. Sci. U S A*, 100(18):10388-10392.

# 8
# Invariant NKT Cells and Immune Regulation in Multiple Sclerosis

Takashi Yamamura

## 1    Introduction

Among the major questions in immunology, "self-tolerance" is unquestionably most fundamental. Therefore countless research papers dealing with autoimmunity and autoimmune diseases are questioning how tolerance to self is maintained in healthy conditions and what could be a key trigger for disrupting the protective mechanism for autoimmune diseases. As also discussed in other chapters, "central tolerance" taking place in the thymus, where self-reactive T cells are vigorously eliminated, is an imperfect mechanism to avoid autoimmunity, which allows dissemination of dangerous lymphocytes in the periphery (Kyewski and Derbinski, 2004; Kyewski and Klein, 2006). As a result, all individuals possess potentially harmful autoimmune T cells in the periphery, which may cause health problems. However, most of us do not suffer from autoimmune diseases thanks to the "peripheral tolerance" mechanisms that are in charge of controlling the dangerous autoimmune responses. Reestablishment of peripheral tolerance is therefore an important goal for immunologists working in the field of autoimmune disease research.

Results of T cell vaccination experiments conducted by Cohen and his colleagues (Ben-Nun et al., 1981), to be extensively described in other chapters of this book, were the first to show that pathological autoimmune T cells are controlled in the context of immunoregulatory network, involving the interactions of counterregulatory T cells with pathogenic autoimmune T cells. Furthermore, it has been firmly established that CD25+ CD4+ regulatory T cells (CD25+ Tregs) play an active role in maintaining "self-tolerance" (Hori et al., 2003; Sakaguchi 2004). In fact, in vivo elimination of the CD25+ T-reg cells, expressing the transcription factor FoxP3 (Hori et *al.*, 2003), was shown to induce or augment various forms of autoimmune diseases in rodents and humans (Kohm et al., 2002; Gambineri et al., 2003). Considering the importance of T cell control of autoimmunity, it is quite reasonable that the research efforts in autoimmunity have focused on seeking the way to empower the counterregulatory T cells that may be useful for treating autoimmune diseases. Along this line, peptide, antibody, and DNA vaccines that could strengthen the power of the immunoregulatory cells have been developed and have proven effective for inducing peripheral tolerance and treating autoimmune diseases.

In this chapter, I do not touch on major histocompatibility complex (MHC)-restricted regulatory T cells but focus on natural killer T cells with invariant T cell receptor (TCR) α-chains, that are specific for glycolipid antigen bound to MHC class I-like CD1d protein (Godfrey and Kronenberg, 2004; Godfrey et al., 2004; Kronenberg and Gapin, 2002; Miyake and Yamamura, 2005; Taniguchi et al., 2003; Wilson and Delovitch 2003; Yamamura et al., 2004). It has been demonstrated that autoimmune diseases such as multiple sclerosis (MS) are associated with functional and numerical changes of the invariant natural killer T cells (NKT cells) (Araki et al., 2004; Illés et al., 2000; Sumida et al., 1995), and that glycolipid ligands stimulating NKT cells could protect or treat the development of the prototypical model of MS experimental autoimmune encephalomyelitis (EAE) (Jahng et al., 2001; Miyamoto et al., 2001). This chapter provides essential information on the invariant NKT cells that may be relevant for understanding their role in MS.

## 2    Invariant NKT Cells

Invariant NKT (iNKT) cell are unique T lymphocytes reactive to CD1d-bound α-galactosylceramide (α-GalCer) with an outstanding ability to produce copious amounts of cytokines within hours after stimulation via TCR. The prototypical ligand for iNKT cells, α-GalCer, was identified from components of marine sponge (Kawano et al., 1997) with a potential to eradicate metastatic tumors. As is the case for other sphingolipids with α-linked sugar, α-GalCer is not present in the mammalian bodies and therefore would not serve as a natural ligand for iNKT cells in rodents or humans. More recent works have identified the self-glycolipid isoglobotrihexosylceramide (iGb3) as a natural ligand for iNKT cells (Mattner et al., 2005; Zhou et al., 2004). Furthermore, studies have demonstrated that some bacterial glycosphingolipids also exhibit an ability to stimulate iNKT cells (Kinjno et al., 2005). As such, iNKT cells would respond to a variety of endogenous and exogenous ligands. However, their response to α-GalCer is particularly robust with regard to induction of cell proliferation and cytokine production.

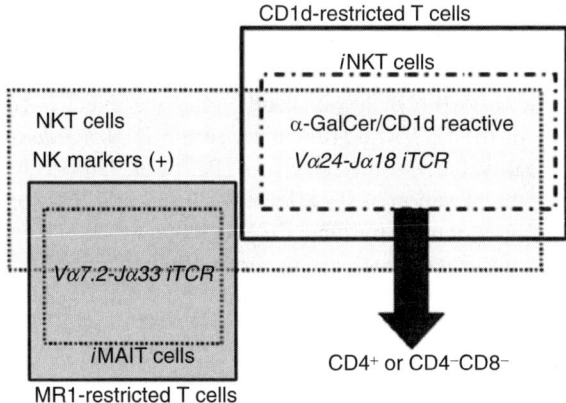

FIGURE 8.1. **Mutual relationship of NK marker-positive T cells in human.** As described in the text, T cells expressing NK cell markers (CD161 etc.) were classically referred to as NKT cells ( ..... ). However, they actually contain conventional MHC-restricted T cells, CD1d-restricted T cells and MR1-restricted T cells. It is currently advised that T cells reactive to α-GalCer bound to CD1d should be called as NKT cells or iNKT cells (–·–·–·–·), regardless of NK cell marker expression. iNKT cells express the Vα24-Jα18 iTCR in human and are restricted by CD1d on definition. Though the majority of CD1d-restricted T cells (———) are iNKT cells, they also contain non-iNKT cells. Likewise, not all MR1-restricted T cells express the Vα7.2-Jα33 iTCR expressed by iMAIT cells (Treiner et al. 2003).

Reflecting the homogeneous specificity for α-GalCer bound to CD1d, iNKT cells uniquely express hemi-invariant TCRs composed of invariant TCR α-chain (Vα14-Jα18 in mice; Vα24-Jα18 in human) paired with TCR β-chain using a biased V gene segments (Vβ8.2 in mice; Vβ11 in humans). It is of note that the glycolipid recognition by iNKT cells is remarkably conserved among the species, and the same ligand α-GalCer is stimulatory for both human and rodent iNKT cells (Spada et al., 1998).

In the past, T cells expressing NK cell markers were simply called "NKT cells." However, "NKT cells" by this definition comprise a variety of lymphocyte subpopulations, including MHC-restricted T cells. To avoid confusion and misunderstanding, it is currently recommended that T cells bearing the iTCR (Vα14-Jα18 in mice; Vα24-Jα18 in humans) that are reactive to α-GalCer bound to CD1d should be referred to as NKT cells or iNKT cells (Godfrey et al., 2004) (Fig. 8.1). With this definition of NKT cells, expression of NK cell markers is not important (in fact, some iNKT cells are missing NK cell markers), but the antigen specificity and expression of the invariant TCR are critical. Notably, MR1-restricted T cells (Treiner et al., 2003), CD1d-restricted Vβ3.2+Vβ9+ nonclassical NKT cells (Duarte et al., 2004), and other NKT-like populations are clearly distinguished from iNKT cells by this definition.

Although the iNKT cells are a numerically minor population in the immune system, they play a critical role in various physiological and pathological conditions. Importantly, iNKT cells express memory T cell markers on the surface. As a reflection of this, they would respond rapidly to TCR stimulation as if they were innate immune cells. That is, shortly after TCR stimulation with α-GalCer they explosively secrete large amounts of regulatory cytokines such as interleukin-4 (IL-4) and interferon-γ (IFNγ). In our hands, after we inject α-GalCer into the peritoneal cavity of the C57/BL6 (B6) mice, serum IL-4 markedly rises within 2 hours to the peak value and then declines (Pál et al., 2001; Oki et al., 2004, 2005). Following the elevation of IL-4, levels of IL-12 and IFNγ increased in the serum, which would stimulate bystander NK cells and CD8+ T cells to produce IFNγ and further augment the increase of serum IFNγ (Carnaud et al., 1999; Oki et al., 2005). These results imply that activation of a small number of iNKT cells would trigger the cascade of the lymphocyte network, leading to a marked shift of the functional Th1/Th2 balance in vivo. As such, iNKT cells are regarded as a unique lymphocyte population bridging innate and acquired immunity.

## 3     Invariant NKT Cells and Autoimmune/Inflammatory Disease Models

Given the remarkable ability to produce IL-4, it was once believed that iNKT cells are the critical source of IL-4 for inducing Th2 cell generation in vivo. However, this hypothesis was abandoned because mice lacking iNKT cells (CD1d KO) could normally develop Th2-dependent immune responses (Smiley et al., 1997). Whereas mice depleted for CD25+ T-reg cells would exhibit a variety of autoimmune disease phenotypes, iNKT cell-deficient mice (B6 mice crossed with CD1d KO or with Jα18 TCR KO) did not develop autoimmune diseases spontaneously. However, they tended to develop an earlier onset of EAE when immunized with an encephalitogenic peptide (our unpublished observations). Although the effect of iNKT cell deficiency is controversial in type 1 diabetes in NOD mice, invariant Vα14+ TCR-transgenic NOD mice, which overexpress iNKT cells, exhibit a milder form of type 1 diabetes (Lehuen et al., 1998). Moreover, EAE induced in the invariant Vα14+ TCR transgenic NOD mice developed a milder form of EAE compared with littermates after immunization with encephalitogenic antigen (Mars et al., 2002). On the other hand, administration of proper glycolipid ligand for iNKT cells would prevent the onset of type 1 diabetes in NOD mice (Hong et al., 2001; Sharif et al., 2001) and EAE (Jahng et al., 2001; Miyamoto et al., 2001) by stimulating iNKT cells. Therefore in classical EAE and NOD diabetes models, iNKT cells appeared to be "potentially good" lymphocytes that could properly deal with harmful autoimmune responses. The mechanism for the iNKT cell-mediated suppression of type 1 diabetes and EAE is still a matter of controversy but probably involve iNKT cell-derived cytokines as key mediators. However, by comparing the severity of diseases induced in wild-type and iNKT cell-deficient mice, studies have demonstrated that iNKT cells would augment the local inflammation in the model of bronchial asthma (Akbari et al., 2003) and arthritis (Chiba et al., 2005; Kim et al., 2005; Ohnishi et al., 2005). In these settings, iNKT cells are obviously "bad" lymphocytes that would endanger the health. In the asthma model (Akbari et al., 2003), Th2 cytokines (IL-4 and IL-13) locally produced by iNKT cells appear to play a pivotal role for provoking airway inflammation. In contrast, local iNKT cells may inhibit production of transforming growth factor-$\beta$ (TGF$\beta$) that would normally down-modulate joint inflammation in the arthritis model (Kim et al., 2005). It is now apparent that iNKT cells modulate autoimmune/inflammatory disease processes using a variety of molecular strategies, which may depend on the local milieu where iNKT cells interact with other cellular components. Cytokine or co-stimultory signals probably play a key role in modulating iNKT cells that would probably recognize the endogenous ligand presented by local antigen-presenting cells (APCs).

## 4   iNKT Glycolipid Ligands and Autoimmune Disease Models

As briefly described above, ligand stimulation of iNKT cells would lead to the suppression of NOD type 1 diabetes and EAE (Hong et al., 2001; Jahng et al., 2001; Miyamoto et al., 2001; Sharif et al., 2001). Because we speculated that iNKT cells should be woven into the natural regulatory network in the immune system, we have been exploring the possibility for use of $\alpha$-GalCer to treat EAE. However, at least in our hands, the protective effect of $\alpha$-GalCer was not demonstrated in EAE induced in B6 mice (Pál et al., 2001). However, EAE induced in IFN$\gamma$ knockout (KO) mice was suppressed by $\alpha$-GalCer (Pál et al., 2001), whereas EAE in IL-4 KO mice was augmented. This indicates that iNKT cell-derived IL-4 is protective for EAE but a simultaneously secreted IFN$\gamma$. might antagonize the effect of the IL-4 triggered by $\alpha$-GalCer. Of note, in a previous clinical trial, administration of IFN$\gamma$. was found to exacerbate MS (Panitch et al., 1987). Inspired by the presence of an altered peptide ligand for myelin basic protein (MBP) able to induce an immune deviation toward Th2 (Bielekova et al., 2000), we decided to synthesize $\alpha$-GalCer analogues and challenged the possibility for discovering a therapeutic glycolipid ligand that would induce IL-4 but not IFN$\gamma$ production by iNKT cells. By screening the synthetic glycolipids for the ability to induce Th1 and Th2 cytokines from iNKT cells, we found that a lipid tail-truncated analogue (referred to as OCH) met our requirements. In fact, OCH compound was found to induce IL-4 selectively in vitro and in vivo (Miyamoto et al., 2001). However, this kind of selective IL-4 production was not seen when lower doses of $\alpha$-GalCer were used for stimulating iNKT cells. This implied that OCH is a qualitatively different ligand from $\alpha$-GalCer. Consistent with the selective IL-4 production, oral or intraperitoneal injection of OCH was found to inhibit the development of EAE induced in B6 mice (Miyamoto et al., 2001), type 1 diabetes in NOD mice (Mizuno et al., 2004), and collagen-induced arthritis (Chiba et al., 2004) in a manner dependent on iNKT cells. Furthermore, a model of Th1-induced colitis was also responsive to oral OCH treatment (Ueno et al., 2005). In all the models tested, the effect of OCH was superior to that of α-GalCer. It appears that the selective induction of Th2 cytokines unique to OCH can account for the superior effect of OCH on the autoimmune disease models. Although much remains to be learned, it is now clear that iNKT cells are lymphocytes of special interest that could play a key role in autoimmune diseases and should serve as a target for immune intervention.

## 5    Human iNKT Cells

As described above, human iNKT cells express an invariant Vα4-Jα18 α-chain that is highly homologous to the Vα14-Jα18 invariant chain of mouse iNKT cells. Because the number of the Vα24+ iNKT cells is much lower in humans than in rodents (~1/1000 blood lymphocytes), it was questioned if such a small population of the cells might play some role in humans. However, a recent study has clearly shown that they are truly involved in human bronchial asthma (Akbari et al., 2006). Although the pulmonary CD4+CD3+ cells in patients with persistent asthma were believed to be MHC II-restricted conventional CD4+ T cells, 60% of the cells were actually iNKT cells, according to the report by Akbari et al. (2006). Furthermore, injection of human dendritic cells pulsed with α-GalCer induced a marked expansion of iNKT cells in patients with malignancy (Chang et al., 2005). Hereafter, I describe the properties of human iNKT cells with some emphasis on functional differences between CD4+ and CD4−CD8− double-negative (DN) iNKT cells. This dichotomy was revealed by using CD1d tetramers loaded with α-GalCer (Gumperz et al., 2002; Lee et al., 2002a). In the literature, production of Th2 cytokines has been described as a cardinal feature of iNKT cells, but the authors did not clarify whether all of the cells are responsible for it or only a subpopulation. Now it is clear that CD4+ iNKT cells would mainly produce antiinflammatory Th2 cytokines, although they produce proinflammatory cytokines as well. On the other hand, DN iNKT cells produce only a trace of IL-4 or IL-5 but secrete a great amount of the proinflammatory cytokines IFNγ and TNFα (upper panel in Fig. 8.2). DN iNKT cells are also characterized by expression of NKG2d, a marker associated with cytolysis, and by upregulation of perforin after exposure to IL-2 or IL-12 (Gumperz et al., 2002). These results indicate that CD4+ iNKT cells would serve as regulatory cells, inducing an immune bias for either Th1 or Th2, whereas DN iNKT cells may resemble NK cells in that they could eliminate virally infected cells and tumor cells. It is also possible that the DN cells may promote pathogenic autoimmune responses by secreting the proinflammatory cytokines IFNγ and TNFα.

## 6    Analyzing iNKT Cells in Human Diseases

After iNKT cells were identified as a distinct lymphocyte population, several reports have documented that the number of iNKT cells in the peripheral blood may be markedly reduced in autoimmune diseases (Illes et al., 2000; Sumida et al., 1995;

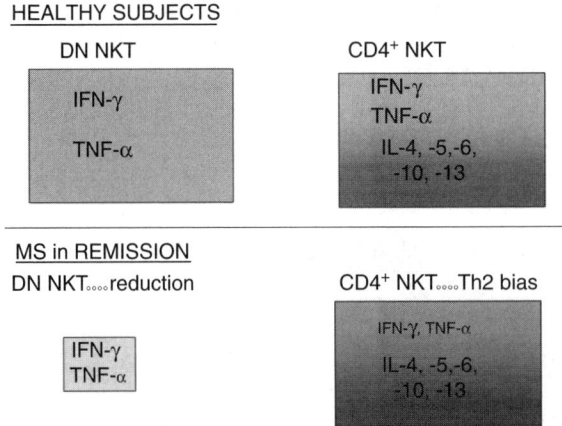

FIGURE 8.2. **DN and CD4+ NKT cells are differentially altered in healthy subjects and MS in remission.** Compared with healthy subjects, the number of DN NKT cells is significantly reduced in MS in remission. As the DN NKT cells mainly produce proinflammatory cytokines such as IFN-γ and TNF-α, this change should help stabilize the remission state of inflammatory process found in MS. In contrast, CD4+ NKT cells are not significantly reduced in MS in remission compared with healthy subjects. However, in the remission of MS, they are significantly biased for producing more Th2 cytokines, which should also contribute to maintaining the remission state.

van der Vliet et al., 2001). Although these earlier studies did not pay attention to the difference between CD4+ and CD4– iNKT cells, they properly suggested that the numerical changes could be associated with some autoimmune diseases. Unfortunately, these observations were sometimes misinterpreted that the reduction of iNKT cells might be a primary cause of the autoimmune disease. Afterward, however, the association of a reduction of iNKT cells with autoimmune disease was challenged in a human disease. Although there were two reports showing a reduction of iNKT cells in insulin-dependent diabetes mellitus (IDDM) (Kukreja et al., 2002; Wilson et al., 1998), one conflicting report argued that iNKT cell frequency is conserved during the course of IDDM (Lee et al., 2002b). However, Oikawa et al. (2002) reported that the number of iNKT cells could be increased in recent-onset IDDM. Furthermore, they showed an inverse correlation between iNKT cell frequency and disease duration (Oikawa et al., 2002). With regard to this discrepancy, there were discussions about differences in the disease phenotypes, duration of disease, disease activity, genetic background of the patients, confounding factors such as medications, and the methods of analysis. Regarding the method for identifying and measuring iNKT cells, the quantitative polymerase chain reaction (PCR), a combination of anti-TCR antibodies (anti-Vα24 and anti-Vβ11), anti-invariant TCR antibody (6B11), and CD1d tetramer bound with α-GalCer were used in the previous studies. The 6B11 antibody and CD1d tetramers are currently recommended as the most reliable means. However, when we analyzed the same peripheral blood samples using anti-Vα24/anti-Vβ11 double-staining and CD1d tetramer loaded with α-GalCer in parallel, the numbers of iNKT cells estimated by both methods showed an excellent correlation ($r = 0.990$; $p < 0.0001$) (Araki et al., 2004). Therefore, we speculated that sample differences, rather than the use of different methods, probably account for the discrepancy. The relevance of this speculation was confirmed when we studied samples from MS.

## 7    iNKT Cells Are Reduced in MS

Although we demonstrated the reduction of total iNKT cells in the blood of MS using the reverse transcription PCR (RT-PCR) SSCP method (Illés et al., 2000), we reanalyzed this issue using a combination of fluorescent antibodies reactive to Vα24, Vβ11, CD4, or CD8, enabling us to distinguish CD4+ and DN iNKT cells (Araki et al., 2004). Using this method, we estimated the mean frequency of total iNKT cells (Vα24+Vβ11+) to be 0.118% in peripheral lymphocytes isolated from healthy Japanese subjects. We were soon able to confirm that the total number of iNKT cells is significantly reduced in the peripheral blood of patients with MS in remission compared with healthy subjects (−86%; $p < 0.01$). We also obtained essentially the same results using CD1d tetramer loaded with α-GalCer (Araki et al., 2004). However, the reduction of iNKT cells was much less clear in the patients during relapse (−53%; $p = 0.196$), indicating that the iNKT cell number would increase in the active state of MS, as has been observed in IDDM (Oikawa et al., 2002). By further analyzing CD4+ and DN cells separately, we noted that DN iNKT cells are also markedly reduced in the patients in remission (−89%; $p < 0.01$). However, the reduction of CD4+ iNKT cells was not significant ($p = 0.18$). These results imply that a reduction of iNKT cells in the peripheral blood is an immunological hallmark of MS, and it is particularly evident during remission. It was also indicated that DN and CD4+ iNKT cells are differentially altered in MS. Provided that DN iNKT cells would preferentially produce proinflammatory cytokines, we hypothesized that the marked reduction of DN iNKT cells may be beneficial for the patients by maintaining the clinical remission.

## 8    CD4+ iNKT Cells Are Th2 Biased in Remission of MS

A next important question is if the iNKT cells are functionally altered in MS. If an immune network problem can lead to the development of autoimmune diseases, we may speculate that iNKT cells are functionally defective. However, when we measured the primary proliferative response of iNKT cells to α-GalCer, they looked quite normal in MS (Araki et al., 2004). To evaluate the functional changes of iNKT cells more precisely, we generated short-term iNKT cell lines for conducting cytokine analysis. In brief, we stimulated peripheral blood monocytic cells (PBMCs) from MS or healthy subjects with α-GalCer and expanded the iNKT cells in culture; we then and sorted them into CD4+ or CD4− fractions.

The sorted CD4+ or CD4− iNKT cells were then assessed for the ability to produce IL-4 and IFNγ after stimulation with anti-CD3 and anti-CD28 coated beads. When we looked at the cytokine profiles of DN lines, those from MS in remission appeared to produce less IL-4 and less IFNγ than those from healthy subjects, indicating that DN iNKT cells are both numerically and functionally attenuated in MS. This is, however, not so bad for MS patients, given that DN iNKT cells have the potential to promote autoimmune inflammation by producing TNFα and IFNγ. The results of CD4+ iNKT cell lines were biologically more remarkable. Those from patients with MS in remission produced significantly more IL-4 than those from healthy individuals, although CD4+ iNKT cell lines from both MS and healthy subjects produced similar amounts of IFNγ. This result indicates that CD4+ iNKT cell lines are functionally biased toward Th2 in remission of MS (lower panel in Fig. 8.2).

# 9   Implications and Future Research

A study has indicated that CD25+ Treg cells are functionally defective in MS (Viglietta et al., 2004), implying that a disease-promoting change is likely to occur in the master regulatory T cells. In contrast, the changes of iNKT cells in remission of MS seem not to be deleterious but, rather, protective. Given that disease-promoting DN cells are reduced and regulatory CD4+ cells are Th2 biased (Fig. 8.2), we assume that these changes are beneficial for the maintenance of clinical remission. Interestingly, NK cells in the peripheral blood are biased for secreting IL-5 in the remission of MS (Takahashi et al., 2001). Therefore, it could be speculated that NK and NKT cells are cooperative in preventing the pathogenic Th1 responses by producing Th2 cytokines. If this cooperation is perfect, the patients may not develop relapses, which is a cure of MS. However, because this is not the case in most patients, a meaningful therapeutic strategy could be to stimulate iNKT cells with glycolipid ligands such as OCH. As described above, glycolipid OCH is potentially an attractive remedy for autoimmune diseases for the following reasons: (1) It is stimulatory for human iNKT cells and would induce a Th2 predominant response by human CD4+ iNKT cells (Araki et al., unpublished data). (2) Because it is presented to the iTCR in the context of monomorphic CD1d, there is no need to consider genetic variations of individuals such as MHC polymorphism. (3) It is orally effective and preventive for a number of Th1-mediated diseases. (4) The effect and safety of α-GalCer is already reported. However, as is the case for other potential therapeutics for MS, we cannot directly extrapolate the promising results obtained in animal models to humans. Obviously, both basic and clinical studies are required for promoting the advance of the glycolipid treatment.

Regarding the specificities of iNKT cells, there is now evidence that they would recognize both endogenous and exogenous ligands. It is of note that similar invariant T cells (MAIT cells) that are restricted by a class 1b-like MR1 molecule are enriched in the gut, and their development appears to depend on microbial flora (Treiner et al., 2003). The increase of MS is evident in Japan over the last 30 years, which may result from environmental changes. However, there was no clue to challenging this problem. Now we propose that changes in the gut flora due to the change of life style somehow alter the functions of gut-derived regulatory cells such as MAIT cells, which leads to the increase in autoimmune and allergic conditions. If this is indeed the case, not only MAIT cells but iNKT cells should be research targets that cross-recognize self and non-self and present in the gut.

*Acknowledgments.* I am grateful to all my colleagues who have collaborated on this study at the National Institute of Neuroscience. I particularly thank Drs. Sachiko Miyake, Ludovic Croxford, Zsolt Illés, Katsuichi Miyamoto, Manabu Araki, Toshimasa Aranami, and Shinji Oki for their discussions and valuable contributions.

# References

Akbari, O., Stock, P., Meyer, E., et al. (2003) Essential role of NKT cells producing IL-4 and IL-13 in the development of allergen-induced airway hyperreactivity. *Nat. Med.* 9:582-588.

Akbari, O., Faul, J. L., Hoyte, E. G., et al. (2006) CD4+ invariant T-cell-receptor+ natural killer T cells in bronchial asthma. *N. Engl. J. Med.* 354:1117-1129.

Araki, M., Kondo, T., Gumperz, J. E., et al. (2003) Th2 bias of CD4+ NKT cells derived from multiple sclerosis in remission. *Int. Immunol.* 15: 279-288.

Ben-Nun, A., Wekerle, H., Cohen, I. R. (1981) Vaccination against autoimmune encephalomyelitis with T-lymphocyte line cells against myelin basic protein. *Nature* 292:60-61.

Bielekova, B., Goodwin, B., Richert, N., et al. (2000) Encephalitogenic potential of the myelin basic protein peptide (amino acids 83-99) in multiple sclerosis: results of a phase II clinical trial with an altered peptide ligand. *Nat. Med.* 6:1167-1175.

Carnaud, C., Lee, D., Donnars, O., et al. (1999) Cross-Talk between cells of the innate immune system: NKT cells rapidly activate NK cells. *J. Immunol.* 163:4647-4650.

Chang, D. H., Osman, K., Connolly, J., et al. (2005) Sustained expansion of NKT cells and antigen-specific T cells after injection of alpha-galactosyl-ceramide loaded mature dendritic cells in cancer patients. *J. Exp. Med.* 201:1503-1517.

Chiba, A., Oki, S., Miyamoto, K., et al. (2004) Natural killer T-cell activation by OCH, a sphingosine truncated analogue of α-galactosylceramide, prevents collagen-induced arthritis. *Arthritis Rheumatol.* 50:305-313.

Chiba, A., Kaieda, S., Oki, S., et al. (2005) The involvement of Vα14 natural killer T cells in the pathogenesis of arthritis in murine models. *Arthritis Rheum.* 52:1941-1948.

Duarte, N., Stenstrom, M., Campino, S., et al. (2004) Prevention of diabetes in nonobese diabetic mice mediated by CD1d-restricted nonclassical NKT cells. *J. Immunol.* 173:3112-3118.

Gambineri, E., Torgerson, T. R., Ochs, H.D. (2003) Immune dysregulation, polyendocrinopathy, enteropathy, and X-linked inheritance (IPEX), a syndrome of systemic autoimmunity caused by mutations of FOXP3, a critical regulator of T-cell homeostasis. *Curr. Opin. Rheumatol.* 15:430-435.

Godfrey, D. I., Kronenberg, M. (2004). Going both ways: immune regulation via CD1d-dependent NKT cells. *J. Clin. Invest.* 114:1379-1388.

Godfrey, D. I., MacDonald, H. R., Kronenberg, M., et al. (2004) NKT cells: what's in a name? *Nat. Rev. Immunol.* 4:231-237.

Gumperz, J. E., Miyake, S., Yamamura, T., Brenner, M. B. (2002) Functionally distinct subsets of CD1d-restricted natural killer T cells revealed by CD1d tetramer staining. *J. Exp. Med.* 195:625-636.

Hong, S., Wilson, M. T., Serizawa, I., et al. (2001) The natural killer T-cell ligand $\alpha$-galactosylceramide prevents autoimmune diabetes in non-obese diabetic mice. *Nat. Med.* 7:1052-1056.

Hori, S., Nomura, T., Sakaguchi, T. (2003) Control of regulatory T cell development by the transcription factor Foxp3. *Science* 299:1030-1031.

Illés, Z., Kondo, T., Newcombe, J., et al. (2000) Differential expression of natural killer T cell $V\alpha24J\alpha Q$ invariant TCR chain in the lesions of multiple sclerosis and chronic inflammatory demyelinating polyneuropathy. *J. Immunol.* 164:4375-4381.

Jahng, A. W., Maricic, I., Pedersen, B., et al. (2001) Activation of natural killer T cells potentiates or prevents experimental autoimmune encephalomyelitis. *J. Exp. Med.* 194:1789-1799.

Kawano, T., Cui, J., Koezuka, Y., et al. (1997) CD1d-restricted and TCR-mediated activation of $V\alpha14$ NKT cells by glycosylceramides. *Science* 278:1626-1629.

Kim, H. Y., Kim, H. J., Min, H. S., et al. (2005) NKT cells promote antibody-induced joint inflammation by suppressing transforming growth factor beta1 production. *J. Exp. Med.* 201:41-47.

Kinjo, Y., Wu, D., Kim, G., et al. (2005) Recognition of bacterial glycosphingolipids by natural killer T cells. *Nature* 434:520-525.

Kohm, A. P., Carpentier, P. A., Anger, H. A., Miller, S. D. (2002) CD4$^+$CD25$^+$ regulatory T cells suppress antigen-specific autoreactive immune responses and central nervous system inflammation during active experimental autoimmune encephalomyelitis. *J. Immunol.* 169:4712-4716.

Kronenberg, M., Gapin, L. (2002) The unconventional lifestyle of NKT cells. *Nat. Immunol.* 2:557-568.

Kukreja, A., Cost, G., Marker, J., et al. (2002) Multiple immuno-regulatory defects in type-1 diabetes. *J. Clin. Invest.* 109:131-140.

Kyewski, B., Derbinski, J. (2004) Self-presentation in the thymus: an extended view. *Nat. Rev. Immunol.* 4:688-698.

Kyewski, B., Klein, L. (2006) A central role for central tolerance. *Annu. Rev. Immunol.* 24:571-606.

Lee, P. T., Benlagha, K., Teyton, L., Bendelac, A. (2002a) Distinct functional lineages of human $V\alpha24$ natural killer T cells. *J. Exp. Med.* 195:637-641.

Lee, P. T., Putnam, A., Benlagha, K., et al. (2002b) Testing the NKT cell hypothesis of human IDDM pathogenesis. *J. Clin. Invest.* 110:793-800.

Lehuen, A., Lantz, O., Beaudoin, L., et al. (1998) Overexpression of natural killer T cells protects $V\alpha14$-J$\alpha281$ transgenic nonobese diabetic mice against diabetes. *J. Exp. Med.* 188:1831-1839.

Mars, L. T., Laloux, V., Goude, K., et al. (2002) $V\alpha14$-J$\alpha281$ NKT cells naturally regulate experimental autoimmune encephalomyelitis in nonobese diabetic mice. *J. Immunol.* 168:6007-6011.

Mattner, J., Debord, K. L., Ismail, N., et al. (2005) Exogenous and endogenous glycolipid antigens activate NKT cells during microbial infections. *Nature* 434:525-529.

Miyake, S., Yamamura T. (2005) Therapeutic potential of glycolipid ligands for natural killer (NK) T cells in the suppression of autoimmune diseases. *Curr. Drug Targets Immune Endocr. Metabol. Disord.* 5:315-322.

Miyamoto, K., Miyake, S., Yamamura, T. (2001) A synthetic glycolipid prevents autoimmune encephalomyelitis by inducing TH2 bias of natural killer T cells. *Nature* 413: 531-534.

Mizuno, M., Masumura, M., Tomi, C., et al. (2004) Synthetic glycolipid OCH prevents insulitis and diabetes in NOD mice. *J. Autoimmun.* 23:293-300.

Ohnishi, Y., Tsutsumi, A., Goto, D., et al. (2005) TCR Vα14 natural killer T cells function as effector T cells in mice with collagen-induced arthritis. *Clin. Exp. Immunol.* 141:47-53.

Oikawa, Y., Shimada, A., Yamada, S., et al. (2002) High frequency of Vα24$^+$Vβ11$^+$ T-cells observed in type 1 diabetes. *Diabetes Care* 25:1818-1823.

Oki, S., Chiba, A., Yamamura, T., Miyake, S. (2004) The clinical implication and molecular mechanism of preferential IL-4 production by modified glycolipid-stimulated NKT cells. *J. Clin. Invest.* 113:1631-1640.

Oki, S., Tomi, C., Yamamura, T., Miyake, S. (2005) Preferential Th2 polarization by OCH is supported by incompetent NKT cell induction of CD40L and following production of inflammatory cytokines by bystander cells in vivo. *Int. Immunol.* 17:1619-1629.

Pál, E., Tabira, T., Kawano, T., et al. (2001) Costimulation-dependent modulation of experimental autoimmune encephalomyelitis by ligand stimulation of Vα14 NK T cells. *J. Immunol.* 166:662-668.

Panitch, H. S., Hirsch, R. L., Schindler, J., Johnson, K. P. (1987) Treatment of multiple sclerosis with gamma interferon: exacerbations associated with activation of the immune system. *Neurology* 37:1097-1102.

Sakaguchi, S. (2004) Naturally arising CD4$^+$ regulatory T cells for immunologic self-tolerance and negative control of immune responses. *Annu. Rev. Immunol.* 22:531-562.

Sharif, S., Arreaza, G. A., Zucker, P., et al. (2001) Activation of natural killer T cells by α-galactosylceramide treatment prevents the onset and recurrence o autoimmune type 1 diabetes. *Nat. Med.* 7:1057-1062.

Smiley, S. T., Kaplan, M. H., Grusby, M. J. (1997) Immunoglobulin E production in the absence of interleukin-4-secreting CD1-dependent cells. *Science* 275:977-979.

Spada, F. M., Koezuka, Y., Porcelli, S. A. (1998) CD1d-restricted recognition of synthetic glycolipid antigens by human natural killer T cells. *J. Exp. Med.* 188:1529-1534.

Sumida, T., Sakamoto, A., Murata, H., et al. (1995) Selective reduction of T cells bearing invariant Vα24 J αQ antigen receptor in patients with systemic sclerosis. *J. Exp. Med.* 182:1163-1168.

Takahashi, K., Miyake, S., Kondo, T., et al. (2001) Natural killer type 2 (NK2) bias in remission of multiple sclerosis. *J. Clin. Invest.* 107:R23-R29.

Taniguchi, M., Harada, M., Kojo, S., et al. (2003) The regulatory role of Vα14 NKT cells in innate and acquired immune response. *Annu. Rev. Immunol.* 21:483-513.

Treiner, E., Duban, L., Bahram, S., et al. (2003) Selection of evolutionarily conserved mucosal-associated invariant T cells by MR1. *Nature* 422:164-169.

Ueno, Y., Tanaka, S., Sumii, M., et al. (2005) Single dose of OCH improves mucosal T helper type 1/T helper type 2 cytokine balance and prevents experimental colitis in the presence of valpha14 natural killer T cells in mice. *Inflamm. Bowel Dis.* 11:35-41.

Van der Vliet, H. J., von Blomberg, B. M., Nishi, N., et al. (2001) Circulating Vα24$^+$Vβ11$^+$ NKT cell numbers are decreased in a wide variety of diseases that are characterized by autoreactive tissue damage. *Clin. Immunol.* 100 :144-148.

Viglietta, V., Baecher-Allan, C., Weiner, H. L., Hafler, D. A. (2004) Loss of functional suppression by CD4+CD25+ regulatory T cells in patients with multiple sclerosis. *J. Exp. Med.* 199:971-979.

Wilson, S. B., Delovitch, T. L. (2003). Janus-like role of regulatory iNKT cells in autoimmune disease and tumor immunity. *Nat. Rev. Immunol.* 3:211-222.

Wilson, S. B., Kent, S. C., Patton, K. T., et al. (1998). Extreme Th1 bias of invariant Vα24JαQ T cells in type 1 diabetes. *Nature* 391:177-181.

Yamamura, T., Miyamoto, K., Illes, Z., et al. (2004) NKT cell-stimulating synthetic glycolipids as potential therapeutics for autoimmune disease. *Curr. Top. Med. Chem.* 4:561-567.

Zhou, D., Mattner, J., Cantu, C., 3rd, et al. (2004) Lysosomal glycosphingolipid recognition by NKT cells. *Science* 306:1786-1789.

# 9
# CD4+CD25+ Regulatory T Cells in Autoimmune Disease

Adam P. Kohm and Stephen D. Miller

## 1   A Need for Regulation

As integral members of the adaptive immune system, CD4+ T cells are key mediators in multiple phases of the protective immune response by recognizing foreign antigens via their antigen-specific T cell receptor (TCR) complex during cognate interactions with antigen-presenting cells (APCs) displaying peptide/major histocompatibility complex class II (MHC II) complexes. Thus, an essential characteristic of intrathymic T cell development is the generation of TCR diversity enabling T cells to respond to an unlimited number of foreign antigens. However, one inevitable consequence of TCR diversity is the generation of self-reactive TCRs creating the potential for autoimmune disease. To balance this, the immune system has developed regulatory checkpoints that govern lymphocyte development, including the biphasic processes of central tolerance, which permits only generation of T cells with a functional TCR while deleting populations of T cells that express TCRs specific for self-peptides. Thus, when functioning properly, the process of central tolerance ensures the selective generation of functional, non-self-reactive T cells. However, many tissue antigens are not expressed at sufficient levels in the thymus to ensure that central tolerance is totally effective in culling the repertoire of all self-reactive T cells. Therefore, autoreactive T cells persist in the mature T cell repertoire with the potential of being activated by various means to mediate autoimmune disease thus creating a requirement for additional endogenous peripheral regulatory mechanisms. One critical component of peripheral tolerance is the presence of "natural" CD4+CD25+ regulatory or suppressor T cells.

## 2    Discovery and Rediscovery of Regulatory/Suppressor T Cells

The concept of a suppressor T cell population was first proposed during the early 1970s, but technical limitations limited the progress of this research. Throughout the 1980s, a contingent of researchers persevered and continued to investigate the role of suppressor T cells in maintaining self-tolerance and preventing autoimmune disease; however, this belief was not widely accepted. Depletion experiments provided the strongest rationale for a potential role for suppressor T cells in preventing autoimmunity. Examples of these types of support include studies in which either thymectomy (Kojima and Prehn, 1981) or selective depletion of Lyt-1+ T cells (Gallo et al., 1989) induced autoimmunity in otherwise normal mice; importantly, these outcomes were reversed by the adoptive transfer of normal splenic T cells (Sakaguchi and Sakaguchi, 1989). These and other similar studies provided further support for the hypothesis that subpopulations of T cells inhibited, not promoted, immune responses.

Studies by Hall et al. (1990) added a piece of the puzzle by identifying the suppressor subpopulation of CD4+ T cells as CD4+CD25+ T cells, but it was not until 1995 that findings from Sakaguchi et al. (1995) invigorated the field of suppressor T cell research. Technological advances provided increased capability for phenotyping T cell populations, and it was observed that depletion of the CD4+CD25+ T cell population in normal mice resulted in autoimmune disease in a manner similar to that used in the earlier T cell depletion experiments (Sakaguchi et al., 1995). The ensuing autoimmune disease observed in these studies covered a wide spectrum of visceral organ systems such that afflicted mice suffered from numerous pathologies including thyroiditis, insulitis, gastritis, sialoadenitis, adrenalitis, oophoritis, glomerulonephritis, polyarthritis, and graft-versus-host wasting disease in the absence of a graft. Importantly, reconstitution of the CD4+CD25+ T cell population prevented autoimmune disease in these mice, indicating a critical regulatory role for the CD4+CD25+ T cell population. Thus, the concept of natural suppressor T cells was reborn, and the cells were renamed CD4+CD25+ regulatory T (Treg) cells.

One of the most enduring and controversial questions concerning Treg cells is how exactly does one identify a Treg; and more importantly, how do we differentiate Tregs from activated effector or memory T cells? The underlying problem is that by almost all measures the cell surface phenotype of a regulatory T cell is relatively indistinguishable from that of an activated T cell. Thus, although it may be fairly simple to discriminate Tregs from activated effector cells in naive transgenic mice, it is quite a different matter to make this same discrimination in normal mice and/or humans, especially in the presence of ongoing immune responses. In light of this, the "holy grail" of regulatory T cell research has been the quest for a true phenotypic surface marker to discriminate regulatory T cells from other populations of CD4+ T cells.

## 3   Identifying a Regulatory T Cell

Ever since the initial suggestion was raised during the early 1970s that a subpopulation of T cells may regulate/suppress the activity of other immune cells, a significant effort has been made to define this regulatory cell population definitively. Since their "rediscovery" during the mid-1990s, the definition of a regulatory T cell has undergone significant evolution and setbacks. The mechanism by which regulatory T cells suppress immune responses continues to be hotly disputed, and efforts to identify this cellular population phenotypically have been equally complicated and controversial (Fig. 9.1). Remarkably, at first glance it appears that 35+ years of research has yet to provide a dependable set of phenotypic markers to discriminate regulatory T cells from other T cell subpopulations.

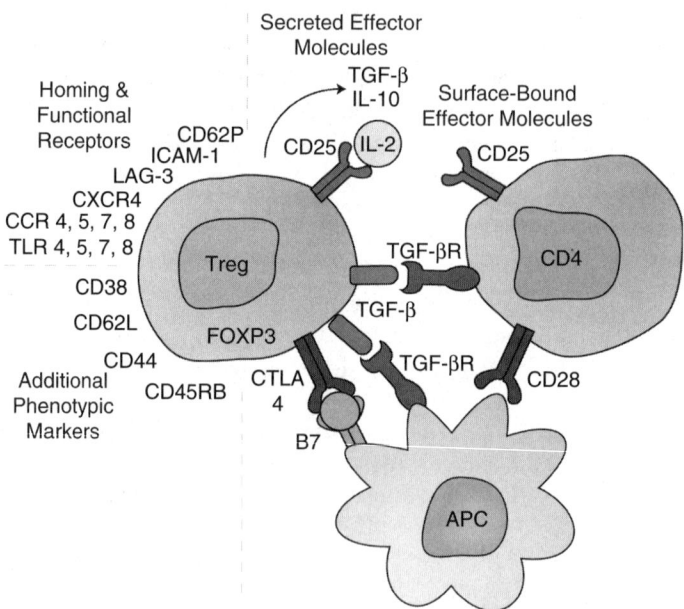

FIGURE 9.1. Phenotypic and potential functional markers of CD4+CD25+ Treg cells.

At a basic level, Tregs are defined as CD4+ T cells that constitutively express elevated levels of CD25 on their surface, in comparison to naive CD4+ T cells (Sakaguchi et al., 1995). It is important to note that effector CD4+ T cells upregulate CD25 so it may trimerize with CD122 interleukin-2 receptor (IL-2R) β-chain and the common γ-chain to form the functional IL-2R during a normal immune response. Upon ligand binding, IL-2R-mediated signaling promotes the proliferation, survival, and effector function of CD4+ T cells. It was the observation that a subpopulation of CD4+ T cells constitutively co-expressed high levels of CD25 in naive mice and that depletion of this population promoted autoimmune disease that led to the rediscovery of regulatory T cells and redefining this population as CD4+CD25+ Tregs (Sakaguchi et al., 1995; Suri-Payer et al., 1998). The practice of defining Treg cells by their co-expression of CD4 and CD25 has been widely adopted and appears to be sufficient in a number of animal model systems. However, because nonregulatory CD4+ effector T cells upregulate/gain CD25 expression as a critical step during the normal process of T cell activation and effector function, the feasibility of using CD4+CD25+ expression as a marker of regulatory cell populations is limited in conditions wherein normal immune responses are ongoing. Thus, over the past few years a number of Treg-specific phenotypic markers have been proposed to distinguish more appropriately the Treg cells from activated effector T cells. In most cases these markers have not proven useful. However, regardless of the need for additional surface markers, elevated CD25 expression remains a defining characteristic of Treg cells and is required for their functionality, as is discussed later. With that said, it is becoming evident that not even CD25 may stand as a true marker of Tregs, as recent reports have documented that some populations of CD4+CD25− T cells are just as effective at suppressing T cell responses as CD25+ T cells (Furtado et al., 2001; Gonzalez et al., 2001). Thus, although CD25 expression was key to the rediscovery of suppressor T cells, it appears that CD25 expression may not be critically required for Treg cell function and that multiple populations of CD4+ Treg cells may exist that may or may not constitutively express high levels of CD25.

## 3.1  GITR

Of all of the potential surface markers proposed to identify CD4+CD25+ Treg cells uniquely, the glucocorticoid-induced tumor necrosis factor (TNF) receptor family-related gene (GITR or TNFRSF18) was the first surface marker significantly adopted as a CD4+CD25+ Treg cell lineage determinant. GITR was first characterized by Nocentini and colleagues (Nocentini et al., 1997) as a member of the TNF growth factor receptor family that includes CD40, CD27, 4-1BB, and OX40. During this original characterization, it was concluded that GITR was expressed on T cells and functioned to inhibit T cell receptor-induced apoptosis. Several years later, GITR expression was linked to Treg cells by the finding that GITR was "predominantly" expressed on CD4+CD25+ T cells from naive mice (Shimizu et al., 2002). These findings, coupled with the observation that GITR ligation abrogated the suppressive function of CD4+CD25+ Treg cells (McHugh et al., 2002; Shimizu et al., 2002), energized the regulatory T cell field and provided a potential marker of Treg cells.

However, as quickly as GITR surfaced and was adopted by the field, its potential as a unique marker of CD4+CD25+ Treg cells was disproved. A number of findings supported another explanation for the apparent GITR-mediated abrogation of CD4+CD25+ Treg cell suppressive function. In contrast to early reports, GITR was not a unique marker of CD4+CD25+ Treg cells and was found to be quickly upregulated on normal responder CD4+ T cells upon activation (Kohm et al., 2004; Ronchetti et al., 2004; Tone et al., 2003). In addition, GITR ligation appeared to mediate direct effects on responder CD4+ T cells in a manner similar to the B7 co-stimulatory family such that ligation of GITR lowered the activation threshold of responder CD4+ T cells in the absence of CD4+CD25+ Treg cells (Kohm et al., 2004). In combination with the observations that high levels of T cell activation confers resistance of CD4+ responder cells to CD4+CD25+ Treg cell suppressive function, it appears that GITR ligation does not serve to inactivate CD4+CD25+ Treg cells but, instead, serves to increase CD4+ responder T cell activation to levels that result in resistance to Treg cell regulation.

## 3.2   FoxP3

At about the same time GITR's role in identifying and regulating CD4+CD25+ Treg cell activity was being disproved, another molecule was proposed as a unique identifier of CD4 + CD25 + Treg cells. The discovery that CD4+CD25+ T cells express the forkhead transcription factor FoxP3 (Fontenot et al., 2003; Khattri et al., 2003) stands as one of the most important Treg cell-related findings to date. It was the linkage of a defect in Foxp3 to the autoimmune-prone scurfy mouse that initiated the investigations of the role of FoxP3 in Treg cell development and function. It is currently well accepted that FoxP3 does indeed stand as a true linage marker of a population of CD4 + CD25 + Treg cells. However, the discovery of FoxP3 does not represent an ideal marker of CD4 + CD25 + Treg cells, as FoxP3 is an intracellular transcription factor, thus making its utility as a phenotypic marker in live cells limited; moreover, FoxP3 mRNA can be expressed in both CD4 + CD25 – and CD8 + T cells upon stimulation in humans (Morgan et al., 2005). Thus, the search for a true surface marker to distinguish CD4 + CD25 + Treg cells from activated responder CD4 + T cells continues.

Regardless, the discovery of FoxP3 has not gone the way of GITR. Interestingly, Foxp3 may contribute to the suppressive mechanism of Treg cells, as ectopic expression of Foxp3 in CD4 + CD25 – T cells confers a Treg cell-like suppressive phenotype to these cells (Fontenot et al., 2003; Khattri et al., 2003). One interesting observation is that TGFβ induces FoxP3 expression in what appear to be normal responder CD4 + T cells (Fantini et al., 2004; Marie et al., 2005) and induces these cells to express a suppressive phenotype. In addition, FoxP3 expression may always correspond with a suppressive phenotype, but FoxP3 expression does not always correspond with CD25 expression. In fact, populations of T cells that express any of the following combinations—CD4+

CD25−FoxP3−, CD4+CD25+FoxP3−, CD4+CD25−FoxP3+, and CD4+CD25+ FOXP3+—are detected in normal mice. With any of these combinations, FoxP3 expression appears to identify populations of T cells with suppressive capacity dependably. In addition, CD4+FoxP3+ T cells appear to gain/regain CD25 expression upon homeostatic proliferation (Zelenay et al., 2005), suggesting that these cells serve as either precursors or a reservoir of potential CD4+CD25+ Treg cells. Thus, the above populations of CD4+ T cells may not represent distinct lineages but, instead, different phenotypic and potentially functional states of regulatory T cells.

The mechanism(s) by which FoxP3 confers suppressive/regulatory capacity on CD4+ T cells is currently unknown. One possibility is that FoxP3 interacts with other transcription factors such as NF-κB to suppress the transcription of cytokine genes necessary for CD4+ T cell proliferation and effector function, including interleukin-2 (IL-2), interferon-γ (IFNγ), and IL-4 (Bettelli et al., 2005). However, the relevance of these observations is yet to be determined. Regardless, it is apparent that FoxP3 plays a significant role in the generation and/or effector function of CD4+ regulatory T cells.

# 4    Origin and Specificity of CD4+CD25+ Treg Cells

Rivaling the complexity of determining a definitive phenotypic marker for CD4+CD25+ Treg cells is the question of Treg cell origin. A major complicating factor in determining the origin of CD4+CD25+ Treg cells is the evolving knowledge that multiple subpopulations of CD4+ regulatory T cells may exist. First and foremost, it is now accepted that "natural" and "inducible" populations of regulatory T cells exist and express CD4/CD25.

It is believed that "natural" Treg cells (nTregs) diverge from the CD4+ T cell developmental pathway earlier such that nTreg cells are not thymically derived from CD4+CD8+ precursor cells (Papiernik et al., 1998). In contrast, nTreg cells are found in mice as early as the 10th day of life and expand in an IL-2-dependent manner. Considering previous findings that day 3 thymectomy results in autoimmune disease, it appears that Treg cells emigrate from the thymus between days 3 and 10 of life and then expand in the periphery.

In light of their ability to quench responder CD4+ T cell effector function effectively, reason dictates that the Treg cells would tend to be specific for autoantigens such that their activation would lead to the prevention of autoimmune responses, whereas it may be detrimental to the host for non-self CD4+CD25+ Treg cells to exist and be activated, as this may impede the clearance of virus or bacteria. In light of this, there has long been speculation that the CD4+CD25+ Treg cell population expresses self-specific TCRs. In the case of CD4+CD25− effector T cells, the process of negative selection is believed to be the cornerstone of central tolerance and the deletion of CD4+ T cells expressing TCRs that recognize self-peptide/MHC complexes. For this to happen, CD4+CD25+ Treg cells must be resistant to the process of negative selection, but this does not appear to be the case. Treg cells appear to be susceptible to negative selection in a manner similar to that of non-Treg cells (Romagnoli et al., 2002; unpublished data from our laboratory and others). Pacholczyk et al. (2002) suggested that the Vβ repertoire of CD4+CD25+ Treg cells resembles that of non-Treg cells. However, at the same time, there are various reports that the CD4+CD25+ Treg cell population is indeed enriched for autoreactive TCRs (Caton et al., 2004; Kawahata et al., 2002; Romagnoli et al., 2002), suggesting that prior to their emigration from the thymus CD4+CD25+ Treg cells escape the process of negative selection (Caton et al., 2004; Kawahata et al., 2002). In support of this, some groups believe that Treg cells must express a TCR that has a high affinity for self-peptide during the early stages of Treg cell development, (Jordan et al., 2001). One theory states that it is the combination of the interaction between high-affinity autoreactive TCR and self-peptide with the constitutive high IL-2R expression that diverts CD4+ T cells into the Treg cell differentiation pathway (Malek et al., 2002). This is in contrast to normal CD4+ T cell differentiation in which high-affinity interactions between TCR and peptide often lead to thymic deletion.

On the surface, the findings that Treg cells undergo negative selection and that peripheral CD4+CD25+ Treg cells express a high frequency of self-reactive TCRs seem to contradict one another. However, a number of explanations may account for the prevalence of autoreactive Treg cells in the presence of an intact process of negative selection. For example, it was originally proposed that constitutive expression of CD25 by CD4+ Treg cells conferred resistance to negative selection, but it is equally possible that high levels of CD25 expression provide a mechanism for lower activation thresholds and increased homeostatic proliferation when autoreactive Treg cells encounter self-antigens. Thus, although negative selection may equally regulate Treg cells and non-Treg cells in a similar manner, postselection events may skew each population in different directions.

Self-recognition may explain a number of observed features of Treg cells. For one, Treg cells appear similar to activated T cells phenotypically. Thus, Treg cells may indeed be the by-product of repeated activation of a self-reactive T cell population. Not only may continuous activation of Treg cells be an essential component to their generation, but self-antigen recognition and activation may also lead to maintenance of the Treg cell population (Cozzo et al., 2003; Fisson et al., 2003). Although the jury is still out on the true specificity of Treg cells, support is growing for a role of Treg cells in preventing autoimmune responses. If autoimmune disease prevention is a primary functional role of Treg cells, future studies may reveal that Treg cell dysfunction is a contributing factor to the susceptibility/onset of autoimmune disease.

## 5  Mechanisms of Regulation

One of the earliest proposed mechanisms (Fig. 9.2) by which CD4+CD25+ Treg cells mediate their suppressive function is centered on their constitutive expression of high levels of CD25. In normal CD4+ T cells, IL-2/IL-2R signaling serves to

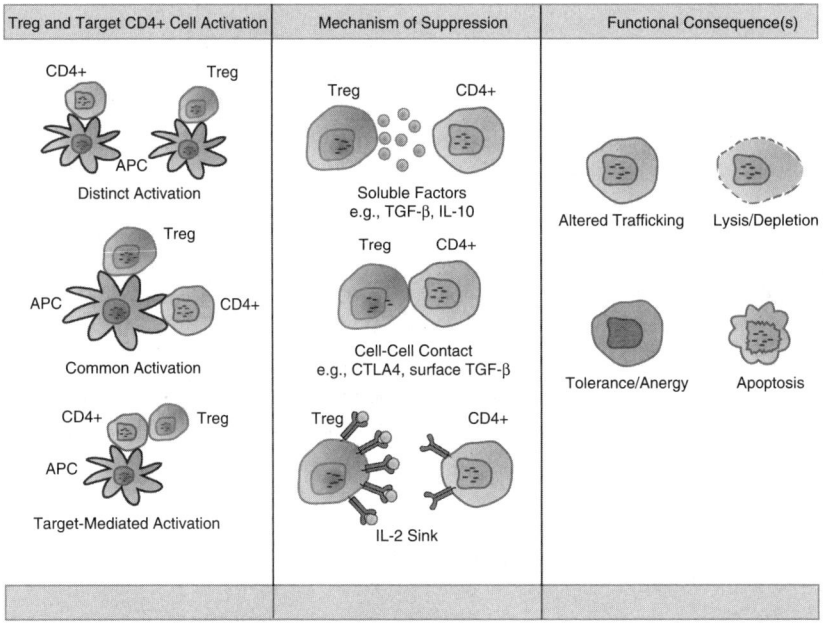

FIGURE 9.2. Proposed mechanisms of Treg cell activation, mechanisms of suppression, and functional consequences. *A.* Three models of Treg cell activation. *Distinct activation*: Responder CD4+ T cells and Treg cells are each activated by cognate interactions with distinct antigen-presenting cells (APC). *Common activation*: Responder CD4+ T cell and Treg cells are each activated by cognate interactions with a common APC. *Target-mediated activation*: Reports of major histocompatibility complex class II (MHC II) expression by CD4+ T cells raises the possibility that Treg cells may be activated by cognate interactions with the responder CD4+ T cell. *B. Mechanism of suppression*: A number of potential mechanisms of suppression have been proposed and may be classified into three general groups: suppression via soluble factors such as the production of TGFβ and/or IL-10; via cell contact and surface protein expression such as CTLA-4 and surface-bound TGFβ; or via IL-2 sequestration. *C. Functional consequences*: Following interaction with Treg cells, CD4+ T cells exhibit decreased effector function. Potential mechanisms by which responder effector function is decreased include altered trafficking patterns, lysis/depletion, tolerance/anergy, and apoptosis.

promote survival and proliferation and to initiate activation-induced cell death. However, despite constitutive high expression the IL-2R, CD4+CD25+ Treg cells are hypoproliferative upon TCR stimulation. This may be due to activation of differential signaling pathways upon IL-2 binding in comparison to that of CD4+CD25− T cells (Bensinger et al., 2004). However, because Treg cells do proliferate in the presence of higher levels of IL-2, it is possible that the constitutive expression of CD25 on Treg cells results in partial desensitization of the IL-2R signaling pathway that may be reversed by higher levels of signaling. Interestingly, the hypoproliferative nature of CD4+CD25+ Treg cells may permit

IL-2R signaling to promote the survival of Treg cells in the absence of continued expansion. At the same time, because CD4+CD25+ Treg cells do not produce IL-2 but do express high levels of CD25, Treg cells may serve as an IL-2 "sink" by competing for the IL-2 produced by CD4+ responder cells (de la Rosa et al., 2004). With this mechanism, Treg cells may inhibit responder cell proliferation by consuming the limited supply of IL-2; and, in turn, this IL-2 may also promote the survival of Treg cells to perpetuate their inhibitory function. Such a theory is supported by the observation that exogenous IL-2 blocks Treg cell suppressive function in vitro (de la Rosa et al., 2004), but the exogenous IL-2 may also serve a direct function on the CD4+ responder cells, permitting them to overcome the suppressive influence of the Treg cells. Also, other cytokines, such as IL-12, confer immunity against CD4+CD25+ Treg cell suppression to CD4+ responder cells (King and Segal, 2005), thus dampening the possibility that CD4+CD25+ Treg cells functioning as an IL-2 "sink" is a major mechanism by which Treg cells inhibit CD4+ T cell effector function.

Early findings suggested that CD4+CD25− T cells mediate their suppressive influence in a contact-dependent, not cytokine-mediated, fashion (Thornton and Shevach, 1998). Over time, this contact-dependent hypothesis of CD4+CD25+ Treg cell effector function was modified. It was later realized that the impression of contact dependence may in fact be due to the requirement of Treg cells to be activated via their TCRs prior to acquisition of suppressive effector function. However, once activated, Treg cells appear to mediate their suppressive phenotype in both a nonspecific and contact-independent manner. This interpretation is controversial, however; and both of these points would undoubtedly be contested by some research groups.

Regardless, significant focus was placed on surface proteins that might serve a regulatory role, such as CTLA-4 and surface transforming growth factor-β (TGFβ). Indeed, TGFβ has continued to remain at the forefront of potential mechanisms by which Treg cells may suppress CD4+ T cell responses. This has been fueled by a number of observations, including those reporting that two of the most prominent cytokines produced by Treg cells are TGFβ and IL-10 (Maloy et al., 2003; Nakamura et al., 2001), both of which are known suppressive factors for CD4+ T cell responses. Skewed focus on TGFβ is justified by the findings that the blockade of TGFβ completely abolishes the suppressive phenotype of Treg cells (Maloy et al., 2003; Nakamura et al., 2001, 2004; Powrie et al., 1996). In addition, CD4+ T cells that are incapable of responding to TGFβ are resistant to the suppressive properties of CD4+CD25+ Treg cells (Fahlen et al., 2005). However, Treg cells develop normally in the absence of TGFβ, suggesting that TGFβ is required for the effector function but not for the development of Treg cells. Although TGFβ may not be required for the initial development of CD4+CD25+ Treg cells, another potential role for TGFβ is regulating the expansion/homeostasis of the Treg cell population because TGFβ appears to induce Treg cell expansion in vivo (Peng et al., 2004).

Other studies have contested the role of TGFβ in mediating the suppressive phenotype of CD4+CD25+ Treg cells (Piccirillo et al., 2002). Rivaling the number of publications reporting a role for TGFβ in mediating CD4+CD25+ Treg cell function, an equal number of reports have failed to observe this in a variety of model systems. One potential complicating factor to the lack of agreement concerning the role of TGFβ in mediating CD4+CD25+ Treg cell effector function is the probable existence of subpopulations of CD4+ regulatory T cells, as discussed earlier. At present, there is more of a consensus in regard to the role of TGFβ in induced Treg cell effector function (Akbar et al., 2003; Zheng et al., 2002) in contrast to natural Treg cells.

In light of the acceptance of cytotoxic T lymphocyte-associated antigen 4 (CTLA-4) as a negative suppressor of CD4+ T cell effector function and the fact that CTLA-4 is a surface-bound protein expressed on activated CD4+ T cells, it is not surprising that a minor focus centered on this molecule as the potential mediator of the contact-dependent mechanisms by which Treg cells attenuate T cell responses. Treg cells constitutively express CTLA-4 (Takahashi et al., 2000), thus providing the rationale to focus of CTLA-4. In some studies, CTLA-4 appears to play a dominant role in the suppressive phenotype of Treg cells (Birebent et al., 2004). However, because the major mechanism by which CTLA-4 inhibits T cell responses is via competition with CD28 for B7 binding, and the observation that CD28 stimulation does not reverse/block the suppressive phenotype of Treg cells, it appears that CTLA-4 is not the sole mechanism by which Treg cells function (Liu et al., 2001). Further studies are required to determine whether CTLA-4 does indeed play a pivotal role in the effector suppressive function of Treg cells.

Currently, there is little agreement concerning the effector suppressive mechanism(s) utilized by CD4+CD25+ Treg cells. As discussed above, IL-2, TGFβ, and CTLA-4 represent the three most championed mechanisms by which CD4+CD25+ Treg cells suppress responder CD4+ T cell responses, but none of these proposed mechanisms has proven to be the critical key to Treg cell effector function. In addition, recent reports continue to identify additional potential mediators of Treg cell effector function, including granzyme B (Gondek et al., 2005), which is a known inducer of apoptosis. However, as is the case with the search for a true phenotypic marker of CD4+CD25+ Treg cells, continued research is required before a complete understanding of the effector suppressive mechanisms of CD4+CD25+ Treg cells is revealed.

# 6    Target Populations of Treg Cells

Until now, we have focused our discussion of CD4+CD25+ Treg cells to their effects on CD4+ responder T cells. Indeed, it was initially believed that Treg cells exclusively exerted their regulatory influence directly on other populations of CD4+ T cells. However, reports have suggested a number of other potential target populations that are also affected by Treg cells.

By far, the most studied secondary target of Treg cells are CD8+ T cells. A number of similarities exist between CD4+ and CD8+ effector cells in regard to their regulation by CD4+CD25+ Treg cells, and both populations appear susceptible to regulation (Green et al., 2003; Piccirillo and Shevach, 2001). For example, similar to the findings concerning effects on CD4+ T cells, Treg cells must be also activated via their TCRs either by antigen or polyclonal stimulation prior to gaining the capacity to downregulate CD8+ T cell effector function (Piccirillo and Shevach, 2001). In addition and potentially providing insight into the general regulatory mechanisms of Treg cells, IL-2 also appears to be a critical participant in the mechanism by which Treg cells attenuate CD8+ T cell responses as Treg cells inhibit the proliferation and the production of IFNγ and IL-2 as well as CD25 expression on CD8+ T cells (Piccirillo and Shevach, 2001). Thus, components of the IL-2/IL-2R pathway appear to be targets of Treg cells regardless of whether the target population is CD4+ or CD8+.

The effects of CD4+CD25+ Treg cells are not limited to T cells. For example, B cells also appear to be a target of CD4+CD25+ Treg cells. It is assumed that B cells are indirectly affected by Treg cells in response to the effects on T-helper (Th) cell cytokine production, although some reports also suggest a direct interaction between Treg cells and B cells potentially via Fas/FasL interactions (Janssens et al., 2003). Although Treg cells may inhibit B cell responses via effects on CD4+ Th cells, they also may directly affect B cells to block immunoglobulin recombination (Lim et al., 2005). In addition, CD4+CD25+ Treg cells are believed to influence co-stimulatory molecule expression on dendritic cells (Cederbom et al., 2000) and inhibit natural killer (NK) cell function (Ghiringhelli et al., 2005), presumably both via production of TGFβ, in addition to perhaps inhibiting CD4+ T cell function indirectly via effects on dendritic cells (Tang et al., 2006). In time, it would not be surprising if reports of other targets of Treg cells surface, and these reports may be stimulated by the increased characterization of the mechanisms by which Treg cells mediate their suppressive effector function.

# 7    Clinical Relevance

## 7.1    EAE/Multiple Sclerosis

Multiple sclerosis (MS) is an immune-mediated disease of the central nervous system (CNS) characterized by perivascular CD4+ T cell and mononuclear cell infiltration with subsequent primary demyelination of axonal tracks leading to progressive paralysis (Wekerle, 1991). MS is generally considered to be an autoimmune disease characterized by T cell responses to myelin basic protein (MBP), proteolipid protein (PLP), and/or myelin-oligodendrocyte glycoprotein (MOG) (Bernard and de Rosbo, 1991; de Rosbo et al., 1997; Ota et al., 1990); however, a clear-cut cause-and-effect relation between myelin reactivity and disease pathology has yet to be demonstrated. In addition to a significant genetic component (Ebers et al., 1995), epidemiological studies provide strong circumstantial evidence for an environmental trigger, most likely viral, in the induction of MS (Kurtzke, 1993; Olson et al., 2001; Waksman, 1995). The CNS pathology may therefore result from bystander myelin damage mediated via T cell targeting of virus persisting in the CNS and/or from direct activation of autoreactive T cells secondary to an encounter with a pathogen via molecular mimicry (Fujinami and Oldstone, 1985; Wucherpfennig and Strominger, 1995); it may also occur indirectly by epitope spreading due to the release of sequestered antigens secondary to virus-specific T cell-initiated myelin damage (McRae et al., 1995; Vanderlugt and Miller, 2002). Despite years of intensive research, the inducing antigen(s) and precise immunological mechanisms involved in the induction and chronic course of MS are still poorly understood, and there are limited therapeutic options available for managing this disease.

As mentioned above, one fundamental question is whether CD4+CD25+ Treg cells influence the onset of autoimmune disease, and, more importantly, does Treg cell dysfunction/deficiency lead to increased susceptibility to autoimmune disease? These are important and hotly contested questions. There is a substantial body of evidence supporting a role for CD4+CD25+ Treg cells in regulating autoimmune T cell function during EAE in mice and potentially MS in humans. Early observations noted that the occurrence of spontaneous EAE in MBP TCR transgenic mice can be efficiently prevented via the transfer of purified, syngeneic CD4+ T cells (reviewed in Furtado et al., 2001). However, the subpopulation of CD4+ T cells responsible for protection in these studies remains unclear, as protection was obtained by the transfer of either CD4+CD25+ and CD4+CD25– T cells.

More recently, studies using the actively induced models of both chronic and relapsing-remitting EAE (R-EAE) provide strong support for a potential contribution of CD4+CD25+ Treg cells in regulating the onset, progression, and spontaneous recovery from EAE (Kohm et al., 2002, 2003). At a basic level, supplementation of the normal population of CD4+CD25+ Treg cells via adoptive transfer attenuates the clinical progression of EAE. One of the most common approaches to determine if Treg cells influence disease progression is to "deplete" this population of cells at various times either prior to or following disease induction (McHugh and Shevach, 2002). To achieve this, animals are injected with one of the two most common clones of anti-CD25 monoclonal antibody (mAb) (PC61/rIgG1 or 7D4/rIgM). Injection of mice with either clone leads to what appears to be a rapid depletion of Treg cells as determined by flow cytometry using the other clone as a detecting antibody. However, our group and others have noted that although anti-CD25 mAb injection leads to rapid depletion of Treg cells the Treg cell population rapidly recovers and returns to normal levels within 7 to 14 days following depletion. Thus, interpretations from previous data resulting from anti-CD25 mAb injection and presumed CD4+CD25+ Treg cell depletion may need to be reevaluated. Accordingly, Treg cell inactivation by the injection of anti-CD25 mAb exacerbates suboptimal clinical EAE (Kohm et al., 2006). These findings are in agreement with those others, who observed that the clinical disease course of EAE is exacerbated in mice previously depleted of CD4+CD25+ Treg cells and that passive transfer of CD4+CD25+ Treg cells prevents EAE (McGeachy et al., 2005). In addition, injection of anti-CD25 mAb to deplete CD4+CD25+ Treg cells converts mice that are normally resistant to EAE

(B10.S) to susceptible to disease induction (Reddy et al., 2004). Thus, these basic findings demonstrate the powerful regulatory role of CD4+CD25+ Treg cells in affecting the clinical disease course of EAE.

A question that arises with any model of autoimmune disease and CD4+CD25+ Treg cells is the suppressive site of action. The two obvious possibilities are either the secondary lymphoid organs or the target organ. Interestingly, the passive transfer of Treg cells does not appear to result in significant effects on the peripheral immune response during EAE. However, the possibility exists that CD4+CD25+ Treg cells mediate their effects in the target organ. As discussed above, supplementation of CD4+CD25+ Treg cells attenuates EAE clinical disease progression. Of interest, CD4+CD25+ Treg cells can be found in the CNS (Kohm et al., 2006; McGeachy et al., 2005) and appear to be localized directly in CNS lesions. The presence of CD4+CD25+ Treg cells in the lesions suggests that regulatory T cells may prevent the continued activation of autoreactive CD4+ responder T cells in the target organ, resulting in decreased levels of CNS inflammation. This may explain the observations that CD4+CD25+ Treg cells appear to prevent clinical disease progression in the absence of observable effects on the peripheral autoreactive immune response (Kohm et al., 2002).

Findings from clinical studies provide additional support for the potential contribution of CD4+CD25+ Treg cells in preventing autoimmune disease in humans. Most notably, it appears that CD4+CD25+ Treg cells isolated from the peripheral blood of MS patients have a reduced capacity to suppress responder CD4+ T cell effector function (Viglietta et al., 2004). There also appears to be a lower frequency of CD4+CD25+ Treg cell in sick versus healthy individuals, arguing against the potential alternative explanation that the CD4+CD25+ T cells isolated from sick individuals were in fact activated responder CD4+ T cells rather than Treg cells. This is a legitimate concern, as a high percentage of circulating CD4+ T cells in autoimmune patients are previously activated and therefore may express elevated levels of CD25. Thus, similar to animal studies, research concerning the role of human CD4+CD25+ Treg cells in regulating autoimmune disease would also benefit significantly from the discovery of a definitive phenotypic marker specifying the regulatory cell population.

Of great interest is the potential role of CD4+CD25+ Treg cells in mediating spontaneous remission from EAE (or other autoimmune diseases). Because of the increased frequency of Treg cells that express self-reactive TCRs, it is conceivable that CD4+CD25+ Treg cells represent an endogenous mechanism by which the immune system discourages self-directed immune responses. Current studies are actively investigating these questions and may provide additional information concerning the role of CD4+CD25+ Treg cells in affecting various components of the pathophysiology of MS including susceptibility, progression, and spontaneous remission.

## 7.2   Type 1 Diabetes

Insulin-dependent diabetes mellitus (IDDM), or type 1 diabetes (T1D), is an autoimmune disease resulting in the progressive loss of insulin-producing β cells in the pancreatic islets of Langerhans (Wong and Janeway, 1999). This tissue destruction appears to be the direct result of a breakdown in central tolerance and subsequent activation of autoreactive CD4+ and CD8+ T cells. In light of the increasing incidence of disease, as well as the contribution of antigen-specific CD4+ and CD8+ T cells to disease pathogenesis, diabetes is a popular model for the general study of autoimmunity and treatment strategy design. Because IDDM is believed to be the direct result of central tolerance breakdown, depletion, inactivation, and/or altered effector function of –cell antigen-specific T cells are efficient techniques to restore tolerance and relieve clinical symptoms. In light of this, a significant effort has been made to understand the endogenous mechanisms that regulate CD4+ T cell effector function and dysfunction.

It is believed that the autoreactive immune responses during T1D are directed against islet antigens (Kent et al., 2005; Nakayama et al., 2005), so one natural mechanism for preventing diabetes would the deletion of islet-reactive T cells during negative selection. However, we now know that negative selection is either not dependable or that deficiencies in the protective process are ultimate indicators of who will develop diabetes because islet-reactive T cells are found in the circulation of healthy individuals (Lohmann et al., 1996). Thus, there appears to be a role for active regulatory mechanisms in preventing the activation of these autoreactive T cells and that it is a loss of central tolerance that leads to IDDM onset and progression. At the center of these investigations is the role of Treg cells in regulatory IDDM onset and progression.

Numerous studies have investigated the role of CD4+CD25+ Treg cells in regulating the onset and progression of IDDM (reviewed in Lan et al., 2005). Not surprisingly, these studies have reached the general consensus that supplementation of the endogenous CD4+CD25+ Treg cell population confers resistance to disease, whereas injection of anti-CD25 mAb exacerbates disease progression (Alyanakian et al., 2003; Anderson and Bluestone, 2005). In addition to modulating diabetes in the diabetes-prone NOD model system, CD4+CD25+ Treg cells also appear to control the spontaneous diabetes that occurs in other model systems, such as B7- and CD28-deficient mice (Salomon et al., 2000).

Interestingly, unlike in EAE (Kohm et al., 2005), TGFβ production by CD4+CD25+ Treg cells appears to play a major role in the mechanisms by which this regulatory T cell population influences the progression of IDDM. For example, TGFβ induces the expansion of Treg cells, which prevents diabetes (Peng et al., 2004), and TGFβ can increase the CD4+CD25+ Treg cell population to approximately 50% of all islet CD4+ T cells. This accumulation of Treg cells is believed to be a direct result of TGFβ-induced proliferation of CD4+CD25+FoxP3+ Treg cells. However, another possibility is the TGFβ-induced conversion of CD4+CD25– T cells to CD4+FoxP3+CD25+ T cells (Fantini et al., 2004). Regardless, TGFβ appears to be a critical component in CD4+CD25+ Treg cell-mediated regulation of IDDM, and this regulation may include both CD4+ and CD8+ autoreactive T cells (Green et al., 2003).

# 8    Conclusion

It appears that CD4+CD25+ Treg cells play a significant role in the body's defense against and control of autoimmune disease. As discussed above, CD4+CD25+ Treg cells appear to play major roles in regulating both EAE/MS and IDDM; and importantly, Treg cells have also been reported to influence a number of other autoimmune diseases such as colitis (Fahlen et al., 2005) and graft-versus-host disease (Ermann et al., 2005; Taylor et al., 2004). By preferentially expressing TCRs that recognize self-peptide/MHC complexes and by efficiently quenching both CD4+ and CD8+ T cell effector function upon activation, CD4+CD25+ Treg cells appear to be a well suited defense against the rogue activation of self-reactive T cells. However, at this point, there appear to be more questions than answers in regard to the origin, phenotype, and mechanism of action of this regulatory cell population. To date, despite an intensive effort, identification of a definitive phenotypic surface marker of CD4+CD25+ Treg cells remains elusive. The realization that multiple subpopulations of regulatory T cells appear to cooperate in regulating the immune response may contribute to future success in defining both the phenotype and the mechanism of action of each of these regulatory cell populations. In addition, the future ability to isolate and propagate self-antigen-specific Tregs may provide a therapeutic option for the therapy of a variety of autoimmune diseases.

## *References*

Akbar, A. N., Taams, L. S., Salmon, M., Vukmanovic-Stejic, M. (2003) The peripheral generation of CD4+ CD25+ regulatory T cells. *Immunology* 109:319-325.

Alyanakian, M. A., You, S., Damotte, D., et al. (2003) Diversity of regulatory CD4+T cells controlling distinct organ-specific autoimmune diseases. *Proc. Natl. Acad. Sci. U S A* 100:15806-15811.

Anderson, M. S., Bluestone, J. A. (2005) The NOD mouse: a model of immune dysregulation. *Annu. Rev. Immunol.* 23:447-485.

Bensinger, S. J., Walsh, P. T., Zhang, J., et al. (2004) Distinct IL-2 receptor signaling pattern in CD4+CD25+ regulatory T cells. *J. Immunol.* 172:5287-5296.

Bernard, C. C., de Rosbo, N. K. (1991) Immunopathological recognition of autoantigens in multiple sclerosis. *Acta Neurol.* 13:171-178.

Bettelli, E., Dastrange, M., Oukka, M. (2005) Foxp3 interacts with nuclear factor of activated T cells and NF-kappa B to repress cytokine gene expression and effector functions of T helper cells. *Proc. Natl. Acad. Sci. U S A* 102:5138-5143.

Birebent, B., Lorho, R., Lechartier, H., et al. (2004) Suppressive properties of human CD4+CD25+ regulatory T cells are dependent on CTLA-4 expression. *Eur. J. Immunol.* 34:3485-3496.

Caton, A. J., Cozzo, C., Larkin, J., 3rd, et al. (2004) CD4(+) CD25(+) regulatory T cell selection. *Ann. N. Y. Acad. Sci.* 1029:101-114.

Cederbom, L., Hall, H., Ivars, F. (2000) CD4+CD25+ regulatory T cells down-regulate co-stimulatory molecules on antigen-presenting cells. *Eur. J. Immunol.* 30:1538-1543.

Cozzo, C., Larkin, J., 3rd, Caton, A. J. (2003) Cutting edge: self-peptides drive the peripheral expansion of CD4+CD25+ regulatory T cells. *J. Immunol.* 171:5678-5682.

De la Rosa, M., Rutz, S., Dorninger, H., Scheffold, A. (2004) Interleukin-2 is essential for CD4+CD25+ regulatory T cell function. *Eur. J. Immunol.* 34:2480-2488.

De Rosbo, N. K., Hoffman, M., Mendel, I., et al. (1997) Predominance of the autoimmune response to myelin oligodendrocyte glycoprotein (MOG) in multiple sclerosis: reactivity to the extracellular domain of MOG is directed against three main regions. *Eur. J. Immunol.* 27:3059-3069.

Ebers, G. C., Sadovnick, A. D., Risch, N. J. (1995) A genetic basis for familial aggregation in multiple sclerosis. *Nature* 377:150-151.

Ermann, J., Hoffmann, P., Edinger, M., et al. (2005) Only the CD62L+ subpopulation of CD4+CD25+ regulatory T cells protects from lethal acute GVHD. *Blood* 105:2220-2226.

Fahlen, L., Read, S., Gorelik, L., et al. (2005) T cells that cannot respond to TGF-beta escape control by CD4(+)CD25(+) regulatory T cells. *J. Exp. Med.* 201:737-746.

Fantini, M. C., Becker, C., Monteleone, G., et al. (2004) Cutting edge: TGF-beta induces a regulatory phenotype in CD4+CD25- T cells through Foxp3 induction and downregulation of Smad7. *J. Immunol.* 172:5149-5153.

Fisson, S., Darrasse-Jeze, G., Litvinova, E., et al. (2003) Continuous activation of autoreactive CD4+ CD25+ regulatory T cells in the steady state. *J. Exp. Med.* 198:737-746.

Fontenot, J. D., Gavin, M. A., Rudensky, A. Y. (2003) Foxp3 programs the development and function of CD4+CD25+ regulatory T cells. *Nat. Immunol.* 4:330-336.

Fujinami, R. S., Oldstone, M. B. (1985) Amino acid homology between the encephalitogenic site of myelin basic protein and virus: mechanism for autoimmunity. *Science* 230:1043-1045.

Furtado, G. C., Olivares-Villagomez, D., Curotto de Lafaille, M. A., et al. (2001) Regulatory T cells in spontaneous autoimmune encephalomyelitis. *Immunol. Rev.* 182:122-134.

Gallo, P., Cupic, D., Bracco, F., et al. (1989) Experimental allergic encephalomyelitis in the monkey: humoral immunity and blood-brain barrier function. *Ital. J. Neurol. Sci.* 10:561-565.

Ghiringhelli, F., Menard, C., Terme, M., et al. (2005) CD4+CD25+ regulatory T cells inhibit natural killer cell functions in a transforming growth factor-beta-dependent manner. *J. Exp. Med.* 202:1075-1085.

Gondek, D. C., Lu, L. F., Quezada, S. A., et al. (2005) Cutting edge: contact-mediated suppression by CD4+CD25+ regulatory cells involves a granzyme B-dependent, perforin-independent mechanism. *J. Immunol.* 174:1783-1786.

Gonzalez, A., Andre-Schmutz, I., Carnaud, C., et al. (2001) Damage control, rather than unresponsiveness, effected by protective DX5+ T cells in autoimmune diabetes. *Nat. Immunol.* 2:1117-1125.

Green, E. A., Gorelik, L., McGregor, C. M., et al. (2003). CD4+CD25+ T regulatory cells control anti-islet CD8+ T cells through TGF-beta-TGF-beta receptor interactions in type 1 diabetes. *Proc. Natl. Acad. Sci. U S A* 100:10878-10883.

Hall, B. M., Pearce, N. W., Gurley, K. E., Dorsch, S. E. (1990). Specific unresponsiveness in rats with prolonged cardiac allograft survival after treatment with cyclosporine. III. Further characterization of the CD4+ suppressor cell and its mechanisms of action. *J. Exp. Med.* 171:141-157.

Janssens, W., Carlier, V., Wu, B., et al. (2003). CD4+CD25+ T cells lyse antigen-presenting B cells by Fas-Fas ligand interaction in an epitope-specific manner. *J. Immunol.* 171:4604-4612.

Jordan, M. S., Boesteanu, A., Reed, A. J., et al. (2001). Thymic selection of CD4+CD25+ regulatory T cells induced by an agonist self-peptide. *Nat. Immunol.* 2:301-306.

Kawahata, K., Misaki, Y., Yamauchi, M., et al. (2002) Generation of CD4(+)CD25(+) regulatory T cells from autoreactive T cells simultaneously with their negative selection in the thymus and from nonautoreactive T cells by endogenous TCR expression. *J. Immunol.* 168:4399-4405.

Kent, S. C., Chen, Y., Bregoli, L., et al. (2005) Expanded T cells from pancreatic lymph nodes of type 1 diabetic subjects recognize an insulin epitope. *Nature* 435:224-228.

Khattri, R., Cox, T., Yasayko, S. A., Ramsdell, F. (2003) An essential role for Scurfin in CD4+CD25+ T regulatory cells. *Nat. Immunol.* 4:337-342.

King, I. L., Segal, B. M. (2005) Cutting edge: IL-12 induces CD4+CD25- T cell activation in the presence of T regulatory cells. *J. Immunol.* 175:641-645.

Kohm, A. P., Carpentier, P. A., Anger, H. A., Miller, S. D. (2002) Cutting edge: CD4(+)CD25(+) regulatory T cells suppress antigen-specific autoreactive immune responses and central nervous system inflammation during active experimental autoimmune encephalomyelitis. *J. Immunol.* 169:4712-4716.

Kohm, A. P., Carpentier, P. A., Miller, S. D. (2003) Regulation of experimental autoimmune encephalomyelitis (EAE) by CD4+CD25+ regulatory T cells. *Novartis Found. Symp.* 252:45-52.

Kohm, A. P., McMahon, J. S., Podojil, J. R., et al. (2006) Anti-CD25 mAb injection results in the functional inactivation, not depletion of CD4+CD25+ Treg cells. *J. Immunol.* 176(6):3301-3305.

Kohm, A. P., Williams, J. S., Bickford, A. L., et al. (2005) Treatment with nonmitogenic anti-CD3 monoclonal antibody induces CD4+ T cell unresponsiveness and functional reversal of established experimental autoimmune encephalomyelitis. *J. Immunol.* 174:4525-4534.

Kohm, A. P., Williams, J. S., Miller, S. D. (2004) Cutting edge : ligation of the glucocorticoid-induced TNF receptor enhances autoreactive CD4+ T cell activation and experimental autoimmune encephalomyelitis. *J. Immunol.* 172:4686-4690.

Kojima, A., Prehn, R. T. (1981) Genetic susceptibility to post-thymectomy autoimmune diseases in mice. *Immunogenetics* 14:15-27.

Kurtzke, J. F. (1993) Epidemiologic evidence for multiple sclerosis as an infection. *Clin. Microbiol. Rev.* 6:382-427.

Lan, R. Y., Ansari, A. A., Lian, Z. X., Gershwin, M. E. (2005) Regulatory T cells: development, function and role in autoimmunity. *Autoimmun. Rev.* 4:351-363.

Lim, H. W., Hillsamer, P., Banham, A. H., Kim, C. H. (2005) Cutting edge: direct suppression of B cells by CD4+ CD25+ regulatory T cells. *J. Immunol.* 175:4180-4183.

Liu, Z., Geboes, K., Hellings, P., et al. (2001) B7 interactions with CD28 and CTLA-4 control tolerance or induction of mucosal inflammation in chronic experimental colitis. *J. Immunol.* 167:1830-1838.

Lohmann, T., Leslie, R. D., Londei, M. (1996) T cell clones to epitopes of glutamic acid decarboxylase 65 raised from normal subjects and patients with insulin-dependent diabetes. *J. Autoimmun.* 9:385-389.

Malek, T. R., Yu, A., Vincek, V., et al. (2002) CD4 regulatory T cells prevent lethal autoimmunity in IL-2Rbeta-deficient mice. Implications for the nonredundant function of IL-2. *Immunity* 17:167-178.

Maloy, K. J., Salaun, L., Cahill, R., et al. (2003) CD4+CD25+ T(R) cells suppress innate immune pathology through cytokine-dependent mechanisms. *J. Exp. Med.* 197:111-119.

Marie, J. C., Letterio, J. J., Gavin, M., Rudensky, A. Y. (2005) TGF-beta1 maintains suppressor function and Foxp3 expression in CD4+CD25+ regulatory T cells. *J. Exp. Med.* 201:1061-1067.

McGeachy, M. J., Stephens, L. A., Anderton, S. M. (2005) Natural recovery and protection from autoimmune encephalomyelitis: contribution of CD4+CD25+ regulatory cells within the central nervous system. *J. Immunol.* 175:3025-3032.

McHugh, R. S., Shevach, E. M. (2002) Cutting edge: depletion of CD4+CD25+ regulatory T cells is necessary, but not sufficient, for induction of organ-specific autoimmune disease. *J. Immunol.* 168:5979-5983.

McHugh, R. S., Whitters, M. J., Piccirillo, C. A., et al. (2002) CD4(+)CD25(+) immunoregulatory T cells: gene expression analysis reveals a functional role for the glucocorticoid-induced TNF receptor. *Immunity* 16:311-323.

McRae, B. L., Vanderlugt, C. L., Dal Canto, M. C., Miller, S. D. (1995) Functional evidence for epitope spreading in the relapsing pathology of experimental autoimmune encephalomyelitis. *J. Exp. Med.* 182:75-85.

Morgan, M. E., van Bilsen, J. H., Bakker, A. M., et al. (2005) Expression of FOXP3 mRNA is not confined to CD4+CD25+ T regulatory cells in humans. *Hum. Immunol.* 66:13-20.

Nakamura, K., Kitani, A., Fuss, I., et al. (2004) TGF-beta 1 plays an important role in the mechanism of CD4+CD25+ regulatory T cell activity in both humans and mice. *J. Immunol.* 172:834-842.

Nakamura, K., Kitani, A., Strober, W. (2001) Cell contact-dependent immunosuppression by CD4(+)CD25(+) regulatory T cells is mediated by cell surface-bound transforming growth factor beta. *J. Exp. Med.* 194:629-644.

Nakayama, M., Abiru, N., Moriyama, H., et al. (2005) Prime role for an insulin epitope in the development of type 1 diabetes in NOD mice. *Nature* 435:220-223.

Nocentini, G., Giunchi, L., Ronchetti, S., et al. (1997) A new member of the tumor necrosis factor/nerve growth factor receptor family inhibits T cell receptor-induced apoptosis. *Proc. Natl. Acad. Sci. U S A* 94:6216-6221.

Olson, J. K., Croxford, J. L., Miller, S. D. (2001) Virus-induced autoimmunity: potential role of viruses in initiation, perpetuation, and progression of T cell-mediated autoimmune diseases *Viral Immunol.* 14:227-250.

Ota, K., Matsui, M., Milford, E. L., et al. (1990) T-cell recognition of an immunodominant myelin basic protein epitope in multiple sclerosis. *Nature* 346:183-187.

Pacholczyk, R., Kraj, P., Ignatowicz, L. (2002) Peptide specificity of thymic selection of CD4+CD25+ T cells. *J. Immunol.* 168:613-620.

Papiernik, M., de Moraes, M. L., Pontoux, C., et al. (1998) Regulatory CD4 T cells: expression of IL-2R alpha chain, resistance to clonal deletion and IL-2 dependency. *Int. Immunol.* 10:371-378.

Peng, Y., Laouar, Y., Li, M. O., et al. (2004) TGF-beta regulates in vivo expansion of Foxp3-expressing CD4+CD25+ regulatory T cells responsible for protection against diabetes. *Proc. Natl. Acad. Sci. U S A* 101:4572-4577.

Piccirillo, C. A., Letterio, J. J., Thornton, A. M., et al. (2002) CD4(+)CD25(+) regulatory T cells can mediate suppressor function in the absence of transforming growth factor beta1 production and responsiveness. *J. Exp. Med.* 196:237-246.

Piccirillo, C. A., Shevach, E. M. (2001) Cutting edge: control of CD8+ T cell activation by CD4+CD25+ immunoregulatory cells. *J. Immunol.* 167:1137-1140.

Powrie, F., Carlino, J., Leach, M. W., et al. (1996) A critical role for transforming growth factor-beta but not interleukin 4 in the suppression of T helper type 1-mediated colitis by CD45RB(low) CD4+ T cells. *J. Exp. Med.* 183:2669-2674.

Reddy, J., Illes, Z., Zhang, X., et al. (2004) Myelin proteolipid protein-specific CD4+CD25+ regulatory cells mediate genetic resistance to experimental autoimmune encephalomyelitis. *Proc. Natl. Acad. Sci. U S A* 101:15434-15439.

Romagnoli, P., Hudrisier, D., van Meerwijk, J. P. (2002) Preferential recognition of self antigens despite normal thymic deletion of CD4(+)CD25(+) regulatory T cells. *J. Immunol.* 168:1644-1648.

Ronchetti, S., Zollo, O., Bruscoli, S., et al. (2004) GITR, a member of the TNF receptor superfamily, is costimulatory to mouse T lymphocyte subpopulations. *Eur. J. Immunol.* 34:613-622.

Sakaguchi, S., Sakaguchi, N. (1989) Organ-specific autoimmune disease induced in mice by elimination of T cell subsets. V. Neonatal administration of cyclosporin A causes autoimmune disease. *J. Immunol.* 142:471-480.

Sakaguchi, S., Sakaguchi, N., Asano, M., et al. (1995) Immunologic self-tolerance maintained by activated T cells expressing IL-2 receptor alpha-chains (CD25): breakdown of a single mechanism of self-tolerance causes various autoimmune diseases. *J. Immunol.* 155:1151-1164.

Salomon, B., Lenschow, D. J., Rhee, L., et al. (2000) B7/CD28 costimulation is essential for the homeostasis of the CD4+CD25+ immunoregulatory T cells that control autoimmune diabetes. *Immunity* 12:431-440.

Shimizu, J., Yamazaki, S., Takahashi, T., et al. (2002) Stimulation of CD25(+)CD4(+) regulatory T cells through GITR breaks immunological self-tolerance. *Nat. Immunol.* 3:135-142.

Suri-Payer, E., Amar, A. Z., Thornton, A. M., Shevach, E. M. (1998) CD4+CD25+ T cells inhibit both the induction and effector function of autoreactive T cells and represent a unique lineage of immunoregulatory cells. *J. Immunol.* 160:1212-1218.

Takahashi, T., Tagami, T., Yamazaki, S., et al. (2000) Immunologic self-tolerance maintained by CD25(+)CD4(+) regulatory T cells constitutively expressing cytotoxic T lymphocyte-associated antigen 4. *J. Exp. Med.* 192:303-310.

Tang, Q., Adams, J. Y., Tooley, A. J., et al. (2006) Visualizing regulatory T cell control of autoimmune responses in nonobese diabetic mice. *Nat. Immunol.* 7:83-92.

Taylor, P. A., Panoskaltsis-Mortari, A., Swedin, J. M., et al. (2004) L-Selectin(hi) but not the L-selectin(lo) CD4+25+ T-regulatory cells are potent inhibitors of GVHD and BM graft rejection. *Blood* 104:3804-3812.

Thornton, A. M., Shevach, E. M. (1998) CD4+CD25+ immunoregulatory T cells suppress polyclonal T cell activation in vitro by inhibiting interleukin 2 production. *J. Exp. Med.* 188:287-296.

Tone, M., Tone, Y., Adams, E., et al. (2003). Mouse glucocorticoid-induced tumor necrosis factor receptor ligand is costimulatory for T cells. *Proc. Natl. Acad. Sci. U S A* 100:15059-15064.

Vanderlugt, C. L., Miller, S. D. (2002). Epitope spreading in immune-mediated diseases: implications for immunotherapy. *Nat. Rev. Immunol.* 2:85-95.

Viglietta, V., Baecher-Allan, C., Weiner, H. L., Hafler, D. A. (2004). Loss of functional suppression by CD4+CD25+ regulatory T cells in patients with multiple sclerosis. *J. Exp. Med.* 199:971-979.

Waksman, B. H. (1995) Multiple sclerosis: More genes versus environment. *Nature* 377: 105-106.

Wekerle, H. (1991) Immunopathogenesis of multiple sclerosis. *Acta Neurol.* 13:197-204.

Wong, F. S., Janeway, C. A. (1999) Insulin-dependent diabetes mellitus and its animal models. *Curr. Opin. Immunol.* 11:643-647.

Wucherpfennig, K. W., Strominger, J. L. (1995) Molecular mimicry in T cell-mediated autoimmunity: viral peptides activate human T cell clones specific for myelin basic protein. *Cell* 80:695-705.

Zelenay, S., Lopes-Carvalho, T., Caramalho, I., et al. (2005) Foxp3+ CD25-CD4 T cells constitute a reservoir of committed regulatory cells that regain CD25 expression upon homeostatic expansion. *Proc. Natl. Acad. Sci. U S A* 102:4091-4096.

Zheng, S. G., Gray, J. D., Ohtsuka, K., et al. (2002) Generation ex vivo of TGF-beta-producing regulatory T cells from CD4+CD25-precursors. *J. Immunol.* 169: 4183-4189.

# 10
# Immunopathogenesis of Multiple Sclerosis: Overview

Til Menge, Bernhard Hemmer, Stefan Nessler, Dun Zhou,
Bernd C. Kieseier, and Hans-Peter Hartung

## 1   Introduction

Multiple sclerosis (MS) is a chronic inflammatory disease of the central nervous system (CNS). With a prevalence of 120 in 100,000 and approximately 1.2 millions affected worldwide, it is a common neurological disease. Its incidence peaks during early adulthood, presents clinically rarely before the age of 15 and after the age of 60, and women are more often affected than men. It represents one of the most frequent causes of neurological disability in the young adult (Compston and Coles, 2002).

The pathology is heterogeneous and is characterized by areas of inflammation, demyelination, remyelination, and axonal degeneration (Lucchinetti et al., 2000). Distinct phenotypes can be differentiated based on the presence or absence of T cell invasion, B cells and immunoglobulin deposits, and damage to the oligodendrocytes. To make the diagnosis the disease has to evolve in time and space, as observed on clinical grounds and supported by diagnostic tools, such as magnetic resonance imaging (MRI), cerebrospinal fluid (CSF) analysis with demonstration of oligoclonal bands and pathological evoked potentials (Poser et al., 1983). Revised diagnostic criteria have put more emphasis on clinically silent lesion development seen by MRI allowing an earlier diagnosis of MS (McDonald et al., 2001; Polman et al., 2005). The etiology of MS remains mostly elusive, although the trio of genetic and environmental factors and autoimmunity seems to influence the risk of disease development. A large body of evidence stems from the animal model of MS, experimental allergic encephalomyelitis (EAE) that was first describe some 70 years ago (Rivers et al., 1933) and has since evolved to one of the most powerful tools to study the pathogenesis of MS (Gold et al., 2000; Gold et al., 2006; Steinman, 1999), although criticism has been leveled regarding its artificial induction, its limited reflection of disease pathology, and different lesion development (Sriram and Steiner, 2005).

## 2   Genetic Factors Influencing MS Susceptibility

The lifetime risk in the general population to acquire MS is approximately 0.25% (Compston and Coles, 2002), but it increases to up to 30% in monozygotic twins and 3% in siblings of patients with MS (Dyment et al., 2004; Oksenberg et al., 1999). Multinational family and genetic linkage studies indicate a highly polygenetic mode of inheritance, with the human leukocyte antigen (HLA) region being to date the only major gene locus. In Caucasian populations the HLA alleles DRB1*1501, DRB5*0101, and DQB1*0602 are associated with a two- to fourfold increased risk of developing MS (Oksenberg et al., 1999). Despite the emergence of some promising chromosomal regions of interest, no other MS-susceptibility genes have been unequivocally identified as yet (Kenealy et al., 2006; Sawcer et al., 2005). Compared to that Caucasians, MS in African Americans is a rather rare, yet more aggressive disease, which had led us to speculate on an additional susceptibility gene locus in the African ancestry (Oksenberg et al., 2004).

## 3   Environmental Factors Contributing to MS Risk

Significant regional shifts in MS prevalence, several migration studies—including decreased MS susceptibility in separated monozygotic twins and increased MS prevalence among Africans who migrated to Europe—a few apparent MS epidemics, and the association between clinical relapses and viral infections may point toward pathogenic agents as the predominant environmental factor in the pathogenesis of MS (Lang et al., 2002; Miller et al., 2001; Wucherpfennig and Strominger, 1995). However, no single infectious organism has been consistently associated with MS so far. Numerous studies have provided data, some of it controversial, on the possible involvement of infectious agents (Christen and von Herrath, 2005), including *Chlamydia pneumoniae* (Goverman et al., 1993; Swanborg et al., 2002), Epstein-Barr virus (Buljevac et al., 2005; Cepok et al., 2005b; Thacker et al., 2006), and human herpesvirus 6 (Moore and Wolfson, 2002), despite the high prevalence of these infectious diseases. Large epidemiological studies have suggested that reinfection with the same agent or initial infection with certain pathogens during late rather than early childhood are predominantly found in MS cases as compared to controls (Casetta and Granieri, 2000). However, regional factors appear to be more powerful in promoting MS susceptibility than local ones (Ebers et al., 1995).

## 4   Autoimmunity as the Pathogenic Effector

Multiple sclerosis is considered an autoimmune disease. Adaptive cellular and humoral autoreactivity can be readily detected in MS patients, and the disease can be elicited in the animal model EAE. Induction of disease by transfer of human-derived antibodies to animals is currently being established. EAE shares many similarities with MS but does not mirror the entire spectrum of disease pathology or course. It is a T cell-mediated inflammatory disease of the CNS with variable degrees of demyelination and axonal damage. A variety of rodents, primates, and even zebrafish are susceptible to disease induction by immunization with myelin antigens (e.g., MOG, MBP and PLP) and adjuvant. The resulting CD4+ T helper-1 (Th1) cell response attacks the myelinated areas of the CNS (Gold et al., 2006; Gold et al., 2000; Steinman, 1999). T cells, recruited monocytes, and activated microglial cells mediate inflammation and demyelination (Kieseier et al., 1999). B cells and antibodies are not essential for the induction of EAE; however, passive transfer of antibodies specific for structural epitopes of MOG enhance demyelination in some models (Schluesener et al., 1987). CD8+ T cells play only a minor role in commonly used EAE models, although in particular circumstances they might also be encephalitogenic (Huseby et al., 2001).

## 5   Interaction of Genes, Environment, and Autoimmunity

Several concepts have been developed to explain the induction of autoimmunity to myelin antigens in EAE and may serve as a model for MS. Cross-reactivity between non-self (viral, bacterial) proteins and self proteins, termed molecular mimicry, has been discussed as a possible mechanism for the onset of autoreactive T cell responses (Fujinami and Oldstone, 1985; Fujinami et al., 2006; Hemmer et al., 1997; McCoy et al., 2006; von Herrath et al., 2003). This is underscored by a recent demonstration that a T cell receptor (TCR) from an MS patient promiscuously recognized both a DRB1*1501-restricted MBP peptide and a DRB5*0101-restricted, however structurally unrelated, Epstein-Barr virus (EBV) peptide (Wucherpfennig and Strominger, 1995). Once self-tolerance is broken, the repeated release of self-antigens from the brain might promote additional autoreactive T cells that respond to additional myelin epitopes. This process, termed epitope spreading, is crucial in the chronic EAE model (Lehmann et al., 1993; Vanderlugt and Miller, 2002) and is, at least in the animal models, dependent on the genetic background.

Comparable CNS pathology and epitope spreading to structurally unrelated antigens are also seen in infectious diseases models. CNS infection with neurotropic viruses results in acute or chronic CNS inflammation and demyelination (Stohlman and Hinton, 2001; Tsunoda et al., 2005; Vanderlugt and Miller, 2002). Depending on the features of the pathogen, the immune response might be beneficial or detrimental. Paradoxically, low neurotoxic viral infections (e.g., Borna virus) may induce limited pathology in the brain per se but result in severe CNS damage secondary to a detrimental cross-reactive immune response to the pathogen; in contrast, death of an animal infected by a highly neurotoxic virus may be prevented by rapid clearing of the pathogen from the brain through a highly specific adaptive immune response (Richt et al., 1994). In most infectious disease models, CD8+ and CD4+ T cells are essential for virus control but also significantly contribute to tissue destruction (Stohlman and Hinton, 2001). B cells appear to have a role primarily in virus control during subacute or chronic CNS infection (Ramakrishna et al., 2002).

## 6    Immune Surveillance of the CNS

The brain was originally considered to be a highly immunoprivileged organ and secluded from immune surveillance, a view that has been challenged by many other studies (Cserr and Knopf, 1992). Even under physiological conditions, the CNS compartment is patrolled by minute numbers of the circulating immune cell repertoire via the intact blood-brain barrier (BBB) (Hickey, 2001). Any damage to CNS tissue leads to leakage and breakdown of the BBB, activation of CNS resident immune cells, in particular microglia, and extravasation of hematogenous T cells into the CNS, irrespective of their antigen specificity (Engelhardt and Ransohoff, 2005). Cellular release of cytokines and chemokines attracts additional monocytes, lymphocytes, and cells with a dendritic cell phenotype into the lesion (i.e., the area of CNS damage). Microglial cells are important for generating and maintaining the inflammatory milieu, whereas dendritic cells seem to play a central role in antigen presentation to invading T cells and their polarization into Th1 or Th2 phenotypes (Greter et al., 2005; Heppner et al., 2005). They can be identified in MS lesions (Serafini et al., 2006).

In parallel with the influx of immune cells into the CNS lesion, antigens from the lesion gain access to the periphery, a phenomenon that is best demonstrated in CNS infection. Non-self antigens introduced into the CNS are rapidly detected in cervical or paraspinal lymph nodes, although it is unclear whether the antigens diffuse passively over the leaky BBB or are transported actively by phagocytic cells. In the lymph node environment, dendritic cells process the proteins and present the resulting peptides bound to major histocompatibility complex class I and II (MHC I and II) molecules to incoming T cells, initiating an adaptive immune response. CD8+ T cells recognize short peptides in the context of MHC I, whereas CD4+ T cells recognize them in the context of MHC II molecules. High-affinity T cells are propagated, clonally expanded, and guided across the

BBB by chemoattractants. They infiltrate the lesion and home to sites where their respective antigen is exposed in the context of relevant MHC expression (Fig. 10.1). Whereas MHC class II is sufficiently displayed only on professional antigen-presenting cells (APCs) (e.g., activated microglial cells, macrophages, dendritic cells), MHC class I can be expressed by all CNS cells in the inflammatory milieu (Dandekar et al., 2001; Neumann et al., 1995). Accordingly, CD4+ T cells are predominantly found in perivascular cuffs and the meninges, whereas CD8+ T cells also seem to invade the parenchyma of the inflamed lesion. Upon contact with their driving antigen, the cells arrest and locally mediate their effector functions (Kawakami et al., 2005). CD4+ T cells recruit monocytes, which release proinflammatory cytokines and toxic molecules, such as nitric oxide, interleukin (IL)-1, IL-6, tumor necrosis factor-$\alpha$ (TNF$\alpha$), and matrix metalloproteinases). In contrast, CD8 T cells might also elicit cytotoxic effects against MHC I-expressing cells directly, such as oligodendrocytes and neurons.

B cell responses are also initiated in the peripheral lymph nodes by antigen presentation through dendritic cells and provision of co-stimulatory signals from antigen-specific T cells. In contrast to T cells, B cells recognize, by virtue of their B cell receptors (BCRs), conformational or linear determinants of proteins displayed on dendritic cells. Analogous to T cell maturation, highly affine B cpells are propagated and clonally expanded in the germinal centers of lymph nodes, where they differentiate either into memory B cells or antibody-secreting plasma cells. Like T cells, activated B cells and plasma cells can pass through the BBB and infiltrate the perivascular space and meninges (Alter et al., 2003). Locally, they release soluble antigen-specific immunoglobulins, exert their effector function through cytokine release, or act as APCs through their BCRs (Fig. 10.1). The question as to whether terminal differentiation of T and B cells may also take place in the brain or exclusively in the lymph nodes is still a matter for debate (Uccelli et al., 2005). Studies, however have pointed toward a local immune response; germinal center-like structures, homeostatic chemokines and B cell activating factor belonging to the TNF family (BAFF), which are produced by activated astrocytes and necessary for B cell survival and homeostasis, have been identified within MS lesion (Kalled, 2005; Krumbholz et al., 2005; Uccelli et al., 2005).

# 7   Immunology of MS

## 7.1   *Immunological Changes in MS Lesions*

Only a few studies have addressed the difficult issue of lesion evolution in MS (Gay et al., 1997). During the first days of lesion development, activation of microglia and macrophages with surface expression of HLA II molecules and complement receptor C3d/immunoglobulin complexes seem to represent the earliest changes that occur. At that time, the BBB appears to remain tightly intact, and only few cell infiltrates can be detected. Demyelination and astrogliosis are largely absent. Lesions between 6 and 20 weeks consist of cell infiltrates, demyelination, leakage of the

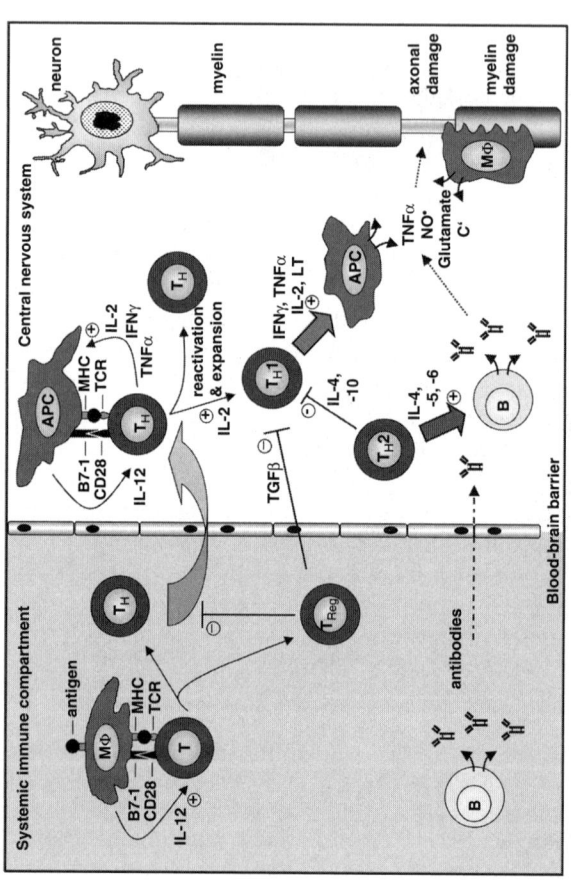

FIGURE 10.1. Immunopathogenetic concept of multiple sclerosis. Autoreactive T cells in the systemic circulation, upon activation and guided by adhesion molecules on the blood-brain-barrier, can (attracted by chemokines) migrate into the central nervous system (CNS). In the CNS, autoreactive T cells when recognising their antigen on an antigen-presenting cell displaying additional recognition and co-stimulatory molecules in a permissive cytokine microenvironment are reactivated, proliferate, and secrete various cytokines. These can in turn activate microglia, macrophages, or astrocytes to elaborate noxious inflammatory molecules. Cytokines can also impede impulse propagation. From EAE models, evidence is being gathered that Th17 cells may play a critical role in regulating this autoimmune response. Autoreactive B cells and autoantibodies can also be detected in blood and cerebrospinal fluid. They can pass through a damaged blood-brain barrier and, upon activating the complement system with ensuing formation of the membrane attack complex, damage myelin. Antibodies can also bind to macrophages and stimulate them to antibody-dependent cytotoxicity. APC, antigen-presenting cell; B, B lymphocyte; T, T lymphocyte (T$_H$, T helper cell; T$_{Reg}$, regulatory T cell; T$_C$, cytotoxic T cell; MΦ, macrophage; C′, complement. See text for further explanation.

BBB, and reactive astrocytes, along with proliferating oligodendroglial cells at the lesion border. These lesions represent the most active phase of the disease. Beyond week 20, strong inflammatory infiltrates decrease in the center of the lesion and later at the lesion border. Many cytokines and chemokines are released in the lesion, including Th1 and Th2 cytokines (Cannella and Raine, 1995; Trebst and Ransohoff, 2001). Axonal damage and demyelination is seen during all phases of the disease (Trapp et al., 1998) but appears to be most pronounced early during the disease course, correlating with the extent of cellular infiltrates (Kuhlmann et al., 2002). Proliferation of oligodendrocytes and remyelination is detectable in many lesions but appears to result only in incomplete myelination of axons. Overall, the extent of inflammation, neurodegeneration, and neuroregeneration is rather heterogeneous in individual patients. Most likely this is due to a temporally and spatially orchestrated sequence of detrimental and beneficial immune agents, such as cytokines or tissue repair markers, as was shown in a relapsing EAE model (Espejo et al., 2005). A systematic analysis of MS lesion pathology revealed four subgroups on the basis of relative quantity and quality of inflammation, antibody deposition, and oligodendrocyte dystrophy (Lucchinetti et al., 2000). Among those subgroups, two are characterized by minor inflammatory changes despite prominent oligodendrocyte pathology. This observation is further supported by a histopathological investigation of very early MS lesions; in some of the patients, no inflammatory infiltrates were observed despite the presence of extensive oligodendrocyte apoptosis and microglia activation (Barnett and Prineas, 2004). Both studies suggest that in a subgroup of patients or at certain disease stages the immune system may play a less dominant role. At present, it remains uncertain which factors trigger oligodendrocyte death in these patients, be it primary neurotoxicity or secondary apoptosis.

## 7.2  T Cells in MS

Because of the key role of CD4+ autoreactive T cells in EAE, MS has been considered primarily a T cell-mediated disease. Many studies have investigated the role of autoreactive, myelin-specific CD4+ T cells in MS (Sospedra and Martin, 2005), but their precise functions in the pathogenesis of the disease remain elusive so far. Recently a new subset of T cells, TH17 cells, has been identified as a highly potent pro-inflammatory T cell component in EAE (Langrish et al., 2005). This subset is exclusively activated by IL-23 and facilitates its detrimental action mainly through IL-17 (hence the denotation), but not classic TH1 cytokines; in fact, rather TH17 than TH1 cells may be the sole effectors of EAE (Langrish et al., 2005; Park et al., 2005; Weaver et al., 2006). Although human myelin-specific T cells were shown to be encephalitogenic in a transgenic animal model (Madsen et al., 1999), no consistent differences have been observed with respect to the frequency or phenotype of these cells between patients and controls, except for a higher level of activation in MS (Ota et al., 1990; Zhang et al., 1994). More and more evidence is being gathered, however, that lend support towards dysfunctional regulatory T cells (Tregs), that is, in MS Tregs are less efficient to suppress autoreactive T cells (Haas et al., 2005; Viglietta et al., 2004).

Peripherally activated immune cells adhere to the activated endothelia of the BBB through a panel of adhesion molecules and integrins (Fig. 10.1) (Archelos et al., 1999; Engelhardt and Ransohoff, 2005), extravasate after disruption of the BBB basement membrane by metalloproteinases (Parks et al., 2004) into the CNS, migrate to the site of inflammation by a fine-tuned gradient of chemokines and chemoattractants (Ubogu et al., 2006), and home at the site of inflammation if specific for the local antigen environment. Importantly, the factors listed here exert a variety of interactions, such as additional chemoattractive effects or further upregulating one another (Archelos et al., 1999; Parks et al., 2004). Interestingly, even immune system components physiologically associated with anti-microbial combat, e.g. toll-like receptors or osteopontin, seem to contribute substantially to the development of EAE as well as MS (Chabas et al., 2001; Prinz et al., 2006).

Both CD4+ and CD8+ T cells are present in MS lesions, with CD4+ T cells being found predominantly in the perivascular cuff and CD8+ T cells (higher in number) being more prevalent in the center and border zone of the lesion (Gay et al., 1997). Several studies indicated that some of the lesion-infiltrating T cells, mostly CD8+ T cells, are clonotypically accumulated in an antigen-driven fashion originating from the same precursor cell (Babbe et al., 2000; Friese and Fugger, 2005; Jacobsen et al., 2002; Oksenberg et al., 1990). The low abundance of these clonotypes in the blood implied specific enrichment of the clonotypes in the diseased organ compartment (Jacobsen et al., 2002; Skulina et al., 2004). Persistence in the local compartment was verified by longitudinal studies and is in marked contrast to the self-limiting of other inflammatory diseases of the CNS.

## 7.3   Humoral Response in MS

The occurrence of intrathecal (i.e., local) antibody synthesis in cerebrospinal fluid (CSF) demonstrated by the presence of oligoclonal bands is still the only diagnostic laboratory marker in MS of high sensitivity but low specificity. Despite conferment of demyelination through anti-myelin immunoglobulin G (IgG) in passive transfer EAE models (Schluesener et al., 1987), B cells were neglected in MS research for decades owing to their dispensable role in EAE and the lack of suitable technology (Archelos et al., 2000; Meinl et al., 2006). Only recently have they regained scientific attention in regards to their effects on MS pathogenesis both through antibody production and antigen presentation.

B cells from MS lesions or CSF comprise dominant clonotypes containing replacement mutations in their B cell receptor genes, compatible with an antigen-driven selection process (Baranzini et al., 1999; Colombo et al., 2000; Qin et al., 1998). The same B cell clonotypes are found in the CSF during the course of the disease, implying that they are periodically recruited or that they persist in the CNS compartment (Colombo et al., 2003). This model is compatible with other CSF parameters, such as intrathecal immunoglobulin (predominantly IgG1) production in the CSF, IgG deposition in MS lesions, and B cell-derived cytokines such as BAFF, that can be detected in MS lesions (Kalled, 2005; Krumbholz et al., 2005; Lucchinetti et al., 2000; Raine et al., 1999). Plasma blasts and plasma cells—terminally differentiated B cells that are usually only found in acute infectious diseases—can be identified in the CSF compartment of MS patients (Cepok et al., 2005a; Corcione et al., 2004). Their number correlates not only with local IgG synthesis but with the extent of CNS inflammation (Cepok et al., 2005a). The CNS provides for a B cell fostering microenvironment, that attracts B cells, and in particular long-lived plasma cells through specific chemokines to settle in survival niches (Meinl et al., 2006; Uccelli et al., 2005; Radbruch et al., 2006). Immunoglobulins against linear MOG epitopes can be specifically detected in demyelinating lesions (Raine et al., 1999), and IgG directed against conformational MOG epitopes can be eluted from MS lesions (O'Connor et al., 2005). Anti-myelin antibodies have been proposed as a marker for early disease progression (Berger et al., 2003). Despite numerous efforts, serum anti-myelin antibodies could not be established as biomarkers for MS diagnosis or prognosis as of yet. This is mainly due to the low discriminatory power between MS and controls for the available assays (Lampasona et al., 2004) and the ongoing quest for the gold standard assay and antigenic preparation; it may also be limited by specific intrathecal antibody production with leakage into the peripheral blood in a fraction of patients.

## 7.4    Axonal Degeneration in MS

A paradigm shift has occurred based on pathological and MRI studies that collectively indicate the importance of axonal damage and loss as the most important determinant of permanent neurological disability (Bjartmar and Trapp, 2001). To date, axonal loss can be assessed noninvasively and correlated to functional loss (De Stefano et al., 1998; Trip et al., 2005). Several promising compounds in serum and CSF, e.g. neurofilament light chains, have been identified, but could not be established as biomarkers as of yet (Teunissen et al., 2005). Axonal degeneration occurs even in early stages of the disease (Bjartmar and Trapp, 2001). The pathogenetic pathways of axonal damage have been considered multifactorial, as the pathological picture appears quite heterogeneous (Deluca et al., 2006). There is controversy as to whether axonal loss is exclusively associated with inflammatory demyelination, or may occur as "collateral" damage secondary to myelin degeneration (Deluca et al., 2006; Kuhlmann et al., 2002; Zipp and Aktas, 2006). Loss of axons could also be demonstrated for normal-appearing white matter, cortical lesions with little inflammation (Peterson et al., 2001), and inactive demyelinated lesions (Deluca et al., 2006). Axonal loss in conjunction with neuroinflammation may be caused by aberrant inflammatory responses in the CNS, such as CD8+ T cells that directly target neurons, vigorous CD4+ T cell responses that recruit macrophages with the release of inflammatory mediators and toxic molecules, or binding of antibodies to neuronal surface antigens with complement activation or antibody-mediated phagocytosis of axons. Indirect mechanisms, such as loss of protective myelin, mitochondrial dysfunction, release of glutamate or nitric oxide, or antibodies against neuronal or para/juxta-nodal antigens may also contribute to axonal damage (Mead et al., 2002; Pitt et al., 2000; Zhang et al., 2005). These mechanisms may be neuron-specific or may occur non-specifically (Zipp and Aktas, 2006). These considerations have been corroborated by addressing the role of inflammation in animal models of primary neurodegeneration. The impact of the immune system in these models is quite heterogeneous and depends on the genetic background of the strain and the timing and quality of the immune response. Although a detrimental effect of the immune response has been well established in most models, immune cells are also capable of producing neurotrophic factors, such as brain-derived neurotrophic factor (BDNF), which might be important for neuroregeneration (Kerschensteiner et al., 2003). Whether an autoimmune response that generates neurotrophic factors has beneficial effects on neurodegeneration in vivo remains controversial (Jones et al., 2004; Moalem et al., 1999). Axonal damage without neuroinflammation may reflect a "dying back" neuropathy or Wallerian degeneration remote from the actual axonal transection at the site of neuroinflammation, or may be related to dysregulation of neuronal ion homeostasis and mitochondrial failure due to upregulation of voltage-gated sodium channels and excitotoxic calcium influx, as demonstrated for MS lesions (Dutta et al., 2006; Stys, 2005; Waxman, 2005; Waxman, 2006). However, the exact molecular events that underlie the axonal damage in MS and that may even facilitate a primary axonopathy remain elusive.

## 8   Immunopathogenetic Concept

Given all the current data available, the following scenario is currently accepted by most investigators as a concept for the immunopathogenesis of MS: In a genetically susceptible individual (i.e., presence of a susceptibility HLA haplotype in conjunction with additional, as yet unidentified alleles), certain environmental factors, such as repeated (re)infections with neurotropic pathogens in specific patterns, result in prolonged, sustained activation of the immune system with insufficient suppression of autoreactive T cells by dysfunctional Tregs. Upon encountering myelin antigens—conceivably in the context of physiological surveillance of the CNS or BBB leakage during an infection—molecular mimicry results in the first autoaggressive attack, causing shedding of more myelin antigens. This results in epitope spreading of the immune response, leading to perpetuation of this vicious cycle. These waves of neuroinflammation cause subclinical damage (i.e., demyelination and axonal degeneration) that cumulatively lead to relapses (demyelination) and sustained disability (axonal damage) (Hemmer et al., 2003; Hemmer et al., 2006).

B cells and CD8+ T cells are thought to play an important role in MS but are not relevant in most EAE models. This suggests that immunological pathways and target antigens differ between human disease and the EAE animal model. The involvement of B cells and CD8+ T cells strongly supports the primary inflammatory nature of this disease in most MS patients.

At present, it remains unclear whether the invasion of the CNS by T and B cells is the initiating event of MS, or it is secondary to the activation of the microglia/macrophage system and the local release of self or foreign antigens. Most probably, however, the highly focused and persisting acquired immune response in MS is driven by a small number of antigens (the identity of which is not yet known) that are presented in the CNS. The involvement of B cells with a dominant IgG1 antibody response suggests that these targets are proteins that are either released or displayed on the surface of CNS cells. Among the possible candidates are myelin, neuronal antigens, and antigens of cross-reactive infectious epitopes.

Although the nature of the target antigens of the immune response is still unknown, most studies suggest a detrimental effect of CNS inflammation in MS. This is supported not only by many findings from experimental models but also by the association between inflammation, demyelination, and axonal damage in MS lesions and the occurrence of irreversible axonal damage early in the disease course. It should be kept in mind that such a detrimental effect is not proof of the autoimmune nature of the disease, yet it may also be observed in chronic CNS infection with microbes that have an inherently low pathogenic potential but elicit a detrimental immune response.

## References

Alter A., Duddy M., Hebert S. et al. (2003) Determinants of human B cell migration across brain endothelial cells. *J. Immunol.* 170:4497-4505.

Archelos J. J., Previtali S. C., Hartung H. P. (1999) The role of integrins in immune-mediated diseases of the nervous system. *Trends Neurosci.* 22:30-38.

Archelos J. J., Storch M. K., Hartung H. P. (2000) The role of B cells and autoantibodies in multiple sclerosis. *Ann. Neurol.* 47:694-706.

Babbe H., Roers A., Waisman A. et al. (2000) Clonal expansions of CD8(+) T cells dominate the T cell infiltrate in active multiple sclerosis lesions as shown by micromanipulation and single cell polymerase chain reaction. *J. Exp. Med.* 192:393-404.

Baranzini S. E., Jeong M. C., Butunoi C. et al. (1999) B cell repertoire diversity and clonal expansion in multiple sclerosis brain lesions. *J. Immunol.* 163:5133-5144.

Barnett M. H., Prineas J. W. (2004) Relapsing and remitting multiple sclerosis: pathology of the newly forming lesion. *Ann. Neurol.* 55:458-468.

Berger T., Rubner P., Schautzer F. et al. (2003) Antimyelin antibodies as a predictor of clinically definite multiple sclerosis after a first demyelinating event. *N. Engl. J. Med.* 349:139-145.

Bjartmar C., Trapp B. D. (2001) Axonal and neuronal degeneration in multiple sclerosis: mechanisms and functional consequences. *Curr. Opin. Neurol.* 14:271-278.

Buljevac D., van Doornum G. J., Flach H. Z. et al. (2005) Epstein-Barr virus and disease activity in multiple sclerosis. *J. Neurol. Neurosurg. Psychiatry* 76:1377-1381.

Cannella B., Raine C. S. (1995) The adhesion molecule and cytokine profile of multiple sclerosis lesions. *Ann. Neurol.* 37:424-435.

Casetta I., Granieri E. (2000) Clinical infections and multiple sclerosis: contribution from analytical epidemiology. *J. Neurovirol.* 6 Suppl 2:S147-S151.

Cepok S., Rosche B., Grummel V. et al. (2005a) Short-lived plasma blasts are the main B cell effector subset during the course of multiple sclerosis. *Brain* 128:1667-1676.

Cepok S., Zhou D., Srivastava R. et al. (2005b) Identification of Epstein-Barr virus proteins as putative targets of the immune response in multiple sclerosis. *J. Clin. Invest* 115:1352-1360.

Chabas D., Baranzini S. E., Mitchell D. et al. (2001) The influence of the proinflammatory cytokine, osteopontin, on autoimmune demyelinating disease. *Science* 294:1731-1735.

Christen U., von Herrath M. G. (2005) Infections and autoimmunity–good or bad? *J. Immunol.* 174:7481-7486.

Colombo M., Dono M., Gazzola P. et al. (2003) Maintenance of B lymphocyte-related clones in the cerebrospinal fluid of multiple sclerosis patients. *Eur. J. Immunol.* 33:3433-3438.

Colombo M., Dono M., Gazzola P. et al. (2000) Accumulation of clonally related B lymphocytes in the cerebrospinal fluid of multiple sclerosis patients. *J. Immunol.* 164:2782-2789.

Compston A., Coles A. (2002) Multiple sclerosis. *Lancet* 359:1221-1231.

Corcione A., Casazza S., Ferretti E. et al. (2004) Recapitulation of B cell differentiation in the central nervous system of patients with multiple sclerosis. *Proc. Natl. Acad. Sci. U. S. A* 101:11064-11069.

Cserr H. F., Knopf P. M. (1992) Cervical lymphatics, the blood-brain barrier and the immunoreactivity of the brain: a new view. *Immunol. Today* 13:507-512.

Dandekar A. A., Wu G. F., Pewe L. et al. (2001) Axonal damage is T cell mediated and occurs concomitantly with demyelination in mice infected with a neurotropic coronavirus. *J. Virol.* 75:6115-6120.

De Stefano N., Matthews P. M., Fu L. et al. (1998) Axonal damage correlates with disability in patients with relapsing-remitting multiple sclerosis. Results of a longitudinal magnetic resonance spectroscopy study. *Brain* 121:1469-1477.

Deluca G. C., Williams K., Evangelou N. et al. (2006) The contribution of demyelination to axonal loss in multiple sclerosis. *Brain* 129:1507-1516.

Dutta R., McDonough J., Yin X. et al. (2006) Mitochondrial dysfunction as a cause of axonal degeneration in multiple sclerosis patients. *Ann. Neurol.* 59:478-489.

Dyment D. A., Ebers G. C., Sadovnick A. D. (2004) Genetics of multiple sclerosis. *Lancet Neurol.* 3:104-110.

Ebers G. C., Sadovnick A. D., Risch N. J. (1995) A genetic basis for familial aggregation in multiple sclerosis. Canadian Collaborative Study Group. *Nature* 377:150-151.

Engelhardt B., Ransohoff R. M. (2005) The ins and outs of T-lymphocyte trafficking to the CNS: anatomical sites and molecular mechanisms. *Trends Immunol.* 26:485-495.

Espejo C., Penkowa M., Demestre M. et al. (2005) Time-course expression of CNS inflammatory, neurodegenerative tissue repair markers and metallothioneins during experimental autoimmune encephalomyelitis. *Neuroscience* 132:1135-1149.

Friese M. A., Fugger L. (2005) Autoreactive CD8+ T cells in multiple sclerosis: a new target for therapy? *Brain* 128:1747-1763.

Fujinami R. S., Oldstone M. B. (1985) Amino acid homology between the encephalitogenic site of myelin basic protein and virus: mechanism for autoimmunity. *Science* 230:1043-1045.

Fujinami R. S., von Herrath M. G., Christen U. et al. (2006) Molecular mimicry, bystander activation, or viral persistence: infections and autoimmune disease. *Clin Microbiol. Rev.* 19:80-94.

Gay F. W., Drye T. J., Dick G. W. et al. (1997) The application of multifactorial cluster analysis in the staging of plaques in early multiple sclerosis. Identification and characterization of the primary demyelinating lesion. *Brain* 120:1461-1483.

Gold R., Hartung H. P., Toyka K. V. (2000) Animal models for autoimmune demyelinating disorders of the nervous system. *Mol. Med Today* 6:88-91.

Gold R., Linington C., Lassmann H. (2006) Understanding pathogenesis and therapy of multiple sclerosis via animal models: 70 years of merits and culprits in experimental autoimmune encephalomyelitis research. *Brain* 129:1953-1971.

Goverman J., Woods A., Larson L. et al. (1993) Transgenic mice that express a myelin basic protein-specific T cell receptor develop spontaneous autoimmunity. *Cell* 72: 551-560.

Greter M., Heppner F. L., Lemos M. P. et al. (2005) Dendritic cells permit immune invasion of the CNS in an animal model of multiple sclerosis. *Nat. Med.* 11:328-334.

Haas J., Hug A., Viehover A. et al. (2005) Reduced suppressive effect of CD4+ CD25high regulatory T cells on the T cell immune response against myelin oligodendrocyte glycoprotein in patients with multiple sclerosis. *Eur. J. Immunol.* 35: 3343-3352.

Hemmer B., Fleckenstein B. T., Vergelli M. et al. (1997) Identification of high potency microbial and self ligands for a human autoreactive class II-restricted T cell clone. *J. Exp. Med.* 185:1651-1659.

Hemmer B., Kieseier B., Cepok S. et al. (2003) New immunopathologic insights into multiple sclerosis. *Curr. Neurol. Neurosci. Rep.* 3:246-255.

Hemmer B., Nessler S., Zhou D. et al. (2006) Immunopathogenesis and immunotherapy of multiple sclerosis. *Nat Clin Pract Neurol* 2:201-211.

Heppner F. L., Greter M., Marino D. et al. (2005) Experimental autoimmune encephalomyelitis repressed by microglial paralysis. *Nat. Med.* 11:146-152.

Hickey W. F. (2001) Basic principles of immunological surveillance of the normal central nervous system. *Glia* 36:118-124.

Huseby E. S., Liggitt D., Brabb T. et al. (2001) A pathogenic role for myelin-specific CD8(+) T cells in a model for multiple sclerosis. *J. Exp. Med.* 194:669-676.

Jacobsen M., Cepok S., Quak E. et al. (2002) Oligoclonal expansion of memory CD8+ T cells in cerebrospinal fluid from multiple sclerosis patients. *Brain* 125:538-550.

Jones T. B., Ankeny D. P., Guan Z. et al. (2004) Passive or active immunization with myelin basic protein impairs neurological function and exacerbates neuropathology after spinal cord injury in rats. *J. Neurosci.* 24:3752-3761.

Kalled S. L. (2005) The role of BAFF in immune function and implications for autoimmunity. *Immunol. Rev.* 204:43-54.

Kawakami N., Nagerl U. V., Odoardi F. et al. (2005) Live imaging of effector cell trafficking and autoantigen recognition within the unfolding autoimmune encephalomyelitis lesion. *J. Exp. Med.* 201:1805-1814.

Kenealy S. J., Herrel L. A., Bradford Y. et al. (2006) Examination of seven candidate regions for multiple sclerosis: strong evidence of linkage to chromosome 1q44. *Genes Immun.* 7:73-76.

Kerschensteiner M., Stadelmann C., Dechant G. et al. (2003) Neurotrophic cross-talk between the nervous and immune systems: implications for neurological diseases. *Ann. Neurol.* 53:292-304.

Kieseier B. C., Storch M. K., Archelos J. J. et al. (1999) Effector pathways in immune mediated central nervous system demyelination. *Curr. Opin. Neurol.* 12:323-336.

Krumbholz M., Theil D., Derfuss T. et al. (2005) BAFF is produced by astrocytes and up-regulated in multiple sclerosis lesions and primary central nervous system lymphoma. *J. Exp. Med.* 201:195-200.

Kuhlmann T., Lingfeld G., Bitsch A. et al. (2002) Acute axonal damage in multiple sclerosis is most extensive in early disease stages and decreases over time. *Brain* 125:2202-2212.

Lampasona V., Franciotta D., Furlan R. et al. (2004) Similar low frequency of anti-MOG IgG and IgM in MS patients and healthy subjects. *Neurology* 62:2092-2094.

Lang H. L., Jacobsen H., Ikemizu S. et al. (2002) A functional and structural basis for TCR cross-reactivity in multiple sclerosis. *Nat. Immunol.* 3:940-943.

Langrish C. L., Chen Y., Blumenschein W. M. et al. (2005) IL-23 drives a pathogenic T cell population that induces autoimmune inflammation. *J. Exp. Med.* 201:233-240.

Lehmann P. V., Sercarz E. E., Forsthuber T. et al. (1993) Determinant spreading and the dynamics of the autoimmune T-cell repertoire. *Immunol. Today* 14:203-208.

Lucchinetti C., Bruck W., Parisi J. et al. (2000) Heterogeneity of multiple sclerosis lesions: implications for the pathogenesis of demyelination. *Ann. Neurol.* 47:707-717.

Madsen L. S., Andersson E. C., Jansson L. et al. (1999) A humanized model for multiple sclerosis using HLA-DR2 and a human T-cell receptor. *Nat. Genet.* 23:343-347.

McCoy L., Tsunoda I., Fujinami R. S. (2006) Multiple sclerosis and virus induced immune responses: autoimmunity can be primed by molecular mimicry and augmented by bystander activation. *Autoimmunity* 39:9-19.

McDonald W. I., Compston A., Edan G. et al. (2001) Recommended diagnostic criteria for multiple sclerosis: guidelines from the International Panel on the diagnosis of multiple sclerosis. *Ann. Neurol.* 50:121-127.

Mead R. J., Singhrao S. K., Neal J. W. et al. (2002) The membrane attack complex of complement causes severe demyelination associated with acute axonal injury. *J. Immunol.* 168:458-465.

Meinl E., Krumbholz M., Hohlfeld R. (2006) B lineage cells in the inflammatory central nervous system environment: migration, maintenance, local antibody production, and therapeutic modulation. *Ann. Neurol.* 59:880-892.

Miller S. D., Olson J. K., Croxford J. L. (2001) Multiple pathways to induction of virus-induced autoimmune demyelination: lessons from Theiler's virus infection. *J. Autoimmun.* 16:219-227.

Moalem G., Leibowitz-Amit R., Yoles E. et al. (1999) Autoimmune T cells protect neurons from secondary degeneration after central nervous system axotomy. *Nat. Med.* 5:49-55.

Moore F. G., Wolfson C. (2002) Human herpes virus 6 and multiple sclerosis. *Acta Neurol. Scand.* 106:63-83.

Neumann H., Cavalie A., Jenne D. E. et al. (1995) Induction of MHC class I genes in neurons. *Science* 269:549-552.

O'Connor K. C., Appel H., Bregoli L. et al. (2005) Antibodies from inflamed central nervous system tissue recognize myelin oligodendrocyte glycoprotein. *J. Immunol.* 175:1974-1982.

Oksenberg J. R., Barcellos L. F., Cree B. A. et al. (2004) Mapping multiple sclerosis susceptibility to the HLA-DR locus in African Americans. *Am. J. Hum. Genet.* 74:160-167.

Oksenberg J. R., Barcellos L. F., Hauser S. L. (1999) Genetic aspects of multiple sclerosis. *Semin. Neurol.* 19:281-288.

Oksenberg J. R., Stuart S., Begovich A. B. et al. (1990) Limited heterogeneity of rearranged T-cell receptor V alpha transcripts in brains of multiple sclerosis patients. *Nature* 345:344-346.

Ota K., Matsui M., Milford E. L. et al. (1990) T-cell recognition of an immunodominant myelin basic protein epitope in multiple sclerosis. *Nature* 346:183-187.

Park H., Li Z., Yang X. O. et al. (2005) A distinct lineage of CD4 T cells regulates tissue inflammation by producing interleukin 17. *Nat. Immunol.* 6:1133-1141.

Parks W. C., Wilson C. L., Lopez-Boado Y. S. (2004) Matrix metalloproteinases as modulators of inflammation and innate immunity. *Nat Rev. Immunol.* 4:617-629.

Peterson J. W., Bo L., Mork S. et al. (2001) Transected neurites, apoptotic neurons, and reduced inflammation in cortical multiple sclerosis lesions. *Ann. Neurol.* 50:389-400.

Pitt D., Werner P., Raine C. S. (2000) Glutamate excitotoxicity in a model of multiple sclerosis. *Nat. Med* 6:67-70.

Polman C. H., Reingold S. C., Edan G. et al. (2005) Diagnostic criteria for multiple sclerosis: 2005 revisions to the "McDonald Criteria". *Ann. Neurol.* 58:840-846.

Poser C. M., Paty D. W., Scheinberg L. et al. (1983) New diagnostic criteria for multiple sclerosis: guidelines for research protocols. *Ann. Neurol.* 13:227-231.

Prinz M., Garbe F., Schmidt H. et al. (2006) Innate immunity mediated by TLR9 modulates pathogenicity in an animal model of multiple sclerosis. *J. Clin. Invest* 116:456-464.

Qin Y., Duquette P., Zhang Y. et al. (1998) Clonal expansion and somatic hypermutation of V(H) genes of B cells from cerebrospinal fluid in multiple sclerosis. *J. Clin. Invest* 102:1045-1050.

Radbruch A., Muehlinghaus G., Luger E. O. et al. (2006) Competence and competition: the challenge of becoming a long-lived plasma cell. *Nat. Rev. Immunol.* 6:741-750.

Raine C. S., Cannella B., Hauser S. L. et al. (1999) Demyelination in primate autoimmune encephalomyelitis and acute multiple sclerosis lesions: a case for antigen-specific antibody mediation. *Ann. Neurol.* 46:144-160.

Ramakrishna C., Stohlman S. A., Atkinson R. D. et al. (2002) Mechanisms of central nervous system viral persistence: the critical role of antibody and B cells. *J. Immunol.* 168:1204-1211.

Richt J. A., Schmeel A., Frese K. et al. (1994) Borna disease virus-specific T cells protect against or cause immunopathological Borna disease. *J. Exp. Med.* 179:1467-1473.

Rivers T. M., Sprunt D. H., Berry G. P. (1933) Observations on the attempts to produce acute disseminated allergic encephalomyelitis in primates. *J. Exp. Med* 58:39-53.

Sawcer S., Ban M., Maranian M. et al. (2005) A high-density screen for linkage in multiple sclerosis. *Am. J. Hum. Genet.* 77:454-467.

Schluesener H. J., Sobel R. A., Linington C. et al. (1987) A monoclonal antibody against a myelin oligodendrocyte glycoprotein induces relapses and demyelination in central nervous system autoimmune disease. *J. Immunol.* 139:4016-4021.

Serafini B., Rosicarelli B., Magliozzi R. et al. (2006) Dendritic cells in multiple sclerosis lesions: maturation stage, myelin uptake, and interaction with proliferating T cells. *J. Neuropathol. Exp. Neurol.* 65:124-141.

Skulina C., Schmidt S., Dornmair K. et al. (2004) Multiple sclerosis: brain-infiltrating CD8+ T cells persist as clonal expansions in the cerebrospinal fluid and blood. *Proc. Natl. Acad. Sci. U. S. A* 101:2428-2433.

Sospedra M., Martin R. (2005) Immunology of multiple sclerosis. *Annu. Rev. Immunol.* 23:683-747.

Sriram S., Steiner I. (2005) Experimental allergic encephalomyelitis: a misleading model of multiple sclerosis. *Ann. Neurol.* 58:939-945.

Steinman L. (1999) Assessment of animal models for MS and demyelinating disease in the design of rational therapy. *Neuron* 24:511-514.

Stohlman S. A., Hinton D. R. (2001) Viral induced demyelination. *Brain Pathol.* 11:92-106.

Stys P. K. (2005) General mechanisms of axonal damage and its prevention. *J. Neurol. Sci.* 233:3-13.

Swanborg R. H., Whittum-Hudson J. A., Hudson A. P. (2002) Human herpesvirus 6 and Chlamydia pneumoniae as etiologic agents in multiple sclerosis -a critical review. *Microbes. Infect.* 4:1327-1333.

Teunissen C. E., Dijkstra C., Polman C. (2005) Biological markers in CSF and blood for axonal degeneration in multiple sclerosis. *Lancet Neurol.* 4:32-41.

Thacker E. L., Mirzaei F., Ascherio A. (2006) Infectious mononucleosis and risk for multiple sclerosis: a meta-analysis. *Ann. Neurol.* 59:499-503.

Trapp B. D., Peterson J., Ransohoff R. M. et al. (1998) Axonal transection in the lesions of multiple sclerosis. *N. Engl. J. Med.* 338:278-285.

Trebst C., Ransohoff R. M. (2001) Investigating chemokines and chemokine receptors in patients with multiple sclerosis: opportunities and challenges. *Arch. Neurol.* 58:1975-1980.

Trip S. A., Schlottmann P. G., Jones S. J. et al. (2005) Retinal nerve fiber layer axonal loss and visual dysfunction in optic neuritis. *Ann. Neurol.* 58:383-391.

Tsunoda I., Kuang L. Q., Kobayashi-Warren M. et al. (2005) Central nervous system pathology caused by autoreactive CD8+ T-cell clones following virus infection. *J. Virol.* 79:14640-14646.

Ubogu E. E., Cossoy M. B., Ransohoff R. M. (2006) The expression and function of chemokines involved in CNS inflammation. *Trends Pharmacol. Sci.* 27:48-55.

Uccelli A., Aloisi F., Pistoia V. (2005) Unveiling the enigma of the CNS as a B-cell fostering environment. *Trends Immunol.* 26:254-259.

Vanderlugt C. L., Miller S. D. (2002) Epitope spreading in immune-mediated diseases: implications for immunotherapy. *Nat. Rev. Immunol.* 2:85-95.

Viglietta V., Baecher-Allan C., Weiner H. L. et al. (2004) Loss of functional suppression by CD4+CD25+ regulatory T cells in patients with multiple sclerosis. *J. Exp. Med.* 199:971-979.

von Herrath M. G., Fujinami R. S., Whitton J. L. (2003) Microorganisms and autoimmunity: making the barren field fertile? *Nat Rev. Microbiol.* 1:151-157.

Waxman S. G. (2005) Sodium channel blockers and axonal protection in neuroinflammatory disease. *Brain* 128:5-6.

Waxman S. G. (2006) Ions, energy and axonal injury: towards a molecular neurology of multiple sclerosis. *Trends Mol. Med* 12:192-195.

Weaver C. T., Harrington L. E., Mangan P. R. et al. (2006) Th17: an effector CD4 T cell lineage with regulatory T cell ties. *Immunity.* 24:677-688.

Wucherpfennig K. W., Strominger J. L. (1995) Molecular mimicry in T cell-mediated autoimmunity: viral peptides activate human T cell clones specific for myelin basic protein. *Cell* 80:695-705.

Zhang J., Markovic-Plese S., Lacet B. et al. (1994) Increased frequency of interleukin 2-responsive T cells specific for myelin basic protein and proteolipid protein in peripheral blood and cerebrospinal fluid of patients with multiple sclerosis. *J. Exp. Med.* 179:973-984.

Zhang Y., Da R. R., Guo W. et al. (2005) Axon reactive B cells clonally expanded in the cerebrospinal fluid of patients with multiple sclerosis. *J. Clin. Immunol.* 25: 254-264.

Zipp F., Aktas O. (2006) The brain as a target of inflammation: common pathways link inflammatory and neurodegenerative diseases. *Trends Neurosci.* 29:518-527.

# 11
# Viral Infection and Multiple Sclerosis

Elizabeth L. Williams and Steven Jacobson

## 1   Introduction

Studies during the 1960s were the first to discover that persistent viral infections can cause chronic neurological disease (Gilden, 2005). Since that time, many viruses have been positively associated with numerous neurological and chronic diseases. The characterization of virus in affected tissues, increased risk of disease in immunosuppressed patients, and response of patients to antiviral therapy all point investigators toward an infectious etiology of the disease. Multiple viruses have been shown to produce demyelination, and a temporal association of postinfectious encephalomyelitis has been shown after smallpox vaccination and measles, varicella, or rubella infection. Viruses are capable of establishing persistent, latent infection in host organisms leading to continuous viral replication over time without killing the host. Reactivation of latent virus can produce clinical disease, although the stimulus of reactivation has yet to be characterized.

The association of viruses with numerous neurological disorders in animals and humans has led to the theory that multiple sclerosis (MS) may also be the result of infectious or virus-triggered neuropathology (Gilden, 2005). One such example is the correlation of JC virus with progressive multifocal leukoencephalopathy (PML), the only human demyelinating disease with a proven viral cause. Another example is the temporal association of viruses with postinfectious encephalomyelitis, a demyelinating disorder that occurs as a complication of either smallpox vaccination or measles infection. Theiler's murine encephalomyelitis virus (TMEV) leading to central nervous system (CNS) demyelination is another example that supports a role for virus in MS. Although a viral association with MS may be the result of primary infection, more likely it is the reactivation of virus years after the primary infection that leads to the development of MS, similar to JC virus in PML and TMEV, both of which result from reactivation of latent virus in the CNS that may lyse oligodendrocytes, initiating an immunopathological process that leads to demyelination (Gilden, 2005).

Although the etiology of MS is still unknown, much research has been done to gain a better understanding of the factors involved in the development of disease. Epidemiological studies have supported both a genetic component and an environmental trigger in MS. These environmental triggers have been suggested to be transmissible agents such as a viral or bacterial infection. Although the immunology of MS is complex, further support for infectious triggers is indicated by the role of Th1-type CD4+ cells in disease pathogenesis. Th1 cells dominate viral infections and allow for the proliferation of CD8+ cells. In addition, studies have demonstrated increased CD8+ cytotoxic T cells in MS lesions, which suggests a role for these effector T cells in the pathogenesis of this disorder. The observation of increased immunoglobulin G (IgG) responses in MS further supports a role for infection in disease, as most diseases that show high concentrations of IgG are inflammatory and infectious in nature. The involvement of viral infections in MS coupled with the long-held view of MS as an autoimmune disease suggests that molecular mimicry may play a role in MS disease pathogenesis, as studies have found that myelin basic protein (MBP)-specific T cells can be effectively activated by MBP homologue from numerous viruses. This review highlights a number of these diverse observations that lend support to a viral etiology in MS.

## 2   Multiple Sclerosis

Multiple sclerosis is a demyelinating disease affecting the CNS, with its onset occurring between 20 and 40 years of age. The pathological marker of disease is white matter lesions or plaques resulting from loss of axonal myelination due to inflammation, with noticeable lymphocyte infiltration and oligodendrocyte loss (Ffrench-Constant, 1994). Lesions are characterized by demyelination, edema, and disruption of the blood-brain barrier (Hemmer et al., 2002). Symptoms of disease begin with recurring inflammatory attacks involving significant neurological impairment, ranging from vision problems and difficulty walking to paralysis (Steinman and Zamvil, 2003). Although the cause of MS is unknown, it has typically been considered a CD4+ Th1 cell-mediated disease. Disease begins when the blood-brain barrier is breached, and activated immune cells attack components of the myelin sheath insulating axons. Epidemiological studies have suggested both a genetic and environmental component in MS, leaving open the possibility of a viral trigger in MS.

## 2.1  Epidemiology and Etiology

The prevalence of MS in specific ethnic groups residing in the same environment supports a role for genetic susceptibility (Haines et al., 1998). The familial aggregation of disease, observed in population and family studies, shows an increased risk in first-, second-, and third-degree relatives over the general population (Dyment et al., 2004). Adoption studies show an increased risk of developing MS only in biologically related individuals (Ebers et al., 1995). In twin studies, monozygotic twins have a higher concordance rate than dizygotic twins (Haines et al., 1998). Although the monozygotic twin concordance rate of 20% to 30% indicates that genetics can predispose an individual to MS, nongenetic factors such as environmental triggers and immune responses must play a role in determining the overall etiology of disease (Dyment et al., 2004). The genetic component most often observed is the HLA-DR2 allele on chromosome 6p21, and this gene could contribute 10% to 60% of the genetic risk factor (Haines et al., 1998). Allelic variants of chromosomal regions are linked to increased disease risk, and these MHC genes could predispose disease through thymic selection or presentation of antigen to T cells, or both (Ermann and Fathman, 2001).

Multiple sclerosis affects approximately 1 million people worldwide, although it is much more prevalent in Caucasians, and 350,000 of people affected reside in North America (Steinman, 2001a). A geographical gradient of north to south has been found, with the Northern Hemisphere having an increased prevalence of disease. The geographical distribution is not related to genetics alone, as the prevalence of disease in Caucasians who migrate outside of Europe or North America is one-half that of those living in many parts of the Northern Hemisphere. Migration studies found that the geographical risk is acquired by the age of 15, as migration from an area of high risk to one of low risk during adolescence confers a reduced risk of developing disease. One explanation for the geographical distribution observed is the reduced sunlight exposure at higher latitudes, as ultraviolet (UV) radiation may exert an effect on vitamin D or cause an excess of melatonin, which enhances Th1 responses (Sospedra and Martin, 2005).

Twice as many women as men develop MS, a phenomenon observed with other autoimmune diseases, including systemic lupus erythematosus and rheumatoid arthritis (Steinman, 2001a). This finding suggests a role for hormones as risk factors in MS and is supported by the finding that relapse rates are decreased during pregnancy and increase afterward. Additionally, disease often worsens during menstruation; high estradiol and low progesterone have been correlated with increased magnetic resonance imaging (MRI) disease activity; and estriol has shown a therapeutic effect in RR-MS (Sospedra and Martin, 2005). The mechanism of hormonal action is unknown, although the stimulatory effect of estrogen on proinflammatory cytokine secretion is a likely mechanism.

## 2.2  Environmental Factors

The mechanisms of action for the observed geographical distribution could be numerous, with the "hygiene hypothesis" frequently mentioned as a possible mechanism. The "hygiene hypothesis" states that less hygienic environments, which predispose children to infections, helps protect against later disease by driving the immune system toward Th1 responsiveness. In contrast, children in a more hygienic environment with low exposure to infectious disease, as is the case in developed countries, tend toward Th2 responsiveness, and these are the children who are thought to be more prone to developing atopic allergic disease. Although this hypothesis is difficult to substantiate, it elucidates a possible mechanism as to the increased prevalence of disease in the Northern Hemisphere.

Interestingly, epidemics of MS have been observed, suggesting that a transmissible agent may be associated with disease pathogenesis. The most well known epidemic is that of the Farøe Islands during World War II, where MS was unknown until 1940 when British soldiers arrived and an epidemic broke out shortly thereafter (Kurtzke, 2000). Prospective studies have also shown that MS relapses often follow viral infections, and seasonal variation in the incidence of new MS cases exists. Experimental autoimmune encephalomyelitis (EAE) studies have also suggested a transmissible agent, as almost 100% of transgenic mice expressing T cell receptors (TCRs) specific for an encephalitogenic peptide of MBP develop EAE when housed in non-pathogen-free conditions, whereas those in pathogen-free conditions do not (Goverman et al., 1993). Additionally, common viral infections are temporally associated with MS exacerbations in some patients (Sibley et al., 1985). Many groups have searched for bacteria and viruses that may be associated with disease, and to date about 20 organisms have been associated, although none has gained acceptance as the causal agent of MS (Table 11.1).

Although the etiology of MS is unknown, epidemiological studies point to a role for both genetics and environmental triggers in the development of disease. Figure 11.1 depicts a visualization of how these overlapping criteria may combine to lead to autoimmune disease.

## 2.3  Immunology of Disease

### 2.3.1  T Helper Lymphocytes

Like many other chronic inflammatory diseases, MS has traditionally been thought to be a CD4+ Th1-mediated disease. Naive T cells differentiate into Th1 cells in the presence of interferon-$\gamma$ (IFN$\gamma$) and interleukin-12 (IL-12) and into

TABLE 11.1. Viruses Implicated in Multiple Sclerosis.

| Family | Virus | Year |
|---|---|---|
| Bornaviridae | Bornea disease virus (BDV) | 1998 |
| Coronaviridae | Human coronavirus (HCV) 229E | 1976 |
| Herpesviridae | Herpes simplex virus 1 and 2 (HSV-1, HSV-2) | 1981 |
| | Epstein-Barr virus (EBV) | 1983 |
| | Cytomegalovirus (CMV) | 1979 |
| | Varicella-zoster virus (VZV) | 1975 |
| | Human herpesvirus-6 (HHV-6) | 1994 |
| Papovaviridae | JC virus | 1998 |
| Paramyxoviridae | Measles virus | 1972 |
| | Mumps virus | 1972 |
| | Parainfluenza virus type 1 | 1972 |
| | Simian virus 5 | 1978 |
| | Canine distemper virus (CDV) | 1979 |
| Poxviridae | Vaccinia virus | 1975 |
| Retroviridae | Human T cell leukemia virus (HTLV-1) | 1986 |
| | Human endogenous retrovirus (HERVs) | 1989 |
| Togaviridae | Rubella virus | 1976 |

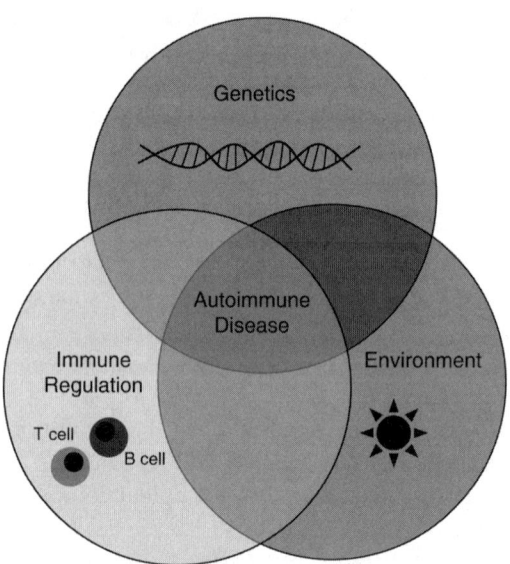

FIGURE 11.1. Development of autoimmune disease. (Adapted from Ermann and Fathman, 2001.)

Th2 cells in the presence of IL-4 and/or IL-6. The presence of Th1- or Th2-type cytokines also function to inhibit the generation of the other response. Th1 cells dominate viral and bacterial infections owing to IL-12 (produced by dendritic cells and macrophages) and IFNγ, produced by natural killer (NK) cells and CD8+ T cells. The two subsets of CD4+ cells have very different functions: Th1 cells activate macrophages and allow proliferation of CD8+ cells and upregulation of major histocompatibility complex class I (MHC I) molecules, and Th2 cells are the most effective activators of B cells, binding to antigen-specific B cells, which leads to the secretion of B cell stimulatory cytokines and drives the proliferation and differentiation of B cells into antibody-secreting plasma cells.

### 2.3.2   Cytotoxic T Lymphocytes

CD8+ cytotoxic T lymphocytes (CTLs) have come to the forefront of MS research after it was found that CD8+ CTL for MBP can induce severe EAE via adoptive transfer (Huseby et al., 2001; Steinman, 2001b). Previous to this, the focus of EAE research was on the role of myelin-specific CD4+ T cells because of their ability to induce EAE predominantly and lead to cytokine activation similar to that seen in MS (Steinman, 2001b). Additionally, lesions in EAE induced by CD8+ CTLs for MBP closely resemble MS lesions and exhibit extensive demyelination, indicating that CD8+ CTLs are capable of damaging MHC I-expressing oligodendrocytes (Huseby et al., 2001).

CD8+ CTLs are the primary effector T cells that lyse virus-infected cells by recognizing short peptide fragments (usually nine amino acids in length) in association with HLA class I molecules (Neumann et al., 2002; Shresta et al., 1998; Trapani et al., 2000). They are associated with antigen-specific apoptosis in the CNS, as their specificity for virus-infected cells prevents widespread tissue damage (Barry and Bleackley, 2002; Russell and Ley, 2002). CTLs kill target cells by a number of mechanisms, including the $Ca^{2+}$-dependent release of lytic granules and the $Ca^{2+}$-independent binding of Fas to Fas-L on the CTLs leading to the activation of caspases (Barry et al., 2000; Green et al., 2003; Medana et al., 2000).

CD8+ CTLs accumulate around lesions in infectious and autoimmune neurological disorders including MS (Neumann et al., 2002). It has come to be appreciated that MHC I-restricted CD8+ CTLs have been found to outnumber CD4+ T cells in MS lesions almost 10:1 (Booss et al., 1983). CD8+ CTLs have been shown to be enriched at the site of actively demyelinating lesions in MS (Monteiro et al., 1995) and the clonal proliferation of CD8+ CTL cells has been observed to be much greater than that of CD4+ cells in the T cell infiltrate in inflammatory lesions (Babbe et al., 2000). CD8+ CTL responses to MHC I peptides of MBP are elevated in MS patients compared to healthy individuals, suggesting that CTLs specific for MBP may become activated in MS patients (Crawford et al., 2004). Neuronal cells do not constitutively express the MHC I molecule necessary for CD8+ activation (Neumann et al., 2002), but invading CD8+ CTLs release proinflammatory cytokines including IFN-γ, tumor necrosis factor-α (TNFα), and TNFβ (Medana et al., 2000; Woodland and Dutton, 2003). IFNγ inhibits viral replication, activates macrophages, and induces the express of MHC I molecules on glial cells and neurons (Vass and Lassmann, 1990).

Recognition of the involvement of CTLs in the pathogenesis of MS is suggestive of a role for virus in lesions, as CD8+ CTLs are implicated in the direct killing of resident cells during infection. The preponderance of CTLs in inflammatory lesions supports a role for CTLs as important effectors in MS, and further study of these cells is crucial to our understanding of their involvement in inflammation and MS.

### 2.3.3    Antibody in Brain and CSF

The response of IgG to autoantigens on cell surfaces leads to the rapid destruction of these cells and inflammatory injury. About 90% of MS patients have high concentrations of IgG in brain and cerebrospinal fluid (CSF), further suggesting an infectious etiology for MS. Other diseases that show high concentrations of IgG are inflammatory and most are infectious (Gilden et al., 1996). Attempts to bind IgG extracted from MS brain and CSF to that of healthy controls and MS patients has been relatively unsuccessful, and points to a possible flaw in the hypothesis of autoantigen involvement in MS (Gilden, 2005). Research on IgG oligoclonal bands from other studies have suggested that oligoclonal IgG in MS is antibody-directed against the infectious cause of MS (Swanborg et al., 2003).

## 2.4    Mechanism of Immunopathology

Viral infections could lead to autoimmune diseases, including MS, via three possible mechanisms: molecular mimicry, bystander activation, and viral persistence. Each of these mechanisms involves the interaction of the virus with the immune system, "priming" the host for subsequent immunopathology by initiating the immunoreactivity that leads to autoimmune disease (Fujinami et al., 2006). In molecular mimicry, T cells react in the periphery to foreign molecules with enough similarity to self-antigen, causing the T cells to become activated and allowing them to cross the blood-brain barrier (Wucherpfennig and Strominger, 1995). Resting autoreactive T cells cannot cross the blood-brain barrier and therefore must be activated in the periphery. Many microbial proteins share sequence homologies with components of the myelin sheath, most notably myelin oligodendroglial glycoprotein (MOG), MBP, and proteolipid protein (PLP); and this homology allows activated T cells to attack myelin (Steinman, 2001a).

Viruses have been shown to cross-react with host proteins, activating MOG- or MBP-specific T cells (Fujinami et al, 2006; Guggenmos et al., 2004). MBP-specific T cells can be effectively activated by MBP homologues from various viruses, including herpes simplex, human papillomavirus, adenovirus type 12, and Epstein-Barr virus (Wucherpfennig and Strominger, 1995). Cross-reaction between viruses and hosts, although common, must occur in a disease-related epitope for autoimmune disease to occur; and this epitope must include peptides that can be presented by MHC II molecules on antigen-presenting cells (APCs) to CD4+ cells (Fujinami et al., 2006). Viruses such as those of the herpesvirus family are good candidates for molecular mimics, as they can establish latent and persistent infection; and chronic infection may permit constant antigenic stimulation of autoreactive T cell clones (Wucherpfennig and Strominger, 1995). Persistent viral infections could damage the immune system as a result of the continual presence of viral antigen (Fujinami et al., 2006).

# 3  Making a Viral Association

Although many pieces of evidence point toward a role for viral infection in the development of MS, making a positive association between a virus and neurological disorders such as MS is challenging, especially when the infection may be ubiquitous in the population. A number of chronic human illnesses have been suggested to be "triggered" by microorganisms, and proving the infectious etiology of these diseases has required extensive support from multiple sources (Carbone, 2005). The traditional rules for proof of pathogenicity, Koch's postulates, are often not practical when examining viral triggers, as it is rare that any single line of evidence is sufficient to prove an infectious nature of chronic disease. Proof is often realized only with the accumulation of sufficient and supportive data by multiple sources and techniques. Although it is extremely important to detect the suspected pathogen in diseased tissue, detection does not necessarily prove causality, as the organism could be an innocent bystander or endogenous flora—present but not a factor in the disease. Additionally, a lack of detection does not necessarily mean an association does not occur, as the method used may not be sensitive enough to detect low levels of virus. Any association must include evidence from multiple sources, including genetic, epidemiological, microbiological, and pathological components (Carbone et al., 2005).

Animal models can be helpful in generating hypotheses and can be useful to support viral associations with disease, but proving a link between an organism and a human chronic illness in an animal model again does not prove causality. Therefore animal models cannot always be used to draw conclusions regarding viral association. Obviously, viral associations in human disease would be more compelling, using methods such as the polymerase chain reaction (PCR) to detect viral nucleic acid in human tissue or by measuring antibody levels in serum, plasma, or CSF. Often, prospective and retrospective serological studies are used to help elucidate the cause of chronic illness; but here too lie problems, as evidence of past exposure does not imply current infection, seronegative individuals could have past infection, antibody levels could decline over time, and often tests are not sensitive enough to detect low levels of virus present in chronic infection.

A viral trigger of disease is any virus that sets into motion or expedites the disease process and can be involved in the development of disease in a number of ways, including persistence as a chronic infection and induction of destructive host immune responses in genetically susceptible individuals (Carbone et al., 2005).

# 4  Viruses Implicated in Multiple Sclerosis

The large body of evidence that points to a role for viral infection(s) in MS pathogenesis has led to numerous possible viral candidates for association with this disease. Viral triggers have been suggested to be involved in MS for more than 100 years (Marie, 1884), and since then numerous viruses have been suggested as causative agents in the pathogenesis of MS (Table 11.1). Some viruses that were initially thought to be associated with MS have, upon further analysis, shown less likelihood of association. This long-list of viruses can be interpreted in one of three ways: (1) viruses have nothing to do with MS pathogenesis, as none of these observations have stood the test of time; (2) the list is incomplete and the "MS virus" has yet to be discovered; (3) some but not all of these viruses are associated with disease in a subset of MS patients. This review argues for the latter interpretation by focusing on three ubiquitous candidate agents: human endogenous retroviruses and the herpesviruses Epstein-Barr virus (EBV) and human herpesvirus-6 (HHV-6), whose association with MS is currently being extensively explored.

## 4.1   Human Endogenous Retroviruses

Retroviruses were named for the presence of the viral enzyme reverse transcriptase (RT), an RNA-dependent DNA polymerase; the assessment of RT activity is the main criterion for distinguishing retroviruses. Most human endogenous retroviruses (HERVs) are endogenized exogenous retroviruses; they were incorporated into our genome during primate evolution and now make up 8% of the human genome. Additionally, retroviruses have been implicated in immunodeficiency and neurological disorders, and RNA for HERVs may be upregulated during inflammation (Christensen, 2005). As HERVs are ubiquitous, their pathogenicity must be modulated by genetic and environmental factors. The diversity of HERVs in the human genome is the result of homologous recombinations, as seen in the MHC region (Christensen, 2005).

Several human endogenous retroviruses have been implicated in MS based on the presence of activated HERVs in MS blood, increased RNA expression of HERV in brain tissue from MS patients, and elevated levels of HERV antibodies in the serum and CSF of MS patients (Brudek et al., 2004). Some studies have found that MS patients have increased concentrations of antibody to RT from HERVs compared to control patients, although other studies have had contradictory findings (Gilden, 2005). Retrovirus-like particles (RVLPs) have been found in T cell cultures from patients with MS, and cultures also express EB virus-encoded proteins (Haahr et al., 1991). Multiple sclerosis-related virus (MSRV, previously known as LM7), a HERV, was first isolated from MS patients, and its expression has been associated with disease severity and duration, although increased expression has also been observed in other inflammatory neurological diseases (Dolei et al., 2002; Zawada et al., 2003). It is unknown if HERVs are a causal factor in MSl; and if not causal they could be involved as a synergistic player, an immune activator or suppressor; alternatively, HERVs could be an epiphenomenon (Christensen, 2005). HERVs or HERV-activated proteins could act as insertional mutagens, as regulators of gene expression, or they could interact with exogenous viral RNA or elicit abnormalities as viruses.

HERVs could also be transactivated by human herpesviruses, as it has been found that herpesviruses and retroviruses are multiplicatively synergistic in stimulating cell-mediated immune responses (Brudek et al., 2004). The combination of HERV-H and either HHV-6 or HSV-1 increased cell proliferation in MS patient cell cultures over that of healthy controls, although the increase was not significant (Brudek et al., 2004). Regardless of the significance, the cell-mediated immune response of HERV and herpesvirus antigens in conjunction has a stimulatory effect (Christensen., 2005). The interaction of HERVs and herpesviruses is similar to situations found in vivo, as HERVs are ubiquitous in the genome and herpesviruses are highly prevalent throughout the human population; therefore an interaction between them is plausible. The herpesviruses could also function to induce aberrant expression of HERVs, as it has been found that the retroviruses human immunodeficiency virus (HIV) and human lymphtropic virus (HTLV) in the presence of herpesviruses encode transactivating factors that can exacerbate disease (Christensen, 2005). It is clear that further studies are needed to better elucidate the role of HERVs in the development of MS.

## 4.2   Epstein-Barr Virus

Epstein-Barr virus (EBV) is a herpesvirus believed to infect 95% of the world's population. EBV is trophic for B cells and can establish latency and persist throughout life. Primary infection can be asymptomatic or develop into infectious mononucleosis (IM), which is associated with lymphoproliferation in an immuno-compromised host.

Whether EBV has a role in MS has yet to be elucidated, but virtually all patients with MS have antibodies against EBV versus 86% to 95% of healthy individuals (Gilden, 2005). Individuals who had IM in the past have an increased risk of developing MS (Lindberg et al, 1991). Prospective samples from MS patients have been shown to be positive for EBV antigens EBNA-1 (Epstein-Barr nuclear antigen-1) and VCA (viral capsid antigen); high activity to EBNA-1 significantly increased the risk for MS, whereas VCA was not associated with MS development (Sundstrom et al, 2004). Antibodies to EBNA-1 have been found in 85% of MS patients compared to only 13% of EBV-seropositive controls (Bray et al, 1992). Increases in serum titers of EBV antibodies have been found before the onset of MS in prospective studies, with the strongest predictor of MS the serum levels of IgG to EBNA-1, supporting a role for EBV as a risk factor in MS (Ascherio et al., 2001; Levin et al., 2005). While oligoclonal IgG bands in brain and CSF of MS patients have been well documented, but there is no indication that these bands are directed against EBV (Gilden, 2005).

Although many studies have pointed toward a role of EBV in MS, EBV RNA and DNA have not been found in MS brain tissue (Challoner et al., 1995; Hilton et al., 1994; Morre et al., 2001; Sanders et al., 1996) or in CSF of MS patients (Martin et al., 1997; Morre et al., 2001), and studies examining EBV DNA in blood and serum have produced mixed results. EBV could lead to MS though molecular mimicry, as EBNA-1 has been shown to have homologies with MBP, and EBV peptides may activate HLA-DR2-restricted MBP-specific T cells (Sundstrom et al., 2004).

## 4.3   Human Herpesvirus-6

Human herpesvirus-6 was first isolated in 1986 from immunosuppressed patients with lymphoproliferative disorders and HIV infection (Salahuddin et al., 1986). HHV-6 is classified as a hematotropic or beta-herpesvirus, along with cytomegalovirus (CMV) and human herpesvirus-7 (HHV-7); HHV-6 bears 67% sequence homology with CMV (Lawrence et al., 1990). HHV-6 infects most individuals between 6 and 12 months of age (Okuno et al., 1989), and more than 90% of the general population is seropositive for HHV-6 (Kimberlin, 1998). Primary HHV-6 infection has been identified as the causative agent in exanthum subitum (roseola infantum), a febrile illness sometimes resulting in seizures and neurological complications including meningitis and encephalitis (Hall et al., 1994; Yamanishi et al., 1988; Yoshikawa and Asano, 2000).

Two variants of the virus have been identified, variant A and variant B, exhibiting a nucleotide sequence homology of between 88% and 96%. HHV-6A has been implicated in multiple sclerosis and is associated with viral persistence and reactivation in the CNS (Akhyani et al, 2000; Hall et al, 1998). The HHV-6B variant is primarily associated with symptomatic infections during infancy and limbic encephalitis and is the variant implicated in exanthem subitum. HHV-6B appears to be the predominant strain of the virus, as it is detected more frequently, although the 6A variant has been suggested to have a greater neurotropic potential than 6B (Dewhurst et al., 1993; McCullers et al., 1995; Wainwright et al., 2001).

### 4.3.1   Latent Infection

After primary infection, HHV-6 can establish a lifelong latent infection, with the viral genome persisting in peripheral blood mononuclear cells (PBMCs) and in the salivary glands (Campadelli-Fiume et al., 1999). HHV-6 DNA has been observed in the CSF of children not only during primary infection but also subsequent to infection (Caserta et al., 1994). Reactivation of latent virus is often seen in immunocompromised patients, as half of all patients who undergo stem cell or bone marrow transplant develop active HHV-6 infection (Caserta et al., 2001; Singh et al., 2000) and may contribute to disease in HIV infection and chronic fatigue syndrome. HHV-6 is suggested to be an immunosuppressive agent, although the mechanism of immunosuppression is unknown. Even in immunocompetent subjects HHV-6 has been shown to invade the CNS directly, creating a persistent, latent infection (Saito et al., 1995).

### 4.3.2   HHV-6 in Neurological Diseases

HHV-6 has been indicated as a cofactor in neurological diseases, most notably demyelinating diseases including multiple sclerosis (Challoner et al., 1995; Soldan et al., 1997). Progressive multifocal leukoencephalopathy (PML), another demyelinating disease of the CNS, is generally thought to be caused by reactivation of the human polyoma virus, JC virus, although it is now suggested that HHV-6 activation, in conjunction with the JC virus, is associated with demyelination in PML (Mock et al., 1999). In addition to demyelinating diseases, HHV-6 has also been implicated as a cause of epilepsy, encephalitis, encephalomyelitis, and meningitis. HHV-6 has been shown to infect numerous CNS cell types, including T cells, oligodendrocytes, and astrocytes, and has been shown to infect primary human fetal astrocytes in vitro (Albright et al., 1998; He et al., 1996; Lusso et al., 1988). HHV-6 DNA has been detected in CSF of patients with limbic encephalitis, and astrocytes positive for HHV-6B DNA were the most commonly infected cell type in the hippocampus (Wainwright et al., 2001).

# 5   Human Herpesvirus-6 and Multiple Sclerosis

The case for HHV-6 as one of many potential viral triggers in MS is supported by a large number of studies that are consistent with the following observations: First, the initial HHV-6 infection during infancy is compatible with the epidemiology of MS that suggests exposure to a microbial agent during childhood. Second, HHV-6 has been shown to infect numerous cell types in the CNS, including lymphocytes and glial cells (He et al., 1996). Third, herpesviruses are highly neurotropic and neuroinvasive, and they are often implicated in neurological diseases and other CNS complications (Wilborn et al., 1994). Fourth, the high seroprevalence of HHV-6 throughout the world is consistent with the high prevalence of MS. Lastly, herpesviruses are able to establish persistent, latent infections in the CNS and can be reactivated as a result of stress or other infections, the same stressors that have also been associated with MS exacerbations. Although an association between HHV-6 and MS remains to be definitively proven, the large numbers of reports that demonstrate a role of this virus in MS pathogenesis (Fig. 11.2) are compelling evidence that suggest that this may be an important area of investigation.

Supportive proof for an infectious etiology of disease can be realized only through the accumulation of data by multiple approaches. For example, the association of HHV-6 with MS has been made based on: (1) immunological detection: differential antibody and virus-specific lymphoproliferative responses between patients and controls; (2) molecular detection: demonstration of virus-specific DNA and RNA sequences in cell-free and cell-associated compartments; (3) pathological detection: expression of viral antigen and/or viral DNA in affected brain tissue from diseased patients; (4) clinical and radiological correlations: temporal association of virus specific responses with clinical and MRI parameters of MS disease. This review highlights the findings from these diverse approaches that have been used to correlate HHV-6 infection and MS as an example of how an association of a ubiquitous virus with MS is currently being explored.

## 5.1   Immunological Detection

Viral associations in MS have been suggested for over 50 years with numerous representations from many viral families including both DNA and RNA viruses

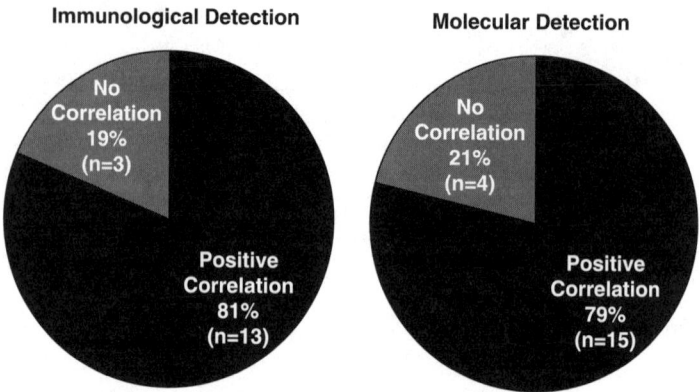

FIGURE 11.2. Studies correlating HHV-6 infection and multiple sclerosis (MS).
*Immunological detection—positive correlation* (Ablashi et al., 1998, 2000; Caselli et al., 2002; Derfuss et al., 2005; Friedman et al., 1999; Liedtke et al., 1995; Ongradi et al., 1999; Sola et al., 1993; Soldan et al., 1997, 2000; Tejada-Simon et al., 2002; Villoslada et al., 2003; Wilborn et al., 1994). *No correlation* (Enbom et al., 1999; Nielsen et al., 1997; Xu et al., 2002).
*Molecular detection—positive correlation* (Ablashi et al., 1998; Akhyani et al., 2000; Alvarez-Lafuente et al., 2002b, 2004b; Berti et al., 2002; Chapenko et al., 2003; Clark, 2004; Fillet et al., 1998; Goldberg et al., 1999; Liedtke et al., 1995; Rotola et al., 2004; Soldan et al., 1997; Tejada-Simon et al., 2002; Tomsone et al., 2001; Wilborn et al., 1994). *No correlation* (Al-Shammari et al., 2003; Martin et al., 1997; Mirandola et al., 1999; Taus et al., 2000).

(Berti et al., 2000; Cermelli and Jacobson, 2000; Johnson, 1994) largely based on increased antibody detection in patients versus controls. These studies typically have measured long-term IgG titers in either serum or CSF as a measure of past exposure. By contrast, a number of compelling studies (Ablashi et al., 1998, 2000; Friedman et al., 1999; Ongradi et al., 1999; Soldan et al., 1997; Villoslada et al., 2003) have demonstrated significant elevations of HHV-6-specific IgM in the serum of MS patients compared to patients with other neurological diseases, other inflammatory diseases, and healthy controls (Soldan et al., 1997). Detection of an early antibody response (IgM) to an early antigen of HHV-6 (p41/38) has suggested that active HHV-6 replication may be occurring in a subset of MS patients. Consistent with these HHV-6-specific IgM responses detected in MS sera, other studies have found an increased percentage of MS patients with anti-HHV-6 IgM in CSF. For example, Ongradi et al. demonstrated anti-HHV-6 IgM in 57% of MS patients and 0% of controls (Ongradi et al., 1999). In addition to IgM responses, increased IgG antibodies to HHV-6 has been one of the most consistent findings in most of the studies (Fig. 11.2) examining the role of HHV-6 in MS serum (Ablashi et al., 1998, 2000; Caselli et al., 2002; Liedtke et al., 1995; Ongradi et al., 1999; Sola et al., 1993, 1997; Wilborn et al., 1994) or CSF

(Ablashi et al., 1998; Derfuss et al., 2005; Ongradi et al., 1999). Although some reports have failed to support these findings (Enbom et al., 1999; Nielsen et al., 1997; Taus et al., 2000; Xu et al., 2002) the wide variety of assay platforms, use of different viral preparations, and variations in selection of patients and controls make it difficult to compare results among studies. There is a clear need to define a "gold standard" for HHV-6-specific IgG and IgM responses in both serum and CSF to better characterize the antibody response to HHV-6 in MS.

Immune responses to HHV-6 have also been investigated by examining virus-specific T cell proliferation. In a study from our own group, although there was no difference between MS patients and controls in response to HHV-6B or HHV-7, a significantly higher percentage of MS patients demonstrated proliferative responses to HHV-6A (Soldan et al., 2000). Increased frequency of HHV-6A specific T cells in MS is of interest because the HHV-6A variant has been suggested to be more neurotropic (Hall et al., 1998) and to have a greater propensity for latency and reactivation (Dewhurst et al., 1993), and HHV-6A sequences are more often detected in MS sera (Akhyani et al., 2000; Alvarez-Lafuente et al., 2002b) and CSF (Rotola et al., 2004) than variant B. More recently, T cells recognizing the recombinant 101-kDa protein of HHV-6 that corresponds to an immunodominant region of the virus occurred at a significantly lower precursor frequency in MS patients than controls (Tejada-Simon et al., 2002). These responses were associated with a skewed cytokine profile characterized by the inability to produce IL-4 and IL-10. The authors concluded that the diminished T cell response to HHV-6 and skewed Th2 cytokine profile was associated with ineffective clearance of HHV-6 in MS, suggesting a role for this virus in MS disease pathogenesis (Tejada-Simon, et al., 2002).

Although not universally accepted (Nielsen et al., 1997; Xu et al., 2002), most reports employing immunological detection methods consistently demonstrate HHV-6-specific responses in MS patients (Fig. 11.2). Collectively, these studies continue to support a role for HHV-6 as a reasonable candidate for an etiological agent in MS.

## 5.2   Molecular Detection

Detection of HHV-6 DNA and RNA sequences by primary and nested PCR has been used to demonstrate the presence of HHV-6 in MS patients compared to controls. The detection of HHV-6 DNA in cell-free compartments (i.e., serum, CSF, urine) has been suggested to reflect potentially active HHV-6 replication, whereas detection of cell-associated (e,g. PBMCs) HHV-6 sequences may not be able to distinguish latent from active virus. Moreover, using HHV-6 variant-specific primers and probes, it is possible to distinguish HHV-6A from HHV-6B infection. Recently, real-time quantitative PCR methods have been used to measure HHV-6 viral loads accurately from PBMCs and CSF lymphocytes; and RT-PCR has been used to amplify HHV-6 mRNA sequences. Demonstration of HHV-6-specific RNA in PBMCs is also suggestive of active HHV-6 replication (Alvarez-Lafuente et al., 2004b).

The results from these molecular analyses support the HHV-6 immunological observations found in MS. In both serum and CSF, HHV-6 DNA has been found in significantly more MS patients than controls (Ablashi et al., 1998; Akhyani et al., 2000; Alvarez-Lafuente et al., 2002b; Berti et al., 2002; Chapenko et al., 2003; Clark, 2004; Fillet et al., 1998; Goldberg et al., 1999; Liedtke et al., 1995; Rotola et al., 2004; Soldan et al., 1997; Tejada-Simon et al., 2002; Tomsone et al., 2001; Wilborn et al., 1994). We have demonstrated HHV-6 DNA in serum from approximately 25% of MS patients compared to 0% of controls including patients with other inflammatory diseases, other neurological diseases, and healthy subjects (Akhyani et al., 2000; Berti et al., 2002; Soldan et al., 1997). Although we find that most MS patients and controls have detectable HHV-6 DNA in the PBMCs, others have demonstrated HHV-6 DNA sequences in PBMCs more frequently in MS patients (Fig. 11.3) (Chapenko et al., 2003; Tomsone et al., 2001). Importantly, the increased frequency of detection in MS patients was HHV-6-specific, as no differences were observed between MS patients and controls to seven other human herpesviruses tested (Fig. 11.3).

More recently a number of studies have focused on HHV-6 RNA in PBMCs of MS patients. Alvarez-Lafuente and colleagues demonstrated HHV-6 mRNA for three immediate early (IE) genes by quantitative real-time RT-PCR in a substantial number of RRMS patients and not in healthy blood donors (Alvarez-Lafuente et al., 2002b). As a method of distinguishing active from latent infection, this study compared mRNA expression of the IE genes U89/90, U16/17, and U94 with the expression of U94 alone. Presence of U94 in the absence of other IE gene transcripts has been associated with latent HHV-6 infection (Mirandola et al., 1998). More MS patients were found to have mRNA for all three IE genes

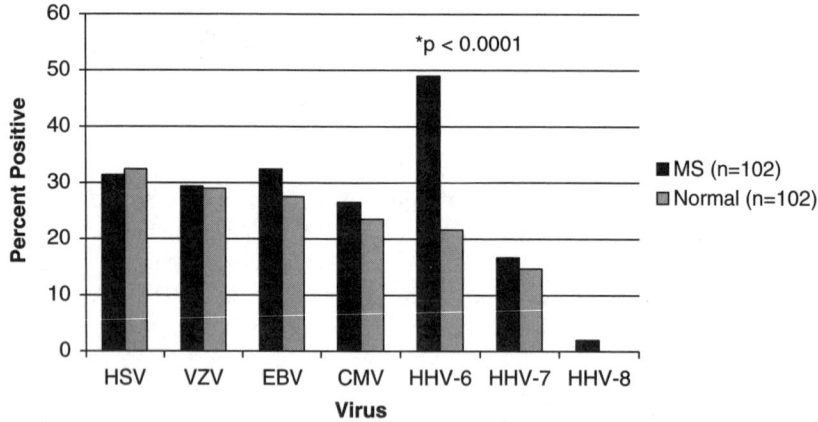

FIGURE 11.3. Prevalence of herpesvirus DNA in peripheral blood monocytic cells of MS patients. HSV, herpes simplex virus; VZV, varicella-zoster virus; EBV, Epstein-Barr virus; CMV, cytomegalovirus; HHV, human herpesvirus. (Adapted from Alvarez-Lafuente et al., 2002a.).

than U94 alone, indicating active HHV-6 infection in this subset of MS patients (Alvarez-Lafuente et al., 2002b). No significant difference in U94 expression was found between MS patients and controls.

Similar to the observations using immunological detection methods that suggest a role of HHV-6 in MS, most reports based on molecular methods support these findings (Fig. 11.2), although with greater variability among these studies (Al-Shammari et al., 2003; Martin et al., 1997; Mirandola et al., 1999; Taus et al., 2000). This is not unexpected as it is well appreciated in molecular PCR-based assays that there is considerable variability in the use of these methods. Different regions of the virus are amplified with different sets of primers and probes having varying degrees of sensitivity and specificity that are used in either primary or nested PCR conditions. Only through standardized PCR assays can these difficulties be overcome.

## 5.3   Pathological Detection

Detection of HHV-6 by immunological and molecular means are important observations in associating this virus with MS. However, the demonstration of this (or any infectious) agent in diseased MS brain material would be even more compelling. As access to MS brain tissue is limited, only a few studies demonstrating HHV-6 in MS brains have been reported. Indeed, the first report of an association of HHV-6 with MS was based on the immunohistochemical detection of HHV-6 antigen in oligodendrocytes from MS plaques (Challoner). This landmark study using an unbiased subtractive hybridization approach (representational differential analysis) demonstrated HHV-6-specific sequences in MS plaque material compared to controls. A more recent report also localized HHV-6 to oligodendrocytes in MS brains (Opsahl and Kennedy, 2005). Although HHV-6 mRNA was detected in both MS and control brain tissue, higher levels of HHV-6 viral activity as determined by percentage of HHV-6 mRNA-positive oligodendrocytes were demonstrated in MS patients compared to controls (Opsahl and Kennedy, 2005). In this study, quantitatively more HHV-6 mRNA to both immediate early and late genes was detected in MS lesions versus controls, suggestive of an active HHV-6 infection (Opsahl and Kennedy, 2005).

The demonstration of HHV-6 in MS brain is also supported by studies examining HHV-6 DNA in brain tissue by PCR. Cermelli et al., using laser microdissection, found statistically more HHV-6 DNA in active MS plaques than in normal-appearing white matter (NAWM) from the same MS patients and brain material from patients with other neurological diseases including inflammation (Cermelli et al., 2003). These findings are consistent with previous studies that demonstrated HHV-6 DNA by PCR more often in brain tissue from MS patients than in controls (Friedman et al., 1999; Sanders et al., 1996). In a case report of brain biopsies from five MS patients, all sections demonstrated high levels of HHV-6 DNA-positive cells by in situ PCR, most of which were oligodendrocytes (Goodman et al., 2003). Although HHV-6 DNA was detected in oligodendrocytes, HHV-6 antigen was not, using immunohistochemical analysis for HHV-6 p41, p101, or gp116. However, HHV-6 antigen was detected in hypertrophic astrocytes to the HHV-6 gp116 protein staining in two of the five patients (Goodman et al., 2003). The authors concluded that the prevalence of HHV-6 genome-containing cells in MS lesions support the hypothesis that HHV-6 plays a role in the demyelinative pathogenesis of MS. The demonstration of HHV-6 in pathological material is crucial to support the association of this agent in this disorder. Although studies using different antibodies have found HHV-6 antigen more often in MS patients than controls (Friedman et al., 1999; Knox et al., 2000; Virtanen et al., 2005), others have not (Blumberg et al., 2000; Coates and Bell, 1998). Clearly, more studies are needed to qualitatively and quantitatively detect both HHV-6 genome and HHV-6 antigen(s) in the CNS of MS patients.

## 5.4    Clinical Correlation

Immunological, molecular, and pathological detection of HHV-6 infection in MS have supported an associative role of HHV-6 in this disease. Even more compelling are the clinical correlative studies between virus and MS disease development or progression. As most MS patients are clinically defined by relapses and remissions, a number of reports have investigated whether HHV-6 can be differentially detected during these phases of disease. A significant correlation was demonstrated between serum HHV-6 DNA and the number of MS patients with clinical exacerbations, diagnosed by patient complaints and neurological examination (Berti et al., 2002). The detection of HHV-6 DNA more often in patient serum during exacerbation than remission suggests that active HHV-6 infection may play a role in the development and/or progression of MS (Berti et al., 2002). These findings were supported by others who found a higher viral load of HHV-6 in MS patients during exacerbations than during remissions for three different IE HHV-6 genes: U89/90, U16/17, and U94 (Alvarez-Lafuente et al., 2004b). Similarly, Chapenko et al. identified periods of HHV-6 viremia by detecting HHV-6 DNA in plasma only during periods of new MS activity (exacerbations), as indicated by the presence of gadolinium (Gd)-enhanced lesions, and not during periods of relative remission, as indicated by an absence of Gd-enhanced lesions (Chapenko et al., 2003). The study concluded that the risk of exacerbation was 2.5 times greater in patients with active HHV-6 infection than in those with latent infection. We have also observed that serum HHV-6 levels cycle over time by a longitudinal analysis of monthly serum samples (24-month time course) (Berti et al., 2002). As it is known that Gd-enhanced MRI lesions also cycle, more patients must be evaluated to determine if there is a correlation between HHV-6 and these lesions.

Of interest is the use of IFNβ, an established therapy for MS based on its ability to reduce the frequency and severity of exacerbations, disability, and brain lesions in patients with MS (IFNB, 1993). IFNβ has become one of the most commonly used treatments for RRMS, and the success of IFNβ in MS is thought to be not only a result of its antiinflammatory properties but may also be due in part to its antiviral activity (Alvarez-Lafuente et al., 2004a; Hong et al., 2002).

IFNβ has been demonstrated to inhibit significantly the viral replication of HHV-6, as treatment with IFNβ decreased the amount of HHV-6 DNA in the serum of MS patients as compared to untreated MS patients (Hong et al., 2002). In another study, IFNβ was not found to diminish HHV-6 DNA in the serum of MS patients during either relapse or remission (Alvarez-Lafuente et al., 2004a). Although DNA in serum was not reduced, they did find that the viral load in the serum of patients undergoing an acute attack was significantly lower in the IFNβ RRMS group than in the untreated RRMS group (Alvarez-Lafuente et al., 2004a).

Collectively, these clinically correlative studies, together with the immunological, molecular, and pathological detection of HHV-6 in MS, continue to support a role for the involvement of HHV-6 in this disorder.

# 6   Conclusion and Future Directions

Further research is needed to associate HHV-6 infection with the immunopathogenesis of MS definitively. Although more data are needed to make an association, it is clear from the breadth of research in the field that most studies find a positive correlation between infection and disease. However, more studies focusing on associations with HHV-6 and MS may get us no closer to proving a causal role for this agent. In all of the studies to date demonstrating the presence of HHV-6 in MS patients, it is difficult to conclude whether the virus is the "cause" of the disease or a mere bystander that results from other immunological events that nonspecifically reactivate the virus. Although some studies have controlled for this (Fig. 11.3), the argument of HHV-6 (or any infectious agent) as an epiphenomenon with little to do with the pathogenesis of MS is valid.

To address this crucial issue, a growing number of investigators believe that only through well controlled interventional clinical trials with effective and safe antiviral agents can a causal role be made for any infectious agent in MS. To date, only a handful of reports have attempted to intervene in MS with anti-beta herpesvirus drugs. Although no compound has been formally approved as an antiviral for the treatment of HHV-6 infection, antiviral agents used for CMV infection or other herpesvirus treatment, including ganciclovir, acyclovir, cidofovir, and foscarnet, are often used (De Bolle et al., 2005). Antiviral drugs used for the treatment of herpesvirus infections act by targeting virus-specific kinases and inhibiting viral DNA polymerases (Bech et al., 2002).

Several case studies of the successful use of ganciclovir for HHV-6 encephalitis in bone marrow transplant (BMT) patients have been published, and prophylactic therapy with ganciclovir has also been shown to be effective in preventing HHV-6 reactivation in BMT patients (Johnston et al., 1999; Mookerjee and Vogelsang, 1997; Rapaport et al., 2002; Rieux et al., 1998; Tokimasa et al., 2002; Wang et al., 1999; Yoshida et al., 2002). Case studies of foscarnet for the treatment of HHV-6 encephalitis have been mixed, with some showing successful results (Bethge et al., 1999; Zerr et al., 2002) and others yielding less success (Rossi et al., 2001; Tiacci et al., 2000). Clinical reports of cidofovir are more limited owing to the risk of nephrotoxicity; and one report found ganciclovir more successful than cidofovir in treating HHV-6 encephalomyelitis in an immunocompromised patient (Denes et al., 2004).

There have been limited studies in the use of antivirals in MS. One study found that valacyclovir, the valine ester of acyclovir with increased bioavailability, did not reduce the formation of active lesions over the 24-week course of treatment (Bech et al., 2002). However, a subgroup of MS patients with high disease activity, as measured by more than one active MRI lesion, valacyclovir was found to have reduced numbers of new active MRI lesions (Bech et al., 2002). In a more recent clinical trial in MS patients, valacyclovir, though the results were not statistically significant, was found to have a stabilizing effect on clinical progression of disease (Friedman et al., 2005).

These clinical trials serve to highlight the challenges in designing and interpreting an antiviral trial in MS. First, the choice of drug is critical, particularly with respect to HHV-6, as it is not clear what is the most effective anti-HHV-6 compound to use. Is the intent to interfere with HHV-6 replication in the periphery or CNS—and in which cell type (e.g., virus-infected lymphocytes or glial cells)? Second, which group of MS patients should one select? If HHV-6 plays a role in only a subset of patients, how is this group to be selected and what assay(s) should be used to monitor patients? Lastly, what clinical and/or radiological measures are to be used as a primary outcome measure of treatment efficacy? The MS trial design has made significant advances over the years with a number of drugs approved for the treatment and many more in the pipeline (McFarland and Reingold, 2005; Mouzaki et al., 2004). If antiviral drugs are to be part of the armamentarium for MS, these drugs must be shown to interfere with virus growth or replication in patients with detectable levels of virus. If these criteria are met, coupled with clinical and/or radiological improvement with antiviral therapy, there can be confidence for an etiological role of a virus in MS.

## References

Ablashi, D. V., Eastman, H. B., Owen, C. B., et al. (2000) Frequent HHV-6 reactivation in multiple sclerosis (MS) and chronic fatigue syndrome (CFS) patients. *J. Clin. Virol.* 16:179-191.

Ablashi, D. V., Lapps, W., Kaplan, M., et al. (1998) Human herpesvirus-6 (HHV-6) infection in multiple sclerosis: a preliminary report. *Mult. Scler.* 4:490-496.

Akhyani, N., Berti, R., Brennan, M. B., et al. (2000) Tissue distribution and variant characterization of human herpesvirus (HHV)-6: increased prevalence of HHV-6A in patients with multiple sclerosis. *J. Infect. Dis.* 182:1321-1325.

Al-Shammari, S., Nelson, R. F., Voevodin, A. (2003) HHV-6 DNAaemia in patients with multiple sclerosis in Kuwait. *Acta Neurol. Scand.* 107:122-124.

Albright, A. V., Lavi, E., Black, J. B., et al. (1998) The effect of human herpesvirus-6 (HHV-6) on cultured human neural cells: oligodendrocytes and microglia. *J. Neurovirol.* 4:486-494.

Alvarez-Lafuente, R., De Las Heras, V., Bartolome, M., et al. (2004a) Beta-interferon treatment reduces human herpesvirus-6 viral load in multiple sclerosis relapses but not in remission. *Eur. Neurol.* 52:87-91.

Alvarez-Lafuente, R., De las Heras, V., Bartolome, M., et al. (2004b) Relapsing-remitting multiple sclerosis and human herpesvirus 6 active infection. *Arch. Neurol.* 61: 1523-1527.

Alvarez-Lafuente, R., Martin-Estefania, C., de las Heras, V., et al. (2002a) Prevalence of herpesvirus DNA in MS patients and healthy blood donors. *Acta Neurol. Scand.* 105: 95-99.

Alvarez-Lafuente, R., Martin-Estefania, C., de Las Heras, V., et al. (2002b) Active human herpesvirus 6 infection in patients with multiple sclerosis. *Arch. Neurol.* 59:929-933.

Ascherio, A., Munger, K. L., Lennette, E. T., et al. (2001) Epstein-Barr virus antibodies and risk of multiple sclerosis: a prospective study. *J.A.M.A.* 286:3083-3088.

Babbe, H., Roers, A., Waisman, A., et al. (2000) Clonal expansions of CD8(+) T cells dominate the T cell infiltrate in active multiple sclerosis lesions as shown by micromanipulation and single cell polymerase chain reaction. *J. Exp. Med.* 192:393-404.

Barry, M., Bleackley, R. C. (2002) Cytotoxic T lymphocytes: all roads lead to death. *Nat. Rev. Immunol.* 2:401-409.

Barry, M., Heibein, J. A., Pinkoski, M. J., et al. (2000) Granzyme B short-circuits the need for caspase 8 activity during granule-mediated cytotoxic T-lymphocyte killing by directly cleaving Bid. *Mol. Cell. Biol.* 20:3781-3794.

Bech, E., Lycke, J., Gadeberg, P., et al. (2002) A randomized, double-blind, placebo-controlled MRI study of anti-herpes virus therapy in MS. *Neurology* 58:31-36.

Berti, R., Brennan, M. B., Soldan, S. S., et al. (2002) Increased detection of serum HHV-6 DNA sequences during multiple sclerosis (MS) exacerbations and correlation with parameters of MS disease progression. *J. Neurovirol.* 8:250-256.

Berti, R., Soldan, S. S., Akhyani, N., et al. (2000) Extended observations on the association of HHV-6 and multiple sclerosis. *J. Neurovirol.* 6(Suppl 2):S85-S87.

Bethge, W., Beck, R., Jahn, G., et al. (1999) Successful treatment of human herpesvirus-6 encephalitis after bone marrow transplantation. *Bone Marrow Transplant* 24:1245-1248.

Blumberg, B. M., Mock, D. J., Powers, J. M., et al. (2000) The HHV6 paradox: ubiquitous commensal or insidious pathogen? A two-step in situ PCR approach. *J. Clin. Virol.* 16:159-178.

Booss, J., Esiri, M. M., Tourtellotte, W. W., Mason, D. Y. (1983) Immunohistological analysis of T lymphocyte subsets in the central nervous system in chronic progressive multiple sclerosis. *J. Neurol. Sci.* 62:219-232.

Bray, P. F., Luka, J., Bray, P. F., et al. (1992) Antibodies against Epstein-Barr nuclear antigen (EBNA) in multiple sclerosis CSF, and two pentapeptide sequence identities between EBNA and myelin basic protein. *Neurology* 42:1798-1804.

Brudek, T., Christensen, T., Hansen, H. J., et al. (2004) Simultaneous presence of endogenous retrovirus and herpes virus antigens has profound effect on cell-mediated immune responses: implications for multiple sclerosis. *AIDS Res. Hum. Retroviruses* 20: 415-423.

Campadelli-Fiume, G., Mirandola, P., Menotti, L. (1999) Human herpesvirus 6: an emerging pathogen. *Emerg. Infect. Dis.* 5:353-366.

Carbone, K. M., Luftig, D. B., Buckley, M. R. (2005) Microbial triggers of chronic human illness. In: *American Academy of Microbiology Critical Issues Colloquia.*

Caselli, E., Boni, M., Bracci, A., et al. (2002) Detection of antibodies directed against human herpesvirus 6 U94/REP in sera of patients affected by multiple sclerosis. *J. Clin. Microbiol.* 40:4131-4137.

Caserta, M. T., Hall, C. B., Schnabel, K., et al. (1994) Neuroinvasion and persistence of human herpesvirus 6 in children. *J. Infect. Dis.* 170:1586-1589.

Caserta, M. T., Mock, D. J., Dewhurst, S. (2001) Human herpesvirus 6. *Clin. Infect. Dis.* 33:829-833.

Cermelli, C., Berti, R., Soldan, S. S., et al. (2003) High frequency of human herpesvirus 6 DNA in multiple sclerosis plaques isolated by laser microdissection. *J. Infect. Dis.* 187:1377-1387.

Cermelli, C., Jacobson, S. (2000) Viruses and multiple sclerosis. *Viral Immunol.* 13:255-267.

Challoner, P. B., Smith, K. T., Parker, J. D., et al. (1995) Plaque-associated expression of human herpesvirus 6 in multiple sclerosis. *Proc. Natl. Acad. Sci. U S A* 92:7440-7444.

Chapenko, S., Millers, A., Nora, Z., et al. (2003) Correlation between HHV-6 reactivation and multiple sclerosis disease activity. *J. Med. Virol.* 69:111-117.

Christensen, T. (2005) Association of human endogenous retroviruses with multiple sclerosis and possible interactions with herpes viruses. *Rev. Med. Virol.* 15:179-211.

Clark, D. (2004) Human herpesvirus type 6 and multiple sclerosis. *Herpes* 11(Suppl 2): 112A-119A.

Coates, A. R., Bell, J. (1998) HHV-6 and multiple sclerosis. *Nat. Med.* 4:537-538.

Crawford, M. P., Yan, S. X., Ortega, S. B., et al. (2004) High prevalence of autoreactive, neuroantigen-specific CD8+ T cells in multiple sclerosis revealed by novel flow cytometric assay. *Blood* 103:4222-4231.

De Bolle, L., Naesens, L., De Clercq, E. (2005) Update on human herpesvirus 6 biology, clinical features, and therapy. *Clin. Microbiol. Rev.* 18:217-245.

Denes, E., Magy, L., Pradeau, K., et al. (2004) Successful treatment of human herpesvirus 6 encephalomyelitis in immunocompetent patient. *Emerg. Infect. Dis.* 10:729-731.

Derfuss, T., Hohlfeld, R., Meinl, E. (2005) Intrathecal antibody (IgG) production against human herpesvirus type 6 occurs in about 20% of multiple sclerosis patients and might be linked to a polyspecific B-cell response. *J. Neurol.* 252:968-971.

Dewhurst, S., McIntyre, K., Schnabel, K., Hall, C. B. (1993) Human herpesvirus 6 (HHV-6) variant B accounts for the majority of symptomatic primary HHV-6 infections in a population of U.S. infants. *J. Clin. Microbiol.* 31:416-418.

Dolei, A., Serra, C., Mameli, G., et al. (2002) Multiple sclerosis-associated retrovirus (MSRV) in Sardinian MS patients. *Neurology* 58:471-473.

Dyment, D. A., Ebers, G. C., Sadovnick, A. D. (2004) Genetics of multiple sclerosis. *Lancet Neurol* 3:104-110.

Ebers, G. C., Sadovnick, A. D., Risch, N. J. (1995) A genetic basis for familial aggregation in multiple sclerosis: Canadian Collaborative Study Group. *Nature* 377:150-151.

Enbom, M., Wang, F. Z., Fredrikson, S., et al. (1999) Similar humoral and cellular immunological reactivities to human herpesvirus 6 in patients with multiple sclerosis and controls. *Clin. Diagn. Lab. Immunol.* 6:545-549.

Ermann, J., Fathman, C. G. (2001) Autoimmune diseases: genes, bugs and failed regulation. *Nat. Immunol.* 2:759-761.

Ffrench-Constant, C. (1994) Pathogenesis of multiple sclerosis. *Lancet* 343:271-275.

Fillet, A. M., Lozeron, P., Agut, H., et al. (1998) HHV-6 and multiple sclerosis. *Nat. Med.* 4:537, author reply 538.

Friedman, J. E., Lyons, M. J., Cu, G., et al. (1999) The association of the human herpesvirus-6 and MS. *Mult. Scler.* 5:355-362.

Friedman, J. E., Zabriskie, J. B., Plank, C., et al. (2005) A randomized clinical trial of valacyclovir in multiple sclerosis. *Mult. Scler.* 11:286-295.

Fujinami, R. S., von Herrath, M. G., Christen, U., Whitton, J. L. (2006) Molecular mimicry, bystander activation, or viral persistence: infections and autoimmune disease. *Clin. Microbiol. Rev.* 19:80-94.

Gilden, D. H. (2005) Infectious causes of multiple sclerosis. *Lancet Neurol* 4:195-202.

Gilden, D. H., Devlin, M. E., Burgoon, M. P., Owens, G. P. (1996).The search for virus in multiple sclerosis brain. *Mult. Scler.* 2:179-183.

Goldberg, S. H., Albright, A. V., Lisak, R. P., Gonzalez-Scarano, F. (1999) Polymerase chain reaction analysis of human herpesvirus-6 sequences in the sera and cerebrospinal fluid of patients with multiple sclerosis. *J. Neurovirol.* 5:134-139.

Goodman, A. D., Mock, D. J., Powers, J. M., et al. (2003) Human herpesvirus 6 genome and antigen in acute multiple sclerosis lesions. *J. Infect. Dis.* 187:1365-1376.

Goverman, J., Woods, A., Larson, L., et al. (1993) Transgenic mice that express a myelin basic protein-specific T cell receptor develop spontaneous autoimmunity. *Cell* 72: 551-560.

Green, D. R., Droin, N., Pinkoski, M. (2003) Activation-induced cell death in T cells. *Immunol. Rev.* 193:70-81.

Guggenmos, J., Schubart, A. S., Ogg, S., et al. (2004) Antibody cross-reactivity between myelin oligodendrocyte glycoprotein and the milk protein butyrophilin in multiple sclerosis. *J. Immunol.* 172:661-668.

Haahr, S., Sommerlund, M., Moller-Larsen, A., et al. (1991) Just another dubious virus in cells from a patient with multiple sclerosis? *Lancet* 337:863-864.

Haines, J. L., Terwedow, H. A., Burgess, K., et al. (1998) Linkage of the MHC to familial multiple sclerosis suggests genetic heterogeneity: the Multiple Sclerosis Genetics Group. *Hum Mol Genet* 7:1229-1234.

Hall, C. B., Caserta, M. T., Schnabel, K. C., et al. (1998) Persistence of human herpesvirus 6 according to site and variant: possible greater neurotropism of variant A. *Clin. Infect. Dis.* 26:132-137.

Hall, C. B., Long, C. E., Schnabel, K. C, et al. (1994) Human herpesvirus-6 infection in children: a prospective study of complications and reactivation. *N. Engl. J. Med.* 331:432-438.

He, J., McCarthy, M., Zhou, Y., et al. (1996) Infection of primary human fetal astrocytes by human herpesvirus 6. *J. Virol.* 70:1296-1300.

Hemmer, B., Archelos, J. J., Hartung, H. P. (2002) New concepts in the immunopathogenesis of multiple sclerosis. *Nat. Rev. Neurosci.* 3:291-301.

Hilton, D. A., Love, S., Fletcher, A., Pringle, J. H. (1994) Absence of Epstein-Barr virus RNA in multiple sclerosis as assessed by in situ hybridisation. *J. Neurol. Neurosurg. Psychiatry* 57:975-976.

Hong, J., Tejada-Simon, M. V., Rivera, V. M., et al. (2002) Anti-viral properties of interferon beta treatment in patients with multiple sclerosis. *Mult. Scler.* 8:237-242.

Huseby, E. S., Liggitt, D., Brabb, T., et al. (2001) A pathogenic role for myelin-specific CD8(+) T cells in a model for multiple sclerosis. *J. Exp. Med.* 194:669-676.

IFNB (1993) Interferon beta-1b is effective in relapsing-remitting multiple sclerosis. I. Clinical results of a multicenter, randomized, double-blind, placebo-controlled trial; the IFNB Multiple Sclerosis Study Group. *Neurology* 43:655-661.

Johnson, R. T. (1994) The virology of demyelinating diseases. *Ann. Neurol.* 36(Suppl):S54-S60.

Johnston, R. E., Geretti, A. M., Prentice, H. G., et al. (1999) HHV-6-related secondary graft failure following allogeneic bone marrow transplantation. *Br. J. Haematol.* 105:1041-1043.

Kimberlin, D. W. (1998) Human herpesviruses 6 and 7: identification of newly recognized viral pathogens and their association with human disease. *Pediatr. Infect. Dis. J.* 17: 59-67; quiz 68.

Knox, K. K., Brewer, J. H., Henry, J. M., et al. (2000) Human herpesvirus 6 and multiple sclerosis: systemic active infections in patients with early disease. *Clin. Infect. Dis.* 31:894-903.

Kurtzke, J. F. (2000) Epidemiology of multiple sclerosis: does this really point toward an etiology? Lectio doctoralis. *Neurol. Sci.* 21:383-403.

Lawrence, G. L., Chee, M., Craxton, M. A., et al. (1990) Human herpesvirus 6 is closely related to human cytomegalovirus. *J. Virol.* 64:287-299.

Levin, L. I., Munger, K. L., Rubertone, M. V., et al. (2005) Temporal relationship between elevation of Epstein-Barr virus antibody titers and initial onset of neurological symptoms in multiple sclerosis. *J.A.M.A.* 293:2496-2500.

Liedtke, W., Malessa, R., Faustmann, P. M., Eis-Hubinger, A. M. (1995) Human herpesvirus 6 polymerase chain reaction findings in human immunodeficiency virus associated neurological disease and multiple sclerosis. *J. Neurovirol.* 1:253-258.

Lindberg, C., Andersen, O., Vahlne, A., et al. (1991) Epidemiological investigation of the association between infectious mononucleosis and multiple sclerosis. *Neuroepidemiology* 10:62-65.

Lusso, P., Markham, P. D., Tschachler, E., et al. (1988) In vitro cellular tropism of human B-lymphotropic virus (human herpesvirus-6). *J. Exp. Med.* 167:1659-1670.

Marie, P. (1884) Sclerose en plaques et maladies infectieuses. *Prog. Med. Paris* 12: 287-289.

Martin, C., Enbom, M., Soderstrom, M., et al. (1997) Absence of seven human herpesviruses, including HHV-6, by polymerase chain reaction in CSF and blood from patients with multiple sclerosis and optic neuritis. *Acta Neurol. Scand.* 95:280-283.

McCullers, J. A., Lakeman, F. D., Whitley, R. J. (1995) Human herpesvirus 6 is associated with focal encephalitis. *Clin. Infect. Dis.* 21:571-576.

McFarland, H. F., Reingold, S. C. (2005) The future of multiple sclerosis therapies: redesigning multiple sclerosis clinical trials in a new therapeutic era. *Mult. Scler.* 11:669-676.

Medana, I. M., Gallimore, A., Oxenius, A., et al. (2000) MHC class I-restricted killing of neurons by virus-specific CD8+ T lymphocytes is effected through the Fas/FasL, but not the perforin pathway. *Eur. J. Immunol.* 30:3623-3633.

Mirandola, P., Menegazzi, P., Merighi, S., et al. (1998) Temporal mapping of transcripts in herpesvirus 6 variants. *J. Virol.* 72:3837-844.

Mirandola, P., Stefan, A., Brambilla, E., et al. (1999) Absence of human herpesvirus 6 and 7 from spinal fluid and serum of multiple sclerosis patients. *Neurology* 53:1367-1368.

Mock, D. J., Powers, J. M., Goodman, A. D., et al. (1999) Association of human herpesvirus 6 with the demyelinative lesions of progressive multifocal leukoencephalopathy. *J. Neurovirol.* 5:363-373.

Monteiro, J., Hingorani, R, Pergolizzi, R., et al. (1995) Clonal dominance of CD8+ T-cell in multiple sclerosis. *Ann. N.Y. Acad. Sci.* 756:310-312.

Mookerjee, B. P., Vogelsang, G. (1997) Human herpes virus-6 encephalitis after bone marrow transplantation: successful treatment with ganciclovir. *Bone Marrow Transplant.* 20:905-906.

Morre, S. A., van Beek, J., De Groot, C. J., et al. (2001) Is Epstein-Barr virus present in the CNS of patients with MS? *Neurology* 56:692.

Mouzaki, A., Tselios, T., Papathanassopoulos, P., et al. (2004) Immunotherapy for multiple sclerosis: basic insights for new clinical strategies. *Curr. Neurovasc. Res.* 1:325-340.

Neumann, H., Medana, I. M., Bauer, J., Lassmann, H. (2002) Cytotoxic T lymphocytes in autoimmune and degenerative CNS diseases. *Trends Neurosci.* 25:313-319.

Nielsen, L., Larsen, A. M., Munk, M., Vestergaard, B. F. (1997) Human herpesvirus-6 immunoglobulin G antibodies in patients with multiple sclerosis. *Acta Neurol. Scand. Suppl.* 169:76-78.

Okuno, T., Takahashi, K., Balachandra, K., et al. (1989) Seroepidemiology of human herpesvirus 6 infection in normal children and adults. *J. Clin. Microbiol.* 27:651-653.

Ongradi, J., Rajda, C., Marodi, CL., et al. (1999) A pilot study on the antibodies to HHV-6 variants and HHV-7 in CSF of MS patients. *J. Neurovirol.* 5:529-532.

Opsahl, M. L., Kennedy, P. G. (2005) Early and late HHV-6 gene transcripts in multiple sclerosis lesions and normal appearing white matter. *Brain* 128:516-527.

Rapaport, D., Engelhard, D., Tagger, G., et al. (2002) Antiviral prophylaxis may prevent human herpesvirus-6 reactivation in bone marrow transplant recipients. *Transpl. Infect. Dis.* 4:10-16.

Rieux, C., Gautheret-Dejean, A., Challine-Lehmann, D., et al. (1998) Human herpesvirus-6 meningoencephalitis in a recipient of an unrelated allogeneic bone marrow transplantation. *Transplantation* 65:1408-1411.

Rossi, C., Delforge, M. L., Jacobs, F., et al. (2001) Fatal primary infection due to human herpesvirus 6 variant A in a renal transplant recipient. *Transplantation* 71:288-292.

Rotola, A., Merlotti, I., Caniatti, L., et al. (2004) Human herpesvirus 6 infects the central nervous system of multiple sclerosis patients in the early stages of the disease. *Mult. Scler.* 10:348-354.

Russell, J. H., Ley, T. J. (2002) Lymphocyte-mediated cytotoxicity. *Annu. Rev. Immunol.* 20:323-370.

Saito, Y., Sharer, L. R., Dewhurst, S., et al. (1995) Cellular localization of human herpesvirus-6 in the brains of children with AIDS encephalopathy. *J. Neurovirol.* 1:30-39.

Salahuddin, S. Z., Ablashi, D. V., Markham, P. D., et al. (1986) Isolation of a new virus, HBLV, in patients with lymphoproliferative disorders. *Science* 234:596-601.

Sanders, V. J., Felisan, S., Waddell, A., Tourtellotte, W. W. (1996) Detection of herpesviridae in postmortem multiple sclerosis brain tissue and controls by polymerase chain reaction. *J. Neurovirol.* 2:249-258.

Shresta, S., Pham, C. T., Thomas, D. A., et al. (1998) How do cytotoxic lymphocytes kill their targets? *Curr. Opin. Immunol.* 10:581-587.

Sibley, W. A., Bamford, C. R., Clark, K. (1985) Clinical viral infections and multiple sclerosis. *Lancet* 1:1313-1315.

Singh, N., Bonham, A., Fukui, M. (2000) Immunosuppressive-associated leukoencephalopathy in organ transplant recipients. *Transplantation* 69:467-472.

Sola, P., Merelli, E., Marasca, R., et al. (1993) Human herpesvirus 6 and multiple sclerosis: survey of anti-HHV-6 antibodies by immunofluorescence analysis and of viral sequences by polymerase chain reaction. *J. Neurol. Neurosurg. Psychiatry* 56:917-919.

Soldan, S. S., Berti, R., Salem, N., et al. (1997) Association of human herpes virus 6 (HHV-6) with multiple sclerosis: increased IgM response to HHV-6 early antigen and detection of serum HHV-6 DNA. *Nat. Med.* 3:1394-1397.

Soldan, S. S., Leist, T. P., Juhng, K. N., et al. (2000) Increased lymphoproliferative response to human herpesvirus type 6A variant in multiple sclerosis patients. *Ann. Neurol.* 47:306-313.

Sospedra, M., Martin, R. (2005) Immunology of multiple sclerosis. *Annu. Rev. Immunol.* 23:683-747.

Steinman, L. (2001a) Multiple sclerosis: a two-stage disease. *Nat. Immunol.* 2:762-764.

Steinman, L. (2001b). Myelin-specific CD8 T cells in the pathogenesis of experimental allergic encephalitis and multiple sclerosis. *J. Exp. Med.* 194:F27-F30.

Steinman, L., Zamvil, S. (2003) Transcriptional analysis of targets in multiple sclerosis. *Nat. Rev. Immunol.* 3:483-492.

Sundstrom, P., Juto, P., Wadell, G., et al. (2004) An altered immune response to Epstein-Barr virus in multiple sclerosis: a prospective study. *Neurology* 62:2277-2282.

Swanborg, R. H., Whittum-Hudson, J. A., Hudson, A. P. (2003) Infectious agents and multiple sclerosis—are Chlamydia pneumoniae and human herpes virus 6 involved? *J. Neuroimmunol.* 136:1-8.

Taus, C., Pucci, E., Cartechini, E., et al. (2000) Absence of HHV-6 and HHV-7 in cerebrospinal fluid in relapsing-remitting multiple sclerosis. *Acta Neurol. Scand.* 101:224-228.

Tejada-Simon, M. V., Zang, Y. C., Hong, J., et al. (2002) Detection of viral DNA and immune responses to the human herpesvirus 6 101-kilodalton virion protein in patients with multiple sclerosis and in controls. *J. Virol.* 76:6147-6154.

Tiacci, E., Luppi, M., Barozzi, P., et al. (2000) Fatal herpesvirus-6 encephalitis in a recipient of a T-cell-depleted peripheral blood stem cell transplant from a 3-loci mismatched related donor. *Haematologica* 85:94-97.

Tokimasa, S., Hara, J., Osugi, Y., et al. (2002) Ganciclovir is effective for prophylaxis and treatment of human herpesvirus-6 in allogeneic stem cell transplantation. *Bone Marrow Transplant* 29:595-598.

Tomsone, V., Logina, I., Millers, A., et al. (2001) Association of human herpesvirus 6 and human herpesvirus 7 with demyelinating diseases of the nervous system. *J. Neurovirol.* 7:564-569.

Trapani, J. A., Davis, J., Sutton, V. R., Smyth, M. J. (2000) Proapoptotic functions of cytotoxic lymphocyte granule constituents in vitro and in vivo. *Curr. Opin. Immunol.* 12:323-329.

Vass, K., Lassmann, H. (1990) Intrathecal application of interferon gamma: progressive appearance of MHC antigens within the rat nervous system. *Am. J. Pathol.* 137:789-800.

Villoslada, P., Juste, C., Tintore, M., et al. (2003) The immune response against herpesvirus is more prominent in the early stages of MS. *Neurology* 60:1944-1948.

Virtanen, J. O., Zabriskie, J. B., Siren, V., et al. (2005) Co-localization of human herpes virus 6 and tissue plasminogen activator in multiple sclerosis brain tissue. *Med. Sci. Monit.* 11:BR84-BR87.

Wainwright, M. S., Martin, P. L., Morse, R. P., et al. (2001) Human herpesvirus 6 limbic encephalitis after stem cell transplantation. *Ann. Neurol.* 50:612-619.

Wang, F. Z., Linde, A., Hagglund, H., et al. (1999) Human herpesvirus 6 DNA in cerebrospinal fluid specimens from allogeneic bone marrow transplant patients: does it have clinical significance? *Clin. Infect. Dis.* 28:562-568.

Wilborn, F., Schmidt, C. A., Brinkmann, V., et al. (1994) A potential role for human herpesvirus type 6 in nervous system disease. *J. Neuroimmunol.* 49:213-214.

Woodland, D. L., Dutton, R. W. (2003) Heterogeneity of CD4(+) and CD8(+) T cells. *Curr. Opin. Immunol.* 15:336-342.

Wucherpfennig, K. W., Strominger, J. L. (1995) Molecular mimicry in T cell-mediated autoimmunity: viral peptides activate human T cell clones specific for myelin basic protein. *Cell* 80:695-705.

Xu, Y., Linde, A., Fredrikson, S., et al. (2002) HHV-6 A- or B-specific P41 antigens do not reveal virus variant-specific IgG or IgM responses in human serum. *J. Med. Virol.* 66:394-399.

Yamanishi, K., Okuno, T., Shiraki, K., et al. (1988) Identification of human herpesvirus-6 as a causal agent for exanthem subitum. *Lancet* 1:1065-1067.

Yoshida, H., Matsunaga, K., Ueda, T., et al. (2002) Human herpesvirus 6 meningoen-cephalitis successfully treated with ganciclovir in a patient who underwent allogeneic bone marrow transplantation from an HLA-identical sibling. *Int. J. Hematol.* 75: 421-425.

Yoshikawa, T., Asano, Y. (2000) Central nervous system complications in human herpesvirus-6 infection. *Brain Dev.* 22:307-314.

Zawada, M., Liwien, I., Pernak, M., et al. (2003) MSRV pol sequence copy number as a potential marker of multiple sclerosis. *Pol. J. Pharmacol.* 55:869-875.

Zerr, D. M., Gupta, D., Huang, M. L., et al. (2002) Effect of antivirals on human herpesvirus 6 replication in hematopoietic stem cell transplant recipients. *Clin. Infect. Dis.* 34:309-317.

# 12
# Multiple Sclerosis Pathology During Early and Late Disease Phases: Pathogenic and Clinical Relevance

Claudia F. Lucchinetti

## 1   Introduction

Multiple sclerosis (MS) is an inflammatory demyelinating disorder of the central nervous system (CNS) with a complex pathology that varies with respect to the extent and character of inflammation, demyelination, gliosis, axonal injury, and remyelination. These factors, in turn, depend on the stage of demyelinating activity in the lesion and the clinical phase of the disease. In early disease phases, MS is characterized by acute exacerbations of neurologic dysfunction owing to a combination of inflammation, edema, and focal demyelination resulting in conduction block. Resolution of the inflammation coupled with early remyelination typically leads to recovery. However, after an initial relapsing course, most patients enter a progressive disease phase characterized by continuous neurologic decline. During these later disease phases, most of the irreversible damage does not depend on the formation of new inflammatory demyelinating lesions. Axonal degeneration, cortical pathology, and diffuse alterations in the normal-appearing white matter (NAWM) likely contribute to disease progression. Although MS pathogenesis is often referred to as a biphasic disease with an inflammatory phase early that leads to a subsequent neurodegenerative phase late, this may inadvertently deemphasize the potential for chronic ongoing inflammatory processes contributing to disease evolution and progression. This review discusses the pathologic hallmarks of MS in relation to early and late disease. A better appreciation of these pathologic differences provides greater insight into the underlying mechanisms of disease initiation, evolution, and progression. Attempts to understand the relative contributions and complex sequence of events related to inflammation, demyelination, gliosis, axonal degeneration, and remyelination in MS will hopefully lead to more effective therapeutic strategies to target these diverse processes.

## 2    MS Lesions in Relation to Lesion Activity and Disease Duration

MS lesions demonstrate different neuropathologic features with disease evolution Furthermore, during early and late disease phases different stages and types of demyelinating activity can be identified.

### 2.1    Stage of Demyelinating Activity

Magnetic resonance imaging (MRI) studies rely on evidence of blood-brain barrier (BBB) leakage, defined by the presence of gadolinium-DTPA leakage, as an indicator of an active lesion (Grossman et al., 1988; Miller et al., 1988). This may not reliably differentiate active from inactive MS plaques, as both can be associated with variable degrees of BBB leakage. MRI sensitivity may not be sufficient to detect potentially small quantitative differences in BBB dysfunction that distinguish active from inactive plaques. Therefore, investigations on pathogenic mechanisms involved in lesion formation must rely on a precise neuropathologic definition of lesional activity and staging. The presence of major histocompatibility class II (MHC II) antigens, adhesion molecules, and cytokines as well as the extent and activation state of lymphocytes and macrophages in MS lesions have all been used to define MS lesional activity (Bruck et al., 1995; Ozawa et al., 1994;

Raine et al., 1990; Sobel et al., 1990; Traughott, 1983). However these approaches do not distinguish demyelinating activity from inflammatory activity, which may be present in the lesion, in the absence of ongoing active demyelination.

A more stringent definition of demyelinating activity in a plaque can be obtained by studying the structural profile and chemical composition of myelin degradation products in macrophages in correlation with the expression of macrophage differentiation markers (Fig. 12.1) (Bruck et al., 1995). The time sequence of myelin degradation in macrophages is based on the evaluation of experimental autoimmune encephalomyelitis (EAE) lesions (Lassmann and Wisniewski, 1979), as well as in vitro studies analyzing the sequential breakdown of myelin by human monocytes (van der Goes, et al., 2005). Whenever myelin sheaths are destroyed, their remnants are taken up by macrophages or microglia cells. Minor myelin proteins, such as myelin oligodendrocyte glycoprotein (MOG) and myelin-associated glycoprotein (MAG), are rapidly degraded within macrophages within 1 to 2 days after phagocytosis. In contrast, major myelin proteins, such as myelin basic protein (MBP) and proteolipid protein (PLP), may persist in macrophages for 6 to 10 days. In later stages, the macrophages contain sudanophilic and periodic acid-Schiff (PAS)-positive "granular lipids" that may persist in the lesion up to several months. In early active lesions, macrophages expressing acute stage inflammatory markers MRP8 and MRP14 contain myelin degradation products immunoreactive for all myelin proteins including minor myelin proteins [MOG, 2′,3′-cyclic nucleotide-3-phosphodiesterase (CNPase)], whereas in late active lesions, only major myelin proteins (MBP and PLP) are immunoreactive in macrophages expressing 27E10. Inactive areas may still contain PAS+ macrophages, but they no longer contain any minor or major myelin proteins, and the chronic stage inflammatory marker 25F9 shows increasing expression. Early remyelinating lesions are characterized by clusters of short,

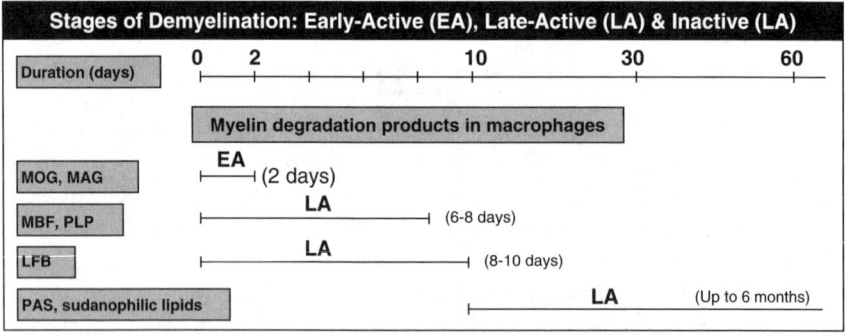

FIGURE 12.1. Staging of demyelinating activity. MS lesions can be classified into early active (EA), late active (LA), and inactive (IA) stages based on the type of myelin degradation products present in the macrophage cytoplasm. MOG, myelin-oligodendrocyte glycoprotein; MAG, myelin-associated glycoprotein; MBP, myelin basic protein; PLP, proteolipid protein; LFB, luxol fast blue; PAS, periodic acid Schiff.

thin, irregularly organized myelin sheaths with greater MAG or CNPase reactivity relative to MOG or PLP. PAS+ macrophages may be present.

A variety of lesions at different stages of demyelinating activity may be present in any MS brain. When these stringent criteria are used, the incidence of active lesions in MS brains is low, especially in classic cases sampled during the chronic phase of the disease. Therefore, as a first step it is critical to define the stage of demyelinating activity in individual MS lesions derived from brain back material before performing biochemical, immunologic, or molecular pathogenic studies. It is also important to recognize that this classification scheme does not capture events in lesion evolution that may have preceded myelin degradation. An alternative method of evaluating tissue has been developed based on a multifactorial cluster analysis, which includes parameters of clinical history, different aspects of inflammation and microglial activation, demyelination, and deposition of immunoglobulins and complement (Gay et al., 1997). Unfortunately, the necessary clinical information is not always available, but the approach may help to identify stages of lesions that precede the structural dissolution of myelin sheaths.

## 2.2    MS Plaque Types

### 2.2.1    Inactive Plaques

Lesions from late chronic MS cases are mainly characterized by the presence of multiple sharply demarcated inactive plaques of demyelination typically ranging from < 1 mm to several centimeters in size, present both in white and gray matter, with a predilection for the periventricular white matter, optic nerves, brain stem, cerebellum, and spinal cord (Lumsden, 1970). By gross inspection, the inactive plaque appears as a well circumscribed, slightly depressed, gray area with increased tissue texture (Fig. 12.2). The lesions may be round or oval but frequently show finger-like extensions that may follow the path of small or medium-sized vessels (Dawson, 1916). Microscopically, the chronic inactive MS plaque demonstrates no evidence for ongoing myelin breakdown and appears as a sharply circumscribed, relatively hypocellular, pale area with marked myelin loss, prominent fibrillary astrocytosis, variably reduced axonal density, and loss of mature oligodendrocytes (Fig. 12.2). Despite the absence of ongoing myelin destruction, inflammation may still be present, with few lymphocytes or macrophages present mainly in the perivascular area.

### 2.2.2    Active Plaques

On gross inspection, the active MS plaque appears as a soft area of irregular pink or gray color. Microscopically, active inflammatory demyelination is characterized by an intimate admixture of lipid-laden macrophages and large reactive astrocytes accompanied by variable perivascular inflammation (Fig. 12.3). The involved areas demonstrate marked pallor of myelin staining with "relative" preservation of axons, although where the damage is most severe axons may be

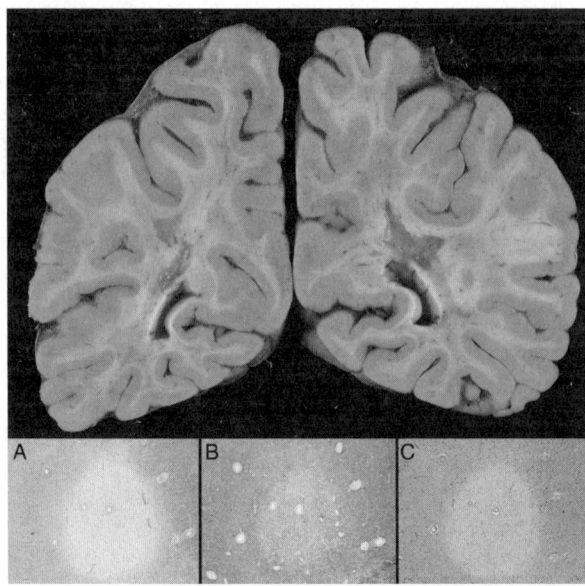

FIGURE 12.2. Chronic MS. Cerebrum, gross coronal section. Plaques appear as well circumscribed, slightly depressed, gray areas with increased tissue texture. The chronic inactive plaque microscopically appears as a sharply circumscribed area of myelin pallor (**A**, LFB-PAS) with variable reduction in axonal density (**B**, neurofilament). The lesions are hypocellular and lack macrophages containing myelin debris (**C**, KiM1P macrophage marker). (© *Handbook of Multiple Sclerosis*. Reprinted with permission.)

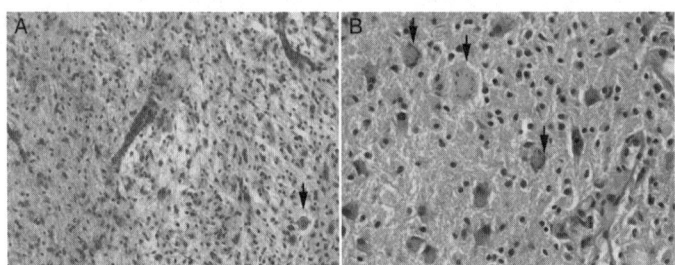

FIGURE 12.3. Active MS Lesion. Active lesions are hypercellular and characterized by an admixture of macrophages and reactive astrocytes associated with myelin loss (**A**, LFB-PAS). Creutzfeld-Peters cells represent astrocytes containing fragmented nuclei that can resemble astrocytic mitoses (arrows in **A** and **B**). (**B**, H&E) (© *Handbook of Multiple Sclerosis*. Reprinted with permission.) (*See Color Plates between pages 256-257.*)

lost or fragmented and display irregular tortuous and clubbed profiles. Many macrophages become engorged with phagocytosed myelin remnants and debris and assume the appearance of classic "gitter cells," with abundant vacuolated cytoplasm. Intimately intermingled are enlarged (reactive) astrocytes with prominent, somewhat polymorphic nuclei and conspicuous eosinophilic cytoplasm (Fig. 12.3). The "granular mitosis" (also referred to as a Creutzfeld-Peters cell), is an unusual finding in some reactive astrocytes.

Based on the topographic distribution of macrophages and the type of myelin degradation products present in the macrophage, several types of active plaques can be distinguished (Fig. 12.4): The acute plaque is characterized by the synchronous destruction of myelin, with all the macrophages containing early and late myelin degradation products distributed evenly throughout the extent of the lesion. Chronic active plaques consist of an accumulation of numerous macrophages containing both early and late myelin degradation products clustered at the radially expanding plaque edge and diminishing in number toward the inactive plaque center. The smoldering active plaque consists of an inactive lesion center surrounded by a rim of macrophages and activated microglia, few of which contain early myelin degradation products; and myelin digestion is completed in most (Prineas et al., 2001).

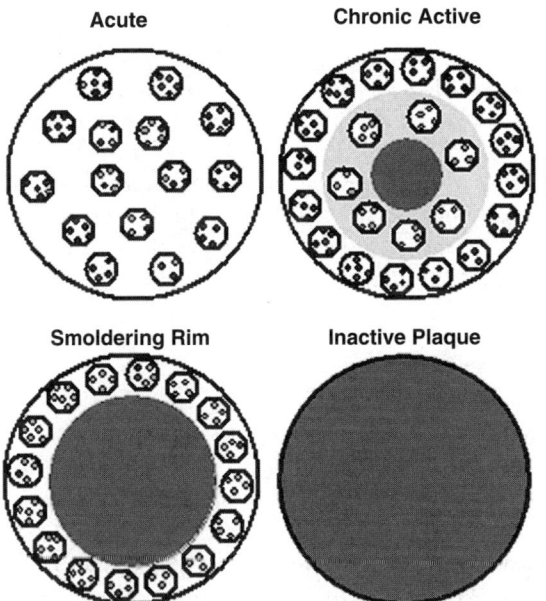

FIGURE 12.4. MS plaque types. The acute active plaque is characterized by the presence of macrophages containing early (red) and late (green) myelin degradation products distributed throughout the extent of the lesion. The chronic active plaque shows the accumulation of numerous macrophages, containing both early (red) and late (green) myelin degradation products clustered at the advancing plaque edge and diminishing in number toward the inactive plaque center (blue). The smoldering rim is defined by the presence of very few macrophages restricted to the plaque edge containing early and late myelin degradation products, whereas the inactive plaque contains no early or late myelin degradation products in the macrophages. (*See Color Plates between pages 256-257.*)

Acute active and chronic active plaques are mainly found in patients with acute or early MS or in secondary progressive MS (SPMS) patients with ongoing clinical exacerbations. These plaques are usually associated with inflammation. Active plaques are rare during late phases of MS or in patients with chronic progressive disease, when inactive plaques typically predominate. However, if there is any ongoing demyelinating activity in the brain, it usually takes the form of a smoldering active plaque in which low-grade demyelination may be observed at the plaque edge. Such lesions are mainly found during late phases of the disease, particularly in patients with primary progressive MS (PPMS), or SPMS not associated with ongoing relapses (Prineas et al., 2001). These plaques typically are associated with less inflammation compared to acute and chronic active MS plaques.

### 2.2.3   Destructive Plaques

Destructive plaques can be found in fulminant variants of MS and are characterized by the presence of demyelination associated with pronounced destruction of axons and astrocytes, often resulting in cystic lesions (Youl et al., 1991). Encephalomalacia associated with prominent gray and white matter atrophy may occur. The destructive plaque is a hallmark of Marburg's acute MS, which is typically characterized by a rapidly progressive fatal course leading to death within 1 year from the onset (Marburg, 1906). Destructive plaques can also be found in patients with neuromyelitis optica, an inflammatory demyelinating disorder mainly affecting the optic nerves and spinal cord; it is characterized by necrotic lesions involving both the gray and white matter, often resulting in cystic cavitation of the spinal cord (Lucchinetti et al., 2002b).

### 2.2.4   Shadow Plaques

Shadow plaques are sharply demarcated plaques in a typical MS distribution, with only moderately reduced axonal density, and a uniform presence of nerve fibers with proportionally thin myelin sheaths. Immunocytochemical and ultrastructural data suggest that shadow plaques represent complete remyelination of previously demyelinated plaques (Lassmann, 1983; Prineas, 1985); they are characterized by reduced staining of myelin (myelin pallor) owing to a decreased ratio between myelin thickness and axonal diameter (Fig. 12.5). However, because a reduction of myelin density may also be seen with active plaques or in areas of secondary wallerian degeneration, a precise definition is needed. Areas of wallerian degeneration have ill-defined borders and preserved nerve fibers in the lesions and show a broad range of thick to thin myelin sheaths, whereas areas of active demyelination are clearly defined by the presence of macrophages containing myelin degradation products. Shadow plaques typically contain few macrophages and are associated with pronounced fibrilllary gliosis. These remyelinated lesions may subsequently become targets of new demyelinating attacks (Prineas et al., 1992b).

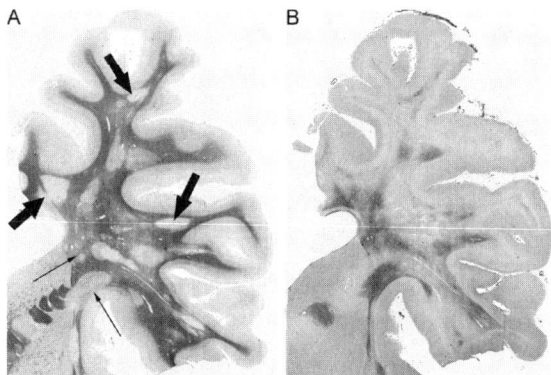

FIGURE 12.5. Remyelination in MS lesion. **A.** Chronic MS. Multiple lesions are in the white matter; some are completely demyelinated (thick arrows), whereas others are shadow plaques (thin arrows) characterized by regions of myelin pallor; X1 (Luxol fast blue myelin stain). **B.** Gliosis is prominent in the shadow plaques (Holzer stain; X1). (© *McAlpines Multiple Sclerosis,* A. Compston, editor. Modified with permission.)

## 3   Complex Role of Inflammation in MS

The presence of variable but persistent inflammation in MS lesions as well as nonlesion areas (i.e., gray matter, NAWM, meninges) is a consistent observation in most descriptions of MS pathology. However, whether inflammation plays a primary or secondary role in disease initiation, evolution, and progression remains disputed. Furthermore, studies have suggested a potential benficial role of inflammation in promoting tissue repair in MS lesions.

### 3.1   Composition

The extent and nature of the inflammatory response in MS lesions may vary depending on the age and stage of the lesion (Fig. 12.6). In active lesions analyzed from early disease phases, the infiltrate is mainly composed of activated macrophages mainly derived from blood-borne monocytes. However, resident microglia may also contribute to lesion development; and ramified activated microglia expressing histocompatibility antigens, adhesion molecules, or markers of activated peripheral macrophages are present mainly in the white or gray matter that surrounds actively demyelinating as well as inactive lesions. Some studies suggest that microglial activation and demyelination may precede recruitment of most hematogeneous cells into the lesions (Gay et al., 1997; Trebst et al., 2001).

Most infiltrating lymphocytes are T cells (Traugott, 1983), with a predominance of MHC I restricted, CD8+ T lymphocytes (Booss et al., 1983). This is also the cell population that shows preferential clonal expansion (Babbe et al., 2000)

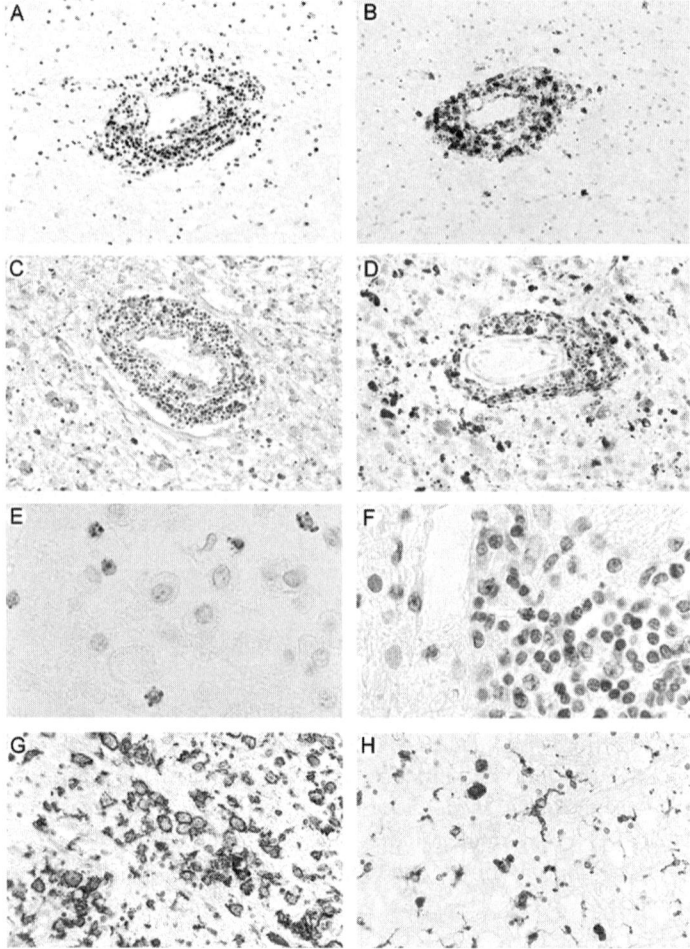

FIGURE 12.6. Acute MS. Inflammation in MS lesions. **A**. Perivascular inflammatory infiltrates with CD3+ T cells. **B**. CD8+ T cells. **C**. IgG-containing plasma cells. **D**. CD68+ macrophages. (×100) **E**. Acute MS with granzyme B-reactive cytotoxic T cells in the lesion parenchyma. **F**. A perivascular cuff. (×600) **G**. Primary progressive MS lesion with CD68+ macrophages in an active plaque. **H**. CD68+ microglia in the normal appearing white matter. (×300) (© *McAlpines Multiple Sclerosis*, A. Compston, editor. Reprinted with permission.) (*See Color Plates between pages 256-257.*)

and close apposition of activated cytotoxic T cells with degenerating oligodendrocytes and axons in some lesions of fulminant acute MS (Neumann et al., 2002). A variable but low number of B cells are also found in MS plaques. The number of antibody-producing plasma cells is reportedly low in active lesions derived from either acute MS cases or from patients during early initial bouts of the disease. However, their absolute and relative numbers increase with disease chronicity (Ozawa et al., 1994). As with all other inflammatory cells, B lymphocytes are more evident in active than inactive lesions (Esiri, 1977). In active MS lesions, inflammatory cells are present in the perivascular space and are dispersed throughout the CNS parenchyma. This holds true for lymphocytes as well as macrophages (Prineas and Wright, 1978). However, B lymphocytes and plasma cells are more concentrated in the perivascular space and meninges, and parenchymal infiltration is relatively rare. In fulminant MS variants, such as Marburg MS or NMO, neutrophils and eosinophils may also be present (Lucchinetti et al., 2002b).

## 3.2    Inflammation and Disease Initiation

There is an ongoing debate as to whether the inflammatory reaction in MS is initiated in the immune system or in response to primary events affecting the neural cells.

### 3.2.1    Primary Role

The immune initiation concept involves the generation of autoreactive T cells in the systemic circulation that gain access to the CNS where they persist and induce an inflammatory cascade that results in CNS tissue injury. A limited trafficking of antigen-nonspecific T cells does occur across an intact BBB via the interaction of adhesion molecules expressed on the surface of lymphocytes and integrins present on the endothelial surface of blood vessels. Rolling, adherence, and diapedesis of T lymphocytes are mediated by VCAM/VLA-4 and ICAM/LFA-1 interactions. Once autoreactive T cells have entered the CNS, matrix metalloproteinases (MMPs), especially MMP-9, are thought to contribute to degrading extracellular matrix macromolecules. Within the CNS, proinflammatory cytokines activate resident and hematogeneous macrophages. Recruitment and attraction of these cells occurs via integrins and chemokines and are believed to contribute to tissue injury and demyelination. Following the formation of the trimolecular complex (consisting of T cell receptor, antigenic peptide, and MHC II molecules) together with the appropriate co-stimulatory molecules, T lymphocytes are activated and secrete various cytokines. These cytokines may cause the surrounding immune cells and glia to produce chemokines, which are chemoattractant substances leading to the recruitment of additional immune cells into the CNS, thereby amplifying the inflammatory response.

Pathologic support for the primary role of the inflammatory response in MS is in part based on evidence that the inflammatory reaction in active lesions is associated with the up-regulation of a variety of Th1 cytokines, including interleukin-2 (IL-2), interferon-γ (IFNγ), and tumor necrosis factor-α (TNFα), which are also found in the cerebrospinal fluid (CSF) of MS patients with active disease. Activated endothelial cells in active lesions also express adhesion molecules, fibronectin, urokinase plasmin activator receptor, MHC II molecules, chemokines and their receptors, and stress proteins. This pathology is similar to that found in

EAE, an experimental paradigm of Th-1-mediated autoimmune disease and an animal model of MS that can be induced in susceptible animals by active sensitization with CNS tissue, myelin, or myelin proteins. Macrophages or microglia cells in MS lesions may express antigenic peptides from CNS proteins on their surface (Prineas and Wright, 1978). Furthermore, clonal expansion of T cells in the lesions suggests their antigen-driven proliferation (Babbe et al., 2000; Skulina et al., 2004). T cells in the MS brains also reveal the phenotype and cytokine spectrum of central memory cells (Kivisakk et al., 2003), and most of these cells are in an activated state, at least during the early stages of active lesions (Hayashi et al., 1988). This view is further supported by serial brain biopsies performed in a single patient with the Marburg type of acute MS in which the first biopsy showed a purely inflammatory disease of the white matter, followed by a second biopsy of the same lesion performed 76 days later revealing large confluent demyelinated lesions (Bitsch et al., 1999).

### 3.2.2    Secondary Role

An alternative view proposed regarding the role of inflammation in MS pathogenesis suggests that events in the CNS initiate the disease process. Such events could include an acquired or persistent infection of neural cells, resulting in the release of tissue antigens provoking an autoimmune response. However, a role for direct infection-mediated cytotoxic injury seems unlikely given the beneficial effects of immunosuppression on MS lesion formation (Prat and Antel, 2005). Infections could also trigger molecular mimicry responses resulting in immune dysregulation or viral persistence. Human herpesvirus 6 and human endogenous retroviruses (HERVs) have been detected in MS tissues (Antony et al., 2004; Rotola et al., 2004). Increased expression of HERVs in astrocytes can result in the release of mediators cytotoxic to oligodendrocytes.

Barnett and Prineas, on the basis of finding extensive oligodendrocyte apoptosis in the absence of inflammation in a pediatric MS patient who died 9 months after disease onset and 17 hours after presentation with acute pulmonary edema, suggested that primary oligodendrogliopathy represents the initial lesion in relapsing-remitting disease (RRMS), preceding inflammation and active myelin breakdown (Barnett and Prineas, 2004). The basis for the oligodendrocyte apoptosis was not defined in this study but could reflect a primary cell injury. This study raised the question as to whether inflammation is a prerequisite in MS pathogenesis. It is important to note, however, that because lymphocyte subsets were not examined a role for inflammation in these lesions cannot be completely excluded. Other confounding features of this case include treatment with high-dose corticosteroids prior to death, which may have dampened the inflammatory response. Corticosteroid administration reduces the number of T cells in the CNS of EAE rats within 4 hours of injection (McCombe et al., 1996). Furthermore, the presence of probable hypoxia related to the patient's known perimortem pulmonary edema is known to result in an identical pattern of myelin and oligodendrocyte pathology (Aboul-Enein et al., 2003).

## 3.3    Pathogenic Role of Inflammation

In most experimental animal models (rats, pigs, primates) of EAE, T cell-mediated immune responses against brain antigens result in brain inflammation but only limited demyelination. This more closely resembles the pathology of acute disseminated encephalomyelitis (ADEM) in which perivascular inflammation dominates, with minimal if any perivenular demyelination. Furthermore, often therapeutic strategies beneficial in EAE have yielded ineffective or at times unexpected aggravation of MS (Hohlfeld, 1997). One possible reason for this discrepancy is that the pathogenesis of MS lesions is more complex than that of a pure Th1-mediated CNS autoimmune disease. These observations suggest that additional pathogenic factors are necessary to produce the widespread demyelination classically seen in MS. These factors may include demyelinating antibodies, cytotoxic T cells, cytokines and other soluble mediators, proteases, reactive oxygen and nitrogen species, excitotoxic mechanisms, and hypoxia (Dawson, 1916).

### 3.3.1    B Cells/Antibody/Complement

Pathology studies have generated renewed interest in the possible contribution of B cells to MS lesion formation. B cells, plasma cells, myelin antibodies, and immunoglobulin transcripts have all been identified in MS lesions (Esiri, 1977; Lock et al., 2002; Mehta et al., 1981; Prineas and Wright, 1978). Molecular studies of B cells and plasma cells in MS CNS tissue and CSF suggest that these cells have undergone T cell-mediated antigen-driven clonal expansion (Baranzini et al., 1999; Cross et al., 2001). Furthermore, the H chain variable (V) regions of immunoglobulin (Ig) expressed in MS plaques and CSF reveal a limited repertoire. The VH sequences were noted to be oligoclonal, extensively mutated, and derived in part from clonally expanded B cell populations, all features suggestive of antigenic stimulation (Baranzini et al., 1999; Owens et al., 1998, 2003; Qin et al., 1998; Smith-Jensen et al., 2000; Williamson et al., 2001). The capping of surface IgG on macrophages involved in myelin breakdown provides further support for a pathogenic role of antibodies in MS (Prineas and Graham, 1981). Autoantibodies to MOG were also found bound to disintegrating myelin (Genain et al., 1999). IgG isolated from CNS tissues was found to bind to MBP in solid-phase assays, and an MBP peptide (83-97) was capable of inhibiting this binding. In addition, Ig isolated from MS lesions recognized folded MOG. This was the first study demonstrating binding of MOG in its native conformation, including glycosylation by Ig isolated from the lesions (Hafler, 2004). More recent studies have described the co-deposition of IgG and activated terminal lytic complement complex in active MS lesions (Lucchinetti et al., 2000; Storch et al., 1998). These findings, coupled with the presence in CSF of membrane attack complex-enriched membrane vesicles, also indicate a potential role for complement mediated injury in MS (Hafler et al., 2005).

### 3.3.2    Cytotoxic CD8 T Cells

Several lines of evidence suggest that MHC I-restricted CD8+ T cell responses may play a role in MS pathogenesis. CD8+ lymphocytes are present in MS lesions and predominate in many lesions. In fact, the CD8 T cell repertoire in MS lesions appears more antigen-driven than that of the CD4 T cells, and studies have shown the propensity of CD8 T cells to mediate demyelinating pathology (Haring et al., 2002). Another pathway of damage to oligodendrocytes, neurons, or axons could be a direct MHC I-dependent attack by CD8+ cytotoxic T cells. This requires MHC I expression on the target cells. Hoftberger et al. examined MHC I expression on microglia and endothelial cells, as well as astrocytes, oligodendrocytes, neurons, and axons, and found that all these cells upregulate MHC I in active lesions and may therefore become targets of a MHC I-restricted immune response (Hoftberger et al., 2004). Furthermore, CD8+ T cells, which express granzyme B as a marker for cytotoxic activation, are sometimes found in close contact to oligodendrocytes in acute MS lesions (Neumann et al., 2002). However, the number of granzyme B+ CD8 T cells diminishes with disease chronicity; therefore it is uncertain whether this mechanism plays a role during later disease phases. CD8+ T cells can also destroy neurons by antigen-independent mechanisms (Nitsch et al., 2004), possibly involving death receptors of the TNF family (Atkas et al., 2005). Such death receptors are also expressed in actively demyelinating MS lesions, although their cellular location is still controversial (Bonetti et al., 1997; Dowling et al., 1996; D'Souza et al., 1996).

### 3.3.3    Cytokines

Cytokines are soluble molecules whose functions include mediating "proinflammatory" and "antiinflammatory" effects on the immune system. Two subsets of T cells exist based on the cytokine profile they produce. The Th1 cells are thought to be proinflammatory by activating cytotoxic T cells and macrophages, whereas Th2 cells induce B cell proliferation and immunoglobulin production. Th2 responses can downregulate Th1 cell responses and are therefore considered "antiinflammatory" in their function. However, it must be emphasized that this is an oversimplification because the cytokine cascade is extremely complex, and it is likely that each cytokine affects the expression and function of another. In addition to exerting nonspecific tissue damage, cytokines can also directly target oligodendrocytes and myelin via the production of lymphotoxin (Selmaj et al., 1991c), perforin (Scolding et al., 1990), and TNFα (Selmaj and Raine, 1988). Conversely, some cytokines play an important role in promoting tissue repair in the CNS. As an example, ciliary neurotrophic factor (CNTF) belongs to the α-helical superfamily of cytokines and is a survival and differentiation factor for a variety of neuronal cell types, including motor, sensory, and sympathetic neurons (Sendtner et al., 1994). In the CNS, CNTF is mainly expressed by astrocytes and promotes maturation of oligodendrocyte precursor cells as well as production of myelin proteins in oligodendrocytes (Barres et al., 1993; Mayer et al., 1994). CNTF protects oligodendrocytes from TNF α-induced apoptotic cell death

(Louis et al., 1993). CNTF exerts trophic and mitogenic effects on oligodendro-cytes by enhancing their sensitivity to insulin-like growth factor-1 (IGF-1) and fibroblast growth factor-2 (FGF-2) (Jiang et al., 1999).

Immunocytochemistry, in situ hybridization, and gene microarray studies reveal that many different cytokines may be involved in MS pathogenesis. Although they appear to be more involved in active than inactive MS lesions, no specific pattern of cytokine expression has emerged that clearly differentiates different lesion stages or disease phases of MS.

### 3.3.4  Proteases

Matrix metalloproteinases are produced and secreted by inflammatory cells to degrade the extracellular matrix, thereby facilitating their migration into the CNS. In MS, MMPs are found in active lesions, where they may directly cause axonal transection (Anthony et al., 1997; Hendricks et al., 2005; Linberg et al., 2001). MMP-12 was shown to be expressed highly selectively by macrophages during active demyelination in MS plaques (Vos et al., 2003). This proteinase cleaves a broad spectrum of molecules, including elastin, MBP, and collagen type IV. Its expression in active MS lesions suggests that it may contribute to demyelination as well as leukocyte migration across the BBB.

The protease tissue plasminogen activator (tPA), secreted by macrophages, is also reportedly increased in MS and EAE, and it induces neuronal apoptosis in vitro (Cuzner et al., 1996; Flavin et al., 1997, 2000; Lu et al., 2002; Onodera et al., 1999). tPA knockout mice demonstrate delayed demyelination and axon degeneration after EAE induction (Lu et al., 2002). Therefore, proteases may contribute to axonal damage in MS.

### 3.3.5  Excitotoxicity

Glutamate, the most abundant excitatory neurotransmitter in the CNS, is secreted in large amounts by macrophages. An excess of glutamate causes excitotoxicity, which can lead to neuronal and oligodendrocyte damage and degeneration during CNS inflammation. Prolonged activation of neurons by glutamate may be dam-aging via production of nitric oxide (NO), reactive oxygen species (ROS), arachi-donic acid, phospholipase $A_2$, and proteases such as calpain causing calcium influx (Matute et al., 2001). An excess of extracellular glutamate may be caused by impaired glutamate clearance and degradation by astrocytes and oligodendro-cytes, as well as inhibition of clearance by proinflammatory cytokines (Hendricks et al., 2005). The expression of glutamate receptors in the CNS of MS patients was reported to be altered, and the presence of a specific glutamate receptor cor-related with, the presence of axonal damage (Geurts et al., 2003). In MS, macrophages and microglia were immunoreactive for the glutamate-producing enzyme glutaminase, which co-localized with dystrophic axons (Werner et al., 2001). Furthermore, a pathogenic role for glutamate excitotoxicity was demon-strated in EAE, where treatment with glutamate receptor antagonists resulted in reduced clinical symptoms and axonal damage (Pitt et al., 2000; Smith et al.,

2000). In addition, an inhibitor of glutamate transmission was noted to have similar protective effects in EAE (Gilgun-Sherki et al., 2003).

### 3.3.6    Nitric Oxide

Nitric oxide (NO) is a signal transducer involved in glutamate-induced excitation. NO has direct effects on BBB permeability, clearance of CNS inflammation possibly via induction of encephalitogenic T cell apoptosis, and mediating demyelination, oligodendrocyte destruction, and injury to axons (Smith and Lassman, 2002). NO is expressed by activated microglia, astrocytes, and macrophages during inflammation in the CNS (Bo et al., 1994) and is induced by several cytokines including IFNγ, TNFα, and IL-1ß (Forstermann and Kleinert, 1995). CSF NO levels correlate with disease activity in MS (Yamashita et al., 1997). NO is synthesized from L-arginine and oxygen by NO synthase (NOS). Although inducible NOS (iNOS) is not detected in the normal human CNS, it is abundantly expressed by astrocytes in plaque areas during autopsy (Giovannoni et al., 1998) and in a biopsy taken 33 days after onset of disseminated symptoms in a patient with fulminant MS (Bitsch et al., 1999). Inducible NOS labeling decreases as lesions age and become less active (De Groot et al., 1997; Liu et al., 2001). A study of brain biopsies from two acute cases of MS detected a signal for iNOS in both reactive astrocytes and perivascular monocytes/macrophages, whereas no signal was found in chronic MS cases (Oleszak et al., 1998). These data suggest that the extent of lesional activity may critically affect which cell type expresses iNOS in the lesion and may account for some previously reported discrepant results.

Confocal microscopy of MS lesions examined the distribution of iNOS and the function of cells producing iNOS. One study demonstrated that iNOS expression was present at the active edge of lesions, in and around perivascular lesions, and in ependymal cells involved in periventricular lesions. iNOS expression in the chronic active plaques was prominent in macrophages and microglial cells with some expression in astrocytes. Further characterization of microglia/macrophages in the chronic active plaques demonstrated that iNOS was frequently found in cells with phagocytic potential including the CD64+ (high-affinity Fcγ receptor) and in a subpopulation of CD14+ scavenger cells. However, iNOS was absent in mature macrophages. Some iNOS-positive microligal/macrophage cells contained intracellular MBP fragments, which is consistent with recent damage, demyelination, and phagocytosis. These cells were also associated with detectable levels of nitrotyrosine in the lesions, consistent with peroxynitirie-mediated damage (Hill et al., 2004). These observations support an important role for iNOS in the pathogenesis of MS lesions.

The effect of inhibition of iNOS in EAE is not clear. Some studies demonstrate inhibition of the disease, whereas others found NOS inhibition can be deleterious (Brenner et al., 1997; Cowden et al., 1998; Zhao et al., 1996). Both oligodendrocytes and axons are highly vulnerable to NO. Demyelinated axons are more vulnerable to NO damage than myelinated axons. NO induces a functional, hence reversible, conduction block in demyelinated axons (Redford et al., 1997), whereas axons that are additionally exposed to high metabolic stress by repetitive stimulation degenerate (Smith et al., 2001). Prolonged exposure to NO can result in irreversible inhibition of enzymes in the respiratory chain in macrophages/microglia in EAE (Bolanos et al., 1997; Brown et al., 1995). NO impairs mitochondrial electron transport and therefore oxidative phosphorylation (Brown and Borutaite, 2002). Exposing CNS white matter to NO causes ATP depletion and irreversible injury (Garthwaite et al., 2002).

These observations suggest that irreversible axonal damage by NO results from the combined effect of direct toxicity and energy failure.

### 3.3.7   Hypoxia

As previously described, inflammatory mediators such as reactive oxygen and nitrogen species are able to mediate mitochondrial dysfunction, and when they are excessively liberated they can result in a state of histiotoxic hypoxia (Aboul-Enein and Lassman, 2005). Hypoxic brain damage leads to the destruction of glial cells and neurons in the lesions. If the hypoxia is incomplete, a cascade of events ensues that increases the resistance of the tissue to subsequent hypoxic damage and thus limits structural damage from the insult ("hypoxic preconditioning") (Sharp and Bernaudin, 2004). One master switch in the induction of hypoxic preconditioning is the expression of hypoxia-inducible factors (HIFs) $\alpha$ and $\beta$ (Sharp and Bernaudin, 2004). They act as transcription factors that induce gene expression of downstream molecules involved in neuroprotection, vasomotor control, angiogenesis, cell growth, and energy metabolism (Aboul-Enein and Lassmann, 2005). These proteins render the tissue more resistant to further hypoxia-induced injury.

Sublethal hypoxia at the border of a stroke lesion also induces the expression of stress proteins, especially heat shock protein-70 (HSP70), a molecular chaperone that helps to renature damaged proteins and protects the tissue against further insult (Christians et al., 2002; Vass et al., 1988).

The expression of these survival proteins is not restricted to hypoxic or ischemic lesions. HIF expression can be induced or increased by pro- or antiinflammatory cytokines, and protein denaturation in injured tissue induces expression of HSPs (Aboul-Enein and Lassman, 2005; Sharp and Bernaudin, 2004). All of these proteins are expressed at the border between active inflammatory lesions and the adjacent normal tissue. Whereas HSP70 is found at the edges of all inflammatory lesions in the CNS, the expression of HIF-1$\alpha$ is restricted to lesions that follow hypoxia-like tissue injury (Aboul-Enein et al., 2003). In a systematic study of more than 80 cases of inflammatory and degenerative CNS white matter diseases and 20 controls, HIF-1$\alpha$ expression was significantly associated ($p < 0.0000001$) with a subset of MS lesions, various viral encephalitides, metabolic encephalopathy, and acute stroke lesions characterized by a preferential loss of MAG and distal oligodendrogliopathy with apoptotic oligodendrocyte cell death. The similar patterns of tissue injury and HIF-1$\alpha$ expression shared by these viral, demyelinating, and ischemic disorders suggests a shared pathogenesis related to energy failure.

## 3.4   Beneficial Role of Inflammation

Pathological analysis of MS lesions have documented that remyelination occurs even in the presence of ongoing inflammation and active demyelination, raising the possibility that the inflammatory response has both pathogenic and reparative roles (Fig. 12.7). The abundance of inflammation in inactive cases, together with observations on the local production of neutrotrophic factors such as brain-derived neurotrophic factor (BDNF) by leukocytes, support a potentially important role for inflammation in the repair of MS lesions (Kerschensteiner et al., 1999). Furthermore, neurotrophin receptors are expressed on glial cells and neurons in or near actively demyelinating MS lesions (Stadelmann et al., 2002). Autoimmune T cells can protect optic nerve neurons after crush injury (Moalem et al., 1999). Macrophages stimulate remyelination in tissue culture (Diemel et al., 1998), and their depletion is associated with diminished remyelination (Kotter et al., 2001). Some studies have revealed two distinct populations of macrophages: M1, or classically activated macrophages; and M2, alternatively activated macrophages (Mosser, 2003). The M1 phenotype is characterized as proinflammatory, whereas the M2 phenotype is associated with Th2 responses, scavenging of debris, promotion of tissue remodeling, and repair, as well as the expression of antiinflammatory markers such as IL-1ra (Mantovani et al., 2004). Active MS lesions contain numerous foamy macrophages, reflecting ingestion and accumulation of myelin lipids. One MS immunocytochemical and in vitro study suggested that foamy macrophages in MS lesions demonstrate a phenotype resembling that of antiinflammatory M2 macrophages and therefore likely contribute to the resolution of inflammation, limit further lesion development, and possibly promote tissue repair (Boven et al., 2006).

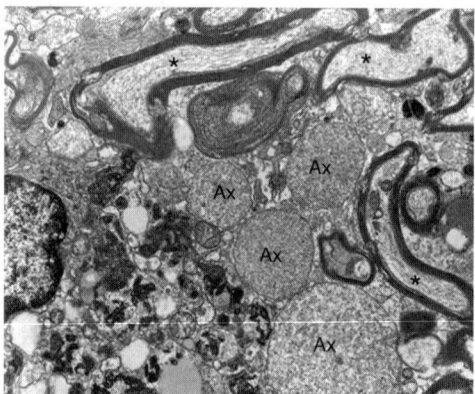

FIGURE 12.7. Electron microscopy of an early MS lesions reveals active demyelination occurring simultaneously with remyelination. A macrophage containing myelin debris is in the proximity of a field of both denuded completely demyelinated axons (Ax), as well as thinly remyelinated axons (*). (Courtesy of Moses Rodriguez, MD, Mayo Clinic, Rochester, MN, USA. © *Handbook of Multiple Sclerosis*. Reprinted with permission.)

It is also significant that a nonremyelinating experimental situation can be transformed to a remyelinating one by the induction of acute inflammation. Although transplanted oligodendrocyte progenitor cells (OPCs) can be established in OPC-depleted areas of chronic demyelination, remyelination is dramatically enhanced only when the environment is made conducive to OPC expansion and differentiation. This is achieved by establishing acute inflammation, thereby supporting a clear relation between acute inflammation and successful remyelination (Foote and Blakemore, 2005; Franklin, 2002). It seems plausible that the limited extent of remyelination during the later phases of MS may in part be the consequence of a lack of sufficient inflammatory signals required to activate OPCs to generate remyelinating oligodendroctes. There is likely a delicate balance between pathogenic and reparative inflammatory mechanisms that determine the final outcome of the MS lesion. Conceivably, complete blockage of all inflammatory responses in the MS lesion could be counterproductive.

## 3.5  Immunopathologic Heterogeneity in MS

Multiple inflammatory mechanisms can potentially lead to demyelination and tissue damage in MS (Fig. 12.8). Whether there is a dominant immune effector pathway of lesion formation or multiple immune effector pathways occur in parallel or sequentially has been a matter of debate. Prior MS immunopathologic studies were largely based on a few cases in which the staging of lesions was often inadequate and the interpretation of the findings was in part biased by immunologic concepts derived from EAE studies (Lassmann, 1979). Over the last few years, the systematic analysis of MS lesions has provided major new insights into the immunopathogenesis of the disease. Detailed immunopathologic studies have revealed profound heterogeneity in the patterns of demyelination and tissue

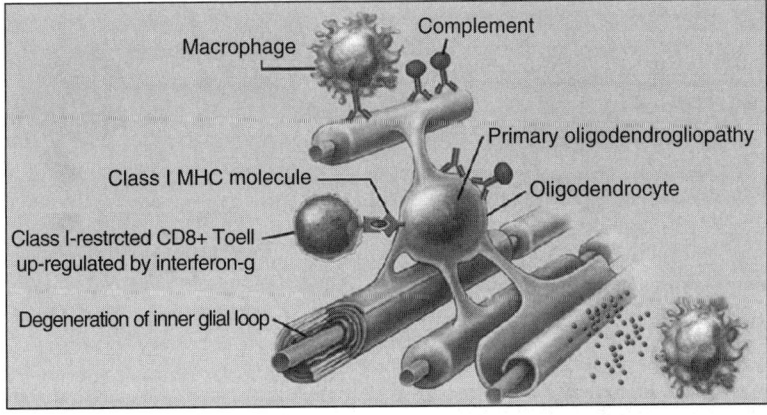

FIGURE 12.8. Potential effector mechanisms of demyelination in MS.

injury observed among active MS lesions and, in particular, among active lesions from different patients, suggesting that the pathogenesis of the disease may differ among patient subgroups. Immunopathogenic heterogeneity may in part underlie the variable treatment response observed among MS patients. However, because such studies require analysis of lesions demonstrating ongoing early active demyelination, they are mainly based on on materials from patients who either died during a fulminant course of the disease or who required brain biopsy during the initial clinical presentation, when the diagnosis was still in question. Detailed clinical and radiographic follow-up studies, however, confirm that most of these biopsied patients subsequently developed a clinical course and disability profile typical of MS, which did not substantially differ from an MS population-based cohort (Pittock et al., 2005). Therefore, the conclusions drawn from immunopathologic studies largely based on biopsied material can be extrapolated to the general MS population.

## 3.6   Immunopathologic Classification of Early Active MS Lesions

A systematic ongoing study of more than 250 MS cases reveals that although all active lesions occur on a background of an inflammatory process, composed mainly of T lymphocytes and macrophages, the lesions segregate into four dominant patterns of demyelination based on plaque geography, extent and pattern of oligodendrocyte pathology, evidence for immunoglobulin deposition and complement activation, and pattern of myelin protein loss (Fig. 12.9) (Lucchinetti et al., 2000, 2004).

In both patterns I and II, macrophages and T cells predominate in well demarcated plaques that surround small veins and venules; however, pattern II lesions demonstrate local precipitation of immunoglobulin and activated complement in regions of active myelin breakdown. The expression of all the myelin proteins (MBP, PLP, MAG, MOG) are reduced similarly. Oligodendrocytes are reduced in number at the active edge but reappear in the plaque's center. Remyelination is often extensive. Pattern I (macrophage-associated demyelination) closely resembles myelin destruction in mouse models of autoimmune encephalomyelitis in which mainly toxic products of activated macrophages such as TNFα (Probert et al., 1995) and nitric oxide (Griot et al., 1990) mediate destruction of myelin sheaths. Lesions similar to pattern II (antibody/complement-associated demyelination) are found in models of EAE, induced by sensitization with MOG. In this model, demyelination is induced by cooperation between encephalitogenic T cells and demyelinating anti-MOG antibodies (Linington et al., 1988). Although pattern II lesions suggest antibody (Ab)- and complement-mediated mechanisms may contribute to demyelination and tissue injury, definite proof is still lacking. A study describing deposition of MOG-reactive Igs on degenerating myelin sheaths in an active MS lesion (Genain et al., 1999) provided some support for this notion.

Pattern III lesions are defined by oligodendrocyte apoptosis, a marked reduction in oligodendrocytes, minimal remyelination, and early loss of MAG and CNPase

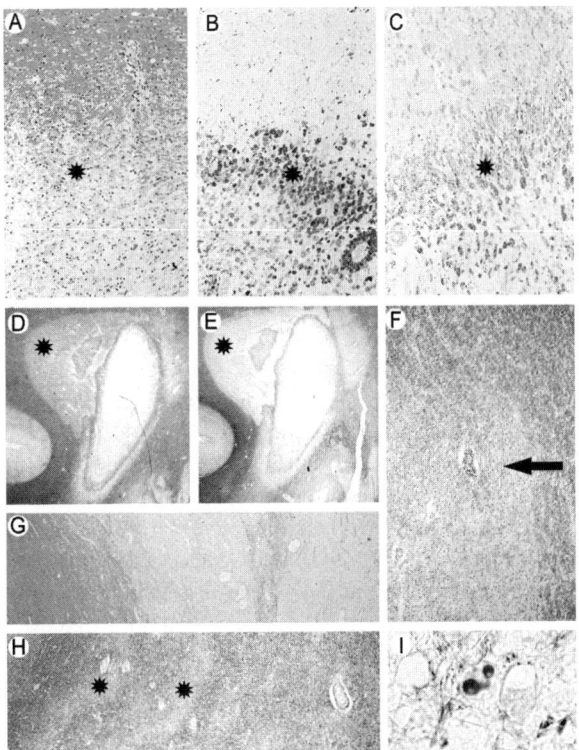

FIGURE 12.9. Immunopathologic heterogeneity of active MS lesions. **A–C.** Antibody/complement-associated demyelination (pattern II). **A.** The lesion is sharply demarcated, and the zone of active myelin destruction (*) is infiltrated by LFB+ containing macrophages. (×100) **B.** The macrophages at the plaque edge form a sharp border. (×100) **C.** Myelin sheaths and myelin degradation products in macrophages at the active plaque edge are immunoreactive for C9neo antigen. (×200) **D–I.** Pattern III. **D, E.** The area of active demyelination (*) still contains MOG (**D**), but has completely lost MAG (**E**) immunoreactivity. (×2) **F.** The lesion edge is ill-defined and contains a broad zone of partial demyelination. Rims of preserved myelin surround blood vessels (arrow). (×100) **G.** The edge of the lesion shows pronounced reduction of myelin density. (LFB-PAS stain, ×30). **H.** Alternating rims of demyelinated (*) and myelinated tissue are seen when stained for MOG. ×30. **I.** Numerous oligodendrocytes show the morphologic changes of apoptosis. (×1000) (© *McAlpines Multiple Sclerosis*, A. Compston, editor. Reprinted with permission.) (*See Color Plates between pages 256–257.*)

myelin proteins. The pronounced reduction in the expression of MAG and CNPase, myelin proteins localized to the most distal extension of the oligodendrocyte cell body—the periaxonal region—has been described in some MS lesions since the early 1980s (Gendelman et al., 1985; Itoyama et al., 1980). A similar selective loss of MAG has been described at the periphery of progressive multifocal leukoencephalopathy (PML) lesions, a known viral infection of glial cells, particularly oligodendrocytes, in which the infected cells are unable to maintain their myelin sheaths (Itoyama et al., 1982). Thus, a pattern of demyelination in which the destruction of MAG precedes that of one of the major myelin proteins (MBP, PLP) suggests a process at the level of the oligodendroglial cell body and is consistent with a distal dying-back oligodendrogliopathy, in which the cell body is unable to support the metabolic demands necessary to maintain the distal axon. Ultrastructurally, this pattern is characterized by alterations in the distalmost extensions of the oligodendroglial processes, the periaxonal region, with uniform widening of inner myelin lamellae and degeneration of inner glial loops antedating destruction of the myelin sheaths. These pathologic alterations have been described in certain experimental models of toxin- and virus-induced demyelination, as well as in several stereotactic brain biopsies obtained for diagnosis in cases of early MS (Ludwin and Johnson, 1981; Rodriguez et al., 1985, 1994).

The preferential loss of MAG, a hallmark of pattern III lesions, is also observed in acute white matter ischemia (Aboul-Enein et al., 2003). Prominent nuclear expression of HIF-$1\alpha$, a specific and sensitive marker for hypoxia-like metabolic injury, also occurs in pattern III. This shared expression of HIF-$1\alpha$ in acute ischemic lesions and pattern III MS lesions suggests that a hypoxia-like metabolic injury may contribute to the pathogenesis of inflammatory white matter damage in a subset of MS patients (Lassmann, 2003a). Microarray analysis of NAWM from 10 postmortem MS brains revealed upregulation of genes, such as HIF-$1\alpha$, known to be involved in neuroprotective mechanisms induced by hypoxic preconditioning (Graumann et al., 2003). Whether this upregulation reflects an adaptation of cells to the chronic progressive pathophysiology of MS or the activation of neuroprotective mechanisms in response to ischemic preconditioning in a subset of patients remains to be determined. Pattern III lesions are ill-defined, typically do not surround blood vessels, and do not demonstrate evidence for immunoglobulin or activated complement. The lesions are invariably associated with T cell-dominated inflammation and microglial activation. However, the quality of microglia activation is different from that observed in other MS lesions Pattern III microglia highly express iNOS but lack other activation markers (e.g., CCR5 and CD14) (Mahad et al., 2004). Because both the pattern of demyelination and microglial activation resemble that found in acute white matter infarction, pattern III lesions may develop on a background of histotoxic hypoxia, perhaps owing to mitochondrial damage induced by oxygen or NO radicals (Aboul-Enein et al., 2005). The recognition of this pathway of tissue damage may have major therapeutic implications, as neuroprotective strategies developed for stroke lesions may also have a beneficial effect in MS.

Pattern IV lesions, the least common, are associated with profound nonapoptotic oligodendrocyte death in the periplaque white matter. Because these lesions are rare and are identified only in autopsy cases, their pathogenesis is unclear.

The frequency of immunopathologic patterns in 286 demyelinating disease cases (238 biopsies, 48 autopsies) analyzed to date reveal a distribution similar to previously published studies (15% pattern I, 58% pattern II, 26% pattern III, 1% pattern IV) (Lucchinetti et al., 2004).

The concept of pathologic heterogeneity in MS lesions is further supported by immunocytochemical studies quantifying chemokines, including cellular expression of CCR1 and CCR5 in pattern II ($n = 21$) and pattern III ($n = 17$) lesions relative to demyelinating activity (Mahad et al., 2004). Infiltrating monocytes in lesions of all patterns coexpress CCR1 and CCR5. In pattern II the number of CCR1 cells decreases, whereas the number of CCR5-expressing cells increases during late active versus early active regions. In contrast, CCR1 and CCR5 cells are equal in all regions of pattern III lesions, resembling the expression pattern seen with acute stroke. These data support the notion of distinct inflammatory microenvironments in pattern II and III lesions and suggest pathologic heterogeneity in MS lesions.

## 3.7   Immunopathologic Heterogeneity: Stage or Patient-Dependent?

Confirming the presence of pathologic heterogeneity in MS lesions must be distinguished from proving that immunopathogenic heterogeneity exists in MS. Whereas many agree MS lesions are pathologically heterogeneous, there is no consensus as to whether it exists in the same patient and is stage-dependent or, alternatively, patient-dependent and reflects pathogenic heterogeneity with a dominant effector mechanism of tissue injury in a given patient.

Barnett and Prineas proposed an alternative hypothesis for the observed immunopathologic heterogeneity in early MS lesions based on their pathologic observations largely derived from a pediatric autopsy case of fulminant MS (Barnett and Prineas, 2004). The presence of oligodendrocyte apoptosis in the absence of overt inflammation in some areas, as well as complement activation with evidence of remyelination in others, was interpreted as evidence of an overlap of features typically associated with patterns III and II, respectively, within a single patient, thus challenging the hypothesis of MS immunopathogenic heterogeneity. The authors suggested that these pattern III-like apoptotic lesions are prephagocytic and represent an early stage in the formation of most if not all lesions in RRMS (starter lesion). They further proposed that these apoptotic lesions either evolve directly into demyelinated lesions or, once remyelinated, become the target of additional demyelinating episodes. However, in all pattern III cases identified in the Lucchinetti et al. series, inflammation and myelin phagocytosis were seen in association with oligodendrocyte apoptosis, suggesting that apoptosis is a secondary event (rather than a primary event) in lesion formation. Furthermore, the limited extent of oligodendrocyte survival, remyelination, or shadow plaques in typical pattern III MS lesions argues against the sequence of pathologic events as suggested by Barnett and Prineas. If all pattern II cases represented preexisting remyelinating lesions that had exhibited new demyelinating activity, one would expect to find remnants of prior lesions (fibrillary gliosis or remyelinated shadow plaques) surrounding areas of active demyelination in all pattern II cases. This has not been true in more than 150 pattern II cases analyzed to date (Lucchinetti et al., 2004). Most importantly, the lack of overlap of patterns among all active lesions or lesion areas from a given patient in more than 280 cases analyzed continues to support a dominant pathway of active lesion formation operating in a given individual patient (Lucchinetti et al., 2004). Furthermore, therapeutic responses associated with specific pathologic subtypes (Keegan et al., 2005) and radiologic differences among immunopathologic patterns continue to suggest that pathologic heterogeneity reflects pathogeneic heterogeneity in the formation of new active MS lesions during early MS phases (Lucchinetti et al., 2004).

How long these patterns persist into disease chronicity is unknown, as the mechanisms of MS lesion formation in chronic active plaques during late phases of the disease are not well understood. Although slowly expanding rims of active demyelination with microglial activation and limited inflammation have been described (Prineas et al., 2001), whether immunopathologic heterogeneity, as seen in early MS lesions, persists in these phases is uncertain. It is possible that once the active demyelinating phase of the disease subsides there is a common mechanism underlying the disease progression among all patients. Studies to address the long-term impact of these various pathologic subtypes on long-term clinical and radiographic progression are ongoing.

## 3.8 Immunopathologic Heterogeneity Among MS Variants

### 3.8.1 Balo Concentric Sclerosis

Considered a variant of inflammatory demyelinating disease closely related MS, Baló concentric sclerosis (BCS) is characterized pathologically by large demyelinating lesions with a peculiar pattern of alternating layers of preserved and destroyed myelin, mimicking the rings of a tree trunk (Fig. 12.10). Clinically, BCS resembles Marburg MS with their similar acute fulminant onsets followed by rapid progression to major disability and death within months (Courville, 1970; Kuroiwa, 1982; Marburg, 1906). Of interest, one of the cases in Marburg's original series (case 3) contained extensive concentric lesions (Marburg, 1906). Reports of less fulminant disease have been described (Korte et al., 1994; Revel et al., 1993), and smaller concentric rims of demyelination have been observed in lesions from some MS patients with a more classic acute or chronic disease course. T2-weighted MRI may reveal a distinct pattern of hypo/isointense and hyperintense rings corresponding to bands of preserved and destroyed myelin, permitting antemortem diagnosis (Chen et al., 1996; Garbern et al., 1986; Hanemann et al., 1993; Karaarslan et al., 2001; Korte et al., 1994; Spiegel et al., 1989).

The etiology of the concentric demyelination in this variant of MS is unknown. Pathologic evaluation of 12 autopsied patients with Balo-type concentric lesions demonstrated expression of iNOS in macrophages and microglia in all active concentric lesions. A role for hypoxia in mediating tissue injury and contributing to lesion concentricity in BCS was suggested by the expression of HIF-1 and HSP70 mainly in oligodendrocytes and to a lesser extent in astrocytes and macrophages at the edge of active lesions and in the outermost layer of preserved myelin

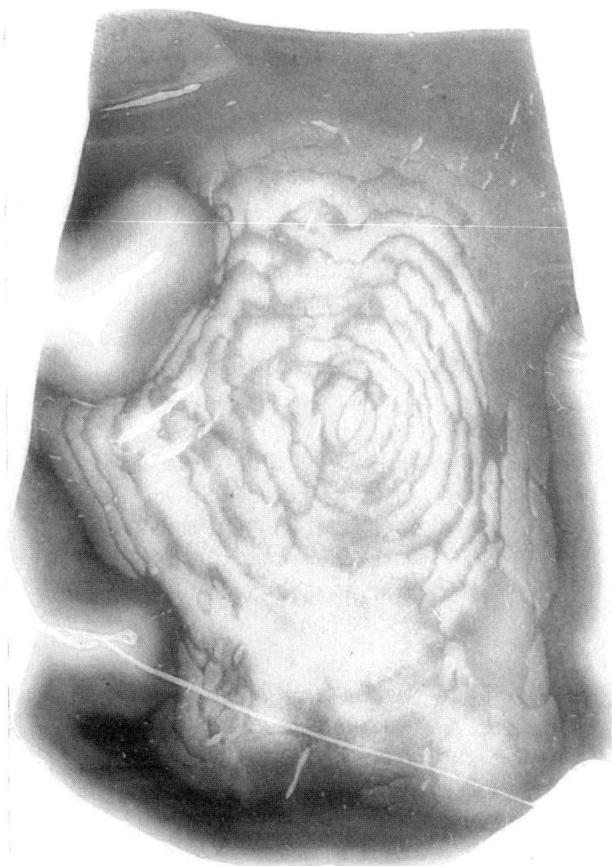

FIGURE 12.10. Balo's concentric sclerosis. This MS variant is characterized by alternating rings of demyelinated and myelinated tissue (MOG immunocytochemistry).

(Stadelmann et al., 2004). Because of their neuroprotective effects, the rim of periplaque tissue expressing these proteins may be resistant to further hypoxia-like injury in an expanding lesion and therefore may remain as a rim of preserved myelinated tissue.

### 3.8.2  Neuromyelitis Optica (Devic's Disease)

Neuromyelitis optica (NMO) is an idiopathic inflammatory CNS demyelinating disease characterized by monophasic or relapsing attacks of optic neuritis and myelitis. Despite traditional views that the lesions of NMO are restricted to the optic nerves and spinal cord, MRI studies of NMO have revealed evidence of brain lesions in 60% of patients, fulfilling the Wingerchuk criteria

(Wingerchuck et al., 2005) for the diagnosis of NMO (except for brain MRI findings) (Pittock et al., 2006).

Pathologically, NMO lesions demonstrate extensive demyelination across multiple spinal cord levels associated with necrosis and cavitation, as well as acute axonal damage in both gray and white matter (Fig. 12.11). There is a pronounced loss of oligodendrocytes in the lesions, and inflammatory infiltrates are comprised of large numbers of macrophages associated with large numbers of perivascular granulocytes and eosinophils, as well as rare CD3+ and CD8+ T cells. A pronounced vasculocentric deposition of immunoglobulin and complement C9neo antigen is associated with prominent vascular fibrosis and hyalinization in both active and inactive lesions (Fig. 12.11) (Lucchinetti et al., 2002a). These findings implicate a potential role for specific autoantibody and local activation of complement in this disorder's pathogenesis. This hypothesis is supported by serologic and clinical evidence of B cell autoimmunity in a large proportion of patients with NMO (Wingerchuck et al., 1999). Furthermore, an NMO IgG-specific marker autoantibody of NMO has been identified that binds at or near the BBB and outlines CNS microvessels, pia, subpia, and Virchow-Robin space; it distinguishes NMO from MS (Fig. 12.12) (Lennon et al., 2004). The staining pattern of patients' serum IgG binding to mouse spinal cord is remarkably similar to the vasculocentric pathologic pattern of immunoglobulin and complement

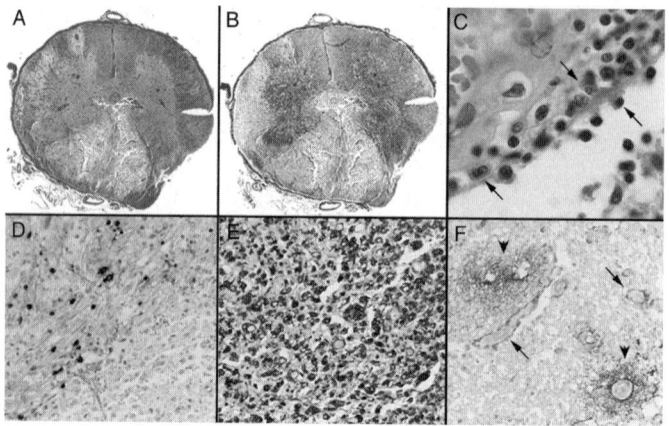

FIGURE 12.11. The pathology of neuromyelitis optica lesions reveals extensive demyelination across multiple spinal cord levels, associated with acute axonal injury, necrosis, and cavitation, involving both gray and white matter (**A, B**). Inflammatory infiltrates of active lesions are associated with large numbers of perivascular granulocytes and eosinophils (**C**), as well as extensive macrophage infiltration (**E**) and rare CD3+ and CD8+ T lymphocytes. There is also a pronounced loss of oligodendrocytes in the lesions (**D**). Actively demyelinating lesions demonstrate a characteristic rim and rosette pattern of immunoglobulin deposition and complement activation (**F**) surrounding blood vessels. (From Coles et al., 2005. Reprinted and modified with permission.) (*See Color Plates between pages 256-257.*)

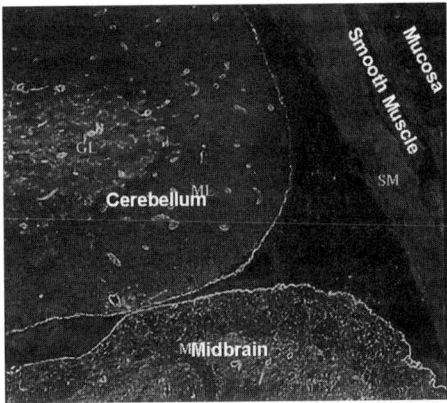

FIGURE 12.12. Immunofluorescence reveals that NMO-IgG stains microvessels, pia, and subpia. (From Lennon et al., 2004. Reprinted and modified with permission.)

deposition. The sensitivity and specificity for this autoantibody are 73% (95% CI = 60–86%) and 91% (95% CI = 79–100%), respectively, for NMO in North American patients and 58% (95% CI = 30–86%) and 100%, respectively, for optic/spinal MS in Japanese patients (Lennon et al., 2004). NMO IgG binds selectively to the mercurial-insensitive water channel protein aquaporin-4 (AQP4), which is concentrated in astrocytic foot processes at the BBB (Lennon et al., 2005). AQP4 is the predominant water channel in the brain (Amiry-Moghaddam et al., 2003). It is also expressed to a limited extent in stomach, kidney, lung, skeletal muscle, and inner ear (Amiry-Moghaddam et al, 2003). In the brain, AQP4 is abundant in the optic nerve and spinal cord, but it is also found throughout the brain (Amiry-Moghaddam et al, 2003). AQP4 has an important role in brain water homeostasis; and consistent with its location in the CNS, it is involved in the development, function, and integrity of the interface between the brain and blood and the brain and CSF (Amiry-Moghaddam et al., 2003; Jung et al., 1994). Studies to determine whether the antibody is pathogenic are ongoing. The careful immunopathologic characterization of NMO lesions reflects a successful paradigm demonstrating the power of clinical-pathologic-serologic approaches to define disease mechanisms and identify surrogates of pathology.

## 4    Oligodendrocyte Pathology and Remyelination in Relation to Lesion Activity and Disease Phase

Oligodendrocytes are susceptible to damage via a number of immune or toxic mechanisms present in the active plaque. They include cytokines such as TNFα (Bitsch et al., 2000a), reactive oxygen or nitrogen species, excitatory amino acids such as glutamate (Pitt et al., 2000), complement components, proteolytic and lipolytic enzymes, T cell-mediated injury via T cell products (perforin/lymphotoxin)

(Selmaj et al., 1991b), the interaction of Fas antigen with Fas ligand (D'Souza et al., 1996), CD8+ MHC I-mediated cytotoxicity (Jurewicz et al., 1998), or persistent viral infection (Merrill and Scolding, 1999).

Most neuropathologic studies agree that mature oligodendrocytes are largely absent from the chronic long-standing MS plaque. However, the situation is less clear with regard to the fate of the oligodendrocyte in actively demyelinating MS lesions sampled during early disease phases (Fig. 12.13). Some studies describe an abundance of oligodendrocytes in the early active MS lesion with a reduction in older lesions, suggesting that oligodendrocytes may escape the primary attack but become the target of additional immune mechanisms during lesion evolution (Raine et al., 1981; Selmaj et al., 1991a, 1992). Prineas et al. described partial reduction of oligodendrocytes at the active plaque edge with reappearance in the plaque center, which was thought to reflect initial destruction followed by progenitor recruitment (Prineas et al., 1992a). Our group has reported that the density of oligodendrocytes in actively demyelinating lesions varies with respect to stage of demyelinating activity and among patients (Bruck et al., 1994; Ozawa et al.,

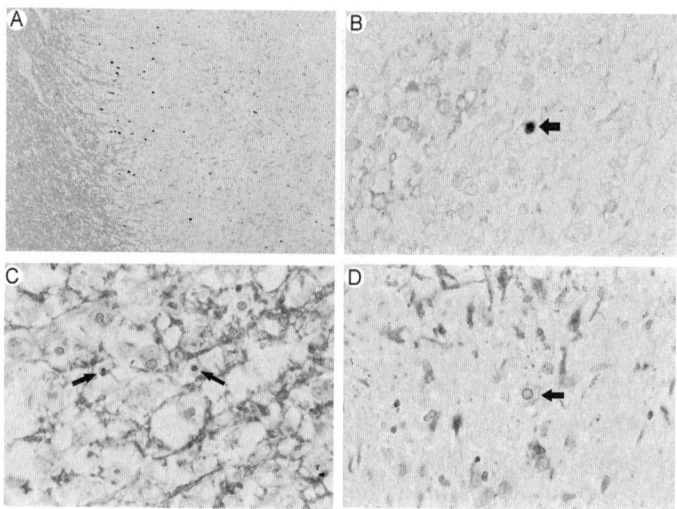

FIGURE 12.13. Oligodendrocyte death in MS lesions. **A.** Chronic MS. Detection of DNA fragmentation by in situ tailing (black) immunocytochemistry with anti-myelin oligodendrocyte glycoprotein (red); a radially expanding lesion. Numerous cells with DNA fragmentation are seen at the lesion border. (×50) **B.** Dying oligodendrocyte identified by immunoreactivity with MOG antibody (red). DNA fragmentation is visualized by in situ tailing (black; arrow). (×500) **C.** Acute MS. Immunocytochemistry with anti-MOG reveals macrophages containing early active myelin debris as well as numerous MOG+ oligodendrocytes with condensed nuclei typical of apoptosis (arrows). (×350) **D.** Active MS lesion reveals a swollen MOG+ oligodendrocyte suggestive of necrosis (arrow). (×350) (© *McAlpines Multiple Sclerosis*, A. Compston, editor. Reprinted with permission.) (*See Color Plates between pages 256-257.*)

1994). These varied observations and interpretations may be explained in part by the prior lack of effective oligodendrocyte markers available for use in paraffin tissue (Prineas et al., 1992a; Yao et al., 1994) and the limited number of well staged lesions with differing demyelinating activity included in the series. Therefore the ability to draw dynamic conclusions based on static pathologic observations was limited.

However, the more recent availability of markers to label oligodendrocytes in paraffin-embedded formalin-fixed tissue led to a systematic analysis of the density of oligodendrocytes in more than 300 lesions from 113 patients with MS during the early phase of the disease (Luccinetti et al., 1999). The numbers were correlated with stages of myelin degradation products in macrophages, thereby providing a snapshot of the temporal evolution of the lesion. Oligodendrocytes were labeled with PLP mRNA, an early marker of oligodendrocytes actively engaged in myelin synthesis and maintenance but not present in surviving oligodendrocytes that have lost their myelin sheaths. Cells were also stained with antibodies directed against MOG, which is expressed on the surface of myelin sheaths and terminally differentiated oligodendrocytes late in myelination. MOG is detectable on oligodendrocytes that have survived demyelination following wallerian degeneration (Ludwin, 1990).

Two principal groups of oligodendrocyte pathology were identified in these early MS lesions (Lucchinetti et al., 1999). Group I (70% of the cases) was characterized by a variable (minor to moderate) reduction of oligodendrocytes at the active demyelinating plaque edge, with reappearance of these cells in inactive or remyelinated regions. These lesions were associated with prominent remyelination. Although markers for the identification of immature oligodendrocytes were not used, the presence of cells expressing PLP mRNA, but not MOG, suggests that these oligodendrocytes may have been derived from the progenitor pool. Group II (30% of cases) was characterized by extensive destruction of oligodendrocytes at active sites of demyelination in the absence of increased oligodendrocyte numbers in inactive plaque areas. In these lesions, remyelination was sparse or absent. Although there was profound heterogeneity of oligodendrocyte damage among patients, lesions from a single individual exactly matched for stage of demyelinating activity showed similar oligodendrocyte densities. Furthermore, the extent of early remyelination correlated with oligodendrocyte numbers in the lesion.

These data suggest that oligodendrocyte survival undergoes dynamic changes with the evolution of activity, onset of remyelination, and disease duration. Early in MS, remyelination may be extensive and may occur simultaneously with demyelination. During the early stage of remyelination (myelin sheath formation), inflammation with prominent macrophage infiltration may be prominent in the lesion. The extent of remyelination at these early stages appears to depend on the availability of oligodendrocytes or their progenitor cells in the lesion.

The recruitment of new oligodendrocytes is typically more efficient during the early stages of relapsing remitting MS, where extensive remyelination is seen in about two-thirds of cases, whereas in the remaining one-third of cases few new oligodendrocytes appear in inactive lesions and remyelination is low or absent (Luccinetti et al., 1999). The profound heterogeneity in extent and topography of oligodendrocyte destruction in active demyelinating lesions suggests that myelin, mature oligodendrocytes, and possibly oligodendrocyte progenitors are differentially affected in subsets of MS patients. Different mechanisms of myelin and/or oligodendrocyte injury may be operating in an individual MS patient and may thereby influence the likelihood of effective remyelination in the MS lesion. These patterns of oligodendrocyte reaction correlate with different immunopathologic patterns of demyelination (Luccinetti et al., 2000).

The pattern is fundamentally different after several years of the disease and in patients with chronic progressive disease, where mature oligodendrocyte numbers in the plaques are, in general, very low (Mews et al., 1998), and remyelination is generally less extensive. The failure of remyelination during these late phases may contribute to disease progression. Although remyelination in chronic lesions is often restricted to the plaque edge, it may also extend throughout the lesion and form a shadow plaque, representing a completely remyelinated lesion. These late remyelinating lesions usually contain few macrophages and are typically associated with profound fibrillary gliosis. The inflammatory response may therefore provide a key stimulus promoting endogenous remyelination in early active MS lesions, which raises the issue of whether, paradoxically, antiinflammatory therapy might contribute to the failure of remyelination during later progressive phases of the disease.

The presence of cells in very early stages of oligodendrocyte development identified in completely demyelinated plaques devoid of mature oligodendrocytes and in chronic lesions devoid of remyelination (Wolswijk, 1998) suggests that the failure of remyelination during these later disease phases is not due to a lack of oligodendrocyte progenitors, as is the case in early remyelinating MS lesions; rather, the lesion microenvironment may not be receptive to remyelination signals (Chang et al., 2002). Whether this is due to an imbalance of growth factors, abnormal composition of axons, glial scarring, or impaired axon–oligodendrocyte interaction is uncertain. To what extent progenitor cells already present in chronic MS lesions can be stimulated to divide, repopulate the lesion, and initiate remyelination must still be determined. Observations on the retina show that axons that have been demyelinated a considerable length of time can still be myelinated following transplantation, and thus axonal changes resulting from demyelination are unlikely responsible for remyelination failure (Franklin, 2002). In addition, oligodendrocyte progenitor cell (OPC) transplant studies have demonstrated that OPCs can repopulate OPC-depleted, chronically demyelinated astrocytosed lesions, and the induction of inflammation stimulates remyelination. Therefore, axonal changes induced by chronic demyelination are unlikely to contribute to remyeliation failure. Rather, remyelination fails either because OPCs fail to repopulate areas of demyelination or because they require the induction of inflammation to enable their proliferation and subsequent differentiation into remyelinating cells (Foot and Blakemore, 2005).

## 5    Axonal Pathology and Mechanisms of Injury During Early and Late Disease

The earliest pathologic descriptions of MS recognized that axonal injury and damage occurred in MS lesions (Kornek and Lassmann, 1999). However the correlation of axonal injury and irreversible disability in MS has refocused attention on axonal pathology in MS. MS patients may become irreversibly disabled in a stepwise fashion due to incomplete recovery from a relapse (i.e., relapse-related disability) (Lublin et al., 2003) or as a result of gradual slow progression (i.e., progression-related disability) that occurs independent of clinical exacerbations or MRI evidence of lesion activity. Attack-related disability may occur in patients with relapsing MS (i.e., RRMS or SPMS with ongoing exacerbations), whereas progression-related disability may occur in patients with progressive forms of the disease (i.e., PPMS or SPMS with or without superimposed exacerbations).

Although in relation to the myelin sheath axons are relatively spared in MS lesions, the degree of axonal destruction can be profound. Axonal loss in MS is widespread, tract-specific, and size-specific (DeLuca et al., 2004). In MS plaques, axon density is reduced by 20% to 80% relative to the periplaque white matter and is highly variable among patients (Bitsch et al., 2000b). Detailed analysis of axonal and neuronal loss in the optic system suggests that small axons are preferentially susceptible to injury in MS patients, supporting a role for differential axonal vulnerability between small and large axons (Evangelou et al., 2001). This may be due to different expression levels of ion channels or excitotoxin receptors between large- and small-diameter axons or that energy failure preferentially affects small-diameter axons owing to their large surface/cytoplasmic volume ratio (Aboul-Enein and Lassmann, 2005).

The causes and pathogenesis of axonal damage in MS are still uncertain. It is likely that inflammation and demyelination play an important role, although the pathogenic mechanisms of injury may differ depending on the type of disability (relapse-related versus progression-related), the stage of demyelinating activity in the lesion, and the phase of the disease phase (early or late). Furthermore, axonal injury is not restricted to white matter plaques. MR spectroscopy studies demonstrate reductions in NAA, a surrogate marker of axonal injury, not only in the MS lesion but also involving the normal-appearing white matter.

During acute demyelination axons are likely damaged owing to the action of toxic inflammatory mediators such as proteases, cytokines, excitotoxins, and free radicals (Kornek et al., 2000). Markers of acute axonal damage in MS lesions include the dephosphorylation of neurofilaments, disturbances of axonal transport, the expression of specific calcium channels, and axonal transaction (Fig. 12.14). When axons are transected, the proximal segment still connected to the neuronal cell body survives, whereas the distal axonal segment undergoes wallerian degeneration. Anterograde transport from the neuronal perikarya continues leading to the accumulation of organelles, proteins, and other molecules at the ends of the proximal axonal segment and results in the formation of axonal

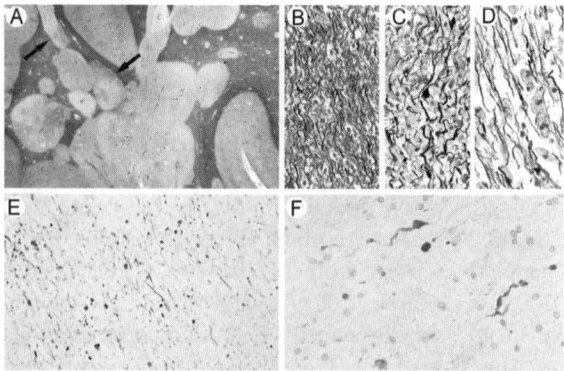

FIGURE 12.14. Axonal pathology in MS lesions. **A.** Multiple demyelinated plaques with variable axonal density in the white matter of a patient with long-standing disease. Arrows identify new demyelianted areas in previously existing plaques and show reduced axonal density compared with the remained of the preexisting lesion (Bielschowsky silver impregnation for axons, ×2). **B.** Axonal density in the normal-appearing white matter (NAWM). Axonal density is reduced mildly (**C**) and moderately (**D**) in two different plaques from the same MS patent (Bielschowsky silver impregnation for axons; ×400). **E, F.** APP expression illustrates dystrophic axons in an active MS lesion reflecting acute axonal injury (**E**, ×300; **F**, ×600). © *McAlpines Multiple Sclerosis*, A. Compston, editor. Reprinted with permission.) (*See Color Plates between pages 256-257.*)

spheroids. These axonal profiles may degenerate or recover axonal function. Disturbances of axonal transport or axonal transection are best detected by the accumulation of amyloid precursor protein (APP), produced by neurons, at sites of impaired transport or axonal damage. Positive correlations between the activity of the demyelinating process and the level of axonal injury (i.e., APP expression) supports a role for inflammatory mediators in contributing to the axonal damage (Fergusson et al., 1997; Trapp et al., 1998). An association between the numbers of CD8+ T cells and the extent of axonal damage suggests T cell-mediated cytotoxicty may be involved, possibly via the release of perforin and/or granzymes from the cytotoxic T cell (Bitsch et al., 2000b). The close contact of CD8+ T cells with a polar orientation of cytotoxic granules toward the axon can be found in actively demyelinating lesions (Neumann et al., 2002). A CD8–MHC I-mediated pathway of axon destruction has also been suggested from experimental studies ( Rivera-Quinones ert al., 1998). Mice deficient for the MHC I light chain $\beta_2$-microglobulin develop inflammatory demyelination but no early axonal damage or clinical deficit following Theiler's virus infection (Murray et al., 1998; Rivera-Quinones et al., 1998). These data demonstrate that the mechanisms of inflammatory demyelination and axonal injury can be dissociated, and functional CD8+ T cells are needed to mediate axonal damage.

Inducible NOS is another molecule mediating functional and structural axon damage (Smith et al., 2001). Macrophages and activated microglia are reported in close contact with degenerating axons. These cells produce inflammatory mediators such as NO and iNOS and are found to damage mainly the small-diameter electrically active axons in vitro (Garthwaite et al., 2002; Smith et al., 2001). NO inhibits the mitochondrial respiratory chain and may lead to energy failure, resulting in failure of the energy-dependent $Na^+/K^+$-ATPase. As a consequence, there is a secondary increase in intraaxonal $Na^+$ as well as an increase in intra-axonal $Ca^{2+}$ due to reversal of the $Na^+/Ca^{2+}$ exchanger (Kapoor et al., 2003). Alternations in $Na^+$ channel subtype and expression and redistribution have been demonstrated in demyelinated axons of EAE (Craner et al., 2003). A postmortem study on spinal cord and optic nerve tissue from patients with severe SPMS revealed a pattern of Na channel expression in acute MS lesions that is similar to the pattern observed in EAE. Control white matter from normal patients displayed abundant MBP and the expected pattern of focal expression of $Na_v1.6$ that was confined to the nodes of Ranvier. However, acute MS lesions demonstrated that $Na_v1.6$ and $Na_v1.2$ were expressed along the demyelinated axons (Craner et al, 2004). A large number of APP-positive axons were present in these plaques, a marker indicative of impaired axonal transport and acute axonal injury. Almost all APP+ axons in the MS lesions co-expressed $Na_v1.6$ channels and the $Na^+–Ca^{2+}$ exchanger was expressed over extended regions. The presence of $Na_v1.6$ was thought to predispose the axons to injury.

The phase of massive acute axonal injury in early active MS lesions typically lasts only a few days to weeks and is most likely the substrate for relapse-related disability predominantly seen during inflammatory phases of the disease (in early relapsing-remitting or active inflammatory SPMS). However, the magnitude of axonal loss in chronic lesions suggests that mechanisms other than inflammatory demyelination may contribute to axonal damage during these later disease phases, when gradual progression appears to occur independent of measurable clinical or radiographic inflammatory activity. During early disease phases, axonal injury convincingly correlates with inflammation, whereas during later phases this correlation is less evident. This might explain the benefit of antiinflammatory and immunomodulatory agents on relapsing MS, with limited if any benefit on gradual disease progression. This is further supported by the effect of Campath-1H, a humanized anti-leukocyte (CD52) monoclonal antibody in a subgroup of SPMS patients in whom the treatment dramatically suppressed MRI markers of cerebral inflammation and limited clinical relapses, but the patients continued to clinically progress, with little effect on MRI markers of axon degeneration (Coles et al., 2004).

Pathology studies reveal that slow, ongoing axonal destruction is present in inactive MS lesions that lack inflammation (Kornek et al., 2000). Although only a few axons are destroyed at a given time point, such lesions may persist in the CNS for years. In addition, repeated demyelination in previously remyelinated lesions may contribute to additional axonal loss in chronic MS (Prineas et al., 1993b). Chronically demyelinated axons may also degenerate owing to the lack of trophic support from myelin and oligodendrocytes. Mice lacking certain myelin proteins (MAG and PLP) demonstrate late-onset axonal pathology and evidence for an increased incidence of wallerian degeneration in the absence of inflammation (Sadahiro et al., 2000; Yoshikawa, 2001). Denuded axons are also more vulnerable to damage to soluble or cellular immune factors that may still be present in the inactive plaque. This is suggested by a study demonstrating that remyelinated axons are protected from ongoing acute axonal damage (Kornek et al., 2000). Secondary (wallerian) degeneration also contributes to diffuse axonal loss (Lassmann, 2003b).

The situation is further aggravated by the fact that demyelinated axons suffer large current leaks through newly exposed $K^+$ channels and the large capacitance of the naked internodal axolemma. In an attempt to restore conduction, these fibers express Na channels along their surface. Impulse traffic under these conditions is inefficient compared to salutatory conduction. This results in increased energy demand as ion gradients are restored by ATPase-consuming pumps (Stys, 2004). Mitochondria are recruited to demyelinated regions to meet the increased energy requirements necessary to maintain conduction. Therefore the mitochondria are functioning at full capacity. The axon may be able to support its function for several years via antioxidant defenses, but free radical damage continues to accumulate and eventually mitochondrial function is compromised (Andrews et al., 2005). As a result, ATP concentration in the axon decreases, resulting in irreversible damage to the axon. Studies have shown reduced mitochondrial complex I and III activity in the brain and spinal cord of MS patients, raising the possibility that in addition to extrinsic inhibitors of mitochondrial respiration such as NO there may be inherent defects in the organelles in MS that may further compromise energy-producing capacity (Lu et al., 2000). This damage could develop initially in association with inflammation but could also occur late in the absence of inflammation, whereby the mitochondria themselves produce excess ROS, driving the oxidative damage that leads to eventual cell degeneration.

Another study based on global transcript profiles, biochemical analysis, and morphologic studies of motor cortex samples suggested that motor neurons in chronic MS patients have significantly impaired mitochondrial function and decreased inhibitory innervations (Dutta et al., 2006). The authors described ultrastructural changes that support $Ca^{2+}$-mediated destruction of chronically demyelinated axons in MS patients. They compared expression levels of 33,000 characterized genes in postmortem cortex from six control and six MS brains matched for age, sex, and postmortem interval and focused on changes in oxidative phosphorylation and inhibitory transmissions. Compared with controls, 488 transcripts were decreased and 67 were increased in the MS cortex. Altogether, 26 nuclear encoded mitochondrial genes and the functional activities of mitochondrial respiratory chain complexes I and III were decreased in the MS motor cortex. Reduced mitochondrial gene expression was specific for neurons. In addition, presynaptic and postsynaptic components of $\gamma$-aminobutyric acid (GABA)ergic neurotransmission, as well as the density of inhibitory interneuron processes, were decreased in the MS motor cortex. These data support increased excitability of upper motor neurons that have reduced capacity to produce ATP. This mismatch between energy demand and ATP supply could cause further degeneration of chronically demyelinated axons in MS and contribute to neurologic disability.[184]

Once axonal injury has been triggered, a cascade of downstream mechanisms leading ultimately to axonal disintegration occurs (Lassmann, 2003b). These mechanisms are similar in a variety of pathologic conditions including inflammation, ischemia, and trauma. Acute axonal injury leads to a disturbance in the axoplasmic membrane permeability and subsequent energy failure, leading to uncontrolled sodium influx into the axoplasm; this, in turn, reverses the sodium/calcium exchanger and results in excess intraxonal calcium. $Ca^{2+}$-dependent proteases, which degrade cytoskeletal proteins, are then activated, further impairing axonal transport. Voltage-gated calcium channels (VGCCs) accumulate at sites of disturbed axonal transport, leading to further $Ca^{2+}$ influx and eventually dissolution of the axonal cytoskeleton and axonal disintegration. Whereas the window for neuroprotection is brief after trauma and during ischemia, MS offers an opportunity to treat patients and limit the subsequent injury before too much axonal damage has occurred. Therapeutic strategies that inhibit the various steps in this execution phase of axonal destruction—such as $Na^+$ channel blockers, inhibitors of the $Na^+$–$Ca^{2+}$ exchanger, blockade of VGCCs, inhibition of calcium-dependent proteases—may help limit axonal destruction in MS. Clinical trials are needed with these agents to determine if they slow disease progression.

# 6    Gray and Nonlesional White Matter Pathology in MS

Although pathologic alterations have previously been described in both the gray matter and so-called normal appearing white matter (NAWM) of MS patients, they have received relatively little attention until recently. By concentrating on focal white matter lesions, previous neuropathologic studies failed to find major differences between patients with relapsing or progressive disease (Luccinetti and Bruck, 2004). However more recent studies focusing on gray and diffuse white matter MS pathology and its relation to focal white matter lesions have suggested that involvement of these regions may contribute to disease progression.

## 6.1   Gray Matter Pathology

MS may involve the gray matter, either as a classically demyelinated plaque or as neuronal loss and atrophy following retrograde degeneration from white matter lesions (Lumsden, 1970). Demyelinated plaques may also be found in deep cerebral nuclei (Cifelli et al., 2002) or in the cerebral cortex (Lumsden, 1970). Although MR studies demonstrate atrophy and intrinsic NAWM and normal-appearing gray matter (NAGM) abnormalities are present from the earliest stages of the disease, intracortical lesions are largely undetected by current imaging techniques (Filippi, 2003; Kidd et al., 1999). As a result, the pathology of gray matter lesions in MS has received little attention in comparison to white matter lesions. Although cortical plaques have been described in MS, little is known regarding their impact on neurologic disability. Most MRI studies have correlated cortical lesions to depression, cognitive deficits, or epilepsy.

Various classification schemes for cortical plaques have been described (Bo et al., 2003b). Bo et al. proposed that type I lesions represent mixed gray matter/white matter lesions, whereas intracortical lesions represent types II to IV, with type II lesions being small intracortical lesions, type III extending from the pia into the deeper cortical layers, and type IV affecting the entire cortex from the pial surface to the underlying white matter. These cortical lesions have a predilection for the cortical sulci as well as the cingulate, temporal, insular, and cerebellar cortex. Subpial cortical lesions may also be associated with meningeal inflammation. Previous studies suggest that extensive subpial demyelination is a prominent feature of progression, mainly during the late stages of MS (Fig. 12.15) (Kutzelnigg et al., 2005).

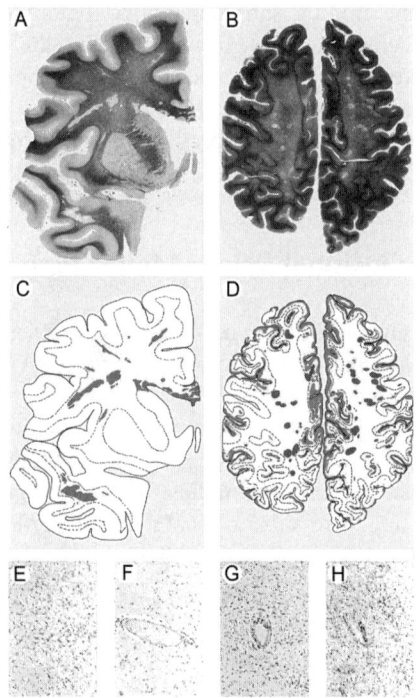

FIGURE 12.15. Comparison of acute and primary progressive MS. **A, C.** Acute MS (7 months disease duration) reveals multiple plaques of demyelination in the white matter (green lesions in **C**). **B, D.** Primary progressive MS with multiple plaques in the white matter (green lesions in **D**) and massive diffuse white matter injury (light myelin staining in **B**). There is also extensive cortical demyelination in the cerebral cortex (red lesions in **D**). (**A, B.** Kluver myelin stain, ×0.3) **E, F.** NAWM in acute MS with HLA-D expression in microglia (**E**) and no expression of iNOS (**F**). **G, H.** NAWM in PPMS with massive expression of HLA-D (**G**) and iNOS (**H**) in the microglia. (×50). (© *McAlpines Multiple Sclerosis*, A. Compston, editor. Reprinted with permission.) (*See Color Plates between pages 256–257.*)

The mechanisms of cortical demyelination in MS are largely unknown, nor is it known whether the same therapies used to limit white matter demyelination influence gray matter pathology. Prior neuropathologic studies have emphasized the paucity of leukocytes in cortical demyelinated lesions compared to white matter lesions, suggesting that the regulation of leukocyte migration may differ in the cortex relative to that in the white matter, and the trafficking of leukocytes into the white matter is related mainly to myelin phagocytosis rather than myelin destruction (Bo et al., 2003a; Peterson et al., 2001). A quantitative study of T cell and B cell infiltrates showed no significant differences between the normal cortex of control patients and those with MS or demyelinated lesions in the cortex (Bo et al., 2003a). However, cortical plaques are associated with massive activation of cortical microglia (Bo et al., 2003a) and high expression levels of iNOS. Cortical lesions tend to be associated with less tissue destruction, likely due to the limited amount of myelin coupled with the limited axonal and neuronal injury.

However, most prior studies describing the features of cortical plaques are derived from chronic long-standing MS cases. A study that analyzed cortical pathology in early MS cases revealed that cortical demyelination is frequent and may be both inflammatory and destructive, suggesting that the lack of inflammation reported in prior studies likely reflects a stage-dependent effect (unpublished observations).

The degree of cortical involvement and whether it correlates with the clinical course or disability in MS is uncertain. Cortical demyelination could affect neuronal, dendritic and axonal function, viability, and survival. A study demonstrated the presence of apoptotic neurons in the demyelinated cortex (Peterson et al., 2001). Another study reported that neuronal density was significantly reduced in cortical lesions compared to the adjacent normal cortex (approximately 20%) from two MS cases showing extensive cortical demyelination (Vercellino et al., 2005). This may be relevant to the pathogenesis of neurologic and cognitive disability in MS and could, in part, explain why the disease progresses in PPMS in the absence of extensive white matter abnormalities. Degeneration of cortical neurons could also partly explain the diffuse NAA loss observed in the NAWM and cortex of PPMS patients. Synaptic loss as determined by optical density of synaptophysin staining does not appear to be a main feature of cortical lesions (Vercellino et al., 2005).

Cortical damage could also lead to secondary tract degeneration, which may account for some of the diffuse spinal cord changes observed in PPMS. In addition to demyelination, the cerebral cortex of MS patients may also be affected by tissue loss and atrophy, particularly at sites of severe focal or diffuse white matter injury. Neurons in such lesions may show signs of retrograde reaction, such as central chromatolysis. Quantitative MRI analyses show that cortical atrophy may occur early and to some extent predicts the clinical course and the development of cognitive impairment (Bozzali et al., 2002). Certain observations suggest that patients with SPMS and PPMS have more widespread cortical demyelination than RRMS patients (Kutzelnigg et al., 2005; Vercellino et al., 2005). These observations could account for the poor correlation between clinical disability and white matter lesion borden in progressive MS, as determined by conventional MRI (Ingle et al., 2003; Rovaris et al., 2001).

## 6.2   "Normal Appearing White Matter" Pathology

Although diffuse NAWM injury is in part due to axonal transection in plaques, leading to secondary (wallerian) degeneration, MRI data indicate that the extent of tissue damage in focal plaques does not fully explain the degree of diffuse white matter changes (Filippi et al., 1995,1998) but suggest that global permanent neurologic deficit may to a major degree be determined by global and diffuse changes in NAWM (Ciccarelli et al., 2001; Dehmeshki et al., 2003; Evangelou, 2003; Rovaris et al., 2002).

There are few pathology studies of the NAWM regarding MS. Most have described mild inflammation (mainly CD8+ T cells), microglial activation, gliosis, increased expression of proteolytic enzymes in astrocytes and microglia, diffuse axonal injury, and nerve fiber degeneration (Allen et al., 1979, 1981, 2001; Evangelou et al., 2000; Kornek et al., 2000; McKeown and Allen, 1978). One study compared the global brain damage in acute, relapsing, and progressive MS and found a diffuse inflammatory process characterized by perivascular and paranechymal inflammatory infiltrates in progressive, but not acute or relapsing, disease (Fig. 12.15) (Kutzelnigg et al., 2005). The extent of inflammation was distributed globally throughout the brain and was associated with widespread microglial activation characterized by CD68 expression, a marker for phagocyitc activity, as well as MHC II antigen and iNOS expression. Despite the lack of primary demyelination in the "normal" white matter, axonal spheroids and terminal axonal swellings were variably present throughout the tissue. The extent of inflammation and axonal injury in the NAWM, as well as the degree and character of cortical demyelination, did not correlate with the number, distribution, activity, or destructiveness of focal white matter lesions (Kutzelnigg et al., 2005).

## 7   Inflammation/Neurodegeneration Paradox: Why Does It Exist in MS?

The apparent dissociation between early inflammation and subsequent disease progression is a well recognized feature of MS and is referred to as the inflammation/neurodegeneration paradox. Understanding the basis of this paradox is critical to develop more effective therapeutic strategies. Clinical and radiographic observations underscore this apparent paradox. Relapse frequency during the early phase of the disease influences the time to onset of progression; however, once a threshold of disability is reached, the rate of progression of disability is not affected by relapses either before the onset of the progressive phase or during this phase (Confavreux et al., 2000; Pittock et al., 2004). During the progression phase, the rate of clinical deterioration is similar for SPMS and PPMS patients

(Confavreux et al., 2000; Pittock et al., 2004). Therapeutic trials have evaluated the effect of three immunomodulatory agents in SPMS: interferon-β, Campath-1H, and cladribine. Although all three suppressed inflammatory MRI lesions, there was no significant slowing in the rate of ongoing cerebral atrophy (Coles et al., 1999). This paradox is also highlighted by the refractory state of progressive MS to antiinflammatory therapies (IFNB Multiple Sclerosis Study Group, 1995; Molyneux et al., 2000).

Imaging studies also suggest a dissociation between inflammation and disease progression. Whereas gadolinium (Gd)-enhancing lesions correlate with relapses, they correlate poorly with disease progression. New Gd-enhancing lesions mainly occur during the early relapsing phase of the disease, whereas Gd-enhancing lesions occur less frequently during this slowly progressive phase of the disease, which is characteristic of PPMS and SPMS without relapses (Thompson et al., 1997). Furthermore, brain atrophy in MS is, in part, independent of T2 lesion load (Pelletier et al., 2003), and diffuse white matter damage and axonal loss can be severe despite few white matter lesions (Rocca et al., 2003; Rovaris et al., 2001).

These clinical and radiographic observations coupled with MRS studies reporting a reduction in NAWM and NAGM NAA, which is contained almost exclusively in neurons and axons, has suggested that brain damage in MS may be mediated by two independent events: an inflammatory reaction that drives the formation of white matter lesions; and neurodegeneration, which is responsible for diffuse and progressive brain damage. The observation of Barnett and Prineas describing oligodendrocyte apoptosis in the absence of ongoing inflammation in an acute lesion from a patient with early RRMS further strengthened this concept (Barnett and Prineas, 2004).

However, from the pathologic perspective, neurodegeneration is defined as degeneration of nervous system cells (neurons or glia) due to a genetic, metabolic, or toxic effect. Inflammatory diseases are distinguished from neurodegenerative diseases by the fact that inflammation is thought to drive this destruction. In contrast to classic neurodegenerative diseases, all MS lesions, regardless of the stage and type of the disease, are associated with inflammation. The apparent lack of obvious correlation between early clinical or radiographic markers of inflammation and later progression does not necessarily imply that the two processes are completely dissociated. The focus on the neurodegenerative aspects of MS may underestimate the pathogenic importance of persistent global inflammation in contributing to ongoing tissue injury. It is likely that the inflammatory response is both qualitatively and quantitatively different during the early versus late phases of the disease.

A careful assessment of MS pathology ultimately reveals that there are three basic pathologic processes. The hallmark of acute or relapsing MS is the focal inflammatory demyelinated white matter lesion, whereas the hallmark of chronic progressive MS additionally includes diffuse "NAWM" damage and cortical demyelination. These three pathologic processes occur in parallel, as well as independently from one another, as supported by the lack of correlation between plaque load in the white matter and the extent and character of cortical demyelination or NAWM injury (Kutzelnigg et al., 2005). These pathologic observations appear consistent with MRI studies that suggest a dissociation between the white matter lesion load and the diffuse global pathology in MS patients. The substrate of disability in MS likely varies in relation to the phase of the disease. However, regardless of the course or phase of the disease, neurodegeneration in MS appears to occur on a background of inflammation. Early axonal loss in the MS lesion contributes to relapse-related disability. This injury correlates with the degree of inflammation in the lesion. Late axonal loss subsequently occurs distal to the lesion as a consequence of wallerian degeneration. This secondary wallerian degeneration is a slow process that may continue for months to years after the acute inflammatory attack has already cleared. This secondary tract degeneration may contribute to the gradual slow progression seen in most MS patients. Furthermore, the presence of global brain injury involving the cortex and "NAWM," occurring diffusely and independent of focal white matter pathology, may also contribute to the gradual progression of disability in MS.

Focal new white matter lesions are associated with BBB damage, inflammation, and acute axonal injury in both the lesion and distal to the lesion owing to wallerian degeneration. This type of injury is likely to be limited by immunomodulatory and immunosuppressant drugs. However, diffuse global brain injury is associated with a compartmentalized inflammatory response, which occurs typically behind an intact BBB in the absence of ongoing focal white matter demyelination. Brain inflammation in slowly progressive MS is typically not associated with significant BBB damage. There is no expression of BBB disturbance markers on endothelial cells, and MRI studies typically demonstrate an absence of Gd-enhancing lesions in PPMS or nonrelapsing SPMS (Thompson et al., 1991). The limited benefit of antiinflammatory or immunomodulatory therapy during the chronic, slowly progressive phase of MS may in part be explained by the compartmentalization of this inflammatory reaction in the CNS.

# 8    Conclusions

The pathogenesis of MS is complex and varies based on the stage of demyelinating activity in the plaque and the clinical phase of the disease, as summarized. An appreciation of this complex pathology is necessary to develop effective therapeutic strategies, which take into account the potential targets of tissue injury (e.g., myelin, oligodendrocyte, axon), the heterogeneous immune effector mechanisms involved in lesion formation (e.g., antibodies, complement, hypoxia, cytotoxic T cells), and the recognition that MS pathology is not restricted to the focal white matter lesion but includes diffuse cortical and white matter damage. Although the inflammatory response is a key feature of MS pathology, what drives the inflammatory response remains to be determined.

# References

Aboul-Enein, F., Lassmann, H. (2005) Mitochondrial damage and histiotoxic hypoxia: a pathway of tissue injury in inflammatory brain disease. *Acta Neuropathol. (Berl.)* 109:49-55.

Aboul-Enein, F., Rauschka, H., Kornek, B., et al. (2003) Preferential loss of myelin-associated glycoprotein reflects hypoxia-like white matter damage in stroke and inflammatory brain diseases. *J. Neuropathol. Exp. Neurol.* 62(1):25-33.

Aktas, O., Smorodchenko, A., Brocke, S., Infante-Duarte, C., Topphoff, US., Vogt, J., Prozorovski, T., Meier, S., Osmanova, V., Pohl, E., Bechmann, I., Nitsch, R., Zipp, F. (2005) Neuronal damage in autoimmune neuroinflammation mediated by the death ligand TRAIL. *Neuron* 46(3):421-432.

Allen, I., McKeown, S. (1979) A histological, histochemical and biochemical study of the macroscopically normal white matter in multiple sclerosis. *J. Neurosci.* 41:81-91.

Allen, I. V., Glover, G., Anderson, R. (1981) Abnormalities in the macroscopically normal white matter in cases of mild or spinal multiple sclerosis. *Acta Neuropathol. (Berl.) Suppl* 7:176-181.

Allen, I. V., McQuaid, S., Mirakhur, M., Nevin, G. (2001) Pathological abnormalities in the normal-appearing white matter in multiple sclerosis. *Neurol. Sci.* 22(2):141-144.

Amiry-Moghaddam, M., Otsuka, T., Hurn, P. D., et al. (2003) An alpha-syntrophin-dependent pool of AQP4 in astroglial end-feet confers bidirectional water flow between blood and brain. *Proc. Natl. Acad. Sci. U S A* 100(4):2106-2111.

Andrews, H. E., Nichols, P. P., Bates, D., Turnbull, D. M. (2005) Mitochondrial dysfunction plays a key role in progressive axonal loss in multiple sclerosis. *Med. Hypotheses* 64:669-677.

Anthony, D. C., Ferguson, M. K., Matyzak, K. M., et al. (1997) Differential matrix metalloproteinase expression in cases of MS and stroke. *Neuropathol. Appl. Neurobiol.* 23:406-415.

Antony, J. M., van Marle, G., Opii, W., et al. (2004) Human endogenous retrovirus glycoprotein-mediated induction of redox reactants causes oligodendorcyte death and demyelination. *Nat. Neurosci.* 7:1088-1095.

Babbe, H., Roers, A., Waisman, A., et al. (2000) Clonal expansion of CD8(+) T cells dominate the T cell infiltrate in active multiple sclerosis lesions as shown by micromanipulation and single cell polymerase chain reaction. *J. Exp. Med.* 192:393-404.

Baranzini, S. E., Jeongk, M. C., Butunol C., et al. (1999) B cell repertoire diversity and clonal expansion in multiple sclerosis brain lesions. *J. Immunol.* 163:5133-5144.

Barnett, M. H., Prineas, J. W. (2004) Relapsing and remitting multiple sclerosis: pathology of the newly forming lesion. *Ann. Neurol.* 55(4).458-468.

Barres, B., Schmid, R., Sendtner, M., Raff, M. (1993) Multiple extracellular signals are required for long-term oligodendrocyte survival. *Development* 118:283-295.

Bitsch, A., Bruhn, H., Vougioukas, V., et al. (1999) Inflammatory CNS demyelination: histopathologic correlation with in vivo quantitative proton MR spectroscopy. *AJNR Am. J. Neuroradiol.* 20(9):1619-1627.

Bitsch, A., Kuhlmann, T., Da Costa, C., et al. (2000a) Tumour necrosis factor alpha mRNA expression in early multiple sclerosis lesions: correlation with demyelinating activity and oligodendrocyte pathology. *Glia* 29(4):366-375.

Bitsch, A., Schuchardt, J., Bunkowski, S., et al. (2000b) Acute axonal injury in multiple sclerosis: correlation with demyelination and inflammation. *Brain* 123(Pt 6):1174-1183.

Bo, L., Dawson, T., Wesselingh, S., et al. (1994) Induction of nitric oxide synthase in demyelinating regions of multiple sclerosis brains. *Ann. Neurol.* 36:778-786.

Bo, L., Vedeler, C., Nyland, H., et al. (2003a) Intracortical multiple sclerosis lesions are not associated with increased lymphocyte infiltration. *Mult. Scler.* 9(4):323-331.

Bo, L., Vedeler, C., Nyland, H., et al. (2003b) Subpial demyelination in the cerebral cortex of multiple sclerosis patients. *J. Neuropathol. Exp. Neurol.* 62(7):723-732.

Bolanos, J., Almeida, A., Stewart, V., et al. (1997) Nitric oxide-mediated mitochondrial damage in the brain: mechanisms and implications for neurodegenerative diseases. *J. Neurochem.* 68:2227-2240.

Bonetti, B., Pohl, J., Gao, Y. L., Raine, C. S. (1997) Cell death during autoimmune demyelination: effector but not target cells are eliminated by apoptosis. *J. Immunol.* 159(11):5733-5741.

Booss, J., Esiri, M. M., Tourtellotte, W. W., Mason, D. Y. (1983) Immunohistological analysis of T lymphocyte subsets in the central nervous system in chronic progressive multiple sclerosis. *J. Neurol. Sci.* 62:219-232.

Boven, L. A., Van Meurs, M., Van Zwam, M., et al. (2006) Mylin-laden macropahges are anti-inflammatory consistent with foam cells in multiple sclerosis. *Brain* 129:517-526.

Bozzali, M., Cercignani, M., Sormani, M. P., et al. (2002) Quantification of brain gray matter damage in different MS phenotypes by use of diffusion tensor MR imaging. *AJNR Am. J. Neuroradiol.* 23:985-988.

Brenner, T., Brocke, S., Szafer, F., et al. (1997) Inhibition of nitric oxide synthase for treatment of experimental autoimmune encephalomyelitis. *J. Immunol.* 158:2940-2946.

Brown, G., Bolanos, J., Heales, S., Clark, J. (1995) Nitric oxide produced by activated astrocytes rapidly and reversibly inhibits cellular respiration. *Neurosci. Lett.* 193:201-204.

Brown, G. C., Borutaite, V. (2002) Nitric oxide inhibition of mitochondrial respiration and its role in cell death. *Free Radic. Biol. Med.* 33:1440-1450.

Bruck, W., Porada, P., Poser, S., et al. (1995) Monocyte-macrophage differentiation in early multiple sclerosis lesions. *Ann. Neurol.* 38:788-796.

Brueck, W., Schmied, M., Suchanek, G., et al. (1994) Oligodendrocytes in the early course of multiple sclerosis. *Ann. Neurol.* 35:65-73.

Chang, A., Tourtellotte, W. W., Rudick, R., Trapp, B. D. (2002) Premyelinating oligodendrocytes in chronic lesions of multiple sclerosis. *N. Engl. J. Med.* 346(3):165-173.

Chen, C., Ro, L., Chang, C., et al. (1996) Serial MRI studies in pathologically verified Balo's concentric sclerosis. *J. Comput. Assist. Tomogr.* 20:732-735.

Christians, E. S., Yan, L. J., Benjamin, I. J. (2002) Heat shock factor 1 and heat shock proteins: critial partners in portection against acute cell injury. *Crit. Care Med.* 30:S43-S50.

Cifelli, A., Arridge, M., Jezzard, P., et al. (2002) Thalamic neurodegeneration in multiple sclerosis. *Ann. Neurol.* 52(5):650-653.

Ciccarelli, O., Werring, D. J., Wheeler-Kingshott, C. A., et al. (2001) Investigation of MS normal-appearing brain using diffusion tensor MRI with clinical correlations. *Neurology* 56:926-933.

Coles, A. J., Wing, M. G., Molyneux, P., et al. (1999) Monoclonal antibody treatment exposes three mechanisms underlying the clinical course of multiple sclerosis. *Ann. Neurol.* 46(3):296-304.

Coles, A., Deans, J., Compston, A. (2004) Campath-1H treatment of multiple sclerosis: lessons from the bedside for the bench. *Clin. Neurol. Neurosurg.* 106(3):270-274.

Confavreux, C., Vukusic, S., Moreau, T., Adeleine, P. (2000) Relapses and progression of disability in multiple sclerosis. *N. Engl. J. Med.* 343:1430-1438.

Courville, C. (1970) Concentric sclerosis. In: Bruyn PVaG (ed) *Handbood of Clinical Neurology.* Amsterdam, Elsevier, pp 437-451.

Cowden, W., Cullen, F., Staykova, M., Willenborg, D. (1998) Nitric oxide is a potential down-regulating molecule in autoimmune disease: inhibition of nitric oxide production renders PVG rats highly susceptible to EAE. *J. Neuroimmunol.* 88:1-8.

Cox, A., Coles, A., Antoun, N., Malik, O., Lucchinnetti, C., Compston, A. (2005) Recurrent myelitis and optic neuritis in a 29-year-old woman. [Case Reports]. Clinical Conference. *Lancet Neurology* 4(8):510-516.

Craner, M. J., Lo, A. C., Blcak, P. A., Wazxman, S. G. (2003) Abnormal sodium channel distribution in optic nerve axons in a model of inflammatory demyelination. *Brain* 126:1552-1561.

Craner, M. J., Newcombe J., Black J. A., et al. (2004) Molecular changes in neurons in MS: altered axonal expression of Nav1.2 and Nav1.6 sodium channels and $Na^+/Ca^{++}$ exchanges. *Proc. Natl. Am. Soc. USA* 101:8168-8173.

Cross, A., Trotter, J., Lyons, J. (2001) B cells and antibodies in CNS demyelinating disease. *J. Neuroimmunol.* 112:1-14.

Cuzner, M., Gveric, D., Strand, C. (1996) The expression of tissue-type plasminogen activator, matrix metalloproteases and endogenous inhibitors in the central nervous system in multiple sclerosis: comparison of stages in lesion evolution. *J. Neuropathol. Exp. Neurol.* 55:1194-1204.

Dawson, J. (1916) The histology of disseminated sclerosis. *Trans. R. Soc. Edinb.* 50: 517-740.

De Groot, C., Ruuls, S., Theeuwes, J., et al. (1997) Immunocytochemical characterization of the expression of inducible and constitutive isoforms of nitric oxide synthase in demyelinating multiple sclerosis lesions. *J. Neuropathol. Exp. Neurol.* 56:10-20.

Dehmeshki, J., Chard, D. T., Leary, S. M., et al. (2003) The normal appearing grey matter in primary progressive multiple sclerosis: a magnetisation transfer imaging study. *J. Neurol.* 250:67-74.

DeLuca, G. C., Ebers, G. C., Esiris, M. M. (2004) Axonal loss in multiple sclerosis: a pathological survey of the corticospinal and sensory tracts. *Brain* 127:1009-1018.

Diemel, L. T., Copelman, C. A., Cuzner, M. L. (1998) Macrophages in CNS remyelination: friend or foe? *Neurochem. Res.* 23(3):341-347.

Dowling, P., Shang, G., Raval, S., et al. (1996) Involvement of the CD95 (APO-1/Fas) receptor/ligand system in multiple sclerosis brain. *J. Exp. Med.* 184(4):1513-1518.

D'Souza, S., Bonetti, B., Balasingam, V., et al. (1996) Multiple sclerosis: Fas signaling in oligodendrocyte cell death. *J. Exp. Med.* 184:2361-2370.

Dutta, R., McDonough, J., Yin, X., et al. (2006) Mitochondrial dysfunction as a cause of axonal degeneration in multiple sclerosis patients. *Ann. Neurol.* 59:178-189.

Esiri, M. M. (1977) Immunoglobulin-containing cells in multiple sclerosis plaques. *Lancet* 2:478.

Evangelou, N. (2003) Regional axonal loss in the corpus callosum correlates with cerebral white matter lesion volume and distribution in multiple sclerosis. *Brain* 2003;123:1845-1849.

Evangelou, N., Esiri, M. M., Smith, S., et al. (2000) Quantitative pathological evidence for axonal loss in normal appearing white matter in multiple sclerosis. *Ann. Neurol.* 47(3):391-395.

Evangelou, N., Konz, D., Esiri, M. M., et al. (2001) Size-selective neuronal changes in the anterior optic pathways suggest a differential susceptibility to injury in multiple sclerosis. *Brain* 124:1813-1820.

Ferguson, B., Matyszak, M. K., Esiri, M. M., Petty, V. H. (1997) Axonal damage in acute multiple sclerosis lesions. *Brain* 120:393-399.

Filippi, M. (2003) Evidence for widespread axonal damage at the earliest clinical stage of multiple sclerosis. *Brain* 126:433-437.

Filippi, M., Campi, A., Dousset, V., et al. (1995) A magnetization transfer imaging study of normal-appearing white matter in multiple sclerosis. *Neurology* 45:478-482.

Filippi, M., Rocca, M., Martino, G., et al. (1998) Magnetization transfer changes in the normal appearing white matter precede the appearance of enhancing lesions in patients with multiple sclerosis. *Ann. Neurol.* 43:809-814.

Flavin, M. P., Coughlin, K., Ho, L. T. (1997) Soluble macrophage factors trigger apoptosis in cultured hippocampal neurons. *Neuroscience* 80:437-448.

Flavin, M. P., Zhao, G., Ho, L. T. (2000) Microglial tissue plasminogen activator (tPA) triggers neuronal apoptosis in vitro. *Glia* 29:347-354.

Foote, A. K., Blakemore, W. F. (2005) Inflammation stimulates remyelination in areas of chronic demyelination. *Brain* 128:528-539.

Forstermann, U., Kleinert, H. (1995) Nitric oxide synthase: expression and expressional control of the three isoforms. *Naunyn Schmiedebergs Arch. Pharmacol.* 352:351-364.

Franklin, R. J. (2002) Why does remyelination fail in multiple sclerosis? *Nat. Rev. Neurosci.* 3(9):705-714.

Garbern, J., Spence, A., Alvord, E. (1986) Balo's concentric demyelination diagnosed premortem. *Neurology* 36:1610-1614.

Garthwaite, G., Goodwin, D. A., Batchelor, A. M., et al. (2002) Nitric oxide toxicity in CNS white matter: an in vitro study using rat optic nerve. *Neuroscience* 109:145-155.

Gay, F., Drye, T., Dick, G., Asiti, N. (1997) The application of multifactorial cluster analysis in the staging of plaques in early multiple sclerosis: identification and characterization of primary demyelinating lesion. *Brain* 120:1461-1483.

Genain, C. P., Cannella, B., Hauser, S. L., Raine, C. S. (1999) Identification of autoantibodies associated with myelin damage in multiple sclerosis. *Nat. Med.* 5(2):170-175.

Gendelman, H. E., Pezeshkpour, G. H., Pressman, N. J., et al. (1985) A quantitation of myelin-associated glycoprotein and myelin basic protein loss in different demyelinating diseases. *Ann. Neurol.* 18(3):324-328.

Geurts, J. J., Wolswijk, G., Bo, L., et al. (2003) Altered expression patterns of group I and II metabotropic glutamate receptors in multiple sclerosis. *Brain* 126:1755-1766.

Gharagozloo, A., Poe, L., Collins, G. (1994) Antemortem diagnosis of Balo concentric sclerosis: correlative MR imaging and pathologic features. *Radiology* 191:817-819.

Gilgun-Sherki, Y., Panet, H., Melamed, E., Offen, D. (2003) Riluzole suppresses experimental autoimmune encephalomyelitis: implications for the treatment of multiple sclerosis. *Brain Res.* 989:196-204.

Giovannoni, G., Heales, S., Land, J., Thompson, E. (1998) The potential role of nitric oxide in multiple sclerosis. *Mult. Scler.* 4:212-216.

Graumann, U., Reynolds, R., Steck, A. J., Schaeren-Wiemers, N. (2003) Molecular changes in normal appearing white matter in multiple sclerosis are characteristic of neuroprotective mechanisms against hypoxic insult. *Brain Pathol.* 13(4):554-573.

Griot, C., Vandevelde, M., Richard, A., et al. (1990) Selective degeneration of oligodendrocytes mediated by reactive oxygen species. *Free Radic. Res. Commun.* 11(4-5):181-193.

Grossman, R. I., Braffman, B. H., Brorson, J. R., et al. (1988) Multiple sclerosis: serial study of gadolinium-enhanced MR imaging. *Radiology* 169:117-122.

Hafler, D. A. (2004) Multiple sclerosis. *J. Clin. Invest.* 113(6):788-794.

# Color Plates

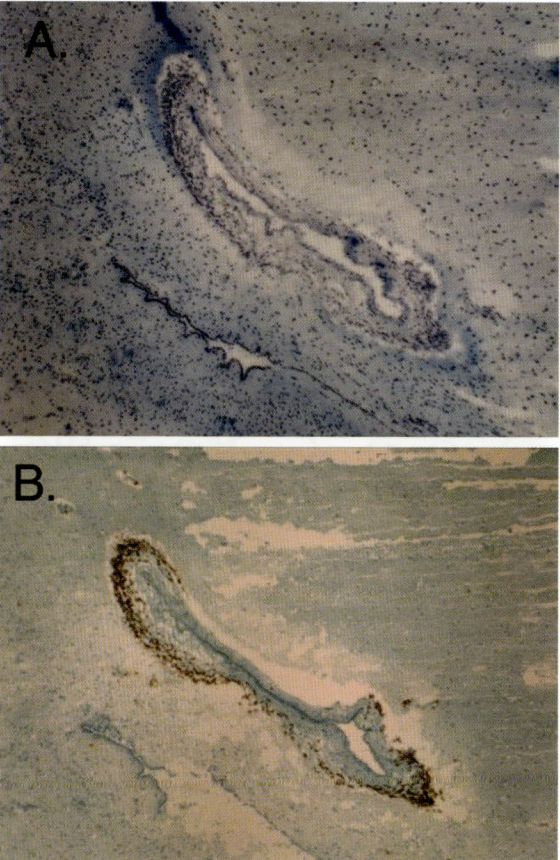

FIGURE 5.1. The perivascular infiltrate proximal to demyelinated CNS tissue includes CD19+ B cells. Perivascular infiltration of CD-19-positive B cells in a demyelinated area of an MS brain. A. An MS lesion stained with luxol fast blue indicates a large area of demyelination proximal to the vasculature. B. Anti-CD19 staining indicates the presence of B cells in the infiltrate. (Histopathology provided by Gerald P. Bailey, MD, PhD and Kevin C. O'Connor, PhD, Brigham and Women's Hospital and Harvard Medical School, Boston, MA.)

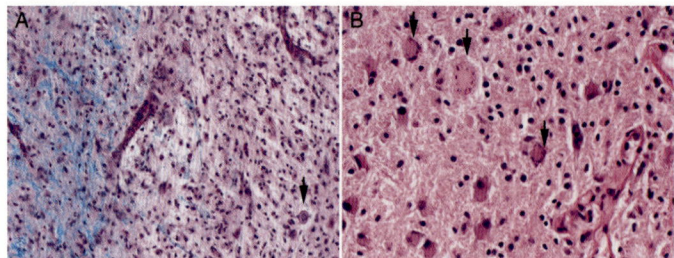

FIGURE 12.3. Active MS Lesion. Active lesions are hypercellular and characterized by an admixture of macrophages and reactive astrocytes associated with myelin loss (**A**, LFB-PAS). Creutzfeld-Peters cells represent astrocytes containing fragmented nuclei that can resemble astrocytic mitoses (arrows in **A** and **B**). (**B**, H&E) (© *Handbook of Multiple Sclerosis*. Reprinted with permission.)

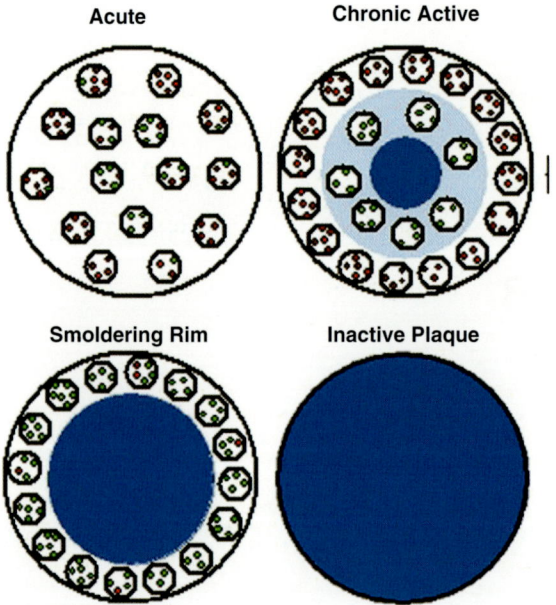

FIGURE 12.4. MS plaque types. The acute active plaque is characterized by the presence of macrophages containing early (red) and late (green) myelin degradation products distributed throughout the extent of the lesion. The chronic active plaque shows the accumulation of numerous macrophages, containing both early (red) and late (green) myelin degradation products clustered at the advancing plaque edge and diminishing in number toward the inactive plaque center (blue). The smoldering rim is defined by the presence of very few macrophages restricted to the plaque edge containing early and late myelin degradation products, whereas the inactive plaque contains no early or late myelin degradation products in the macrophages.

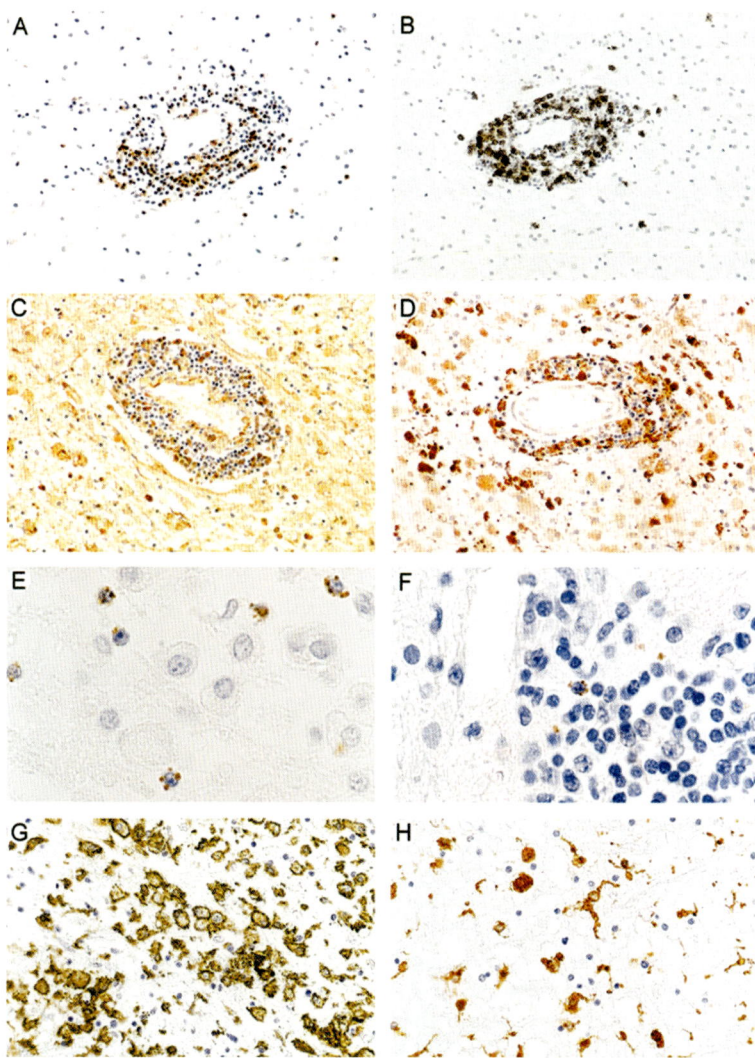

FIGURE 12.6. Acute MS. Inflammation in MS lesions. **A**. Perivascular inflammatory infiltrates with CD3+ T cells. **B**. CD8+ T cells. **C**. IgG containing plasma cells. **D**. CD68+ macrophages. (×100) **E**. Acute MS with granzyme B-reactive cytotoxic T cells in the lesion parenchyma. **F**. A perivascular cuff. (×600) **G**. Primary progressive MS lesion with CD68+ macrophages in an active plaque. **H**. CD68+ microglia in the normal appearing white matter. (×300) (© *McAlpines Multiple Sclerosis*, A. Compston, editor. Reprinted with permission.)

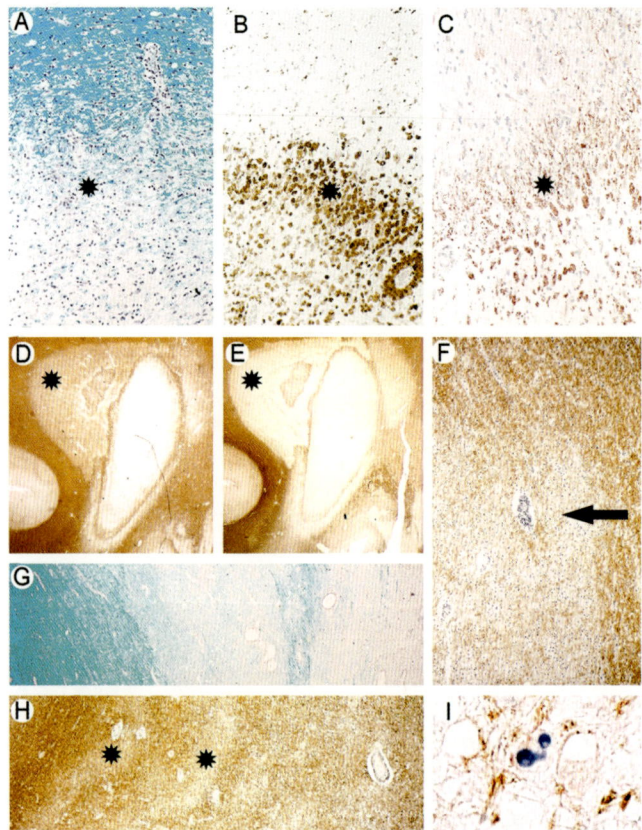

FIGURE 12.9. Immunopathologic heterogeneity of active MS lesions. **A–C.** Antibody/complement-associated demyelination (pattern II). **A.** The lesion is sharply demarcated, and the zone of active myelin destruction (*) is infiltrated by LFB+ containing macrophages. (×100) **B.** The macrophages at the plaque edge form a sharp border. (×100) **C.** Myelin sheaths and myelin degradation products in macrophages at the active plaque edge are immunoreactive for C9neo antigen. (×200) **D–I.** Pattern III. **D, E.** The area of active demyelination (*) still contains MOG (**D**), but has completely lost MAG (**E**) immunoreactivity. (×2) **F.** The lesion edge is ill-defined and contains a broad zone of partial demyelination. Rims of preserved myelin surround blood vessels (arrow). (×100) **G.** The edge of the lesion shows pronounced reduction of myelin density. (LFB-PAS stain, ×30). **H.** Alternating rims of demyelinated (*) and myelinated tissue are seen when stained for MOG. ×30. **I.** Numerous oligodendrocytes show the morphologic changes of apoptosis. (×1000) (© *McAlpines Multiple Sclerosis,* A. Compston, editor. Reprinted with permission.)

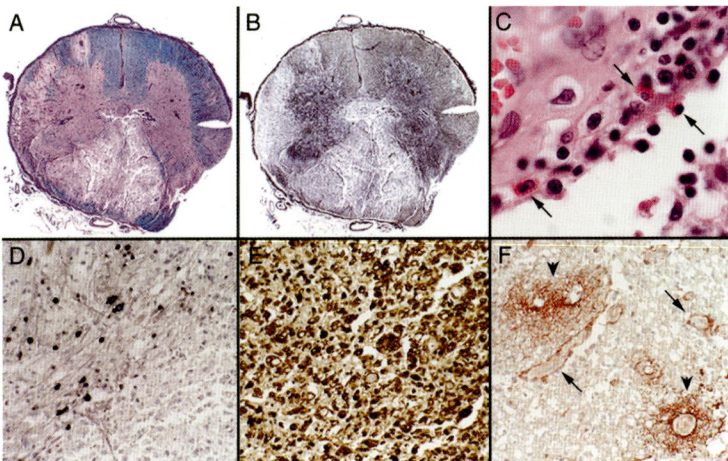

FIGURE 12.11. The pathology of neuromyelitis optica lesions reveals extensive demyelination across multiple spinal cord levels, associated with acute axonal injury, necrosis, and cavitation, involving both gray and white matter (**A, B**). Inflammatory infiltrates of active lesions are associated with large numbers of perivascular granulocytes and eosinophils (**C**), as well as extensive macrophage infiltration (**E**) and rare CD3+ and CD8+ T lymphocytes. There is also a pronounced loss of oligodendrocytes in the lesions (**D**). Actively demyelinating lesions demonstrate a characteristic rim and rosette pattern of immunoglobulin deposition and complement activation (**F**) surrounding blood vessels. (From Coles et al., 2005. Reprinted and modified with permission.)

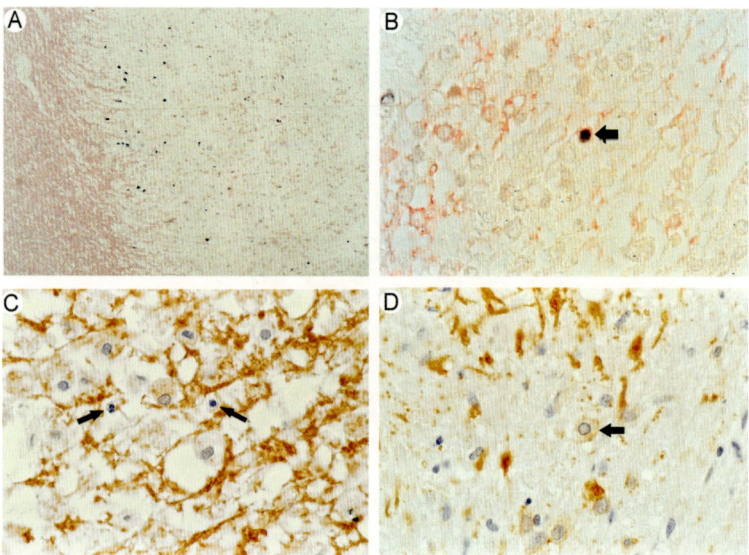

FIGURE 12.13. Oligodendrocyte death in MS lesions. **A.** Chronic MS. Detection of DNA fragmentation by in situ tailing (black) immunocytochemistry with anti-myelin oligodendrocyte glycoprotein (red); a radially expanding lesion. Numerous cells with DNA fragmentation are seen at the lesion border. (×50) **B.** Dying oligodendrocyte identified by immunoreactivity with MOG antibody (red). DNA fragmentation is visualized by in situ tailing (black; arrow). (×500) **C.** Acute MS. Immunocytochemistry with anti-MOG reveals macrophages containing early active myelin debris as well as numerous MOG+ oligodendrocytes with condensed nuclei typical of apoptosis (arrows). (×350) **D.** Active MS lesion reveals a swollen MOG+ oligodendrocyte suggestive of necrosis (arrow). (×350) (© *McAlpines Multiple Sclerosis*, A. Compston, editor. Reprinted with permission.)

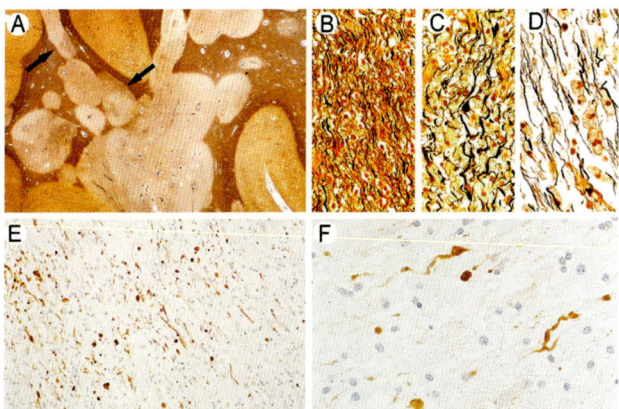

FIGURE 12.14. Axonal pathology in MS lesions. **A.** Multiple demyelinated plaques with variable axonal density in the white matter of a patient with long-standing disease. Arrows identify new demyelianted areas in previously existing plaques and show reduced axonal density compared with the remained of the preexisting lesion (Bielschowsky silver impregnation for axons, ×2). **B.** Axonal density in the normal-appearing white matter (NAWM). Axonal density is reduced mildly (**C**) and moderately (**D**) in two different plaques from the same MS patent (Bielschowsky silver impregnation for axons; ×400). **E, F.** APP expression illustrates dystrophic axons in an active MS lesion reflecting acute axonal injury (**E,** ×300; **F,** ×600). © *McAlpines Multiple Sclerosis*, A. Compston, editor. Reprinted with permission.)

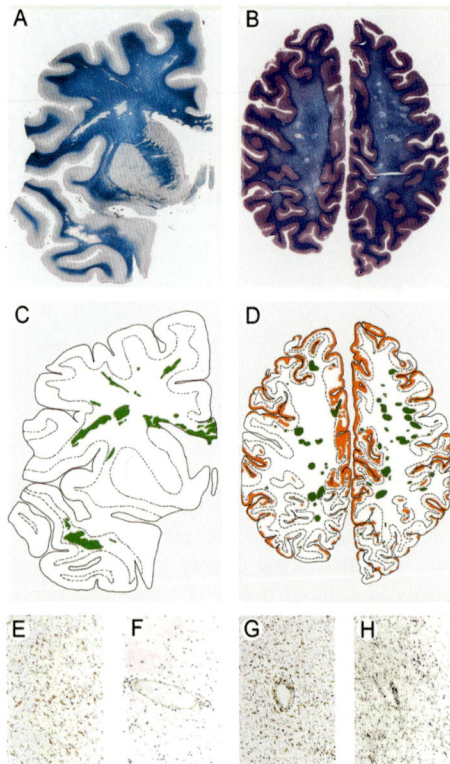

FIGURE 12.15. Comparison of acute and primary progressive MS. **A, C.** Acute MS (7 months disease duration) reveals multiple plaques of demyelination in the white matter (green lesions in **C**). **B, D.** Primary progressive MS with multiple plaques in the white matter (green lesions in **D**) and massive diffuse white matter injury (light myelin staining in **B**). There is also extensive cortical demyelination in the cerebral cortex (red lesions in **D**). (**A, B.** Kluver myelin stain, ×0.3) **E, F.** NAWM in acute MS with HLA-D expression in microglia (**E**) and no expression of iNOS (**F**). **G, H.** NAWM in PPMS with massive expression of HLA-D (**G**) and iNOS (**H**) in the microglia. (×50). (© *McAlpines Multiple Sclerosis*, A. Compston, editor. Reprinted with permission.)

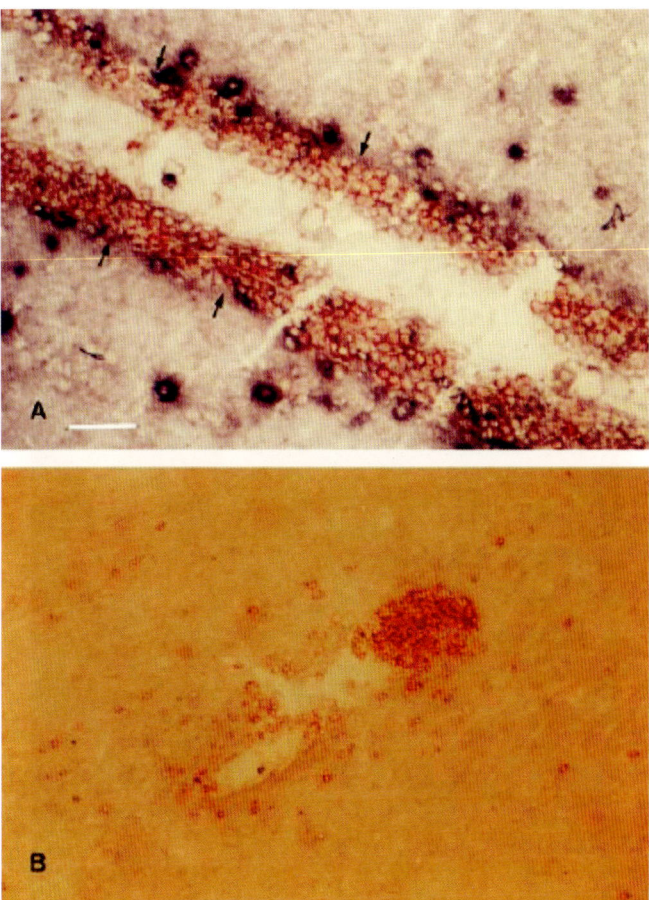

FIGURE 13.1. Concurrent single-field analysis of lymphocytes. **A.** Double labeling for CD4 and CD8 T lymphocytes. Cryostat section of a venule in an advanced type I multiple sclerosis lesion double-labeled for CD4 (brown-red) and CD8 (blue-purple) T cells. Biotinylated anti-CD4, MT.310 (IgG) with streptavidin–HRP–AEC, and anti-CD8, RFT8 (IgM) with anti-mouse IgM, phosphatase-NBT. Brown-red cells (CD4) are concentrated typically in the perivascular space, whereas the blue-purple cells (CD8) are distributed along the glia limitans (arrows) and in the perivascular parenchyma. The bar represents 20 mm. **B.** Technique to increase the sensitivity of CD4 T cell detection. After labeling for both CD4 and CD8 T cells, preliminary CD4 (AEC) development and counting in a wet preparation

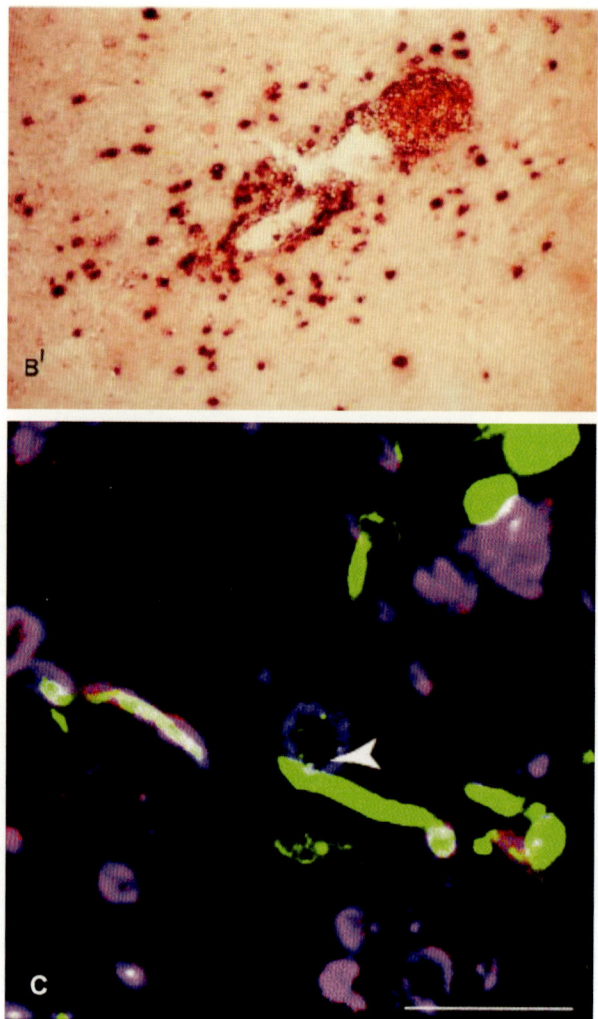

FIGURE 13.1. (*Continued*) (**B′**) was followed by the development.(**B**) of NBT to detect CD8 cells. (From Gay et al., 1997. Reprinted with permission from Oxford University Press.) **C.** A cytotoxic T lymphocyte attached to a demyelinated axon in an acute multiple sclerosis lesion. The section was quadruple-stained for CD3 (blue) and proteolipid protein (pink), SMI31 recognizing axonal neurofilaments (green) and granzyme-B (GrB) (green dots) and visualized using confocal microscopy. GrB+ granules are polarized, facing the surface of a demyelinated axon (arrowhead). Bar = 50 μm. (From Neumann et al., 2002. Reprinted with permission from Elsevier.)

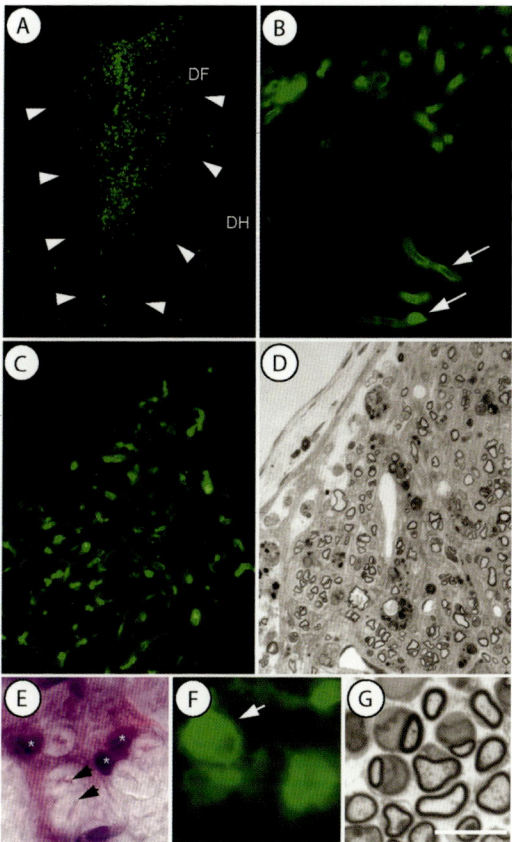

FIGURE 21.1. **A–G.** Identification of donor bone marrow cells in vivo. **A.** Dorsal funiculus of coronally cut spinal cord from a rat that was injected with bone marrow stromal cells from green fluorescent protein (GFP)-expressing mice 3 weeks after transplantation. *Arrows* indicate the lateral margins of the dorsal funiculus. Numerous GFP-positive cells are observed in the remyelinated region. **B.** Higher-power image of same field showing profiles reminiscent of myelinated axons. **C, D.** Frozen and plastic embedded sections, respectively, from the same animal showing co-localization of GFP fluorescence and more clearly defined myelination in the plastic section. **E, F.** H&E-stained frozen section and a fluorescent unstained image with GFP fluorescence at the same high power. Note that in the frozen H&E section the axon cylinder is collapsed (*arrows*) and the myelin is "puffy" as is typical with this staining technique. **G.** Comparable semithin plastic section from the same animal showing myelinated axons. The myelin is better preserved and the tissue more shrunken form dehydration protocols required for plastic embedding. Bar in **G** corresponds to: 250 μm in **A**, 50 μm in **B**, 40 μm in **C** and **D**, 12 μm in **E** and **F**, and 10 μm in **G**. (Modified from Akiyama et al., 2002.)

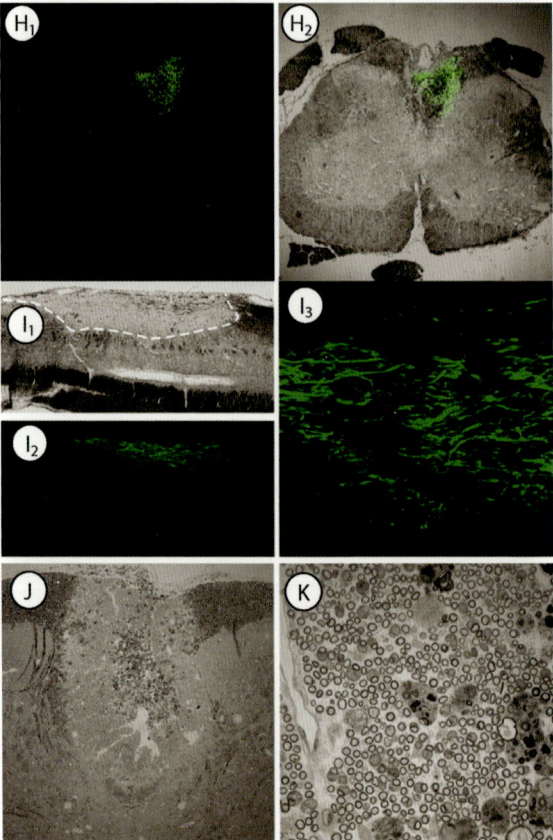

FIGURE 21.1. (*Continued*) **H–K.** Distribution of eGFP-expressing OECs transplanted into the demyelinated dorsal funiculus. **H₁**. Coronal spinal cord section from rat that was transplanted with Schwann cells from GFP-expressing mice 3 weeks after transplantation. **H₂**. Superimposition of the GFP fluorescent and DIC images. Note that donor Schwann cells expressing GFP are localized only in the dorsal funiculus. **I.** Sagittal sections through the lesion site showing the distribution of transplanted OECs. The lesion site is within the stippled area in **I₁**. **I₂**. The transplanted cells extend throughout the lesion site. **I₃**. Higher-power micrograph showing clusters of transplanted cells. **J, K.** Low- and high-power micrographs of plastic semithin sections of the lesion with remyelinated axons. (Modified from Akiyama et al., 2004.)

Hafler, D. A., Slavik, J. M., Anderson, D. E., et al. (2005) Multiple sclerosis. *Immunol. Rev.* 204:208-231.

Hanemann, C., Kleinschmidt, A., Reifenberger, G., et al. (1993) Balo concentric sclerosis followed by MRI and positron emission tomography. *Neuroradiology* 35:578-580.

Haring, J. S., Pewe, L. L., Perlman, S. (2002) Bystander CD8 T cell-mediated demyelination after viral infection of the central nervous system. *J. Immunol.* 169:1550-1555.

Hayashi, T., Burks, J. S., Hauser, S. L. (1988) Expression and cellular localization of major histocompatibility complex antigens in active multiple sclerosis lesions. *Ann. N. Y. Acad. Sci.* 540:301-305.

Hendricks, J. J., Teunissen, C. E., de Vries, H. E., Dijkstra, C. D. (2005) Macrophages and neurodegeneration. *Brain Res. Rev.* 48:185-195.

Hill, K. E., Zollinger, L. V., Watt, H. E., et al. (2004) Inducible nitric oxide synthase in chronic active multiple sclerosis plaques: distribution, cellular expression, and association with myelin damage. *J. Neuroimmunol.* 151:171-179.

Hoftberger, R., Aboul-Enein, F., Brueck, W., et al. (2004) Expression of major histocompatibility complex class I molecules on the different cell types in multiple sclerosis lesions. *Brain Pathol.* 14:43-50.

Hohlfeld, R. (1997) Biotechnological agents for the immunotherapy of multiple sclerosis: principles, problems and perspectives. *Brain* 120:865-916.

IFNB Multiple Sclerosis Study Group and the University of British Columbia MS/MRI Analysis Group. (1995) Interferon b-1b in the treatment of MS: final outcome of the randomized controlled trial. *Neurology* 45:1277-1285.

Ingle, G. T., Stevenson, V. L., Miller, D. H., Thompson, A. J. (2003) Primary progressive multiple sclerosis: a 5-year clinical and MR study. *Brain* 126:2528-2536.

Itoyama, Y., Sternberger, N. H., Webster, H. D., et al. (1980) Immunocytochemical observations on the distribution of myelin-associated glycoprotein and myelin basic protein in multiple sclerosis lesions. *Ann. Neurol.* 7:167-177.

Itoyama, Y., Webster, H. D., Sternberger, N. H., et al. (1982) Distribution of papovavirus, myelin-associated glycoprotein, and myelin basic protein in progressive multifocal leukoencephalopathy lesions. *Neurology* 11(4):396-407.

Jiang, F., Levison, S., Wood, T. (1999) Ciliary neurotrophic factor induces expression of the IGF type I receptor and FGF receptor 1 mRNAs in adult rat brain oligodendrocytes. *J. Neurosci. Res.* 57:447-457.

Jung, J., Bhat, R., Preston, G., et al. (1994) Molecular characterization of an aquaporin cDNA from brain: candidate osmoreceptor and regulator of water balance. *Proc. Natl. Acad. Sci. USA* 91:13052-13056.

Jurewicz, A., Biddison, W. E., Antel, J. P. (1998) MHC class I-restricted lysis of human oligodendrocytes by myelin basic protein peptide-specific CD8 T lymphocytes. *J. Immunol.* 160:3056-3059.

Kapoor, R., Davies, M., Blaker, P., et al. (2003) Blockers of sodium and calcium entry protect axons from nitric-oxide medaited degeneration. *Ann. Neurol.* 53(2):174-180.

Karaarslan, E., Altintas, A., Senol, U., et al. (2001) Balo's concentric sclerosis: clinical and radiologic features of five cases. *AJNR Am. J. Neuroradiol.* 22(7):1362-1367.

Keegan, M., Konig, F., McClelland, R., et al. (2005) Relation between humoral pathological changes in multiple sclerosis and response to therapeutic plasma exchange. *Lancet* 2005;366(9485):579-582.

Kerschensteiner, M., Gallmeier, E., Behrens, L., et al. (1999) Activated human T cells, B cells and monocytes produce brain-derived neurotrophic factor (BDNF) in vitro and in brain lesions: a neuroprotective role for inflammation? *J. Exp. Med.* 189:865-870.

Kidd, D., Barkhof, F., McConnell, R., et al. (1999) Cortical lesions in multiple sclerosis. *Brain* 122(1):17-26.

Kivisakk, P., Mahad, D., Callahan, M. K., et al. (2003) Cerebrospinal fluid central memory T-cells traffic through choroid plexus and meninges. *Proc. Natl. Acad. Sci. U S A* 100:8389-8394.

Kornek, B., Lassmann, H. (1999) Axonal pathology in multiple sclerosis: a historical note. *Brain Pathol.* 9:651-656.

Kornek, B., Storch, M. K., Weissert, R., et al. (2000) Multiple sclerosis and chronic autoimmune encephalomyelitis: a comparative quantitative study of axonal injury in active, inactive, and remyelinated lesions. *Am. J. Pathol.* 157(1):267-276.

Korte, J., Born, E., Vos, L., et al. (1994) Balo concentric sclerosis: MR diagnosis. *AJNR Am. J. Neuroradiol.* 15:1284-1285.

Kotter, M. R., Setzu, A., Sim, F. J., et al. (2001) Macrophage depletion impairs oligodendrocyte remyelination following lysolecithin-induced demyelination. *Glia* 35(3):204-212.

Kuroiwa, Y. (1982) Clinical and epidemiological aspects of multiple sclerosis in Japan. *Jpn. J. Med.* 21(2):135-140.

Kutzelnigg, A., Lucchinetti, C., Stadelmann, C., et al. (2005) Cortical demyelination and diffuse white matter injury in multiple sclerosis. *Brain* 128(11):2705-2712.

Lassmann, H. (1983) Comparative neuropathology of chronic experimental allergic encephalomyelitis and mulitple sclerosis. *Springer Schriftenr. Neurol.* 25:1-135.

Lassmann, H. (2003a) Hypoxia-like tissue injury as a component of multiple sclerosis lesions. *J. Neurol. Sci.* 206(2):187-191.

Lassmann, H. (2003b) Axonal injury in multiple sclerosis. *J. Neurol. Neurosurg. Psychiatry* 74:695-697.

Lassmann, H., Wisniewski, H. M. (1979) Chronic relapsing experimental allergic encephalomyelitis: morphological sequence of myelin degradation. *Brain Res.* 169:357-368.

Lennon, V. A., Wingerchuk, D. M., Kryzer, T. J., et al. (2004) A serum autoantibody marker of neuromyelitis optica: distinction from multiple sclerosis. *Lancet* 364:2106-2112.

Lennon, V. A., Kryzer, T. J., Pittock, S. J., et al. (2005) IgG marker of optic-spinal MS binds to the aquaporin 4 water channel. *J. Exp. Med.* 202(4):473-477.

Linberg, R. L., De Groot, C. J., Montagne, L., et al. (2001) The expression profile of matrix metalloproteinases and their inhibitors in lesions and normal appearing white matter of multiple sclerosis. *Brain* 124:1743-1753.

Linington, C., Bradl, M., Lassmann, H., et al. (1988) Augmentation of demyelination in rat acute allergic encephalomyelitis by circulating mouse monoclonal antibodies directed against a myelin/oligodendrocyte glycoprotein. *Am. J. Pathol.* 130:443-454.

Liu, J., Zhao, M., Brosnan, C., Lee, S. (2001) Expression of inducible nitric oxide synthase and nitrotyrosine in multiple sclerosis lesions. *Am. J. Pathol.* 158:2057-2066.

Lock, C., Hermans, G., Pedotti, R., et al. (2002) Gene-microarray analysis of multiple sclerosis lesions yields new targets validated in autoimmune encephalomyelitis. *Nat. Med.* 8:500-508.

Louis, J., Magal, E., Takayama, S., Varon, S. (1993) CNTF protection of oligodendrocytes against natural and tumor necrosis factor-induced death. *Science* 259:689-692.

Lu, F., Selak, M., O'Connor, J., et al. (2000) Oxidative damage to mitochondrial DNA and activity of mitochondrial enzymes in chronic active lesions of multiple sclerosis. *J. Neurol. Sci.* 177:95-103.

Lu, W., Bhasin, M., Tsirka, S. E. (2002) Involvment of tissue plasminogen activator in onset and effector phases of experimental allergic encpehalomyelitis. *J. Neurosci.* 22:10781-10789.

Lublin, F. D., Baier, M., Cutter, G. (2003) Effect of relapses on development of residual deficit in multiple sclerosis. *Neurology* 61:1528-1532.

Lucchinetti C., Bruck W. (2004) The pathology of primary progressive multiple sclerosis. *Mult. Scler.* 10(Suppl 1):S23-S30.

Lucchinetti, C. F., Brueck, W., Rodriguez, M., et al. (1999) A quantitative study on the fate of the oligodendrocyte in multiple sclerosis lesions: a study of 113 cases. *Brain* 122:2279-2295.

Lucchinetti, C. F., Bruck, W., Parisi, J., et al. (2000) Heterogeneity of multiple sclerosis lesions: implications for the pathogenesis of demyelination. *Ann. Neurol.* 47: 707-717.

Lucchinetti, C. F., Mandler, R. N., McGavern, D., et al. (2002a) A role for humoral mechanisms in the pathogenesis of Devic's neuromyelitis optica. *Brain* 125(Pt 7):1450-1461.

Lucchinetti, C., Mandler, R., Weinshenker, B., et al. (2002b) Humoral mechanisms in the pathogenesis of Devic's neuromyelitis optica. *Neurology* 54(Suppl 3):A259.

Lucchinetti, C. F., Bruck, W., Lassmann, H. (2004) Evidence for pathogenic heterogeneity in multiple sclerosis. *Ann. Neurol.* 56:308.

Ludwin, S., Johnson, E. (1981) Evidence of a "dying back" gliopathy in demyelinating disease. *Ann. Neurol.* 9:301-305.

Ludwin, S. K. (1990) Oligodendrocyte survival in wallerian degeneration. *Acta Neuropathol. (Berl.)* 80:184-191.

Lumsden, C. (1970) The neuropathology of multiple sclerosis. In: Vinken P, Bruyn G (eds) *Handbook of Clinical Neurology*. New York, Elsevier, pp 217-309.

Mahad, D. J., Trebst, C., Kivisakk, P., et al. (2004) Expression of chemokine receptors CCR1 and CCR5 reflects differential activation of mononuclear phagocytes in pattern II and pattern III multiple sclerosis lesions. *J. Neuropathol. Exp. Neurol.* 63(3):262-273.

Mantovani, A., Sica, A., Sozzani, S., et al. (2004) The chemokine system in diverse forms of macrophage activation and polarization. *Trends Immunol.* 25:677-686.

Marburg, O. (1906) Die sogenannte "akute Multiple Sklerose." *J. Psychiatr. Neurol.* 27:211-312.

Matute, C., Alberdi, E., Domercq, M., et al. (2001) The link between excitotoxic oligodendroglial death and demyelinating diseases. *Trends Neurosci.* 24(4):224-230.

Mayer, M., Bhakoo, K., Noble, M. (1994) Ciliary Neurotrophic factor and leukemia inhibitory factor promote the generation, maturation and survival of oligodendrocytes in vitro. *Development* 120:143-153.

McCombe, P. A., Nickson, I., Tabi, Z., Pender, M. P. (1996) Corticosteroid treatment of experimental autoimmune encephalomyelitis in the Lewis rat results in loss of V beta 8.2+ and myelin basic protein-reactive cells from the spinal cord, with increased total T-cell apoptosis but reduced apoptosis of V beta 8.2+ cells. *J. Neuroimmunol.* 70(2): 93-101.

McKeown, S. R., Allen, V. (1978) The cellular origin of lysosomal enzymes in the plaque of multiple sclerosis: a combined histological and biochemical study. Neuropathol. *Appl. Neurobiol.* 4:471-482.

Mehta, P. D., et al. (1981) Bound antibody in multiple sclerosis brains. *J. Neurol. Sci.* 49:91-98.

Merrill, J. E., Scolding, N. J. (1999) Mechanisms of damage to myelin and oligodendrocytes and their relevance to disease. Neuropathol. *Appl. Neurobiol.* 25(6):435-458.

Mews, I., Bergmann, M., Bunkowski, S., et al. (1998) Oligodendrocyte and axon pathology in clinically silent multiple sclerosis lesions. *Mult. Scler.* 4(2):55-62.

Miller, D. H., Rudge, P., Johnson, G., et al. (1988) Serial gadolinium enhanced magnetic resonance imaging in multiple sclerosis. *Brain* 111:927-939.

Moalem, G., Leibowitz-Amit, R., Yoles, E., et al. (1999) Autoimmune T cells protect neurons from secondary degeneration after central nervous system axotomy. *Nat. Med.* 5(1):49-55.

Molyneux, P. D., Kappos, L., Polman, C., et al. (2000) The effect of interferon beta-1b treatment on MRI measures of cerebral atrophy in secondary progressive multiple sclerosis: European Study Group on interferon beta-1b in secondary progressive multiple sclerosis. *Brain* 123(Pt 11):2256-2263.

Mosser, D. M. (2003) The many faces of macrophage activation. *J. Leukoc. Biol.* 73: 209-212.

Murray, P., McGavern, D., Lin, X., et al. (1998) Perforin-dependent neurologic injury in a viral model of multiple sclerosis. *J. Neurosci.* 18:7306-7314.

Neumann, H., Medana, I. M., Bauer, J., Lassmann, H. (2002) Cytotoxic T lymphocytes in autoimmune and degenerative CNS diseases. *Trends Neurosci.* 25(6):313-319.

Nitsch, R., Pohl, E. E., Smorodchenko, A., Infante-Duarte, C., Aktas, O., Zipp, F. (2004) Direct impact of T cells on neurons revealed by two-photon microscopy in living brain tissue. *Journal of Neuroscience* 24(10):2458-2464.

Oleszak, E. L., Zaczynska, E., Bhattacharjee, M., et al. (1998) Indcible nitric oxide synthase and nitrotyrosine are dound in monocytes/macropahges and/or astrocytes in acute, but not in chronic multiple sclerosis. *Clin. Diagn. Lab. Immunol.* 5: 438-445.

Onodera, H., Nakashima, I., Fujihara, K., et al. (1999) Elevated plasma level of plasminogen activator inhibitor-1 (PAI01) in patients with relapsing-remitting multiple sclerosis. *Tohoku J. Exp. Med.* 189:259-265.

Owens, G. P., Kraus H., Burgoon, M. P., et al. (1998) Restricted use of VH4 germline segments in an acute multiple sclerosis brain. *Ann. Neurol.* 43:236-243.

Owens, G. P., Ritchie, A. M., Burgoon, M. P., et al. (2003) Single-cell repertoire analysis demonstrates that clonal expansion is a prominent feature of the B cell response in multiple sclerosis cerebrospinal flu. *J. Immunol.* 171:2725-2733.

Ozawa, K., Suchanek, G., Breitschopf, H., et al. (1994) Patterns of oligodendroglia pathology in multiple sclerosis. *Brain* 117:1311-1322.

Pelletier, D., Nelson, S. J., Oh, J., et al. (2003) MRI lesion volume heterogeneity in primary progressive MS in relation with axonal damage and brain atrophy. *J. Neurol. Neurosurg. Psychiatry* 74:950-952.

Peterson, J. W., Bo, L., Mork, S., et al. (2001) Transected neurites, apoptotic neurons, and reduced inflammation in cortical multiple sclerosis lesions. *Ann. Neurol.* 50(3): 389-400.

Pitt, D., Werner, P., Raine, C. S. (2000) Glutamate excitotoxicity in a model of multiple sclerosis. *Nat. Med.* 6(1):67-70.

Pittock, S. J., Mayr, W. T., McClelland, R. L., et al. (2004) Change in MS-related disability in a population-based cohort: a 10-year follow-up study. *Neurology* 62(1):51-59.

Pittock, S. J., McClelland, R. L., Achenbach, S. J. The neuroimmunology of multiple sclerosis: possible roles of T and B lymphocytes in immunopathogenesis, et al. (2005) Clinical course, pathologic correlations and outcome of biopsy proven inflammatory demyelinating disease. *J. Neurol. Neurosurg. Psychiatry* 767(12):1693-1697.

Pittock, S. J., Lennon, V. A., Krecke, K., et al. (2006) Brain abnormalities in patients with neuromyelitis optica. *Arch. Neurol.* 63:390-396.

Prat, A., Antel, J. (2005) Pathogenesis of multiple sclerosis. *Curr. Opin. Neurol.* 18: 225-230.

Prineas, J. (1985) The neuropathology of multiple sclerosis. In: Vinken P, Bruyn G, Klawans H (eds) *Handbook of Clinical Neurology.* New York, Elsevier Science, pp 213-257.

Prineas, J. W., Graham, J. S. (1981) Multiple sclerosis: capping of surface immunoglobulin G on macrophages engaged in myelin breakdown. *Ann. Neurol.* 10:149-158.

Prineas, J. W., Wright, R. G. (1978) Macrophages, lymphocytes, and plasma cells in the perivascular compartment in chronic multiple sclerosis. *Lab. Invest.* 38(4):409-421.

Prineas, J. W., Barnard, R. O., Kwon, E. E., et al. (1992a) Multiple sclerosis: remyelination of nascent lesions. *Ann. Neurol.* 33:137-151.

Prineas, J. W., Barnard, R. O., Revesz, T., et al. (1992b) Multiple sclerosis: pathology of recurrent lesions. *Brain* 116(Pt 3):681-693.

Prineas, J. W., Kwon, E. E., Cho, E. S., et al. (2001) Immunopathology of secondary-progressive multiple sclerosis. *Ann. Neurol.* 50(5):646-657.

Probert, L., Akassoglou, K., Pasparakis, M., et al. (1995) Spontaneous inflammatory demyelinating disease in transgenic mice showing central nervous system-specific expression of tumor necrosis factor alpha. *Proc. Natl. Acad. Sci. USA* 92:11294-11298.

Qin, Y., et al. (1998) Clonal expansion and somatic hypermutation of V (H) genes of B cells from cerebrospinal fluid in multiple sclerosis. *J. Clin. Invest.* 102:1045-1050.

Raine, C. S., Scheinberg, L., Waltz, J. M. (1981) Multiple sclerosis: oligodendrocyte survival and proliferation in an active established lesion. *Lab. Invest.* 45:534-546.

Raine, C. S., Cannella, B., Duijvestijn, A. M., Cross, A. H. (1990) Homing to central nervous system vasculature by antigen-specific lymphocytes. II. Lymphocyte/endothelial cell adhesion during the initial stages of autoimmune demyelination. *Lab. Invest.* 63:476-489.

Redford, E. J., Kapoor, R., Smith, K. J. (1997) Nitric oxide donors reversibly block axonal conduction: demyelinated axons are especially susceptible. *Brain* 120: 2149-2157.

Revel, M., Valiente, E., Gray, F., et al. (1993) Concentric MR patterns in multiple sclerosis: report of two cases. *J. Neuroradiol.* 20(4):252-257.

Rivera-Quinones, C., McGavern, D., Schmelzer, J. D., et al. (1998) Absence of neurological deficits following extensive demyelination in a class I-deficient murine model of multiple sclerosis. *Nat. Med.* 4(2):187-193.

Rocca, M. A., Iannucci, G., Rovaris, M., et al. (2003) Occult tissue damage in patients with primary progressive multiple sclerosis is independent of T2-visible lesions: a diffusion tensor MR study. *J. Neurol.* 250:456-460.

Rodriguez, M. (1985) Virus-induced demyelination in mice: "dying back" of oligodendrocytes. *Mayo Clin. Proc.* 60:433-438.

Rodriguez, M., Sheithauer, B. (1994) Ultrastructure of multiple sclerosis. *Ultrastruct. Pathol.* 18:3-13.

Rotola, A., Merlotti, I., Caniatti, L., et al. (2004) Human herpesvirus 6 infects the central nervous system of multiple sclerosis patients in the early stages of the disease. *Mult. Scler.* 10:348-354.

Rovaris, M., Bozzali, M., Santuccio, G., et al. (2001) In vivo assessment of the brain and cervical cord pathology of patients with primary progressive multiple sclerosis. *Brain* 124:2540-2549.

Rovaris, M., Bozzali, M., Iannucci, G., et al. (2002) Assessment of normal-appearing white and gray matter in patients with primary progressive multiple sclerosis: a diffusion-tensor magnetic resonance imaging study. *Arch. Neurol.* 59:1406-1412.

Sadahiro, S., Yoshikawa, H., Yagi, N., et al. (2000) Morphometric analysis of the myelin-associated oligodendrocytic basic protein-deficient mouse reveals a possible role for myelin-associated oligodendrocytic basic protein in regulating axonal diameter. *Neuroscience* 98(2):361-367.

Scolding, N. J., Jones, J., Compston, D. A., Morgan, B. P. (1990) Oligodendrocyte susceptibility to injury by T-cell perforin. *Immunology* 70:6-10.

Selmaj, K. W., Raine, C. S. (1988) Tumor necrosis factor mediates myelin and oligodendrocyte damage in vitro. *Ann. Neurol.* 23:339-346.

Selmaj, K., Brosnan, C. F., Raine, C. S. (1991a) Colocalization of lymphocytes bearing gamma delta T-cell receptor and heat shock protein hsp65+ oligodendrocytes in multiple sclerosis. *Proc. Natl. Acad. Sci. USA* 88:6452-6456.

Selmaj, K., Cross, A. H., Farooq, M., et al. (1991b) Non-specific oligodendrocyte cytotoxicity mediated by soluble products of activated T cell lines. *J. Neuroimmunol.* 35:261-271.

Selmaj, K., Raine, C. S., Cannella, B., Brosnan, C. F. (1991c) Identification of lymphotoxin and tumor necrosis factor in multiple sclerosis lesions. *J. Clin. Invest.* 87:949-954.

Selmaj, K., Brosnan, C. F., Raine, C. S. (1992) Expression of heat shock protein-65 by oligodendrocytes in vivo and in vitro: implications for multiple sclerosis. *Neurology* 42:795-800.

Sendtner, M., Carroll, P., Holtmann, B., et al. (1994) Ciliary neurotrophic factor. *J. Neurobiol.* 25:1436-1453.

Sharp, F. R., Bernaudin, M. (2004) HIF1 and oxygen sensing in the brain. *Nat. Rev. Neurosci.* 5:437-448.

Skulina, C., Schmidt, S., Dornmair, K., et al. (2004) Multiple sclerosis: brain-infiltrating CD8+ T cells persist as clonal expansions in the cerebrospinal fluid and blood. *Proc. Natl. Acad. Sci. USA* 101:2428.

Smith, K. J., Lassmann, H. (2002) The role of nitric oxide in multiple sclerosis. *Lancet Neurol.* 1(4):232-241.

Smith, K. J., Kapoor, R., Hall, S. M., Davies, M. (2001) Electrically active axons degenerate when exposed to nitric oxide. *Ann. Neurol.* 49(4):470-476.

Smith, T., Groom, A., Zhu, B., Turski, L. (2000) Autoimmune encephalomyelitis ameliorated by AMPA antagonists. *Nat. Med.* 6(1):62-66.

Smith-Jensen, T., Burgoon, M. P., Anthony, J., et al. (2000) Comparison of immunoglobulin G heavy-chain sequences in MS and SSPE brains reveals an antigen-driven response. *Neurology* 54:1227-1232.

Sobel, R., Mitchell, M., Fondren, G. (1990) Intercellular adhesion molecule-1 (ICAM-1) in cellular immune reactions in the human central nervous system. *Am. J. Pathol.* 136:1309-1316.

Spiegel, M., Kruger, H., Hofmann, E., Kappos, L. (1989) MRI study of Balo's concentric sclerosis before and after immunosuppressant therapy. *J. Neurol.* 236:487-488.

Stadelmann, C., Kerschensteiner, M., Misgeld, T., et al. (2002) BDNF and gp145trkB in multiple sclerosis brain lesions: neuroprotective interactions between immune cells and neuronal cells. *Brain* 125:75-85.

Stadelmann, C., Ludwin, S. K., Tabira, T., et al. (2004) Hypoxic preconditioning explains concentric lesions in Balo's type of multiple sclerosis. Submitted.

Stadelmann, C., Ludwin, S., Tabira, T., Guseo, A., Lucchinetti, C. F., Leel-Ossy, L., Ordinario, A. T., Bruck, W., Lassmann, H. (2005) Tissue preconditioning may explain concentric lesion in Balo's type of multiple sclerosis. *Brain* 128(Pt 5):979-987.

Storch, M. K., Piddlesden, S., Haltia, M., et al. (1998) Multiple sclerosis: in situ evidence for antibody- and complement-mediated demyelination. *Ann. Neurol.* 43(4): 465-471.

Stys, P. K. (2004) Axonal degeneration in multiple sclerosis: is it time for neuroprotective strategies? *Ann. Neurol.* 55:601-603.

Thompson, A., Kermode, A., Wicks, D., et al. (1991) Major differences in the dynamics of primary and secondary progressive multiple sclerosis. *Ann. Neurol.* 29:53-62.

Thompson, A. J., Polman, C. H., Miller, D. H., et al. (1997) Primary progressive multiple sclerosis. *Brain* 120:1085-1096.

Trapp, B. D., Peterson, J., Ransohoff, R. M., et al. (1998) Axonal transection in the lesions of multiple sclerosis. *N. Engl. J. Med.* 338:278-285.

Traugott, U. (1983) Multiple sclerosis: relevance of class I and class II MHC-expressing cells to lesion development. *J. Neuroimmunol.* 16:283-302.

Trebst, C., Sorensen, T., Kivisakk, P., et al. (2001) CCR1+/CCRf+ mononuclear phagocytes accumulate in the central nervous system of patients with multiple sclerosis. *Am. J. Pathol.* 159:1701-1710.

Van der Goes, A., Boorsma, W., Hoekstra, K., et al. (2005) Determination of the sequential degradation of myelin proteins by macrophages. *J. Neuroimmunol.* 161:12-20.

Vass, K., Welch, W. J., Nowak, T. S. (1988) Localization of 70-kDa stress protein induction in gerbil brain after ischemia. *Acta Neuropathol. (Berl.)* 77:128-135.

Vercellino, M., Plano, F., Votta, B., et al. (2005) Grey matter pathology in multiple sclerosis. *J. Neuropathol. Exp. Neurol.* 64(12):1101-1107.

Vos, C. M., van Haastert, E. S., De Groot, C. J., et al. (2003) Matrix metalloproteinase-12 is expressed in phagocytotic macrophages in active multiple sclerosis lesions. *J. Neuroimmunol.* 138:106-114.

Werner, P., Pitt, D., Raine, C. S. (2001) Multiple sclerosis: altered glutamate homeostasis in lesions correlates with oligodendrocyte and axonal damage. Ann. Neurol. 50(2): 169-180.

Williamson, R. A., Burgoon, M. P., Owens, G. P., et al. (2001) Anti-DNA antibodies are a major component of the intrathecal B cell response in multiple sclerosis. *Proc. Natl. Acad. Sci. USA* 2001;98:1793-1798.

Wingerchuk, D. M., Hogancamp, W. F., O'Brien, P. C., Weinshenker, B. G. (1999) The clinical course of neuromyelitis optica (Devic's syndrome). *Neurology* 53(5):1107-1114.

Wingerchuk, D., Pittock, S., Lennon, V., et al. (2005) Neuromyelitis optica diagnostic criteria revisited: validation and incorporation of the NMO-IgG serum autoantibody. *Neurology* 64:A38.

Wolswijk, G. (1998) Chronic stage multiple sclerosis lesions contain a relatively quiescent population of oligodendrocyte precursor cells. *J. Neurosci.* 18(2):601-609.

Yamashita, T., Ando, Y., Obayashi, K., et al. (1997) Changes in nitrite and nitrate ($NO_2^-$ /$NO_3^-$) levels in cerebrospinal fluid of patients with multiple sclerosis. *J. Neurol. Sci.* 153:32-34.

Yao, D. L., Webster, H., Hudson, L. D., et al. (1994) Concentric sclerosis (Balo): morphometric and in situ hybridization study of lesions in six patients. *Ann. Neurol.* 35:18-30.

Yoshikawa, H. (2001) Myelin-associated oligodendrocytic basic protein modulates the arrangement of radial growth of the axon and the radial component of myelin. *Med. Electron Microsc.* 34(3):160-164.

Youl, B. D., Kermode, A. G., Thompson, A. J., et al. (1991) Destructive lesions in demyelinating disease. *J. Neurol. Neurosurg. Psychiatry* 54:288-292.

Zhao, W., Tilton, R. G., Corbett, J. A., et al. (1996) Experimental allergic encephalomyelitis in the rat is inhibited by aminoguanidine, an inhibitor of nitric oxide synthase. *J. Neuroimmunol.* 64:123-133.

# 13
# CD8+ T Cells in Multiple Sclerosis

Manuel A. Friese and Lars Fugger

## 1   Genetic Basis for an Association of MS
## With Different T Cell Subsets

Autoimmune diseases such as multiple sclerosis (MS) are common and are thought to result from complex interactions of susceptibility genes at multiple loci, environmental factors, and stochastic events. During the last decade there has been great interest in testing candidate genomic regions or genes for associations with particular autoimmune diseases. This is complicated by the difficulties of distinguishing between true and false associations and demonstrating causality. Furthermore, for complex diseases many genes usually influence the disease in only a modest way, underlining that autoimmune diseases occur owing to a combination of certain genes. Therefore, identification of these genes ultimately relies on functional studies that link the gene or gene combinations and phenotype. The most extensively studied genetic region is the major histocompatibility complex (MHC), which shows associations in almost all autoimmune diseases (Rioux and Abbas, 2005). Insulin-dependent diabetes mellitus is associated with alleles belonging to the HLA-DR3 and HLA-DR4 haplotypes (Nepom and Erlich, 1991), rheumatoid arthritis with HLA-DR4 alleles (Stastny, 1978), and celiac disease with HLA-DQ2 and HLA-DQ8 alleles (Sollid et al., 1989; Sollid and Thorsby, 1993). Familial aggregation of MS indicates that this disease also has a significant genetic component (Ebers et al., 1995). In particular, it associates with the HLA-DR2 haplotype, which contains three alleles: DRB1*1501, DRB5*0101, and DQB1*0602. HLA-DR2 confers a fourfold relative risk in northern European Caucasian patients, of whom about two-thirds are HLA-DR2+ (Olerup and Hillert, 1991). Although MS most commonly associates with the HLA-DR2 haplotype, this can vary among populations. In the Sardinian population, for example, it is associated with the HLA-DR3 (DRB1*0301– DQA1*0501–DQB1*0201) and HLA-DR4 (DRB1*0405–DQA1*0501–DQB1*0301) haplotypes (Marrosu et al., 1997) in addition to HLA-DR2 (Marrosu et al., 2001).

There is also evidence implicating MHC loci other than DRB1 and DQB1; in the Sardinian MS population, there are additional associations with the HLA class II DPB1 locus and a microsatellite in the HLA class I region (Marrosu et al., 2001). Other studies have also documented associations with MHC class I (MHC I) microsatellites (Ebers et al., 1996; Rubio et al., 2002), and two recent studies have shown that HLA-A*0301 increases the risk of MS independently of HLA-DR2 (Fogdell-Hahn et al., 2000; Harbo et al., 2004). Indeed, the first reported associations were with this allele (Naito et al., 1972) or possibly with the linked HLA-B7 (Jersild et al., 1972). Whereas the previously recorded association with HLA-B7 (B*0702) proved to be secondary to that with the linked HLA-DR2 allele, more refined methods in larger series have now convincingly shown that HLA-A3 (A*0301) roughly doubles the risk of developing MS and does so independently of the HLA-DR2 alleles (Fogdell-Hahn et al., 2000; Harbo et al., 2004). Indeed, subjects with both HLA-A*0301 and HLA-DR2 have a more than additive risk of developing MS. In contrast, HLA-A2 (A*0201) confers some protection against MS, approximately halving the relative risk (Fogdell-Hahn et al., 2000; Harbo et al., 2004). Another study supports these findings (Boon et al., 2001), although the protective MHC I region locus has not yet been mapped precisely.

Therefore, MS is positively and independently associated with HLA-A*0301 and HLA-DR2 genes. The relative risk is increased substantially by HLA-A3 and HLA-DR2 but not by HLA-B7. This association can be partly overridden by the protective allele HLA-A*0201.

## 2   Evidence from Animal Models of MS for Contributions by Different T Cell Subsets

Although these association studies imply an important role for CD4+ helper T cells (which interact with MHC II molecules), they also suggest that CD8+ T cells (interacting with MHC I molecules) could be involved in the disease process. Strong support for the prominent role of HLA class II (HLA II) molecules in MS came from the detection of CD4+ T cell receptors (TCRs) specific for myelin basic protein (MBP) in MS brains (Oksenberg et al., 1993). In addition, MS patients show an immunodominant MBP peptide complexed with HLA-DR2 on antigen-presenting cells (APCs) at sites of demyelination (Krogsgaard et al., 2000). However, one should remember that occasional CD4+ T cells can interact with MHC I molecules (Huseby et al., 2005; Logunova et al., 2005) and CD8+ T cells with MHC II molecules (Tyznik et al., 2004). Though observed with only a few highly cross-reactive T cell clones in autoimmune animal models so far, this phenomenon might question conclusions from MHC association studies in humans. In addition, although studying genetic associations can lead to valuable concepts in pathogenesis, it is important to realize that in complex diseases such as MS many genes usually influence disease in only a modest way and that many affected individuals do not carry any of the so far identified genes. Therefore, the relevance of these genes relies on functional studies and hence on studies in animal models.

Experimental autoimmune encephalomyelitis (EAE) is a widely used model for MS; it is induced by immunizing mice (Steinman, 1999) with whole spinal cord, myelin proteins, or their encephalitogenic epitopes, usually in complete Freund's adjuvant (CFA). This induction results in an autoimmune response that mimics MS pathologically, with a similar composition of inflammatory lesions (T cells and macrophages) in the central nervous system (CNS) (Lassmann, 1983).

It has been clearly established that in EAE myelin proteins such as MBP, proteolipid protein (PLP), and myelin oligodendrocyte oligoprotein (MOG) can be potent activators of encephalitogenic CD4+ T cells, and they have been isolated and propagated as oligoclonal lines (Steinman, 1999). Such cloned CD4+ T cells can transfer EAE to healthy recipients, emphasizing that CD4+ T cells are sufficient to induce EAE (Ben-Nun et al., 1981; Zamvil et al., 1985a,b). These early observations led to an almost exclusive focus on CD4+ T cells, which was reinforced by the observation of spontaneous EAE in *rag2* gene-deleted mice colonized solely with CD4+ T cells transgenic for a TCR from a human MBP-specific CD4+ T cell clone (Madsen et al., 1999). The widely accepted view that CD4+ T cells dominate in pathogenesis was reinforced by the early evidence that CD8+ T cells in EAE were predominantly suppressive, with minimal effector activity (Jiang et al., 1992; Koh et al., 1992).

These findings led to the neglect of CD8+ T cells in EAE—which was almost certainly premature; the apparent "dominance" of CD4+ T cells and of their Th1 subset in EAE is largely based on a self-fulfilling prophecy resulting from the mode of immunization against myelin. Almost all EAE induction protocols use CFA as adjuvant; crucially, it includes killed mycobacteria. These are potent activators of certain Toll-like receptors and inevitably bias toward CD4+ Th1 rather than CD8+ T cell responses (Su et al., 2005). Optimal priming regimens for the latter are quite different (Bevan, 2004), so it seems premature to ignore their potential involvement in any of the varied subgroups of MS patients.

More convincing evidence in the favor of CD8+ T cells in EAE has recently emerged from two animal models where disease can be induced without any CD4+ T cell help by adoptively transferring CD8+ T cells (Huseby et al. 2001; Sun et al. 2001), as was shown during the 1980s for CD4+ T cells. Surprisingly, C3H mice immunized with MBP in CFA show clonal expansion of CD8+ T cells specific for the $MBP_{79-87}$ peptide presented by the murine MHC I molecule $H-2K^k$ (Huseby et al., 1999, 2001). These CD8+ T cells induce severe EAE when isolated and transferred into naive mice. Interestingly, the disease pattern resembles MS more closely than does conventional EAE, with prominent ataxia, spasticity, and higher mortality. Histopathologic analysis revealed a predominance of lesions in the brain rather than the spinal cord. These lesions showed severe demyelination and perivascular cell death. Co-injection of anti-interferon-$\gamma$ (IFN$\gamma$) antibodies markedly reduced disease severity, whereas blocking tumor necrosis factor-$\alpha$ (TNF$\alpha$) with a TNF receptor-Fc fusion protein had no such effect (Huseby et al., 2001).

A parallel study showed a similar pathogenic MHC I-dependent CD8+ T cell response to standard $MOG_{35-55}$ immunization (restricted to $H-2D^b$ in C57BL/6 mice), likewise with CFA as adjuvant (Sun et al., 2001). After adoptive transfer of CD8+ T cells reactive to $MOG_{35-55}/H-2D^b$, disease was severe in wild-type recipients but absent if they were $\beta_2$-microglobulin ($\beta$2m) gene-deleted (i.e., MHC I-deficient). Histopathology again showed a more MS-like pattern, with massive infiltration by CD8+ T cells and macrophages/microglial cells and demyelination, mainly in the brain rather than in the spinal cord. Interestingly, the disease followed a more relapsing and remitting course instead of the chronic progressive pattern seen in conventional MOG-induced models. In fact, the $MOG_{35-55}$ sequence includes two closely overlapping epitopes: $MOG_{37-46}$, restricted to MHC I ($H-2D^b$) (Ford and Evavold, 2005; Sun et al., 2003) and $MOG_{40-48}$ to MHC II ($IA^b$) (Mendel et al., 1996). Although the latter may well contribute to the pathology (Mendel et al., 1996), the minimal epitope ($MOG_{37-46}$) that binds only to $H-2D^b$ nevertheless induced EAE in its own right in C57BL/6 mice (Ford and Evavold, 2005; Sun et al., 2003). Moreover, labeling with peptide:$H2D^b$-tetramers identified MOG-specific CD8+ T cells early in the disease process and detected them in the CNS prior to the onset of neurologic signs. Furthermore, these cells secreted IFN$\gamma$ but no detectable transforming growth factor-$\beta$ (TGF$\beta$) or interleukin-10 (IL-10), thus showing an effector rather than a regulatory phenotype (Ford and Evavold, 2005).

# 3   Clinical Trials Emphasizing a Possible Role of CD8+ T Cells in MS Patients

Although EAE can now be induced by either T cell subset, the dominance of CD4+ T cells in the pathogenesis of human MS has been questioned only recently. Depletion of CD4+ T cells caused no improvement in relapse rates or inflammatory activity as shown by magnetic resonance imaging (MRI) in a carefully controlled phase II trial with the chimeric anti-CD4 monoclonal antibody (mAb) (Hohlfeld and Wiendl, 2001; van Oosten et al., 1997), even though such depletion is highly effective in CD4+-mediated EAE models (Steinman, 1999). In contrast, depletion of both CD8+ and CD4+ T cells with mAbs directed against CD52 (Campath-1H) substantially reduced MS relapses and new lesion formation, although with little improvement in long-term neurologic deficits (Coles et al., 1999; Paolillo et al., 1999). Comparing these two trials forces one to conclude that CD4+ T cells are not the only pathogenic cells, at least at the ongoing chronic stage of the human disease where treatment is normally introduced, and that CD8+ cytotoxic T cells deserve more attention as possible effectors in MS (Friese and Fugger, 2005; Goverman et al., 2005; Hemmer et al., 2002; Lassmann and Ransohoff, 2004; Neumann et al., 2002; Steinman, 2001).

## 4    CD8+ T Cell Infiltration in MS Lesions

In fact, a predominance of CD8+ T cells has long been observed in MS lesions (Babbe et al., 2000; Booss et al., 1983; Gay et al., 1997; Hauser et al., 1986) (Fig. 13.1). Although CD4+ T cells outweigh CD8+ T cells by 1.5- to 3-fold in the blood, and by 3- to 6-fold in the normal cerebrospinal fluid (CSF) (Kivisakk et al., 2003), CD8+ T cells exceed CD4+ T cells in chronically inflamed MS plaques (Hauser et al., 1986), and by 3- to 10-fold in regions of demyelination and axonal damage (Babbe et al., 2000; Booss et al., 1983). This observation suggests that CD8+ T cells actively migrate into MS lesions and are not just passive bystanders. Furthermore, the amount of axonal damage correlates better with the numbers of CD8+ T cells (Bitsch et al., 2000; Kuhlmann et al., 2002) and macrophages/microglial cells (Ferguson et al., 1997; Kornek et al., 2000; Trapp et al., 1998) than the number of CD4+ T cells. Axonal damage is most prominent in acute MS lesions, where it again correlates with the numbers of infiltrating CD8+ T cells, both of which decline later (Kuhlmann et al., 2002). Recently, CD8+ T cells have been found diffusely infiltrating the relatively normal-appearing white matter in primary (PPMS) and secondary (SPMS) progressive MS brains, which is correlated with diffuse signs of neurodegeneration (Kutzelnigg et al., 2005). These histopathological observations argue strongly for an important role for this subset in MS pathogenesis.

## 5    Preferential Clonal Expansion of CD8+ T Cells in MS

These inferences are further supported by the clonal expansions of CD8+ rather than of CD4+ T cells in MS. The clones were identified by their specific TCRs or their TCR-β variable gene usage, and were confined to CD8+ T cells in studies of MS patients' brains, blood (Babbe et al., 2000; Skulina et al., 2004), or CSF (Jacobsen et al., 2002). The same clones persisted for 2 to 5 years, implying

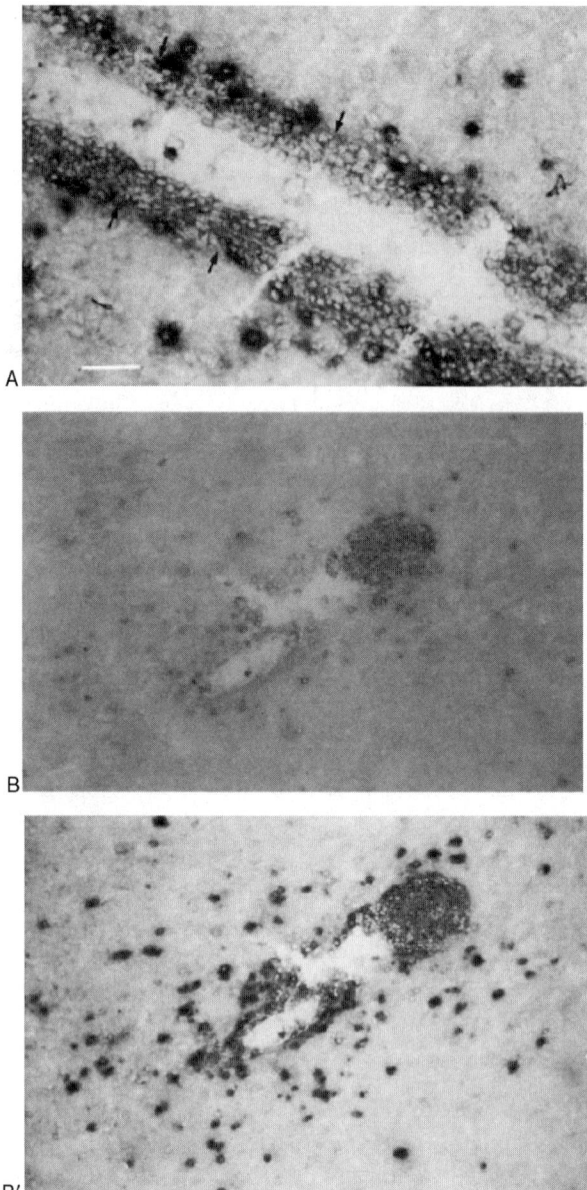

FIGURE 13.1. Concurrent single-field analysis of lymphocytes. **A.** Double labeling for CD4 and CD8 T lymphocytes. Cryostat section of a venule in an advanced type I multiple sclerosis lesion double-labeled for CD4 (brown-red) and CD8 (blue-purple) T cells. Biotinylated anti-CD4, MT.310 (IgG) with streptavidin–HRP–AEC, and anti-CD8, RFT8 (IgM) with anti-mouse IgM, phosphatase-NBT. Brown-red cells (CD4) are concentrated typically in the perivascular space, whereas the blue-purple cells (CD8) are distributed along the glia limitans (arrows) and in the perivascular parenchyma. The bar represents 20 mm. **B.** Technique to increase the sensitivity of CD4 T cell detection. After labeling for both CD4 and CD8 T cells, preliminary CD4 (AEC) development and counting in a wet preparation (**B′**) was followed by the development.(**B**) of NBT to detect CD8 cells. (From Gay et al., 1997. Reprinted with permission from Oxford University Press.)

FIGURE 13.1. (*continued*) **C.** A cytotoxic T lymphocyte attached to a demyelinated axon in an acute multiple sclerosis lesion. The section was quadruple-stained for CD3 (blue) and proteolipid protein (pink), SMI31 recognizing axonal neurofilaments (green) and granzyme-B (GrB) (green dots) and visualized using confocal microscopy. GrB+ granules are polarized, facing the surface of a demyelinated axon (arrowhead). Bar = 50 μm. (From Neumann et al., 2002. Reprinted with permission from Elsevier.) (*See Color Plates between pages 256-257.*)

chronic antigen stimulation, as does their memory phenotype (Babbe et al., 2000; Skulina et al., 2004). Blood T cells isolated from the blood of MS patients also showed a significantly skewed T cell repertoire, especially in the CD8+ T cell subset, which also exhibited increased expression of mRNAs for proinflammatory cytokines (Laplaud et al., 2004).

# 6    MHC Class I Expression in the CNS

Even an invasion dominated by CD8+ T cells would be harmless in the brain if MHC I expression was as low or absent as previously believed (Joly et al., 1991; Rall et al., 1995). In fact, it is now widely accepted that MHC I is constitutively expressed in the normal human brain on endothelial cells, perivascular macrophages, and some microglial cells (Hoftberger et al., 2004) and that all glial and neuronal cells can express MHC I after stimulation with appropriate cytokines in vitro, especially IFNγ (Neumann et al., 1995, 1997, 2002; Rensing-Ehl et al., 1996), whereas neuronal cells never express MHC II.

In MS lesions, MHC I is clearly upregulated, primarily on endothelial and microglial cells, as documented by several studies (Gobin et al., 2001; Hayashi et al., 1988; Ransohoff and Estes, 1991). Indeed, it is expressed by all cell types (microglial cells, astrocytes, oligodendrocytes, neurons) in the human CNS at certain stages of the disease (Hoftberger et al., 2004) (Fig. 13.2) Periplaque white matter and inactive demyelinated plaques in MS brains upregulate MHC I on microglial cells. In contrast, MHC I expression in active lesions in acute or chronic MS was reported on all cell types (Hoftberger et al., 2004). These observations underline that all cell types in the CNS are potential targets of a cytotoxic

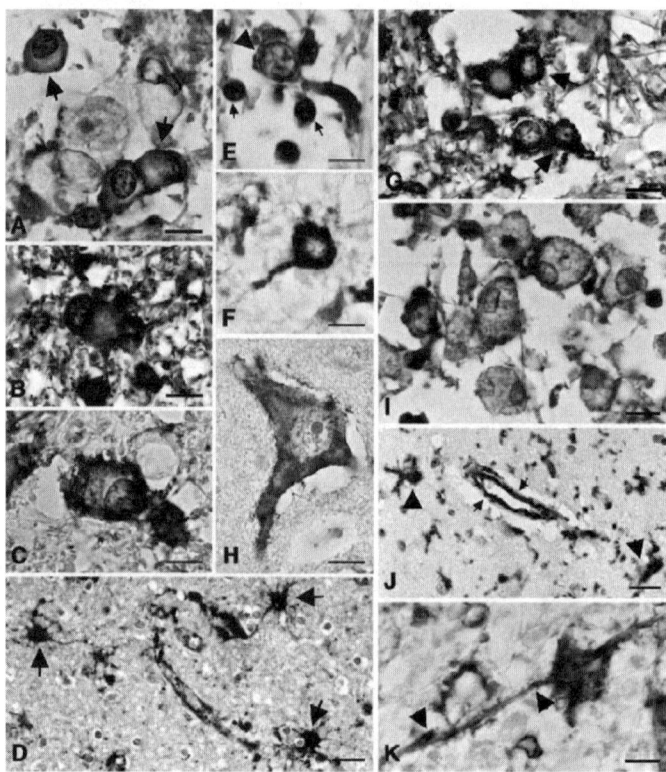

Figure 13.2. Immunohistochemistry for major histocompatibility complex class I (MHC I) molecules on different cell types in MS lesions. **A.** Plasma cells ($\alpha$-chain of MHC I) (arrows). **B.** Astrocytes ($\beta_2$-microglobulin). **C.** Astrocytes ($\alpha$-chain of MHC I). **D.** Astrocytes ($\beta_2$-microglobulin). **E, F.** Oligodendrocytes ($\alpha$-chain of MHC I) (arrowhead). The small arrows in **E** indicate adjacent lymphocytes positive for $\alpha$-chain as well. **G.** Oligodendrocytes ($\alpha_2$-microglobulin) (arrows). **H.** Neuron ($\alpha$-chain of MHC I). **I.** Foamy macrophages ($\alpha$-chain of MHC I). **J.** Microglia (arrowheads) and endothelial cells (arrows). **K.** Axon in the MS lesion positive for $\alpha$-chain of MHC I complex (arrowhead). Bars **A–C, E–I, K** = 10 $\mu$m; **D, J** = 30 $\mu$m. (From Hoftberger et al., 2004. Reprinted with permission from ISN Neuropathology Press.)

CD8+ T cell attack. However, MHC I expression in MS does not necessarily lead to a CD8+ T cell response, as neurons that overexpress MHC I can be resistant to CD8+ T cell-mediated lysis following viral infection despite increased infiltration into brain and enhanced viral clearance from MHC-overexpressing infected cells (Rall et al., 1995). Electrically active and intact neurons usually suppress the induction of MHC I expression both on their glial neighbors and their own surface, but the loss of integrity and electrical activity allows MHC I induction by proinflammatory cytokines (Neumann et al. 1998).

# 7    Cytotoxic Response of CD8+ T Cells on CNS Resident Cells

How CD8+ T cells kill their target cells in the CNS is still poorly understood. It is generally accepted that they can use two mechanisms. Engagement (by the TCRs) of the cognate peptide class I complexes on target cells induces transloca-tion of Fas ligand (FasL) to the immune synapse, the ligation of Fas/CD95 on the target cell, and its subsequent death. CD8+ T cells can also kill by releasing the contents of their cytotoxic granules, including perforin and granzymes, to the effector-target cell contact zone. The relative contribution of these two pathways is poorly understood in MS. In vitro, CD8+ T cells can form direct, stable adhe-sions with neurons when they recognize MHC I/peptide complexes, leading to selective transsection of neurites (Medana et al., 2001b). Analysis of this killing process showed that neurons are protected against perforin but susceptible to Fas/CD95-mediated apoptosis (Medana et al., 2001a). However, CD8+ T cells in MS brains are seen in close contact with oligodendrocytes and demyelinated axons, toward which cytolytic granules were polarized (Neumann et al., 2002) (Fig. 13.1C). In addition to neurons, human oligodendrocytes can be effectively lysed by blood-derived CD8+ T cells, whether specific for MHC I-presented pep-tides or alloantigens (Jurewicz et al.,1998; Ruijs et al., 1990). IFNγ-treated human oligodendrocytes express Fas/CD95 and are susceptible to FasL-induced apoptosis (Pouly et al., 2000). In addition, activated CD8+ T cells express and secrete TNFα and IFNγ, among other cytokines. In combination with IL-1, these cytokines alone can induce neuronal cell death (Allan and Rothwell, 2001). However, Fas-mediated signaling does not appear to be essential for CNS cell death; in one of the MHC I-restricted models, EAE was just as severe when the pathogenic CD8+ T cells were transferred into *fas* gene-deleted mice (Huseby et al., 2001).

# 8    Peripheral Activation of Autoreactive CD8+ T Cells

Given that CD8+ T cells are the predominant cell type in MS lesions and can kill any cell type in the CNS, one next needs to understand how and why they enter the brain. Apparently, only primed memory T cells do so, as naive T cells were not detected in a thorough study on CSF from healthy donors (Kivisakk et al., 2003). Thus it seems likely that most of the T cells in MS lesions are first activated in the periphery and then migrate into the CNS, where they reactivate when they encounter their target antigens, expand clonally, and cause cytotoxic damage (Kivisakk et al., 2004).

It is still unknown where and when CD8+ T cells initially recognize MHC I-self-antigen complexes in MS. Oligodendrocytes and neurons seem to be insufficient APCs for priming and generating memory because they lack the essential co-stimulatory molecules (Odeberg et al., 2005). Far likelier culprits are any professional APCs expressing surface peptide/MHC I complexes and co-stimulatory molecules after migrating to peripheral lymph nodes (Seder and Ahmed, 2003).

Direct transfer of exogenous antigens from antigen-expressing cells to APCs is called cross-presentation and is well established for both MHC I and MHC II pathways in experimental models (Ackerman and Cresswell, 2004; Heath and Carbone, 2001); indeed, it has been observed for CNS antigens in brain-derived APCs (Calzascia et al., 2003).

Whether it results in activation or unresponsiveness reflects the quality and quantity of cross-presentation, which depends on the activation state of the APCs, the cytokine milieu, the number of peptide/MHC complexes (Kurts et al., 1999), and the affinity of the TCRs for them (Gronski et al., 2004); autoreactive CD8+ T cells with intermediate to low avidity might also require help from activated CD4+ T cells. The latter express CD154 (CD40L) and promote both maturation of CD40+ APCs and CD8+ T cell cross-priming, shifting the outcome from tolerance to immunization (Hernandez et al., 2002). Such CD40 expression on CNS resident APCs is also required for local reactivation and progression of EAE and again acts through CD154 on the encephalitogenic T cells (Becher et al., 2001; Howard et al., 1999). Likewise in MS, CD4+ and CD8+ T cells in the blood of patients with SPMS constitutively express CD154, which is not seen in controls (Jensen et al., 2001).

Which APCs might be responsible for cross-presentation in CNS? The perivascular spaces surrounding cerebral and meningeal blood vessels contain a population of myeloid cells that are thought to play an important role as APCs (Hickey and Kimura, 1988; Weller, 1998). They are necessary for the initiation of EAE (Greter et al., 2005; Hickey and Kimura, 1988), and they are thought to be the only APCs in the CNS capable of efficiently stimulating naive myelin-specific cells (McMahon et al., 2005). Dendritic cells, which in most tissues endocytose, process, and transport potential antigens to lymphoid organs, have long been thought not to be present in the CNS. However, they have now been identified in normal human meninges, choroid plexus, and CSF (Pashenkov et al., 2001; Pashenkov and Link, 2002). They are also present in the murine brain and initiate T cell-mediated immune responses after migrating into cervical lymph nodes (Karman et al., 2004; Serafini et al., 2000). Indeed, CNS-antigen-specific CD8+ T cells are primed in the cervical and lumbar lymph nodes, where they acquire an activated phenotype, which allows them to migrate specifically into the CNS in a tumor model (Calzascia et al., 2005).

# 9    Mechanisms of CD8+ T Cell Infiltration of the CNS

Activated T cells can only invade into the CNS if they express certain adhesion molecules that allow them to attach to blood vessels (Weninger et al., 2002). Important clues to the players involved came from injection of MS patient T cells directly into meningeal vessels of living mice. Although vascular cell adhesion molecule-1 (VCAM-1) was upregulated on CD4+ T cells during acute attacks, it did not increase their adherence to the murine brain venules. In striking contrast, the CD8+ T cells from MS patients showed increased adhesiveness in this in vivo model (Battistini et al., 2003); it depended on expression of P-selectin glycoprotein ligand-1 (PSGL-1) by these memory-effector T cells. Importantly, its partner, P-selectin, is constitutively expressed in microvessels of the choroid plexus and may mediate T cell trafficking into the brain and CSF (Ransohoff et al., 2003). Moreover, CD8+ T cells activated in the cervical or lumbar lymph nodes were imprinted preferentially to home into the CNS, most likely by expressing characteristic adhesion molecules such as the $\alpha_4\beta_1$ integrin (very late antigen-4, VLA-4) (Calzascia et al., 2005), which is critical for T cell homing to the CNS in EAE (Yednock et al., 1992) and MS (Miller et al., 2003). A humanized blocking anti-$\alpha_4$ integrin mAb (natalizumab) was highly effective in preclinical EAE studies and MS patients (Miller et al., 2003). This remarkable result was attributed to the blockade of CD4+ T cells, as seen in the EAE model (Yednock et al., 1992), although CD8+ T cells express the same integrin.

Tragically, three patients (among more than 3000 total) under natalizumab treatment developed progressive multifocal leukoencephalopathy (PML), an opportunistic infection of the CNS by polyomavirus JC (Kleinschmidt-DeMasters and Tyler, 2005; Langer-Gould et al., 2005; Van Assche et al., 2005). PML is almost exclusively observed in immunosuppressed individuals and is normally controlled by memory CD8+ T cells (Monaco et al., 1998). Exclusion of CD8+ T cells from the CNS by this mAb might be one explanation for this opportunistic infection, as well as for the generally beneficial effects on the inflammatory course in MS.

The preferential recruitment and accumulation of CD8+ T cells inside the CNS has been reported not only in MS but also in two EAE models (Brabb et al., 2000; Carson et al., 1999). Even after using specific MHC II-restricted antigens for T cell priming, a greater percentage of CD8+ than CD4+ T cells entered the CNS (Carson et al., 1999), possibly because the memory T cell population (CD44[high]) in the periphery contains a higher percentage of CD8+ T cells that preferentially cross the blood-brain barrier (Brabb et al. 2000; Hickey, 1999).

In addition, chemokines are the key regulators of lymphocyte trafficking and recruitment and an attractive target in drug development (Ubogu et al., 2005). Indeed, virus-specific CD8+ T cells need to express CXCR3 to migrate from the meninges into the brain parenchyma (Christensen et al., 2004), and its expression is increased in progressive and relapsing/remitting (RR) MS (Balashov et al., 1999). It is the receptor for CXCL10, an IFNγ-inducible protein of 10 kDa (IP-10) whose levels are increased in peripheral blood monocytic cells (PBMCs), serum, and CSF of MS patients (Muller et al., 2004; Scarpini et al., 2002) and especially in CSF during relapses in RRMS (Ubogu et al., 2005). CXCL10 is mainly expressed by astrocytes in MS and EAE lesions (Balashov et al., 1999; Ubogu et al., 2005), but some PLP-specific CD8+ T cell lines derived from MS patients also secrete such chemokines as MIP-1α, MIP-1β, IL-16, CXCL10, and matrix metalloproteinases (Biddison et al., 1997; Honma et al., 1997).

## 10    Reactivation of Autoreactive CD8+ T Cells in the CNS

Having entered the CNS, activated CD8+ T cells survey cell surfaces for their specific epitopes. After recognizing them on MHC I+ cells, they rapidly divide and expand clonally and attack them (Ramakrishna et al., 2004). In EAE, this local reactivation and the consequent inflammation depend on the co-stimulatory molecules CD80 (B7-1) and CD86 (B7-2), as implied by the reduced inflammation and preferential meningeal localization of T cells in mice deficient for these molecules (Chang et al., 2003). Conversely, transgenic mice constitutively expressing CD86 on microglia spontaneously develop autoimmune demyelinating disease. Interestingly, the CNS infiltrates show a predominance of CD8+ memory-effector T cells; evidently, myelin-specific autoreactive CD8+ T cells are prevalent in the normal repertoire and can be specifically and productively activated in situ by CD86-expressing microglial cells (Zehntner et al., 2003).

## 11    Proteins and Epitopes Recognized by Autoreactive CD8+ T Cells in the CNS

What MHC I-restricted epitopes are presented by APCs in the brain? Most efforts have focused on the predominant proteins in the myelin sheath of the CNS, which include MBP and PLP (~30% and ~50% of CNS myelin protein, respectively) and the minor component MOG (Baumann and Pham-Dinh, 2001). Studies have identified the ubiquitously expressed transaldolase (TAL), which is highly expressed in oligodendrocytes, another potential target for CD8+ T cells in MS (Niland et al., 2005). Many MHC I-restricted epitopes have been identified by prediction algorithms, and their actual recognition and relevance in vivo in MS lesions await confirmation (Table 13.1). Indeed, most of the CD8+ T cell responses studied have focused on HLA-A2, partly because it is so frequent; even though it is clearly protective in genetic studies, all of the HLA-A2-restricted CD8+ T cell responses so far studied appear to be cytotoxic. Several predicted epitopes in MBP, PLP, and myelin-associated glycoprotein (MAG) have been screened in MS patients, but no healthy controls were tested for comparison. Cytotoxic T cell lines and clones were isolated against: $MBP_{110-118}$, $PLP_{80-88}$, $MAG_{287-295}$, $MAG_{509-517}$, and $MAG_{556-564}$; and they secreted IFNγ and TNFα after stimulation by phytohemagglutinin (Tsuchida et al., 1994). The $MBP_{110-118}$/HLA-A2-restricted epitope was evidently processed naturally in freshly isolated

TABLE 13.1. MHC class I immunodominant epitopes in humans and mice.

| Protein | Human | Mouse |
|---|---|---|
| MBP | HLA-A2: 87-95, 110-118 | H-2K$^k$: 79-87 |
| PLP | HLA-A2: 80-88 | None known |
| | HLA-A3: 45-53 | |
| MOG | None known | H-2D$^b$: 37-46 |
| MAG | HLA-A2: 287-295, 509-517, 556-564 | None known |
| TAL | HLA-A2: 168-176 | None known |

MHC, major histocompatibility complex; MBP, myelin basic protein; PLP, proteolipid protein; MOG, myelin-oligodendrocyte glycoprotein; MAG, myelin-associated glycoprotein; TAL, transaldolase.

HLA-A2+ oligodendrocytes, which were killed by the specific T cell lines. Although MS patients and controls showed no difference in cytotoxic activity (Jurewicz et al., 1998), MBP$_{110\text{-}118}$/HLA-A2-restricted CD8+ T cells were found to be more frequently expanded in patients (Zang et al., 2004).

Other epitopes recognized by MS patient-derived CD8+ T cell lines include PLP$_{80\text{-}88}$/HLA-A2 or PLP$_{45\text{-}53}$/HLA-A3 (Biddison et al., 1997; Honma et al., 1997). Again, PLP$_{80\text{-}88}$/HLA-A2-restricted T cell lines showed cytolytic activity regardless of whether the donor was an MS patient or a healthy control (Dressel et al., 1997). One study identified the HLA-A2-resticted TAL$_{168\text{-}176}$ epitope, which is recognized more frequently in HLA-A2+ MS patients than in HLA-A2+ controls. It is again processed endogenously, rendering HLA-A2+ oligodendrocytes susceptible to killing (Niland et al., 2005). A different approach used peptide mixtures from several CNS proteins, including MBP, PLP, MAG, MOG, S100β glycoprotein, oligodendrocyte-myelin glycoprotein, myelin-associated oligodendrocytic basic protein, αB-crystallin, and 2'-3'-cyclic nucleotide 3'-phosphodiesterase. Interestingly, there was no difference in the number of positive responses between MS and normal donors in the CD4+ T cell compartment. However, the CD8+ T cells showed more frequent responses in the MS patients (Crawford et al., 2004). Although the pathogenic antigens/epitopes in these mixtures await identification or mapping, these studies clearly show that CD8+ T cells can, in principle, recognize MHC I-restricted myelin autoantigens.

## 12   Conclusions and Perspective

CD8+ T cells are highly potent killers with an array of cytotoxic weapons and are vital for defense against viral or intracellular bacterial infections. As memory cells, they can enter the CNS and sensitively detect signs of viral replication and infection; when activated, they limit its spread by killing infected cells. Therefore, they need to be tightly controlled to prevent them from mistakenly targeting normal self-tissues. However, it is now clear that CD8+ T cells can escape such controls to participate in autoimmune diseases just like CD4+ T cells, as clearly shown in type I diabetes and its animal model (DiLorenzo and Serreze, 2005). Although they have long been underestimated, they could be decisive players in tissue destruction. Although their role in MS is far from clear, several observations implicate them in its pathogenesis and should prompt a reappraisal. CD8+ T cells are the dominant cell type in the inflammatory infiltrates in MS lesions and are clonally expanded in the brain, CSF, and blood of MS patients. They apparently use PSGL-1, $\alpha_4\beta_1$-integrin, and CXCR3 to invade the CNS. HLA-I is highly upregulated in these MS lesions, and CD8+ T cells are able to kill oligodendrocytes and neurons by recognizing myelin protein-derived epitopes presented by MHC I on these cells. Moreover, the MHC class I HLA-A*0301 allele is positively and independently associated with MS, conferring a relative twofold risk.

However, it is still controversial whether CD4+ T cell help is required for initiating CD8+ T cell responses (Bevan, 2004), though it clearly is essential for CD8+ memory generation (Sun et al., 2004). In addition, with certain viral infections (e.g., herpes simplex virus-1) and in situations with only mild inflammatory reactions (i.e., autoimmune responses), "licensing" of APCs by CD4+ T cells is needed to prime CD8+ T cell responses (Smith et al., 2004). Therefore, the key triggering event in MS pathogenesis is probably still the initial activation of CD4+ T cells, though it remains a mystery.

Because we cannot identify presymptomatic patients for prophylactic interventions, therapies currently need to be tailored for the chronic phase of MS. By then, CD8+ memory T cells have already expanded and no longer require CD4+ T cell help. They may be crucial targets for novel therapies to prevent them from perpetuating the inflammatory process in chronic MS patients. Particularly attractive candidate targets might include the chemokines, cytokines, adhesion molecules, and cytotoxic pathways that are preferentially used by autoreactive CD8+ T cells, though they clearly need sharper definition. The conventional CD4+-mediated EAE models are valuable tools for investigating how the disease process is initiated. However, they have so far revealed little about the chronic phases of MS—during which most patients are currently treated (Friese et al., 2006) (Sriram and Steiner, 2005) and for which better animal models are sorely needed. These chronic models must involve both CD4+ and CD8+ T cells and allow their characterization at different phases (Friese et al., 2006).

Finally, in addition to their cytotoxic activity, CD8+ T cells can adopt the phenotype of a regulatory T cell, which suppresses the autoimmune response. In certain EAE models, these CD8+ regulatory cells have an important role in dampening the immune response. We have focused here on the cytotoxic effector role; their suppressor functions are discussed in detail in Chapter 3.

## References

Ackerman, A. L., Cresswell, P. (2004) Cellular mechanisms governing cross-presentation of exogenous antigens. *Nat. Immunol.* 5:678-684.

Allan, S. M., Rothwell, N. J. (2001) Cytokines and acute neurodegeneration. *Nat. Rev. Neurosci.* 2:734-744.

Babbe, H., Roers, A., Waisman, A., et al. (2000) Clonal expansions of CD8(+) T cells dominate the T cell infiltrate in active multiple sclerosis lesions as shown by micromanipulation and single cell polymerase chain reaction. *J. Exp. Med.* 192:393-404.

Balashov, K. E., Rottman, J. B., Weiner, H. L., Hancock, W. W. (1999) CCR5(+) and CXCR3(+) T cells are increased in multiple sclerosis and their ligands MIP-1alpha and IP-10 are expressed in demyelinating brain lesions. *Proc. Natl. Acad. Sci. U S A* 96:6873-6878.

Battistini, L., Piccio, L., Rossi, B., et al. (2003) CD8+ T cells from patients with acute multiple sclerosis display selective increase of adhesiveness in brain venules: a critical role for P-selectin glycoprotein ligand-1. *Blood* 101:4775-4782.

Baumann, N., Pham-Dinh, D. (2001) Biology of oligodendrocyte and myelin in the mammalian central nervous system. *Physiol. Rev.* 81:871-927.

Becher, B., Durell, B. G., Miga, A. V., et al. (2001) The clinical course of experimental autoimmune encephalomyelitis and inflammation is controlled by the expression of CD40 within the central nervous system. *J. Exp. Med.* 193:967-974.

Ben-Nun, A., Wekerle, H., Cohen, I. R. (1981) The rapid isolation of clonable antigen-specific T lymphocyte lines capable of mediating autoimmune encephalomyelitis. *Eur. J. Immunol.* 11:195-199.

Bevan, M. J. (2004) Helping the CD8(+) T-cell response. *Nat. Rev. Immunol.* 4:595-602.

Biddison, W. E., Taub, D. D., Cruikshank, W. W., et al. (1997) Chemokine and matrix metalloproteinase secretion by myelin proteolipid protein-specific CD8+ T cells: potential roles in inflammation. *J. Immunol.* 158:3046-3053.

Bitsch, A., Schuchardt, J., Bunkowski, S., et al. (2000) Acute axonal injury in multiple sclerosis: correlation with demyelination and inflammation. *Brain* 123(Pt 6):1174-1183.

Boon, M., Nolte, I. M., Bruinenberg, M., et al. (2001) Mapping of a susceptibility gene for multiple sclerosis to the 51 kb interval between G511525 and D6S1666 using a new method of haplotype sharing analysis. *Neurogenetics* 3:221-230.

Booss, J., Esiri, M. M., Tourtellotte, W. W., Mason, D. Y. (1983) Immunohistological analysis of T lymphocyte subsets in the central nervous system in chronic progressive multiple sclerosis. *J. Neurol. Sci.* 62:219-232.

Brabb, T., von Dassow, P., Ordonez, N., et al. (2000) In situ tolerance within the central nervous system as a mechanism for preventing autoimmunity. *J. Exp. Med.* 192:871-880.

Calzascia, T., Di Berardino-Besson, W., Wilmotte, R., et al. (2003) Cutting edge: cross-presentation as a mechanism for efficient recruitment of tumor-specific CTL to the brain. *J. Immunol.* 171:2187-2191.

Calzascia, T., Masson, F., De Berardino-Besson, W., et al. (2005) Homing phenotypes of tumor-specific CD8 T cells are predetermined at the tumor site by crosspresenting APCs. *Immunity* 22:175-184.

Carson, M. J., Reilly, C. R., Sutcliffe, J. G., Lo, D. (1999) Disproportionate recruitment of CD8+ T cells into the central nervous system by professional antigen-presenting cells. *Am. J. Pathol.* 154:481-494.

Chang, T. T., Sobel, R. A., Wei, T., et al. (2003) Recovery from EAE is associated with decreased survival of encephalitogenic T cells in the CNS of B7-1/B7-2-deficient mice. *Eur. J. Immunol.* 33:2022-2032.

Christensen, J. E., Nansen, A., Moos, T., et al. (2004) Efficient T-cell surveillance of the CNS requires expression of the CXC chemokine receptor 3. *J. Neurosci.* 24:4849-4858.

Coles, A. J., Wing, M. G., Molyneux, P., et al. (1999) Monoclonal antibody treatment exposes three mechanisms underlying the clinical course of multiple sclerosis. *Ann. Neurol.* 46:296-304.

Crawford, M. P., Yan, S. X., Ortega, S. B., et al. (2004) High prevalence of autoreactive, neuroantigen-specific CD8+ T cells in multiple sclerosis revealed by novel flow cytometric assay. *Blood* 103:4222-4231.

DiLorenzo, T. P., Serreze, D. V. (2005) The good turned ugly: immunopathogenic basis for dianetogenic CD8+ T cells in NOD mice. *Immunol. Rev.* 204:250-263.

Dressel, A., Chin, J. L., Sette, A., et al. (1997) Autoantigen recognition by human CD8 T cell clones: enhanced agonist response induced by altered peptide ligands. *J. Immunol.* 159:4943-4951.

Ebers, G. C., Sadovnick, A. D., Risch, N. J. (1995) A genetic basis for familial aggregation in multiple sclerosis: Canadian Collaborative Study Group. *Nature* 377:150-151.

Ebers, G. C., Kukay, K., Bulman, D. E., et al. (1996) A full genome search in multiple sclerosis. *Nat. Genet.* 13:472-476.

Ferguson, B., Matyszak, M. K., Esiri, M. M., Perry, V. H. (1997) Axonal damage in acute multiple sclerosis lesions. *Brain* 120(Pt 3):393-399.

Fogdell-Hahn, A., Ligers, A., Gronning, M., et al. (2000) Multiple sclerosis: a modifying influence of HLA class I genes in an HLA class II associated autoimmune disease. *Tissue Antigens* 55:140-148.

Ford, M. L., Evavold, B. D. (2005) Specificity, magnitude, and kinetics of MOG-specific CD8(+) T cell responses during experimental autoimmune encephalomyelitis. *Eur. J. Immunol.* 35:76-85.

Friese, M. A., Fugger, L. (2005) Autoreactive CD8+ T cells in multiple sclerosis: a new target for therapy? *Brain* 128(Pt 8):1747-1763. Erratum in *Brain* 2005;128 (Pt 9):2215.

Friese, M. A., Montalban, X., Wilcox, N., Bell, J. I., Martin, R., Fugger, L. (2006) The value of animal models for drug development in multiple sclerosis, *Brain* 129: 1940-1952.

Gay, F. W., Drye, T. J., Dick, G. W., Esiri, M. M. (1997) The application of multifactorial cluster analysis in the staging of plaques in early multiple sclerosis: identification and characterization of the primary demyelinating lesion. *Brain* 120(Pt 8):1461-1483.

Gobin, S. J., Montagne, L., Van Zutphen, M., et al. (2001) Upregulation of transcription factors controlling MHC expression in multiple sclerosis lesions. *Glia* 36:68-77.

Goverman, J., Perchellet, A., Huseby, E. S. (2005) The role of CD8(+) T cells in multiple sclerosis and its animal models. *Curr. Drug Targets Inflamm. Allergy* 4:239-245.

Greter, M., Heppner, F. L., Lemos, M. P., et al. (2005) Dendritic cells permit immune invasion of the CNS in an animal model of multiple sclerosis. *Nat. Med.* 11:328-334.

Gronski, M. A., Boulter, J. M., Moskophidis, D., et al. (2004) TCR affinity and negative regulation limit autoimmunity. *Nat. Med.* 10:1234-1239.

Harbo, H. F., Lie, B. A., Sawcer, S., et al. (2004) Genes in the HLA class I region may contribute to the HLA class II-associated genetic susceptibility to multiple sclerosis. *Tissue Antigens* 63:237-247.

Hauser, S. L., Bhan, A. K., Gilles, F., et al. (1986) Immunohistochemical analysis of the cellular infiltrate in multiple sclerosis lesions. *Ann. Neurol.* 19:578-587.

Hayashi, T., Morimoto, C., Burks, J. S., et al. (1988) Dual-label immunocytochemistry of the active multiple sclerosis lesion: major histocompatibility complex and activation antigens. *Ann. Neurol.* 24:523-531.

Heath, W. R., Carbone, F. R. (2001) Cross-presentation, dendritic cells, tolerance and immunity. *Annu. Rev. Immunol.* 19:47-64.

Hemmer, B., Archelos, J. J., Hartung, H. P. (2002) New concepts in the immunopathogenesis of multiple sclerosis. *Nat. Rev. Neurosci.* 3:291-301.

Hernandez, J., Aung, S., Marquardt, K., Sherman, L. A. (2002) Uncoupling of proliferative potential and gain of effector function by CD8(+) T cells responding to self-antigens. *J. Exp. Med.* 196:323-333.

Hickey, W. F. (1999) Leukocyte traffic in the central nervous system: the participants and their roles. *Semin. Immunol.* 11:125-137.

Hickey, W. F., Kimura, H. (1988) Perivascular microglial cells of the CNS are bone marrow-derived and present antigen in vivo. *Science* 239:290-292.

Hoftberger, R., Aboul-Enein, F., Brueck, W., et al. (2004) Expression of major histocompatibility complex class I molecules on the different cell types in multiple sclerosis lesions. *Brain Pathol.* 14:43-50.

Hohlfeld, R., Wiendl, H. (2001) The ups and downs of multiple sclerosis therapeutics. *Ann. Neurol* 49:281-284.

Honma, K., Parker, K. C., Becker, K. G., et al. (1997) Identification of an epitope derived from human proteolipid protein that can induce autoreactive CD8+ cytotoxic T lymphocytes restricted by HLA-A3: evidence for cross-reactivity with an environmental microorganism. *J. Neuroimmunol.* 73:7-14.

Howard, L. M. Miga, A. J., Vanderlugt, C. L., et al. (1999) Mechanisms of immunotherapeutic intervention by anti-CD40L (CD154) antibody in an animal model of multiple sclerosis. *J. Clin. Invest.* 103:281-290.

Huseby, E. S., Ohlen, C., Goverman, J. (1999) Cutting edge: myelin basic protein-specific cytotoxic T cell tolerance is maintained in vivo by a single dominant epitope in H-2k mice. *J. Immunol.* 163:1115-1118.

Huseby, E. S., Liggitt, D., Brabb, T., et al. (2001) A pathogenic role for myelin-specific CD8(+) T cells in a model for multiple sclerosis. *J. Exp. Med.* 194:669-676.

Huseby, E. S., White, J., Crawford, F., et al. (2005) How the T cell repertoire becomes peptide and MHC specific. *Cell* 122:247-260.

Jacobsen, M., Cepok, S., Quak, E., et al. (2002) Oligoclonal expansion of memory CD8+ T cells in cerebrospinal fluid from multiple sclerosis patients. *Brain* 125:538-550.

Jensen, J., Krakauer, M., Sellebjerg, F. (2001) Increased T cell expression of CD154 (CD40-ligand) in multiple sclerosis. *Eur. J. Neurol.* 8:321-328.

Jersild, C., Svejgaard, A., Fog, T. (1972) HL-A antigens and multiple sclerosis. *Lancet* 1:1240-1241.

Jiang, H., Zhang, S. I., Pernis, B. (1992) Role of CD8+ T cells in murine experimental allergic encephalomyelitis. *Science* 256:1213-1215.

Joly, E., Mucke, L., Oldstone, M. B. (1991) Viral persistence in neurons explained by lack of major histocompatibility class I expression. *Science* 253:1283-1285.

Jurewicz, A., Biddison, W. E., Antel, J. P. (1998) MHC class I-restricted lysis of human oligodendrocytes by myelin basic protein peptide-specific CD8 T lymphocytes. *J. Immunol.* 160:3056-3059.

Karman, J., Ling, C., Sandor, M., Fabry, Z. (2004) Initiation of immune responses in brain is promoted by local dendritic cells. *J. Immunol.* 173:2353-2361.

Kivisakk, P., Mahad D. J., Callahan, M. K., et al. (2003) Human cerebrospinal fluid central memory CD4+ T cells: evidence for trafficking through choroid plexus and meninges via P-selectin. *Proc. Natl. Acad. Sci. U S A* 100:8389-8394.

Kivisakk, P., Mahad, D. J., Callahan, M. K., et al. (2004) Expression of CCR7 in multiple sclerosis: implications for CNS immunity. *Ann. Neurol.* 55:627-638.

Kleinschmidt-DeMasters, B. K., Tyler, K. L. (2005) Progressive multifocal leukoencephalopathy complicating treatment with natalizumab and interferon beta-1a for multiple sclerosis. *N. Engl. J. Med.* 353:369-374.

Koh, D. R., Fung-Leung, W. P., Ho, A., et al. (1992) Less mortality but more relapses in experimental allergic encephalomyelitis in CD8–/– mice. *Science* 256:1210-1213.

Kornek, B., Storch, M. K., Weissert, R., et al. (2000) Multiple sclerosis and chronic autoimmune encephalomyelitis: a comparative quantitative study of axonal injury in active, inactive, and remyelinated lesions. *Am. J. Pathol.* 157:267-276.

Krogsgaard, M., Wucherpfennig, K. W., Cannella, B., et al. (2000) Visualization of myelin basic protein (MBP) T cell epitopes in multiple sclerosis lesions using a monoclonal antibody specific for the human histocompatibility leukocyte antigen (HLA)-DR2-MBP 85-99 complex. *J. Exp. Med.* 191:1395-1412.

Kuhlmann, T., Lingfeld, G., Bitsch, A., et al. (2002) Acute axonal damage in multiple sclerosis is most extensive in early disease stages and decreases over time. *Brain* 125:2202-2212.

Kurts, C., Sutherland, R. M., Davey, G., et al. (1999) CD8 T cell ignorance or tolerance to islet antigens depends on antigen dose. *Proc. Natl. Acad. Sci. U S A* 96:12703-12707.

Kutzelnigg, A., Lucchinetti, C. F., Stadelmann, C., et al. (2005) Cortical demyelination and diffuse white matter injury in multiple sclerosis. *Brain* 128:2705-2712.

Langer-Gould, A., Atlas, S. W., Green, A. J., et al. (2005) Progressive multifocal leukoen-cephalopathy in a patient treated with natalizumab. *N. Engl. J. Med.* 353:375-381.

Laplaud, D. A., Ruiz, C., Wiertleewski, S., et al. (2004) Blood T-cell receptor beta chain transcriptome in multiple sclerosis: characterization of the T cells with altered CDR3 length distribution. *Brain* 127:981-995.

Lassmann, H. (1983) Comparative neuropathology of chronic experimental allergic encephalomyelitis and multiple sclerosis. *Schriftenr. Neurol.* 25:1-135.

Lassmann, H., Ransohoff, R. M. (2004) The CD4-Th1 model for multiple sclerosis: a critical [correction of crucial] re-appraisal. *Trends Immunol.* 25:132-137.

Logunova, N. N., Viret, C., Pobezinsky, L. A., et al. (2005) Restricted MHC-peptide repertoire predisposes to autoimmunity. *J. Exp. Med.* 202:73-84.

Madsen, L. S., Andersson, E. C., Jansson, L., et al. (1999) A humanized model for multiple sclerosis using HLA-DR2 and a human T-cell receptor. *Nat. Genet.* 23:343-347.

Marrosu, M., Murru, M. R., Costa, G., et al. (1997) Multiple sclerosis in Sardinia is associated and in linkage disequilibrium with HLA-DR3 and -DR4 alleles. *Am. J. Hum. Genet.* 61:454-457.

Marrosu, M. G., Murru, R., Murru, M. R., et al. (2001) Dissection of the HLA association with multiple sclerosis in the founder isolated population of Sardinia. *Hum. Mol. Genet.* 10:2907-2916.

McMahon, E. J., Bailey, S. L., Castenada, C. V., et al. (2005) Epitope spreading initiates in the CNS in two mouse models of multiple sclerosis. *Nat. Med.* 11:335-339.

Medana, I., Li, Z., Flugel, A., et al. (2001a) Fas ligand (CD95L) protects neurons against perforin-mediated T lymphocyte cytotoxicity. *J. Immunol.* 167:674-681.

Medana, I., Martinic, M. A., Wekerle, H., Neumann, H. (2001b) Transection of major histocompatibility complex class I-induced neurites by cytotoxic T lymphocytes. *Am. J. Pathol.* 159:809-815.

Mendel, I., Kerlero de Rosbo, N., Ben-Nun, A. (1996) Delineation of the minimal encephalitogenic epitope within the immunodominant region of myelin oligodendrocyte glycoprotein: diverse V beta gene usage by T cells recognizing the core epitope encephalitogenic for T cell receptor V beta b and T cell receptor V beta a H-2b mice. *Eur. J. Immunol.* 26:2470-2479

Miller, D. H., Khan, O. A., Sheremata, W. A., et al. (2003) A controlled trial of natalizumab for relapsing multiple sclerosis. *N. Engl. J. Med.* 348:15-23.

Monaco, M. C., Shin, J., Major, E. O. (1998) JC virus infection in cells from lymphoid tissue. *Dev. Biol. Stand.* 94:115-122.

Muller, D. M., Pender, M. P., Greer, J. M. (2004) Chemokines and chemokine receptors: potential therapeutic targets in multiple sclerosis. *Curr. Drug Targets Inflamm. Allergy* 3:279-290.

Naito, S., Namerow, N., Mickey, M. R., Terasaki, P. I. (1972) Multiple sclerosis: association with HL-A3. *Tissue Antigens* 2:1-4.

Nepom, G. T., Erlich, H. (1991) MHC class-II molecules and autoimmunity. *Annu. Rev. Immunol.* 9:493-525.

Neumann, H., Cavalie, A., Jenne, D. E., Wekerle, H. (1995) Induction of MHC class I genes in neurons. *Science* 269:549-552.

Neumann, H., Schmidt, H., Cavalie, A., et al. (1997) Major histocompatibility complex (MHC) class I gene expression in single neurons of the central nervous system: differential regulation by interferon (IFN)-gamma and tumor necrosis factor (TNF)-alpha. *J. Exp. Med.* 185:305-316.

Neumann, H., Misgeld, T., Matsumuro, K., Wekerle, H. (1998) Neurotrophins inhibit major histocompatibility class II inducibility of microglia: involvement of the p75 neurotrophin receptor. *Proc. Natl. Acad. Sci. U S A* 95:5779-5784.

Neumann, H., Medana, I. M., Bauer, J., Lassmann, H. (2002) Cytotoxic T lymphocytes in autoimmune and degenerative CNS diseases. *Trends Neurosci.* 25:313-319.

Niland, B., Banki, K., Biddison, W. E., Perl, A. (2005) CD8+ T cell-mediated HLA-A*0201-restricted cytotoxicity to transaldolase peptide 168-176 in patients with multiple sclerosis. *J. Immunol.* 175:8365-8378.

Odeberg, J., Piao, J. H., Samuelsson, E. B., et al. (2005) Low immunogenicity of in vitro-expanded human neural cells despite high MHC expression. *J. Neuroimmunol.* 161:1-11.

Oksenberg, J. R., Panzara, M. A., Begovich, A. B., et al. (1993) Selection for T-cell receptor V beta-D beta-J beta gene rearrangements with specificity for a myelin basic protein peptide in brain lesions of multiple sclerosis. *Nature* 362:68-70.

Olerup, O., Hillert, J. (1991) HLA class II-associated genetic susceptibility in multiple sclerosis: a critical evaluation. *Tissue Antigens* 38:1-15.

Paolillo, A., Coles, A. J., Molyneux, P. D., et al. (1999) Quantitative MRI in patients with secondary progressive MS treated with monoclonal antibody Campath 1H. *Neurology* 53:751-757.

Pashenkov, M., Link, H. (2002) Dendritic cells and immune responses in the central nervous system. *Trends Immunol.* 23:69-70; author reply 70.

Pashenkov, M., Huang, Y. M., Kostulas, V., et al. (2001) Two subsets of dendritic cells are present in human cerebrospinal fluid. *Brain* 124:480-492.

Pouly, S., Becher, B., Blain, M., Antel, J. P. (2000) Interferon-gamma modulates human oligodendrocyte susceptibility to Fas-mediated apoptosis. *J. Neuropathol. Exp. Neurol.* 59:280-286.

Rall, G. F., Mucke, L., Oldstone, M. B. (1995) Consequences of cytotoxic T lymphocyte interaction with major histocompatibility complex class I-expressing neurons in vivo. *J. Exp. Med.* 182:1201-1212.

Ramakrishna, C., Stohlman, S. A., Atkinson, R. A., et al. (2004) Differential regulation of primary and secondary CD8+ T cells in the central nervous system. *J. Immunol.* 173:6265-6273.

Ransohoff, R. M., Estes, M. L. (1991) Astrocyte expression of major histocompatibility complex gene products in multiple sclerosis brain tissue obtained by stereotactic biopsy. *Arch. Neurol.* 48:1244-1246.

Ransohoff, R. M., Kivisakk, P., Kidd, G. (2003) Three or more routes for leukocyte migration into the central nervous system. *Nat. Rev. Immunol.* 3:569-581.

Rensing-Ehl, A., Malipiero, U., Irmler, M., et al. (1996) Neurons induced to express major histocompatibility complex class I antigen are killed via the perforin and not the Fas (APO-1/CD95) pathway. *Eur. J. Immunol.* 26:2271-2274.

Rioux, J. D., Abbas, A. K. (2005) Paths to understanding the genetic basis of autoimmune disease. *Nature* 435:584-589.

Rubio, J. P., Bahio, M., Butzkueven, H., et al. (2002) Genetic dissection of the human leukocyte antigen region by use of haplotypes of Tasmanians with multiple sclerosis. *Am. J. Hum. Genet.* 70:1125-1137.

Ruijs, T. C., Freedman, M. S., Grenier, Y. G., et al. (1990) Human oligodendrocytes are susceptible to cytolysis by major histocompatibility complex class I-restricted lymphocytes. *J. Neuroimmunol.* 27:89-97.

Scarpini, E., Galimberti, D., Baron, P., et al. (2002) IP-10 and MCP-1 levels in CSF and serum from multiple sclerosis patients with different clinical subtypes of the disease. *J. Neurol. Sci.* 195:41-46.

Seder, R. A., Ahmed, R. (2003) Similarities and differences in CD4+ and CD8+ effector and memory T cell generation. *Nat. Immunol.* 4:835-842.

Serafini, B., Columba-Cabezas, S., Di Rosa, F., Aloisi, F. (2000) Intracerebral recruitment and maturation of dendritic cells in the onset and progression of experimental autoimmune encephalomyelitis. *Am. J. Pathol.* 157:1991-2002.

Skulina, C., Schmidt, S., Dornmair, K., et al. (2004) Multiple sclerosis: brain-infiltrating CD8+ T cells persist as clonal expansions in the cerebrospinal fluid and blood. *Proc. Natl. Acad. Sci. U S A* 101:2428-2433.

Smith, C. M., Wilson, N. S., Waithman, J., et al. (2004) Cognate CD4(+) T cell licensing of dendritic cells in CD8(+) T cell immunity. *Nat. Immunol.* 5:1143-1148.

Sollid, L. M., Thorsby, E. (1993) HLA susceptibility genes in celiac disease: genetic mapping and role in pathogenesis. *Gastroenterology* 105:910-922.

Sollid, L. M., Markussen, G., Ek, J., et al. (1989) Evidence for a primary association of celiac disease to a particular HLA-DQ alpha/beta heterodimer. *J. Exp. Med.* 169: 345-350.

Sriram, S., Steiner, I. (2005) Experimental allergic encephalomyelitis: a misleading model of multiple sclerosis. *Ann. Neurol.* 58:939-945.

Stastny, P. (1978) Association of the B-cell alloantigen DRw4 with rheumatoid arthritis. *N. Engl. J. Med.* 298:869-871.

Steinman, L. (1999) Assessment of animal models for MS and demyelinating disease in the design of rational therapy. *Neuron* 24:511-514.

Steinman, L. (2001) Myelin-specific CD8 T cells in the pathogenesis of experimental allergic encephalitis and multiple sclerosis. *J. Exp. Med.* 194:F27-30.

Su, S. B., Silver, P. B., Grajewski, R. S., et al. (2005) Essential role of the MyD88 pathway, but nonessential roles of TLRs 2, 4, and 9, in the adjuvant effect promoting Th1-mediated autoimmunity. *J. Immunol.* 175:6303-6310.

Sun, D., Whitaker, J. N., Huang, Z., et al. (2001) Myelin antigen-specific CD8+ T cells are encephalitogenic and produce severe disease in C57BL/6 mice. *J. Immunol.* 166:7579-7587.

Sun, D., Zhang, Y., Wei, B., et al. (2003) Encephalitogenic activity of truncated myelin oligodendrocyte glycoprotein (MOG) peptides and their recognition by CD8+ MOG-specific T cells on oligomeric MHC class I molecules. *Int. Immunol.* 15:261-268.

Sun, J. C., Williams, M. A., Bevan, M. J. (2004) CD4+ T cells are required for the maintenance, not programming, of memory CD8+ T cells after acute infection. *Nat. Immunol.* 5:927-933.

Trapp, B. D., Peterson, J., Ransohoff, R. M., et al. (1998) Axonal transection in the lesions of multiple sclerosis. *N. Engl. J. Med.* 338:278-285.

Tsuchida, T., Parker, K. C., Turner, R. V., et al. (1994) Autoreactive CD8+ T-cell responses to human myelin protein-derived peptides. *Proc. Natl. Acad. Sci. U S A* 91:10859-10863.

Tyznik, A. J., Sun, J. C., Bevan, M. J. (2004) The CD8 population in CD4-deficient mice is heavily contaminated with MHC class II-restricted T cells. *J. Exp. Med.* 199:559-565.

Ubogu, E. E., Cossoy, M. B., Ransohoff, R. M. (2005) The expression and function of chemokines involved in CNS inflammation. *Trends Pharmacol. Sci.* 27:48-55.

Van Assche, G., Ban Ranst, M., Sciot, R., et al. (2005) Progressive multifocal leukoencephalopathy after natalizumab therapy for Crohn's disease. *N. Engl. J. Med.* 353: 362-368.

Van Oosten, B. W., Lai, M., Hodgkinson, S., et al. (1997) Treatment of multiple sclerosis with the monoclonal anti-CD4 antibody cM-T412: results of a randomized, double-blind, placebo-controlled, MR-monitored phase II trial. *Neurology* 49:351-357.

Weller, R. O. (1998) Pathology of cerebrospinal fluid and interstitial fluid of the CNS: significance for Alzheimer disease, prion disorders and multiple sclerosis. *J. Neuropathol. Exp. Neurol.* 57:885-894.

Weninger, W., Manjunath, N., von Andrian, U. H. (2002) Migration and differentiation of CD8+ T cells. *Immunol. Rev.* 186:221-233.

Yednock, T. A., Cannon, C., Fritz, L. C., et al. (1992) Prevention of experimental autoimmune encephalomyelitis by antibodies against alpha 4 beta 1 integrin. *Nature* 356: 63-66.

Zamvil, S., Nelson, P., Trotter, J., et al. (1985a) T-cell clones specific for myelin basic protein induce chronic relapsing paralysis and demyelination. *Nature* 317:355-358.

Zamvil, S. S., Nelson, P. A., Mitchell, D. J., et al. (1985b) Encephalitogenic T cell clones specific for myelin basic protein: an unusual bias in antigen recognition. *J. Exp. Med.* 162:2107-2124.

Zang, Y. C., Li, S., Rivera, V. M., et al. (2004) Increased CD8+ cytotoxic T cell responses to myelin basic protein in multiple sclerosis. *J. Immunol.* 172:5120-5127.

Zehntner, S. P., Brisebois, M., Tran, E., et al. (2003) Constitutive expression of a costimulatory ligand on antigen-presenting cells in the nervous system drives demyelinating disease. *FASEB J.* 17:1910-1912.

# Part II
## Novel Immunotherapeutic Strategies and Emerging Treatments

# 14
# Review of Novel Immunotherapeutic Strategies for MS

Heinz Wiendl and Reinhard Hohlfeld

Novel insights in the immunopathologic processes, advances in biotechnology, development of powerful magnetic resonance imaging (MRI) technologies together with improvements in clinical trial design has led to a variety of evaluable therapeutic approaches in multiple sclerosis (MS). Therapy has changed dramatically over the past decade, yielding significant progress for the treatment of relapsing/remitting and secondary progressive MS. A substantial number of pivotal and preliminary reports continue to demonstrate encouraging new evidence that advances are being made in the care of MS patients. This chapter provides a compilation of novel immunotherapeutic strategies or new aspects of known immunotherapeutic agents that have evolved in recent years. Their immunopathogenetic rationale as well as the accompanying preclinical and clinical data to support their potential to endorse the currently available disease modifying drugs are highlighted.

## 1   Introduction

Based on the growing immunopathogenetic understanding of MS and with the assistance of modern biotechnology, a growing arsenal of potential therapeutic drugs has been developed. There has been a vast amount of interest in MS clinical research and trial conduction, accompanied by a dramatic change of therapy over the past decade (Hohlfeld and Wiendl, 2001) (Fig. 14.1). Several agents have been approved by the Food and Drug Administration (FDA) and the European Agency for the Evaluation of Medicinal Products (EMEA) and are now being widely used. Because of the increasing knowledge about the cellular and molecular mechanisms of immune cell migration and activation, emphasis was partly shifted to more selective therapies (Fox and Ransohoff, 2004; Hohlfeld and Wekerle, 2004). However, recent years have taught us that selectivity does not necessarily imply more efficacy (Wiendl and Hohlfeld, 2002). This notion is fueled by the recognition of MS disease heterogeneity at the clinical, immunologic, and pathomorphologic levels. Thus, in parallel with the sophisticated and elegant immune-selective strategies, concepts of more general immunosuppression and immunomodulation in MS therapy are receiving increased attention. Furthermore, emerging treatment strategies take into account the various pathologic mechanisms, particularly strategies to protect neurons against axonal damage and loss. Neurodegeneration and lack of significant regenerative mechanisms clearly seem to predominate over inflammatory damage during the later stages of the disease course, emphasizing the emerging trend for therapeutic intervention.

The following compilation of novel immunotherapeutic strategies in MS summarizes the most important agents under clinical investigation with respect to its pathophysiologic rationale, comprising also those strategies that are already officially reported as clinical trials and listed on the National MS Society web page (http://nationalmssociety.org).

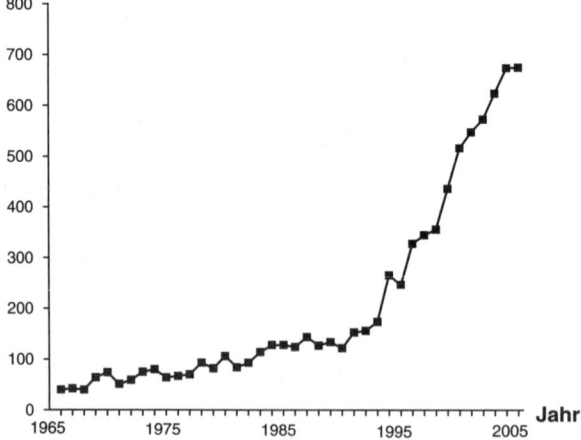

FIGURE 14.1. Publications per year on multiple sclerosis (MS) therapy as searched in PubMed.

# 2   Immunopathogenetic Rationale for Strategies in MS

Current concepts assume that MS occurs as a consequence of immune tolerance breakdown in genetically susceptible hosts. The major contributing factors thus include genetics, the environment, and immune dysregulation (Frohmann, 2006; Hafler, 2005; Hemmer et al., 2002; Noseworthy et al., 2000a, Sospedra and Martin, 2005a). T helper cells are considered to play a pivotal role in orchestrating self-reactive immune responses in the central nervous system (CNS), primarily hallmarked by inflammatory demyelination. Although not yet formally proven, MS pathogenesis is thought to comprise an initial inflammatory phase, which fulfils the criteria for an autoimmune disease. This is followed by a phase of selective demyelination and, last, a neurodegenerative phase (Hemmer et al., 2002; Hohlfeld, 1997; Noseworthy et al., 2000a,b; Steinman, 2001). Crucial immunopathogenetic steps and their most important mediators are shown in Figure 14.2. These mechanisms provide the rationale for specific and nonspecific therapeutic approaches currently applied or envisaged. Potentially self-reactive T cells, acknowledged to exist in any individual, are activated in the periphery, probably by molecular mimicry (i.e., recognition of epitopes that are common to autoantigens and microbial structures as exogenous triggers (Benoist and Mathis, 2001; Lang et al., 2002) or by self-antigens (Fig. 14.2, **1, 4**). The regulative interfaces between antigen-presenting cells (APCs) and T cells, including those molecules critically modulating APC–T cell interactions, represent potential targets. T cells may trigger B cell activation and antibody formation, the latter eventually exerting detrimental effector functions at the myelin sheath (Fig. 14.2, **2, 4**) (Archelos and Hartung, 2000). T cell activation enables transmigration through the blood-brain barrier to the sites of inflammation, a cascade of events influenced by adhesion molecules, chemotactic factors, and migration promotors (Fig. 14.2, **3**). Reactivated in the CNS, these T cells of both CD4 T helper and CD8 cytotoxic phenotype (Babbe et al., 2000) release proinflammatory Th1 cytokines and orchestrate the destruction of the myelin sheath via various types of immune cells. Destruction follows either of two pathologic patterns (Lassmann et al., 2001): (1) T cell- and macrophage-mediated demyelination or (2) antibody-mediated demyelination that involves complement activation (Fig. 14.2, **4**) (Archelos and Hartung, 2000). In addition to this autoaggressive inflammatory phase, axonal damage and loss occurs early in the disease course, causing irreversible disability. It is unclear whether axonal damage is the consequence of a primary active destructive process executed by, for example, macrophages and cytotoxic molecules derived from CD8 T cells or a pathophysiologic response that occurs secondarily to demyelination based on increased vulnerability and loss of trophic support (Fig. 14.2, **5**) (Bjartmar and Trapp, 2001). Finally, dysregulation in T cells' apoptotic mechanisms would lead to accumulation of autoreactive cells in the periphery or in the CNS and perpetuate the undesirable and detrimental lymphocytic effector functions (Fig. 14.2, **6**) (Zipp et al., 1999).

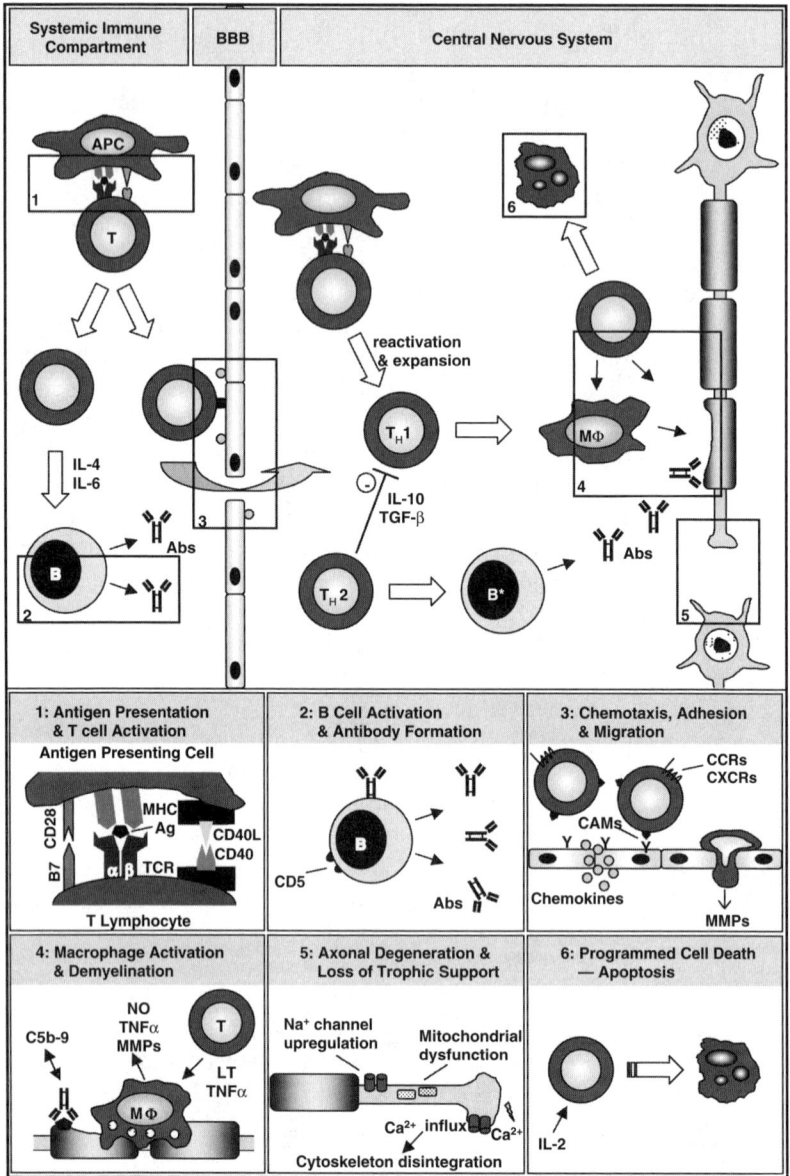

FIGURE 14.2. Synoptic view of the immune response in the pathogenesis of MS. Autoreactive T cells (T) recognize with their T cell receptor (TCR) a specific autoantigen (Ag) presented by major histocompatibility complex class II molecules (MHC) and the simultaneous delivery of co-stimulatory signals (CD28, B7, CD40, CD40L) on the cell surface of antigen-presenting cells (APCs), such as macrophages, in the systemic immune compartment (highlighted in 1). Activated T lymphocytes can cross the blood-brain barrier (BBB) to enter the central nervous system (CNS). The mechanisms of transendothelial migration is mediated by the complex interplay of cellular adhesion molecules (CAMs),

# 3   Novel Immunotherapeutic Strategies

According to the National MS Society, more than 160 trials in MS have recently been completed, are in progress or are planned (http://www.nationalms society.org). Table 14.1 has been created to assist in understanding possible sites of action of various treatment strategies currently evaluated or displaying future directions. Major groupings of agents or strategies, including some of their clinical evaluations, are described below.

## 3.1   Immunosuppressants and Immunomodulators

For a variety of reasons, the concept of immunosuppression as a strategy to treat autoimmune disorders (AIDs) has recently attracted considerable reattention. The "resurrection" of this classic concept (in contrast to various strategies of highly selective intervention and immunomodulation, listed below) is reasonable for a number of possible explanations, including: (1) the heterogeneity and complexity of disease mechanisms including its interindividual differences in MS; (2) the lack of a clear (auto)antigen identification in most AIDs, including MS; (3) the recent "failures" of highly (antigen)-specific and selective immune therapies in MS; (4) the availability of novel drugs with expected good safety profile and potential for longer-term use; (5) the possibility of drugs that can be applied orally; and last but not least (6) novel immunologic concepts as the view of "tolerogenic immunosuppression."

It becomes increasingly difficult to make a clear distinction between a group of agents classified as "immunosuppressants" and those classified as "immunomodulatory" or "immune-specific." The World Health Organization (WHO) has

---

FIGURE 14.2. (*continued*) chemokines and their receptors (CCRs, CXCRs), and matrix metalloproteinases (MMPs) (highlighted in **3**). Within the CNS the T cells, once reactivated and expanded, activate microglia cells/macrophages (MF) to enhance phagocytic activity, with the production of cytokines such as tumor necrosis factor (TNFα) and lymphotoxin (LT), and the release of toxic mediators such as nitric oxide (NO), thereby propagating demyelination and axonal loss. Autoantibodies (Abs), crossing the BBB or locally produced by B cells or mast cells (B*) contribute to this process. Autoantigens activate the complement cascade, resulting in the formation of the membrane-attack complex (C5b-9) and subsequent lysis target structure (highlighted in **2** and **4**). The upregulation of $Na^+$ and $Ca^{2+}$ channels on the axon as well as mitochondrial dysfunction and loss of trophic support contribute to axonal disintegration and degeneration (highlighted in **5**). The inflammatory response is regulated by anti-inflammatory cytokines, such as interleukin-10 (IL-10) or transforming growth factor-β (TGFβ), as well as IL-2, inducing programmed cell death (apoptosis) in immunoreactive T lymphocytes (highlighted in **6**).

TABLE 14.1. Immunotherapeutic Strategies Based on Putative Immunopathologic Mechanisms in Multiple Sclerosis.

| Process | Target and mediators | Therapeutic approaches |
|---|---|---|
| Antigen presentation and T cell activation Nonselective activation of T cells | T cells | Bone marrow transplantation Extracorporeal photopheresis |
| Selective activation of myelin-reactive T cells APC–T cell interaction | Molecular mimicry? Superantigens? Other factors? | T cell vaccination T cell receptor peptide vaccination Antigen-derived therapies |
| Interaction of the trimolecular complex | MHC class II T cell receptor (TCR) | MHC-peptide vaccine Anti-MHC-mAbs TCR-peptide vaccination TCR-mAbs T cell vaccination Antigen-derived therapies (e.g., altered peptide ligands) |
| Activation of "second signals" | Co-stimulatory molecules B7-1, B7-2, CD40-CD40-L, CTLA-4, other CD molecules | Anti-B7 mAbs CD40/CD40-L interaction CTLA-4-Ig fusion protein Anti-T-cell mAbs (e.g., CD2) T cell vaccines |
| B cell activation and antibody formation Humorally mediated damage | B-cells Antibodies Complement system | Immunosuppression Cell-specific mAbs Autologous stem-cell transplantation Anti-autoantibody mAbs Intravenous immunoglobulins (IVIg) Plasmapheresis Complement inhibitors MAC (C5b-C9 membrane attack complex inhibitors, CD59, CD46, DAF) |
| Chemotaxis, adhesion, and migration into the CNS Adherence and diapedesis | Selectins (L-, E-, P-) Integrins (VLA-4, VLA-5, VLA-6, LFA-1, MAC-1, CR4) Ig superfamily (ICAM-1, ICAM-2, ICAM-3, VCAM, LFA-2, LFA-3) | Selectin inhibitors Anti-selectin mAbs Integrin/ligand inhibitors Anti-adhesion molecule mAbs (anti-VLA-4 mAb) |
| Blood-brain barrier disruption | MMPs Cytokines ? Infection (viral, bacterial) ? Reactive metabolites | MMP inhibitors Anti-MMP mAbs Cytokine modulators (anti-cytokine drugs, anti-cytokine-mAbs, cytokines) Antiviral/antimicrobial therapies |

*(Continued)*

TABLE 14.1. Immunotherapeutic Strategies Based on Putative Immunopathologic Mechanisms in Multiple Sclerosis—Cont'd

| Process | Target and mediators | Therapeutic approaches |
|---|---|---|
| Cellular CNS invasion | MMPs TIMPs | Free radical scavengers Nitric oxide synthase inhibitors MMP inhibitors Anti-MMP Mabs TIMP promotors |
| Cell recruitment in the CNS | Integrins (VLA-4, VLA-5, VLA-6, LFA-1, MAC-1, CR4) Chemokines | Integrin/ligand inhibitors Anti-integrin mAbs Chemokine inhibitors Anti-chemokine mAbs Cytokine modulators (anti-cytokine drugs, anti-cytokine mAbs, cytokines) Thalidomide, pentoxifylline |
| Macrophage activation and demyelination | | |
| Soluble toxic mediators | TNFα, IFNγ, lymphotoxin MMPs/proteases | Cytokine modulators (anti-cytokine drugs, anti-cytokine-mAbs, cytokines) MMP inhibitors Anti-MMP Mabs TIMP promoters |
| Cytokine network | Th1 predominance Th1/Th2 dysbalance | Anti-Th1 cytokines Th1 cytokine inhibitors (IL-1 inhibitors; IL-1R blockade, anti-IL-12) Anti-CD52 (Campath) Anti-IL-2R-α Th2 cytokines (IL-4, IL-10, TGFβ) Th2 cytokine promoters (e.g., GA, estradiol, statins) Shift Th1 to Th2 (e.g., GA, statins, manipulation of co-stimulation) |
| Cell-mediated damage | CD4+ and CD8+ T cells Macrophages Microglia Astrocytes | Immunosuppression Cell-specific Mabs inhibitors of excitotoxin products (arachidonic acid, platelet-activating factor, free radicals, glutamate, quinolinate, cysteine, amines, chemokines) Protein kinase inhibitors, caspase signaling inhibitors, Neuroimmunophilins Autologous stem-cell transplantation |
| Humorally mediated damage | Antibodies Complement system | Anti-autoantibody mAbs Intravenous immunoglobulins (IVIg) |

*(Continued)*

TABLE 14.1. Immunotherapeutic Strategies Based on Putative Immunopathologic Mechanisms in Multiple Sclerosis—Cont'd

| Process | Target and mediators | Therapeutic approaches |
|---|---|---|
| | | Plasmapheresis |
| | | Complement inhibitors MAC (C5b-C9 membrane attack complex inhibitors, CD59, CD46, DAF) |
| Axonal degeneration and loss of trophic support | Survival and maturation of oligodendrocyte precursors | BDNF, GDNF, CNTF, IGF-1, IL-6 |
| | Availability of oligodendrocyte precursors | Leukemia inhibitory factor |
| | | Neurotrophin-3 |
| | Axon regeneration | PDGF |
| | | Gene therapy |
| | | Stem cell transplantation |
| | | Precursor cell implantation |
| | | Immortalized cell transplantation |
| | | Schwann cell transplantation |
| | | Xenotransplantation |
| | | Axon guidance (Semaphorin-neurophilin-plexin) |
| | | Therapeutic antibodies (IF-1 antibody, anti-Nogo-Ab) |
| | | Protective factors: NGF, BDNF, basic fibroblast GF, IL-10, neurotrophin-3 |
| Other damage mechanisms | Glutamate | AMPA/kainate inhibitors, GABA stimulation |
| | Reactive oxygen species | |
| | Reactive nitrogen specie | Free radical scavengers |
| | Channels | Nitric oxide synthase inhibitors |
| | Viral infection | Channel blockers (Na+, Ca+) |
| | Microbial infection | |
| Programmed cell death—apoptosis | T and B cell overdrive | Fas/Fas-L, TRAIL, and other receptor modulations |
| | Neuronal and glial demise | Anti-apoptosis strategies |

See also Figure 14.2.
APC, antigen presenting cell; VLA, very late antigen; LFA, lymphocyte function-associated antigen; ICAM, intercellular adhesion molecule; VCAM, vascular cell adhesion molecule; MMP, matrix metalloproteinase; TIMP, tissue inhibitor of MMP; MHC, major histocompatibility complex; CD, cluster of differentiation; TNFα, tumor necrosis factor α; IFNγ, interferon-γ; mAb, monoclonal antibody; DAF, decay-accelerating factor; BDNF, brain-derived growth factor; GDNF, glial-cell derived growth factor; CNTF, ciliary neurotrophic factor; IGF-1, insulin-like growth factor-1; PDGF, platelet-derived growth factor.

proposed a classification of immunosuppressants based on distinguishing agents that target (I) "intracellular ligands," (II) cell surface ligands, and (III) soluble factors (e.g., "anticytokines"). Figure 14.3 illustrates this classification, including important examples of drugs that currently are in use or under investigation for its use in MS therapy.

Classification of "immunosuppressants" according to the WHO

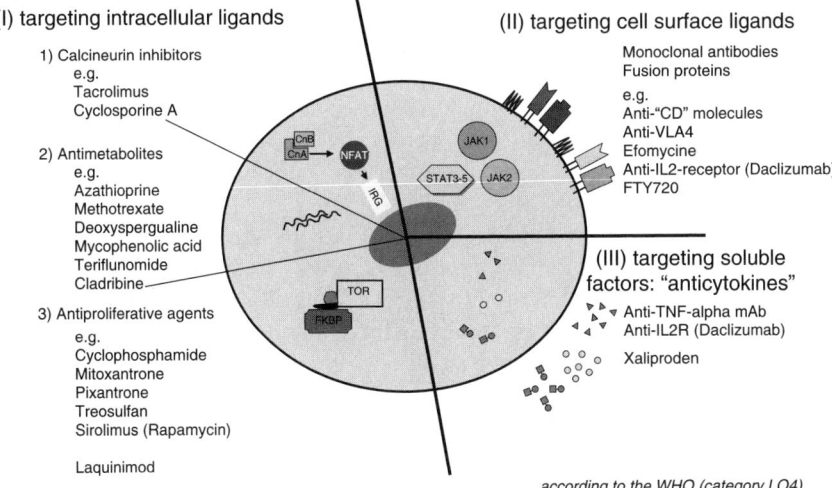

(I) targeting intracellular ligands

1) Calcineurin inhibitors
   e.g.
   Tacrolimus
   Cyclosporine A

2) Antimetabolites
   e.g.
   Azathioprine
   Methotrexate
   Deoxyspergualine
   Mycophenolic acid
   Teriflunomide
   Cladribine

3) Antiproliferative agents
   e.g.
   Cyclophosphamide
   Mitoxantrone
   Pixantrone
   Treosulfan
   Sirolimus (Rapamycin)

   Laquinimod

(II) targeting cell surface ligands

Monoclonal antibodies
Fusion proteins

e.g.
Anti-"CD" molecules
Anti-VLA4
Efomycine
Anti-IL2-receptor (Daclizumab)
FTY720

(III) targeting soluble
factors: "anticytokines"

Anti-TNF-alpha mAb
Anti-IL2R (Daclizumab)

Xaliproden

*according to the WHO (category LO4)*

FIGURE 14.3. Classification of immunosuppressants according to the World Health Organization (category LO4). Immunosuppressants are separated into three families according to their respective targets: intracellular ligands, cell surface ligands, and "anticytokines". Drugs acting on intracellular ligands are subdivided into three groups: calcineurin inhibitors, antimetabolites, and antiproliferative agents. Immunosuppressants targeting cell surface ligands include mainly monoclonal antibodies (mAb) and fusion proteins. "Anticytokines" are directed against TNFα and several interleukins. Examples are mentioned in the figure.

Among the group of nonselective immunosuppressants, only one agent, mitoxantrone, has so far been approved for the treatment of progressive MS. Potentially new immunosuppressive substances that would come into question for clinical trials in MS should generally fulfill the following conditions: (1) application should be safe and associated with little acute or chronic toxicity; (2) longer-term usage should be tolerated without major cumulative toxicity; and (3) mechanisms of action should be rational with respect to a putative therapeutic effect on MS. Hitherto, no "perfect drug" exists that meets all these criteria.

### 3.1.1   Immunosuppressants: Classic and Novel

*Mitoxantrone.* The anthracenedione mitoxantrone has profound antiinflammatory and immunosuppressive effects. The alkylating immunosuppressant is widely used for both progressive relapsing/remitting (RR) and secondary progressive (SP) MS (Edan et al. 1997; Hartung et al., 2002), mainly in patients with a suboptimal response to first-line disease-modifying drugs. Significant adverse effects that limit mitoxantrone therapy are cardiotoxicity and hematotoxicity (Brassat et al., 2002; Ghalie et al., 2002). The total cumulative mitoxantrone dosage per patient should not exceed 100 mg/m$^2$ body surface area, which is an important limitation in terms of treatment duration (normally 2–3 years with applications every 3 months). With a total dose above 100 mg/m$^2$, not only does the risk of cumulative cardiotoxicity increase but so does the risk of breast cancer and treatment-related acute myelogenous leukemia (TRAL) in patients receiving additional chemotherapy (Edan, 2005).

Cardiotoxicity is one of the major concerns with mitoxantrone. One possible way to overcome this limitation is to employ cardioprotective agents in parallel to the administration of mitoxantrone. Dexrazoxane is an iron-chelating cardioprotective agent that is approved for combined use with doxorubicin (Swain, 1998). The results from a small open-label study exploring the potential of dexrazoxane to prevent cardiotoxicity in MS patients receiving mitoxantrone have recently been published (Bernitsas et al., 2006). Taken together, no patient experienced symptoms of heart failure after at least four quarterly mitoxantrone infusions, but subjects in the dexrazoxane group exhibited a significantly lower decline in left ventricular ejection fraction, which may indicate a cardioprotective effect of dexrazoxane in the mitoxantrone-treated MS patients.

Because of the inherent limitations of mitoxantrone monotherapy, the question of long-term safety remains important. A phase IV study of long-term safety and tolerability is currently under way in 509 MS patients aged 18 to 65 years who have been treated with mitoxantrone. Patients were enrolled by December 2002 and are being followed for 5 years, including treatment and posttreatment periods. Fifty participating MS treatment centers are involved in the study. The results are reported to the FDA and study investigators every 3 months for the first 2 years and every 6 months for the remaining 3 years (Cohen and Mikol, 2004).

The question of how the combination of mitoxantrone with any disease-modifying drug could be dose-sparing or enhance the efficacy of the treatment has raised considerable interest (Jeffery, 2004). A small pilot study examined the combination of mitoxantrone plus interferon-β-1b (IFNβ-1b) in active RRMS or SPMS over a period of 6 months. In this study, 10 patients under therapy with high-dose, high-frequency IFNβ and signs of active disease by both clinical and MRI criteria were enrolled. Safety analysis showed no serious adverse events. In addition to the expected neutropenia after mitoxantrone infusion, upper respiratory and urinary tract infections occurred more frequently. After addition of mitoxantrone, the annualized relapse rate declined by 70% compared to baseline conditions, and gadolinium-enhancing MRI lesions were reduced by 90% after 6 months (Jeffery, 2004). Although the combination of mitoxantrone and IFNβ-1b seems to be safe in view of this small, timely restricted pilot study, there is clearly a need for larger trials to address the questions of long-term side effects and sustained efficacy. A small study assessing the combination of mitoxantrone with glatiramer acetate has been performed, documenting safety and potential efficacy with this combination (Ramtahal et al., 2005).

*Cyclophosphamide.* Cyclophosphamide is a cytotoxic alkylating agent that is widely used for the treatment of neoplastic and autoimmune disorders. It has also long been applied as a therapeutic agent in MS, especially in the group of patients with progressive disease. Frequent side effects include nausea, alopecia, infertility, and infection. One key reason why long-term use in MS is avoided is cumulative bladder toxicity (e.g., hemorrhagic cystitis and bladder cancer) (Gauthier et al., 2005; Weiner and Cohen, 2002). Data for its use in progressive MS are not unequivocally positive. Although intravenous pulses of cyclophosphamide have been shown to be effective in patients with early, aggressive inflammatory MS (Hauser et al., 1983), other studies could not demonstrate clear efficacy in primary progressive (PP) MS and SPMS (Canadian Cooperative MS Study Group, 1991; Likosky et al., 1991). One trial investigated the combination of cyclophosphamide plus IFNβ and found a significant clinical improvement during treatment and in the 24-month follow-up period (Reggio et al., 2005). A double-blind, placebo-controlled study of combination therapy with IFNβ is now in progress (Patti et al., 2004). In general, because cyclophosphamide is not patented, large clinical trials to obtain formal approval are not expected and its off-label use has widely been replaced by mitoxantrone.

*Cyclosporine A.* Another rather old immunosuppressant tentatively employed in MS is the calcineurin inhibitor cyclosporine. Mechanistically, the substance blocks the production of IL-2 and IL-4 from T cells. Common serious side effects associated with the administration of cyclosporine are nephrotoxicity and neurotoxicity, which affect 25% to 60% of transplant patients (Serkova et al., 2004). Others include hypertension, diabetes, and hyperlipidemia. Therefore, the overall risk/benefit ratio seems unfavorable. Analysis of a double-blind German multicenter trial of cyclosporine versus azathioprine in MS revealed that patients in the cyclosporine group were more than twice as likely to report side effects than those in the azathioprine arm (Kappos et al., 1988). In 1990 the results from a 2-year randomized, double-blind, placebo-controlled study that included 547 patients with chronic progressive MS were published (MS Study Group, 1990). Patients in the verum group experienced a statistically significant but modest delay of disease progression, but the overall lesion load was not affected, as revealed by a subsequent MRI study (Zhao et al., 1997). As the small benefits of cyclosporine are overbalanced by its markedly toxicity, the Therapeutics and Technology Assessment Subcommittee of the American Academy of Neurology advised against the use of cyclosporine in patients with MS (Goodin et al., 2002).

*Azathioprine.* Azathioprine is an antimetabolite that interferes with purine pathways used in DNA synthesis and cell division. The substance is well tolerated, and its application is convenient (oral). Albeit available studies have not unequivocally demonstrated significant efficacy of azathioprin in MS, its therapeutic use has a long tradition. When applied as a monotherapy, the drug only marginally affected the disease course in MS (Yudkin et al., 1991). A post hoc meta-analysis, however, concluded that the effects of azathioprine might be comparable to those seen in the trials with interferons, glatiramer acetate, and intravenous immunoglobulin (IVIg) (Palace and Rothwell, 1997). Massacesi et al. (2005) provided an open labeled study to demonstrate the efficacy of azathioprine on paraclinical markers of MS disease activity. In the 14 patients included in the study, azathioprine at lymphocyte-suppressing dosage, showed a reduction in the development of new inflammatory brain lesions as assessed by MRI.

Because azathioprin is well tolerated, it may be the prototypic agent for usage in combination therapy together with the available disease-modifying drugs. To date, several studies have examined the effects of combined administration of IFNβ and azathioprine (Fernandez et al., 2002; Lus et al., 2004). Azathioprine was added to IFNβ-1b in six patients who were clinically refractory to therapy (active RRMS at a dose of 2.0 mg/kg/day over 6 months) (Markovic-Plese et al., 2003). After combination with azathioprine, the frequency of gadolinium-enhancing MRI lesions was reduced by 69%. However, the overall lesion burden on T2-weighted MRI continued to increase (Markovic-Plese et al., 2003), and no effect on the relapse rate was found. In contrast, in another small study that included 10 patients with SPMS, relapse frequency in the group receiving combination therapy was reduced by 50% compared to patients on IFNβ alone (Fernandez et al., 2002). The total lesion load on T2-weighted MRI scans was also significantly smaller at 12 and 24 months, and a discrete improvement in neuropsychological tests was observed. Other than lymphopenia, no notable side effects occurred.

Azathioprine is a reasonable candidate for testing the safety and efficacy of combination therapy in controlled studies. In theory, combination therapies seem particularly promising owing to assumed synergistic or complementary mechanisms of action. Treatment regimens in a variety of other autoimmune disorders indicate that this assumption is meaningful in clinical practice. However, the goal is to find optimal combinations that simultaneously address different targets, such as relapse inhibition, delay of disease progression, prevention of axonal loss, and improved remyelination. At present, all of theses approaches are still at an "experimental" state and (except for statements on safety and tolerability) none of them has achieved a level that warrants recommendation or even approval based on the studies' evidence. It has to be kept in mind, however, that treatment combinations might produce unpredictable adverse effects, and therefore patients and physicians who may wish to try combinations that have not been tested adequately should be vigilant about unexpected clinical or laboratory toxicity (e.g., the experiences from the SENTINEL trial with natalizumab anti-VLA-4, where two cases of PML occurred in the group treated with the combination of IFNß-1a and Natalizumab; see below).

*Methotrexate.* Methotrexate is another classic immunosupressive agent with the potential to be studied in combination therapy. Like azathioprine, methotrexate belongs to the group of antimetabolites. Some studies with methotrexate as a monotherapy exist, but only one showed significant effects, with a trend toward increasing the time to a state of confirmed progression of disability in PPMS and SPMS patients was reached (Goodkin, 1995). Because of a preferential effect to improve the function of the upper extremities, methotrexate is assumed by some MS therapists to be a potential alternative treatment for stabilizing or improving upper extremity function. The agent has also been studied in combination with interferon therapy (IFNβ-1a). In a preliminary open-label study, addition of 20 mg methotrexate once weekly reduced the number of active MRI lesions by 44% without any significant effect on relapse frequency. The combination was well tolerated, with nausea as the only major side effect (Calabresi et al., 2002).

*Cladribine.* Cladribine is an adenosine deaminase-resistant nucleoside analogue with selective lymphotoxic specificity. After phosphorylation into the active triphosphate deoxynucleotide, the substance accumulates in lymphocytes and monocytes causing DNA damage and subsequent cell death (Beutler, 1992; Sipe, 2005). Its long-lasting lymphocytotoxic activity suggests that it could be useful in modulating conditions that involve lymphocyte abnormalities. Thus, cladribine has been tested extensively for the treatment of lymphoid neoplasms and autoimmune disorders, especially MS. The substance has been approved for the treatment of hairy cell leukemia since 1993.

Evidence of the efficacy of cladribine in delaying disease progression mainly results from smaller placebo-controlled trials in chronic progressive MS (Beutler et al., 1996; Sipe et al., 1994) and RRMS (Romine et al., 1999; Sipe et al., 1997) patients. The clinical observations were underlined by remarkable MRI effects (e.g., nearly complete elimination of enhanced T1 lesions and stabilization of T2 lesion volume (Sipe et al., 1997). However, albeit Phase I and II studies raised high expectations, a multicenter, double-blind, placebo-controlled study of cladribine in 159 patients with SPMS and PPMS failed to show significant clinical benefits after 1 year. No significant effect on whole-brain volume changes and T1 "black holes" were observed (Filippi et al., 2000a,b; Rice et al., 2000). In these studies, the substance was administered subcutaneously in two dose levels (0.7 or 2.1 mg/kg, respectively, versus placebo). The most common adverse events included upper respiratory tract infections, muscle weakness, and injection site reactions. The study showed favorable effects on the presence, number, and volume of brain lesions by gadolinium-enhanced T1-weighted MRI and on the lesion load by T2-weighted MRI; however, there is a discrepancy between those MRI endpoints and the clinical effects (Rice et al., 2000). Evaluation of the MRI data led to the conclusion that cladribine triggers strong, prolonged antiinflammatory effects by MRI criteria but does not influence the mechanisms of continuous tissue destruction (Filippi et al., 2000a,b).

Cladribine is also available as an oral formulation. Phase I and II trials in MS have provided encouraging data concerning safety and potential efficacy, leading to the initiation of a randomized, double-blind, placebo-controlled Phase III study that is including 1290 patients with active inflammatory RRMS. This trial was launched in April 2005 to test the safety and effectiveness of this oral immuno-suppressant further. According to the study protocol, 10 mg cladribine or placebo is given orally over 5 days per month, administered in two to four cycles per year. The outcome parameters include relapse rate, Expanded Disability Status Scale score progression, and MRI activity.

*Treosulfan.* Treosulfan (dihydroxybusulfane, or DHB; L-threitol-1,4-bis[methane sulfonate]) is a cytostatic alkylating agent approved for the treatment of advanced ovarian cancer. The drug is characterized by overall good patient tolerability and a favorable safety profile concerning acute and long-term toxicity. Treosulfan seems to have superior immediate and long-term tolerance compared to cyclophos-phamide (Meden et al., 1997). It was successfully applied in an animal model of MS, myelin-oligodendrocyte-glycoprotein (MOG)-induced experimental autoim-mune encephalomyelitis (EAE) (Weissert et al., 2003). The severity of acute EAE and the long-term disease outcome were improved under pretherapeutic as well as therapeutic conditions. Furthermore, treosulfan induced concentration-dependent apoptosis in human peripheral blood lymphocytes in vitro (Weissert et al., 2003). A recent study demonstrated superior immunosuppressive and myeloablative prop-erties compared to cyclophosphamide and busulfan (Sjöö et al., 2006).

A small Phase I pilot trial investigated the safety and efficacy of treosulfan in active SPMS. Eleven patients with active disease who failed or did not qualify for conventional therapy were treated over 1 year; progression of disability and safety parameters were assessed in parallel with MRI activity and lesion parameters. Treosulfan was well tolerated. Of the 11 patients, 9 remained on treatment and showed clinical stabilization or improvement of the EDSS and MS Function Composite (MSFC) components. No corticosteroid-requiring relapses were observed. Treosulfan reduced MRI activity and stabilized measures of the disease burden (Wiendl et al., 2005). The development of this agent is currently pursued in a Phase II trial.

*Mycophenolate Mofetil.* Another interesting member of the immunosuppressants is mycophenolate mofetil (Frohman et al., 2004). It belongs to the antimetabolite group and is a prodrug of the active metabolite mycophenolic acid. The drug is applied orally at a usual dose of 2 g per day, has a good adverse effect profile, and thus is convenient for the patient. Its use in neuroimmunologic autoimmune dis-orders is becoming increasingly popular, albeit almost all applications are "off label" (Schneider-Gold, 2006). For its use in MS, currently only a few reports or open-label studies exist. Two small open-label trials using mycophenolate mofetil as monotherapy in various progressive MS cases suggest efficacy in this patient population, and application seems safe (Ahrens et al., 2001; Ducray et al., 2004). 30 patients suffering from RRMS were enrolled in an open-label Phase II clini-cal trial testing the combination of mycophenolate mofetil and IFNβ-1a (Vermersch et al., 2004). Overall, few patients complained about side effects (e.g., diarrhea). Preliminary efficacy data suggest that the combination may be effective in decreasing the relapse rate (Vermersch et al., 2004). To estimate its role either as monotherapy or an agent to combine with other disease-modifying drugs, controlled, larger studies are necessary.

*Sirolimus and Temsirolimus.* Sirolimus and temsirolimus (formulated for improved oral absorption) are new, potent antiproliferative agents with a specific mechanism of action. Sirolimus forms a complex with the intracellular protein FKBP12, which blocks activation of a kinase cascade essential for cell–cycle progression. This novel macrocyclic compound is derived from the bacterium *Streptomyces hygroscopius* and specifically inhibits mTOR (mammalian target of rapamycin) kinase, an enzyme required to control a cell's life cycle, preventing cell division into new cells. Sirolimus thus inhibits antigen-induced proliferation of T and B cells and antibody production (Sehgal, 2003). In addition, the agent probably induces persisting regulatory T cells (Chen et al., 2000). The substance has already been used to prevent organ rejection and is currently being tested in a Phase I/II open-label trial in RRMS patients. One major side effect is the consistent elevation of blood lipids (hypercholesterolemia) due to administration.

Temsirolimus (CCI-779, rapamycin) is a newly developed ester analogue of sirolimus (Anonymous, 2004). Several studies are currently investigating the safety and effectiveness of this substance in regard to breast and renal cancer and rheumatoid arthritis. A fast track approval has been received for the treatment of renal cancer. In addition, a Phase II study has been completed in MS. The substance was tested with three different doses against placebo in 296 individuals with RRMS and active SPMS for a period of 9 months (Kappos et al., 2005b). At the end of the study patients on the highest dose had significantly less contrast-medium enhancing MRI lesions (48% reduction) and 51% fewer relapses compared to those on placebo. Side effects occurred more frequently in the high-dose subgroup and included menstrual dysfunction, mouth ulceration or inflammation, hyperlipidemia, and rashes.

*Pixantrone.* Pixantrone (BBR2778) is a novel analogue derived from mitoxantrone. Structurally, the 5,8-dihydroxyphenyl ring of mitoxantrone, which has been noted to cause cardiotoxicity, was replaced by a pyridine ring (Gonsette, 2004). In animal experiments and clinical pilot studies of cancer, pixantrone did not induce toxic effects on cardiac tissue, and the pharmacokinetic and immunosuppressive properties were similar to those of mitoxantrone (Beggiolin et al., 2001; Cavaletti et al., 2004; Gonsette, 2004). Because the substance has proven to be as effective as mitoxantrone in MS animal models, an open-label study in 20 RRMS patients has been initiated to investigate the safety and MRI effects of intravenous pixantrone (Gonsette and Dubois, 2004).

### 3.1.2 Novel Immunomodulatory Agents

*Teriflunomide.* Teriflunomide is an analogue of leflunomide (Arava), which is a known therapy for rheumatoid arthritis. Teriflunomide belongs to the group of malononitrilamide agents that block the mitochondrial enzyme dihydroorotate dehydrogenase and inhibit T and B cell proliferation (Korn et al., 2004; Nakajima et al., 2003). The oral prodrug is rapidly converted to its active metabolite A771726. Leflunomide has been successfully used to treat rheumatoid arthritis. Teriflunomide has been found to suppress EAE in Lewis rats probably via the suppression of tumor necrosis factor-$\alpha$ (TNF$\alpha$) and interleukin-2 (IL-2) production (Korn et al., 2001; Smolen et al., 2004). The results from a clinical Phase II study testing teriflunomide in patients with relapsing RRMS and SPMS have recently been published (O'Connor et al., 2006a). Two different teriflunomide dose regimens (7 and 14 mg once daily) were compared to placebo over an observation period of 36 weeks. The primary endpoint of the study was met because subjects receiving verum (either dose) had significantly less active MS lesions and reduced numbers of new lesions on MRI. EDSS progression was delayed in the high-dose arm and a trend toward a reduced number of relapses was observed. Overall, teriflunomide was well tolerated, with upper respiratory tract infections and headache being the most common adverse effects. However, serious adverse events, including toxic liver necrosis and pancytopenia, have been described in relation to rheumatoid arthritis. The safety and efficacy of the drug in MS is being further investigated in an ongoing Phase III trial.

Thus, teriflunomide belongs to an interesting category of new oral immunomodulatory agents that are currently tested in Phase III trials. Given that teriflunomide is able to produce superior or at least noninferior safety and efficacy in comparison to available disease-modifying drugs, these compounds could represent an important addition to the available armamentarium of drugs in MS therapy.

*Laquinimod.* Laquinimod (ABR-215062, SAIK-MS) is a new, orally active immunomodulator that was shown to be approximately 20 times more potent than its "ancestor" roquinimex (linomide) in EAE (Brunmark et al., 2002). The synthetic compound has excellent oral bioavailability and serves as an immunoregulatory drug without general immunosuppressive properties. Its sustained inhibitory activity has been shown in other autoimmune and inflammatory diseases in several animal models (Jönsson et al., 2004; Runstrom et al., 2006; Yang et al., 2004). Roquinimex (linomide) efficiently reduced active MRI lesions in Phase II and III clinical studies of MS, although a Phase III trial in MS had to be stopped prematurely owing to unexpected severe inflammatory side effects (serositis, myocardial infarction) (Andersen et al., 1996; Karussis et al., 1996; Noseworthy et al., 2000b).

Two clinical Phase I trials with laquinimod demonstrated that the drug is well tolerated by healthy volunteers and patients with MS. The results from a double-blind, randomized, multicenter proof-of-concept study testing two different doses of oral laquinimod (0.1 and 0.3 mg/day) versus placebo in 180 patients with RRMS have been published (Polman et al., 2005). The duration of the study was 24 weeks. Taken together, a significant difference between 0.3 mg laquinimod/day and placebo was observed for the primary outcome measure: that is, the mean cumulative number of active MRI lesions (5.24 vs. 9.44, respectively; 44% reduction). This effect was even more pronounced in patients with at least one active lesion at baseline (52% reduction). Clinical outcome parameters (relapse rate, disability) were not different between the groups. The overall safety profile was favorable with no signs of undesired tissue inflammation. Laquinimod has considerable potential to become a safe, effective oral treatment of MS (Polman et al., 2005) and is currently being tested in a Phase III trial.

*Phosphodiesterase Inhibitors.* Phosphodiesterase (PDE) inhibitors downregulate inflammatory responses by changing the levels of intracellular cyclic adenosine monophosphate (cAMP) and cyclic guanosine monophosphate (cGMP) (Dyke, 2002). In addition, their capacity to skew the immune balance toward a more Th2-driven response is mediated by influencing the local cytokine milieu, T cell priming, and co-stimulatory signaling (Bielekova et al., 2000b). Treatment with PDE inhibitors led to clinical and histomorphologic improvement in several EAE models (Fujimoto et al., 1999; Jung et al., 1996; Sommer et al., 1995). The non-specific PDE inhibitor ibudilast was found to induce a favorable shift in T cell subtype and natural killer (NK) cells in MS (Feng et al., 2004) However, an open-label, crossover study with 18 MS patients testing the orally applied PDE-4 specific inhibitor rolipram has recently been terminated owing to lack of efficacy on the primary outcome measure; further results regarding clinical effectiveness are awaited.

*PPARγ Agonists: Thiazolidinedions (Pioglitazone, Rosiglitazone).* Thiazolidinedions are insulin-sensitizing compounds originally approved for the therapy of type 2 diabetes; they act via potent activation of the nuclear transcription factor peroxisome proliferator-activated receptor-gamma (PPARγ) (Pershadsingh, 2004). In addition to their glucose-lowering properties, thiazolidinedions have significant antiinflammatory and antiproliferative effects: They inhibit T cell activation, reduce the production of proinflammatory cytokines, and increase the resistance to cytotoxicity (Dello Russo et al., 2003; Dormandy et al., 2005; Storer et al., 2005). Observed adverse effects caused by this substance class are hepatotoxicity, peripheral edema, and weight gain. Several studies have demonstrated that thiazolidinedions can ameliorate the clinical course and histopathologic features of EAE (Diab et al., 2002; Feinstein et al., 2002; Lovett-Racke et al., 2004; Niino et al., 2001; Storer et al., 2005). In humans, proinflammatory stimulation and pioglitazone treatment regulates peroxisome proliferator-activated receptor gamma levels in peripheral blood mononuclear cells from healthy controls and MS patients (Klotz, 2005). In addition, preliminary evidence suggested that thiazolidinedions might also influence the disease course of human MS (Pershadsingh et al., 2004). A Phase I/II proof-of-principle study including 15 patients with RRMS was launched in April 2004 to test the safety, tolerability, and efficacy of pioglitazone (30 mg/day) versus placebo. The observation period was 2 years, and the final results are still pending. Results of another study investigating the safety and effectiveness of another thiazolidinedion, rosiglitazone, have been made public. A total of 51 individuals with RRMS were treated with rosiglitazone. In comparison to placebo, no differences were found on MRI measures of tissue damage or clinical or immunologic responses.

*FTY720.* The compound FTY720 is derived from the fungus *Isaria sinclairii* and exhibits profound and unique immunoregulatory effects (Brinkmann et al., 2002; Chiba, 2005). After in vivo phosphorylation, FTY720 forms FTY720P, a high-affinity nonselective mimetic of the sphingosine 1-phosphate receptor 1 (S1P1). The mechanisms underlying FTY action remain somewhat controversial. Two alternative hypotheses have been proposed involving different modes of action and sites of action. The "functional antagonism" hypothesis postulates that some agonists bind directly to S1P1 receptors on lymphocytes. This would cause receptor internalization and degradation, thus preventing lymphocyte egress from lymph nodes along an endogenous chemotactic gradient of S1P (Matloubian et al., 2004). FTY 720 therefore entraps CD4+ and CD8+ T cells and B cells in secondary lymphatic organs, preventing them from being recruited to possible sites of inflammation (Brinkmann et al., 2002; Cyster, 2005). This "entrapped" homing of lymphocytes primarily seems to depend on the expression of various chemokines (Cyster, 2005; Yopp et al., 2004). Furthermore, some studies imply a direct impact of FTY720 on dendritic cells (Muller et al., 2005). The second hypothesis states that ligand binding to S1P1 receptors on endothelial cells results in alterations of the endothelial barrier functions, thereby blocking lymphocytic egress from the medullary parenchyma to draining sinuses (Mandala et al., 2002; Wei et al., 2005). According to the second hypothesis, S1P1 receptor agonism on endothelial cells, not in lymphocytes, is the crucial mechanism leading to sequestration of lymphocytes in the nodes. For a review of sphingosine 1-phosphate and its receptors, readers are referred to Rosen and Goetzl (2005). Because of its mechanism of action, FTY720 induces marked lymphopenia in peripheral blood counts but does not provoke general immunosuppression, as neither the activation of T cells nor the memory of T and B cell responses is impaired.

The potency of this agent has already been demonstrated in human organ transplantation (Budde et al., 2002). Furthermore, preclinical studies in various autoimmune EAE models demonstrated its efficacy (Fujino et al., 2003; Rausch et al., 2004; Webb et al., 2004). The results of an international, double-blind, placebo-controlled Phase II study involving 281 subjects with active RRMS have been made public (Kappos et al., 2006). The participants received one of two doses of oral FTY720 (1.25 mg or 5 mg, respectively) or placebo daily for 6 months. At the end of this observation period, the total number of enhancing MS lesions on monthly MRI scans (primary outcome) was significantly reduced in both groups receiving treatment versus those on placebo. Similar results favoring the verum groups were seen on MRI measures of disease activity. Volumes of enhancing lesions and new T2 lesions were significantly diminished by verum treatment; in addition, significantly more patients receiving FTY720 remained relapse-free (86% in the FTY720 group versus 70% in the placebo group). The relapse rate under FTY720 treatment was reduced 53% to 55%. Results of the extension phase (18 months) with switch of the placebo group to verum demonstrate persistent efficacy of the agent (O'Connor et al., 2006b). Although in general the substance was safe and no severe adverse events peaked in either of the verum groups, frequently reported side effects included common colds, sinusitis, mild headaches, and gastrointestinal disorders. Adverse events were more frequent in the high-dose group. The underlying mechanism of action, however, warrants extensive safety analysis, especially with longer periods of administration. In the transplant program, several cases of macular edema appeared. Furthermore, due to the expression of the sphingosine 1-phosphate receptor 1 (S1P1) in the cardiovascular system, alterations in blood pressure and cardiac rhythm must be monitored. Two large Phase III studies of FTY720 in MS are currently being launched (2006). One study is testing the safety and efficacy of two doses (1.25 mg and 0.5 mg) against placebo, the other is assessing the efficacy of FTY720 against an active comparator (IFNß-1a, Avonex).

FTY720 is clearly one of the most interesting novel immunomodulatory agents being studied, both from an "immune-mechanistic" and a practical point of view (oral drug). However, a number of caveats must be taken into account. Phase III studies will demonstrate whether FTY720 is able to demonstrate long-term efficacy and—even more important—safety in larger numbers of patients. Thus, it is currently not known how the influence with lymphocyte homing and migration might affect immune surveillance of parenchymal organs including the CNS during long-term treatment.

*Fumaric Acid.* Another recent approach to induce beneficial immune deviation in MS is the oral application of fumaric acid esters (BG12, fumarate). Fumaric acid is an immunomodulator that also upregulates Th2 responses (Anonymous, 2005). BiogenIdec is currently evaluating the product in clinical trials as an oral treatment for MS.

An exploratory, prospective, open-label study of fumaric acid esters (FAE) (Fumaderm) was conducted in patients with RRMS. In this small, open-label pilot study of 10 MS patients, a reduction of relapse rate and volume of gadolinium-enhancing lesions assessed by MRI was observed after 12 weeks (Schimrigk et al., 2005). The study consisted of the following four phases: 6-week baseline, 18-week treatment (target dose of 720 mg/day), 4-week washout, and a second 48-week treatment phase (target dose of 360 mg/day). Ten patients with an EDSS score of 2.0 to 6.0 and at least one gadolinium-enhancing (Gd+) lesion on T1-weighted MRI brain scans participated in the study. However, three patients discontinued during the first 3 weeks of treatment, and several patients experienced mild to moderate gastrointestinal discomfort.

A double-blind, placebo-controlled Phase II study was launched in November 2005 to test the drug further regarding its safety and effectiveness in controlling the disease course and development of MS brain lesions. In January 2006, the product sponsor announced that the study met its primary endpoint by reducing the number of active MRI lesions with 6 months of treatment (http://wwww.FDA.org). Details about the secondary outcome measures have not been released so far.

### 3.1.3    Other Agents With "Secondary" Immunomodulatory Properties

*Lipid-Lowering Agents: Statins.* Statins are lipid-lowering drugs approved for the therapy of dyslipidemia. They act on the enzyme 3-hydroxy-3-methylglutaryl coenzyme A (HMG-CoA) reductase, which converts HMG-CoA to L-meval-onate, a key intermediate in cholesterol biosynthesis. Evidence has accumulated indicating that statins have potent immunomodulatory properties (Kwak et al., 2000; Menge et al., 2005). Accordingly, statins have demonstrated beneficial effects in different EAE models (Aktas et al., 2003; Paintlia et al., 2005; Youssef et al., 2002). Relevant immunomodulatory properties that might be of benefit in the treatment of MS include (1) inhibition of the proliferative activity of T and B cells; (2) reduction of the secretion of matrix metalloproteinase-9 (MMP-9); (3) reduction of the expression of activation-induced adhesion molecules on T cells; (4) downregulation of chemokine receptors on B and T cells; (5) inhibition of IFNγ-inducible major histocompatibility class II (MHC II) expression on various types of nonprofessional antigen-presenting cells (APCs); (6) suppression of the upregulation of co-stimulatory molecules required for antigen presentation; (7) inhibition of secretion of proinflammatory cytokines; and (8) promotion of the secretion of anti-inflammatory cytokines (Rizvi and Bashir, 2004; Weber et al., 2005). Dunn et al. (2006) presented evidence that specific isoprenoid intermedi-ates are linked to Th1 differentiation. By depleting these isoprenoid intermedi-ates, atorvastatin seems to inhibit extracellular signal-regulated kinase (ERK) and p38 signaling pathways, thereby inducing a Th2 cytokine shift.

The beneficial effects of oral simvastatin (80 mg daily) in MS were first demonstrated in a small, multicenter, open-label 6-month cohort study involving 28 patients with RRMS, where the number and volume of gadolinium-enhancing lesions were reduced by 44% and 41%, respectively (Vollmer at al., 2004). Although these results are promising, the lack of a placebo arm in the study lim-its interpretation. Currently, a larger placebo-controlled trial including 152 patients with carcinoma in situ is currently being conducted in 14 centers in North America (treatment for 12 months with 80 mg atorvastatin or placebo). In exper-imental systems, the combination of atorvastatin with glatiramer acetate shows an interesting immunomodulatory synergy (Stüve et al., 2006), thus making the combination attractive for clinical trials.

Both the pharmacological profile and its assumed mechanisms of action make statins attractive candidates for combination studies. An in vitro study by Neuhaus and colleagues has demonstrated an additional effect of IFNβ and statins in inhibiting the activation of T cells (Neuhaus et al., 2002). A small clin-ical trial is currently testing the combination of high-dose IFNβ-1a (44 μg three times a week) with atorvastatin in RRMS (Birnbaum and Altafullah, 2005). In addition, a trial testing atorvastatin in combination with glatiramer acetate is being planned (Weber et al., 2005).

*Sex Hormones.* There is a long-standing notion on the important influence of sex hormones in MS: (1) Females have a higher susceptibility to various autoimmune disorders; (2) pregnancy has a protective effect regarding the relapse rate; (3) EAE can be ameliorated with estrogens in vivo; and (4) estrogens have clear immunomodulatory effects such as induction of an immune shift from Th1 to Th2 in vitro (Farrell et al., 2005). These findings triggered an open pilot study with

12 female patients suffering from RRMS or SPMS (Sicotte et al., 2002). "Pregnancy" doses of estriol were administered orally to 6 patients over 6 months, and all patients were followed for a further 6 months. Although there were no significant changes in either the EDSS or the clinical relapse rate, a significant reduction in new and enhancing MRI lesions was seen up to 3 months after treatment. Although the treatment with sex hormones might have some beneficial effect on MS patients, the study is limited because of the small number of patients. Furthermore, there is a considerable risk of vascular side effects of high-dose oral estriols, and the application itself is an inherent restriction in that it can be used only in female patients (Farrell et al., 2005).

In another study funded by the National MS Society, the efficacy of testosterone gel (AndroGel), 100 mg daily over 6 months, on immune function was tested by the same group of researchers. Ten men suffering from RRMS were treated. After 12 months of treatment, patients showed stable disease activity, significant improvement in cognitive function, and a lesser extent of brain atrophy (67%) (Sicotte et al., 2006). Furthermore blood tests indicated increased levels of brain-derived neurotrophic factor (BDNF) (about twofold) in testosterone gel-treated males. No adverse events were reported during the study.

Although these first clinical results of sex hormones are promising, further studies involving larger numbers of patients are necessary to allow any firm conclusions concerning their potential role in future MS patient care.

*Uric Acid Precursors.* Uric acid (UA) is a purine metabolite that selectively inhibits peroxynitrite-mediated reactions and is implicated in the pathogenesis of MS. MS patients have low serum UA levels, and individuals with hyperuricemia (gout) rarely develop the disease. Moreover, the administration of UA has been shown to be of therapeutic benefit in EAE (Scott, 2002). Thus, raising serum UA levels in MS patients by oral administration of a UA precursor such as inosine may have therapeutic value. A double-blind, placebo-controlled trial in 30 subjects with RRMS is currently being performed.

## 3.2 Pathogenetically Oriented Immune Therapies

### 3.2.1 Antigens, T Cells, and the Immune Synapse

*Antigen-Based Immune Therapy.* Conceptually, antigen-specific therapy of autoimmune disorders would still be one of the most desirable therapeutic approaches (Hohlfeld and Wekerle, 2004; Sospedra and Martin, 2005b). An elegant way to induce antigen-specific tolerance is the tolerogenic application of putative autoantigens (e.g., myelin proteins or myelin basic protein via the mucosal surfaces of the gut ("oral tolerance") or respiratory tract ("nasal tolerance"). The application of antigens leads to the induction of antigen-specific suppressor or regulatory T cells in the periphery. After recognizing their antigen in the target organ, they release antiinflammatory cytokines and trigger bystander suppression (Faria and Weiner, 2005; Weiner et al., 1993). Albeit this concept has attracted considerable attention during the last few years, the clinical success in humans is still limited. Application of oral myelin led to impressive amelioration of EAE and suggested favorable effects in a small Phase I pilot study (Weiner et al., 1993). However, a larger Phase III multicenter trial testing the oral administration of myelin in MS showed no effect (Francis et al., 1997; Panitch et al., 1997; Wiendl and Hohlfeld, 2002).

An alternative antigen-based strategy utilizes so-called altered peptide ligands (APLs) (Sette et al., 1994). APLs are peptides whose structure is similar but not identical to the expected autoantigen. Although some effects are incompletely understood, APLs have become interesting candidates for antigen-selective immunotherapy. Acting as agonists, partial agonists, or antagonists, APLs can still bind to T cell receptors (TCRs), but they significantly alter the outcome of the TCR-induced T cell response. Thus they may modulate the cytokine pattern of T cells and can induce T cell anergy (Germain and Stefanova, 1999). Administration of APLs led to disease inhibition in various EAE models (Nicholson et al., 1995; Smilek et al., 1991). However, clinical trials testing APLs derived from immunodominant peptides of myelin basic protein (MBP) thus far have been disappointing. One trial demonstrated that the MBP-derived APL (amino acids 83-99; CGP77116) was able to induce considerable disease exacerbation accompanied by a drastic increase in the frequency of MBP-specific T cells (Bielekova et al., 2000a), suggesting that APLs can cross-stimulate encephalitogenic, MBP-specific T cells under certain circumstances (Bielekova et al., 2000a; Kim et al., 2003). These inadvertent effects indicate that in certain patients the APLs can cross-stimulate and activate encephalitogenic, MBP (83-99)-specific T cells. Furthermore, it must be a concern that the autoimmune reaction eventually evades the effects of the APL therapy (e.g., by epitope spreading). The results of the second multicenter trial indicated that low-dose APL treatment induces a shift of the T cell response from Th1 to Th2 (Kappos et al., 2000), which is reminiscent of the immunologic effect of glatiramer acetate. Albeit this trial was suspended because of hypersensitivity reactions in 9% of the patients (Kappos et al., 2000), it clearly demonstrated that the MBP APL was capable of exerting antiinflammatory effects. The lessons to be learned from these first clinical experiences of an antigen-selective and antigen-specific therapeutic approach may be that (1) MBP can be a real target in MS; (2) peripheral modulation of antigen-specific T cells can translate into changes in the CNS compartment; and (3) effects can persist for years after therapy is discontinued. Based on these experiences, a new trial is currently testing the safety and efficacy of the same MBP APL but using a different dosing regimen.

Another antigen-based approach based on MBP is the repetitive application of MBP8298, a MBP-derived peptide containing MBP residues 85-96, the minimal epitope identified for the dominant autoantibody response in the cerebrospinal fluid (CSF) of MS patients. Residues 85-96 were found to represent the critical contact residues also for T cell recognition as assessed by systematic amino acid substitution. MBP 8298 includes amino acid extensions at both ends of the minimal B cell epitope and serves to optimize the potential for interaction with T cells. The assumed mechanism of action of the intravenously applied peptide is the induction of immunologic tolerance. Data from subjects treated in open protocols (advanced, progressive MS) suggested that intravenous injections of 500 mg MBP 8298 as single or repeated injections induced long-lasting immunologic tolerance, confirmed by low or undetectable CSF anti-MBP in some subjects with progressive MS. Interestingly, HLA-DR haplotypes with high affinity for the administered peptide (DR2, DR4, DR7) had more durable tolerization effects (Warren et al., 1997). A Phase II study of chronic progressive MS (2 year placebo-controlled, double-blind, matched-pair phase followed up by a 24-month open-label phase) showed that fewer subjects progressed (based on confirmed EDSS progression: 31% vs. 56%) compared to placebo group. The effect of MBP was most impressive in the subgroup of HLA-DR2 and/or DR4 patients. A 2-year Phase II/III study in 553 patients with SPMS (comprising stratification for DR2/4) is currently being launched in multiple centers in the United States and Europe.

Another antigen-selective approach is a soluble complex of MBP peptide bound to the MS-associated HLA molecule DR2 (Goodkin et al., 2000). This Phase I trial had only a short observation time. Apart from safety aspects over this period, available clinical data do not allow firm conclusions on its clinical potential.

Vaccination using naked DNA encoding self-antigen protects and even reverses established disease in several autoimmune disease animal models for various diseases, including rheumatoid arthritis, insulin-dependent diabetes mellitus, and MS (reviewed in Fontoura et al., 2005). The convincing preclinical data on the immunologic effects of a plasmid encoding full-length MBP led to a Phase I/II trial currently investigating the effect of an MBP encoding plasmid (BHT-3009) in 30 RRMS/SPMS patients (Bar-Or et al., 2006). Intramuscular injections of three different doses at weeks 1, 3, 5, and 9 are being tested against placebo.

Although previous attempts toward antigen-specific immunomodulation have often been disappointing, these difficulties have led to renewed interest in therapies that aim to re-establish tolerance to autoantigens at the level of either T cell-mediated or antibody-mediated immune responses, or both. Such antigen-specific immunotherapies offer the prospect of correcting pathologic immune reactivity against autoantigens in a highly specific, effective manner and achieve this goal with relatively few side effects (Sospedra and Martin, 2005b; Steinman and Zamvil, 2005). There are, however, a number of caveats when judging the future perspectives of these approaches. The T cell response in the human system toward various CNS-candidate autoantigens is much more complex and diverse than in the EAE model using inbred animal strains. The heterogeneity and individuality of the TCR repertoire implies that such selective immune therapy might work only if the therapy can be "individualized" or "tailored" by developing specific therapies for individual patients or groups of patients with similar immunologic features.

*T Cell and T Cell Receptor Vaccination.* T cell as well as T cell receptor vaccination therapy is covered as a separate chapter (see Chapter 15).

*Leukocyte Differentiation Molecules.* Among the "biologicals" for the treatment of autoimmune disorders, monoclonal antibodies (mAbs) are considered especially promising. Initially, the goal of antibody treatment was to deplete certain subsets of immune cells that were suspected to be pathogenetic. In MS, this was primarily based on the assumption that pathogenesis is mainly T cell-mediated. Therefore, early studies targeted T cell differentiation molecules such as CD3 and CD4 (for review, see Hohlfeld and Wekerle, 2005). However, evidence accumulated that nondepleting mAbs can have subtle immunoregulatory effects and in some instances might even induce lasting immune tolerance (Waldmann, 1989).

Muromonab (OKT3) is a murine anti-human CD3 monoclonal antibody that exhibits both T cell activation and suppression properties. A decade and a half ago muronomab was tested in an open-label trial in 16 MS patients. However, acute side effects (fever, chills, nausea) were associated with a surge of TNFα and IFNγ production (Weinshenker et al., 1991). Furthermore, application was associated with significant lymphopenia and hypotension. First-dose reactions probably resulted from significant T cell activation induced by muromonab and concomitant cytokine release. Although MRI activity remained stable, disability scores worsened. Only two patients showed clinical improvement at the 1-year follow-up; and because of the attendant toxicity, anti-CD3 is considered to be an unattractive therapy.

Nevertheless, CD3-specific mAbs have recently elicited renewed interest, largely for two reasons: First, humanized anti-CD3 mAbs are now available and second, nondepleting anti-CD3 mAbs seem to be capable of inducing long-lasting immune tolerance (in contrast to the expected complications of long-lasting T cell depletion). Those non-FcR binding anti-CD3 antibodies are structurally modified and deliver only a partial signal to the TCRs. This modification decreases not only their T cell activation capacity but also their immunogenicity and toxicity (Alegre et al., 1994; Chatenoud, 2003; Li et al., 2005). Thus, non-FcR binding anti-CD3 antibodies are promising candidates for the treatment of autoimmune diseases such as diabetes (Bisikirska et al., 2005; Keymeulen et al., 2005), arthritis (Utset et al., 2002), or MS (Kohm et al., 2005).

In several trials, anti-CD4 mAbs were relatively well tolerated by MS patients (Llewellyn-Smith et al., 1997l; Rumbach et al., 1996; van Oosten et al., 1997). An open trial of a chimeric anti-CD4 mAb, priliximab, induced long-lasting suppression of CD4+ T cells without causing major toxicity (Lindsey et al., 1994a,b). A randomized, double-blind, placebo-controlled exploratory trial of the chimeric monoclonal anti-CD4 antibody cM-T412 in 71 patients with active RRMS and SPMS revealed that the antibody produced a long-lasting reduction of circulating CD4 T cells (van Oosten et al., 1997). However, clinical and paraclinical measures were essentially negative and yielded no effect of anti-CD4 treatment on MRI activity or disease progression (van Oosten et al., 1997). Using CD4 counts as a covariate, however, there was a statistically significant effect on the number of active lesions over 18 months and a significant reduction of 41% in the number of clinical relapses (a secondary efficacy parameter) after 9 months. It was concluded by the investigators that cM-T412 resulted in a substantial and sustained reduction in the number of circulating CD4 cells but was unlikely to influence MS activity or clinical disease course. Since then, several studies have investigated the differential effects/susceptibility of anti-CD4 mAbs on various T cell subsets and have provided possible explanations of why treatment with anti-CD4 might have produced disappointing results (reviewed in Hohlfeld and Wekerle, 2005).

Aletumzumab (Campath-1) is a humanized mAb directed against the CD52 antigen, which is commonly expressed on T and B cells, monocytes, and eosinophils. The antibody causes rapid, significant, persistent lymphocyte depletion. Potential clinical applications include hematologic malignancies, avoidance of transplant rejection, and treatment of autoimmune diseases. Its application in SPMS has demonstrated an interesting dissociation between (1) the observed suppression of T cells and inflammatory lesions but (2) evidence of progressive CNS tissue loss and disability (Coles et al., 1999a; Paolillo et al., 1999). At least in patients with SPMS this might indicate that nonspecific T cell downregulation alone at this stage of the disease may be ineffective. It also raises the possibility that by increasing B cell activity Campath-1 treatment led to increased antibody-mediated CNS damage with resultant progression of disability. For unknown reasons, about on-third of Campath-1-treated patients develop antibodies against the thyrotropin receptor as well as carbimazole-responsive autoimmune hyperthyroidism (Grave's disease) (Coles et al., 1999b). A Phase II clinical trial designed to compare the safety and efficacy of aletumzumab with subcutaneous IFNβ-1a three times weekly in patients with RRMS showed impressive effects on relapse rates and disease progression. However, the trial was suspended because there was evidence of severe toxicity (idiopathic thrombocytopenic purpura, or ITP) including two cases of death (see http://www.schering.de/scripts/en/index.php). The distributing companies have notified participants and regulatory authorities about the risk of ITP and are working on the development of tests that would help to identify patients who might have a higher risk for serious adverse events (http://www.FDA.org).

Taken together, therapeutic approaches targeting leukocyte differentiation molecules decisively depend on the expression of the targeted CD molecule. However, immunologic knowledge concerning the in vivo effects of mAbs and the differential influences of certain mAbs in various immune subsets sharing the same leukocyte differentiation molecules have opened new views. In general, one still might assume that antibodies aiming at widely distributed leukocyte antigens are less likely to become important in the treatment of MS, as the potential of serious side effects and the risk of compromising immune defense mechanisms may counteract their efficacy. It is conceivable, however, that intelligent combinations of antibodies might induce a lasting state of self-tolerance in T cells, provided they are applied at an optimal time point of the disease.

*Co-stimulatory Molecules.* Therapeutic blockade of co-stimulation as a treatment for autoimmune disorders has attracted considerable attention (Howard et al., 2005). T and B cells require two distinct types of signal for effective activation. One signal originates from ligation of the TCR complex and its co-receptors (CD4 and CD8) to an antigenic peptide bound to the presenting MHC molecule ("trimolecular complex"). The second signal is dependent on either soluble factors such as IL-2 or the ligation of cell surface molecules that provide essential co-stimulatory signals complementary to TCR engagement (Frauwirth and Thompson, 2002). It is assumed that co-stimulatory signals are critically relevant for the regulation of T cell activation as well as for keeping the balance between Th1 and Th2 (T helper) cell differentiation (Kobata et al., 2000).

1. *CTLA4-Ig*. Two co-stimulatory pathways, the B7/CD28-CTLA-4 and CD40-CD40L systems, have attracted attention also for the treatment of MS (Carreno and Collins, 2002; Keir and Sharpe, 2005; Stuart and Racke, 2002). Genetic alterations or agents interacting with the CD28-CD80/CD86 network decisively influence the development and progression of many autoimmune disease models, including EAE (e.g., Anderson et al., 1999; Salomon and Bluestone, 2001; Sharpe and Freeman, 2002) and, in addition, contribute to the (dys)regulation of autoreactive T cells in MS (e.g.. Anderson et al., 2000; Lovett-Racke et al., 1998; Maurer et al., 2002).

Probably the most important modulator of the B7/CD28-CTLA-4 pathway currently under investigation is CTLA4-Ig, a chimeric protein consisting of the extracellular domain of human CD152 fused to the Fc region of human IgG1 (hinge, CH2, and CH3) (Bluestone et al., 2006; Linsley et al., 1991). CTLA4 binds with higher avidity to CD80 (B7.1) and CD86 (B7.2) than to CD28 and thus interferes with the most important second signal of co-stimulation.

In rheumatoid arthritis patients refractory to TNFα inhibition, Abatacept (Orencia; BMS 188667; Orencia, Bristol-Myers Squibb Company) produced significant clinical and functional benefits (Genovese et al., 2005), thus receiving approval from the FDA in December 2005. Smaller Phase I/II studies already demonstrated clear proof of the concept concerning safety and clinical effectiveness in psoriasis (Abrams et al., 1999, 2000). Importantly, CTLA4Ig therapy was proving to have an acceptable toxicity profile. In parallel, advances have been made on the transplant front. LEA29Y is a higher-affinity form of CTLA-4-Ig, which—by two amino acid mutations—binds with a 10 times higher avidity for CD80 and CD86 compared to CTLA4Ig. LEA29Y (Belatacept) showed significantly greater efficacy in the transplant setting in nonhuman primates (Larsen et al., 2006) and proved not inferior to cyclosporine in renal transplants (Vincenti et al., 2005).

A total of 330 patients with RRMS were intended for enrollment in a double-blind, placebo-controlled multicenter study testing two doses of MBS18667 (2 and 10 mg/kg) against placebo. However, after inclusion of 130 patients, the study was prematurely halted by the safety board because of an accumulation of relapses and MR activity in the 2 mg/kg verum group. Although the investigators were worried about unexpected effects of the verum treatment, unblinding of the patient groups revealed that the 2 mg/kg group already had higher relapse rates at inclusion. Therefore, a randomization problem, rather than a true treatment failure, was suggested. The final interpretation is still pending. The results from a small pilot study testing the safety and immune mechanisms of another CTLA-4-Ig (RG2077) in MS have been reported (Khoury et al., 2005): Taken together, no major adverse events were observed, and immunologic and mechanistic analysis revealed evidence of biologic drug activity.

2. *Anti-CD40L*. Blockade of the CD154-CD40 pathway was effective in a variety of autoimmune models, including EAE (Aloisi et al., 1999; Howard et al., 1999). After encouraging preclinical data, the clinical development of a monoclonal antibody to CD40L (CD154; IDC131) was pushed. However, in Phase II clinical trials, anti-CD40L was associated with unanticipated side effects (Sidiropoulos and Boumpas, 2004). A few patients treated with anti-CD40L experienced thromboembolic events most likely due to reactivity of the monoclonal antibody with CD40L expressed on the surface of activated platelets (Kawai et al., 2000). Therefore, all clinical trials underway, including those on MS, were halted (http://www.nationalmssociety.org/research.asp).

3. *Anti-CD25 (IL-2 receptor)*. The interaction of IL-2 and its receptor CD25 mediates a pivotal signal for T cell activation and proliferation. The IL-2 receptor antagonist daclizumab is a humanized mAb that interferes with the $\alpha$-chain of the IL-2 receptor. CD25 treatment aims to limit T cell proliferation by blocking IL-2 signaling via its high-affinity receptor. The antibody has been successfully applied to prevent transplant rejection, and a variety of other studies with this indication are underway (Huston and George, 2001; Vincenti et al., 1998). Two open-label Phase II studies showed that daclizumab is well tolerated and leads to a significant reduction in MRI activity and significant improvement in several clinical outcome measures (Bielekova et al., 2004; Rose et al., 2004). Overall, daclizumab treatment was well tolerated, with a slight increase in infections and rash. At first glance, these results might seem surprising or even paradoxical because CD25 is expressed not only on "activated" putatively pathogenic T cells but also on suppressor T cells (Sakaguchi, 2005). Recent findings indicate that MS is accompanied by dysfunction of these suppressor cells (Baecher-Allan and Hafler, 2004; Haas et al., 2005; Viglietta et al., 2004). This may imply a (theoretical) risk of further compromising the CD25+ suppressor system. However, the network of putative regulatory cells in MS is complex, and other types of regulatory cell are also likely to be involved. Bielekova and coworkers have provided evidence that anti-CD25 treatment leads to an expansion of immunoregulatory CD56[bright] regulatory NK cells, which might contribute to the clinical benefits observed under daclizumab (Bielekova et al., 2006). Currently, several studies are recruiting subjects to test the safety and efficacy of daclizumab in a placebo-controlled, double-blind manner. They include monocenter trials at the National Institutes of Health (NIH) and multicenter trials against placebo in RRMS/SPMS including the question as to how daclizumab might be beneficial as an add-on therapy to incomplete interferon responders.

### 3.2.2    B Cells and Antibody Formation

There is increasing evidence that B cells and antibodies play a decisive role in the development and perpetuation of MS, at least in some forms (Archelos et al., 2000; Lucchinetti et al., 2000; Sospedra and Martin, 2005b). Therefore, inhibition of autoreactive B cells is a logical therapeutic aim. Certain MS variants, such as neuromyelitis optica, seem especially to be predominantly mediated by humoral mechanisms (Bruck et al., 2001; Lennon et al., 2005; Lucchinetti et al., 2002; Scolding, 2005), thus making them potentially accessible to plasma exchange or immunoglobulin administration (Keegan et al., 2002; Wingerchuk and Weinshenker, 2005). In "pattern II" lesions, antibodies and complement are found in vast amounts in parallel with T cells and monocyte/macrophages. Importantly, data correlating the clinical response between humoral pathologic changes in MS and the response to therapeutic plasma exchange provided a first clinicopathologic correlation of this histologically based concept in "classic" MS (Keegan et al., 2005).

Rituximab (mabthera) is a genetically engineered chimeric murine/human mAb against CD20, a differentiation antigen that is found on normal and malignant pre-B and mature B lymphocytes but is absent on hematopoietic stem cells, activated B cells (plasma cells), and in normal tissues. CD20 is vital for regulation of the activation process of cell cycle initiation and differentiation. The mAb is already approved for the treatment of B cell malignancies. In general, the use of rituximab is well tolerated. Known side effects are transfusion reactions with fever, hypotension, and dyspnea. Other events, such as tumor lysis syndrome, have been reported in the literature relating to malignancies but is not applicable to the MS population.

In an open pilot study in patients with neuromyelitis optica, administration of rituximab led to a clear reduction in relapse frequency, thus providing a first proof of the concept study. Meanwhile a number of case series or case reports describing the use of rituximab in "classic" MS are available (Monson et al., 2005; Petereit and Rubbert, 2005; Stüve et al., 2005). At present, two trials are investigating the potential use of rituximab in MS: the rituximab in PPMS (OLYMPUS) trial, which is a Phase II/III trial; and a Phase II trial involving RRMS patients. The safety and efficacy of rituximab in RRMS and PPMS is being investigated in two parallel Phase II/III clinical trials.

### 3.2.3    Cytokine-Based Immunotherapy

For a comprehensive overview on the cytokine network and cytokine-based therapy we refer to Wiendl and Hohlfeld, 2006.

### 3.2.4    Adhesion, Chemotaxis, and Migration

*Anti-Adhesion Molecules*

Leukocyte recruitment into the CNS is regulated by cell adhesion molecules on endothelial cells and chemokines that are released from the site of inflammation to attract hematogenous cells (Archelos et al., 1999; Charo and Ransohoff, 2006).

As a first step, the velocity of leukocytes in the bloodstream is focally reduced ("rolling") as they loosely attach to endothelial cells via selectins (E-selectin, P-selectin, L-selectin). Then, firm adhesion of leukocytes to the endothelium is mediated by integrin interactions. Integrins comprise the $\alpha_4\beta_1$, $\alpha_5\beta_1$, $\alpha_6\beta_1$ iso-forms—also known as very late antigen 4 (VLA-4), VLA-5, and VLA-6, respectively—and leukocyte function antigen 1 (LFA-1), which is also a co-stimulatory molecule in T cell activation. Whereas leukocytes constitutively express the lig-ands (LFA-1, VLA-4) on their surface, the corresponding counterreceptors such as intercellular adhesion molecule-1 (ICAM-1), ICAM-2, ICAM-3, and vascular cell adhesion molecule-1 (VCAM-1) on the endothelial cells require specific induction and thereby determine the location of leukocyte–endothelial cell inter-action. Antibodies directed against single adhesion molecules can potently inhibit crucial steps in the pathogenesis of MS, especially leukocyte migration.

The most promising candidate derived from this substance class is natalizumab (Tysabri). This monoclonal humanized antibody targets $\alpha_4\beta_1$ integrin (VLA-4) on the surface of lymphocytes. It is composed of a human immunoglobulin G4 (IgG4) framework at the complementary determining region (CDR), which is linked to a murine antibody clone to reduce immunogenicity (Keeley et al., 2005). Experiments in EAE established that treatment with anti-$\alpha_4$ integrin mAb inhibits the interaction betweeen the $\alpha_4\beta_1$ integrin (expressed on leukocytes) and its ligands on brain endothelium, preventing the accumulation of leukocytes (especially T cells) in the CNS (Rice et al., 2005; Steinman, 2005). Additional mechanisms, including inhibition of the reactivation of T cells in the CNS, are also conceivable. Because natalizumab also acts on $\alpha_4\beta_7$ integrin, which is involved in leukocyte migration through intestinal endothelium, this treatment is also developed and investigated in Crohn's disease. Infusions of this antibody are given at monthly intervals.

A preceding Phase Ib/II study applying natalizumab in MS patients suggested beneficial effects (Tubridy et al., 1999), but clear proof of this concept was demonstrated in a subsequent three-armed, randomized, double-blind Phase II trial in RRMS and active SPMS. A total of 213 patients received intravenous natalizumab at either 3 or 6 mg/kg body weight or placebo every 4 weeks for 6 months (Miller et al., 2003). Treatment with both dosage regimens significantly reduced the number of new MS brain lesions on monthly gadolinium-enhanced MRI and the relapse frequency. Two Phase III trials have been completed (Polman, 2006; Rudick et al., 2006). The AFFIRM trial tested natalizumab as monotherapy (300 mg i.v. every 28 days for up to 28 months) versus placebo in 942 MS patients who had not received any immunotherapy during the preceding 6 months. Altogether, 96% of subjects in the treatment arm had no new gadolin-ium-enhancing lesions compared to 68% of those on placebo. The rate of clinical relapses was reduced by 67%, and natalizumab significantly delayed the pro-gression of disability after 2 years (Polman et al., 2006). In the SENTINEL study, natalizumab was combined with the intramuscular formulation of IFNβ-1a and tested against IFNβ-1a alone. Subjects continued IFNβ-1a therapy and were ran-domizd to either natalizumab 300 mg ($n = 589$) or placebo ($n = 582$) infusions every 28 days. The natalizumab + IFNβ-1a group had a 54% reduced relapse rate with IFNβ-1a alone, significantly fewer MRI lesions, and a markedly higher proportion of relapse-free patients. The most frequent adverse events in both studies included anaphylactoid reactions and an increased risk of infections, rash, arthralgia, and headache. Approximately every tenth patient receiving natalizumab developed antibodies, which had a clear negative impact on the therapeutic efficacy in subjects who remained persistently antibody-positive (e.g., second testing).

Based on the positive interim analysis of both Phase III trials after 1 year of therapy, natalizumab was approved by the FDA via fast track for the treatment of "relapsing forms of MS" in November 2004. However, in February 2005, the substance was suspended, and ongoing clinical trials were stopped because two patients receiving natalizumab in combination with IFNβ-1a developed progressive multifocal leukencephalopathy (PML), a severe opportunistic infection of the CNS. One of these MS patients finally died (Kleinschmidt-DeMasters and Tyler, 2005; Langer-Gould et al., 2005). In addition, a third person enrolled in a study of Crohn`s disease was classified to have PML based on a histopathologic reexamination that was primarily classified as an astrocytoma. This subject had received natalizumab over an 18-month period but had also been treated with various additional immunosuppressive and immunomodulatory agents (except IFN) prior to and in parallel with natalizumab (van Assche et al., 2005).

As a consequence, extensive reexamination of all patients who had received natalizumab in clinical trials was performed, and the safety profile of natalizumab was collected and reviewed (Yousry et al., 2006). Information from the clinical history, physical examination, brain MRI, and CSF testing for JC virus DNA were collected by an expert panel to evaluate patients for PML. However, there is a number of reasons that vote for remarketing natalizumab despite those serious risks and side effects (Yousry et al., 2006): (1) No additional cases of PML were identified in more than 3000 patients exposed to natalizumab. (2) Highly specific plans for close monitoring of patients receiving natalizumab have been developed, and various prerequisites have to be fulfilled prior to natalizumab administration (e.g., application as a monotherapy in RRMS patients only, "patient checklist," and screening for PML symptoms prior to starting treatment, mandatory registry program, administration restricted to authorized natalizumab infusion centers (for further information see http://wwww.FDA.org). (3) The available efficacy data indicate that natalizumab is a highly effective drug in a devastating disease such as MS, where available treatment options are limited.

Owing to its mechanism of action, natalizumab might impede immune surveillance of the CNS. It is unknown, however, whether the PML reactivation (or de novo infection) that occurred was caused by a general immunosuppressive effect on immune surveillance or by a more specific effect related to the particular mechanism of action of natalizumab (off target effect or direct effect of the mechanism of action) (Ransohoff, 2005). One must first consider $\alpha_4\beta_1$/VCAM-1 interactions in the adult hematopoietic system, beyond those that support inflammation; $\alpha_4\beta_1$ is required for generating T cells and B cells from bone marrow progenitors in adult mice and for some inflammatory lymphocyte trafficking but not for entry into spleen, lymph nodes, or intestinal epithelium (Arroyo et al.,

1996). Patients who received natalizumab exhibited elevated basophil, eosinophil, and lymphocyte counts, with approximately 5% showing circulating nucleated erythrocytes (Tysabri package insert). Natalizumab, unlike many other leukocyte trafficking modulators, potently affects bone marrow physiology by blocking $\alpha_4\beta_1$/VCAM-1 interactions. Thus the following scenario is possible (Ransohoff, 2005): Natalizumab treatment led to PML beginning with its action toward infected bone marrow cells by promoting their premature exit from the marrow, which reduced their ability to control JCV replication owing to non-physiologic deprivation of contact with bone marrow stroma, and it gave these infected cells access to all body compartments. These effects of natalizumab treatment, in combination with reduced surveillance and inflammatory trafficking to the infected CNS, promoted PML pathogenesis. A natalizumab-induced spike of JC viremia in a Crohn's disease patient may have represented release of virus from marrow stores, rather than uncontrolled JCV proliferation due to immuno-suppression. If this hypothesis is correct, agents that do not affect bone marrow lack this complication, whereas those affecting bone marrow and lymphoid organs (anti-CXCR4; anti-$\alpha_4$ integrin; FTY720, for example) might require caution and surveillance.

In conclusion, PML may indeed be regarded as an off-target adverse effect, and it remains possible to consider treating MS patients with agents that address leukocyte trafficking to the CNS. Given the risk of PML and the potential benefits of this class of agents, it is necessary to learn much more about JCV infection without delay. It is not known how many healthy individuals harbor pathogenic, reactivation-competent JCV or if we can reliably detect such persons; further-more, it is not know how often JC viremia occurs. It is imperative to clarify whether there is a reservoir of JCV in the CNS, as some have suggested.

Against the background of the experience with natalizumab, the FDA requested that all clinical trials involving substances that target $\alpha_4\beta_1$/VLA-4 interactions should be suspended until the analysis of the natalizumab safety data is completed. This also affected the recently developed oral VLA-4 integrin antagonist SB683699. The oral application route is expected to satisfy key unmet needs of patients (e.g., lack of flu-like symptoms and injection site reac-tions seen with IFN$\beta$s and glatiramer acetate or simplified dosing. A Phase II study is currently on hold that was designed to analyze the safety and efficacy of SB683699 versus placebo in 260 RRMS patients testing four dosing regimens of SB683699.

The LFA-1/ICAM interaction pathway is another key mediator of cell adhesion between leukocytes and vascular endothelial cells (Simmons and Buckley, 2005). This engagement is mechanistically targeted by Hu23F2G, a humanized anti-LFA1 (CD11/CD18) antibody. Application of this drug in an open Phase I study in 24 patients with MS led to a high saturation of LFA 1 on circulating lympho-cytes with in vivo inhibition of leukocyte migration (Bowen et al., 1998). However, a subsequent Phase II study enrolling 169 MS patients could not demonstrate beneficial effects on MRI activity or clinical measures (Lublin et al., 1999).

Efalizumab is a recombinant humanized IgG1 kappa isotype mAb against the CD11a molecule administered as a weekly subcutaneous injection (Marecki and Kirckpatrick, 2004). Efalizumab selectively targets the α-chain (CD11a) of LFA-1 and is approved for the treatment of moderate to severe psoriasis (Lebwohl et al., 2003). Experience with 19,000 patients (Scheinfeld, 2006) allows the assessment that treatment with efalizumab can be regarded as remarkably safe. Interestingly, treatment with monoclonal anti-CD11a antibodies protected rats from EAE (Willenborg et al., 1996) and diminished the severity of the disease in mice (Gordon et al., 1995) underlining that this approach principally might also be attractive for the therapy of human MS.

*Chemotaxis*

Active MS is characterized by the presence of inflammatory foci disseminated in the CNS. The cellular composition of these inflammatory infiltrates is determined in part by the local spectrum of secreted chemokines, which are key contributors to the directional movement of leukocytes (Baggiolini, 1998; Charo and Ransohoff, 2006; Engelhardt and Ransohoff, 2005). Various studies underline the importance of the interaction of chemokines with their receptors in EAE and MS pathogenesis. For example, the expression of certain chemokines and their respective receptors are upregulated in MS and EAE (Columba-Cabezas et al., 2002; Omari et al., 2005, 2006; Sörensen et al., 2002). A variety of experimental studies have shown that specific interference of genetic ablation of chemokine/ chemokine receptor interactions has the potential to prevent and/or ameliorate CNS inflammation (Charo and Ransohoff, 2006; Gaupp et al., 2003; Ransohoff, 1999; Sörensen et al., 1999; Ubogu et al., 2005). Pharmacologic interference with the chemokine/chemokine receptor system has therefore been regarded as a promising target for effective treatment of MS (Fox and Ransohoff, 2004).

BX-471 (ZK811752) is a recently developed chemokine receptor 1 (CCR1) antagonist for oral application. Positive results from Phase I trials in autoimmune diseases and in MS have been reported (Elices, 2002). A double-blind, placebo-controlled Phase II study failed to demonstrate efficacy on primary or secondary outcome measures (MRI, relapses) in a subsequent Phase II trial (Zipp et al., 2005).

Several other Phase I/II studies addressing chemokine receptor antagonists (e.g., CCR2 and CCR5) are currently underway (Charo and Ransohoff, 2006). One has to take into consideration, however, that the chemokine system, like the complex cytokine network (see above), is characterized by its large redundancy. Thus, a definite assessment to which extent the blockade of a solitary chemokine receptor mediates significant clinical effects remains difficult. Large controlled trials are therefore warranted to address this therapeutic approach in MS.

*Blood-Brain Barrier Disruption and Migration*

There is increasing evidence that various members of the matrix metalloproteinases (MMPs) family mediate fundamental steps in the development of inflammatory demyelinating disorders, such as cell migration, blood-brain barrier (BBB) disruption, demyelination, and cytokine activation (Hartung and Kieseier, 2000; Sellebjerg and Sörensen, 2003; Yong, 2005). Various MMPs were detected in postmortem tissue samples from patients with MS (Lindberg et al., 2001; Maeda and Sobel, 1996), and particularly MMP-7 and MMP-9 seem to play a key role in the pathogenesis of CNS inflammation. Both proteases were found to be increased in EAE, and peak expression levels correlated with the maximum disease activity (Clements et al., 1997; Kieseier et al., 1998). Selective MMP inhibitors prevented or ameliorated inflammatory CNS demyelination when applied in the EAE mode (Brundula et al., 2002; Popovic et al., 2002), suggesting that MMPs appear as suitable targets for the treatment of MS (Kieseier et al., 1999). Various broad-spectrum MMP inhibitors have been developed and tested in different in vivo models or knockout mice, emphasizing the potential impact of MMPs on the pathophysiology of certain diseases. Consequently, clinical trials for these broad-spectrum MMP inhibitors were initiated in various disorders, but not in MS so far.

Based on their chelating properties, tetracyclines and chemically modified tetracyclines are capable of blocking MMP activity. The tetracycline minocycline has attracted increasing interest for the treatment of various neurologic diseases (Yong et al., 2004). Minocycline is capable of inhibiting MMP activity independent of its antimicrobial activity (Paemen et al., 1996) and exhibits potent synergistic effects with glatiramer acetate or IFN in animal models of MS (Giuliani et al., 2005a,b). A small open-label trial in 10 MS patients revealed impressive reduction of the mean total number of active MRI lesions, implying that minocycline inhibited the MMPs that permit transmigration of the BBB by autoreactive T cells (Metz et al., 2004). Consequently, controlled follow-up studies including larger patient numbers are currently underway.

Alpha lipoic acid (ALA) is an endogenous fatty acid that is able to decrease MMP-9 activity and T cell migration in vitro (Marracci et al., 2004). In addition, the substance acts as a free radical scavenger. Oral application of ALA prevented the development of EAE and ameliorated the disease course when administered under therapeutic conditions (Morini et al., 2004). Data from a preliminary study of 37 MS patients revealed favorable safety properties and a significant negative correlation between serum ALA levels and serum MMP-9, at least in the high-dose arm (Yadav et al., 2005). Similar results were obtained from a small study supplementing omega-3 fatty acid which is also known to inhibit MMP-9 (Marrecci et al., 2005). At present, to our knowledge no larger trials addressing ALA or omega-3 fatty acids and MS are ongoing.

### 3.2.5   Macrophage Activation and Macrophage-Mediated Demyelination

Although macrophages are thought to play a pivotal role in the pathogenesis of MS, no treatment strategy is currently available specifically targeting these cells. However, most of the immunosuppressive drugs available exhibit inhibitory effects on macrophages and as such prevent or ameliorate demyelination in the CNS.

### 3.2.6    Autologous Stem Cell Transplantation

The issue on stem cell therapy in MS is discussed in Chapter 17.

## 4    Neuroprotection and "Neurorepair"

Protecting neuronal tissue from degeneration following or accompanied by inflammatory injury is a key challenge for recent and future treatment strategies in MS. Many of the existing therapies that directly target inflammation have marked benefits on the relapse rate (as an indicator of inflammatory activity) but insufficiently influence disability progression and MRI parameters of brain tissue degeneration. Thus, aspects of neuroprotection, neuronal repair, and remyelination have received increasing attention lately. Albeit the number of agents or strategies currently under experimental and clinical investigation is large, they cannot be carried out in more detail, as they are not "immunotherapeutic" approaches. Notwithstanding, to establish the use of neuroprotective agents in clinical practice requires new therapeutic strategies and animal models that are able to help us understand and "image" the interdependence and influence of inflammation and neuronal degeneration. Strategies or targets currently under investigation or debate comprise neurotrophic factors, cannabinoids, erythropoietin, glutamate transmission and excitotoxicity, ion channels (sodium, calcium, potassium), strategies of remyelination, and neuronal stem cells. Although experimental data on the involvement of certain molecular structures or cellular compartments are in many cases fascinating and convincing, the road from "bench to bedside" is anticipated to be particularly long. At this point in time, no agent or strategy complementing daily practice in MS patient care can clearly be anticipated within the next few years.

## 5    Conclusions

A substantial number of pivotal and preliminary reports continue to demonstrate encouraging new evidence that advances are being made in the care of MS patients. Despite the disappointments for a number of large pivotal trials, there is an auspicious array on novel agents and upcoming strategies. At present, it is not possible to foresee which, if any, of the new treatment strategies will supersede or complement the currently approved therapies in the near future. Clearly, strategies to enhance repair and to promote neuroprotection are of immense importance, and the next years will hopefully demonstrate how theoretical concepts, existing experimental data, or already available agents can be transferred successfully to the treatment of MS. The approach of combining agents or strategies remains theoretically promising. At present, however, all of these approaches are still at an "experimental" state; and except for statements on their safety and tolerability, none has achieved the level at which recommendations or even approval can be made based on the evidence.

From the interesting class of monoclonal antibodies to treat MS, natalizumab is anticipated to (re)enter clinical practice soon. However, three cases of PML that occurred in patients exposed to the mAb have realerted us to the intrinsic limitations of monoclonal antibody treatment. Even the latest generation of so-called humanized mAbs are clearly immunogenic; that is, they can induce the production of anti-antibodies, including anti-idiotypic antibodies. These anti-antibodies can neutralize the desired therapeutic effects and induce allergic reactions, postulating the need for close monitoring of "anti-antibody" development. Second, after binding to their target, mAbs may induce a systemic inflammatory response by activating inflammatory cells and mediators. Systemic release of cytokines and inflammatory mediators is most likely responsible for these reactions. Third, apart from the effects based on the postulated mechanism of action, numerous "off-target" effects are possible and should be expected or anticipated. Such an "off-target" effect might have contributed to the cases of PML observed with the anti-VLA4 antibody natalizumab. It should be noted that many "immunomodulatory" mAbs act as nonselective immunosuppressants, even though their molecular targets are precisely known. Therefore, like conventional "chemical" immunosuppressive drugs, these agents can have serious adverse effects, most notably infections (not only opportunistic) and tumors. It should furthermore be noted that completely unexpected side effects cannot be predicted from the preclinical studies in animal models (e.g., the induction of MS-like symptoms in patients treated with anti-TNFα mAbs in rheumathoid arthritis or exacerbation of MS symptoms with anti-TNFα therapy) when patients are treated with mAbs. One sad example in recent history is the first clinical use of a monoclonal superagonistic antibody against CD28, which has shown fascinating effects in rodents and nonhuman primates in modulating the course and severity of experimental inflammatory models. However, in contrast to the preclinical experience, where the superagonistic antibody preferentially induced regulatory cells that counterbalanced the inflammatory reaction in various experimental models, in the human system six individuals treated with the antibody developed severe systemic immune hyperreactions, leading to multiorgan failure. In contrast to the rodent data and the human in vitro data, where anti-CD28 stimulated preferentially regulatory T cells, in humans in vivo TGN1412 probably switched on T-helper cells, which produced cytokines leading to mass activation causing a devastating cytokine storm. Altogether, this leads to the conclusion that albeit part of the attraction of biologicals in general is that the molecular targets of these agents seem to be precisely known, it cannot be predicted what consequences the inhibition (or stimulation) of the molecular targets of the biologicals (or therapeutic mAbs in particular) in the complex organism of a living human organism might induce.

It should clearly be noted that there is a lively refueling of the "older" idea of general immunosuppression or immunomodulation in the treatment of MS, in contrast to the highly immunoselective therapeutic approaches propagated in the class of biologicals. In these categories, a number of agents are currently under Phase II/III clinical development and are already partly approved and on the market for different indications (e.g., statins); they may fulfill some of the key requirements for treating a nonfatal disorder: convenience of drug administration accompanied by long-term efficacy plus safety. It is to be expected that within the next few years oral immunomodulatory or immunosuppressive agents will be available for the treatment of MS patients and complement the currently available armamentarium of disease-modifying drugs.

## References

Abrams, J. R., Lebwohl, M. G., Guzzo, C. A., et al. (1999) CTLA4Ig-mediated blockade of T-cell costimulation in patients with psoriasis vulgaris. *J. Clin. Invest.* 103(9):1243-1252.

Abrams, J. R., Kelley, S. L., Hayes, E., et al. (2000) Blockade of T lymphocyte costimulation with cytotoxic T lymphocyte-associated antigen 4-immunoglobulin (CTLA4Ig) reverses the cellular pathology of psoriatic plaques, including the activation of keratinocytes, dendritic cells, and endothelial cells. *J. Exp. Med.* 192(5): 681-694.

Ahrens, N., Salama, A., Haas, J. (2001). Mycophenolate-mofetil in the treatment of refractory multiple sclerosis. *J. Neurol.* 248(8):713-714.

Aktas, O., Waiczies, S., Smorodchenko, A., et al. (2003) Treatment of relapsing paralysis in experimental encephalomyelitis by targeting Th1 cells through atorvastatin. *J. Exp. Med.* 17;197(6):725-733.

Alegre, M. L., Sattar, H. A., Herold, K. C., et al. (1994) Prevention of the humoral response induced by an anti-CD3 monoclonal antibody by deoxyspergualin in a murine model. *Transplantation* 57(12):1786-1794.

Aloisi, F., Penna, G., Polazzi, E., et al. (1999) CD40-CD154 interaction and IFN-gamma are required for IL-12 but not prostaglandin E2 secretion by microglia during antigen presentation to Th1 cells. *J. Immunol.* 162(3):1384-1391.

Andersen, O., Lycke, J., Tollesson, P. O. (1996) Linomide reduces the rate of active lesions in relapsing-remitting multiple sclerosis. *Neurology* 47:895-900.

Anderson, D. E., Sharpe, A. H., Hafler, D. A. (1999) The B7-CD28/CTLA-4 costimulatory pathways in autoimmune disease of the central nervous system. *Curr. Opin. Immunol.* 11(6):677-683.

Anderson, D. E., Bieganowska, K. D., Bar-Or, A., et al. (2000) Paradoxical inhibition of T-cell function in response to CTLA-4 blockade; heterogeneity within the human T-cell population. *Nat. Med.* 6(2):211-214.

Anonymous. (2004) *Drugs RD* 5(6):363-367.

Archelos, J. J., Hartung, H. P. (2000) Pathogenetic role of autoantibodies in neurological diseases. *Trends Neurosci.* 23:317-327.

Archelos, J. J., Previtali, S. C., Hartung, H. P. (1999) The role of integrins in immune-mediated diseases of the nervous system. *Trends Neurosci.* 22(1):30-38.

Archelos, J. J., Storch, M. K., Hartung, H. P. (2000) The role of B cells and autoantibodies in multiple sclerosis. *Ann. Neurol.* 47(6):694-706.

Arroyo, A. G., Yang, J. T., Rayburn, H., Hynes, R. O. (1996) Differential requirements for alpha4 integrins during fetal and adult hematopoiesis. *Cell* 85(7):997-1008.

Babbe H., Roers A. Waisman A. (2000) Clonal expansions of CD8(+) T cells dominate the T cell infiltrate in active multiple sclerosis lesions as shown by micromanipulation and single cell polymerase chain reaction. *J. Exp. Med.* 1192:393-404.

Baecher-Allan, C., Hafler, D. A. (2004) Suppressor T cells in human diseases. *J. Exp. Med.* 200(3):273-276.

Baggiolini, M. (1998) Chemokines and leukocyte traffic. *Nature* 392(6676):565-568.

Bar-Or, A., Antel, J., Bodner, C. A. (2006) Antigen-specific immunomodulation in multiple sclerosis patients treated with MBP encoding DNA plasmid (BHT-3009) alone or combined with atorvastatin. *Neurology* 66(Suppl.):A62.

Beggiolin, G., Crippa, L., Menta, E., et al. (2001) Bbr 2778, an aza-anthracenedione endowed with preclinical anticancer activity and lack of delayed cardiotoxicity. *Tumori* 87(6):407-416.

Benoist, C., Mathis, D. (2001) Autoimmunity provoked by infection: how good is the case for T cell epitope mimicry? *Nat. Immunol.* 2:797-801.

Bernitsas, E., Wei, W., Mikol, D. D. (2006) Suppression of mitoxantrone cardiotoxicity in multiple sclerosis patients by dexrazoxane. *Ann. Neurol.* 59(1):206-209.

Beutler, E. (1992) Cladribine (2-chlorodeoxyadenosine). *Lancet* 340(8825):952-956.

Beutler, E., Sipe, J. C., Romine, J. S., et al. (1996) The treatment of chronic progressive multiple sclerosis with cladribine. *Proc. Natl. Acad. Sci. U S A* 93(4):1716-1720.

Bielekova, B., Goodwin, B., Richert, N., et al. (2000a) Encephalitogenic potential of the myelin basic protein peptide (amino acids 83-99) in multiple sclerosis: results of a phase II clinical trial with an altered peptide ligand. *Nat. Med.* 6(10):1167-1175.

Bielekova, B., Lincoln, A., McFarland, H., Martin, R. (2000b) Therapeutic potential of phosphodiesterase-4 and -3 inhibitors in Th1-mediated autoimmune diseases. *J. Immunol.* 164(2):1117-1124.

Bielekova, B., Richert, N., Howard, T., et al. (2004) Humanized anti-CD25 (daclizumab) inhibits disease activity in multiple sclerosis patients failing to respond to interferon beta. *Proc. Natl. Acad. Sci. U S A.* 101(23):8705-8708.

Bielekova, B., Catalfamo, M., Reichert-Scrivner, S., et al. (2006) Regulatory CD56bright natural killer cells mediate immunomodulatory effects of IL-2R{alpha}-targeted therapy (daclizumab) in multiple sclerosis. *Proc. Natl. Acad. Sci. U S A.* 103(15):5941-5946.

Birnbaum, G., Altafullah, I. (2005) *Neurology* 64(Suppl. 1):A385, P06.158.

Bisikirska, B., Colgan, J., Luban, J., et al. (2005) TCR stimulation with modified anti-CD3 mAb expands CD8+ T cell population and induces CD8+CD25+ Tregs. *J. Clin. Invest.* 115(10):2904-2913.

Bjartmar, C., Trapp, B. D. (2001) Axonal and neuronal degeneration in multiple sclerosis: mechanisms and functional consequences. *Curr. Opin. Neurol.* 14:271-278.

Bluestone, J. A., St. Clair, E. W., Turka, L. A. (2006) CTLA4Ig: bridging the basic immunology with clinical application. *Immunit.* 24(3):233-238.

Bowen, J. D., Petersdorf, S. H., Richards, T. L., et al. (1998) Phase I study of a humanized anti-CD11/CD18 monoclonal antibody in multiple sclerosis. *Clin. Pharmacol. Ther.* 64(3):339-346.

Brassat, D., Recher, C., Waubant, E., et al. (2002) Therapy-related acute myeloblastic leukemia after mitoxantrone treatment in a patient with MS. *Neurology* 59(6):954-955.

Brinkmann, V., Davis, M. D., Heise, C. E., et al. (2002) The immune modulator FTY720 targets sphingosine 1-phosphate receptors. *J. Biol. Chem.* 277(24):21453-21457.

Bruck, W., Neubert, K., Berger, T., Weber, J. R. (2001) Clinical, radiological, immunological and pathological findings in inflammatory CNS demyelination—possible markers for an antibody-mediated process. *Mult. Scler.* 7(3):173-177.

Brundula, V., Rewcastle, N. B., Metz, L. M., et al. (2002) Targeting leukocyte MMPs and transmigration: minocycline as a potential therapy for multiple sclerosis. *Brain* 125 (Pt 6):1297-1308.

Brunmark, C., Runstrom, A., Ohlsson, L., et al. (2002) The new orally active immunoreg-ulator laquinimod (ABR-215062) effectively inhibits development and relapses of experimental autoimmune encephalomyelitis. *J. Neuroimmunol.* 130(1-2):163-172.

Budde K., Schmouder R. L., Brunkhorst R., et al. (2002) First human trial of FTY720, a novel immunomodulator, in stable renal transplant patients. *J. Am. Soc. Nephrol.* 13(4):1073-1083.

Calabresi, P. A., Wilterdink, J. L., Rogg, J. M., et al. (2002) An open-label trial of combi-nation therapy with interferon beta-1a and oral methotrexate in MS. *Neurology* 58(2):314-317.

Canadian Cooperative Multiple Sclerosis Study Group. (1991) The Canadian cooperative trial of cyclophosphamide and plasma exchange in progressive multiple sclerosis. *Lancet* 337(8739):441-446.

Carreno, B. M., Collins, M. (2002) The B7 family of ligands and its receptors: new path-ways for costimulation and inhibition of immune responses. *Annu. Rev. Immunol.* 20:29-53.

Cavaletti, G., Cavalletti, E., Crippa, L., et al. (2004) Pixantrone (BBR2778) reduces the severity of experimental allergic encephalomyelitis. *J. Neuroimmunol.* 151(1-2):55-65.

Charo, I. F., Ransohoff, R. M. (2006) The many roles of chemokines and chemokine recep-tors in inflammation. *N. Engl. J. Med.* 354(6):610-621.

Chatenoud, L. (2003) CD3-specific antibody-induced active tolerance: from bench to bed-side. *Nat. Rev. Immunol.* 3(2):123-132.

Chen, B. J., Morris, R. E., Chao, N. J. (2000) Graft-versus-host disease prevention by rapamycin: cellular mechanisms. *Biol. Blood Marrow Transplant.* 6(5A):529-536.

Chiba, K. (2005) FTY720, a new class of immunomodulator, inhibits lymphocyte egress from secondary lymphoid tissues and thymus by agonistic activity at sphingosine 1-phosphate receptors. *Pharmacol. Ther.* 108(3):308-319.

Clements, J. M., Cossins, J. A., Wells, G. M., et al. (1997) Matrix metalloproteinase expression during experimental autoimmune encephalomyelitis and effects of a com-bined matrix metalloproteinase and tumour necrosis factor-alpha inhibitor. *J. Neuroimmunol.* 74(1-2):85-94.

Cohen, B. A., Mikol, D. D. (2004) Mitoxantrone treatment of multiple sclerosis: safety considerations. *Neurology* 63(Suppl. 6):S28-S32.

Coles, A. J., Wing, M. G., Molyneux, P., et al. (1999a) Monoclonal antibody treatment exposes three mechanisms underlying the clinical course of multiple sclerosis. *Ann. Neurol.* 46(3):296-304.

Coles, A. J., Wing, M., Smith, S., et al. (1999b) Pulsed monoclonal antibody treatment and autoimmune thyroid disease in multiple sclerosis. *Lancet* 354(9191):1691-1695.

Columba-Cabezas, S., Serafini, B., Ambrosini, E., et al. (2002) Induction of macrophage-derived chemokine/CCL22 expression in experimental autoimmune encephalomyelitis and cultured microglia: implications for disease regulation. *J. Neuroimmunol.* 130 (1-2):10-21.

Cyster, J. G. (2005) Chemokines, sphingosine-1-phosphate, and cell migration in secondary lymphoid organs. *Annu. Rev. Immunol.* 23:127-159.

Dello Russo, C., Gavrilyuk, V., Weinberg, G., et al. (2003) Peroxisome proliferator-activated receptor gamma thiazolidinedione agonists increase glucose metabolism in astrocytes. *J. Biol. Chem.* 278(8):5828-5836.

Diab, A., Deng, C., Smith, J. D., et al. (2002) Peroxisome proliferator-activated receptor-gamma agonist 15-deoxy-delta(12,14)-prostaglandin J(2) ameliorates experimental autoimmune encephalomyelitis. *J. Immunol.* 168(5):2508-2515.

Dinter, H., Tse, J., Halks-Miller, M., Asarnow, D., et al. (2004) The type IV phosphodiesterase specific inhibitor mesopram inhibits experimental autoimmune encephalomyelitis in rodents. *J. Neuroimmunol.* 108(1-2):136-146.

Dormandy, J. A., Charbonnel, B., Eckland, D. J., et al. (2005) Secondary prevention of macrovascular events in patients with type 2 diabetes in the PROactive Study (PROspective pioglitAzone Clinical Trial In macroVascular Events): a randomised controlled trial. *Lancet* 366(9493):1279-1289.

Ducray, F., Vukusic, S., Gignoux, L. (2004) Mycophenolate mofetil: an open-label study in 42 MS patients. *Mult. Scler.* 10(Suppl.):S263-S263.

Dunn, S. E., Youssef, S., Goldstein, M. J., et al. (2006) Isoprenoids determine Th1/Th2 fate in pathogenic T cells, providing a mechanism of modulation of autoimmunity by atorvastatin. *J. Exp. Med.* 203:401-412.

Dyke, H. J., Montana, J. G. (2002) Update on the therapeutic potential of PDE4 inhibitors. *Expert Opin. Invest. Drugs.* 11(1):1-13.

Edan, G. (2005) Immunosuppressive therapy in multiple sclerosis. *Neurol. Sci.* 26(Suppl.)1:S18.

Edan, G., Miller, D., Clanet, M., et al. (1997) Therapeutic effect of mitoxantrone combined with methylprednisolone in multiple sclerosis: a randomised multicentre study of active disease using MRI and clinical criteria. *J. Neurol. Neurosurg. Psychiatry* 62(2):112-118.

Elices, M. J. (2003) Natalizumab: Elan/Biogen. *Curr. Opin. Invest. Drugs* 4(11):1354-1362.

Engelhardt, B., Ransohoff, R. M. (2005) The ins and outs of T-lymphocyte trafficking to the CNS: anatomical sites and molecular mechanisms. *Trends Immunol.* 26(9):485-495.

Faria, A. M., Weiner, H. L. (2005) Oral tolerance. *Immunol. Rev.* 206:232-259.

Farrell, R., Heaney, D., Giovannoni, G., (2005) Emerging therapies in multiple sclerosis. [Review] *Expert. Opin. Emerg. Drugs* 10(4):797-816.

Feinstein, D. L., Galea, E., Gavrilyuk, V., et al. (2002) Peroxisome proliferator-activated receptor-gamma agonists prevent experimental autoimmune encephalomyelitis. *Ann. Neurol.* 51(6):694-702.

Feng, J., Misu, T., Fujihara, K., et al. (2004) Ibudilast, a nonselective phosphodiesterase inhibitor, regulates Th1/Th2 balance and NKT cell subset in multiple sclerosis. *Mult. Scler.* 10(5):494-498.

Fernandez, O., Guerrero, M., Mayorga, C., et al. (2002) Combination therapy with interferon beta-1b and azathioprine in secondary progressive multiple sclerosis: a two-year pilot study. *J. Neurol.* 249(8):1058-1062.

Filippi, M., Rovaris, M., Rice, G. P., et al. (2000a) The effect of cladribine on T(1) 'black hole' changes in progressive MS. *J. Neurol Sci.* 176(1):42-44.

Filippi, M., Rovaris, M., Iannucci, G., et al. (2000b) Whole brain volume changes in patients with progressive MS treated with cladribine. *Neurology* 12;55(11):1714-1718.

Fontoura, P., Garren, H., Steinman, L. (2005) Antigen-specific therapies in multiple sclerosis: going beyond proteins and peptides. *Int. Rev. Immunol.* 24(5-6):415-446.

Fox, R. J., Ransohoff, R. M. (2004) New directions in MS therapeutics: vehicles of hope. *Trends Immunol.* 25(12):632-636.

Francis, G., Evans, A., Panitch, H. (1997) MRI results of a phase III trial of oral myelin in relapsing-remitting multiple sclerosis [abstract]. *Ann. Neurol.* 42:467.

Frauwirth, K. A., Thompson, C. B. (2002) Activation and inhibition of lymphocytes by costimulation. *J. Clin. Invest.* 109(3):295-299.

Frohman, E. M., Brannon, K., Racke, M. K., Hawker, K. (2004) Mycophenolate mofetil in multiple sclerosis. *Clin. Neuropharmacol.* 27(2):80-83.

Frohman, E. M., Racke, M. K., Raine, C. S. (2006) Multiple sclerosis—the plaque and its pathogenesis. *N. Engl. J. Med.* 354(9):942-955.

Fujimoto, T., Sakoda, S., Fujimura, H., Yanagihara, T. (1999) Ibudilast, a phosphodiesterase inhibitor, ameliorates experimental autoimmune encephalomyelitis in Dark August rats. *J. Neuroimmunol.* 95(1-2):35-42.

Fujino, M., Funeshima, N., Kitazawa, Y., et al. (2003) Amelioration of experimental autoimmune encephalomyelitis in Lewis rats by FTY720 treatment. *J. Pharmacol. Exp. Ther.* 305(1):70-77.

Gaupp, S., Pitt, D., Kuziel, W. A., et al. (2003) Experimental autoimmune encephalomyelitis (EAE) in CCR2(-/-) mice: susceptibility in multiple strains. *Am. J. Pathol.* 162(1):139-150.

Gauthier, S. A., Buckle, G. J., Weiner, H. L. (2005) Immunosuppressive therapy for multiple sclerosis. *Neurol. Clin.* 23(1):247-272, viii-ix.

Genovese, M. C., Becker, J. C., Schiff, M., et al. (2005) Abatacept for rheumatoid athritis refractory to tumor necrosis factor alpha inhibition. *N. Engl. J. Med.* 353(11):1114-1123.

Germain, R. N., Stefanova, I. (1999) The dynamics of T cell receptor signaling: complex orchestration and the key roles of tempo and cooperation. *Annu. Rev. Immunol.* 17:467-522.

Ghalie, R. G., Mauch, E., Edan, G., et al. (2002) A study of therapy-related acute leukaemia after mitoxantrone therapy for multiple sclerosis. *Mult. Scler.* 8(5):441-445.

Giuliani, F., Fu, S. A., Metz, L. M., Yong, V. W. (2005a) Effective combination of minocycline and interferon-beta in a model of multiple sclerosis. *J. Neuroimmunol.* 165(1-2):83-91.

Giuliani, F., Metz, L. M., Wilson, T., et al. (2005b) Additive effect of the combination of glatiramer acetate and minocycline in a model of MS. *J. Neuroimmunol.* 158(1-2):213-221.

Gonsette, R. E. (2004) New immunosuppressants with potential implication in multiple sclerosis. *J. Neurol. Sci.* 223(1):87-93.

Gonsette, R. E., Dubois, B. (2004) Pixantrone (BBR2778): a new immunosuppressant in multiple sclerosis with a low cardiotoxicity. *J. Neurol. Sci.* 223(1):81-86. Erratum in: *J. Neurol. Sci.* 2005;235(1-2):79.

Goodin, D. S., Frohman, E. M., Garmany, G. P., Jr., et al. (2002) Disease modifying therapies in multiple sclerosis: report of the Therapeutics and Technology Assessment Subcommittee of the American Academy of Neurology and the MS Council for Clinical Practice Guidelines. *Neurology* 58(2):169-178.

Goodkin, D. E., Rudick, R. A., VanderBrug Medendorp, S., et al. (1995) Low-dose (7.5 mg) oral methotrexate reduces the rate of progression in chronic progressive multiple sclerosis. *Ann. Neurol.* 37(1):30-40.

Goodkin, D. E., Shulman, M., Winkelhake, J., et al. (2000) A phase I trial of solubilized DR2:MBP84-102 (AG284) in multiple sclerosis. *Neurology* 54(7):1414-1420.

Gordon, E. J., Myers, K. J., Dougherty, J. P., et al. (1995) Both anti-CD IIa (LFA-1) and anti-CD IIb (MAC-1) therapy delay the onset and diminish the severity of experimental autoimmune encephalomyelitis. *J. Neuroimmunol.* 62(2):153-160.

Grillo-Lopez, A. J. (2005). $^{90}$Y-ibritumomab tiuxetan: rationale for patient selection in the treatment of indolent non-Hodgkin's lymphoma. *Semin. Oncol.* 32(Suppl. 1):S44-S49.

Haas, J., Hug, A., Viehover, A., et al. (2005) Reduced suppressive effect of CD4+CD25 high regulatory T cells on the T cell immune response against myelin oligodendrocyte glycoprotein in patients with multiple sclerosis. *Eur. J. Immunol.* 35(11):3343-3352.

Hafler, D. A., Slavik, J. M., Anderson, D. E., et al. (2005) Multiple sclerosis. *Immunol. Rev.* 204:208-231.

Hartung, H. P., Kieseier, B. C. (2000) The role of matrix metalloproteinases in autoimmune damage to the central and peripheral nervous system. *J. Neuroimmunol.* 107(2):140-147.

Hartung, H. P., Gonsette, R., König, N., et al. (2002) Mitoxantrone in progressive multiple sclerosis: a placebo-controlled, double-blind, randomised, multicentre trial. *Lancet* 360(9350):2018-2025.

Hauser, S. L., Hauser, S. L., Dawson, D. M., et al. (1983) Intensive immunosuppression in progressive multiple sclerosis: a randomized, three-arm study of high-dose intravenous cyclophosphamide, plasma exchange, and ACTH. *N. Engl. J. Med.* 308(4):173-180.

He, K, Zeng, L, Shi, G, et al. (1997) Bioactive compounds from Taiwania cryptomerioides. *J. Nat. Prod.* 60(1):38-40.

Hemmer, B., Archelos, J. J., Hartung, H. P. (2002) New concepts in the immunopathogenesis of multiple sclerosis. *Nat. Rev. Neurosci.* 3(4):291-301.

Hohlfeld, R. (1997) Biotechnological agents for the immunotherapy of multiple sclerosis: principles, problems and perspectives. *Brain* 120:865-916.

Hohlfeld, R, Wekerle, H. (2004) Autoimmune concepts of multiple sclerosis as a basis for selective immunotherapy: from pipe dreams to (therapeutic) pipelines. *Proc. Natl. Acad. Sci. U S A* 101(Suppl. 2):14599-14606.

Hohlfeld, R., Wekerle, H. (2005) Drug insight: using monoclonal antibodies to treat multiple sclerosis: Nature Clinical Practice. *Neurology* 1:34-44.

Hohlfeld, R., Wiendl, H. (2001) The ups and downs of multiple sclerosis therapeutics. *Ann. Neurol.* 49:281-284.

Howard, L. M., Miga, A. J., Vanderlugt, C. L., et al. (1999) Mechanisms of immunotherapeutic intervention by anti-CD40L (CD154) antibody in an animal model of multiple sclerosis. *J. Clin. Invest.* 103(2):281-290.

Howard, L. M., Kohm, A. P., Castaneda, C. L., Miller S. D. (2005) Therapeutic blockade of TCR signal transduction and co-stimulation in autoimmune disease. *Curr. Drug Targets Inflamm. Allergy* 4(2):205-216.

Huston, J. S., George, A. J. (2001) Engineered antibodies take center stage. *Hum. Antibodies* 10(3-4):127-142.

Jeffery, D. R. (2004) Use of combination therapy with immunomodulators and immunosuppressants in treating multiple sclerosis. *Neurology.* 63(Suppl. 6):S41-S46.

Jönsson, S., Andersson, G., Fex, T., et al. (2004) Synthesis and biological evaluation of new 1,2-dihydro-4-hydroxy-2-oxo-3-quinolinecarboxamides for treatment of autoimmune disorders: structure-activity relationship. *J. Med. Chem.* 47(8):2075-2088.

Jung, S., Zielasek, J., Kollner, G., et al. (1996) Preventive but not therapeutic application of Rolipram ameliorates experimental autoimmune encephalomyelitis in Lewis rats. *J. Neuroimmunol.* 68(1-2):1-11.

Kappos, L., Patzold, U., Dommasch, D., et al. (1988) Cyclosporine versus azathioprine in the long-term treatment of multiple sclerosis—results of the German multicenter study. *Ann. Neurol.* 23(1):56-63.

Kappos, L., Comi, G., Panitch, H., et al. (2000). Induction of a non-encephalitogenic type 2 T helper-cell autoimmune response in multiple sclerosis after administration of an altered peptide ligand in a placebo-controlled, randomized phase II trial: the Altered Peptide Ligand in Relapsing MS Study Group. *Nat. Med.* 6(10):1176-1182.

Kappos, L., Radu, E. W., Antel, J., et al. (2005a) Promising results with a novel oral immunomodulator-FTY720-in relapsing multiple sclerosis. *Mult. Scler.* 11(Suppl.): S13.

Kappos, L., Barkhof, F., Desmet, A. (2005b) The effect of oral temsirolimus on new magnetic resonance imaging scan lesions, brain atrophy, and the number of relapses in multiple sclerosis: results from a randomised, controlled clinical trial. *J. Neurol.* 252(Suppl. 2):46.

Kappos, L., Antel, J., Comi, G., et al. (2006) Oral fingolimod (FTY720) for releasing multiple sclerosis. *N. Engl. J. Med.* 355(11):1124-1140.

Karussis, D. M., Meiner, Z., Lehmann, D., et al. (1996) Treatment of secondary progressive multiple sclerosis with the immunomodulator linomide: a double-blind, placebo-controlled pilot study with monthly magnetic resonance imaging evaluation. *Neurology* 47(2):341-346.

Kawai, T., Andrews, D., Colvin, R. B., et al. (2000) Thromboembolic complications after treatment with monoclonal antibody against CD40 ligand. *Nat. Med.* 6(2):114.

Keegan, M., Pineda, A. A., McClelland, R. L., et al. (2002) Plasma exchange for severe attacks of CNS demyelination: predictors of response. *Neurology* 58(1):143-146.

Keegan, M., Konig, F., McClelland, R., et al. (2005) Relation between humoral pathological changes in multiple sclerosis and response to therapeutic plasma exchange. *Lancet* 366(9485):579-582.

Keeley, K. A., Rivey, M. P., Allington, D. R. (2005) Natalizumab for the treatment of multiple sclerosis and Crohn's disease. *Ann. Pharmacother.* 39(11):1833-1843.

Keir, M. E., Sharpe, A. H. (2005) The B7/CD28 costimulatory family in autoimmunity. *Immunol. Rev.* 204:128-143.

Keymeulen, B., Vandemeulebroucke, E., Ziegler, A. G., et al. (2005) Insulin needs after CD3-antibody therapy in new-onset type 1 diabetes. *N. Engl. J. Med.* 352(25):2598-2608.

Kieseier, B. C., Kiefer, R., Clements, J. M., et al. (1998) Matrix metalloproteinase-9 and -7 are regulated in experimental autoimmune encephalomyelitis. *Brain* 121 (Pt 1):159-166.

Kieseier, B. C., Seifert, T., Giovannoni, G., Hartung, H. P. (1999) Matrix metalloproteinases in inflammatory demyelination: targets for treatment. *Neurology* 53(1): 20-25.

Khoury, J., Viglietta, V., Buckle, G., et al. (2005) *Neurology* 64(Suppl. 1):A393, S46.004.

Kim, K. S., Jacob, N., Stohl, W. (2003) In vitro and in vivo T cell oligoclonality following chronic stimulation with staphylococcal superantigens. *Clin. Immunol.* 108(3):182-189.

Kleinschmidt-DeMasters, B. K., Tyler, K. L. (2005) Progressive multifocal leukoencephalopathy complicating treatment with natalizumab and interferon beta-1a for multiple sclerosis. *N. Engl. J. Med.* 353(4):369-374.

Klotz, L., Schmidt, M., Giese, T., et al. (2005) Proinflammatory stimulation and pioglitazone treatment regulate peroxisome proliferator-activated receptor gamma levels in peripheral blood mononuclear cells from healthy controls and multiple sclerosis patients. *J. Immunol.* 175(8):4948-4955.

Kobata, T., Azuma, M., Yagita, H., Okumura, K. (2000) Role of costimulatory molecules in autoimmunity. *Rev. Immunogenet.* 2(1):74-80.

Kohm, A. P., Williams, J. S., Bickford, A. L., et al. (2005) Treatment with nonmitogenic anti-CD3 monoclonal antibody induces CD4+ T cell unresponsiveness and functional reversal of established experimental autoimmune encephalomyelitis. *J. Immunol.* 174(8):4525-4534.

Korn, T., Toyka K., Hartung, H. P., Jung, S. (2001) Suppression of experimental autoimmune neuritis by leflunomide. *Brain* 124(Pt 9):1791-1802.

Korn, T., Magnus, T., Toyka, K., Jung, S. (2004) Modulation of effector cell functions in experimental autoimmune encephalomyelitis by leflunomide—mechanisms independent of pyrimidine depletion. *J. Leukoc. Biol.* 76(5):950-960.

Kwak, B., Mulhaupt, F., Myit, S., Mach, F. (2000) Statins as a newly recognized type of immunomodulator. *Nat. Med..* 6(12):1399-1402.

Lang, H. L., Jacobsen, H., Ikemizu, S. (2002) Nat. Immunol. 3:940-943.

Langer-Gould, A., Atlas, S. W., Green, A. J., et al. (2005) Progressive multifocal leukoencephalopathy in a patient treated with natalizumab. *N. Engl. J. Med.* 353(4):375-381.

Larsen, C. P., Knechtle, S. J., Adams, A., et al. (2006) A new look at blockade of T-cell costimulation: a therapeutic strategy for long-term maintenance immunosuppression. *Am. J. Transplant.* 6(5):876-883.

Lassmann, H., Bruck, W., Lucchinetti, C. (2001) Heterogeneity of multiple sclerosis pathogenesis: implications for diagnosis and therapy. *Trends Mol. Med.* 14:259-269.

Lebwohl, M., Tyring, S. K., Hamilton, T. K., et al. (2003) A novel targeted T-cell modulator, efalizumab, for plaque psoriasis. *N. Engl. J. Med.* 349(21):2004-2013.

Lennon, V. A., Kryzer, T. J., Pittock, S. J., et al. (2005) IgG marker of optic-spinal multiple sclerosis binds to the aquaporin-4 water channel. *J. Exp. Med.* 202(4):473-477.

Li, B., Wang, H., Dai, J., et al. (2005) Construction and characterization of a humanized anti-human CD3 monoclonal antibody 12F6 with effective immunoregulation functions. *Immunology* 116(4):487-498.

Likosky, W. H., Fireman, B., Elmore, R., et al. (1991) Intense immunosuppression in chronic progressive multiple sclerosis: the Kaiser study. *J. Neurol. Neurosurg. Psychiatry.* 54(12):1055-1060.

Lindberg, R. L., De Groot, C. J., Montagne, L., et al. (2001) The expression profile of matrix metalloproteinases (MMPs) and their inhibitors (TIMPs) in lesions and normal appearing white matter of multiple sclerosis. *Brain* 124(Pt 9):1743-1753.

Lindsey, J. W., Hodgkinson, S., Mehta, R., et al. (1994a) Repeated treatment with chimeric anti-CD4 antibody in multiple sclerosis. *Ann. Neurol.* 36(2):183-189.

Lindsey, J. W., Hodgkinson, S., Mehta, R., et al. (1994b) Phase 1 clinical trial of chimeric monoclonal anti-CD4 antibody in multiple sclerosis. *Neurology* 44(Pt 1):413-419.

Linsley, P. S., Brady, W., Urnes, M., et al. (1991) CTLA-4 is a second receptor for the B cell activation antigen B7. *J. Exp. Med.* 174(3):561-569.

Llewellyn-Smith, N., Lai, M., Miller, D. H., et al. (1997) Effects of anti-CD4 antibody treatment on lymphocyte subsets and stimulated tumor necrosis factor alpha production: a study of 29 multiple sclerosis patients entered into a clinical trial of cM-T412. *Neurology* 48(4):810-816.

Lovett-Racke, A. E., Bittner, P., Cross, A. H., et al. (1998) Regulation of experimental autoimmune encephalomyelitis with insulin-like growth factor (IGF-1) and IGF-1/IGF-binding protein-3 complex (IGF-1/IGFBP3). *J. Clin. Invest.* 101(8):1797-1804.

Lovett-Racke, A. E., Hussain, R. Z., Northrop, S., et al. (2004) Peroxisome proliferator-activated receptor alpha agonists as therapy for autoimmune disease. *J. Immunol.* 172(9):5790-5798.

Lublin, F., et al. (1999) A phase II trial of anti-CD11/CD18 monoclonal antibody in acute exacerbations of MS [abstract 290]. *Neurology* 52, Suppl 2.

Lucchinetti, C., Bruck, W., Parisi, J., et al. (2000) Heterogeneity of multiple sclerosis lesions: implications for the pathogenesis of demyelination. *Ann. Neurol.* 47(6):707-717.

Lucchinetti, C. F., Mandler, R. N., McGavern, D., et al. (2002) A role for humoral mechanisms in the pathogenesis of Devic's neuromyelitis optica. *Brain* 125(Pt 7): 1450-1461.

Lus, G., Romano, F., Scuotto, A., et al. (2004) Azathioprine and interferon beta(1a) in relapsing-remitting multiple sclerosis patients: increasing efficacy of combined treatment. *Eur. Neurol.* 51(1):15-20.

Maeda, A., Sobel, R. A. (1996) Matrix metalloproteinases in the normal human central nervous system, microglial nodules, and multiple sclerosis lesions. *J. Neuropathol. Exp. Neurol.* 55(3):300-309.

Mandala, S., Hajdo, R., Bergstrom, J., et al. (2002) Alteration of lymphocyte trafficking by sphingosine-1-phosphate receptor agonists. *Science* 296(5566): 346-349.

Marecki, S., Kirkpatrick, P. (2004) Efalizumab. *Nat. Rev. Drug Discov.* 3(6):473-474.

Markovic-Plese, S., Bielekova, B., Kadom, N., et al. (2003) Longitudinal MRI study: the effects of azathioprine in MS patients refractory to interferon beta-1b. *Neurology* 60(11):1849-1851.

Marracci, G. H., McKeon, G. P., Marquardt, W. E., et al. (2004) Alpha lipoic acid inhibits human T-cell migration: implications for multiple sclerosis. *J. Neurosci. Res.* 78(3): 362-370.

Marracci, G. H., Shinto, L., Strehlow, A., et al. (2005) Effect of omega-3 fatty acid supplementation on matrix metalloproteinase-9 production in multiple sclerosis. *Neurology* 64(Suppl. 1):A192-A193, P03.119.

Massacesi, L., Parigi, A., Barilaro, A., et al. (2005) Efficacy of azathioprine on multiple sclerosis new brain lesions evaluated using magnetic resonance imaging. *Arch Neurol.* 62(12):1843-7.

Matloubian, M., Lo, C.G., Cinamon, G., et al. (2004) Lymphocyte egress from thymus and peripheral lymphoid organs is dependent on S1P receptor 1. *Nature.* 427(6972): 355-60.

Maurer, M., Loserth S., Kolb-Maurer A., et al. (2002) A polymorphism in the human cytotoxic T-lymphocyte antigen 4 (CTLA4) gene (exon 1 +49) alters T-cell activation. *Immunogenetics.* 54(1):1-8. Epub 2002 Mar 12.

Meden, H., Mielke S., Marx D., et al. (1997) Hormonal treatment with sex steroids in women is associated with lower p105 serum concentrations. *Anticancer Res.* 17(4B):3075-7.

Menge, T., Hartung, H. P., Stüve, O. (2005) Statins-a cure-all for the brain? *Nat Rev Neurosci.* 6(4):325-31.

Metz, L. M., Zhang, Y., Yeung, M., et al. (2004) Minocycline reduces gadolinium-enhancing magnetic resonance imaging lesions in multiple sclerosis. *Ann Neurol.* 55(5):756.

Miller, D. H., Khan, O. A., Sheremata, W. A., et al. (2003) A controlled trial of natalizumab for relapsing multiple sclerosis. *N Engl J Med.* 348(1):15-23.

Monson, N. L., Cravens, P. D., Frohman, E. M., et al. (2005) Effect of rituximab on the peripheral blood and cerebrospinal fluid B cells in patients with primary progressive multiple sclerosis. *Arch Neurol.* 62(2):258-64.

Morini, M., Roccatagliata, L., Dell'Eva, R., et al. (2004) Alpha-lipoic acid is effective in prevention and treatment of experimental autoimmune encephalomyelitis. *J. Neuroimmunol.* 148(1-2):146-153.

Muller, H., Hofer, S., Kaneider, N., et al. (2005) The immunomodulator FTY720 interferes with effector functions of human monocyte-derived dendritic cells. *Eur. J. Immunol.* 35(2):533-545.

Multiple Sclerosis Study Group. (1990) Efficacy and toxicity of cyclosporine in chronic progressive multiple sclerosis: a randomized, double-blinded, placebo-controlled clinical trial. *Ann. Neurol.* 27(6):591-605.

Nakajima, A., Yamanaka, H., Kamatani, N. (2003) Leflunomide: clinical effectiveness and mechanism of action. *Clin. Calcium* 13(6):771-775.

Neuhaus, O., Strasser-Fuchs, S., Fazekas, F., et al. (2002) Statins as immunomodulators: comparison with interferon-beta 1b in MS. *Neurology* 59(7):990-997.

Nicholson, L. B., Greer, J. M., Sobel, R. A., et al. (1995) An altered peptide ligand mediates immune deviation and prevents autoimmune encephalomyelitis. *Immunity* 3(4): 397-405.

Niino, M., Kikuchi, S., Fukazawa, T., et al. (2001) Genetic polymorphisms of IL-1beta and IL-1 receptor antagonist in association with multiple sclerosis in Japanese patients. *J. Neuroimmunol.* 118(2):295-299.

Noseworthy, J. H., Lucchinetti, C., Rodriguez, M., Weinshenker, B. G. (2000a) Multiple sclerosis. *N. Engl. J. Med.* 343:938-952.

Noseworthy, J. H., Wolinsky, J. S., Lublin, F. D., et al. (2000b) Linomide in relapsing and secondary progressive MS. I. Trial design and clinical results; North American Linomide Investigators. *Neurology* 2000;54(9):1726-1733.

Noseworthy, J. H., Lucchinetti, C., Rodriguez, M., Weinshenker, B. G. (2001) Multiple Sclerosis. *N. Engl. J. Med.* 343(13):938-952.

O'Connor, P. W., Li, D., Freedman, M. S., et al. (2006a) Teriflunomide Multiple Sclerosis Trial Group; University of British Columbia MS/MRI Research Group: a phase II study of the safety and efficacy of teriflunomide in multiple sclerosis with relapses. *Neurology* 66(6):894-900.

O'Connor, P., Antel, J., Comi, G. (2006b) Oral FTY720 in relapsing MS: results of the dose-blinded, active drug extension phase of a phase II study. *Neurology* 66(Suppl. 5):A123.

Omari, K. M., John, G. R., Sealfon, S. C., Raine, C. S. (2005) CXC chemokine receptors on human oligodendrocytes: implications for multiple sclerosis. *Brain* 128(Pt 5): 1003-1015.

Omari, K. M., John, G., Lango, R., Raine, C. S. (2006) Role for CXCR2 and CXCL1 on glia in multiple sclerosis. *Glia* 53(1):24-31.

Paemen, L., Martens, E., Norga, K., et al. (1996) The gelatinase inhibitory activity of tetracyclines and chemically modified tetracycline analogues as measured by a novel microtiter assay for inhibitors. *Biochem. Pharmacol.* 52(1):105-111.

Paintlia, A. S., Paintlia, M. K., Khan, M., et al. (2005) HMG-CoA reductase inhibitor augments survival and differentiation of oligodendrocyte progenitors in animal model of multiple sclerosis. *FASEB* J. 19(11):1407-1421.

Palace, J., Rothwell, P. (1997) New treatments and azathioprine in multiple sclerosis. *Lancet* 350(9073):261.

Panitch, H., Francis, G., Oral Myelin Study Group. (1997) Clinical results of a phase III trial of oral myelin in relapsing-remitting multiple sclerosis [abstract]. *Ann. Neurol.* 42:459.

Paolillo, A., Coles, A. J., Molyneux, P. D., et al. (1999) Quantitative MRI in patients with secondary progressive MS treated with monoclonal antibody Campath 1H. *Neurology* 53(4):751-757.

Patti, F., Amato, M. P., Filippi, M., et al. (2004) A double blind, placebo-controlled, phase II, add-on study of cyclophosphamide (CTX) for 24 months in patients affected by multiple sclerosis on a background therapy with interferon-beta study denomination: CYCLIN. *J. Neurol. Sci.* 223(1):69-71.

Pershadsingh, H. A. (2004) Peroxisome proliferator-activated receptor-gamma: therapeutic target for diseases beyond diabetes: quo vadis? *Expert Opin. Invest. Drugs* 13(3): 215-228.

Pershadsingh, H. A., Heneka, M. T., Saini, R., et al. (2004) Effect of pioglitazone treatment in a patient with secondary multiple sclerosis. *J. Neuroinflamm.* 1(1):3.

Petereit, H. F., Rubbert, A. (2005) Effective suppression of cerebrospinal fluid B cells by rituximab and cyclophosphamide in progressive multiple sclerosis. *Arch. Neurol.* 62(10):1641-1642; author reply 1642.

Polman, C., Barkhof, F., Sandberg-Wollheim, M., et al. (2005) Laquinimod in Relapsing MS Study Group: Treatment with laquinimod reduces development of active I lesions in relapsing MS. *Neurology* 64(6):987-991.

Polman, C. H., O'Connor, P. W., Havrdova, E., et al. (2006) A randomized, placebo-controlled trial of natalizumab for relapsing multiple sclerosis. *N. Engl. J. Med.* 354(9):899-910.

Popovic, N., Schubart, A., Goetz, B. D., et al. (2002) Inhibition of autoimmune encephalomyelitis by a tetracycline. *Ann. Neurol.* 51(2):215-223.

Ramtahal, J., Jacob, A., Das, K., Boggild, M. (2005) 5-Year retrospective study of the use of mitoxantrone and glatiramer acetate combination in Patients with very active relapsing remitting multiple sclerosis. *Neurology* 4(Suppl. 1):A386-387, P06.163.

Ransohoff, R. M. (1999) Mechanisms of inflammation in MS tissue: adhesion molecules and chemokines. *J. Neuroimmunol.* 98(1):57-68.

Ransohoff, R. M. (2005) Natalizumab and PML. *Nat. Neurosci.* 8(10):1275.

Rausch, M., Hiestand, P., Foster, C. A., et al. (2004) Predictability of FTY720 efficacy in experimental autoimmune encephalomyelitis by in vivo macrophage tracking: clinical implications for ultrasmall superparamagnetic iron oxide-enhanced magnetic resonance imaging. *J. Magn. Reson. Imaging* 20(1):16-24.

Reggio, E., Nicoletti, A., Fiorilla, T. (2005) The combination of cyclophosphamide plus interferon beta as rescue therapy could be used to treat relapsing-remitting multiple sclerosis patients: twenty-four months follow-up. *J. Neurol.* 252(10):1255-1261.

Rice, G. P., Filippi M., Comi G. (2000) Cladribine and progressive MS: clinical and MRI outcomes of a multicenter controlled trial: Cladribine MRI Study Group. *Neurology* 14;54(5):1145-1155.

Rice G. P, Hartung, H. P., Calabresi, P. A. (2005) Anti-alpha4 integrin therapy for multiple sclerosis: mechanisms and rationale. *Neurology* 64(8):1336-1342.

Rizvi, S. A., Bashir, K. (2004) Other therapy options and future strategies for treating patients with multiple sclerosis. *Neurology* 63(Suppl. 6):S47-S54.

Romine, J. S., Sipe, J. C., Koziol, J. A., et al. (1999) A double-blind, placebo-controlled, randomized trial of cladribine in relapsing-remitting multiple sclerosis. *Proc. Assoc. Am. Physicians* 111(1):35-44.

Rose, J. W., Watt, H. E., White, A. T., Carlson, N. G. (2004) Treatment of multiple sclerosis with an anti-interleukin-2 receptor monoclonal antibody. *Ann. Neurol.* 56(6): 864-867.

Rosen, H., Goetzl, E.J (2005) Spingosine 1-phosphate and its receptors: an autocrine and paracrine network. *Nat. Rev. Immunol.* 5(7):560-570.

Rudick, R. A., Stuart, W. H., Calabresi, P. A., et al. (2006) Natalizumab plus interferon beta-1a for relapsing multiple sclerosis. *N. Engl. J. Med.* 354(9):911-923.

Rumbach, L., Racadot, E., Armspach, J. P., et al. (1996) Biological assessment and MRI monitoring of the therapeutic efficacy of a monoclonal anti-T CD4 antibody in multiple sclerosis patients. *Mult. Scler.* 1(4):207-212.

Runstrom, A., Leanderson, T., Ohlsson, L., Axelsson, B. (2006) Inhibition of the development of chronic experimental autoimmune encephalomyelitis by laquinimod (ABR-215062) in IFN-beta k.o. and wild type mice. *J. Neuroimmunol.* 173(1-2):69-78.

Sakaguchi, S. (2005) Naturally arising Foxp3-expressing CD25+CD4+ regulatory T cells in immunological tolerance to self and non-self. *Nat. Immunol.* 6(4):345-352.

Salomon, B., Bluestone, J. A. (2001) Complexities of CD28/B7: CTLA-4 costimulatory pathways in autoimmunity and transplantation. *Annu. Rev. Immunol.* 19:225-252.

Scheinfeld, N. (2006) Efalizumab: a review of events reported during clinical trials and side effects. *Expert Opin. Drug Saf.* 5(2):197-209.

Schimrigk, K., Brune, N., Hellwig, K., et al. (2005) An open-label, prospective study of oral fumaric acid therapy for the treatment of relapsing-remitting multiple sclerosis (RRMS). *Neurology.* 64(Suppl. 1):A392, S46.003

Scolding, N. (2005) Devic's disease and autoantibodies. *Lancet Neurol.* 4(3):136-137.

Scott, G. S., Spitsin, S. V., Kean, R. B., et al. (2002) Therapeutic intervention in experimental allergic encephalomyelitis by administration of uric acid precursors. *Proc. Natl. Acad. Sci. U S A.* 99(25):16303-16308.

Sehgal, S. N. (2003) Sirolimus: its discovery, biological properties, and mechanism of action. *Transplant Proc.* 35(Suppl):7S-14S.

Sellebjerg, F., Sorensen, T. L. (2003) Chemokines and matrix metalloproteinase-9 in leukocyte recruitment to the central nervous system. *Brain Res. Bull.* 61(3):347-355.

Serkova, N. J., Christians, U., Benet, L. Z. (2004) Biochemical mechanisms of cyclosporine neurotoxicity. *Mol. Interv.* 4(2):97-107.

Sette, A., Alexander, J., Ruppert, J., et al. (1994) Antigen analogs/MHC complexes as specific T cell receptor antagonists. *Annu. Rev. Immunol.* 12:413-431.

Sharpe, A. H., Freeman, G. J. (2002) The B7-CD28 superfamily. *Nat. Rev. Immunol.* 2(2):116-126.

Sicotte, N. L., Liva, S. M., Klutch, R., et al. (2002) Treatment of multiple sclerosis with the pregnancy hormone estriol. *Ann. Neurol.* 52(4):421-428.

Sicotte, N. L., Giesser, B. S., Tandon, V. (2006) A pilot study of testosterone treatment for men with relapsing remitting multiple sclerosis. *Neurology* 66(Suppl.):A30-A30.

Sidiropoulos, P. I., Boumpas, D. T. (2004) Lessons learned from anti-CD40L treatment in systemic lupus erythematosus patients. *Lupus* 13(5):391-397.

Simmons, D. L., Buckley, C. D. (2005) Some new, and not so new, anti-inflammatory targets. *Curr. Opin. Pharmacol.* 5(4):394-397.

Sipe, J. C. (2005) Cladribine for multiple sclerosis: review and current status. *Expert Rev. Neurother.* 5(6):721-727.

Sipe, J. C., Romine, J. S., Koziol, J. A., et al. (1994) Cladribine in treatment of chronic progressive multiple sclerosis. *Lancet* 344(8914):9-13.

Sipe, J. C., Romine, J. S., Koziol, J., et al. (1997) Cladribine improves relapsing-remitting MS: a double blind placebo controlled study. *Neurology* 48(Suppl. 2):A340.

Sjöö, F., Hassan Z., Abedi-Valugerdi, M., et al. (2006) Myeloablative and immunosuppressive properties of treosulfan in mice. *Exp. Hematol.* 34(1):115-121.

Smilek, D. E., Wraith, D. C., Hodgkinson, S., et al. (1991). A single amino acid change in a myelin basic protein peptide confers the capacity to prevent rather than induce experimental autoimmune encephalomyelitis. *Proc. Natl. Acad. Sci. U S A* 88(21):9633-9637.

Smolen, J. S., Emery, P., Kalden, J. R., et al. (2004) The efficacy of leflunomide monotherapy in rheumatoid arthritis: towards the goals of disease modifying antirheumatic drug therapy. *J. Rheumatol.* Suppl. 71:13-20.

Sommer, N., Loschmann, P. A., Northoff, G. H., et al. (1995) The antidepressant rolipram suppresses cytokine production and prevents autoimmune encephalomyelitis. *Nat. Med.* 1(3):244-248.

Sorensen, T. L., Tani, M., Jensen, J., et al. (1999). Expression of specific chemokines and chemokine receptors in the central nervous system of multiple sclerosis patients. *J. Clin. Invest.* 103(6):807-815.

Sorensen, T. L., Trebst, C., Kivisakk, P., et al. (2002) Multiple sclerosis: a study of CXCL10 and CXCR3 co-localization in the inflamed central nervous system. *J. Neuroimmunol.* 127(1-2):59-68.

Sospedra, M., Martin, R. (2005a) Immunology of multiple sclerosis. *Annu. Rev. Immunol.* 23:683-747.

Sospedra, M., Martin, R. (2005b) Antigen-specific therapies in multiple sclerosis. *Int. Rev. Immunol.* 24(5-6):393-413.

Steinman, L. (2001) Multiple sclerosis: a two-stage disease. *Nat. Immunol.* 2:762-764.

Steinman, L. (2005) Blocking adhesion molecules as therapy for multiple sclerosis: natalizumab. *Nat. Rev. Drug Discov.* 4(6):510-518.

Steinman, L., Zamvil, S. S. (2005) Virtues and pitfalls of EAE for the development of therapies for multiple sclerosis. *Trends Immunol.* 26(11):565-571.

Storer, P. D, Xu, J., Chavis, J., Drew, P. D. (2005). Peroxisome proliferator-activated receptor-gamma agonists inhibit the activation of microglia and astrocytes: implications for multiple sclerosis. *J. Neuroimmunol.* 161(1-2):113-122.

Stuart, R. W., Racke, M. K. (2002) Targeting T cell costimulation in autoimmune disease. *Expert Opin. Ther. Targets.* 6(3):275-289.

Stüve, O., Cepok, S., Elias, B., et al. (2005) Clinical stabilization and effective B-lymphocyte depletion in the cerebrospinal fluid and peripheral blood of a patient with fulminant relapsing-remitting multiple sclerosis. *Arch. Neurol.* 62(10):1620-1623.

Stüve, O., Youssef, S., Weber, M.S., et al. (2006) Immunomodulatory synergy by combination of atorvastatin and glatiramer acetate in treatment of CNS autoimmunity. *J. Clin. Invest.* 116:1037-1044.

Swain, S. M. (1998) Adult multicenter trials using dexrazoxane to protect against cardiac toxicity. *Semin. Oncol.* 25(Suppl. 10):43-47.

Tubridy, N., Behan, P. O., Capildeo, R., et al. (1999) The effect of anti-alpha4 integrin antibody on brain lesion activity in MS; the UK Antegren Study Group. *Neurology* 53(3):466-472.

Ubogu, E. E., Cossoy, M. B., Ransohoff, R. M. (2005) The expression and function of chemokines involved in CNS inflammation. *Trends Pharmacol. Sci.* 27(1):48-55.

Utset, T. O., Auger, J. A., Peace, D., et al. (2002) Modified anti-CD3 therapy in psoriatic arthritis: a phase I/II clinical trial. *J. Rheumatol.* 29(9):1907-1913.

Van Assche, G., Van Ranst, M., Sciot, R., et al. (2005) Progressive multifocal leukoencephalopathy after natalizumab therapy for Crohn's disease. *N. Engl. J. Med.* 353(4): 362-368.

Van Oosten, B. W., Uitdehaag, B. M., Barkhof, F., et al. (1997) Interleukin-2 therapy does not exacerbate multiple sclerosis. *Neurology* 49(2):633-634.

Vermersch, P., Waucquier, N., Bourteel, H., et al. (2004) Treatment of multiple sclerosis with a combination of interferon beta-1a (Avonex) and mycophenolate mofetil (Cellcept): results of a phase II clinical trial. *Neurology* 62(Suppl.):A259-A259.

Viglietta, V., Baecher-Allan, C., Weiner, H. L., Hafler, D. A. (2004) Loss of functional suppression by CD4+CD25+ regulatory T cells in patients with multiple sclerosis. *J. Exp. Med.* 199(7):971-979.

Vincenti, F., Kirkman, R., Light, S., et al. (1998) Interleukin-2-receptor blockade with daclizumab to prevent acute rejection in renal transplantation: Daclizumab Triple Therapy Study Group. *N. Engl. J. Med.* 338(3):161-165.

Vollmer, T., Key, L., Durkalski, V., et al. (2004) Oral simvastatin treatment in relapsing-remitting multiple sclerosis. *Lancet* 363(9421):1607-1608.

Waldmann, H. (1989) Manipulation of T-cell responses with monoclonal antibodies. *Annu. Rev. Immunol.* 7:407-444.

Warren, K. G., Catz, I., Wucherpfennig, K. W. (1997) Tolerance induction to myelin basic protein by intravenous synthetic peptides containing epitope P85 VVHFFKNIVTP96 in chronic progressive multiple sclerosis. *J. Neurol. Sci.* 152(1):31-38.

Webb, M., Tham, C. S., Lin, F. F., et al. (2004) Sphingosine 1-phosphate receptor agonists attenuate relapsing-remitting experimental autoimmune encephalitis in SJL mice. *J. Neuroimmunol.* 153(1-2):108-121.

Weber, M. S., Prod'homme, T., Steinman, L., Zamvil, S. S. (2005) Drug Insight: using statins to treat neuroinflammatory disease. *Nat. Clin. Pract. Neurol.* 1:106-112.

Wei, S. H., Rosen, H., Matheu, M. P., et al. (2005) Shingosine-1-phosphate type 1 receptor agonism inhibits transendothelial migration of medullary T cells to lymphatic sinuses. *Nat. Immunol.* 6(12):1228-1235.

Weiner, H. L., Cohen, J. A. (2002) Treatment of multiple sclerosis with cyclophosphamide: critical review of clinical and immunologic effects. *Mult. Scler.* 8(2): 142-154.

Weiner, H. L., Mackin, G. A., Matsui, M., et al. (1993) Double-blind pilot trial of oral tolerization with myelin antigens in multiple sclerosis. *Science* 259(5099): 1321-1324.

Weinshenker, B. G., Bass, B., Karlik, S., et al. (1991) An open trial of OKT3 in patients with multiple sclerosis. *Neurology* 41(7):1047-1052.

Weissert, R., Wiendl, H., Pfrommer, H., et al. (2003) Action of treosulfan in myelin-oligo-dendrocyte-glycoprotein-induced experimental autoimmune encephalomyelitis and human lymphocytes. *J. Neuroimmunol.* 144(1-2):28-37.

Wiendl, H., Hohlfeld, R. (2002). Therapeutic approaches in multiple sclerosis: lessons from failed and interrupted treatment trials. *Biodrugs* 16(3):183-200.

Wiendl, H., Hohlfeld, R. (2006) Cytokine-based therapeutics for intrinsic CNS disease processes. In Ransohoff R. M., Benveniste E. N. (Eds): Cytokines and the CNS, Taylor & Francis, 2nd Edition, 328-351.

Wiendl, H., Kieseier, B. C., Weissert, R., et al. (2005) Treatment of active secondary progressive multiple sclerosis with treosulfan: an open label pilot study. *Mult. Scler.* 11(Suppl.1):181, P680.

Willenborg, D. O., Fordham, S., Bernard, C. C., et al. (1996) IFN-gamma plays a critical down-regulatory role in the induction and effector phase of myelin oligodendrocyte glycoprotein-induced autoimmune encephalomyelitis. *J. Immunol.* 157(8): 3223-3227.

Wingerchuk, D. M., Weinshenker, B. G. (2005) Neuromyelitis optica. *Curr. Treat. Options Neurol.* 7(3):173-182.

Yadav, V., Marracci, G., Lovera, J., et al. (2005) Lipoic acid in multiple sclerosis: a pilot study. *Mult. Scler.* 11(2):159-165.

Yang, J. S., Xu, L. Y., Xiao, B. G., et al. (2004) Laquinimod (ABR-215062) suppresses the development of experimental autoimmune encephalomyelitis, modulates the Th1/Th2 balance and induces the Th3 cytokine TGF-beta in Lewis rats. *J. Neuroimmunol.* 156 (1-2):3-9.

Yong, V. W. (2005). Metalloproteinases: mediators of pathology and regeneration in the CNS. *Nat. Rev. Neurosci.* 6(12):931-944.

Yong, V. W., Wells, J., Giuliani, F., et al. (2004) The promise of minocycline in neurology. *Lancet Neurol.* 3(12):744-751.

Yopp, A. C., Fu, S., Honig, S. M., et al. (2004) FTY720-enhanced T cell homing is dependent on CCR2, CCR5, CCR7, and CXCR4: evidence for distinct chemokine compartments. *J. Immunol.* 173(2):855-865.

Yousry, T. A., Major, E. O., Ryschkewitsch, C., et al. (2006) Evaluation of patients treated with natalizumab for progressive multifocal leukoencephalopathy. *N. Engl. J. Med.* 354(9):924-933.

Youssef, S., Stuve, O., Patarroyo, J. C., et al. (2002) The HMG-CoA reductase inhibitor, atorvastatin, promotes a Th2 bias and reverses paralysis in central nervous system autoimmune disease. *Nature* 420(6911):78-84.

Yudkin, P. L., Ellison, G. W., Ghezzi, A., et al. (1991) Overview of azathioprine treatment in multiple sclerosis. *Lancet* 338(8774):1051-1055.

Zhao, G. J., Li, D. K., Wolinsky, J. S., et al. (1997) Clinical and magnetic resonance imaging changes correlate in a clinical trial monitoring cyclosporine therapy for multiple sclerosis; the MS Study Group. *J. Neuroimaging* 7(1):1-7.

Zipp, F., Krammer, P. H., Weller, M. (1999) Immune (dys)regulation in multiple sclerosis: role of the CD95-CD95 ligand system. *Immunol. Today* 20:550-554.

Zipp, F., Hartung, H. P., Hillert, J., et al. (2005). Blockade of chemokine receptor in multiple sclerosis patients. *Mult. Scler.* 11(Suppl. 1):S13.

# 15
# T Cell Vaccination in Autoimmune Disease

Sheri M. Skinner, Ying C. Q. Zang, Jian Hong, and Jingwu Z. Zhang

## 1  Introduction

Autoreactive T cells that may appear in the normal course of physiologic events are dealt with in the periphery by a wide variety of regulatory mechanisms, keeping them in check and ensuring the continued maintenance of homeostasis in the immune system. However, with dysregulation of such regulatory mechanisms, a variety of autoimmune pathologies can develop. This dysregulation probably takes the form of combinations of genetic, environmental, and other unknown factors that arise in ways and according to time lines not yet understood. The earliest attempts to use animal models to understand autoimmune mechanisms involved the induction of experimental allergic encephalomyelitis (EAE) in susceptible rodent species. This disease was induced by injecting nervous system tissue or protein and peptides isolated from this tissue emulsified with complete Fruend's adjuvant, resulting in perivascular myelin damage that causes symptoms reminiscent of human multiple sclerosis (MS). Investigators were able finally to determine that T cells were the only cells required for the induction of EAE after such an immunization protocol (Ortiz-Ortiz and Weigle, 1976). At this point, it was realized that regardless of whether antibody was produced it was necessary neither for the initiation nor the induction of the disease. Ben-Nun, Cohen, and colleagues undertook a series of studies that confirmed the requirement for T cells in EAE induction and also established these myelin-reactive cells as sufficient to prevent the disease (Ben-Nun et al., 1981a,b; Ben-Nun and Cohen, 1982; Holoshitz et al., 1983).

Initially, the antigenic determinants operative in causing EAE were unknown. However, the T cell lines grown and used by these investigators in the prevention of EAE in susceptible animals were found to be specific to myelin basic protein (MBP), and work proceeded to isolate the peptide determinants in MBP that were responsible. Several such residues arose as possibilities, and further animal studies implicated other myelin-related proteins as candidate antigens. It was found that EAE could be induced in experimental animals by inoculation with proteolipid protein (PLP) (Lees, 1982; Yoshimura et al., 1985). In addition, mouse monoclonal antibodies to myelin oligodendrocyte glycoprotein (MOG) could accelerate MBP-induced EAE and could induce a similar condition (Lassman et al., 1988; Linington and Lassmann, 1987; Linington et al., 1988; Schluesener et al., 1987). These animal studies laid the groundwork for later human studies establishing these proteins as potential autoimmune targets in human MS and spurred work on determining the immunodominant residues in these proteins.

The importance of immunoreactive T cells in both progression and prevention of EAE in animal models led logically to the notion that suppressing the depletion of autoreactive T cells should be possible in human cases of MS if these autoimmune T cells are indeed involved in the pathogenesis of MS. In so doing, it was hoped that some clinical benefits would follow. Thus began more than 10 years of human studies in which a variety of protocols were attempted until consensus was obtained with regard to a few basic strategies. T cell vaccination (TCV) came to be understood as the use of inactivated autoreactive T cells as an immunogen to stimulate the immune system and thus set into motion a network of regulatory mechanisms potentially contributing to the depletion or suppression of subsets of circulating autoreactive effector T cells and potential clinical improvement. In researching the molecular results of such immunization practices, it has been possible to uncover some of the bases of in vivo immune regulatory mechanisms. Human trials extended into areas of other autoimmune disorders, such as rheumatoid arthritis (RA), and protocols continued to be refined. While these were being worked out, it was heartening to discover a continuous record of an excellent safety profile. Additionally, some clinical improvement was noted—enough to encourage continued human trials of TCV. However, the real proof of treatment efficacy lies in larger, controlled studies, several of which are either in planning stages or are ongoing. As these studies add to our assurance of safety and effectiveness of TCV as a potential therapeutic approach, we continue to add also to our understanding of the immunologic bases of immune regulation. Such increased understanding can then lead to more effective therapeutic strategies for treating a variety of autoimmune pathologies and to additional lines of inquiry about the in vivo immune mechanisms that keep us healthy.

Details of the vaccination approaches and of the inroads made in understanding the immune dysfunctions involved are discussed in the chapters to follow. Here, we briefly outline the history, recent progress, and future directions of TCV research with an emphasis on the recent advances seen in the areas of MS and RA.

## 2    Early TCV in Autoimmune Animal Models

The use of T cells specifically focused against a particular pathogen to direct the immune system in an in vivo attack against similarly armed T cells emerged from the early work of Cohen and colleagues. They studied the in vivo function of autoreactive T cell clones/lines specific for the MBP antigen and found not only that these cells were capable of inducing the disease but that use of these same T cells in the context of a vaccine could prevent EAE (Ben-Nun et al., 1981a,b; Naparstek et al., 1983; Zamvil et al., 1985). This T cell vaccination technique was attempted successfully in a number of other rodent autoimmune disease models (Ben-Yeduda et al., 1996; Beraud et al., 1992; Elias et al., 1990; Holoshitz et al., 1983; Kakimoto et al., 1988; Maron et al., 1983). Additionally, Madsen and colleagues (1999) utilized transgenic mice, "humanized" in their expression of particular T cell responses to MBP, to demonstrate physiologic responses that were histologically identical to those seen in human MS patients. These animal experiments set the stage for research into human autoimmune conditions.

Furthermore, it was demonstrated, using the EAE model, that anti-idiotypic T cell responses induced by TCV involve the specific recognition of a target T cell receptor (TCR) (Elias et al., 1999; Lider et al., 1988). These responses involve an internal image formed through recognition of the hypervariable determinants of specific antibodies or T cells in order to regulate immune responses to both foreign and self antigens. Anti-idiotypic T cells have been recognized as part of the normal T cell population and are present in the healthy immune system (Cohen, 1989). Evidence suggests strongly that it is the TCR of the vaccinating T cells that forms the target for such regulation. Animal studies suggest that it is this molecular structure that is identified when anti-idiotypic T cells are presented with a target choice between immunizing T cell clones and T cells expressing differing TCR determinants (Saruhan-Direskeneli et al., 1993). Moreover, sequence diversity in the CDR3 and CDR2 regions make these the likely target residues of anti-idiotypic regulatory processes. Here also, it became established that autoreactive T cells used in TCV must be both immunologically activated and artificially attenuated (by irradiation or chemical treatment) to render them unable to reproduce in vivo. This strategy became a fundamental tenet of TCV in all future studies.

In addition to the responses aimed at recognition of idiotypic determinants of the TCR, TCV can induce another subset of regulatory T cells that recognize cellular markers other than idiotypes. Elegant studies by Lohse and Cohen led to the description of cells that were originally dubbed "antiergotypic" (Lohse et al., 1989). They were thought to respond not to the TCR in the manner described above but, rather, to cell surface activation markers (ergotopes). However, for many years, the true molecular identity of these ergotopes remained a puzzle. Only recently has new evidence emerged that these ergotopes may represent T cell activation markers, such as interleukin-2 (IL-2) receptor and self heat-shock proteins as described below, which rounded out the understanding of these mechanisms.

In addition to the animal studies involving autoimmune demyelinating conditions, several such studies utilized animal models of RA to test TCV approaches. RA models are considered to have good reproducibility of data and have much in common with the human disease. Some differences that must be taken into consideration were reviewed by Bendele (2001) and include (1) the tendency of animal models to progress at a more rapid rate than the human disease, (2) their primary characteristic is acute inflammatory responses, and (3) the tendency for rodents to show marked bone resorption and bone formation. Testing aimed at uncovering human disease responses to treatment are rat adjuvant arthritis, rat type II collagen arthritis (with lesions more analogous to those seen in humans), and antigen-induced arthritis (which can be induced in a number of species).

TCV was applied specifically to RA animal models by the following investigators. The rat collagen type II model was used by Holoshitz and colleagues (1983) to illustrate that a T cell line responsive to this antigen could be used to induce the arthritic condition or could serve as a useful vaccine against it. An effector T cell clone that caused adjuvant arthritis in Lewis rats proved useful as a vaccinogen merely with treatment aimed at disruption of the cellular cytoskeleton. No isolation of autoimmune lines or clones was necessary (Lider et al., 1987). The mouse collagen type II arthritis model was used in experiments by Kakimoto and colleagues (1988) and showed the ability of a mouse T cell line reactive to human type II collagen to induce, as well as protect from, clinical arthritis, suggesting involvement of an anti-TCR response in the vaccination mechanism. Mor and colleagues (1990) showed that expansion of autoimmune T cells either specifically by use of antigen or nonspecifically by mitogen could be used as a vaccinogen to prevent or ameliorate disease in syngeneic animals. Anderton and colleagues (1995) gathered evidence showing that vaccination with a T cell line specific to an epitope of mammalian heat shock protein (HSP) protected against adjuvant arthritis, thus showing that cross-reactivity between bacterial and self HSP might be used to maintain a self-reactive T cell population.

# 3   Human Studies in TCV

## 3.1   Initial Human Trial

Research aimed at use of TCV in humans began in 1991 when approaches utilizing autologous CD4+ T cells derived from cerebrospinal fluid (CSF) of MS patients resulted in short-term immunosuppression with no adverse effects (Hafler et al., 1992). In a pilot study, four patients with chronic progressive MS were chosen based on the desire to arrest or slow the disabling progression of this form of the disease. Patients were given one, two, or three subcutaneous inoculations and were monitored over the next year. Five untreated secondary progressive MS patients were studied as controls. Autologous CSF cells were cloned by a nonspecific T cell stimulus such as phytohemagglutinin (PHA) for use as an immunogen. Given that there was no information at that time defining immunodominant regions of any of the myelin antigens, T cell clones were grown and chosen based simply of their CD4+ nature, their ease of expansion in vitro, and the presence of common TCR gene rearrangements, which indicate a dominant T cell clonotype. PHA stimulation was used, and formaldehyde attenuation of the clones was chosen.

Although this study was too small to prove either safety or effectiveness, no side effects were seen. T cell choice, expansion, and attenuation procedures proved feasible for expansion of TCV to larger groups of patients. Unlike results from animal studies, there was no indication of specific anticlonotypic T cell response in this human trial as gauged by proliferation experiments. Partial, short-term immunosuppression was shown by a decrease of subsequent responses to stimulation via the CD2 pathway and by increases in the autologous mixed lymphocyte response.

## 3.2   Central Human Study Series

Research in MS during the early 1990s had discovered the immunodominant epitopes of MBP and other candidate myelin autoantigens that were implicated in the pathogenesis of MS (Chou et al., 1992; Martin et al., 1991, 1992, 1994; Ota et al., 1990; Pette et al., 1990; Saruhan-Direskeneli et al., 1993; Zhang et al., 1990, 1994). It became obvious that a better protocol was needed and was possible with the increasing knowledge about myelin-reactive T cells in MS, which led to the planning and execution of a more advanced human TCV trial for MS. In addition to the questions suggested by previous studies, several points needed clarifying.

1. Can the downregulation of circulating myelin-responsive T cells achieved by TCV be targeted to an especially relevant population?
2. Is it possible to confirm a clonotypic response to TCV equivalent to that seen in animal studies?
3. Can the immunosuppressive response to TCV be achieved long term in a manner technically feasible for the MS patient and patient care staff?

### 3.2.1 Establishing Fundamental Vaccine Protocols and Parameters

In a limited Phase I trial, Zhang and colleagues chose eight MS patients to participate. Five suffered with remitting/relapsing (RR)-MS, one with primary progressive (PP)-MS, and two with secondary progressive (SP)-MS. They were matched with eight untreated MS control patients. Each of the vaccinated MS patients received three subcutaneous injections of two to four vaccine clones ($15 \times 10^6$ cells/clone) at intervals of 2 to 4 months. All were monitored over 2 to 3 years for changes in a variety of clinical parameters to determine the extent and longevity of symptom improvement. Furthermore, extensive immunologic studies were done to explore the nature and extent of the impact of TCV on the immune regulatory networks. One of the goals of this study was to, for the first time, specifically target populations of autoreactive T cells based on their anticipated myelin antigen specificity. The feasibility of this approach was supported by previous work (Zhang et al., 1994), which indicated that MBP-reactive T cells isolated from the CSF of MS patients possessed the same peptide reactivity to the two immunodominant epitopes as did the T cell clones derived from the peripheral circulation. Therefore, selected MBP-reactive T cell clones were derived from the blood of each patient to be vaccinated. The clones were selected based on their reactivity in vitro to two immunodominant regions of MBP (peptides 84-102 and 143-168). Selected vaccine clones were activated with autologous MBP-pulsed antigen-presenting cells (APCs) in vitro and irradiated to render them incapable of proliferation. Each inoculation consisted of a pool of two or three such clones.

Clinical results were gathering in more detail now. Observations of the vaccine recipients were carried out for 2 to 3 years to capture data useful for predicting a long-term response to this TCV protocol. As seen previously, the vaccine was well tolerated with minimal adverse events, all of which were limited to injection site irritation. Clinical parameters monitored included the exacerbation rate, Expanded Disability Status Score (EDSS), and magnetic resonance imaging (MRI)-determined changes in brain lesions. Paired controls matched for age, sex, clinical characteristics, and disease duration were similarly monitored. The longitudinal evaluation suggested a moderately lower level of clinical exacerbation, EDSS, and brain lesions than that seen in the matched controls. Exacerbations of the group as a whole dropped from a level of 16 during the 2 years prior to treatment to a level of 3 during the same period after vaccination. A slower evolution in the clinical course and a net reduction in the final EDSS in some of the patients contrasted with the expected natural progressive increase for unvaccinated patients over the same time period. A semiquantitative analysis of the MRI scans showed a 3.5% lesion increase compared with 39.5% for the unvaccinated cohort. However, no clinical conclusion could be drawn from this limited Phase I clinical trial.

Immunologic responses provided a basis for future attempts at vaccination. For all patients, this protocol resulted in T cell responses that corresponded reciprocally with a progressive decrease in the frequency of circulating MBP-reactive

T cells. With each vaccination, a boosting effect was seen in the decline of circulating MBP-reactive cells. In most patients this decline dropped below the detectable limit. When circulating responding T cells were isolated from patients and tested for specificity of action against the immunizing vaccine T cells, specificity was found in both CD4+ and CD8+ cells. These could then be termed anticlonotypic for their specific reactivity against a clonotypic structure of the vaccine inoculates. For the cytolytic anticlonotypic response (the major one in this protocol), the frequency increased approximately 10-fold after the second and third vaccination (Zhang et al., 1993, 1995). No attempt was made to identify antiergotypic T cells in this study. Later studies of these patients determined that some experienced declining immunity 1 to 2 years after vaccination. This was evident in decreased anti-idiotypic T cell responses and in a loss of gained clinical benefit (Hermans et al., 2000). In a little more than half of the vaccinated patients 5 years after TCV, the MBP-reactive T cell clones isolated had a different clonal origin from those present before vaccination. These clones were seen to have similar effector functions and could be depleted by additional vaccine boosters with new clone preparations.

This groundbreaking study determined the parameters and the protocol that was effective in depleting autoreactive T cells in the MS patient. After many years of TCV investigation, the guidelines established here remain the ones universally recognized as fundamental to the practice of TCV in MS and have led to effective depletion or suppression of myelin-reactive T cells in MS patients (Achiron et al., 2004; Correale et al., 2000). With the fundamental aspects of the TCV protocol elucidated, two major questions remained: Are reactions important in the TCV response other than the CD8+ anti-idiotypic T cell response illustrated here? Also, is there is a clinical effect on the course of MS? Other questions suggest themselves as well: What is the specific target molecule for these anticlonotypic T cells? Are antiergotypic T cell populations also active in this model?

### 3.2.2   Attempts to Further Determine Clinical Efficacy

The promising results of the foregoing clinical trials indicated, but could not prove, that the consistently demonstrated depletion of circulating myelin-reactive T cells produced clinical improvement in vaccinated patients. Hence, Zhang and colleagues found it necessary to conduct a larger study aimed at better determining such a correlation. The excellent safety profile and the potential clinical benefit of TCV as previously performed presaged continued success with this procedure. A group of 54 MS patients (28 RR-MS, 26 SP-MS) were recruited for this extended Phase I/II open label study. Patients were monitored for immunologic and clinical changes over a period of 24 months, and the results were compared with values collected from prevaccination tests and samples in a self-paired manner. Over the following years, collected samples were submitted to a variety of immunologic tests aimed at exploring the specific effects induced by the TCV procedure. The vaccine was produced in the same manner as described for the pilot studies (Zhang et al., 1993, 1995). Patients received three subcutaneous injections of irradiated autologous MBP-reactive T cell clones ranging from 30 to $60 \times 10^6$ cells per injection (two to four T cell clones per injection) at 2-month intervals.

As indicated in previous studies, this trial confirmed that TCV, in the form of vaccination with autologous MBP-reactive T cells, remains a consistent and powerful procedure for depleting circulating MBP-reactive T cells, an effect that correlates favorably with improved clinical variables. These variables include a prolonged time to progression in both RR-MS and SP-MS as compared with the natural history observed for these conditions. However, it was noted that a trend for a lost clinical stability appeared 12 months after the last injection. It is possible that this may be associated with a gradual decline in immunity or with the development of new and different myelin-reactive T cell populations, as suspected in previous studies. It is possible that some patients experience a clonal shift, or "epitope spreading," in their MBP-reactive T cell repertoire (Tuohy et al., 1999). In this study, some clinical improvement was observed in terms of reduced EDSS, rate of relapse, and MRI lesion burden (Zhang et al., 2002). Again, no clinical conclusion could be drawn from this uncontrolled clinical trial. The observed clinical improvement in this clinical trial setting could be subject to trial bias. Thus, controlled clinical trials with large groups of patients are needed to prove clinical efficacy.

Immunological studies done immediately upon availability of study samples confirmed the depletion of circulating MBP-reactive T cells in vaccinated MS patients. However, much more remained to be done to unveil immune regulatory mechanisms induced by TCV (Fig. 15.1). Also, increased understanding of these mechanisms is necessary for the improved choices of relevant T cell populations to be included in future vaccine elaborations. Consequently, continued studies of these patient samples were instituted over the ensuing years to gain a deeper understanding of the TCV process.

There had been several attempts to detect anti-idiotypic antibodies induced by TCV in serum specimens, but they failed. In one study (Hong et al., 2000), Zhang and colleagues found that although such antibodies could not be detected in sera of vaccinated patients B cells producing such antibodies could be identified at an increased precursor frequency using a more sensitive approach and did indeed produce such antibodies. They were seen to have an inhibitory effect on those immunizing MBP-reactive T cells expressing the CDR3 sequence and had specific regulatory properties, as judged by their inhibition of the proliferation of the vaccinating T cells (Fig. 15.1).

FIGURE 15.1. Regulatory mechanisms induced by T cell vaccination. Autoimmune T cells are known to carry idiotopes characteristic of unique clonotypic sequences in the T cell receptors and ergotopes, a loosely connected class of molecules, comprised of T cell activation markers (e.g., IL-2 receptors). Immunization with selected autologous autoimmune T cells induces the host immune system to develop anti-idiotypic T cell responses involving mainly cytotoxic T cells of the CD8 phenotype and anti-ergotypic T cell responses involving a heterogeneous population of CD4+ regulatory T cells that recognize ergotopes. At least two types of CD4+ anti-ergotypic T cells are identified. One is expanded from a

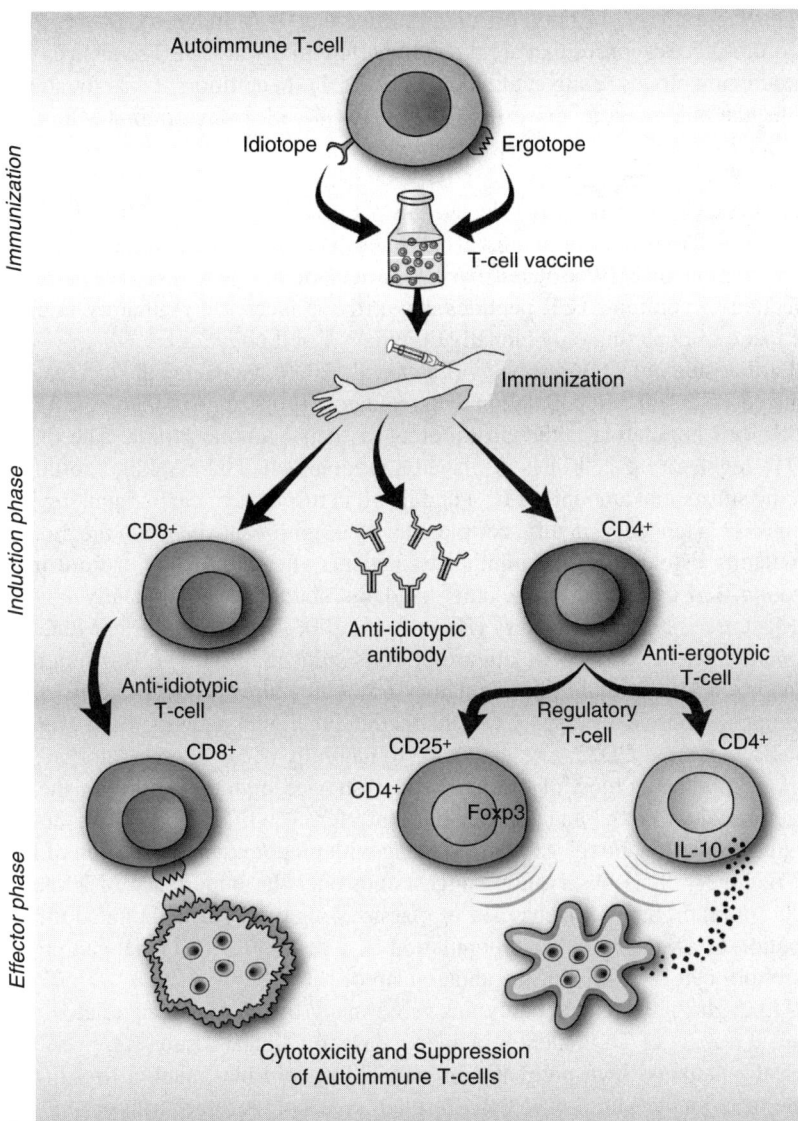

FIGURE 15.1. (*continued*) CD4+CD25+ regulatory T cell pool that expresses high levels of transcription factor Foxp3. The other represents a subset of CD4+ T cells characterized by high production of IL-10. In some studies, anti-idiotypic antibody is found. These antibodies are directed at the T cell receptors of target T cells and can functionally inhibit activation of autoimmune T cells by blocking T cell recognition. These immune regulatory mechanisms identified so far are thought to act in concert and contribute to depletion and suppression of autoimmune T cells selected for immunization and other immunologic effects induced by T cell vaccination.

Our laboratory confirmed for the first time that TCV induces CD8+ cytotoxic anti-idiotypic responses that are directed preferentially at CDR3 sequences of the immunizing clones (Zang et al. 2000) (Fig. 15.1). In addition, it was found that a molecular commonality exists in MBP-reactive T cells from different MS patients that may some day allow development of a more relevant population of vaccinating T cells. There is a common CDR3 sequence motif in peripheral T cells of more than 31% of randomly selected MS patients (Hong et al., 2000). This motif (LGRAGLTY) is present in most T cell lines that recognize the immunodominant 83-99 region of MBP isolated from a variety of MS patients. The purposeful selection of common TCR peptides may provide increased regulatory impact to the tactic of depleting only clonally dominant T cell populations.

Further study of TCV responses (Zang et al., 2000) uncovered the fact that TCV induced a CD4+ T cell population with regulatory properties similar to the induced CD8+ cell population, although different in their cytokine profile. The induced CD4+ regulatory T cell lines displayed a predominant Th2 cytokine profile, producing substantial amounts of IL-4 and IL-10 in response to the immunizing T cell clones. This tendency, in turn, correlated with a significant change in the profile of circulating cytokines in the immunized patients after the second or third immunization. Th1 cytokines, on the other hand, did not change significantly.

Most recently, our laboratory (Hong et al., 2006) uncovered further functional properties of the CD4+ population of T cells induced by TCV. It was found that, in addition to displaying the regulatory properties already identified for such a population, these cells display two distinctive functional patterns. Both groups of T cells expressed CD25, a resemblance to naturally occurring regulatory T cells. However, whereas most of these cells also showed high expression of the transcription factor Foxp3 and secreted both interferon-$\gamma$ (IFN$\gamma$) and IL-10, the other group showed low Foxp3 expression along with predominant production of IL-10 but not IFN$\gamma$. This observation suggests not only the importance of a vaccine-induced antiergotypic mechanism of classic T regulatory cells but also the contribution of a vaccine-induced population of T cells (Fig. 15.1) that can produce inhibitory cytokines free of the antagonism of IFN$\gamma$.

Although this extended study answered many of the questions related to the clinical impact of TCV, there remained a caveat: Clinical results were compared with the patients' own pretreatment status. To avoid bias, such a trial must be done in a double-blind, placebo-controlled protocol. A larger trial on this basis can also answer additional questions: What are the molecular correlates of the eventual accelerated progression observed to occur approximately a year after the last immunization? In what percentage of vaccinated MS patients can one expect a clonal shift to take place, and can this tendency be predicted in the individual patient? What is the optimal timing for booster injections to maintain adequate immunity, and for how long must they continue? Are there better choices of antigenic epitopes that would result in a more potent vaccine? What is the best source (CSF or circulation) of relevant T cells for vaccination purposes? What is the best expansion procedure for vaccination T cells? Why the selective activation of CD8+ cytotoxic anti-idiotypic T cells in TCV? How do anti-idiotypic mechanisms combine with antiergotypic mechanisms in the effect of TCV, and how can these factors be utilized to better advantage in improved TCV formulations?

### 3.2.3    Ongoing Human MS Trial

To answer these questions, the Zhang vaccine is currently being taken into further clinical trials by Loftus and colleagues in Houston, TX. An initial dose escalation study seems to indicate that all dosages are safe and well tolerated and that even the low doses are effective immunologically. Hence, a Phase II, multicenter, randomized, double-blind, placebo-controlled, year-long study was planned to begin in 2006, enrolling150 participants with early RR-MS. It plans to evaluate the efficacy, safety, and tolerability of a new TCV formulation and will evaluate biomarkers of efficacy and effects on epitope spreading.

## 3.3    TCV Studies by Other Investigators

Concurrent with the studies ongoing in the Zhang laboratory, a number of investigators took a variety of independent approaches to TCV in MS.

### 3.3.1    Hermans and Colleagues: T Cell Responses to TCV

This study sought to understand something of the relative contributions of various T cell subsets in response to TCV in MS. Cellular responses of interest here included establishing the cytokine profile of responding circulating T cells and investigation of antibody responses to their vaccination procedure. A group of 49 MS patients of undetermined disease subtype were recruited for the study and were treated with the TCV as described above. Three subcutaneous immunizations at 2-month intervals were performed, and the patients were monitored for a 2-year period after vaccination. A control group of untreated MS patients was recruited. The MBP-reactive T cells isolated from the patients' blood were cloned and analyzed for their TCR V-(D)-J rearrangements and CDR3 sequence and were expanded by alternate rounds of stimulation with MBP pulsed autologous peripheral blood monocytic cells (PBMCs) containing APCs or the irradiated feeders PHA and rIL-2. Again, irradiation was used for inactivation. Although limited clinical results were available, it could be seen that no safety problems were noted. Antivaccine responses were present up to 2 years after the third vaccination, but it is still unknown whether these responses are vaccine-specific. Phenotypically, the profile was mixed, consisting of CD4+ and CD8+ cells as well as a CD4−CD8− population. Most of these were TCR$\alpha\beta$+. Isolation experiments indicated that CD4+ T cells played a large part in both the proliferative response and cytokine production of the antivaccine lines. Analysis of the cytokine profile of responding T cells indicated that stimulation with vaccine clones gave rise predominantly to a Th1-type cytokine response. Finally, antibody responses were confirmed to be much less important in this TCV protocol than would have been predicted from animal studies. Therefore, this study showed that TCR $\alpha\beta$+CD8+ T cells incorporate the most important direct anti-idiotypic effects, whereas CD4+ T cells become the predominant cytokine producers after vaccination (Hermans et al., 2000).

### 3.3.2    Correale and Colleagues: Widening the Antigenic Target

The true autoantigens related to MS remained a mystery. Investigators began to suspect a variety of proteins present in the myelin sheath and wondered which, if any, might stimulate the initial autoimmune response. The reappearance of MBP responses in formerly vaccinated patients during clinical exacerbations drew the attention of many investigators; and the differing epitopes associated with these reappearing MBP-responsive T cell populations suggested epitope spreading and intimated that only a fraction of reactive T cell populations had been deleted in previous TCV protocols. Correale's group therefore took an alternative approach, using autologous T cell lines directed against a broad array of epitopes present in their whole bovine myelin antigen, aiming to affect the maximum number of myelin antigens possible with a single vaccine (Correale et al., 2000). Four patients with SP-MS were recruited for this pilot study and were studied over a 3-year period. They were vaccinated at 3-month intervals over 24 months prior to an additional 12-month follow up. Autologous T cell lines were developed from patient blood against whole bovine myelin. Cells were selected, expanded, and restimulated using autologous APCs with myelin proteins. Each vaccination provided $40 \times 10^6$ cells given subcutaneously.

Clinically, two of the patients showed a stable EDSS score over time, one showed temporary mild improvement, and one continued to progress. No significant side effects were noted. Immunologically, circulating T cells reacting against bovine myelin and against immunodominant regions of human MBP, MOG, and PLP declined over time after vaccination, suggesting that a pan-antigen effect can be achieved with such a protocol. There was also a decline in bovine myelin-specific IL-2- and IFNγ-secreting T cells. A variety of T cell lines were isolated from patients' blood and were characterized with respect to cytolytic activity and cytokine secretion patterns. Human myelin peptide-specific lysis activity was seen, and secretion of IFNγ and tumor necrosis factor (TNF)α/β predominated. Once again, relevant serum antibodies were not detected. It was further determined that the cytolytic mechanism was mediated in part in a Fas-independent manner with an IL-10 immunosuppressive response appearing after vaccination.

### 3.3.3    Van der Aaa and Colleagues: CSF-Derived Vaccinogen

If a number of myelin proteins are found to be viable autoimmune targets, how may one produce a T cell vaccine containing a sufficient variety of activated T cell clones for optimal coverage of the important antigens? Circulating T cells may not provide sufficient coverage, so Van der Aaa and colleagues identified a second likely source of patient-specific activated T cells. They observed that animal studies and earlier human studies had suggested that, in addition to MBP, myelin proteins such as MOG and PLP may contribute to the disease process. Recognizing the difficulty of generating a sufficient variety of specific T cell clones from circulating T cells and noting the higher frequency of activated myelin-reactive T cells present in MS CSF, these investigators initiated a pilot clinical trial in which the latter condition was utilized for a specialized vaccine elaboration (Van der Aaa et al., 2003). Van der Aaa and colleagues utilized earlier optimized expansion T cell techniques to produce a vaccine containing CSF-derived CD4+ activated T cells characterized as being predominantly of the Th1/Th0 TCRαβ+CD4+ phenotype and restricted clonality. Four RR-MS and one CP-MS patients were recruited for this study. Patients were given three immunizations of 10 million cells per inoculation at an interval of 2 months. Although the small size of the trial prevented proof of efficacy, all patients were seen to remain clinically stable or showed reduced EDSS with no relapses during or after the treatment. As with all other TCV studies, no significant adverse events were noted. Immunogenicity of the vaccine used in this study was found to rely on both anticlonotypic and antiergotypic responses (Van der Aaa et al., 2003). They found a significant reduction in circulating MBP/MOG/PLP-reactive T cells after vaccination in two of the five patients, with the myelin-reactive T cells in the other three patients retaining their initial low level or becoming further reduced.

### 3.3.4  Achiron and Colleagues: Tweaking the Protocol to Help Nonresponders

While established TCV parameters were proving themselves in providing clinical relief to most MS patients taking part in TCV clinical trials, a far greater number of MS patients had spent years being treated with conventional pharmacologic drug regimens. Although drug treatment helps a great many in postponing or easing accumulating symptoms of MS, even these established treatment regimens fail in a number of RR-MS patients. For these individuals, defined as "nonresponders," additional interventions are needed to halt ongoing deterioration. To this end, Achiron and colleagues attempted to rescue this patient group with cells aimed at particular peptide targets (Achiron et al., 2004). Altogether, 20 RR-MS patients, shown to be nonresponders to conventional immunomodulatory therapies, were recruited for the study. PBMCs isolated from patient blood samples were cultured and stimulated with immunodominant peptide sequences of MBP and MOG. Cell lines with more than 60% CD4+ content were used in the vaccine at a level of up to $1.5 \times 10^7$ cells per peptide and an overall cell count not exceeding $6 \times 10^7$. Patients were given three subcutaneous injections of irradiated T cell vaccine preparations at 6- to 8-week intervals. Follow-up was done over a 1-year period. The results were compared to the patients' pretreatment status.

Once again, no serious adverse events were noted. MRI analysis demonstrated significant reduction in active brain lesions. There was also a reduction in relapse rate and a decreased rate of progression to further disability. Patient response was tested to five immunodominant myelin peptides. Almost half of the patients responded to all five peptides, 30% responded to four, 20% to three, and 5% responded to two; 90% or more responded to MBP 87-110 and to MOG 34-56. However, there were no correlations found between the clinical responses and any of the specific epitopes tested.

### 3.3.5   Ongoing Clinical Trials

*Medaer and Colleagues: Emphasis on the Brain.* In September 2000, a group of 30 RR-MS couples were recruited into a double-blind, placebo-controlled study by Medaer and colleagues aimed at controlling lesion development. Patients received three subcutaneous inoculations of $50 \times 10^6$ myelin-specific, attenuated, CSF-derived T cells. Patients would be followed for 18 months.

*Karussis and Colleagues: Disease Progression.* In the spring of 2002, Karussis and colleagues began enrollment into a double-blind study in which 30 RR-MS patients were inoculated with myelin-attacking T cells and observed over a period of 1 year. The aim was to control disease progression and lesion development.

*Achiron and Colleagues: Early Intervention.* In July 2002, Achiron and colleagues initiated a Phase III, double-blind, placebo-controlled study. A total of 76 patients exhibiting an acute onset of neurologic symptoms suggestive of a first clinical demyelinating event were enrolled. Half of the patients received three subcutaneous injections of attenuated T cell lines reactive against immunodominant MBP, PLP, and MOG peptides; the other half received placebo injections. Inoculations were given once every 6 to 8 weeks. Boosters were given to those who did not convert to definite MS within the 48-week follow-up period. The study assessed the value of early TCV at disease onset. It was hoped that the timing of TCV would stop the process of epitope spreading associated with exposure to myelin antigens. In so doing, a second attack might be prevented, and the inflammatory process with its ongoing recognition of new self-determinants might be slowed or stopped. Patients were followed using clinical, immunologic, and MRI methods.

## 4   Application of TCV to Other Autoimmune Diseases

T cells chosen as immunogen candidates for TCV in MS are chosen chiefly on the basis of their antigen specificity. However, useful as this is expected to be for treatment of autoimmune disease, this antigen-based approach is not yet possible for TCV treatment of other autoimmune conditions such as RA because the identities of autoantigens for RA are as yet unknown. Hence, the elaboration of TCV vaccine must depend instead on identification of pathogenic T cells on the basis of TCR sequence features.

### 4.1   Pilot Studies

#### 4.1.1   Van Laar and Colleagues: Feasibility and Toxicity

Early work had suggested that TCV might be just as useful in rheumatoid arthritis as it had proven to be in MS. It was now time to take the leap into human studies with this new treatment approach. Animal studies had demonstrated the use of TCV in RA. Moreover, human trials by Hafler (1992) in MS and Lohse et al. (1993a) in RA had demonstrated that downregulation of circulating autoimmune T cells provided clinical benefit. Therefore, van Laar and colleagues initiated a trial to further test the benefits of TCV in human RA (van Laar et al., 1993). A group of 13 patients with definite RA were enrolled in this pilot study to test the feasibility, toxicity, clinical influence, and immunomodulatory effect of TCV on human RA. Patients were followed from 2 weeks prior to 6 months after vaccination. They were inoculated subcutaneously with vaccine preparations consisting of attenuated, CD4+ T cell lines or clones isolated originally from synovial fluid or tissue ($50 \times 106$ cells per inoculation). Activation was accomplished polyclonally by a mitogenic anti-CD3 antibody. Although the pathogenic T cells had not yet been identified in RA, it was assumed that they would be present in the affected joint and could be retrieved and expanded for use in the inoculum. Although time-consuming and labor-intensive, the procedure was accomplished for each patient enrolled. As observed in all previous studies, TCV was well tolerated by these patients with no adverse events noted. All clinical parameters indicated a small decrease in disease activity, which was most obvious at 8 weeks after inoculation. A humoral anti-T cell response could not be observed, nor was specific immune reactivity against the immunizing T cells detected. Immunosuppressive effects were heterogeneous and appeared unrelated to clinical effects. A placebo effect was considered possible. One patient was exceptional by her extreme positive response to TCV, occurring perhaps because of an unintended booster effect and because of particularly short disease duration.

### 4.1.2   Lohse and Colleagues: A Careful Approach in Primates and Humans

Lohse and colleagues' own and others' detailed work on the immunologic bases of response to TCV in animals suggested the feasibility of moving to mammalian species. It seemed wise to try this approach in just a few very ill patients while taking precautions that they should not lose any benefit of past or current treatment they may have had. Based on the successes of animal models and the additional work on anti-idiotypic and antiergotypic vaccine mechanisms in rodent models (Lider et al., 1988; Lohse et al., 1989), Lohse and colleagues (1993a) tested the immunologic effects and toxicity of nonspecific T cell vaccination in primates and humans. In a small pilot study, only two human patients were recruited. Both had severe, longstanding RA, and one was continued on immunosuppressive therapy throughout the study. T cells to use for establishing lines for vaccination were isolated from synovial and circulation sources. Stimulation with anti-CD3 antibody was used, and glutaraldehyde fixation was applied. One patient received one injection of $40 \times 10^6$ cells SQ, and the other received two such injections at a 55-day interval. The patients were followed for approximately 2 months when the study was halted because of failure to obtain sufficient symptomatic relief. At this point, both patients were returned to high-dose steroid therapy. The objective clinical criteria showed little change, although one patient experienced lessening of symptoms for a few weeks following the first injection, but relapsed. Progression of the second patient could not be isolated to influence of the vaccine alone. As before, no adverse events occurred, yet no clear benefit was appreciated in this small study. Proliferative response to the vaccinating cells was marked only for one patient and only after the second vaccination. A weaker response to control T cell blasts suggested an antiergotypic response for this sample. Results for the second patient were confusing and could not be associated with the vaccination. Reactivity against any of the proposed RA target antigens (a 65-kDa heat shock protein, collagen type 2) was not detected in these patients.

## 4.2    Clinical Trials

### 4.2.1    Chen and Colleagues: Establishing TCV Parameters for RA

Problems remain in formulating a TCV protocol that would be useful for all RA patients. Chief among them are the selection, isolation, and expansion of appropriate populations of autologous, autoreactive T cells in the face of still unknown autoantigens. Promising as TCV has become, as demonstrated by animal studies in a variety of autoimmune conditions and in all of the foregoing MS human trials, approaches for use in RA must still be worked out. This is in part because, even now, specific autoantigens involved in the disease process remain unknown, although a few have been proposed (Corrigall and Panayi, 2002; Harris, 1984). Fortunately, two gambits suggest themselves through both clinical and immunochemical observations. First, identification of pathogenic cells should be possible through isolation of infiltrating T cells from synovial lesions—in effect, allowing the in vivo processes to accomplish the selection for us. Second, further selection of pathologic relevance should be possible through examining these cells for related TCR sequence features. Previous studies have provided clues as to the likely TCR V gene usage and relevant CDR3 sequence motifs likely for lesion-derived cells. A group of 16 patients with RA were enrolled in this study (Chen et al., manuscript in preparation). Vaccine was elaborated from autologous synovial T cells preselected for RA relevance on the basis of (1) the in vivo activation state, (2) the Th1 cytokine profile, and (3) TCR V gene expression. Selective expansion of activated T cells was done via PHA and IL-2 and, irradiation was used for attenuation. Patients received five subcutaneous inoculations with 30 to $45 \times 10^6$ cells per inoculation at monthly intervals followed by two boosters at months 6 and 9. Immunologic and clinical parameters were evaluated every 3months and compared with baseline values.

As before, no adverse reactions were noted. Observed clinical improvements were significant; and those measured by ACR20, ACR50, and ACR70 responses also were notable and were associated with a significant reduction in serum markers of inflammation. However, substantial (ACR50 and ACR70) responses were achieved only later in the course of T cell vaccination, a delay that is consistent with the time assumed to be required for the increasing effect of regulatory mechanisms identified for these patients by immunochemical testing. State-of-the-art genetic and immunochemical assays were applied to patient samples to clarify the regulatory mechanisms at work after TCV. Vaccination led to induction of circulating CD4+ regulatory T cells and CD8+ cytotoxic T cells, both specific for the T cell vaccine. Selective expansion was noted for the population of

CD4+Vβ2+ regulatory T cells that produced IL-10 and expressed a high level of Foxp3. Expansion of this population of regulatory T cells was seen to correlate with a depletion of the formerly overexpressed BV14+ T cells in vaccinated patients. CD4+ T cells responded to vaccination by increasing expression of Foxp3 and inhibitory activity of the CD4+CD25+ subset. Finally, antiinflammatory cytokines increased markedly after vaccination. All regulatory changes correlated with the clinical benefit.

A number of questions remain to be answered in the use of TCV for RA: Do the CD4+Vβ2+ regulatory cells seen to expand as a result of TCV actually represent one subset, or do they include both the Foxp3+-expressing cells and the IL-10-secreting cohort? Do the CD4+ IL-10-secreting cells reacting specifically to activated rather than resting T cells represent an anti-ergotypic subset such as those seen in animal models? The delayed clinical response, in which patients achieved ACR50 and ACR70 only later in the vaccination course, suggests a possible correlation with induction and response time to T cell regulatory networks. Can immunochemical details of this timing be further applied to the service of downregulation of inflammatory processes in this and other diseases?

Although no information is yet available, TCV studies are also underway in the area of lupus erythematosus.

# 5    Regulatory Mechanisms Induced by TCV

## 5.1    Anti-idiotypic T Cell Responses

After the discovery of anti-idiotypic T cell mechanisms in earlier animal models, Oksenberg and colleagues (1993) illustrated TCR commonality in human MS brain lesions. Later, Hong and colleagues (1999) illustrated such a commonality among MBP reactive T cells, showing recognition of an immunodominant MBP peptide in a series of MS patients. Anti-idiotypic cell lines isolated from vaccinated patients have been studied extensively and are recognized as being predominantly CD8+ cytotoxic T cells. These appear highly specific to vaccinating T cells (Cohen, 1989; Correale et al., 2000; Tejada-Simon et al., 2000; Zang et al., 2000; Zhang et al., 1993, 1995). Zhang and colleagues have further defined the TCR sequences forming targets of TCV for MS (Zang et al., 2000, 2003), placing them within the CDR3 region with CDR2 cryptic determinants still to be elucidated. Furthermore, studies into TCR gene expression in RA patients have yielded additional information on TCV mechanisms.

A number of laboratories (Alam, et al., 1995; Jenkins et al., 1993; Van der Borght et al., 2000; Williams et al., 1992; Zagon et al., 1994) investigating clonal expansion in RA have noted that in Caucasian populations TCR BV gene usage of T cells derived from RA patients was variably skewed to BV14 and BV17. However, Sun and colleagues (2005) studied a cohort of Chinese patients and found highly skewed BV14 and BV16 usage in synovial T cells in these patients. Detailed sequence analysis confirmed common CDR3 sequences and structural characteristics among these patients. This lack of randomness suggests the presence of as yet unidentified common autoantigens involved in RA pathogenesis.

It remains puzzling that the immune system selectively activates a CD8+ cytotoxic anti-idiotype response as the predominant one to T cell vaccination (Zhang et al., 1993, 1995). A number of aspects peculiar to TCV may account for this, including the route of administration, the size of the inoculum, and altered cell surface properties of the prepared vaccine (Zhang, 2004). The favorable clinical response that follows is closely correlated with the depletion of myelin autoreactive T cells in the patients' circulation. Indeed, this augurs well for future success of protocols aimed at booster vaccinations to extend anti-idiotypic regulation and its clinical benefits.

## 5.2  Anti-idiotypic Antibody Response

As evidenced by earlier studies, anti-idiotypic antibodies are not a dominant response to TCV and may represent only a small fraction of the total serum immunoglobulins, making them essentially impossible to detect in the circulation. However, more recent studies into samples gleaned from a larger human trial (Zhang et al., 2002) did indeed identify the activity of anti-idiotypic antibodies in the regulatory processing of TCV by an alternate, far more sensitive approach to the problem. Hong and colleagues (2000) synthesized a 20-mer TCR peptide that incorporated the CDR3 sequence expressed by the immunizing MBP-reactive T cell clones and used it as a screening agent to identify anti-idiotypic antibodies initially. Then, using limiting dilution methods, Hong et al. isolated and cloned patient circulatory B cells, producing anti-idiotypic antibodies of interest; they obtained and tested these antibodies as produced directly from the B cells in culture. This was the first demonstration that these antibodies appeared at an increased precursor frequency in vaccinated patients and not only reacted to the CDR3 sequence against which they had been screened but also had an inhibitory effect on proliferation of the vaccine T cells expressing this particular sequence. Such an effect is likely to contribute to the suppression of myelin-reactive T cells in vivo and has important implications for understanding the B cell component of the idiotypic/anti-idiotypic network as it is modulated by TCV.

## 5.3  Anti-ergotypic T Cell Responses

Early animal work by Cohen and Lohse on cell surface activation markers that came to be known as ergotopes was followed by later work in which several such markers were studied. One such that has been extensively investigated is a chain of the IL-2 receptor (CD25) (Mimran et al., 2004). Other ergotopes include HSP60 epitopes and even TCRs. Cohen et al. (2004) pointed out that whereas the TCR is expressed on resting cells ergotypic TCR peptides are presented only by activated T cells. A transient response downregulating autoimmune disease has been demonstrated not only in animal models (Lohse et al., 1993b) but also in early primate and human TCV experiments (Lohse et al., 1993a). Induction of the response appeared to be dose-dependent, requiring few vaccinating cells; and it appeared earlier and more strongly than the anti-idiotypic response in the same animals. TCV initiated by Van der Aaa and colleagues (2003) allowed recognition of anti-ergotypic responses contributing to suppression of activated T cells following vaccination.

## 5.4    CD4+ Regulatory T Cells: Responses Induced by TCV

In recent work, Hong et al. (2006) succeeded in elucidating for the first time two subsets of regulatory T cells, both of which inhibit activated T cells and appear to be antiergotypic in their recognition of a specific sequence of the IL-2 receptor α-chain. Chen and colleagues (manuscript in preparation) found similar subsets of T regulatory cells in their recent TCV trial for RA patients. It is important to note that early work of Cohen and colleagues, as well as others working in animal models, demonstrated that only activated (not resting) T cells could stimulate full protection when used as a vaccinogen (Cohen et al., 2002; Lohse et al., 1989). It is suggested (Hong et al., 2006) that these T cells may contribute a slower but longer-lasting series of responses necessary for overall immune balance. An exhaustive discussion of regulatory T cells, the history of their surface marker discovery, and functional attributes can be found in Chapters 3 and 9.

It appears, then, that TCV provides its benefit first through the cytotoxic and anti-idiotypic contribution of the CD8+ T cells, quickly depleting the autoreactive T cells (Achiron et al., 2004; Correale et al., 2000; Medaer et al., 1995; Van der Aaa et al., 2003; Zhang et al., 1993, 1995, 2002). This is then followed, over time, with repair of homeostatic maintenance functions, including reestablishment of an antiinflammatory cytokine environment offered by the CD4+CD25+ Treg contingent. Taken together with the clinical course observed with TCV in both MS and RA patients, it appears that whatever the cause the functional defect seen in Tregs in these autoimmune states may be countered by replacing that regulatory function. Cohen et al. (2004) pointed out that antiergotypic cells activated by whole T effector cells become anergic after producing their initial effects and are thus depleted with no hope of reactivation. However, "cross-presentation" of ergotopes by APCs allows such reactivation. Hence, TCV can achieve symptom relief either by introducing a new cytokine milieu through introduction of a Treg vaccinogen (Hong et al., 2006; Zhang, 2002) or by energizing such a vaccine with ergotopic peptide vaccination added within or in addition to the conventional whole cell vaccine. Such a concept is currently under investigation.

Although years of detailed immunologic study has effectively "proven" the extensive regulatory effects set into motion by TCV, the extent of its clinical effectiveness cannot be considered reliably established until ongoing controlled clinical trials are completed. One of the central targets of such efforts is a greater understanding of the inflammatory condition(s) seen in MS. Inflammation is known to involve more than one physiologic mechanism. One involves specific inflammatory processes driven by myelin-reactive T cells and ameliorated by the elimination of these cells followed by suppression of their inflammatory effects.

A second mode of inflammation is seen to be nonspecific and independent of these T cells. It is noted at relatively late stages of MS and is thought to involve both anti-idiotypic and regulatory T cell processes, both of which can be nonspecific and both may enter the pathologic picture at this stage of disease. During the application of TCV, all the regulatory networks described above act in concert to provide potent regulation of the disease. It should be noted that no current pharmacologic treatment is able to achieve this depth and breadth of immunologic regulation and symptom relief.

## 6    Technical Improvements

When considering technologic aspects of TCV, we would do well to keep in mind the fact that this procedure was never intended to become an off-the-shelf pharmaceutical product. Cohen (2002), who began this long discovery process with his early work in the area, remarked that "TCV is a notable example of the attempt to cure autoimmune disease by inducing active regulation rather than by suppressing the immune system." As more is learned about the immunochemistry of autoimmune conditions and the regulatory interventions possible with TCV (Fig. 15.2), we are increasingly mindful of the need to incorporate this new knowledge into each new attempt at TCV. However, out of these studies, along with ideas for new cellular approaches to vaccination, come a fleet of additional technical questions that must be answered to ensure safety, efficiency, and cost containment for a useful TCV protocol. Some of these areas are discussed below.

### 6.1    Sources of T Cells

Accessing T cells in an efficient, cost-effective way is the first step toward providing a vaccine that would be available to most of the patients requiring it. It is necessary, at the same time, to include in the vaccine T cell populations that are or will give rise to the cell subsets known to provide the immunologic changes necessary for an improved clinical picture. The source of T cell vaccine is critical because it determines the target of immune regulation induced by T cell vaccination. It was also important to determine the best source of T cells for vaccination, taking into consideration accessibility and other factors. Three sources of vaccine T cells (Fig. 15.2) have been investigated: peripheral circulation, cerebrospinal fluid, and lesion sites. Each has its advantages and disadvantages as described in these clinical studies.

---

FIGURE 15.2. T cell vaccination procedure: elements of a fundamental protocol. Pathogenic T cells can be obtained from various sources for vaccine elaboration: cerebrospinal fluid of multiple sclerosis patients, synovial fluid of rheumatoid arthritis patients, or blood as commonly used. Selection and expansion of the pathogenic T cells is undertaken to arrive at cell numbers adequate for therapy. Two main methods are used: stimulation with selected peptides of candidate autoantigens through a T cell cloning procedure or selective

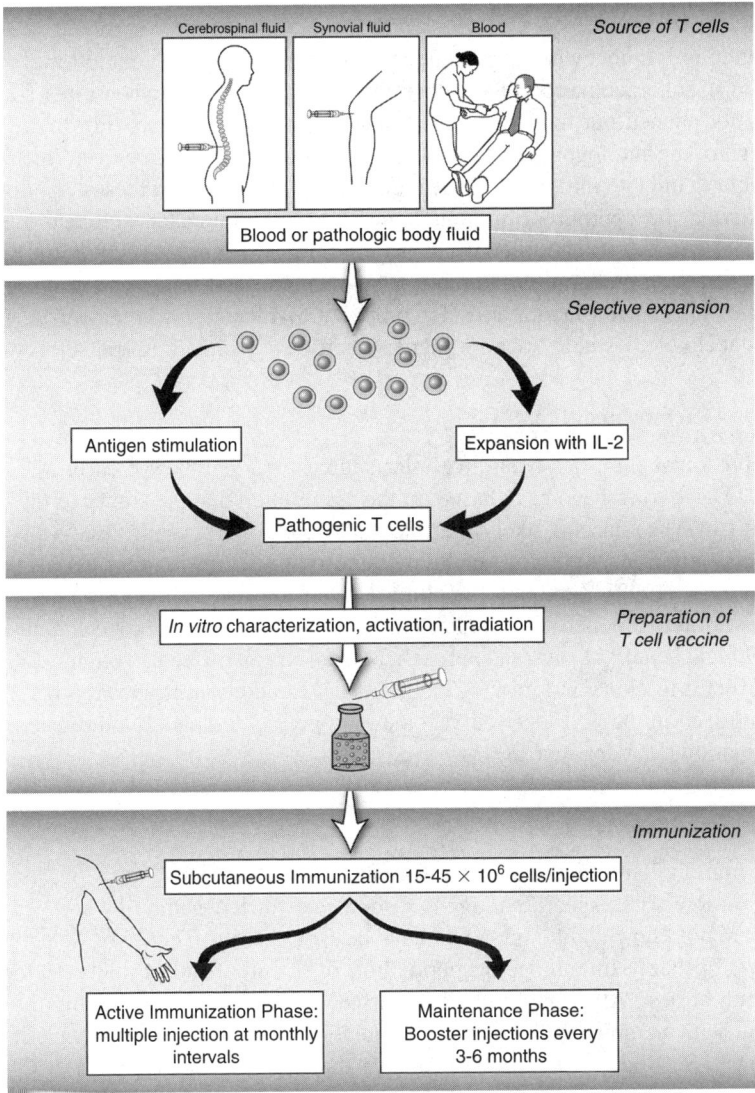

Figure 15.2. (*continued*) expansion of in situ activated T cells by IL-2 (in situ activated T cells expressing IL-2 receptor here are considered autoimmune T cells relevant to the disease process). Attenuation of selected pathogenic T cells, generally by irradiation, is then accomplished to render them unable to propagate, and immunization proceeds. Clinical trials and laboratory research have indicated that cell number, activation state, and relevance to target antigens are all central to the effectiveness of T cell vaccine as judged by immunologic studies. Immunization is best done as active immunization and maintenance or booster immunization. Active immunization phase of four or five injections monthly is followed by a lengthier maintenance phase during which booster injections may be given every 3 to 6 months. Protocol steps continue to be refined and improved as clinical trials proceed.

### 6.1.1   Peripheral Circulation

Circulating lymphocytes have been perhaps the most studied source for elaboration of T cell vaccines and have been the fall-back position when other sources have not panned out for individual patients. The circulation provides more than enough of the autologous cells for each patient's treatment regimen, and such cells are able to migrate into the CNS (Hong et al., 1999). Although these cells do tend to provide the appropriate immunologic milieu for immediate clinical benefit, boosters are often found to be necessary, and not all patients respond. The protocol for vaccine elaboration is highly labor-intensive and time-consuming, and it requires specialized laboratories for production (Zhang, 2001). However, flow-based technologies hold hope for increased efficiency of sorting and selection.

### 6.1.2   Cerebrospinal Fluid

Cerebrospinal fluid has been a cellular source for a number of early MS TCV trials. These trials have been based on the assumption that proximity to the brain lesions of MS makes it likely that the cells found therein will be a population especially relevant to disease pathology and progression (Fujimura et al., 1997; Saeki et al., 1992). Although the clinical data on these trials hint at the truth of this assumption, none of the studies has been large enough to prove conclusively the clinical benefit of such an approach. Myelin-reactive T cells from the CSF can be difficult to clone and grow (Zhang, 2004). In addition, it appears that T cell clones used in the CSF-derived vaccine may not be completely eliminated after vaccination (Van der Aaa, 2003).

### 6.1.3   Lesion Sites

Two immunologic facts combine to make such a source an attractive one for TCV in RA. First, specific antigens responsible for RA pathology and progression are not known, although a few have been suggested. Therefore, selection of an appropriately immunogenic population of T cells from peripheral blood is difficult at best. Allowing the body to do the filtration work, enriching the synovium with pathologically relevant populations of T cells makes a great deal of sense. Second, there is good evidence that the frequencies of CD4+CD25+ T cells found in the synovium of RA patients are elevated over those found in their peripheral circulation (Cao et al., 2003; van Amelsfort et al., 2004). An enhanced population of T regulatory cells are likely to be a plus in restoring the immune balance to the patient's system; and if in vitro growth and expansion problems can be overcome, this may be the most efficient and patient-specific approach to take.

It becomes clear from these studies that lesion-derived T cells, as are used in RA procedures, would be a better source for T cell vaccine. However, in the case of MS, the extremely low number of T cells that can be obtained from CSF through lumbar puncture seems to constrain the use of this T cell source. This is because of the technical diffculty of expanding relevant T cells.

## 6.2   Nature of T Cells

Selection of T cells for vaccine elaboration involves choices based not only on epitope specificity but on the cell number achievable for injection needs. The immunologically relevant optimum is too often at war with the practical necessities of vaccine production.

### 6.2.1   Specificity

T cell relevance to the autoimmune condition to be treated has always been a chief concern. Establishing the antigen of interest is the first concern of scientists working in this area. In the case of MS, MBP has long been identified as central to the pathogenic nature of the disease. Recently, immunodominant epitopes have been identified and T cells specific for these epitopes targeted for vaccine use. Success with monovalent vaccines has led, logically, to the desire to expand the usefulness of the vaccine by adding other target antigens to the selection process, namely MOG and PLP (Zhang, 2002). Trials are ongoing to test whether a wider variety of antigens can provide clinical improvement in a larger number of patients.

In MS, one can run into problems with using T cells from the CSF in that selection of relevant T cell clones from the relatively heterogeneous populations can be difficult. The myelin-reactive cells in the CSF are a small fraction of all T cells and are often difficult to clone and grow. However, selection can be done based on knowledge of in vivo clonal expansion of certain T cells taking place in the autoimmune condition. There is evidence of oligoclonal activation and expansion of T cells in CSF of MS patients, so they are more likely to be relevant to the disease process (Zhang, 2004). Although their mode of expansion is unknown, one can select appropriate populations using TCR CDR3 DNA sequence analysis or immunoscope technology (Sun et al., 2005) and expand in vitro using PHA or anti-CD3 antibody methods.

However, what are the choices when specific pathogenic antigens are not known, as in RA? While relying on the body's own filtration system to provide them with enriched populations of T cells of some antigenic relevance to the disease condition, investigators increase their chances of success by selecting these further on the basis of the known immunologic parameters that are desired for initiation of repair of the dysregulated immune system. Chen and colleagues (manuscript in preparation) did this by examining related TCR sequence features (V gene usage and CDR3 sequence motifs) and choosing candidate clones on the basis of the activation state, Th1 cytokine profile, and characteristic TCR V gene expression, as commonly seen in rheumatoid synovium (Corrigall and Panayi, 2002; Harris, 1984). Expansion with PHA and IL-2 allowed collection of sufficient cell numbers for inoculation.

### 6.2.2   Clones Versus Lines

Whenever possible, one would like to hedge one's bets for antigenic relevance of the vaccinogen by including a number of clones, each specific for one's antigen of choice. This allows several "shots" at the pathogenic effector cell while retaining specificity for the antigen or immunodominant determinant that has been identified. However, clones can be more difficult to grow and expand than T cell lines, which provide a somewhat diluted population of epitope-specific cells. Although a number of epitopes are covered in a T cell line, not all of them may be immunologically useful for targeting appropriate effector cells. Given the realities of in vitro expansion, however, it is often necessary to include cell lines along with the clones in the vaccine to arrive at the necessary T cell number for inoculation purposes.

## 6.3    Methods of T Cell Activation

Once the selection process is done and one has in hand the chosen clones/lines of appropriate antigen specificity or optimal immunogenic potential, it is time to expand these cells to give sufficient numbers for inoculation (Fig. 15.2). Here there are some choices to make. Expansion by antigen stimulation may safeguard the specificity of one's population, but repeated stimulation of this type leads to cellular anergy and loss of antigen specificity. The alternative is mitogen stimulation, a frequently used gambit for TCV expansion. PHA and IL-2 are generally the reagents of choice. This type of stimulation does indeed expand the cells in hand but does so across the board. Although these stimuli are highly effective in expanding T cells in culture, the major concern is related to loss of antigen specificity in T cell lines or other T cell preparations in which T cells of interest are mixed with other T cells of irrelevant specificity. Expansion of T cells with mitogens and IL-2 is appropriate for T cell clones in most cases. Once again, one force must be balanced against another. One approach is to alternate antigen stimulation with mitogen stimulation if several rounds of expansion are needed to arrive at the necessary cell number for inoculation. Another important aspect is activation of T cell vaccine prior to irradiation and use for vaccination. There is experimental evidence generated by Cohen's group indicating that only activated (not resting) T cells are effective in the induction of protective immune responses. It now becomes clear that T cell activation markers expressed on the surface of T cell vaccine are critical to the induction of antiergotypic T cell response. There is consensus that all TCV protocols must require preactivation of T cell vaccine.

## 6.4    Dosage Level and Frequency of Vaccination

As in any clinical trial, one stands on the shoulders of those who have gone before when planning the details of vaccine dosage and inoculation frequency. The nature of the immune processes at work in TCV requires that a period of 1 to 2 months passes between active immunization treatments. Beyond that, it becomes a highly individualized process. TCV is, by its very nature, a highly individualized procedure composed of two arms: testing and inoculation. Both processes are standardized as completely as possible, and they are applied to each patient on a case-by-case basis. Some generalities can be drawn from previous experience. We can assume that, for most MS patients, more than one active immunization inoculation will be needed. For RA patients, this may not be the case. For other autoimmune conditions, only time and trial experience will tell. Before each inoculation it will be necessary to test at least for the presence of effector cells in the circulation. In addition, of course, the necessary clinical testing must be undertaken to chart symptomatic progress. Experience tells us that old clones may in time renew themselves (Hermans et al., 2000), requiring additional injections.

With increasing knowledge, more advanced immunization schemes have been designed in recent clinical trials to include an active immunization phase consisting of multiple monthly injections over 5 to 6 months followed by a maintenance phase requiring booster injections every 3 to 6 months to maintain the immunity (Fig. 15.2). Once in a while clonal shift may take place (Tuohy et al., 1999), requiring a new selection process to maintain symptom relief. Changes in antigen number and choice and/or additional approaches—such as Cohen's suggestion of increasing the number of anti-idiotypic and antiergotypic regulatory cells by adding ergotope peptide vaccines to the basic TCV vaccine strategy (Cohen, 2002)—are expected to change the symptom picture and thus the inoculation needs.

Although the patient is always the best barometer and his or her symptom picture the best guide to inoculation frequency, the immunological success of any TCV protocol rests on the answers to four questions: Is there depletion of autoreactive effector cells? Is there a change of cytokine environment to Th2? Is there upregulation of Tregs? Is there a lengthy clinical benefit? One standardizes all the elements of the TCV process that it is possible to standardize. However, ultimately all of the considerations discussed here constitute a balancing act with the fate of the patient at its center.

## 7  Future Directions

As reviewed above, TCV has an interesting history of development and an excellent example that concepts rising from basic science research can be translated into a potential treatment in humans. There are a number of critical issues that influence the future research and clinical application of TCV. First, the concept of TCV has been proven in both animal models and human clinical trials with respect to its immunologic properties and regulatory mechanisms, as extensively reviewed in this chapter. In human studies, since our landmark publication in *Science* in 1993, we have been continuously exploring and learning new mechanisms about TCV with the help of the guiding work by Cohen and colleagues in this field. It is proven that TCV is a useful tool for probing the immune system to investigate how autoreactive T cells are regulated in vivo by various regulatory networks and what regulatory networks are in operation and can be boosted to achieve a desired effect on the immune system or to correct an aberrant autoimmune reaction. In this regard, we will continue to learn from this research and perhaps be given the opportunity to develop a more advanced approach, if the essential elements of

TCV are fully identified. For example, peptides derived from idiotypic and ergo-typic epitopes of a T cell vaccine may be combined to create a powerful peptide-based approach to achieving the same or most of the immunologic effects induced by whole T cell vaccination. Such an approach is not impossible given our recent findings in the identification of an IL-2 receptor α-chain peptide as an essential ergotope (Hong et al., 2006) and the existence of CDR3 common sequence motifs in TCRs of myelin-reactive T cells recognizing the immunodominant epitope in the context of the same DR2 element (Hong et al., 1999).

Second, we must realize that the current protocol of TCV is still not fully opti-mized for maximum effects and desired practicality. During the last decade or so, a number of investigators designed and tested independently various protocols based on their understanding of TCV and the pathogenesis of MS and RA. This exploration phase was extremely important for us to learn what might be the opti-mal protocol in terms of selecting T cells to make T cell vaccine, dosage, fre-quency, method of activation and subsequent inactivation, and so on, as reviewed in this chapter. TCV is a highly sophisticated technical procedure that has many undetermined elements and technical details. The effectiveness of TCV lies con-siderably in the details. Inappropriate selection or handling of any of these essen-tial elements would lead to a lack of the expected immunologic effects. Over the years, some procedures survived repeated testing in human trials by independent groups to stand out as effective protocols, whereas other protocols failed to demonstrate the expected immunologic effects. We have learned new tricks and many lessons. It was the consensus from a recent international TCV workshop that as a group we must now summarize the technical and clinical data obtained so far and carry on with an effective protocol to the next stage—large controlled clinical trials.

That brings us to the next critical issue related to the clinical efficacy of TCV in MS, RA, and other autoimmune conditions. There have been more than 10 clini-cal trials on MS and RA during the past 15 years. All except one were open-labeled pilot, or Phase I, trials in small groups of patients, ranging from 4 to 54 patients. Given the limitation of the clinical trial design/setting, the results are sug-gestive of various clinical improvements but are not clinically conclusive. With better protocols emerging from our exploration phase, we must now move on to begin multicenter, double-blind, randomized clinical trials to reach a definitive conclusion as to whether TCV is efficacious in the treatment of MS and RA, the two autoimmune conditions for which TCV has been tested extensively. Whether TCV will endure as an effective treatment depends on these pivotal clinical trials on which some groups or entities are currently embarking. Definitive clinical con-clusions may also finally settle the issues related to the role of myelin-reactive T cells in the disease process of MS.

## References

Achiron, A., Lavie, G., Kishner, I., et al. (2004) T cell vaccination in multiple sclerosis relapsing-remitting nonresponders patients. *Clin. Immunol.* 113:155-160.

Alam, A., Lule, J., Coppin, H., et al. (1995) T-cell receptor variable region of the beta-chain gene use in peripheral blood and multiple synovial membranes during rheumatoid arthritis. *Hum. Immunol.* 42:331-339.

Anderton, S. M., van der Zee, R., Prakken, B., et al. (1995) Activation of T cells recognizing self 60-kD heat shock protein can protect against experimental arthritis. *J. Exp. Med.* 181:943-952.

Bendele, A. M. (2001) Animal models of rheumatoid arthritis. *J. Muskuloskel. Neuron. Interact.* 1:377-385.

Ben-Nun, A., Cohen, I. R. (1982) Spontaneous remission and acquired resistance to autoimmune encephalomyelitis (EAE) are associated with suppression of T cell reactivity: suppressed EAE effector T cells recovered as T cell lines. *J. Immunol.* 128(3): 1450-1457.

Ben-Nun, A., Wekerle, H., Cohen, I. R. (1981a) The rapid isolation of clonable antigen-specific T-lymphocyte lines capable of mediating autoimmune encephalomyelitis. *Eur. J. Immunol.* 11:195-199.

Ben-Nun, A., Wekerle, H., Cohen, I. R. (1981b) Vaccination against autoimmune encepahlomyelitis with T-lymphocyte line cells reactive against myelin basic proteins. *Nature* 292:60-61.

Ben-Yeduda, A., Bar-Tana, R., Livoff, A., et al. (1996). Lymph node cell vaccination against the lupus syndrome of MrL/lpr/lpr mice. *Lupus* 5:232-236.

Beraud, E., Kotake, S., Caspi, R. R., et al. (1992) Control of experimental autoimmune uveoretinitis by low dose T-cell vaccination. *Cell Immunol.* 140(1):112-122.

Cao, D., Malmstrom, V., Baecher-Allan, C., et al. (2003) CD4+CD25[bright] regulatory T cells from the target organ of patients with rheumatoid arthritis. *Eur. J. Immunol.* 33:215-223.

Chou, Y. K., Bourdette, N., Offner, H., et al. (1992) Frequency of T cells specific for myelin basic protein and myelin proteolipid protein in blood and cerebrospinal fluid in multiple sclerosis. *J. Neuroimmunol.* 38(1-2):105-113.

Cohen, I. R. (1989) Natural id-anti-id networks and the immunological homunculus. In: Atlan, H., Cohen, I. R. (eds) *Theories of Immune Networks.* Springer-Verlag, Berlin, pp 6-12.

Cohen, I. R. (2002) T-cell vaccination for autoimmune disease: a panorama. *Vaccine* 20:706-710.

Cohen, I. R., Quintana, F. J., Mimran, A. (2004) Tregs in T cell vaccination: exploring the regulation of regulation. *J. Clin. Invest.* 114(9):1227-1232.

Correale, J., Lund, B., McMillan, M., et al. (2000) T cell vaccination in secondary progressive multiple sclerosis. *J. Neuroimmunol.* 107:130-139.

Corrigall, V. M., Panayi, G. S. (2002) Autoantigens and immune pathways in rheumatoid arthritis. *Crit. Rev. Immunol.* 22:281-293.

Elias, D., Markovits, D., Reshef, T., et al. (1990) Induction and therapy of autoimmune diabetes in the non-obese diabetic (NOD/Lt) mouse by a 65-kD heat shock protein. *Proc. Natl. Acad. Sci. USA* 87:1576-1580.

Elias, D., Tikochinsky, Y., Frankel, G., Cohen, I. R. (1999) Regulation of NOD mouse autoimmune diabetes by T cells that recognize a T-cell receptor CDR3 peptide. *Int. Immunol.* 11:957-966.

Fujimura, H., Nakatsuji, Y., Sakoda, S., et al. (1997) Demyelination in severe combined immunodeficient mice by intracisternal injection of cerebrospinal fluid cells from patients with multiple sclerosis: neuropathological investigation. *Acta Neuropathol. Berl.* 93:567-578.

Hafler, D., Cohen, I. R., Benjamin, D., Weiner, H. L. (1992) T-cell vaccination in multiple sclerosis: a preliminary report. *Clin. Immunol. Immunopathol.* 62:307-312.

Harris, E. D. (1984) Pathogenesis of rheumatoid arthritis. *Clin. Orthop.* 182:14-23.

Hermans, G., Medaer, R., Raus, J., Stinissen, P. (2000) Myelin reactive T cells after T cell vaccination in multiple sclerosis: cytokine profile and depletion by additional immunizations. *J. Neuroimmunol.* 102:79-84.

Holoshitz, J., Naparstek, Y., Ben-Nun, A., Cohen, I. R. (1983) Lines of T lymphocytes induce or vaccinate against autoimmune arthritis. *Science* 129:56-58.

Hong, J., Zang, Y. C., Tejada-Simon, M. V., et al. (1999) A common TCR V-D-J-sequence in V beta 13.1 T cells recognizing an immunodominant peptide of myelin basic protein in multiple sclerosis. *J. Immunol.* 163:3530-3538.

Hong, J., Zang, Y. C. Q., Tejada-Simon, M. V., et al. (2000) Reactivity and regulatory properties of human anti-idiotypic antibodies induced by T cell vaccination. *J. Immunol.* 165:6858-6864.

Hong, J., Zang, Y. C. Q., Nie, H., Zhang, J. Z. (2006) CD4+ regulatory T cell responses induced by T cell vaccination in patients with multiple sclerosis. *Proc. Nat. Acad. Sci. USA* 103:5024-5029.

Jenkins, R. N., Nikaein, A., Zimmermann, A., et al. (1993) T cell receptor V beta gene bias in rheumatoid arthritis. *J. Clin. Invest.* 92:2688-2701.

Kakimoto, K., Katsuki, M., Hirofuji, T., et al. (1988) Isolation of T cell line capable of protecting mice against collagen-induced arthritis. *J. Immunol.* 140(1):78-83.

Lassmann, H., Brunner, C., Bradl, M., Linington, C. (1988) Experimental allergic encephalomyelitis: the balance between encephalitogenic T lymphocytes and demyelinating antibodies determines size and structure of demyelinated lesions. *Acta Neuropathol. (Berl.)* 75:566-576.

Lees, M. B. (1982) Proteolipids. *Scand. J. Immunol.* 15:147-166.

Lider, O., Kaarin, N., Shinitzky, M., Cohen, I. R. (1987). Therapeutic vaccination against adjuvant arthritis using autoimmune T cells treated with hydrostatic pressure. *Proc. Nat. Acad. Sci. USA* 84:4577-4580.

Lider, O., Reshef, T., Beraud, E., et al. (1988) Anti-idiotypic network induced by T cell vaccination against experimental autoimmune encephalomyelitis. *Science* 239:181-184.

Linington, C., Lassmann, H. (1987) Antibody responses in chronic relapsing experimental allergic encephalomyelitis: correlation of serum demyelination activity with antibody titre to the myelin/oligodendrocyte glycoprotein (MOG). *J. Neuroimmunol.* 17:61-69.

Linington, C., Bradl, M., Lassmann, H., et al. (1988) Augmentation of demyelination in rat acute allergic encephalomyelitis by circulating mouse monoclonal antibodies directed against a myelin/oligodendrocyte glycoprotein. *Am. J. Pathol.* 130:443-454.

Lohse, A.W., Mor, F., Karin, N., Cohen, I. R. (1989) Control of experimental autoimmune encephalomyelitis by T cells responding to activated T cells. *Science* 244:820-822.

Lohse, A. W., Bakker, N. P. M., Hermann, E., et al. (1993a) Induction of an anti-vaccine response by T cell vaccination in non-human primates and humans. *J. Autoimmun.* 1:121-130.

Lohse, A. W., Spahn, T. W., Wolfel, T., et al. (1993b) Induction of the anti-ergotypic response. *Int. Immunol.* 5(5):533-539.

Madsen, L. S., Andersson, E. C., Jansson, L., et al. (1999) A humanized model for multiple sclerosis using HLA-DR2 and a human T-cell receptor. *Nat. Genet.* 23:343-347.

Maron, R., Zerubavel, R., Friedman, A., Cohen, I. R. (1983) T-lymphocyte line specific for thyroglobulin produces or vaccinates against autoimmune thyroiditis in mice. *J. Immunol.* 131:2316-2322.

Martin, R., Howell, M. D., Jaraquemada, D., et al. (1991) A myelin basic protein peptide is recognized by cytotoxic T cells in the context of four HLA-DR types associated with multiple sclerosis. *J. Exp. Med.* 173(1):19-24.

Martin, R., Utz, U., Coligan, J. E., et al. (1992) Diversity in fine specificity and T cell receptor usage of the human CD4+ cytotoxic T cell response specific for the immunodominant myelin basic protein peptide 87-106. *J. Immunol.* 148(5):1359-1366.

Martin, R., Whitaker, J. N., Rhame, L., et al. (1994) Citrulline-containing myelin basic protein is recognized by T-cell lines derived from multiple sclerosis patients and healthy individuals. *Neurology* 44(1):123-129.

Medaer, R., Stinissen, P., Truyen, L., et al. (1995) Depletion of myelin-basic-protein autoreactive T cells by T-cell vaccination: pilot trial in multiple sclerosis. *Lancet* 346(8978):807-808.

Mimran, A., Mor, F., Carmi, P., et al. (2004) DNA vaccination with CD25 protects rats from adjuvant arthritis and induces an antiergotypic response. *J. Clin. Invest.* 113:924-932.

Mor, F., Lohse, A. W., Karin, N., Cohen, I. R. (1990) Clinical modeling of T cell vaccination against autoimmune disease in rats: Selection of antigen-specific T cells using a mitogen. *J. Clin. Invest.* 85:1594-1598.

Naparstek, Y., Ben-Nun, A., Holoshitz, J., et al. (1983) T-lymphocyte lines producing or vaccinating against autoimmune encephalomyelitis (EAE): functional activation induces PNA receptors and accumulation in the brain and thymus of line cells. *Eur. J. Immunol.* 13:418-423.

Oksenberg, J. R., Panzara, M. A., Begovich, A., et al. (1993) Selection for T–cell receptor V beta-D beta-J beta gene rearrangements with specificity for a myelin basic protein peptide in brain lesions of multiple sclerosis. *Nature* 362:68-70.

Ortiz-Ortiz, L., Weigle, W. O. (1976) Cellular events in the induction of experimental allergic encephalomyelitis in rats. *J. Exp. Med.* 114:604-616.

Ota, K., Matsui, M., Milford, E. L., et al. (1990) T-cell recognition of an immunodominant myelin basic protein epitope in multiple sclerosis. *Nature* 12;346(6280):183-187.

Pette, M., Fujita, K., Wilkinson, D., et al. (1990) Myelin autoreactivity in multiple sclerosis: recognition of myelin basic protein in the context of HLA-DR2 products by T lymphocytes of multiple-sclerosis patients and healthy donors. *Proc. Natl. Acad. Sci. USA* 87(20):7968-7972.

Saeki, Y., Mima, T., Sakoda, S., et al. (1992) Transfer of multiple sclerosis into severe combined immunodeficiency mice by mononuclear cells from cerebrospinal fluid of the patients. *Proc. Natl. Acad. Sci. USA* 89:6157-6161.

Saruhan-Direskeneli, G., Weber, F., Meinl, E., et al. (1993) Human T cell autoimmunity against myelin basic protein: CD4+ cells recognizing epitopes of the T cell receptor β chain from a myelin basic protein-specific T cell clone. *Eur. J. Immunol.* 23:530-540.

Schluesener, H., Sobel, R., Linington, C., Weiner, H. (1987) A monoclonal antibody against a myelin oligodendrocyte glycoprotein induces relapses and demyelination in CNS autoimmune disease. *J. Immunol.* 139:4016-4021.

Sun, W., Nie, H., Li, N., et al. (2005) Skewed T-cell receptor BV14 and BV16 expression and shared CDR3 sequence and common sequence motifs in synovial T cells of rheumatoid arthritis. *Genes Immun.* 6:248-261.

Tejada-Simon, M. V., Yang, D., Zang, Y., et al. (2000) T-cell reactivity to myelin antigens and their structural and functional properties at different clinical stages of multiple sclerosis. *Int. Immunol.* 12:1641-1650.

Tuohy, V. K., Yu, M., Yin, L., et al. (1999) Spontaneous regression of primary autoreactivity during chronic progression of experimental autoimmune encephalomyelitis and multiple sclerosis. *J. Exp. Med.* 189:1033-1042.

Van Amelsfort, J. M., Jacobs, K. M., Bijlsma, J. W., et al. (2004) CD4+CD25+ regulatory T cells in rheumatoid arthritis: differences in the presence, phenotype, and function between peripheral blood and synovial fluid. *Arthitis Rheum.* 50:2775-2785.

Van der Aaa, A., Hellings, N., Medaer, R., et al. (2003) T cell vaccination in multiple sclerosis patients with autologous CSF-derived activated T cells: results from a pilot study. *Clin. Exp. Immunol.* 131:155-168.

Van der Borght, A., Geusens, P., Vandevyver, C., et al. (2000) Skewed T-cell receptor variable gene usage in the synovium of early and chronic rheumatoid arthritis patients and persistence of clonally expanded T cells in a chronic patient. *Rheumatology (Oxford)* 39:1189-1201.

Van Laar, J. M., Miltenburg, A. M. M., Verdonk, M. J. A., et al. (1993) Effects of inoculation with attenuated autologous T cells in patients with rheumatoid arthritis. *J. Autoimmun.* 6:159-167.

Williams, W. V., Fang, Q., Demarco, D., et al. (1992) Restricted heterogeneity of T cell receptor transcripts in rheumatoid synovium. *J. Clin. Invest.* 90:326-333.

Yoshimura, T., Kunishita, T., Sakai, K., et al. (1985) Chronic experimental allergic encephalomyelitis in guinea pig induced by proteolipid protein. *J. Neurol. Sci.* 69:47-58.

Zagon, G., Tumang, J. R., Li, Y., et al. (1994) Increased frequency of V beta 17-positive T cells in patients with rheumatoid arthritis. *Arthritis Rheum.* 37:1431-1440.

Zamvil, S. S., Nelson, P., Trotter, J., et al. (1985) Encephalitogenic T-cell clones specific for myelin basic protein induce chronic relapsing paralysis and demyelination. *J. Exp. Med.* 162(6):2107-2124.

Zang, Y. C. Q., Hong, J., Tejada-Simon, M., et al. (2000) Th2 immune regulation induced by T cell vaccination in patients with multiple sclerosis. *Eur. J. Immunol.* 30:908-913.

Zang, Y. C. Q., Hong, J., Rivera, V. M., et al. (2003) Human anti-idiotypic T cells induced by TCR peptides corresponding to a common CDR3 sequence motif in myelin basic protein-reactive T cells. *Int. Immunol.* 15(9):1073-1080.

Zhang, J. Z. (2001) T-cell vaccination in multiple sclerosis: immunoregulatory mechanism and prospects for therapy. *Crit. Rev. Immunol.* 21:41-55.

Zhang, J. (2002) T-cell vaccination for autoimmune diseases: immunologic lessons and clinical experience in multiple sclerosis. *Expert Rev. Vaccines* 1(3):285-292.

Zhang, J. (2004) T cell vaccination as an immunotherapy for autoimmune diseases. *Cell. Mol. Immunol.* 1(5):321-327.

Zhang, J. W., Chou, S. J., Hashim, G. A., et al. (1990) Preferential peptide and HLA restriction of MBP specific T cell clones derived from MS patients. *Cell. Immunol.* 129:189-198.

Zhang, J., Medaer, R., Stinissen, P., et al. (1993) MHC-restricted depletion of human myelin basic protein-reactive T cells by T cell vaccination. *Science* 261:1451-1454.

Zhang, J., Markovic, S., Lacet, B., et al. (1994) Increased frequency of inerleukin-2-responsive T-cells specific for myelin basic protein and proteolipid protein in peripheral blood and cerebrospinal fluid of patients with multiple sclerosis. *J. Exp. Med.* 179: 973-984.

Zhang, J., Vandevyver, C., Stinissen, P., Raus, J. (1995) In vivo clonotypic regulation of human myelin basic protein-reactive T cells by T cell vaccination. *J. Immunol.* 155: 5868-5877.

Zhang, J. Z., Rivera, V. M., Tejada-Simon, M. V., et al. (2002) T cell vaccination in multiple sclerosis: results of a preliminary study. *J. Neurol.* 249:212-218.

# 16
# Trivalent T Cell Receptor Peptide Vaccine for Treatment of Multiple Sclerosis Targets Predominant V Genes Widely Implicated in Autoimmune Diseases and Allergy

Arthur A. Vandenbark and Nicole E. Culbertson

Over the past one and a half decades, intense effort has been expended to address the question of whether the T cell receptor (TCR) variable (V) gene repertoire differs between patients with disease versus healthy controls. Many individual studies suggested that certain V genes were utilized by T cells more often than expected in the disease state, whereas other studies seemed not to confirm or support this contention. We here present the summarized results of a careful and comprehensive review of the literature, with a focus on which V genes predominate in the healthy repertoire in T cells specific for autoantigens and allergens and on differences that occur in patients with related diseases. Although our meta-analysis had to deal with substantial methodologic differences, our results consistently demonstrated significant disease-associated V gene expression. Surprisingly, however, a relatively small subset of V genes predominated among most of the diseases studied, with only a few distinct V genes being unique for each disease. For the most part, the predominant disease-associated V genes emerged from the most prominent V gene families. Thus, V genes used to form pathogenic TCR chains appear to arise from the most productively rearranged V gene families in the normal repertoire. This implies that the autoreactive and allergic TCR V genes represent normal, positively selected clones that have been further expanded by the disease process in accordance with their respective starting frequencies rather than through other mechanisms such as superantigens, which would produce a skewed pattern of expression. These observations have important ramifications for the design of TCR-based vaccines, which may broadly affect patients with different diseases using a relatively small subset of commonly expressed V genes. Our experience using TCR peptide vaccines in several animal models of experimental encephalomyelitis (EAE) demonstrated that injection of TCR peptides can reverse ongoing paralytic disease by boosting natural regulatory T cells that inhibit target T effector cells by both specific and nonspecific mechanisms. Clinical trials in multiple sclerosis are in progress to evaluate the potential of this therapeutic approach in humans.

# 1   Introduction

The involvement of T cells in the pathogenesis of human autoimmune disease has long been suspected, even though few data are available that directly implicate these T cells. In contrast, antigen-specific CD4+ T cells have clearly been demonstrated to induce autoimmune tissue destruction and clinical disease in experimental animal models. These pathogenic T cells retained the tendency to develop characteristic CDR3 "motifs" specific for target peptides, as well as to utilize common V gene families. This knowledge has led to immune interventions directed at both clonotypic CDR3 determinants present on only the peptide-specific T cells and more broadly represented V gene determinants, typically the CDR1, CDR2, and FW3 loops of the TCR.

The overexpression of BV8S2 genes by pathogenic T cells from both rats and mice led to the "V gene hypothesis," which states that only certain selected V genes may be involved in forming pathogenic TCRs. This provocative idea, coupled with a newfound ability to clone autoreactive T cells from human fluids and tissues and a fairly complete knowledge of TCR V gene sequences, subsequently triggered a tremendous international effort to identify disease-associated patterns of V gene expression in various patient groups. In most cases, the focus of this effort was to determine if V genes expressed by autoreactive T cell clones from patients occurred more often than in healthy controls (HCs) or other disease controls. Several studies demonstrated oligoclonal TCR expansions in individual patients, but no consensus emerged as to common overexpressed V genes in any specific disease. However, the biggest limitation of each individual study was the lack of enough well characterized T cell clones to draw firm conclusions. We have readdressed this issue by carrying out an exhaustive meta-analysis of TCR V gene expression patterns in several autoimmune and allergic diseases thought to involve pathogenic T cells. The pattern of TCR V gene expression in MS patients compared to healthy controls provided key information regarding which V gene families and peptide sequences would be useful targets for immune regulation. This analysis led to the development of a tripeptide TCR vaccine that has now been utilized in more than 60 MS patients to boost expansion of TCR-specific T cells. Our recent studies demonstrated that monthly intramuscular injection of this trivalent vaccine can strongly increase the frequency of TCR-specific T cells that appear to mediate immune suppression through secretion of interleukin-10 (IL-10) and Treg cell inhibition of effector cell activation. In this chapter, we discuss mechanisms through which the TCR V gene repertoire is selected, identify predominant AV and BV genes associated with MS and other diseases, review animal studies using TCR peptides to EAE, and present data from recent MS clinical trials in which our trivalent TCR vaccine was used to induce regulatory TCR-specific T cells.

## 2    Development of the Peripheral T Cell Repertoire

We used the method of reverse transcription-polymerase chain reaction (RT-PCR) with family-specific primers to determine the levels of AV and BV gene transcripts in the peripheral blood monocytic cells (PBMCs) from 10 healthy individuals. Previous experiments showed that amplification of the cDNA was in the linear range (data not shown), and a comparison of the PCR primer sequences against the published V gene sequences (Arden et al., 1995) shows that our set of

PCR primers should amplify most V genes. It has been shown that the level of expression of TCR V genes by RT-PCR is not identical to the level of expression as determined by monoclonal antibody (mAb) staining and fluorescence-activated cell sorter (FACS) analysis (Diu et al., 1993). In some cases the oligonucleotide-driven amplification showed somewhat higher levels of V gene expression than the antibody staining, in others it was sometimes lower, and sometimes there was close agreement between the two methods. In the absence of a complete panel of antibodies that can stain most V genes (including all subfamilies and alleles), we chose RT-PCR as the best method for quickly determining the relative levels of expression for the AV and BV genes.

Our results showed that the level of expression of the V genes varied, and that different healthy individuals had similar expression levels, in agreement with several other studies (Baccala et al., 1991; Choi et al., 1989; Geursen et al., 1993; Theofilopoulos et al., 1993; Usuku et al., 1993). As is shown in Fig. 16.1, we

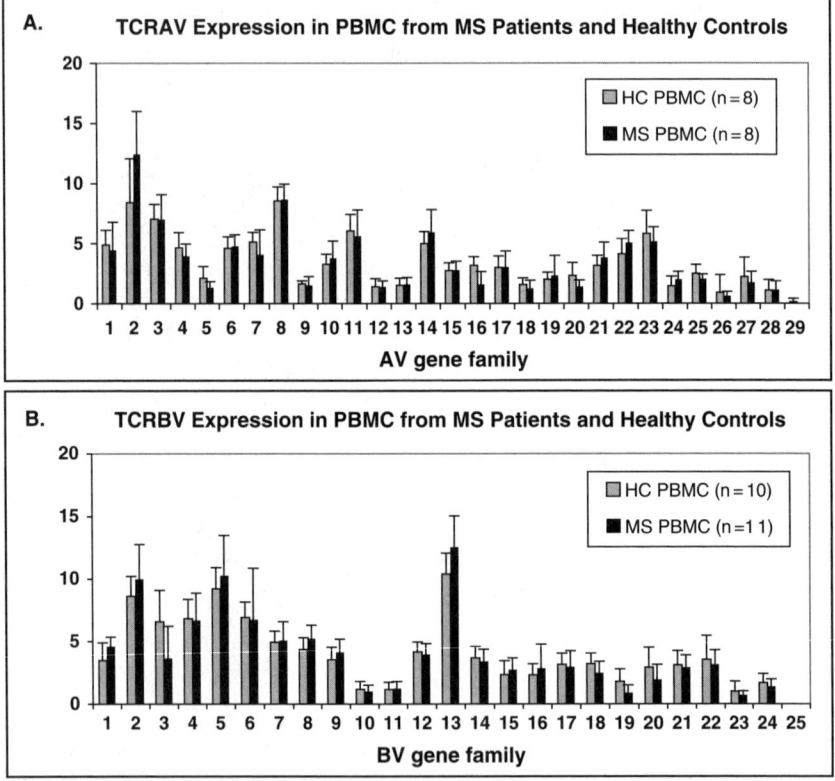

FIGURE 16.1. Expression pattern of T cell receptor (TCR) V genes in the peripheral blood monocytic cells (PBMC) of multiple sclerosis (MS) patients and healthy controls (HC) as assessed by the reverse transcription-polymerase chain reaction (RT-PCR).

found that AV8, AV2, and AV3 were the most common α-chain genes; and BV13, BV5, BV2, and BV6 were the most common β-chain genes. Despite great differences in methodology and the ethnic background of the individuals studied, these genes are often mentioned as most abundant in other reported studies. The expression level for each of these common V genes ranged from 7% to 10%. Other V genes had very low expression levels of < 1% of all T cells, with most having mid-range expression levels of 1% to 4%. Numerous factors are involved in the development of the peripheral TCR V gene repertoire (reviewed in Kay, 1996), and they can be described as either genetic mechanisms or environmental factors.

## 2.1   Genetic Mechanisms

The αβ T cell receptor is a disulfide-linked heterodimer consisting of an α-chain and a β-chain. Similar to immunoglobulins (Ig), the TCR chains are encoded by variable (V), diversity (D, β-chain only), joining (J), and constant (C) region genes, which are spliced together to form a complete α- or β-chain. The TCR AV genes are located on chromosome 14 (14q11-12), and it is estimated that 42 gene segments are functionally expressed, representing 32 AV gene families (Arden et al., 1995). These AV gene families are further subdivided into subfamilies, which share 75% nucleotide sequence identity. AV1, for example, has five subfamily members, AV2 has three subfamily members, AV3 has a single subfamily member, and AV8 has two subfamily members. If the peripheral AV gene repertoire were influenced by the number of subfamily members alone, one would predict high levels of AV1 and low levels of AV3, but that is not the case.

The TCR BV genes are located on chromosome 7 (7p15) and consist of 47 functionally expressed BV genes grouped into 23 functional families (BV10 and BV19, respectively, exist as a pseudogene and an orphon gene) (Arden et al., 1995). Several BV families have numerous subfamily members (BV5 has six functional subfamily members, BV6 and BV13 both have seven functional subfamily members), and all three of these BV gene families are highly expressed in peripheral blood. BV2 is also highly expressed in peripheral blood, yet it consists of only two subfamily members. Previously published results also showed that the usage of V gene segments in a human subject were not proportional to the number of gene subfamily members, with BV1 to BV4 being expressed in the peripheral repertoire at higher levels than expected, but other highly expressed V genes (BV6 and BV13) being used in proportion to the number of subfamily members (Robinson, 1992).

During the development of the T cell, the β-chain is rearranged first, followed by the α-chain. Owing to allelic exclusion, only one β-chain is rearranged on a T cell, although a certain percentage of T cells escape allelic exclusion and do go on to rearrange two β-chains (Triebel et al., 1988). It is estimated that up to 1% of human T cells express two distinct β-chains on their surface (Davodeau et al., 1995; Padovan et al., 1995). Each of the dual receptors on a single T cell is expressed at lower levels than if there were a single receptor (mutual dilution), and there may or may not be equal numbers of each of the two types expressed on the cell's surface. These dual receptors are apparently functional; they occur on single positive thymocytes (CD4+CD8−, CD4−CD8+); they can have distinct superantigen reactivities (Davodeau et al., 1995); and they occur in both the CD45RO+ and CD45RO− subsets, indicating that they survive thymic and peripheral selection (Padovan et al., 1995). Normal mice also show the presence of dual β-chain TCRs, and the number of such cells increases with age (Balomenos et al., 1995).

There is no allelic exclusion for the α-chain, and the proportion of T cells bearing two α–receptors has been estimated to be 25% (Corthay et al., 2001) to 33% (Padovan et al., 1993) in humans. The presence of these dual receptor T cells has important implications for the development of autoimmune disease, as such T cells can have dual specificities. It has been shown in a dual αβ-chain transgenic mouse system that both positive and negative selection is less efficient on dual receptor T cells, presumably because of the lower number of each type of TCR available for engagement (Dave et al., 1999). Another transgenic mouse study (this time with dual α-chains) showed that the low levels of expression of the autoreactive TCR allowed the T cell to escape tolerance and cause autoimmune diabetes. T cells bearing the autoreactive TCR as the sole receptor were deleted (Sarukhan et al., 1998). A third transgenic mouse model with dual αβ-chain TCRs demonstrated cross inhibition when one TCR was engaged by its antagonist and the second TCR was engaged by its agonist (Dittel et al., 1999).

During the rearrangements of the α- and β-chains, the splicing of the germline V-(D)-J segments is not random. One study found preferential pairing of certain BV genes to BJ segments, which would influence the BV repertoire available for thymic selection (Nanki et al., 1998). Another study did not find preferential pairing but did find that some BJ gene segments were preferentially used by all five individuals tested, even in unselected (aberrant—lacking the BV gene) BD-BJ-BC transcripts, indicating that the biased usage of BJ is due to genetic recombination events, not thymic selection (Rosenberg et al., 1992). In mice, the position of the individual TCR AV and AJ gene segments on the chromosome can also affect the likelihood of recombination occurring (Roth et al., 1991). A recent study of $V_H$ gene rearrangements in mouse Ig found that chromosomal position also played an important role in determining the recombination frequency for $V_H$ genes, and allelic polymorphisms of the recombination signal sequences (RSS) played a smaller role. For B cells, it is thought that the gene elements must be transcriptionally active for successful rearrangement to occur. If this is true in T cells as well, the fact that human TCR BV gene promoters display a wide range of activities could lead to different frequencies of recombination for individual BV genes (Deng et al., 1998).

The pairing of individual α- and β-chains is also not random. The expression levels of various αβ pairings can differ markedly (Saito and Germain, 1989). This may be due to structural constraints of the various α-β pairings (Uematsu, 1992) or may be due to self/major histocompatibility complex (MHC) or other individual (nonstructural) genetic constraints (Vacchio et al., 1993).

All of these nonrandom events probably contribute in some way to the V gene biases one sees in the peripheral T cell repertoire, and because many of these mechanisms are operative in all individuals, this may explain why the expression levels of the V gene families are so similar between individuals. Additional skewing in the peripheral repertoire is due to allelic polymorphism in TCR V genes, which would lead to individual variation in the expression of TCRs. Allelic polymorphisms can occur in the coding region of the mature V gene protein and can result in amino acid substitutions, or "silent" mutations. In one study, it was found that the two allelic forms of Vβ6.7 (BV6S5A1 and BV6S5A2, Arden nomenclature) (Arden et al., 1995)) were expressed at very different levels in the peripheral blood despite a difference of only two amino acids. Heterozygous individuals varied regarding the allele they favored, and this was not related to HLA type (Vissinga et al., 1994). The same study showed no such difference in the expression level of BV12 alleles with a silent mutation. Another known coding region allelic polymorphism includes a premature stop codon, which results in a null allele in BV20S1 for some individuals (Charmley et al., 1993; Malhotra et al., 1992).

Allelic polymorphisms of the noncoding regions of TCR V genes can also affect the expression levels in peripheral blood. As one example, Posnett, et al. (1994) found an allelic polymorphism in the spacer region between the heptamer and nonamer sequences of the recombination signal sequence (RSS) of BV3S1 that affected the number of T cells expressing this V gene. Individuals who were homozygous for allele 2 had high levels of circulating BV3 T cells, homozygotes for allele 1 had low levels of BV3 T cells, and heterozygotes had intermediate levels. This polymorphism in the RSS may function to modify recombination efficiency.

Two insertion/deletion-related polymorphisms (IDRPs) of the TCR BV gene locus have also been noted (Seboun et al., 1989). They were found to be widespread, with 49 of 50 healthy individuals positively typed for one or both alleles. At least one of these IDRPs involves the coding region of functional V genes and leads to formation of a new gene segment (BV7S3) and the duplication of another (BV13S2) (Zhao et al., 1994).

## 2.2  Thymic Selection Mechanisms

Once the αβ TCR is present on the surface, the immature T cell is subject to positive and negative thymic selection. Negative selection leads to death of the thymocyte due to apoptosis, whereas positive selection leads to the activation and differentiation of the thymocyte into either CD4+ or CD8+ T cells. The role of the MHC molecule is crucial, as it presents the self peptide to the TCR of the thymocyte during these selection events. According to the differential avidity model of T cell selection (Ashton-Rickardt and Tonegawa, 1994), high-avidity interactions between the TCR and MHC/peptide leads to negative selection, whereas low avidity interactions lead to positive selection. If there is no recognition of the peptide/MHC by the T cell, this also leads to the death of the T cell (death by neglect). Others have postulated that positive and negative selection result from the thymocyte receiving qualitatively different signals during the selection process through interactions with distinct thymic stromal types (Laufer et al., 1999).

## 2.3  Role of the MHC

Allelic differences in the MHC molecules can affect the peptide-binding groove and determine whether a given MHC molecule can bind and present a self peptide efficiently. This would presumably affect the peptides available for selection of thymocyte TCRs and may be one reason that certain MHC haplotypes are associated with various autoimmune disorders. In addition, analysis of the crystal structure of an αβ TCR indicates that the germline-encoded CDR1 and CDR2 loops of the TCR contact the α-helices of the MHC molecule as well as the ends of the bound peptide (Garcia et al., 1996; Reinherz et al., 1999); hence, during positive selection it is possible that the MHC I or MHC II alleles can preferentially select different V genes.

A number of studies have looked at the role of the MHC in selecting the peripheral T cell repertoire. Theophilopoulos et al. (1993) found that the V gene expression in immature thymocytes is essentially the same as for mature single-positive (CD4+CD8–, CD4–CD8+) T cells but at the same time found that certain V genes are skewed toward either the CD4+ or CD8+ subsets; many of these V gene biases were applicable to all individuals tested. In contrast, Baccala et al. (1991) found no correlation between CD4+ and CD8+ thymocytes and V gene usage, although they did report that there was no statistically significant difference between V gene usage in total thymocytes (including single-positives), double-negative thymocytes, and double-positive thymocytes. Usuku et al. (1993) found skewing of the V gene repertoire in CD4+ and CD8+ subsets, with elevated BV2 and BV5S1 in CD4+ cells and elevated BV10, BV14, and BV16 in CD8+ cells. Others have found elevated BV6S7a in the CD4+ cells and elevated BV5.2/5.3 in CD8+ cells (Gulwani-Akolkar et al., 1991). Akolkar et al. (1993) found that HLA-identical siblings had more similar patterns of V gene usage than HLA-mismatched or partially mismatched siblings; they also found elevated use of BV2, BV5S1, BV9, and BV20 in CD4+ cells, with elevated BV7 and BV14 in CD8+ cells. They were unable to detect any relation between HLA haplotype and patterns of V gene usage in totally unrelated individuals, but this may be due to the fact that there were, at most, only four HLA-A, B, DR, or DQ alleles in common among the most similar unrelated individuals. Hawes et al. (1993) also found that V usage in identical twins was more similar than the V gene usage in unrelated individuals; they also found that AV11, AV17, AV22, BV3, BV9, BV12, and BV18 were skewed to the CD4+ subset, and AV2, AV6, AV12, AV15, AV20, BV7, BV14, and BV17 were skewed to the CD8+ subset. These studies of related individuals suggest that the peripheral repertoire is shaped largely through genetic factors, as environmental antigens seem to play a small role in the expression of V genes in identical twins. It appears then, that genetic factors are probably the major contributors to shaping the peripheral repertoire for healthy individuals.

## 2.4   Environmental Factors

Environmental factors also alter the peripheral repertoire and are probably responsible for the smaller individual differences seen in V gene expression between closely related individuals such as homozygous twins. A persistent question concerns the role of superantigens in the development of the human T cell repertoire. Unlike conventional antigens, superantigens bind to the outer surfaces of the MHC II molecule and TCR and are able to stimulate every T cell bearing a given BV gene(s) regardless of the amino acid sequence of the BD, BJ, BC, or α-chain. There is a family of endogenous superantigens in the mouse, called Mls (minor lymphocyte stimulation antigen), that are able to stimulate proliferation of T cells bearing certain BV genes in vitro; when expressed in the thymus, they cause deletions of all T cells bearing these same BV genes. To date, no human studies have shown similar deletions of large subsets of V genes in the human TCR repertoire, so it is unlikely that endogenous superantigens play a large role in human repertoire development. Exogenous superantigens are microbial or viral proteins with potent immunostimulatory effects that cause food poisoning and shock in animals and humans. Staphylococcal enterotoxins (SEs) A, B, C1, C2, C3, D, and E as well as toxic shock syndrome toxin (TSST-1) are produced by various strains of *Staphylococcus aureus*, whereas the streptococcal pyrogenic exotoxins A, B, and C are from *Streptococcus pyogenes*. When cultured with human T cells in vitro, SEB causes the proliferation of virtually all BV3 T cells, SEE causes proliferation of BV6 and BV8 T cells, and TSST-1 causes proliferation of BV2 and BV6 T cells (Choi et al., 1989; reviewed in Herman et al., 1991; Marrack and Kappler, 1990; Papageorgiou and Acharya, 2000). One study found that five of eight patients suffering from toxic shock syndrome expressed significantly elevated frequencies of BV2 T cells; and a longitudinal analysis of two of the patients showed that the BV2 levels returned to normal after 45 to 60 days (Choi et al., 1990). The authors did not sequence the TCR transcripts, so it is unknown if the expansion of BV2 T cells was oligoclonal. Based on this study, it seems that even if there are superantigen-driven expansions of various V genes the levels of expression decline to normal levels once the disease has resolved.

Other viral infections are also known to affect the T cell repertoire. One such study looked at two HLA−B8+ patients with chronic Epstein-Barr virus (EBV) infection and found that a clonally expanded population of BV6S2BJ2S7 T cells with identical CDR3 sequence dominated the CD8+CD45RO+CD45RA− repertoires of both patients (Silins et al., 1998). The role of infection in the development of autoimmune diseases is still unclear, but streptococcal infection can lead to rheumatic fever and heart disease, and there is evidence that in some cases autoimmune disease was preceded by bacterial or viral infections.

In conclusion, it appears that no one factor can explain the overall level of V gene expression. Rather, it is a combination of genetic factors—many of which are common to all individuals—that largely shapes the expression levels of V genes. Superimposed on this innate "potential" repertoire are the individual genetic polymorphisms that lead to the final expression levels seen in the mature thymocytes.

Environmental factors then interact with this pool of available T cells, leading to further differences in expression levels of V genes and longitudinal variation within the repertoire of a given individual.

# 3   V Gene Expression Patterns in Autoimmune Diseases and Allergy

## 3.1   Peripheral TCR Repertoire Does Not Differ in MS Patients Versus Healthy Controls

By far the most effort has been directed at evaluating V gene expression in patients with multiple sclerosis (MS). Using primers and probes specific for most members of each AV and BV gene family, we compared the expressed TCR repertoire of unselected PBMCs from MS patients versus those from HCs. As shown in Figure 16.1 and Table 16.1, the expression of AV and BV genes was quite similar for MS patients and HCs, although levels of expression among different AV and BV gene families varied considerably.

Among AV gene families, AV2, AV8, AV3, AV11, AV14, and AV23 predominated, in descending order, altogether representing about 40% of total AV gene expression (Fig. 16.1A). Among the BV gene families, BV13, BV5, BV2, BV6, and BV4 predominated, in descending order, altogether representing about 45% of total BV gene expression. All other AV and BV gene families were detectable in PBMCs. The overall pattern of expression of AV and BV genes in PBMCs did not differ significantly between MS patients and HCs, as assessed by the Mann-Whitney test (AV, $p = 0.3483$; BV, $p = 0.3745$), nor were there any significant differences in expression of individual V gene families, as assessed by the Wilcoxin test with Bonferroni adjustment. This result demonstrates that the expressed peripheral T cell repertoire of MS patients is not skewed relative to that in HCs.

## 3.2   V Genes Used by MBP-Specific T Cell Clones from MS Patients

We evaluated AV and BV gene usage by T cell clones specific for the 18.6-kDa form of human myelin basic protein (MBP) (Afshar et al., 1998; Ben-Nun et al., 1991; Goebels et al., 2000; Joshi et al., 1993; Kotzin et al., 1991; Martin et al., 1992; Meinl et al., 1993; Muraro et al., 1997; Richert et al., 1995; Shanmugam et al., 1996; Uccelli et al., 1998; Vandevyver et al., 1995; Wucherpfennig et al., 1990, 1994; Zhang et al., 1995a,b). The isolates included in the analysis were all selected from PBMCs using intact MBP and were not segregated into groups according to peptide specificity. A total of 263 T cell isolates from 73 MS patients were evaluated for AV gene expression. Among them, 163 expressed a single AV gene and 101 expressed two different AV genes, resulting in a total of 364 plotted AV genes (Fig. 16.2A). For BV genes, a total of 528 T cell isolates from 99 MS patients were evaluated. Among them, 368 isolates were clonal, expressing one predominant BV gene, and 148 isolates contained two clones with relatively equal expression of two different BV genes, resulting in a total of 676 plotted BV genes (Fig. 16.2B). The isolates from each donor are stacked according to V gene use using different colors and textures.

Figure 16.2 shows that MBP-specific T cells from MS patients used a wide variety of T cell receptors in recognition of MBP epitopes, and that nearly all V genes were used at least once. However, some V genes were utilized much more extensively than others. Figure 16.2A shows that AV8 was by far the most abundant AV gene, and AV1, AV2, AV3, AV4 were also frequently used. Figure 16.2B

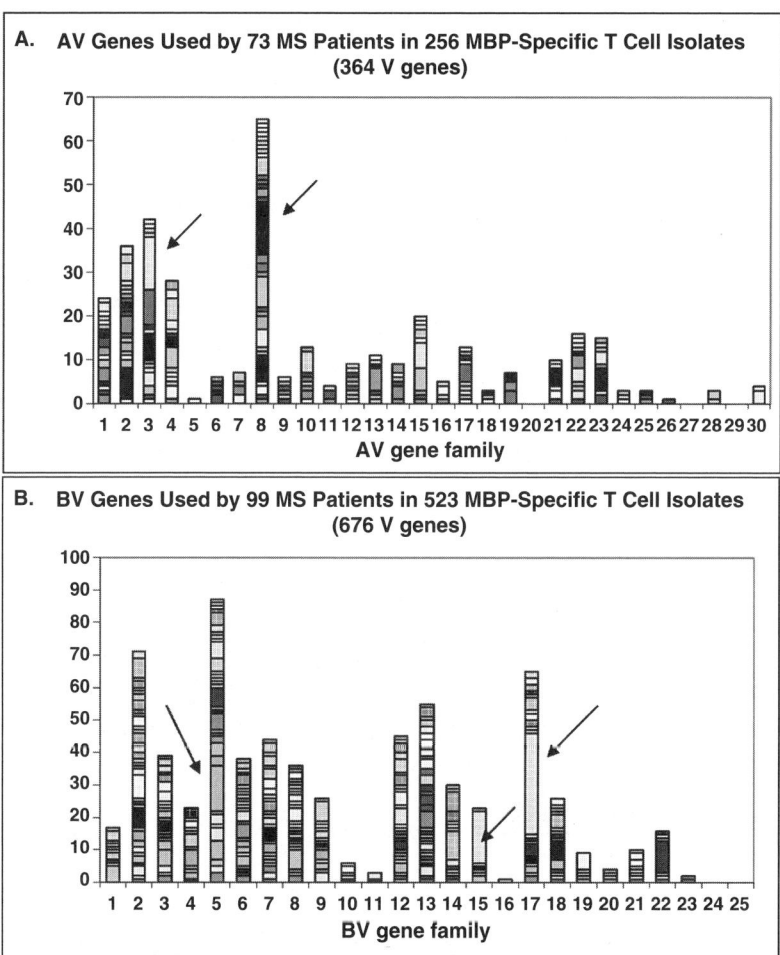

FIGURE 16.2. Number of occurrences of AV and BV genes used by MBP-specific T cell isolates from PBMC of MS patients. Arrows indicate oligoclonal expansions in some MS donors.

shows that BV5 was the most abundant BV gene, and BV2, BV17, and BV13 were also frequent. Each colored segment represents the isolates derived from a single patient, and it is possible to see from these figures that some patients have clonal or oligoclonal expansions (represented by large segments; see arrows indicating biased usage of AV3 or AV8 in Figure 16.2A or BV5, BV15, and BV17 in Figure 16.2B). This biased usage of just a few V genes by isolates from a given patient seems to be the exception, however, as most patients used a variety of V genes.

## 3.3    Pattern of MS V Gene Expression by MBP-Specific T Cells Reflects Pattern in PBMCs

The most commonly used V genes in the MBP-specific isolates appeared to match the most abundant V genes found in PBMCs. As is shown in Figure 16.3, we compared the percent contribution of each V gene seen in the T cell isolates (from Fig. 16.2) to the percent expression of each V gene in PBMC (from Fig. 16.1). These patterns had much similarity and were not significantly different from each other when evaluated for rank order (Mann-Whitney test). It is noteworthy that the three most predominant AV genes in MBP-specific T cells

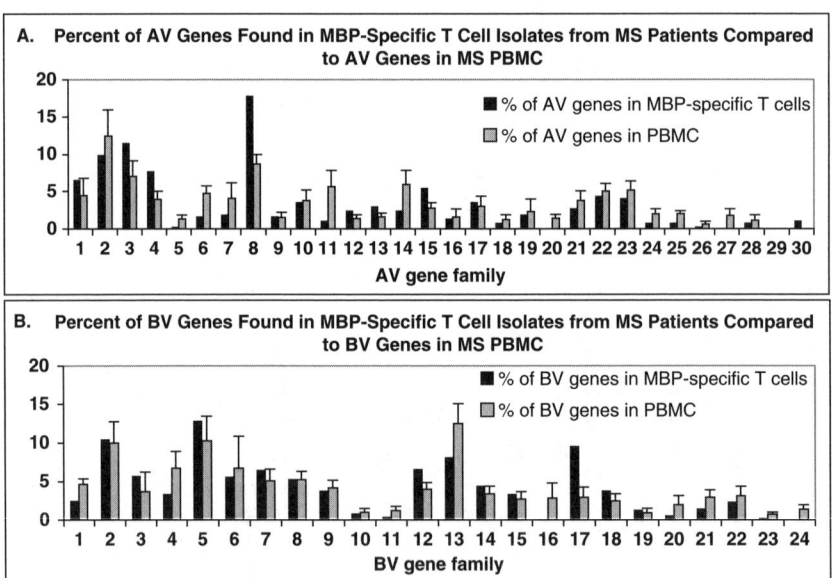

FIGURE 16.3. Comparison of the percent total for each V gene used by myelin basic protein (MBP)-specific T cells with percent total V gene expression in PBMC from MS patients. The overall rank order of expression of V genes by MBP-specific T cells did not differ significantly from the pattern of PBMC (**A**: AV comparison, $p = 0.138$; **B**: BV comparison, $p = 0.405$; Mann-Whitney test).

(AV8, AV3, AV2 = 40% of total) coincided with the three most expressed AV genes from PBMCs (AV2, AV8, AV3 = 27% of total); and three of the four most predominant BV genes from MBP-specific T cells (BV5, BV2, BV13, but not BV17 = 33% of total) coincided with the three most expressed BV genes from PBMCs (BV13, BV5, BV2 = 33% of total). This comparison demonstrates that, for the most part, the TCR V genes expressed most abundantly by MBP-specific T cells from MS patients emerged from the most abundant V gene families present in the circulating TCR repertoire.

## 3.4   V Genes Used by MBP-Specific T Cell Clones from HCs

We carried out a similar evaluation of V genes expressed by MBP-specific isolates derived from the PBMCs of HCs. From the literature we analyzed TCR AV gene data from 129 T cell isolates (154 AV genes plotted) from 32 individuals (Fig. 16.4A) and TCR BV data from 178 T cell isolates (213 BV genes plotted) from 46 individuals (Fig. 16.4B). Similar to the MS patients, the MBP-specific T cell isolates from HCs used a wide variety of T cell receptors, with nearly all V genes being represented. Again, some V genes were utilized much more extensively than others, although some of the predominant V genes were different from those in MS patients. AV2, AV3, AV4, and AV21 were the most abundant AV genes (Fig. 16.4A), and BV6, BV7, BV14, and BV5 were the most abundant BV genes (Fig. 16.4B). As with the MS patients, the HCs also had oligoclonal expansions (such as biased usage of AV4, BV7, or BV14 by single individuals).

## 3.5   Overall Pattern of V Gene Expression by MBP-Specific T Cells from HC Donors Is Generally the Same as for PBMCs

As for the MS patients, we compared the pattern of expression of V genes used in MBP-specific T cells (from Fig. 16.4) to the pattern of expression of V genes in PBMCs for HCs (from Fig. 16.1). Although most of the AV genes appeared to be expressed at levels similar to those in PBMCs, AV3, AV4, AV12, AV15, and AV21 all appear more commonly in the T cell isolates. All of these AV genes were used preferentially by T cell isolates from one or two individuals, which may explain the discrepancy. Overall, however, the differences between the levels of TCR AV genes found in isolates versus PBMCs were not found to be statistically significant ($p = 0.145$, Mann-Whitney test). As with the AV genes, the expression levels of BV genes in the MBP-specific T cells of HCs versus PBMCs were also roughly equivalent ($p = 0.215$, Mann-Whitney test). Again, however, there were apparent differences in expression of some BV gene families, some of which could be attributed to the preferential use of a given BV gene by a single individual.

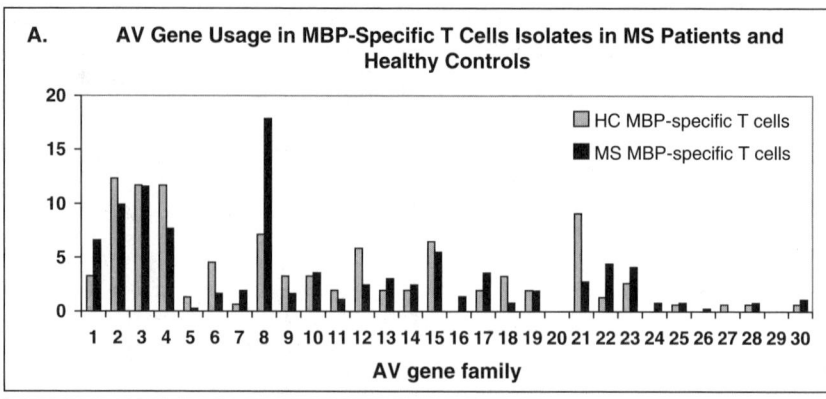

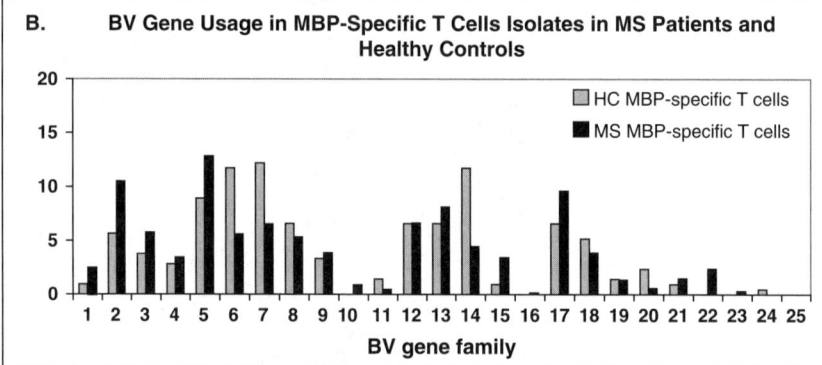

FIGURE 16.4. Comparison of V gene expression by MBP-specific T cells in MS patients and HCs. The overall rank order comparisons between isolates from MS patients and HCs were not significantly different. **A.** AV comparison, $p = 0.500$. **B.** BV comparison, $p = 0.417$; both by the Mann-Whitney test). However, note the significantly increased levels of AV8 and BV2 in MS patients and the increased levels of AV21, BV6, BV7, and BV14 in HCs (*) as assessed by chi-squared analyses.

## 3.6    AV8 and BV2 Genes Are Overexpressed by MBP-Specific T Cells from MS Patients versus HCs

V genes used by MBP-specific T cells were plotted as the percent of the total expressed V genes and compared in MS patients versus HCs (Fig. 16.4). Most V gene families were expressed at comparable levels, and there was no significant difference in the overall pattern of AV and BV genes utilized by MS patients versus HCs, as assessed by the Mann-Whitney test. However, there were highly significant differences in expression of six individual V gene families, as assessed by chi-squared analysis. AV8 ($p < 0.003$) and BV2 ($p < 0.05$) were significantly overexpressed by MBP-specific isolates from MS patients versus HCs (Fig. 16.4).

Other prevalent MBP-specific MS V genes, including BV5, BV13, and BV17, were also expressed at higher levels than in HCs, but the differences were not significant. On the other hand, AV21 ($p < 0.004$), BV6 ($p = 0.004$), BV7 ($p = 0.01$), and BV14 ($p < 0.001$) were significantly overexpressed by HCs versus MS patients (Fig. 16.4).

## 3.7   Differences in Patterns of V Gene Expression Among Contributing Research Groups

Much of the contention regarding V gene expression by MBP-specific T cells arose from profound differences in the patterns reported by various research groups. Among six groups that contributed most of the T cell clones to the literature [Vandenbark, Portland, OR, USA (150 AV genes from 18 patients; 210 BV genes from 23 patients); Raus-Zhang, Diepenbeek, Belgium (88 AV genes from 22 patients; 143 BV genes from 25 patients; Martin-McFarland, Bethesda, MD, USA (76 AV genes from 23 patients); 70 BV genes from 22 patients); Hafler, Boston, MA, USA (42 AV genes from 6 patients); 109 BV genes from 11 patients); Hauser, San Francisco, CA, USA (82 BV genes from 7 patients); and Meinl-Wekerle-Hohlfeld, Munich, Germany (8 AV genes from 4 patients; 44 BV genes from 8 patients)], there was a rough consensus for many of the V gene families but striking differences for others (i.e., AV2, AV15, AV16, AV22; and BV9, BV15, BV17, BV22). Overall, the patterns obtained by the various groups were significantly different from each other ($p < 0.001$ for AV genes; $p < 0.1$ for BV genes; Kreskal Wallace test). Nearly all of these differences could be explained by the contribution of many similar or identical clones from a single donor. For example, the high percentages of AV3 and BV17 reported by Hafler's group represented > 30 clones isolated from one patient. Although some of these isolates clearly were replicate clones (with the same CDR3 sequences), most were oligoclonal, expressing the same AV and BV genes but different CDR3 sequences; and in a few isolates, BV17 paired with a different AV gene. Such oligoclonal T cells specific for MBP were sporadically isolated by all the contributing groups from both MS patients (see arrows in Fig. 16.2) and HCs and clearly are characteristic of the human response to MBP. However, this phenomenon just as clearly skewed the results of individual studies that included relatively few isolates and may well have contributed to the widely held impression that there are regional differences in V gene expression (a contention that cannot be addressed here). The inclusion of larger numbers of isolates from many sources thus provides a more balanced view of V gene expression that is less influenced by individual oligoclonal expansions.

A second factor that may have contributed to the observed differences was variations in the concentration of MBP used to select the T cell isolates in vitro. As shown by Mazzanti et al. (2000), using MBP concentrations ranging from 0.1 μg/ml to 50 μg/ml to select T cells from both MS patients and HC, low MBP concentrations yielded a small number of T cell isolates with limited epitope specificity that recognized the antigen with higher avidity, whereas high doses yielded a larger number of isolates with broader specificity and lower avidity. The MBP-specific T cell isolates evaluated for V gene expression included in our analysis were selected using MBP concentrations ranging from 10 to 100 μg/ml, depending on the research group. Evaluation of numbers of isolates obtained at different MBP concentrations showed a positive trend (more isolates at higher MBP concentrations) for BV2, BV5, and BV6; a flat trend for BV13 and BV17; and a negative trend for BV1, BV3, BV7, BV8, BV12, and BV14 (other BV genes were not evaluated). Thus, at higher MBP concentrations (e.g., in the range of 50 μg/ml used by most groups), isolates expressing BV2, BV5, BV6, BV13, and BV17 would be favored, and isolates expressing the other BV genes would be disfavored. Although this explanation more or less fits the observed data, it would also imply that the choice of a particular V gene segment, without consideration of its CDR3 segment, could influence the avidity of the TCR. Although this possibility is conceivable because of the demonstrated interactions of CDR1 and CDR2 loops with MHC-Ag, there is no clear precedent in the literature that speaks to this issue.

## 3.8    Distribution of MBP-Specific V Genes Among MS and HC Donors

Due to the uneven contribution of isolates from different donors and the tendency for some donors to have oligoclonal expansions of MBP-specific T cells, it is possible that the distribution of V genes from each MS patient and HC donor might differ from the overall expression pattern of the T cell isolates. Thus, we evaluated the percentage of donors who used a given V gene at least once by an MBP-specific isolate. For MS patients, AV8 was the most commonly used V gene, with 32 of the 73 patients (44%) using this particular TCR. Altogether, 20 patients (27%) used AV2, 17 patients (23%) used AV1, 16 patients (22%) used AV3, and 15 patients (20%) used AV4. For BV genes, BV2 was the most commonly used, with 37 (37%) of the 99 patients using this particular receptor family at least once. Additionally, BV5 was present in 34 patients (34%), and BV13 was present in 31 patients (31%). These data clearly show that the distribution of V genes used by MBP-specific T cells among donors was similar to the overall percent expression of total V genes (Fig. 16.2), with the possible exception of BV17, which was less broadly expressed among donors than among total isolates due to oligoclonality (see above).

Similarly, we evaluated the percentage of HCs who used a given AV or BV gene at least once in their MBP-specific T cell isolates. AV2, AV3, AV8, and AV21 and BV5, BV6, and BV12 were present in the highest percentage of HCs. Notably, AV4, BV7, and BV14, which were among the most prevalent V genes in HCs (Fig. 16.4), were not as broadly represented among HCs owing to the oligoclonal expression of these V genes in single HC donors. Interestingly, the percentage of donors expressing any given AV or BV gene did not differ significantly between MS patients and HCs, although there was a strong trend for BV2, which was present in 37% (37/99) of MS patients versus only 19% (9/46) of HCs ($p = 0.051$).

## 3.9    DR2-Associated V Genes in MS Patients and HCs

HLA-DR2 (more precisely, DRB1*1501) is an MS-associated MHC II allele, being present in about 60% of Caucasian MS patients but only about 25% of HCs. Although the contribution of the DR2 allele to the MS disease process is not known, it is conceivable that the ridges of the peptide binding groove for DR2 could contribute to the preferential selection of compatible residues in the CDR1 and CDR2 loops of TCR AV and BV chains that interface with the MHC/Ag complex and thus contribute to V gene selection. To discern the effect of DR2 on V gene use, we analyzed V gene expression patterns in DR2+ versus DR2– MS patients and HCs. For MS patients, the number of occurrences for AV3 ($p = 0.011$), BV4 ($p = 0.020$), and BV17 ($p = 0.033$) was significantly higher in DR2+ donors than in DR2– donors. However, there were no statistically significant differences in the number of DR2+ patients versus DR2– patients that used these or any other particular V gene. This result implies that a given DR2+ patient is more likely to have several clones that expressed AV3, BV4, or BV17, suggesting positive selection and in vivo expansion of these particular V genes in the DR2+ patients. For HC donors, the number of occurrences for BV7 and BV17 was significantly higher in DR2+ donors than in DR2– donors. Again, there were no statistically significant differences in the numbers of DR2+ patients versus DR2– patients that used these or any other particular V gene, indicating oligoclonality within these V gene families among HC donors.

## 3.10    V Gene Expression in MS T Cell Clones Specific for Peptides from MBP, PLP, and MOG

We found five articles in the literature that evaluated V gene expression in T cell clones selected for response to myelin peptides. The V gene use for MBP-peptide-selected clones is presented separately from those selected with whole MBP because of the presence of "cryptic" epitopes in synthetic peptides that may not be present after processing by antigen-presenting cells (APCs) of "natural" epitopes in intact MBP. Moreover, the three studies reporting MBP-83-99 or MBP-85-99 selected T cell clones evaluated only DR2+ MS patients (Ausubel et al., 1999; Hong et al., 1999; Zang et al., 1998). As is shown in Table 16.1, the predominant V genes used by MS MBP-85-99-specific T cell clones included AV3 and AV8 plus BV17, BV13, BV12, and BV2. In particular, AV3, AV8, BV13, and BV17 were mostly expressed on DRB1*1501-restricted clones. The strong representation of AV3 and BV17 in MBP-85-99-specific clones confirmed results from the MBP-selected T cells (see above). The lack of BV4 in these data sets from DR2+ donors may be due to the limitation of clonal responses to a single MBP epitope rather than to the full set of determinants present on the intact MBP molecule. Thus, it is possible that DR2-restricted T cells that express BV4 are specific for a different MBP determinant.

Two articles were also found that evaluated V gene used in T cell clones specific for peptide determinants from myelin oligodendroglial cell glycoprotein (MOG)

TABLE 16.1. TCR V Gene Expression Patterns in Healthy Controls vs. Patients With Various Diseases

| T cells analyzed | | | AV gene expression | | | BV gene expression | | |
|---|---|---|---|---|---|---|---|---|
| Disease | Cell source | T cell specificity | No. of V genes | No. of donors | Dominant AV genes | No. of V genes | No. of donors | Dominant BV genes |
| HC | Blood | Unselected | All | 10 | 8, 2, 3 | All | 10 | 13, 5, 2, 6 |
| | | MBP (whole) | 154 | 32 | 2, 3, 4 | 213 | 46 | 6, 7, 14, 5 |
| MS | Blood | Unselected | All | 11 | 2, 8, 3 | All | 11 | 13, 5, 2, 6 |
| | | MBP (whole) | 364 | 73 | 8, 2, 3, 4 | 676 | 99 | 5, 2, 17, 13 |
| | | MBP 85-99 | 46 | 16 | 3, 8 | 53 | 19 | 17, 13, 12, 2 |
| | | MOG | | | | 8 | 5 | 8, 13, 17 |
| | | PLP peptides | | | | 11 | 1 | 5, 2, 17 |
| | | HPRT- MBP/PLP | | | | 9 | 3 | 13, 2, 5, 6, 8, 14, 17 |
| | CSF/ Blood | Unselected | | | | All | 84 | 6, 2, 5, 13, 7 |
| | | MBP (whole) | 52 | 7 | 2, 10 | 65 | 11 | 7, 5, 18, 13 |
| | Brain | Unselected | All | 21 | 10, 8, 2 | All | 21 | 5, 7, 8, 6, 12 |
| MG | Blood | AChR | 13 | 6 | 8 | 28 | 18 | 6, 7, 5, 2 |
| Lupus | Blood | DNA | 42 | 5 | 8 | 42 | 5 | 6, 2, 5, 8 |
| IDDM | Blood | ISG | | | | 16 | 9 | 8, 2, 19 |
| | | ICA69 | | | | 18 | 10 | 8, 13, 14 |
| | | HPRT- | | | | 164 | 9 | 14, 6, 2, 13 |
| Arthritis | SF | Unselected | All | 51 | 5, 19 | All | 67 | 13, 2, 6 |
| | | CD4+ | | | | All | 7 | 2, 6, 21, 15 |
| | ST | Unselected | | | | All | 25 | 3, 6, 15 |
| | | CD4+ | | | | All | 5 | 6, 2, 13 |
| Allergy | Blood | Der p 1 dust mite | 8 | 7 | 3, 8 | 59 | 6 | 5, 6, 21 |
| | | Bet v 1 birch pollen | 19 | 11 | 8, 2 | 19 | 12 | 2, 6 |

MS, multiple sclerosis; MG, myasthenia gravis; Lupus, lupus nephritis; IDDM, type 1 diabetes (IDDM); CSF, cerebrospinal fluid; SF, synovial fluid; ST, synovial tissue; ISG, insulin secretory granules; ICA69, a recombinant B cell protein; MBP, myelin basic protein; MOG, myelin oligodendroglial cell glycoprotein; PLP, proteolipid protein; AChR, acetylcholine receptor; Der p 1 dust mite, major antigen of *Dermatophagoides pteronyssinus*, Bet v 1 birch pollen, major antigen of *Betula verrucosa*. Gray highlighting indicates the presence of the same dominant AV or BV genes in patient samples as in HC PBMCs.
Dominant V genes are listed in decreasing order of expression.

and proteolipid protein (PLP). These two myelin antigens are highly encephalitogenic in rodents and thus may represent potentially important T cell determinants in MS. Lindert et al. (1999) used a partial panel of antibodies to BV gene products to stain 8 MS T cell lines specific for the extracellular domain of MOG or a pool of MOG peptides. The results from this study indicated that the most frequent V genes included BV8, BV13S1, BV17, BV20, and BV22 (Table 16.1). A similar approach using a partial panel of antibodies to BV genes was used by Correale et al. (1995) to stain T cell clones specific for PLP104-117 or PLP142-153 peptides from eight MS patients. BV genes could be positively identified in about half of the MS T cell clones, and the results indicated a marked diversity of BV genes present. However, evaluation of 11 clones from one particular donor revealed use of BV5 (five clones), BV2 and BV17 (one clone each) (Table 16.1), and four clones that could not be identified.

An additional data set should also be mentioned. Two studies evaluated HPRT-mutant T cell clones specific for MBP or PLP peptides obtained from the PBMCs of MS patients (Lodge et al., 1994; Trotter et al., 1997). The HPRT-mutation indicates that these T cells have been activated in vivo and thus may be relevant to the MS disease process. The results from nine HPRT-clones indicated expression of BV2, BV5, BV6, BV8, BV12, BV13 (2), BV14, and BV17 (Table 16.1). Taken together, these studies suggest that BV2, BV5, BV13, and BV17, which are prominent BV genes present on MS T cell clones specific for MBP, are also prominent BV genes utilized by MOG and PLP peptide-specific T cells.

## 3.11   V Gene Expression in Brain Tissue and CSF T Cells from MS Patients

Several studies looked at the V genes expressed in T cells taken from MS brain lesions (presumably of greatest disease relevance), as well as from the cerebrospinal fluid (CSF) of MS patients. Three studies looked at TCR AV usage in MS brain plaques, and all found a polyclonal T cell response (Oksenberg et al., 1990, 1993; Wucherpfennig et al., 1992). More V gene diversity was seen in the acute plaques than in the chronic plaques (Wucherpfennig, et al., 1992), which may be due to the paucity of TCR RNA found in the chronic plaques. The most common AV genes in plaque tissue were AV10 (100% of 14 patients) and AV8 (93% of patients). Other commonly seen AV genes were AV1 (71% of patients' brains), AV2 (79%), and AV7 (64%).

Three studies looked at TCR BV usage in MS brain tissue (Birnbaum and van Ness, 1992; Oksenberg et al., 1993; Wucherpfennig et al., 1992), and a heterogeneous BV response was also seen (see Fig. 16.5B). One study (Birnbaum and van Ness, 1992) found that BV12 was greatly increased in both of the MS brains tested, but there was no corresponding oligoclonal expansion found in the CSF of these patients. Additionally, there were clear differences between V gene usage in CSF versus blood, and both differed from the V gene usage in the brain lesions. When all three of these brain plaque BV gene studies are combined, 61% of the 23 patients tested used BV5, 52% used BV12, and 43% used BV7 and/or BV8. The absence of BV13 usage is striking and may be due partially to the fact that one of the three studies (Wucherpfennig et al., 1992) apparently did not use a PCR primer that could detect V genes from the BV13 family.

We also looked at CSF-derived T cells described in the literature and found that unselected (ex vivo) CSF T cells utilized V genes that were considerably different than CSF T cells that were either expanded and selected in the presence of MBP

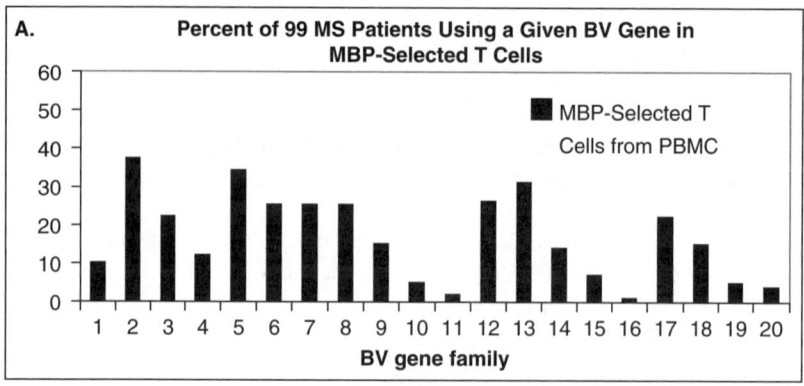

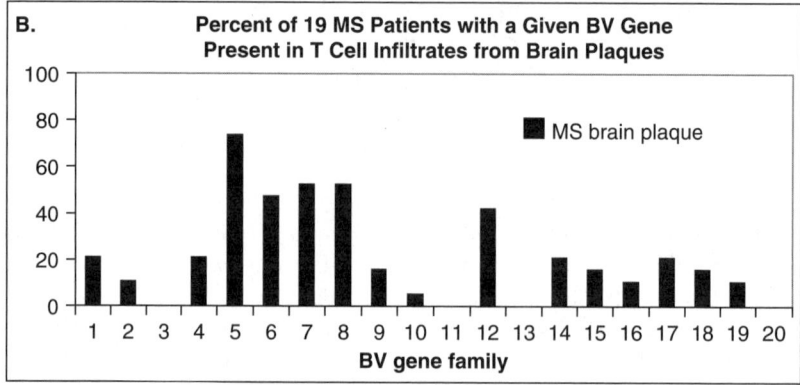

FIGURE 16.5. Percentage of MS patients who used a given BV gene at least once in unselected or MBP-specific T cell isolates from blood, brain, and cerebrospinal fluid. Note that the most broadly represented V genes (BV2, BV5, BV13) were also the most prevalent V genes.

or were found to be MBP-specific after expansion in IL-2 and IL-4. We divided the TCR V gene data from the various CSF studies into: (1) T cells taken directly from the CSF without any manipulation or in vitro expansion; (2) CSF cells that were expanded by culture in IL-2 and IL-4 (presumably activated T cells, expressing the IL-2 receptor); (3) CSF cells that were expanded in IL-2 and IL-4 and also found to be MBP-specific; (4) CSF cells that were expanded and selected in the presence of MBP.

The V gene repertoire of the CSF T cells appears to be strongly affected by the method of culture. Studies that compared V genes in unmanipulated (ex vivo) CSF found a heterogeneous population where most V genes were present (see Fig. 16.5C). Birnbaum (1992) looked at paired samples of ex vivo CSF and peripheral blood from MS patients and concluded that the V gene usage patterns were clearly different, whereas Usuku (1996) thought the V genes in ex vivo CSF

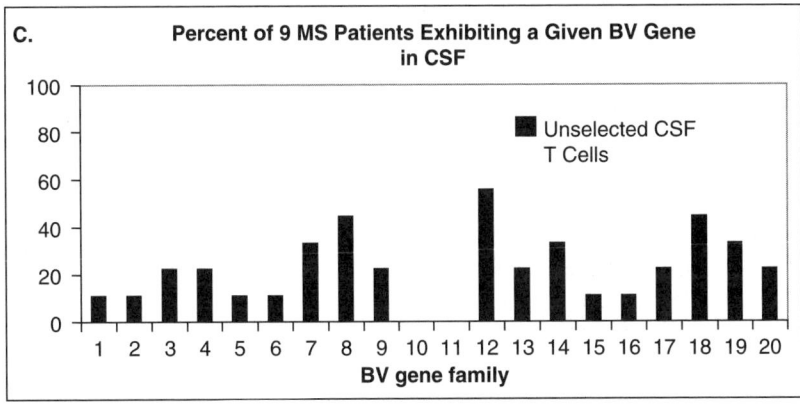

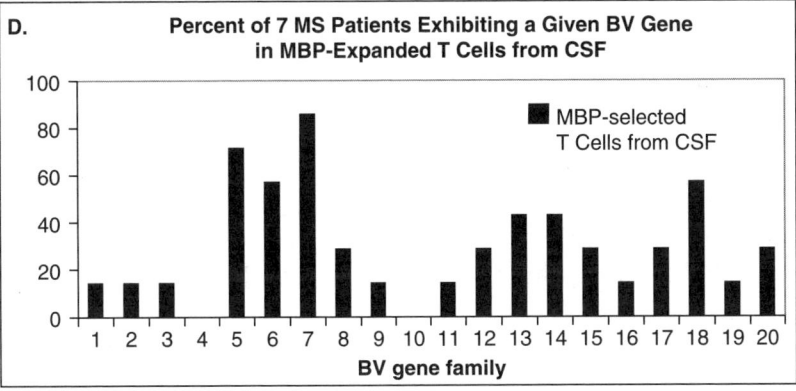

FIGURE 16.5. (*Continued*)

occurred in proportion to their representation in peripheral blood. Anti-CD3 stimulation appeared to expand all T cells indiscriminately, and data on V genes from this method of expansion were not included in our analysis.

One point of agreement in all studies was that culturing CSF cells in IL-2 resulted in a narrowing of the V gene repertoire. Usuku (1996) thought that using IL 2 did not enrich for MBP-specific T cells and may in fact have reduced the ability to detect these cells, but our study using this approach (Chou et al., 1994) found that the MBP reactivity was retained in these IL-2 expanded cells. Probably both studies are correct, and all specificities of activated cells are expanded in IL-2 culture. If MBP-specific cells are present, they are expanded, but they may not always be present in the CSF of all patients at all times. Birebent found that overexpressed T cells in CSF could produce proinflammatory cytokines, whereas others were producing IL-4 (Birebent et al., 1998). They concluded that pathogenic and protective/restorative T cells are simultaneously present in the CSF.

A comparison of the V gene usage in plaques, and the four types of CSF T cells is shown in Figure 16.5. To be able to compare these various data sets more easily, the data are presented as the percent of MS patients expressing a given V gene. The most commonly seen AV genes in MS brain are AV10, AV8, AV2, and AV1. In MBP-reactive cells from CSF, the pattern of AV gene expression shows no AV8, less expression of AV1 and AV10, but higher expression of AV2 and AV19. As mentioned earlier, the most common plaque BV genes are BV5, BV7, BV8, and BV12. Interestingly, BV5 is also one of the most commonly used genes in MBP-specific T cells selected directly from CSF or after initial expansion in IL-2 and IL-4. T cells that were taken directly from the CSF and examined for V gene expression did not demonstrate high usage of BV5 by most patients, and CSF T cells that were expanded in IL-2 and IL-4 but not selected (or even tested) for MBP reactivity showed a similar low level of BV5. When one considers that BV5 is one of the most commonly utilized BV genes for the recognition of MBP in peripheral blood T cells, it seems that BV5 may be particularly suited to the recognition of MBP antigens. BV13 often appeared in MBP-selected or MBP-specific CSF T cells but was not expressed in MS lesions. In general, BV genes used most often in unmanipulated CSF T cells (BV8, BV18, BV7, BV14, BV19) are not representative of the cells seen most often in the brain lesions of MS patients or in their PBMCs. Thus, we may conclude that the CSF compartment probably includes numerous activated T cells that cross the blood-brain barrier, but those that are likely to be disease-relevant become prominent only after selection with MBP, with or without prior expansion in IL-2 and IL-4.

## 3.12    V Gene Expression in Patients with Other Autoimmune Diseases and Allergy

Based on the discussion above, it is apparent that MS-associated V genes expressed by neuroantigen-specific T cells reflected to a large degree the pattern of V genes expressed in the PBMCs, with some additional BV gene expression that appears to be disease-related (e.g., BV17). It seemed likely that the process by which self-reactive T cells are expanded in the periphery would apply to the peripheral repertoire in response to other self antigens that could contribute to the pathogenesis of autoimmune diseases and allergy. Thus, we evaluated the available published literature on V gene expression patterns in myasthenia gravis, lupus, diabetes, arthritis, and allergy, with an eye toward identifying disease-associated BV genes suitable for targeting using the TCR peptide vaccination approach.

In myasthenia gravis (MG), it is well established that the acetylcholine receptor (AChR) is an important target antigen for both T cells and antibodies. In two studies involving 13 T cell lines from five MG patients (Hill et al.,1999; Melms et al.,1993), AV8 was clearly predominant, being present in six of the lines from three patients. For BV genes, BV6, BV7, BV5, and BV2 were prominent in five studies involving 28 T cell lines evaluated from 18 MG patients (Bond et al., 1997. 1998; Hill et al., 1999; Melms et al., 1993; Raju et al., 1997) (Table 16.1).

Again, the pattern suggested an increase in predominant BV genes from PBMCs and a disease-specific V gene, in this case BV7. Similar results were obtained in lupus nephritis patients, in which AV8, BV6, BV5, BV2, and BV8 predominated in T cells that provided help for antibody production against DNA (Desai-Mehta et al., 1995) (Table 16.1) and in psoriasis, in which T cells from dermal/epidermal psoriatic lesions overexpressed BV6 and BV2 genes (Chang et al., 1994; Menssen et al., 1995). In a single study of insulin-dependent diabetes mellitus (IDDM) (Kallan et al., 1997), the pattern of BV gene usage differed somewhat from the examples cited above, with BV8 predominating in T cells specific for insulin secretory granules (ISG) from nine patients, and BV8, BV13, and BV14 predominating in T cells specific for ICA69 (a recombinant B cell protein) from 10 patients (Table 16.1). Interestingly, analysis of 164 HPRT-mutant T cell clones from nine IDDM patients demonstrated clear predominance of BV14, with lesser expression of BV6, BV13, and BV2 (Falta et al., 1999). These clones were not selected for responsiveness to any particular antigen and represented clones that were recently activated in vivo in the donor IDDM patients. Although it is not known if they represent disease-inducing cells, their BV gene profile includes both predominant BV genes from PBMCs (BV13, BV2, BV6) and a relative unique V gene (BV14).

Studies in rheumatoid arthritis (RA) have focused mostly on differences in general AV and BV gene profiles in synovial fluid (SF) and synovial tissue (ST) samples relative to that in PBMCs, without evaluation of T cells specific for putative tissue-derived target antigens. In analyses of AV gene expression from 51 RA patients in three studies (Lunardi et al., 1992; Maruyama et al., 1993; VanderBorght et al., 2000), there was a modest increase in the percentage of donors expressing AV5 and AV19 (Table 16.1), neither of which is predominant in PBMCs. Evaluation of overexpressed BV genes from RA patients from five studies (Davey and Munkirs, 1993; Jenkins et al., 1993; Lunardi et al., 1992; Maruyama et al., 1993; VanderBorght et al., 2000) indicated modest overexpression of the common BV13, BV2, and BV6 genes in unsorted T cells from SF (67 patients) and of BV3, BV6, and BV15 in unsorted T cells from ST (25 patients) (Table 6.1) (Alam et al., 1995; Jenkins et al., 1993; VanderBorght et al., 2000). A second set of studies (Fischer et al., 1996; Sottini et al., 1991) looked at the presence of BV genes (no overexpression compared to PBMCs) and had some similarities in the expression pattern of SF cells, showing a modest increase in expression of BV13, BV6, and BV4 (11 patients), but a highly heterogeneous pattern of V gene expression in ST cells (16 patients) (Bucht et al., 1992; Olive et al., 1991; Struyk et al., 1996). On the other hand, there were pronounced increases in expression of BV2, BV6, BV21, and BV15 in CD4+ T cells from SF (7 patients) (Cooper et al., 1994) and of BV6, BV2, and BV13 in CD4+ T cells from ST (5 patients) (Davey et al.,1997) (Table 6.1). Given the argument that ST may be enriched in disease-relevant cells that promote a local inflammatory environment, these data suggest that RA may also represent an autoimmune disease in which potentially pathogenic CD4+ T cells are expanded from predominant V gene families from PBMCs.

Finally, V gene expression patterns were evaluated in T cell clones specific for the dust-mite Der p 1 allergen or the birch pollen Bet v 1 allergen from atopic patients. As is shown in Table 6.1, Der p 1-specific T cells overexpressed AV8, AV3, BV5, BV6, and BV21 (Bohle et al., 1998; Richards et al., 1997; Wedderburn et al., 1993; Werfel et al., 1996; Yssel et al., 1992), whereas Bet v 1-specific T cells overexpressed AV8, AV2, BV2, and BV6 (Bohle et al., 1998; Breiteneder et al., 1995, 1996; Friedl-Hajek et al., 1999). This pattern again appears to reflect expansion of T cells from predominant V gene families from PBMCs, with occasional BV gene expression of a minority V gene family (BV21).

# 4    Regulation of Autoimmunity With T Cell Receptor Peptides

T cell recognition of self TCR sequences represents a natural autoregulatory network for limiting inflammatory reactions mediated by Th1 cells such as those directed at organ-specific antigens (Cohen and Young, 1991; Howell et al., 1989; Vandenbark et al., 1989). This mechanism, as currently postulated, involves the display of internally processed TCR determinants in association with major histocompatibility complex class II (MHC II) molecules on the surface of Ag-specific T cells or APCs, a complex that can both stimulate and function as a target for anti-TCR-specific T cells (Jiang et al., 1991; Vandenbark et al., 1996b). We and others developed TCR peptide and recombinant vaccines corresponding to the BV8S2 sequence that could induce regulatory T cells and antibodies that prevented or reversed clinical paralysis in EAE (Howell et al., 1989; Kumar et al., 1997; Offner et al., 1991; Vandenbark et al., 1989). Peptides from CDR1, CDR2, FW3, and CDR3 all had protective activity against autoimmune disease (Kumar et al., 1995; Vainiene et al., 1996). Of these, CDR1, CDR2, and FW3 determinants are V gene-specific, whereas CDR3 peptides would be clonotypic. The outcome of T–T interactions directed at TCR epitopes depends on the functional properties of both the responder and target T cells. Thus, CD4+ T helper cells recognize a variety of specific TCR determinants on AV and BV chains in association with MHC II, resulting in secretion of cytokines—including IL-10, IL-4, and to a lesser degree interferon-γ (IFNγ) and transforming growth factor-β (TGFβ)—that can inhibit activation, secretion of inflammatory cytokines, and encephalitogenic activity of neuroantigen-specific Th1 cells (Adlard et al., 1999; Offner et al., 1998, 1999). Because of the probability that suppressive cytokines released by TCR-reactive T cells can also affect bystander Th1 cells bearing different V genes specific for other neuroantigens, we believe that the broader scope of regulation triggered by CDR1, CDR2, and FW3 V gene epitopes might be better than that directed at clonotypic CDR3 determinants. Thus, TCR therapy has the potential to regulate complex autoimmune responses that may involve epitope spreading (Lehmann et al., 1992) and overexpression of multiple V genes (Hafler et al., 1996). Application of this regulatory strategy in humans indicated that TCR peptides are immunogenic (Bourdette et al., 1994, 1998; Gold et al., 1997) and may beneficially affect the clinical course of disease in patients with RA or MS (Moreland et al., 1998; Vandenbark et al., 1996a).

## 4.1   Summary of Work Accomplished in Animal Models

### 4.1.1   Effects of TCR Reactive T cells on Encephalitogenic BV8S2+ T cells in Rats

We have carried out a series of studies in rats that have partially defined the mechanism by which MBP-specific encephalitogenic T cells are inhibited by activated TCR BV8S2-39-59 reactive T cell lines. In our initial studies (Offner et al.,1994), we found that treatment of EAE with the BV8S2-39-59 peptide did not induce DNA fragmentation (apoptosis) in spinal cord cells. Moreover, BV8S2+ T cells with the $Asp_{96}Ser_{97}$ CDR3 motif persisted in spinal cord in clinically well rats (Buenafe et al., 1995, 1996). These findings implicated a regulatory rather than a deletional mechanism. Further studies using T cell co-cultures demonstrated that supernatants from activated BV8S2-39-59 reactive T cells inhibited proliferation, expression of IL-3 and to a lesser extent IL-2 and IFNγ, and transfer of lethal EAE by MBP-reactive T cells. Follow-up studies using allotypic markers to distinguish pathogenic and regulatory T cells demonstrated that BV8S2-but not BV10-specific T cells could recognize cellular determinants and inhibit proliferation, migration into the CNS, and encephalitogenic activity of MBP-specific T cells. Interestingly, both BV8S2 and BV10 reactive T cells migrated into the central nervous system (CNS), but only the BV8S2-reactive T cells prevented entry of pathogenic or recruited CD4+ T cells into the CNS (Offner et al., 1999). Taken together, these studies indicate that regulation of activation and migration of pathogenic T cells occurs locally through cytokines released by TCR-reactive T cells after specific activation.

### 4.1.2   TCR Regulation in a Genetically Restricted Model of EAE

To further explore basic mechanisms, we evaluated whether network regulation could be induced naturally in TCR BV8S2 single Tg mice, in which TCR peptide-specific T cells should be forced to express the same BV chain as the pathogenic T cell targeted for regulation. Using the BV8S2 transgenic B10.PL mouse specific for MBP-Ac1-11 peptide, we found that T cells specific for recombinant rat BV8S2 protein expressed the same BV8S2 transgene as the MBP-Ac1-11-specific T cells (Buenafe et al., 1997). Although the TCR BV repertoire was essentially frozen, presumably any TCR AV chain could be rearranged in response to BV8S2 determinants. Surprisingly, however, the TCR BV8S2-specific T cells also expressed the same AV2S3 gene as the MBP-Ac1-11-specific T cells, differing only in the CDR3 sequence. Additional studies demonstrated that overexpression of the BV8S2 transgene (1) engendered a natural preeffector T cell population specific for MBP-Ac1-11 peptide that acquired encephalitogenic capability only after immunization with this peptide in CFA, and (2) naturally induced a second population of T cells specific for BV8S2 determinants (Offner et al.,1998). Vaccination of male mice with heterologous Rt-BV8S2 protein induced strong cross-reactivity to mouse BV8S2 determinants and was more effective than Mo-BV8S2 for inducing protection against EAE. Moreover, boosting weekly with BV8S2 protein maintained protection against EAE more effectively than an initial vaccination without further boosting. Conversely, neonatal tolerization with BV8S2 protein enhanced severity of EAE (Siklodi et al., 1998). Regulatory T cells secreted IL-4, IL-10 and IFNγ; and supernatants could inhibit activation and transfer of EAE by MBP-Ac1-11-specific T cells (Offner et al., 1998). Using neutralizing anti-cytokine antibodies to reverse inhibitory effects of supernatants from BV8S2-specific T cells, we found that IL-4, IL-10, and to a lesser extent IFNγ and TGFβ were the major regulatory cytokines responsible for inhibiting encephalitogenic activity, proliferation, and IFNγ secretion of MBP-Ac1-11-reactive T cells (Adlard et al., 1999). TCR vaccination also drastically reduced the expression of chemokines and chemokine receptors in the spinal cords of EAE-protected mice (Matejuk et al., 2000).

### 4.1.3    TCR Protection Studies in TCR DTg B10.PL and B6 Mice

We carried out experiments in TCR double Tg B10.PL mice specific for MBP-Ac1-11 obtained previously from Dr. Charles Janeway. Because of the presence of TCR α- and β-chains, this model is even more genetically restricted than the single Tg mice that we explored previously, with the possibility that selection of TCR-specific regulatory T cells might be largely precluded. However, when we immunized male mice with rat BV8S2 protein on days −7 and +3 relative to challenge with MBP-Ac1-11 on day 0, we found that TCR double transgenic mice were fully protected against EAE, and T cells from the vaccinated mice could protect T/R-mice from developing spEAE (Matejuk et al., 2003). This finding firmly establishes the presence of TCR-reactive T cells in these mice and ensures the feasibility of experiments described below using DR2/TCR DTg mice reactive to MBP-85-99 peptide.

The B6 mouse model of EAE induced by the MOG-35-55 peptide is also extremely important for assessing the role of cytokines and chemokines because KO mice have mostly been produced on this background. Encephalitogenic MOG-35-55 T cells have been described as largely BV8S2+, and we confirmed this finding in a MOG-35-55-specific T cell line from B6 mice. Subsequently, we demonstrated that EAE induced with MOG-35-55 peptide could be significantly inhibited by vaccination with rat BV8S2 protein given on days −7 and +3 relative to EAE induction. The ability to induce protective TCR reactive T cells in B6 mice is important because this finding demonstrates that TCR-reactive T cells are present on this background shared with the DR2 and DR2/TCR DTg mice to be used in this proposal. We also recently demonstrated that chronic EAE could be induced in DR2 mice with the MOG-35-55 peptide/CFA/Ptx (Rich et al., 2004).

### 4.1.4    Working Hypothesis and Model

The work outlined above in four rodent model systems strongly supports the contention that regulation induced by TCR determinants involves activation and expansion of a naturally occurring set of CD4+ T cells specific for TCR peptides that are expressed with MHC II molecules on the T cell surface (humans or rats) or that are taken up and presented by professional APCs (mice). Although activation is TCR peptide-specific, the regulatory mechanisms involve secretion of cytokines—including IL-4, IL-10, INFγ, and TGFβ—that may locally inhibit not only the triggering Th1 cell expressing the cognate V gene but also bystander Th1 cells that may be specific for different myelin antigens and that may express different V genes (Vandenbark et al., 1997). Specific activation resulting in local nonspecific inhibition of bystander inflammatory responses is an important aspect inherent in the TCR mechanism that would allow regulation of complex autoimmune diseases such as MS that likely involve multiple neuroantigen T cell specificities that may change during the disease course.

## 4.2   Human Studies

### 4.2.1   Natural Recognition of TCR Determinants in Healthy Human Donors

To evaluate native recognition of TCR determinants, we assessed the frequency of IL-10- and IFNγ-secreting T cells from the blood of five HCs using the ELISPOT assay to detect responses to a comprehensive panel of 113 unique CDR2 peptides representing nearly all of the AV and BV repertoires (Buenafe et al., 2004; Vandenbark et al., 2000). Peptide-specific T cells secreting either IL-10 or IFNγ were detected in response to nearly all of the TCR peptides tested. Frequencies varied considerably by peptide and donor but were not affected by gender. Of interest, the average frequency of IL-10-secreting T cells recognizing BV peptides was > 600 cells/million PBMCs, and for AV peptides > 300 cells/million. IFNγ responses to TCR peptides were less vigorous than IL-10 responses, with an average frequency of 250 cells/million BV-reactive T cells and 182 cells/million AV-reactive T cells.

### 4.2.2   Deficient TCR-Reactive T Cells in MS Patients

In both a previous study (Vandenbark et al., 2001b) and in our expanded analysis (Buenafe et al., 2004), we found striking differences in both the magnitude and pattern of response to the 113 AV and BV CDR2 peptides in the MS patients versus HCs. Overall, the total frequency of MS T cells responding to the panel of CDR2 peptides was significantly reduced ($p = 0.03$) compared to HCs. Moreover, the pattern of response was clearly different in MS patients, showing overall reduced frequencies to most peptides. However, for a few peptides and for concanavalin A (ConA), the MS patients responded as well or better than the HCs, indicating no global deficit in the ability of MS T cells to respond. These unique data demonstrate a broad deficiency of TCR-reactive T cells that may explain why MS patients have oligoclonal expansions of potentially pathogenic neuroreactive Th1 cells.

### 4.2.3   Specific Activation of TCR-Reactive T Cells With Co-cultured MBP-Reactive T Cells Expressing a Cognate TCR BV Chain

Local triggering of TCR-reactive T cells is thought to depend on natural expression of TCR determinants on the Th1 cell surface in association with MHC II molecules. To test this assumption in a rigorous manner, we selected highly char-acterized Th1 cell clones and used these cells to activate T cell lines that had been selected from the same donor after immunization with a TCR CDR2 peptide. Thus, both BV6S1+ and BV2+ Th1 clones specific for myelin basic protein (MBP) were selected from a DR2 (DRB1*1501) homozygous MS patient and were shown to be clonal by expression of a single BV chain. Both clones were activated to enhance expression of MHC II molecules and irradiated to prevent them from proliferating. These attenuated Th1 clones were then incubated at varying ratios with a T cell line specific for BV6S1-38-58 peptide that had been selected previously from the same MS donor after vaccination in vivo with the same TCR peptide. Within 2 hours, clump formation was observed with the BV6S1+ Th1 cells but not the BV2+ Th1 cells in a dose-dependent manner. Moreover, the cell–cell interaction with the BV6S1+ clone but not the BV2+ clone induced proliferation and release of both IL-10 and IFNγ by the TCR-reac-tive T cells, demonstrating cognate peptide specificity. The response was further shown to be DR2-restricted. These are the first data to demonstrate clearly that MHC II-restricted T cell-associated TCR determinants can induce proliferation and cytokine release of TCR peptide-specific T cells, and they lend direct support to the contention that naturally processed TCR peptides can be recognized on the Th1 cell surface. Importantly, these data demonstrate that T cells specific for a CDR2 peptide used in human vaccination trials can be stimulated by contact with whole T cells that express the cognate TCR. Cytokines released from this inter-action included both IFNγ and IL-10. Of note, the 3:1 ratio of stimulator Th1 cells to responder TCR-reactive T cells induced high levels of IL-10 and reduced levels of IFNγ.

### 4.2.4   TCR-Specific T Cells Have Treg Activity

To define the spectrum of immunoregulatory TCR epitopes from a potentially pathogenic Th1 cell, we cloned an MBP-85-99-specific T cell from an HLA-DR2 homozygous MS donor (Burrows et al., 2001). This T cell clone expressed BV6S1 in combination with two AV chains, AV14 and AV23. After sequencing the AV and BV chains, we produced an scTCR representative of the MBP-specific T cell clone (Buenafe et al., 2004). Of importance, we found that T cells from the same MS donors that were selected against the BV6S1 CDR2 peptide could respond specifically to the scTCR molecule, verifying that the BV6S1-CDR2 peptide is preserved during natural processing and MHC presentation of the TCR chain. Moreover, we found that both CD4+ and CD8+ T cell lines selected from the MS donor were highly reactive to the BV6S1:AV23 scTCR molecule. Using overlapping peptides for both BV and AV chains, we found that the CD4+ T cells responded primarily to CDR3 determinants unique to this mol-ecule, with lesser V gene specific responses to CDR1, CDR2, and FW3 peptides. T cell responses to scTCR included proliferation and release of IFNγ and IL-10 but not IL-2 or IL-4, and they were restricted by DRB1*1501 but not DRB5*0101. Most importantly, the scTCR-reactive T cells, which were CD25+, were able to strongly inhibit anti-CD3-induced proliferation of CD25– indicator T cells in a cell–cell contact-dependent manner typical of Treg cells. These studies demonstrated that TCR-reactive T cells are present naturally in MS patients and possess functional Treg activity.

In additional studies, we demonstrated that CD4+CD25+ T cells from healthy donors could respond to a pool of TCR-CDR2 peptides; and T cell lines selected by TCR but not recall antigens could inhibit indicator T cells in the Treg assay (Buenafe et al., 2004). Interestingly, antibodies to IL-10, GITR, CTLA-4, IL-17, and TGFβ could reverse suppression. These data suggest that the TCR-reactive T cell population is heterogeneous, containing classic Treg cells, and possibly IFNγ-producing Th1, IL-10-producing Tr1, and TGFβ-producing Th3 cells but not IL-4-producing Th2 cells. These data are the first to associate TCR-reactive cells with Treg cells and were key to the formation of our current hypothesis that TCR specificities have heterogeneous regulatory activity.

### 4.2.5   Reduced Treg Activity and FOXP3 Expression in MS Patients

We recently developed and refined an in vitro Treg assay for use with human PBMC (Tsaknaridis et al., 2003). Moreover, we demonstrated that both FOXP3 message and protein and Treg functional activity were decreased in MS patients versus paired age- and gender-matched donors, and that FOXP3 levels correlated with Treg suppressive activity in the CD4+CD25+ fraction from PBMCs (Huan et al., 2005). Subsequently, we evaluated Treg responses in 19 MS patients versus 19 age- and gender-matched HCs. MS patients had reduced Treg activity compared to HCs, in concert with a previous report (Viglietta et al., 2004). Moreover, there was a significant reduction in expression of FOXP3 message. These data indicate that MS patients may have a functional lack of Treg-mediated suppression that may be a predisposing risk factor allowing unregulated expression of neuroantigen-reactive T cells.

### 4.2.6   TCR Peptide Vaccination Studies in Patients With MS

Unlike mouse models of EAE, the TCRs of neuroantigen-reactive Th1 cells in MS patients are diverse. To address the V gene issue in MS, we evaluated TCR expression in 150 MBP-specific T cell clones from 24 MS patients. This evaluation revealed that several AV and BV gene families were predominant, including BV2, BV5, BV6, BV13, AV2, and AV8 (Kotzin et al., 1991; Offner and Vandenbark, 1999). Based on this expression pattern and from related studies in CSF, we targeted three BV genes—BV5S2, BV6S5, BV13S1—for use in a trivalent TCR peptide vaccine (Vandenbark et al., 2001c). Evaluation of the patients expressing these three BV genes in our worldwide analysis indicated that nearly 90% of MS patients would have MBP-specific clones expressing at least one of the three V gene families present in the trivalent TCR peptide vaccine (Fig. 16.6) Despite the heightened expression of these BV genes, however, native T cell proliferation responses to the BV peptides were relatively low (< 1 cell/million PBMCs). Vaccination intradermally with a BV5S2 peptide in saline induced significant

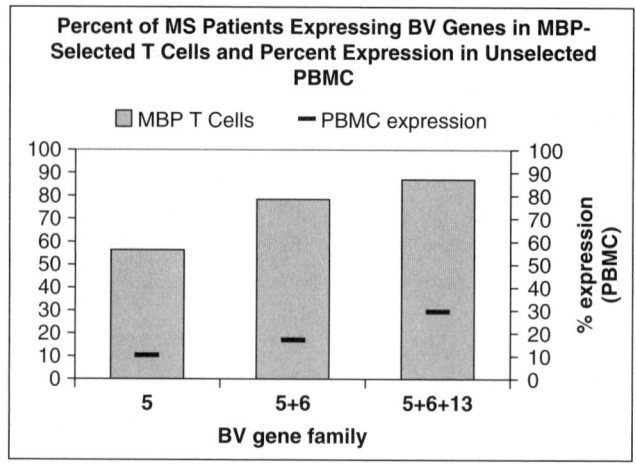

FIGURE 16.6. Evaluation of percentage of MS patients with MBP-specific T cell clones expressing BV5, BV6, and BV13 genes targeted by our trivalent T cell receptor peptide vaccine.

anti-TCR responses in about half of the patients and resulted in reduced response to MBP and a significant trend toward clinical benefit in TCR-responsive patients (Bourdette et al., 1994, 1998; Vandenbark et al., 1996a, 2001a). The release of inhibitory cytokines, mainly IL-10, by TCR-reactive T cells would allow local regulation of target Th1 cells expressing the cognate BV5S2 gene as well as bystander Th1 cells that may express different V genes and respond to different CNS target antigens (e.g., PLP and MOG) (Chou et al., 1994, 1996). In addition, because of the homologies present among V genes in the CDR2, recognition of TCR epitopes is probably degenerate (Vandenbark et al., 2000). These mechanisms account for the broader effects of vaccination observed with BV5S2 peptide than would be predicted solely on the basis of BV5S2 expression by MBP-specific T cells. Injection with our trivalent TCR vaccine containing CDR2 peptides from BV5S2, BV6S5, and BV13S1 genes in incomplete Freund's adjuvant (IFA) induced strong responses to these peptides in 100% of MS patients tested, and responsive patients showed a trend toward reduced MRI activity, thus providing us with an excellent candidate vaccine for future trials (Bourdette et al., 2005). Taken together, the results demonstrate unequivocally the antigenicity of TCR determinants and the ability of TCR vaccines to expand the frequency of TCR-reactive T cells in MS patients.

Preliminary indications show that vaccination with our trivalent TCR vaccine can boost FOXP3+ Treg cells in MS patients (Vandenbark, 2005). FOXP3 expression that was decreased in MS patients prior to TCR vaccination was restored to normal levels within 12 weeks after vaccination. These data indicate that T cell responses to TCR peptides and Treg activity are deficient in MS patients and that low TCR and Treg activity may be corrected by successful vaccination. One MS patient with strongly enhanced Treg activity also exhibited clinical improvement over the course of a 1-year study.

## 4.3    Proposed Mechanism of Action of TCR Peptide Vaccination

Our studies have produced strong evidence that activated Th1 cells expressing self TCR sequences in association with upregulated MHC II molecules are capable of activating TCR-reactive T cells specific for processed determinants present in the cognate TCR. Additionally, uptake and processing of TCR from dead or dying T cells by APCs may also result in activation of TCR-reactive T cells. The result of T–T cell activation or APC presentation of TCR determinants results in secretion of a variety of cytokines, including IL-10, that potentially could inhibit the targeted Th1 cells, as well as "bystander" Th1 cells including those that express BV genes different than the target T cell. Immunoregulation through release of IL-10 is a prominent feature of Tr1 cells (Jonuleit, 2003 #2055), and it is possible that TCR-reactive T cells include a subset of Tr1 cells. The concept of bystander suppression is an attractive feature of regulation induced by TCR peptide vaccination because it allows for localized cytokine-mediated suppression of any number of specificities of potentially encephalitogenic T cells, regardless of their V gene expression, in the vicinity of the target T cell. A critical unknown variable is how many target Th1 cells would be needed in an inflammatory lesion to trigger sufficient regulatory factors to inhibit lesion development. In this regard, the inverse correlation of TCR-reactive T cells versus MBP-reactive T cells observed after TCR vaccination (Vandenbark et al., 1996a) and the generalized reduction of inflammatory cells in CSF after vaccination with a single BV peptide (Wilson et al., 1997) support the notion that bystander suppression is an important component of TCR vaccination.

Another major pathway involves TCR peptide-reactive Treg cells. Treg function is mediated by cel—cell contact and does not involve to any great degree the secretion of regulatory cytokines. The clear demonstration that TCR-reactive T cell lines have classic Treg activity and that CD4+CD25+ T cells from blood respond to TCR peptides strongly suggest that the Treg cell lineage includes TCR specificities. Our ability to key on TCR reactive T cells has allowed us to propose a general developmental pathway for Treg cells specific for TCR. The major elements of this pathway are in accordance with other recently published characteristics of Treg cells, including the ability of Treg cells to expand in vivo (Levings et al., 2001; Walker et al., 2003a; Yamazaki et al., 2003) and the ability of non-activated CD4+CD25− T cells to become activated and to later acquire suppressive activity upon upregulation of *Foxp3* and expression of CD25 (Walker et al., 2003b). TCR-specific Treg cells would undergo intrathymic activation to TCR determinants and seeding of the periphery with TCR-reactive T cells, production of a pool of nonactivated T cells without Treg activity in the absence of recent stimulation by cognate TCRs, reactivation of quiescent TCR-reactive T cells with cell-expressed TCR/MHC, or TCR peptide vaccination leading to early expansion of Treg cells that then mature to a nondividing state capable of exerting cell–cell suppression, migration into the tissues, and finally suppression of cognate and bystander T cells. *Thus, the activity exerted by TCR-reactive T cells may include MHC II recognition of TCR determinants expressed by targeted T cells, cytokine-mediated Tr1-like suppression (involving IL-10), and cell contact–induced Treg inactivation of target and bystander encephalitogenic T cells.*

*Acknowledgments.* Dr. Vandenbark has a significant financial interest in The Immune Response Corporation, a company that has a commercial interest in the results of this research and technology.

The authors wish to acknowledge collaboration with The Immune Response Corporation, Carlsbad, CA, and support from NIH grants NS23221 and NS23444, The National Multiple Sclerosis Society, The Nancy Davis MS Center Without Walls, and the Department of Veterans Affairs Department of Biomedical Research.

## *References*

Adlard, K., Tsaknaridis, L., Beam, A., et al. (1999) Immunoregulation of encephalitogenic MBP-Ac1-11-reactive T cells by CD4+ TCR-specific T cells involves IL-4, IL-10 and IFN-gamma. *Autoimmunity* 31:237-248.

Afshar, G., Muraro, P. A., McFarland, H. F., Martin, R. (1998) Lack of over-expression of T cell receptor Vbeta5.2 in myelin basic protein-specific T cell lines derived from HLA-DR2 positive multiple sclerosis patients and controls. *J. Neuroimmunol.* 84(1):7-13.

Akolkar, P. N., Gulwani-Akolkar, B., Pergolizzi, R., et al. (1993) Influence of HLA genes on T cell receptor V segment frequencies and expression levels in peripheral blood lymphocytes. *J. Immunol.* 150(7):2761-2773.

Alam, A., Lule, J., Coppin, H., et al. (1995) T-cell receptor variable region of the beta-chain gene use in peripheral blood and multiple synovial membranes during rheumatoid arthritis. *Hum. Immunol.* 42(4):331-339.

Arden, B., Clark, S. P., Kabelitz, D., Mak, T. W. (1995) Human T-cell receptor variable gene segment families. *Immunogenetics* 42:455-500.

Ashton-Rickardt, P. G., Tonegawa, S. (1994) A differential-avidity model for T-cell selection. *Immunol. Today* 15(8):362-366.

Ausubel, L. J., Bieganowska, K. D., Hafler, D. A. (1999) Cross-reactivity of T-cell clones specific for altered peptide ligands of myelin basic protein. *Cell. Immunol.* 193(1):99-107.

Baccala, R., Kono, D. H., Walker, S., et al. (1991) Genomically imposed and somatically modified human thymocyte Vb gene repertoires. *Proc. Natl. Acad. Sci. USA* 88:2908-2912.

Balomenos, D., Balderas, R. S., Mulvany, K. P., et al. (1995) Incomplete T cell receptor Vβ allelic exclusion and dual Vβ-expressing cells. *J. Immunol.* 155:3308-3312.

Ben-Nun, A., Liblau, R. S., Cohen, L., et al. (1991) Restricted T-cell receptor V beta gene usage by myelin basic protein-specific T-cell clones in multiple sclerosis: predominant genes vary in individuals. *Proc. Natl. Acad. Sci. USA* 88(6):2466-2470.

Birebent, B., Semana, G., Commeurec, A., et al. (1998) TCR repertoire and cytokine profiles of cerebrospinal fluid- and peripheral blood-derived T lymphocytes from patients with multiple sclerosis. *J. Neurosci. Res.* 51(6):759-770.

Birnbaum, G., van Ness, B. (1992) Quantitation of T-cell receptor V beta chain expression on lymphocytes from blood, brain, and spinal fluid in patients with multiple sclerosis and other neurological diseases. *Ann. Neurol.* 32(1):24-30.

Bohle, B., Schwihla, H., Hu, H. Z., et al. (1998) Long-lived Th2 clones specific for seasonal and perennial allergens can be detected in blood and skin by their TCR-hypervariable regions. *J. Immunol.* 160(4):2022-2027.

Bond, A., Corlett, L., Nagvekar, N., et al. (1997) Heterogeneity and immunotherapy of specific T-cells in myasthenia gravis. *Biochem. Soc. Trans.* 25:665-670.

Bond, A. P., Corlett, L., Curnow, S. J., et al. (1998) Diverse patterns of unresponsiveness in an acetylcholine receptor-specific T-cell clone from a myasthenia gravis patient after engaging the T-cell receptor with three ligands. *J. Neuroimmunol.* 82:182-190.

Bourdette, D. N., Whitham, R. H., Chou, Y. K., et al. (1994) Immunity to T cell receptor peptides in multiple sclerosis. I. Successful immunization of patients with synthetic V 5.2 and V 6.1 CDR2 peptides. *J. Immunol.* 152:2510-2519.

Bourdette, D. N., Chou, Y. K., Whitham, R. H., et al. (1998) Immunity to T cell receptor peptides in multiple sclerosis. III. Preferential immunogenicity of complementarity determining region 2 peptides from disease-associated T cell receptor BV genes. *J. Immunol.* 161:1034-1044.

Bourdette, D. N., Edmonds, E., Smith, C., et al. (2005) A highly immunogenic trivalent T cell receptor peptide vaccine for multiple sclerosis. *Mult. Scler.* 11:1-10.

Breiteneder, H., Scheiner, O., Hajek, R., et al. (1995) Diversity of TCRAV and TCRBV sequences used by human T-cell clones specific for a minimal epitope of Bet v 1, the major birch pollen allergen. *Immunogenetics* 42(1):53-58.

Breiteneder, H., Friedl-Hajek, R., Ebner, C., et al. (1996) Sequence comparisons of the CDR3 hyper-variable loops of human T cell receptors specific for three major T cell epitopes of the birch pollen allergen Bet v 1. *Mol. Immunol.* 33(13):1039-1048.

Bucht, A., Oksenberg, J. R., Lindblad, S., et al. (1992) Characterization of T-cell receptor alpha beta repertoire in synovial tissue from different temporal phases of rheumatoid arthritis. *Scand. J. Immunol.* 35(2):159-165.

Buenafe, A. C., Vainiene, M., Celnik, B., et al. (1995) Analysis of VB8.2 CDR3 sequences from spinal cord T cells of Lewis rats vaccinated or treated with TCR VB8.2-39-59 peptide. *J. Immunol.* 155:1556-1564.

Buenafe, A. C., Weinberg, A. D., Culbertson, N. E., et al. (1996) VB CDR3 motifs associated with BP recognition are enriched in OX-40+ spinal cord T cells of Lewis rats With EAE. *J. Neurosci. Res.* 44:562-567.

Buenafe, A. C., Tsu, R. C., Bebo, B. F., et al. (1997) Myelin basic protein-specific and TCR VB8.2-specific T-cell lines from TCR VB8.2 transgenic mice utilize the same Valpha and Vbeta genes: specificity associated with the Valpha CDR3-Jalpha region. *J. Neurosci. Res.* 47.489-499.

Buenafe, A. C., Tsaknaridis, L. J., Spencer, L., et al. (2004) Specificity of regulatory CD4+CD25+ T cells for self-T cell receptor determinants. *J. Neurosci. Res.* 76:129-140.

Burrows, G. G., Chou, Y. K., Wang, C., et al. (2001) Rudimentary TCR signaling triggers default IL-10 secretion by human Th1 cells. *J. Immunol.* 167:4386-4395.

Chang, J. C., Smith, L. R., Froning, K. J., et al. (1994) CD8+ T cells in psoriatic lesions preferentially use T-cell receptor V beta 3 and/or V beta 13.1 genes. *Proc. Natl. Acad. Sci. USA* 91(20):9282-9286.

Charmley, P., Wang, K., Hood, L., Nickerson, D. A. (1993) Identification and physical mapping of a polymorphic T cell receptor gene with a frequent null allele. *J. Exp. Med.* 177:135.

Choi, Y., Kotzin, B., Herron, L., et al. (1989) Interaction of Staphylococcus aureus toxin "superantigens" with human T cells. *Proc. Natl. Acad. Sci. USA* 86:8941-8945.

Choi, Y., Lafferty, J. A., Clements, J. R., et al. (1990) Selective expansion of T cells expressing Vβ2 in toxic shock syndrome. *J. Exp. Med.* 172:981-984.

Chou, Y. K., Morrison, W. J., Weinberg, A. D., et al. (1994) Immunity to T cell receptor peptides in multiple sclerosis. II. T cell recognition of Vβ5.2 and Vβ6.1 CDR2 peptides. *J. Immunol.* 152:2520-2529.

Chou, Y. K., Weinberg, A. D., Buenafe, A., et al. (1996) MHC-restriction, cytokine profile, and immunoregulatory effects of human T cells specific for TCR VB CDR2 peptides: comparison with myelin basic protein-specific T cells. *J. Neurosci. Res.* 45:838-851.

Cohen, I. R., and Young, D. B. (1991) Autoimmunity, microbial immunity and the immunological homunculus. *Immunol. Today* 12:105-110.

Cooper, S. M., Roessner, K. D., Naito-Hoopes, M., et al. (1994) Increased usage of V beta 2 and V beta 6 in rheumatoid synovial fluid T cells. *Arthritis Rheum.* 37(11):1627-1636.

Correale, J., McMillan, M., McCarthy, K., et al. (1995) Isolation and characterization of autoreactive proteolipid protein-peptide specific T-cell clones from multiple sclerosis patients. *Neurology* 45(7):1370-1378.

Corthay, A., Nandakumar, K. S., Holmdahl, R. (2001) Evaluation of the percentage of peripheral T cells with two different T cell receptor α-chains and of their potential role in autoimmunity. *J. Autoimmun.* 16:423-429.

Dave, V. P., Allman, D., Wiest, D. L., Kappes, D. J. (1999) Limiting TCR expression levels leads to quantitative but not qualitative changes to thymic selection. *J. Immunol.* 162:5764-5774.

Davey, M. P., Munkirs, D. D. (1993) Patterns of T-cell receptor variable beta gene expression by synovial fluid and peripheral blood T-cells in rheumatoid arthritis. *Clin. Immunol. Immunopathol.* 68(1):79-87.

Davey, M. P., Burgoine, G. A., Woody, C. N. (1997) TCRB clonotypes are present in CD4+ T cell populations prepared directly from rheumatoid synovium. *Hum. Immunol.* 55:11-21.

Davodeau, F., Peyrat, M. A., Romagne, F., et al. (1995) Dual T cell receptor beta chain expression on human T lymphocytes. *J. Exp. Med.* 181(4):1391-1398.

Deng, X., Sun, G-R., Zheng, Q., Li, Y. (1998) Characterization of human Vβ gene promoter. *J. Biol. Chem.* 273(37):23709-23715.

Desai-Mehta, A., Mao, C., Rajagopalan, S., et al. (1995) Structure and specificity of T cell receptors expressed by potentially pathogenic anti-DNA autoantibody-inducing T cells in human lupus. *J. Clin. Invest.* 95(2):531-541.

Dittel, B. N., Stefanova, I., Germain, R. N., Janeway, C. A., Jr (1999) Cross-antagonism of a T cell clone expressing two distinct T cell receptors. *Immunity* 11:289-298.

Diu, A., Romagné, F., Genevée, C., et al. (1993) Fine specificity of monoclonal antibodies directed at human T cell receptor variable regions: comparison with oligonucleotide-driven amplification for evaluation of Vβ expression. *Eur. J. Immunol.* 23:1422-1429.

Falta, M. T., Magin, G. K., Allegretta, M., et al. (1999) Selection of HPRT mutant T cells as surrogates for dividing cells reveals a restricted T cell receptor BV repertoire in insulin-dependent diabetes mellitus. *Clin. Immunol.* 90(3):340-351.

Fischer, D. C., Opalka, B., Hoffmann, A., et al. (1996) Limited heterogeneity of rearranged T cell receptor V alpha and V beta transcripts in synovial fluid T cells in early stages of rheumatoid arthritis. *Arthritis Rheum.* 39(3):454-462.

Friedl-Hajek, R., Spangfort, M. D., Schou, C., et al. (1999) Identification of a highly promiscuous and an HLA allele-specific T-cell epitope in the birch major allergen Bet v 1: HLA restriction, epitope mapping and TCR sequence comparisons. *Clin. Exp. Allergy* 29(4):478-487.

Garcia, K. C., Degano, M., Stanfield, R. L., et al. (1996) An αβ T cell receptor structure at 2.5Å and its orientation in the TCR-MHC complex. *Science* 274:209-219.

Geursen, A., Skinner, M. A., Townsend, L. A., et al. (1993) Population study of T cell receptor Vβ gene usage in peripheral blood lymphocytes: differences in ethnic groups. *Clin. Exp. Immunol.* 94:201-207.

Goebels, N., Hofstetter, H., Schmidt, S., et al. (2000) Repertoire dynamics of autoreactive T cells in multiple sclerosis patients and healthy subjects: epitope spreading versus clonal persistence. *Brain* 123(Pt 3):508-518.

Gold, D. P., Smith, R. A., Golding, A. B., et al. (1997) Results of a phase I clinical trial of a T-cell receptor vaccine in patients with multiple sclerosis. II. Comparative analysis of TCR utilization in CSF T-cell populations before and after vaccination with a TCRVβ6 CDR2 peptide. *J. Neuroimmunol.* 76:29.

Gulwani-Akolkar, B., Posnett, D. N., Janson, C. H., et al. (1991) T cell receptor V-segment frequencies in peripheral blood T cells correlate with human leukocyte antigen type. *J. Exp. Med.* 174:1139-1146.

Hafler, D. A., Saadeh, M. G., Kuchroo, V. K., et al. (1996) TCR usage in human and experimental demyelinating disease. *Immunol. Today* 17:152-159.

Hawes, G. E., Struyk, L., van der Elsen, P. J. (1993) Differential usage of T cell receptor V gene segments in CD4+ and CD8+ subsets of T lymphocytes in monozygotic twins. *J. Immunol.* 150(5):2033-2045.

Herman, A., Kappler, J. W., Marrack, P., Pullen, A. M. (1991) Superantigens: mechanism of T-cell stimulation and role in immune response. *Annu. Rev. Immunol.* 9:745-772.

Hill, M., Beeson, D., Moss, P., et al. (1999) Early-onset myasthenia gravis: a recurring T-cell epitope in the adult-specific acetylcholine receptor epsilon subunit presented by the susceptibility allele HLA-DR52a. *Ann. Neurol.* 45(2):224-231.

Hong, J., Zang, Y. C., Tejada-Simon, M. V., et al. (1999) A common TCR V-D-J sequence in V beta 13.1 T cells recognizing an immunodominant peptide of myelin basic protein in multiple sclerosis. *J. Immunol.* 163(6):3530-3538.

Howell, M. D., Winters, S. T., Olee, T., et al. (1989) Vaccination against experimental allergic encephalomyelitis with T cell receptor peptides. *Science (Wash. DC)* 246:668-670.

Huan, J., Culbertson, N., Spencer, L., et al. (2005) Decreased FOXP3 levels in multiple sclerosis patients. *J. Neurosci. Res.* 81:45-52.

Jenkins, R. N., Nikaein, A., Zimmermann, A., et al. (1993) T cell receptor V beta gene bias in rheumatoid arthritis. *J. Clin. Invest.* 92(6):2688-2701.

Jiang, H., Sercarz, E., Nitecki, D., Pernis, B. (1991) The problem of presentation of T cell receptor peptides by activated T cells. *Ann. N.Y. Acad. Sci.* 636:28-32.

Jonuleit, H., Schmitt, E. (2003) The regulatory T cell family: distinct subsets and their interrelations. *J. Immunol.* 171(12):6323-6327.

Joshi, N., Usuku, K., Hauser, S. L. (1993) The T-cell response to myelin basic protein in familial multiple sclerosis: diversity of fine specificity, restricting elements, and T-cell receptor usage. *Ann. Neurol.* 34(3):385-393.

Kallan, A. A., Duinkerken, G., de Jong, R., et al. (1997) Th1-like cytokine production profile and individual specific alterations in TCRBV-gene usage of T cells from newly diagnosed type 1 diabetes patients after stimulation with beta-cell antigens. *J. Autoimmun.* 10(6): 589-598.

Kay, R. A. (1996) TCR gene polymorphisms and autoimmune disease. *Eur. J. Immunogenet.* 23:159-177.

Kotzin, B. L., Karuturi, S., Chou, Y. K., et al. (1991) Preferential T cell receptor ß-chain variable gene use in myelin basic protein-reactive T cell clones from patients with multiple sclerosis. *Proc. Natl. Acad. Sci. USA* 88:9161-9165.

Kumar, V., Tabibiazar, R., Geysen, H. M., Sercarz, E. (1995) Immunodominant framework region 3 peptide from TCR VB8.2 chain controls murine experimental autoimmune encephalomyelitis. *J. Immunol.* 154:1941-1950.

Kumar, V., Coulsell, E., Ober, B., et al. (1997) Recombinant T cell receptor molecules can prevent and reverse experimental autoimmune encephalomyelitis. *J. Immunol.* 159:5150-5156.

Laufer, T. M., Glimcher, L. H., Lo, D. (1999) Using thymus anatomy to dissect T cell repertoire selection. *Semin. Immunol.* 11:65-70.

Lehmann, P. V., Forsthuber, T., Miller, A., Sercarz, E. E. (1992) Spreading of T-cell autoimmunity to cryptic determinants of an autoantigen. *Nature* 358:155-157.

Levings, M. K., Sangregorio, R., Roncarolo, M-G. (2001) Human CD25+CD4+ T regulatory cells suppress naive and memory T cell proliferation and can be expanded in vitro without loss of function. *J. Exp. Med.* 193:1295-1301.

Lindert, R. B., Haase, C. G., Brehm, U., et al. (1999) Multiple sclerosis: B- and T-cell responses to the extracellular domain of the myelin oligodendrocyte glycoprotein. *Brain* 122( Pt 11):2089-2100.

Lodge, P. A., Allegretta, M., Steinman, L., Sriram, S. (1994) Myelin basic protein peptide specificity and T-cell receptor gene usage of HPRT mutant T-cell clones in patients with multiple sclerosis. *Ann. Neurol.* 36(5):734-740.

Lunardi, C., Marguerie, C., So, A. K. (1992) An altered repertoire of T cell receptor V gene expression by rheumatoid synovial fluid T lymphocytes. *Clin. Exp. Immunol.* 90:440-446.

Malhotra, U., Spielman, R., Concannon, P. (1992) Variability in T cell receptor Vβ gene usage in human periphral blood lymphocytes: studies of identical twins, siblings, and insulin-dependent diabetes mellitus patients. *J. Immunol.* 149(5):1802-1808.

Marrack, P., and Kappler, J. (1990) The staphylococcal enterotoxins and their relatives. *Science* 248:705-711.

Martin, R., Utz, U., Coligan, J. E., et al. (1992) Diversity in fine specificity and T cell receptor usage of the human CD4+ cytotoxic T cell response specific for the immunodominant myelin basic protein peptide 87-106. *J. Immunol.* 148(5):1359-1366.

Maruyama, T., Saito, I., Miyake, S., et al. (1993) A possible role of two hydrophobic amino acids in antigen recognition by synovial T cells in rheumatoid arthritis. *Eur. J. Immunol.* 23:2059-2065.

Matejuk, A., Vandenbark, A. A., Burrows, G. G., et al. (2000) Reduced chemokine and chemokine receptor expression in spinal cords of TCR BV8S2 transgenic mice protected against experimental autoimmune encephalomyelitis with BV8S2 protein. *J. Immunol.* 164:3924-3931.

Matejuk, A., Buenafe, A. C., Dwyer, J., et al. (2003) Endogenous CD4+BV8S2– T cells from Tg BV8S2+ donors confer complete protection against spontaneous experimental encephalomyelitis (Sp-EAE) in TCR transgenic, Rag–/– mice. *J. Neurosci. Res.* 71:89-103.

Mazzanti, B., Hemmer, B., Traggiai, E., et al. (2000) Decrypting the spectrum of antigen-specific T-cell responses: the avidity repertoire of MBP-specific T-cells. *J. Neurosci. Res.* 59(1):86-93.

Meinl, E., Weber, F., Drexler, K., et al. (1993) Myelin basic protein-specific T lymphocyte repertoire in multiple sclerosis: complexity of the response and dominance of nested epitopes due to recruitment of multiple T cell clones. *J. Clin. Invest.* 92(6):2633-2643.

Melms, A., Oksenberg, J. R., Malcherek, G., et al. (1993). T-cell receptor gene usage of acetylcholine receptor-specific T-helper cells. *Ann. N. Y. Acad. Sci.* 681:313-314.

Menssen, A., Trommler, P., Vollmer, S., et al. (1995) Evidence for an antigen-specific cellular immune response in skin lesions of patients with psoriasis vulgaris. *J. Immunol.* 155:4078-4083.

Moreland, L. W., Morgan, E. E., Adamson, T. C., et al. (1998)T cell receptor peptide vaccination in rheumatoid arthritis: a placebo-controlled trial using a combination of Vβ3, Vβ14, and Vβ17 peptides. *Arthritis Rheum.* 41(11):1919-1929.

Muraro, P. A., Vergelli, M., Kalbus, M., et al. (1997) Immunodominance of a low-affinity major histocompatibility complex-binding myelin basic protein epitope (residues 111-129) in HLA-DR4 (B1*0401) subjects is associated with a restricted T cell receptor repertoire. *J. Clin. Invest.* 100(2):339-349.

Nanki, T., Kohsaka, H., Miyasaka, N. (1998) Development of human peripheral TCRBJ gene repertoire. *J. Immunol.* 161:228-233.

Offner, H., Vandenbark, A. A. (1999) T cell receptor V genes in multiple sclerosis: increased use of TCRAV8 and TCRBV5 in MBP-specific clones. *Intern. Rev. Immunol.* 18:9-36.

Offner, H., Hashim, G. A., Vandenbark, A. A. (1991) T cell receptor peptide therapy triggers autoregulation of experimental encephalomyelitis. *Science* 251:430-432.

Offner, H., Vainiene, M., Celnik, B., et al. (1994) Co-culture of TCR peptide-specific T cells with basic protein-specific T cells inhibits proliferation, IL-3 mRNA and transfer of experimental autoimmune encephalomyelitis. *J. Immunol.* 153:4988-4996.

Offner, H., Adlard, K, Bebo, B. F., et al. (1998) Vaccination with BV8S2 protein amplifies TCR-specific regulation and protection against experimental autoimmune encephalomyelitis in TCR BV8S2 mice. *J. Immunol.* 161:2178-2186.

Offner, H., Jacobs, R., Bebo, B. F., Vandenbark, A. A. (1999) Treatments targeting the TCR: effects of TCR peptide-specific T cells on activation, migration, and encephalitogenicity of MBP-specific T cells. *Springer Semin. Immunopathol.* 21:77-90.

Oksenberg, J. R., Stuart, S., Begovich, A. B., et al. (1990) Limited heterogeneity of rearranged T-cell receptor V alpha transcripts in brains of multiple sclerosis patients. *Nature* 345(6273):344-346.

Oksenberg, J. R., Panzara, M. A., Begovich, A. B., et al. (1993) Selection for T-cell receptor V beta-D beta-J beta gene rearrangements with specificity for a myelin basic protein peptide in brain lesions of multiple sclerosis. *Nature* 362(6415):68-70.

Olive, C., Gatenby, P. A., Serjeantson, S. W. (1991) Analysis of T cell receptor V alpha and V beta gene usage in synovia of patients with rheumatoid arthritis. *Immunol. Cell Biol.* 69( Pt 5):349-354.

Padovan, E., Casorati, G., Dellabona, P., et al. (1993) Expression of two T cell receptor α chains: dual receptor T cells. *Science* 262:422-424.

Padovan, E., Giachino, C., Cella, M., et al. (1995) Normal T lymphocytes can express two different T cell receptor β chains: implications for the mechanism of allelic exclusion. *J. Exp. Med.* 181:1587-1591.

Papageorgiou, A. C., Acharya, K. R. (2000) Microbial superantigens: from structure to function. *Trends Microbiol.* 8(8):369-375.

Posnett, D. N., Vissinga, C. S., Pambuccian, C., et al. (1994) Level of human TCRBV3S1 (Vβ3) expression correlates with allelic polymorphism in the spacer region of the recombination signal sequence. *J. Exp. Med.* 179:1707.

Raju, R., Navaneetham, D., Protti, M. P., et al. (1997) TCR V beta usage by acetylcholine receptor-specific CD4+ T cells in myasthenia gravis. *J. Autoimmun.* 10(2):203-217.

Reinherz, E. L., Tan, K., Tang, L., et al. (1999) The crystal structure of a T cell receptor in complex with peptide and MHC class II. *Science* 286:1913-1921.

Rich, C., Link, J. M., Zamora, A., et al. (2004) Myelin oligodendrocyte glycoprotein-35-55 peptide induces severe chronic experimental autoimmune encephalomyelitis in HLA-DR2 transgenic mice. *Eur. J. Immunol.* 34:1251-1261.

Richards, D., Chapman, M. D., Sasama, J., et al. (1997) Immune memory in CD4+ CD45RA+ T cells. *Immunology* 91(3):331-339.

Richert, J. R., Robinson, E. D., Camphausen, K., et al. (1995) Diversity of T-cell receptor V alpha, V beta, and CDR3 expression by myelin basic protein-specific human T-cell clones. *Neurology* 45(10):1919-1922.

Robinson, M. A. (1992) Usage of human T-cell receptor Vβ, Jβ, Cβ, and Vα gene segments is not proportional to gene number. *Hum. Immunol.* 35:60-67.

Rosenberg, W. M. C., Moss, P. A., Bell, J. I. (1992) Variation in human T cell receptor Vβ and Jβ repertoire: analysis using anchor polymerase chain reaction. *Eur. J. Immunol.* 22:541-549.

Roth, M. E., Holman, P. O., Kranz, D. M. (1991) Nonrandom use of Jα gene segments: influence of Vα and Jα location. *J. Immunol.* 147:1075-1081.

Saito, T., Germain, R. N. (1989) Marked differences in the efficiency of expression of distinct αβ T cell receptor heterodimers. *J. Immunol.* 143(10):3379-3384.

Sarukhan, A., Garcia, C., Lanoue, A., von Boehmer, H. (1998) Allelic inclusion of T cell receptor α genes poses an autoimmune hazard due to low-level expression of autospecific receptors. *Immunity* 8:563-570.

Seboun, E., Robinson, M. A., Kindt, T. J., Hauser, S. L. (1989) Insertion/deletion-related polymorphisms in the human T cell receptor β gene complex. *J. Exp. Med.* 170:1263-1270.

Shanmugam, A., Copie-Bergman, C., Falissard, B., et al. (1996) TCR alpha beta gene usage for myelin basic protein recognition in healthy monozygous twins. *J. Immunol.* 156(10):3747-3754.

Siklodi, B., Jacobs, R., Burrows, G. G., et al. (1998) Neonatal exposure of TCR BV8S2 transgenic mice to recombinant TCR BV8S2 results in reduced T cell proliferation and elevated antibody response to BV8S2 and increased severity of EAE. *J. Neurosci. Res.* 52:750.

Silins, S. L., Cross, S. M., Krauer, K. G., et al. (1998) A functional link for major TCR expansions in healthy adults caused by persistent Epstein-Barr infection. *J. Clin. Invest.* 102:1551-1558.

Sottini, A., Imberti, L., Gorla, R., et al. (1991) Restricted expression of T cell receptor V beta but not V alpha genes in rheumatoid arthritis. *Eur. J. Immunol.* 21(2):461-466.

Struyk, L., Hawes, G. E., Mikkers, H. M., et al. (1996) Molecular analysis of the T-cell receptor beta-chain repertoire in early rheumatoid arthritis: heterogeneous TCRBV gene usage with shared amino acid profiles in CDR3 regions of T lymphocytes in multiple synovial tissue needle biopsies from the same joint. *Eur. J. Clin. Invest.* 26(12):1092-1102.

Theofilopoulos, A. N., Baccalà, R., González-Quintial, R., et al. (1993) T-cell repertoires in health and disease. *Ann. N.Y. Acad. Sci.* 756:53-65.

Triebel, F., Breathnach, R., Graziani, M., et al. (1988) Evidence for expression of two distinct T cell receptor b-chain transcripts in a human diptheria toxoid-specific T cell clone. *J. Immunol.* 140(1):300-304.

Trotter, J. L., Damico, C. A., Cross, A. H., et al. (1997) HPRT mutant T-cell lines from multiple sclerosis patients recognize myelin proteolipid protein peptides. *J. Neuroimmunol.* 75(1-2):95-103.

Tsaknaridis, L., Spencer, L., Culbertson, N., et al. (2003) Functional assay for human CD4+CD25+ Treg cells reveals an age-dependent loss of suppressive activity. *J. Neurosci. Res.* 74:296-308.

Uccelli, A., Giunti, D., Salvetti, M., et al. (1998) A restricted T cell response to myelin basic protein (MBP) is stable in multiple sclerosis (MS) patients. *Clin. Exp. Immunol.* 111(1):186-192.

Uematsu, Y. (1992) Preferential association of α and β chains of the T cell antigen receptor. *Eur. J. Immunol.* 22:603-606.

Usuku, K., Joshi, N., Hatem, C. J., Jr., et al. (1993)The human T-cell receptor beta-chain repertoire: longitudinal fluctuations and assessment in MHC matched populations. *Immunogenetics* 38:193-198.

Usuku, K., Joshi, N., Hatem, C. J., Jr., Wong, M. A., Stein, M. C., Hauser, S. L. (1996) Biased expression of T cell receptor genes characterizes activated T cells in multiple sclerosis cerebrospinal fluid. *J. Neurosci. Res.* 45(6):829-837.

Vacchio, M. S., Granger, L., Kanagawa, O., et al. (1993) T cell receptor Vα-Vβ combinatorial selection in the expressed T cell repertoire. *J. Immunol.* 151(3):1322-1327.

Vainiene, M., Celnik, B., Hashim, G. A., et al. (1996) Natural immunodominant and EAE-protective determinants within the Lewis rat V 8.2 sequence include CDR2 and framework 3 idiotopes. *J. Neurosci. Res.* 43:137-145.

Vandenbark, A. A. (2005) TCR peptide vaccination in multiple sclerosis: boosting a deficient natural regulatory network that may involve TCR-specific CD4+CD25+ Treg cells. *Curr. Drug Targets Inflamm. Allergy* 4:217-229.

Vandenbark, A. A., Hashim, G., Offner., H. (1989) Immunization with a synthetic T-cell receptor V-region peptide protects against experimental autoimmune encephalomyelitis. *Nature* 341:541-544.

Vandenbark, A. A., Chou, Y. K., Whitham, R., et al. (1996a) Treatment of multiple sclerosis with T cell receptor peptides: results of a double-blind pilot trial. *Nat. Med.* 2:1109-1115.

Vandenbark, A. A., Hashim, G. A., Offner, H. (1996b) T cell receptor peptides in treatment of autoimmune disease: rationale and potential. *J. Neurosci. Res.* 43:391-402.

Vandenbark, A. A., Chou, Y. K., Bourdette, D. N., et al. (1997) Immunogenicity is critical to the therapeutic application of T cell receptor peptide. *Drug News Perspect.* 10(6):341.

Vandenbark, A. A., Culbertson, N., Finn, T., et al. (2000) Human TCR as antigen: homologies and potentially cross-reactive HLA-DR2-restricted epitopes within the AV and BV CDR2 loops. *Crit. Rev. Immunol.* 20(1):57-83.

Vandenbark, A. A., Bourdette, D. N., Group, T. T. P. V. S. (2001a) T cell receptor peptide vaccination in multiple sclerosis: results from a multicenter, double-blind, placebo controlled phase I/II trial. *Neurology* 56(Suppl 3):A75.

Vandenbark, A. A., Finn, T., Barnes, D., et al. (2001b) Diminished frequency of IL-10 secreting, TCR peptide-reactive T-cells in multiple sclerosis patients may allow expansion of activated memory T-cells bearing the cognate BV gene. *J. Neurosci. Res.* 66:171-176.

Vandenbark, A. A., Morgan, E., Bourdette, D., et al. (2001c) TCR peptide therapy in human autoimmune diseases. *Neurochem. Res.* 26:713-730.

VanderBorght, A., Geusens, P., Vandevyver, C., et al. (2000) Skewed T-cell receptor variable gene usage in the synovium of early and chronic rheumatoid arthritis patients and persistence of clonally expanded T cells in a chronic patient. *Rheumatology* 39:1189-1201.

Vandevyver, C., Mertens, N., van den Elsen, P., et al. (1995) Clonal expansion of myelin basic protein-reactive T cells in patients with multiple sclerosis: restricted T cell receptor V gene rearrangements and CDR3 sequence. *Eur. J. Immunol.* 25(4):958-968.

Viglietta, V., Baecher-Allan, C., Weiner, H. L., Hafler, D. A. (2004) Loss of functional suppression by CD4+CD25+ regulatory T cells in patients with multiple sclerosis. *J. Exp. Med.* 199:971-979.

Vissinga, C. S., Charmley, P., Concannon, P. (1994) Influence of coding region polymorphism on the peripheral expression of a human TCR Vb gene. *J. Immunol.* 152:1222-1227.

Walker, L. S. K., Chodos, A., Abbas, A. K. (2003a) Antigen-dependent proliferation of CD4+CD25+ regulatory T cells in vivo. *J. Exp. Med.* 198:249-258.

Walker, M. R., Kasprowicz, D. J., Gersuk, V. H., et al. (2003b) Induction of FoxP3 and acquisition of T regulatory activity by stimulated human CD4+CD25– T cells. *J. Clin. Invest.* 112:1437-1443.

Wedderburn, L. R., O'Hehir, R. E., Hewitt, C. R., et al. (1993) In vivo clonal dominance and limited T-cell receptor usage in human CD4+ T-cell recognition of house dust mite allergens. *Proc. Natl. Acad. Sci. USA* 90(17):8214-8218.

Werfel, T., Morita, A., Grewe, M., et al. (1996) Allergen specificity of skin-infiltrating T cells is not restricted to a type-2 cytokine pattern in chronic skin lesions of atopic dermatitis. *J. Invest. Dermatol.* 107(6):871-876.

Wilson, D. B., Golding, A. B., Smith, R. A., et al. (1997) Results of a phase I clinical trial of a T-cell receptor peptide vaccine in patients with multiple sclerosis. I. Analysis of T-cell receptor utilization in CSF cell populations. *J. Neuroimmunol.* 76:15-28.

Wucherpfennig, K. W., Ota, K., Endo, N., et al. (1990) Shared human T cell receptor V beta usage to immunodominant regions of myelin basic protein. *Science* 248(4958):1016-1019.

Wucherpfennig, K. W., Newcombe, J., Li, H., et al. (1992) T cell receptor V alpha-V beta repertoire and cytokine gene expression in active multiple sclerosis lesions. *J. Exp. Med.* 175(4):993-1002.

Wucherpfennig, K. W., Zhang, J., Witek, C., et al. (1994) Clonal expansion and persistence of human T cells specific for an immunodominant myelin basic protein peptide. *J. Immunol.* 152(11):5581-5592.

Yamazaki, S., Iyoda, T., Tarbell, K., et al. (2003) Direct expansion of functional CD25+CD4+ regulatory T cells by antigen-processing dendritic cells. *J. Exp. Med.* 198:235-247.

Yssel, H., Johnson, K. E., Schneider, P. V., et al. (1992) T cell activation-inducing epitopes of the house dust mite allergen Der p I: proliferation and lymphokine production patterns by Der p I-specific CD4+ T cell clones. *J. Immunol.* 148(3):738-745.

Zang, Y. C., Kozovska, M., Aebischer, I., et al. (1998) Restricted TCR Valpha gene rearrangements in T cells recognizing an immunodominant peptide of myelin basic protein in DR2 patients with multiple sclerosis. *Int. Immunol.* 10(7):991-998.

Zhang, J., Vandevyver, C., Stinissen, P., et al. (1995a) Activation and clonal expansion of human myelin basic protein-reactive T cells by bacterial superantigens. *J. Autoimmun.* 8(4):615-632.

Zhang, J., Vandevyver, C., Stinissen, P., Raus, J. (1995b) In vivo clonotypic regulation of human myelin basic protein-reactive T cells by T cell vaccination. *J. Immunol.* 155(12):5868-5877.

Zhao, T. M., Whitaker, S. E., Robinson, M. A. (1994) A genetically determined insertion/deletion related polymorphism in human T cell receptor b chain (TCRB) includes functional variable gene segments. *J. Exp. Med.* 180:1405.

# 17
# Nonmyeloablative Stem Cell Transplantation for Multiple Sclerosis

Richard K. Burt, Yvonne Loh, and Larissa Verda

## 1   Stem Cell Replacement Therapy

There have been major advancements in our understanding of the biology of stem cells and their potential in clinical applications during the last few years. Stem cells are capable of both self-renewal and lineage commitment and differentiation into more specialized cells; and they are broadly categorized as either embryonic stem cells (ESC) or adult stem cells. ESC can be obtained by culturing the inner cell mass derived from preimplantation embryos (Carpenter et al., 2003; Edwards, 2004; Thompson et al., 1998; Wobus et al., 1984). The propagation of ESC requires a feeder layer of fibroblasts of mouse origin (Edwards, 2004; Thompson et al., 1998). Another disadvantage of ESC use in vivo is a tendency to form teratomas at the site of injection (Burt et al., 2004b; Edwards, 1987). This complication may be overcome by induction of ESC differentiation into lineage-committed progenitor cells with subsequent stringent selection of this desired cell population prior to in vivo application (Burt et al., 2004b). It has been shown that ESCs can be successfully differentiated in vitro into neuronal stem cells, neurons, oligodendrocyte progenitor cells, hematopoietic stem cells, and others (Burt et al., 2004b; Reubinoff et al., 2001; Xian et al., 2003; Zhang et al., 2001). Although preliminary data on ESC-based cell replacement therapy appear promising (Bjorklund et al., 2002; McDonald et al., 1999), numerous unanswered questions and unsolved issues need to be confronted: (1) ethical issues; (2) safety of supplements used for propagation of ESCs; (3) characteristics of ESC-derived cells, including their mechanisms of action in vivo; and (4) more experimental studies to confirm the therapeutic benefit of ESCs or ESC-derived cell use in the clinical situation.

Adult stem cells are obtained from differentiated tissue compartments during or after birth and are lineage-restricted (pluripotent) to differentiate into and replenish a particular tissue or organ system. In experimental autoimmune, chemical, or traumatic models of central nervous system (CNS) demyelination, adult neural stem cells (NSCs) show the ability to reach the areas of tissue damage selectively, differentiate into axon-ensheathing oligodendrocytes, and promote functional recovery while injected intraparenchymally, intracerebroventricularly, or intravenously (Beh-Hur et al., 2003; Pluchino et al., 2003). The greatest disadvantage of NSCs is the difficulty of safely collecting this cell population in clinically relevant numbers. NSCs are located in the periventricular area of the brain, which renders harvesting from a living patient impractical. Although it has been reported that adult NSCs can be expanded and maintained safely in a chemically defined medium for years (Nunes et al., 2003), further studies to assess the safety, efficacy, and in vivo plasticity of these cells are required before any future human applications can be brought into clinical use.

Hematopoietic stem cells (HSCs) may be easily and safely collected in clinically significant numbers from the bone marrow, blood, or umbilical cord blood. HSCs maintain the ability to self-renew and differentiate into cells of all blood lineages throughout adult life. HSCs have been recently shown to have greater plasticity than previously recognized: they can differentiate into specific cells other than hematopoietic cells, including neural cells (Eglitis and Mezey, 1997; Krause et al., 2001; Mezey et al., 2003). Experiments on animal models of CNS demyelination have indicated that injected bone marrow cells could differentiate into physiologically functional nonhematopoietic cells and contribute to remyelination (Inoue et al., 2003). Whether this phenomenon is the result of transdifferentiation or cell fusion remains unclear (Alvarez-Dolado et al., 2003; Wagers et al., 2002). The true identity of cells of such great plasticity is still questionable as most studies have used nonselected bone marrow, which is a known source of hematopoietic stem cells, stromal stem cells, and possibly other stem cells such as germline stem cells (Johnson et al., 2005). The route of administration (direct intralesional, intracerebroventricular, or intravenous) represents another key issue for cell replacement therapy in CNS pathology. Although these findings support the possibility of using bone marrow stem cells (BMSCs) for the repair of CNS pathologies, the actual contribution of BMSCs to this process remains controversial, and it remains unproven that clinically significant levels of transdifferentiation and/or cell fusion occur in vivo.

# 2    Hematopoietic Stem Cell Transplantation in Multiple Sclerosis

Hematopoietic stem cell transplantation (HSCT), discussed below, is an intensive form of immunosuppressive therapy. The patient (recipient of HSCs) is prepared for transplant by potent immunosuppressive or myeloablative treatment. This treatment—referred to as a transplant conditioning regimen—is followed by the transfer of HSCs (autologous if the cells were harvested from the recipient before therapy and allogeneic if the cells were harvested from another individual). The transplant conditioning regimen destroys the aberrant disease-causing immune cells while HSCs regenerate a new, antigen-naive immune system. Therefore, the toxicity and efficacy of an autologous HSCT is likely a consequence of the conditioning regimen. Allogeneic HSCT (disscussed section 2.4) has the additional risk of graft versus host disease (GVHD).

HSCs mobilized into the peripheral blood are collected by apheresis, which is an outpatient procedure performed through a double-lumen catheter inserted into the internal jugular vein. Blood is drawn from one lumen of the catheter; the mononuclear cells are then separated by an external centrifuge and returned to the patient through the second catheter lumen. Approximately 10 to 15 liters of blood are processed in a procedure that requires several hours. The peripheral blood stem cells (PBSCs) are cryopreserved or further processed by immunoselection for an HSC phenotype and then cryopreserved. Purification or enrichment ex vivo for HSCs may be performed using antibodies to select for CD34 or CD133 or by negative selection by using antibodies to remove lymphocytes. In general, a minimum of $2 \times 10^6$ CD34+ cells/kg of recipient weight ensures engraftment (Fig. 17.1).

HSCT was proposed as a treatment for multiple sclerosis (MS) in 1995 based on favorable results in experimental autoimmune encephalomyelitis (EAE), an animal model of MS (Burt et al., 1995). EAE is an autoimmune demyelinating disease of the CNS induced by either in vivo immunization with myelin peptides or adoptive transfer of ex vivo primed CD4+ T cells. HSCs are acquired from a euthanized animal of a different strain (allogeneic HSCT) or the same highly inbred strain (syngeneic HSCT) or from a syngeneic animal with the same stage of disease (pseudoautologous HSCT). Any of three donor HSC sources (allogeneic, syngeneic, pseudoautologous) are capable of alleviating the neurological disability when performed during the acute phase of disease (Burt et al., 1998; Karussis et al., 1992, 1993; van Gelder et al., 1996a,b). In contrast, HSCT does not diminish the neurological impairment when performed during the chronic progressive phase of EAE (Burt et al, 1998).

## 2.1    Results of Autologous HSCT for MS

High-dose chemotherapy with autologous hematopoietic stem cell transplantation is standard treatment for selected hematological malignancies. Initial HSCT

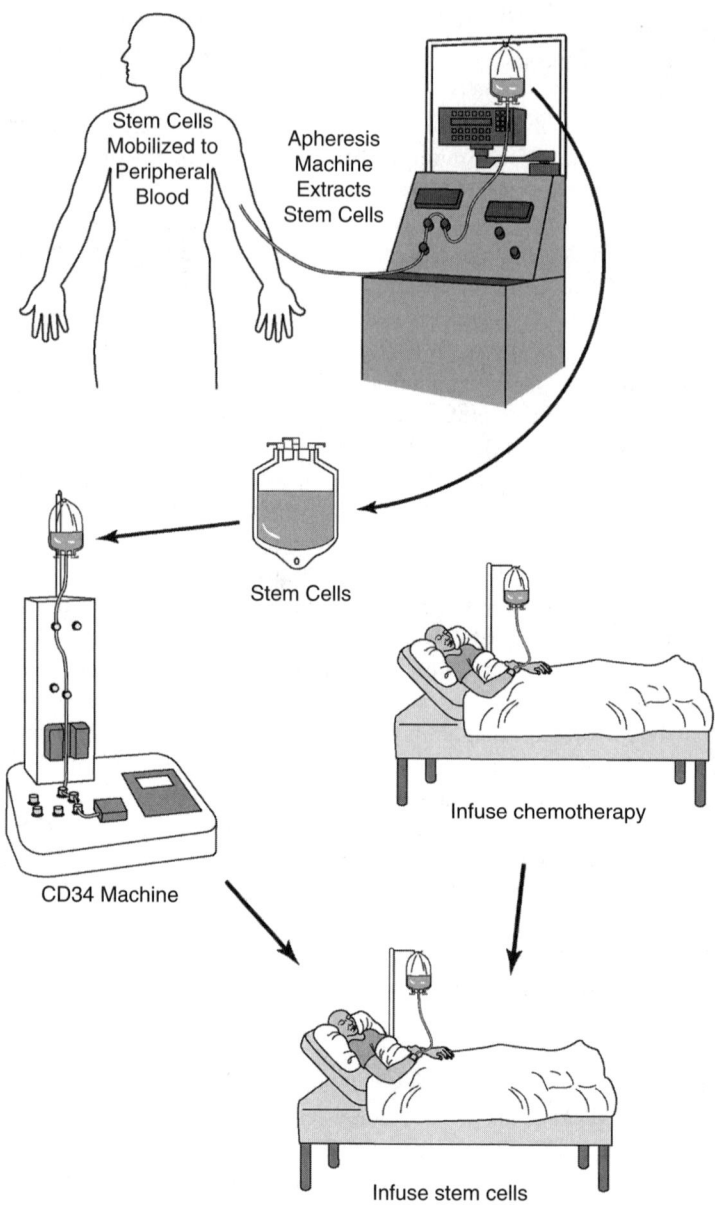

Figure 17.1. Stem cell transplantation.

protocols for the treatment of MS generally employed rather aggressive malignancy-specific myeloablative regimens in patients with progressive MS (Table
17.1). Based on these studies, immunosuppression following autologous HSCT
appears to be an effective therapy to halt inflammation, as seen by magnetic
resonance imaging (MRI). In fact, there is no other therapy that may provide such

TABLE 17.1. Myeloablative Autologous Hematopoietic Stem Cell Transplant Trials for Multiple Sclerosis

| Study | Conditioning regimen | No. of patients (no. of deaths) | Cause of death/time after transplant | Comment |
|---|---|---|---|---|
| Fassas et al. (2000) | BEAM/ATG | 24 (1) | Invasive aspergillosis at day +65 | 9 Patients improved EDSS score by 1.0–3.0 points<br>5 Patients progressed with worsened EDSS scores<br>1 Patient – venooocclussive disease of liver<br>1 Patient – TTP |
| Kozak et al. (2001) | BEAM | 10 (0) | None | 1 Patient improved EDSS score by 1.5 point, 3 patients by 0.5 point<br>1 Patient worsened EDSS score by 1.0 point<br>3 Patients – bacteremia |
| Saiz et al. (2004) | BCNU/CY/ ATG | 14 (0) | None | 4 Patients – improved EDSS score by 0.5–1.5 points<br>2 Patients worsened; EDSS score (1.0 and 1.5 points) |
| Mancardi et al. (2001) | BEAM | 10 (0) | None | 5 Patients – improved EDSS score by 0.5–1.5 points<br>2 Patients – UTI<br>2 Patients – VZV infection<br>1 Patient – CMV infection |
| Burt et al. (2003) | Cy/TBI | 21 (2) | Neurologic disease-related deterioration at 13 and 18 months, respectively | Pretranpslant EDSS ≤ 6 (4 patients): 1 improved by 2.5 points, 6 progressed by 0.5 point<br>Pretransplant EDSS > 6 (12 patients): 10 progressed by 0.5–1.0 point<br>1 Patient – bacteremia<br>2 Patients – VZV infection<br>1 Patient – asymptomatic subdural hematoma |
| Nash et al. (2003) | TBI/CY/ATG | 26 (2) | 1 Patient – EBV-associated PTLD at day +53<br><br>1 Patient – neurologic deterioration with aspiration pneumonia at 23 months | 2 Patients improved EDSS score by 0.5 point<br><br>6 Patients worsened EDSS score by >1.0 point, 3 patients worsened by 0.5 point<br>4 Patients – bacteremia<br>8 Patients – UTI<br>2 Patients – viral upper respiratory and gastrointestinal infections<br>2 Patients – VZV infection<br>1 Patient – HSV infection<br>1 Patient – CMV infection<br>1 Patient – ITP |

(*Continued*)

TABLE 17.1. Myeloablative Autologous Hematopoietic Stem Cell Transplant Trials for Multiple Sclerosis—Cont'd

| Study | Conditioning regimen | No. of patients (no. of deaths) | Cause of death/time after transplant | Comment |
|---|---|---|---|---|
| Openshaw et al. (2000) | Bu/CY/ATG | 5 (2) | 1 Patient – influenza A pneumonia at day +22<br>1 Patient – pneumonia sepsis at 19 months | 1 Patient – worsened EDSS score by 1.0 point |
| Atkins (verbal communi-cation) | Bu/Cy/ATG | 11 (1) | Venoocclussive disease of liver | No follow-up data of neurologic function is available |
| Samijn et al. (2006) | TBI/Cy/ATG | 14 (1) | Radiation-related myelodysplasia | 9 Patients progressed with worsening EDSS score |
| Voltarelli J. (verbal communi-cation) | BEAM/ATG | 25 (3) | 1 Patient – pneumonitis<br>1 Patient – CMV pneumonia<br>1 Patient – alveolar hemorrhage | No follow-up data of neurologic function is available<br>2 Patients – fungal infection<br>2 Patients – sinusitis<br>2 Patients – CMV infection<br>1 Patient – HSV-6 infection<br>2 Patients – alveolar hemorrhage<br>1 Patient – venoocclussive disease of liver<br>1 Patient – ATG anaphylaxis |
| Voltarelli J., off protocol (verbal commun-ication) | BEAM/ATG | 5 (4) | 1 Patient – pneumonia<br>1 Patient – CMV<br>1 Patient – HHV-6<br>1 Patient – gastrointestinal bleed | No follow-up data of neurologic function is available |
| EBMT retrospe-ctive summary – Fassas et al. (2002) | BEAM (16%)<br>BEAM/ATG (47%)<br>CY/ATG/ other (12%)<br>CY/TBI/ATG (6%)<br>Bu/CY/ATG (18%)<br>Flu/ATG (1%) | 85 (7) | 5 Patients – toxicity/infection-related at day +7 to 19 months<br>2 Patients – neurologic deterioration-related within 3 months | 18 Patients (21%) – improved EDSS score by ≥ 1.0 point<br>6 Patients – disease progressed<br>Probability of confirmed disease progression at 3 years 20% ± 11%<br>59% of patients had nonhematologic grade III and IV toxicities (infection, cardiac, and hepatic toxicities; bleeding; TTP) |

a striking and long-term effect on suppressing MRI gadolinium enhanced and new T2-weighted lesions. Saiz et al. (2001, 2004), using a regimen of carmustine (BCNU), cyclophosphamide, antithymocyte globulin, and CD34 selection of the graft, reported no post-HSCT enhancing lesions and a decrease in the mean T2-weighted lesion load by 11.8%. Mancardi et al. (2001), using a regimen of BCNU, etoposide, cytosine arabinoside, and melphalan (BEAM), performed triple-dose gadolinium MRI monthly for 3 months before HSCT and then monthly for 6 months followed by every 3 months after HSCT. Complete and durably suppressed MRI activity was documented following HSCT.

Autologous HSCT also appears to reset the immune system effectively. The mechanism of autologous transplant-induced remission of an immune-mediated disease may be transient immunosuppression-related lymphopenia and/or a more durable "immune reset" due to regeneration of an antigen-naive immune system from the HSCs. By analyzing T cell receptor repertoires with flow cytometry, polymerase chain reaction (PCR) spectratyping, and sequenced-based clonotyping, as well as recent thymic emigrant output by the T cell receptor excision circle (TREC), we have shown in MS patients undergoing HSCT that a new and antigen-naive T cell repertoire arises from the stem cell compartment via thymic regeneration (Muraro et al., 2005). This suggests that intense immunosuppression via HSCT results in long-term immune reset independent of transient immunosuppression-mediated lymphopenia.

Despite suppression of MRI enhancing lesions and encouraging data on immune reconstitution, myeloablative HSCT protocols were complicated by treatment-related deaths (Table 17.1); moreover, the clinical outcome in terms of progressive neurological disability was not obviously better than the natural history of patients with progressive MS. The discordance between promising immune analysis and MRI data continued clinical disability and is most likely due to the selection for transplant of patients with late progressive disease without significant ongoing CNS inflammation. In a European retrospective analysis on 85 patients, the progression-free survival at 3 years was 78% for secondary progressive MS (SPMS) and 66% for primary progressive MS (PPMS) (Fassas et al., 2000). At Northwestern University (Chicago), of 21 patients with SPMS treated using a myeloablative HSCT regimen, disease progression in more disabled patients with a pretreatment Expanded Disability Status Scale (EDSS) score of $\geq 6.0$ was significantly worse than in those with an EDSS score $< 6.0$ (Fig. 17.2) (Burt et al., 2003). In fact, none of 9 patients with an EDSS score $< 6.0$ had disease progression of 1.0 or more EDSS points after more than 2 years of follow-up. The single patient in this study with relapsing-remitting MS (RRMS) not only failed to progress but had a sustained improvement by 2.5 EDSS steps. In a Rotterdam, (The Netherlands) study using a total body irradiation-based myeloablative regimen in patients with SPMS, 9 of 14 patients had continued posttransplant progression of disability by EDSS rating (Samijn et al., 2006).

In retrospect, because autologous HSCT is a form of intense immunosuppression, it is unlikely to have a beneficial impact on the noninflammatory (i.e., degenerative) aspects of MS. This is supported by MRI data in patients with

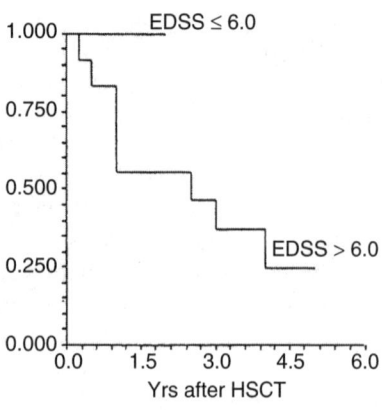

FIGURE 17.2. Difference in the Expanded Disability Status Scale (EDSS) progression between patients with less disability (EDSS ≤ 6.0) versus higher disability (EDSS > 6.0) at the time of autologous hematopoietic stem cell transplantation. (From Blood (Burt et al., 2003, with permission.)

progressive disease undergoing HSCT who have a continued decrease in brain volume, suggesting continued axonal atrophy for the duration of reported follow-up, at least 2 years, after HSCT (Inglese et al., 2004). The importance of selecting patients with inflammatory disease is also supported by data on two patients with pretreatment striking gadolinium-enhancing lesions and severe deficits (nonambulatory with EDSS scores of 7.5 and 8.0) after a short clinical duration of disease (1 and 3 years). After autologous HSCT, they were able to ambulate 100 and 300 meters, respectively, with only unilateral assistance by 6 months after HSCT (Mancardi et al., 2004).

The initial myeloablative regimens have also been associated with treatment-related deaths (Table 17.1). A Phase 1 study performed at the City of Hope (Duarte, California) utilizing a maximum dose myeloablative cancer regimen of busulfan and cyclophosphamide (also known as Big BuCy) along with ATG and CD34+ selection resulted in treatment-related death in two of five patients (Openshaw et al., 2000). A similar Big BuCy regimen performed in Ottawa (Canada) resulted in a treatment-related death due to hepatic venoocclusive disease in 1 of 11 patients (M.S. Freedman, personal communication). A slightly less intense regimen using a leukemia-specific protocol of myeloablative total body irradiation (TBI), cyclophosphamide, ATG, and CD34+ selection performed at the Fred Hutchinson Cancer Center resulted in one reported transplant-related death among 26 patients (Nash et al., 2003). A similar TBI-based regimen performed in Rotterdam that enrolled 14 patients ended with one patient developing radiation-related preleukemic myelodysplasia (Samijn et al., 2006). A retrospective European analysis of 85 patients treated with a lymphoma-specific regimen, BEAM (BCNU, etoposide, cytosine arabinoside (ara-c), melphalan) reported five treatment-related deaths (Fassas et al., 2002). The BEAM regimen, which was also used in a Brazilian MS transplant trial, was recently closed owing to excessive morbidity and mortality (J. Voltarelli, personal communication). Although an Italian trial of the BEAM regimen in MS showed better safety with no deaths in 19 patients (Saccardi et al., 2005), the high overall transplant-related mortality, mostly due to infection but also due to end-organ damage and treatment-related leukemia, has resulted in the termination of trials, dose reduction in treatment (conditioning regimen) drug intensity, enrolling less disabled patients, and/or limiting the procedure to more experienced centers. Nevertheless, significant concerns remain as to whether any of these first-generation myeloablative cancer-specific regimens are capable of achieving equipoise in a disease of low mortality such as MS, especially as the patients likely to benefit are generally not severely disabled.

In contrast, the experience with nonmyeloablative regimens in active inflammatory MS is more encouraging. In Berlin, a nonmyeloablative regimen of cyclophosphamide and rabbit ATG has been used in children with marked improvements in EDSS scores and with little morbidity (R. Arnold, personal communication). A nonmyeloablative regimen of cyclophosphamide and anti-lymphocyte antibodies has been well tolerated with little morbidity at Northwestern University (Chicago). No patients had early or late infections. One-half of the patients never required a red blood cell transfusion, and one-fourth never required a platelet transfusion. In 18 patients with 6 months or more of follow-up, the EDSS has improved by at least 1 point in most (R. Burt, unpublished manuscript in preparation).

## 2.2    Myeloablative versus Nonmyeloablative (Lymphoablative) Autologous HSCT for MS

The rationale for autologous HSCT for an immune-mediated disease is that the disease is not due to a genetic defect of HSCs but, rather, is a disorder triggered by an environmental component. For MS, the logical goal of an autologous transplant conditioning regimen is therefore lymphoablation, not myeloablation. Following a nonmyeloablative conditioning regimen, autologous HSCs are infused to shorten the duration of the conditioning regimen-related cytopenia. Compared to myeloablative regimens, nonmyeloablative regimens have lower treatment-related mortality, which in terms of risk/benefit from treatment is a significant advantage for MS, a disease that has significantly lower mortality than malignancies. Despite the lack of myeloablation, aggressive combinations of lymphoablative agents are highly immunosuppressive, which could also result in lethal opportunistic infections. Therefore, nonmyeloablative transplant regimens must be tailored for the degree of immunosuppression desired. Myeloablative drugs such as etoposide, TBI, busulfan, melphalan, or carmustine that are used in myeloablative first-generation HSCT studies have nothing to do with treating MS but were chosen because of their familiarity to oncologists for treating cancer. The regimen must also avoid further damage to already injured axons and oligodendrocytes. By definition, myeloablative agents are lethal to HSCs and, apart from their myeloablative effect on bone marrow, may be similarly lethal to tissue-specific stem cells such as oligodendrocyte progenitor cells or NSCs. In animal models, cranial irradiation impairs the mechanism of CNS repair by NSC apoptosis, alteration in cell cycle progression, and/or destruction of the NSC niche or milieu through invasion of macrophages and microglia (Monje et al., 2002). This raises concerns about using a total body irradiation-based, or otherwise stem cell-ablative regimen in the treatment of MS. Fever-related deterioration of neural function in MS, termed pseudoexacerbation, due to conduction blocks in marginally functioning demyelinated axons should be avoided during transplant by minimizing the use of pyrogenic agents in the conditioning regimen. Similarly, the risk of infection-related fever should be minimized during transplant by using prophylactic antibiotics. In comparison, nonmyeloablative regimens employ lymphoablative agents (e.g., cyclophosphamide) that neurologists already employ to treat MS and that have little nonlymphopoietic toxicity.

As the success of HSCT in hematologic malignancies depends on early intervention, it is possible that the earlier HSCT is performed for the treatment of MS the better the outcome. The goal of nonmyeloablative HSCT is to prevent inflammation and suppress relapses by intervening before the onset of irreversible progressive axonal degeneration. Studies suggest that axonal damage is a relatively early event in MS lesions and normal-appearing white matter (PRISMS Study Group, 1998). Treatment at an earlier stage of disease, particularly in patients with a relapsing remitting course, might help prevent massive axonal loss associated with permanent deficits and disabilities. Rather than selecting for rapidly progressive disease (i.e., an increase in EDSS score of 1.0 or more points during the preceding 12 months), transplant candidates should be selected for active inflammation. Criteria may include RRMS or RPMS with multiple acute relapses despite interferon treatment and gadolinium-enhanced MRI lesions with less accumulated disability (EDSS score 2.0–6.0). Only the nonmyeloablative approach should be explored in these patients. Patients with higher EDSS scores could be considered if they have rapid clinical deterioration and striking gadolinium enhancement on MRI.

## 2.3   Summary of Autologous HSCT for MS

In limited safety trials, myeloablative HSCT for progressive MS has been complicated by both treatment-related and disease-related deaths. In contrast, Phase I safety trials with nonmyeloablative HSCT in relapsing MS have *not* been fraught with morbidity or mortality; and in most of the patients, nonmyeloablative HSCT results in improved neurologic function with lower disability scores. Although the long-term durability of nonmyeloablative HSCT-induced remission of active inflammation is yet to be determined, it holds promise for patients with active inflammatory disease if performed before the onset of significant irreversible axonal injury. The exact role of nonmyeloablative HSCT in the treatment of MS is currently being explored in the Multiple Sclerosis International Stem Cell Transplant (MIST) Trial (Table 17.2).

TABLE 17.2. MIST—Multiple Sclerosis International Stem Cell Transplant Trial.

**Eligibility**—EDSS 2.0 to 6.0 and two relapses treated with intravenous corticosteroids within the past 12 months despite interferon therapy or one relapse treated with intravenous corticosteroids within the last 12 months despite interferon therapy and gadolinium-enhanced lesions on MRI

**Treatment group**—Nonmyeloablative autologous HSCT

**Control group**—Continued FDA approved therapy (e.g., interferons, copaxone, mitoxantrone)

**Endpoint**—Treatment failure defined as increase in EDSS of at least 1.0 points sustained on two consecutive examinations at least 3 months apart

**Crossover**—Crossover between groups allowed when meeting above endpoint

**Center/Contact**

1. Northwestern University Feinberg School of Medicine, Chicago, IL; Richard Burt, MD: rburt@northwestern.edu

## 2.4   Allogeneic HSCT for MS

Similar to autologous HSCT, an immunosuppressive conditioning regimen is necessary before allogeneic stem cell infusion. The conditioning regimen for autologous HSCT is designed to be lymphoablative to induce disease remission and to facilitate immune regeneration from the patient's own stem cell compartment. For allogeneic HSCT, the conditioning regimen is also designed to be lymphoablative to induce disease remission, allow engraftment of allogeneic or foreign HSCs, and subsequently facilitate immune regeneration from the donor's allogeneic stem cells. The allogeneic immune compartment will have a different and presumably disease-resistant genetic predisposition toward disease recurrence. Compared to autologous HSCT, allogeneic HSCT is therefore more likely to cure or prevent recurrence of immune-mediated diseases such as MS. The traditional complications arising from allogeneic HSCT are more serious than after an autologous transplant due to graft-versus-host disease (GVHD), a disease mediated by donor immune cells that recognize host tissue as foreign and in severe cases is lethal. GVHD may be eliminated by infusing a graft enriched for donor HSCs through aggressive purging of donor lymphocytes. In patients with malignancies, depleting donor lymphocytes to prevent GVHD results in leukemia relapse rates similar to those seen with an autologous HSC transplant. Therefore, in terms of malignancies, donor lymphocytes convey a beneficial graft versus leukemia or graft versus tumor effect as well as a detrimental GVHD. For autoimmune diseases, the risk of trading one immune-initiated disease (MS) for another even more lethal immune-mediated disease, GVHD, would be unacceptable. Thus, for autoimmune diseases, allogeneic HSCT must be performed without risk of GVHD (i.e., by using an HSC-enriched and/or lymphocyte-depleted allogeneic graft).

At Northwestern University, we investigated whether HSC could convey a graft-versus-autoimmune (GVA) effect in an animal model of type 1 diabetes mellitus. Donor chimerism achieved by the transplantation of HSCs lacking hematopoietic-derived immune cells exerted a profound antidiabetic effect without adverse events such as GVHD. This study suggested that, unlike the relation between GVL and GVHD, both of which result from donor lymphocytes, the GVA effect is separable from GVHD and arises from donor HSCs (L. Verda, unpublished work). Clinical trials with lymphocyte-depleted allogeneic donor HSCT have already begun to determine if a GVA effect may occur without GVHD in patients with autoimmune disease. A 52-year-old woman with a poor prognosis and refractory rheumatoid arthritis (RA) was given a transplant of purified lymphocyte-depleted HSCs from her HLA-matched normal sibling. The patient has stable mixed chimerism with both donor and host hematolymphopoiesis; she never developed GVHD, and her RA is in complete remission (Burt et al., 2004a,b). However, more experience with lymphocyte-depleted allogeneic HSCTs in patients with aggressive and potentially lethal autoimmune diseases is necessary before applying this approach to an immune-mediated disease of low mortality such as MS.

## 2.5  Mesenchymal Stem Cell Transplantation for MS

It has been reported that intravenous injection of mesenchymal stem cells (MSCs) can ameliorate EAE, an animal model of MS, without use of any chemotherapy or immunosuppressive medications (Zappia et al., 2005). Human bone marrow stromal cells appear to suppress T lymphocyte proliferation induced by cellular or nonspecific mitogenic stimuli (Di Nicola et al., 2002). Because MSCs can be easily obtained and expanded from bone marrow, their direct immunomodulatory effect independent of chemotherapeutics can be desirable for treatment of immune-mediated diseases. Although the exact mechanism(s) is unclear, MSC-mediated immunosuppression occurs through both direct cell contact and cytokine secretion (Zhao et al., 2003). In fact, treatment of human immune-mediated diseases with intravenous infusion of MSCs has already begun following a report of rapid resolution of severe GVHD after transplant of MSCs from the patient's haplo-identical mother (Le Blanc et al., 2004). Nevertheless, many questions and concerns about MSCs remain unanswered: (1) What are the late complications of MSC infusions? For example, do MSCs lodge in tissue such as the pulmonary vascular bed, resulting in local proliferation and differentiation into fibroblasts with late fibrosis? Can MSCs proliferate in vivo to form tumors when placed outside their normal bone marrow niche? (2) What are the exact mechanisms of MSC-mediated immunosuppression? Do MSCs normally suppress immune reactions in vivo, or is an immunosuppressive phenotype the consequence of extended passage in tissue culture? (3) Because MSCs are a heterogeneous population of cells, is there a unique identifiable MSC marker? (4) What is the best method to isolate, purify, and assay MSCs? (5) Are MSCs true stem cells that can be transplanted, isolated, and retransplanted in serial generations of recipients? (6) Is their immunosuppressive effect durable, or will repeated infusions be required? Will immunologic sensitization or rejection occur with repeated exposure? (7) Will nonspecific MSC-mediated immunosuppression lead to increased opportunistic infections? These questions not withstanding, the application of MSCs for immune modulation may allow MS-directed stem cell therapy that results in favorable immune modulation without exposure to chemotherapy or the use of combined allogeneic HSC and MSC transplantation for engraftment without GVHD.

# 3    Conclusion

There are several types of stem cell that may have potential for treating MS. Due to practical concerns related to good manufacturing procedures required for their culture and expansion, ESCs are not yet available in the United States for clinical trials. Adult neuronal stem cells are, at present, impractical to harvest safely in sufficient quantities for clinical use. Therefore, clinical stem cell trials for MS have focused on hematopoietic stem cells. To date, only autologous HSCT has been performed for treatment of MS. Autologous HSCT is a form of immunosuppression, with the conditioning regimen ablating the immune system while HSCs regenerate a new immune system. Initial autologous HSCT protocols employed malignancy-specific myeloablative regimens, enrolled predominantly late SPMS patients, were complicated by some treatment-related deaths, and failed to stop the continued decline in neurologic disability (increase in EDSS score) despite effective suppression of CNS inflammation. These results helped clarify the pathogenesis of late progressive MS as predominantly an axonal degenerative disease and raised questions about the use of any immunosuppressive medication to treat progressive MS. The current goal of autologous HSCT studies in MS is to intervene earlier in the disease course with safer MS-specific nonmyeloablative regimens to prevent or significantly delay the onset of progressive neurologic disability. In patients with inflammatory MS, defined by frequent active relapses and gadolinium-enhancing lesions on MRI, nonmyeloablative HSCT has diminished the neurologic disability (lowered the EDSS score) in most patients. The duration of autologous nonmyeloablative HSCT-induced MS remission is unknown. Its role in treating MS is being evaluated in an ongoing randomized trial (the Multiple Sclerosis International Stem Cell Transplant Trial, MIST). Allogeneic HSCT or MSC transplants have not yet been performed to treat patients with MS, although such approaches are being evaluated in EAE.

## References

Alvarez-Dolado, M., Pardal, R., Garcia-Verdugo, J. M., et al. (2003) Fusion of bone marrow-derived cells with Purkinje neurons, cardiomyocytes and hepatocytes. *Nature* 425:968-973.

Ben-Hur, T., Einstein, O., Mizrachi-Kol, R., et al. (2003) Transplanted multipotential neural precursor cells migrate into the inflamed white matter in response to experimental autoimmune encephalomyelitis. *Glia* 41:73-80.

Bjorklund, L. M., Sanchez-Pernaute, R., Chung, S., et al. (2002) Embryonic stem cells develop into functional dopaminergic neurons after transplantation in a Parkinson rat model. *Proc. Natl. Acad. Sci. USA* 99:2344-2349.

Burt, R. K., Burns, W., Hess, A. (1995) Bone marrow transplantation for multiple sclerosis. *Bone Marrow Transplant.* 16:1-6.

Burt, R. K., Padilla, J., Begolka, W. S., et al. (1998) Effect of disease stage on clinical outcome after syngeneic bone marrow transplantation for relapsing experimental autoimmune encephalomyelitis. *Blood* 91:2609-2616.

Burt, R. K., Cohen, B. A., Russell, E., et al. (2003) Hematopoietic stem cell transplantation for progressive multiple sclerosis: failure of a total body irradiation-based conditioning regimen to prevent disease progression in patients with high disability scores. *Blood* 102:2373-2378.

Burt, R. K., Oyama, Y., Verda, L., et al. (2004a) Induction of remission of severe and refractory rheumatoid arthritis by allogeneic mixed chimerism. *Arthritis Rheum.* 50:2466-2470.

Burt, R. K., Verda, L., Kim, D. A., et al. (2004b) Embryonic stem cells as an alternate marrow donor source: engraftment without graft-versus-host disease. *J. Exp. Med.* 199:895-904.

Carpenter, M. K., Rosler, E., Rao, M. S. (2003) Characterization and differentiation of human embryonic stem cells. *Cloning Stem Cells* 5:79-88.

Di Nicola, M., Carlo-Stella, C., Magni, M., Milanesi, M., et al. (2002) Human bone marrow stromal cells suppress T-lymphocyte proliferation induced by cellular or nonspecific mitogenic stimuli. *Blood* 99:3838-3843.

Edwards, R. G. (1987) In vitro fertilization: past and future. *Ann. Biol. Clin.* 45:321-329.

Edwards, R. G. (2004) Stem cells today. A. Origin and potential of embryo stem cells. *Reprod. Biomed. Online* 8:275-306.

Eglitis, M. A., Mezey, E. (1997) Hematopoietic cells differentiate into both microglia and macroglia in the brains of adult mice. *Proc. Natl. Acad. Sci. USA* 94:4080-4085.

Fassas, A., Anagnostopoulos, A., Kazis, A., et al. (2000) Autologous stem cell transplantation in progressive multiple sclerosis—an interim analysis of efficacy. *J. Clin. Immunol.* 20:24-30.

Fassas, A., Passweg, J. R., Anagnostopoulos, A., et al. (2002) Hematopoietic stem cell transplantation for multiple sclerosis: a retrospective multicenter study. *J. Neurol.* 249:1088-1097.

Inglese, M., Mancardi, G. L., Pagani, E., et al. (2004) Brain tissue loss occurs after suppression of enhancement in patients with multiple sclerosis treated with autologous haematopoietic stem cell transplantation. *J. Neurol. Neurosurg. Psychiatry* 75:643-644.

Inoue, M., Honmou, O., Oka, S., et al. (2003) Comparative analysis of remyelinating potential of focal and intravenous administration of autologous bone marrow cells into the rat demyelinated spinal cord. *Glia* 44:111-118.

Johnson, J., Bagley, J., Skaznik-Wikiel, M., et al. (2005) Oocyte generation in adult mammalian ovaries by putative germ cells in bone marrow and peripheral blood. *Cell* 122:303-315.

Karussis, D. M., Slavin, S., Lehmann, D., et al. (1992) Prevention of experimental autoimmune encephalomyelitis and induction of tolerance with acute immunosuppression followed by syngeneic bone marrow transplantation. *J. Immunol.* 148:1693-1698.

Karussis, D. M., Vourka-Karussis, U., Lehmann, D., et al. (1993) Prevention and reversal of adoptively transferred, chronic relapsing experimental autoimmune encephalomyelitis with a single high-dose cytoreductive treatment followed by syngeneic bone marrow transplantation. *J. Clin. Invest.* 92:765-772.

Kozak, T., Havrdova, E., Pit'ha, J., et al. (2001) Immunoablative therapy with autologous stem cell transplantation in the treatment of poor risk multiple sclerosis. *Transplant. Proc.* 33:2179-2181.

Krause, D. S., Theise, N. D., Collector, M. I., et al. (2001) Multi-organ, multi-lineage engraftment by a single bone marrow-derived stem cell. *Cell* 105:369-377.

Le Blanc, K., Rasmusson, I., Sundberg, B., et al. (2004) Treatment of severe acute graft-versus-host disease with third party haploidentical mesenchymal cells. *Lancet* 363:1439-1441.

Mancardi, G. L., Saccardi, R., Filippi, M., et al. (2001) Autologous hematopoietic stem cell transplantation suppresses Gd enhanced MRI activity in MS. *Neurology* 57:62-68.

Mancardi, G., Saccardi, R., Murialdo, A., et al. (2004) Intense immunosuppression followed by autologous stem cell transplantation in severe multiple sclerosis cases: MRI and clinical data. In: Burt, R. K., Marmont, A. (eds) *Stem Cell Therapy for Autoimmune Diseases.* Landes Bioscience, Georgetown, pp 302-307.

McDonald, J. W., Liu, X. Z., Qu, Y., et al. (1999) Transplanted embryonic stem cells survive, differentiate and promote recovery in injured rat spinal cord. *Nat. Med.* 5:1410-1412.

Mezey, E., Key, S., Vogelsang, G., et al. (2003) Transplanted bone marrow generates new neurons in human brains. *Proc. Natl. Acad. Sci. USA* 100:1364-1369.

Monje, M. L., Mizumatsu, S., Fike, J. R., Palmer, T. D. (2002) Irradiation induces neural precursor-cell dysfunction. *Nat. Med.* 8:955-962.

Muraro, P. A., Douek, D. C., Packer, A., et al. (2005) Thymic output generates a new and diverse TCR repertoire after autologous stem cell transplantation in multiple sclerosis patients. *J. Exp. Med.* 201:805-816.

Nash, R. A., Bowen, J. D., McSweeney, P. A., et al. (2003) High-dose immunosuppressive therapy and autologous peripheral blood stem cell transplantation for severe multiple sclerosis. *Blood* 102:2364-2372.

Nunes, M. C., Roy, N. S., Keyoung, H. M., et al. (2003) Identification and isolation of multipotential neural progenitor cells from the subcortical white matter of the adult human brain. *Nat. Med.* 9:439-447.

Openshaw, H., Lund, B. T., Kashyap, A., et al. (2000) Peripheral blood stem cell transplantation in multiple sclerosis with busulfan and cyclophosphamide conditioning: report of toxicity and immunological monitoring. *Biol. Blood Marrow Transplant.* 6:563-575.

Pluchino, S., Quattrini, A., Brambilla, E., et al. (2003) Injection of adult neurospheres induces recovery in a chronic model of multiple sclerosis. *Nature* 422:688-694.

PRISMS Study Group. (1998) Randomised, double-blind, placebo-controlled study of interferon-beta 1a in relapsing-remitting multiple sclerosis. *Lancet* 352:1498-1504.

Reubinoff, B. E., Itsykson, P., Turetsky, T., et al. (2001) Neural progenitors from human embryonic stem cells. *Nat. Biotechnol.* 19.1134-1140.

Saccardi, R., Mancardi, G. L., Solari, A., et al. (2005) Autologous HSCT for severe progressive multiple sclerosis in a multicenter trial: impact on disease activity and quality of life. *Blood* 105:2601-2607.

Saiz, A., Carreras, E., Berenguer, J., et al. (2001) MRI and CSF oligoclonal bands after autologous hematopoietic stem cell transplantation in MS. *Neurology* 56:1084-1089.

Saiz, A., Blanco, Y., Carreras, E., et al. (2004) Clinical and MRI outcome after autologous hematopoietic stem cell transplantation in MS. *Neurology* 62:282-284.

Samijn, J. P., te Boekhorst, P. A., Mondria, T., et al. (2006) Intense T cell depletion followed by autologous bone marrow transplantation for severe multiple sclerosis. *J. Neurol. Neurosurg. Psychiatry* 77:46-50.

Thompson, J. A., Itskovitz-Eldor, J., Shapiro, S. S., et al. (1998) Embryonic stem cell lines derived from human blastocysts. *Science* 283:1145-1147.

Van Gelder, M., van Bekkum, D. W. (1996a) Effective treatment of relapsing experimental autoimmune encephalomyelitis with pseudoautologous bone marrow transplantation. *Bone Marrow Transplant.* 18:1029-1034.

Van Gelder, M., Mulder, A. H., van Bekkum, D. W. (1996b) Treatment of relapsing experimental autoimmune encephalomyelitis with largely MHC-mismatched allogeneic bone marrow transplantation. *Transplantation* 62:810-818.

Wagers, A. J., Sherwood, R. I., Christensen, J. L., Weissman, I. L. (2002) Little evidence for developmental plasticity of adult hematopoietic stem cells. *Science* 297:2256-2259.

Wobus, A. M., Holzhausen, H., Jakel, P., Schoneich, J. (1984) Characterization of a pluripotent stem cell line derived from a mouse embryo. *Exp. Cell. Res.* 152:212-219.

Xian, H. Q., McNichols, E., St. Clair, A., Gottlieb, D. I. (2003) A subset of ES-cell-derived neural cells marked by gene targeting. *Stem Cells* 21:41-49.

Zappia, E., Casazza, S., Pedemonte, E., et al. (2005) Mesenchymal stem cells ameliorate experimental autoimmune encephalomyelitis inducing T-cell anergy. *Blood* 106: 1755-1761.

Zhang, S. C., Wernig, M., Duncan, I. D., et al. (2001) In vitro differentiation of transplantable neural precursors from human embryonic stem cells. *Nat. Biotechnol.* 19:1129-1133.

Zhao, R. C., Liao, L., Han, Q. (2003) Mechanisms of and perspectives on the mesenchymal stem cells in immunotherapy. *J. Lab. Clin. Med.* 143:284-291.

# 18
# T Cell-Targeted Therapy
# for Rheumatoid Arthritis

Hendrik Schulze-Koops and Peter E. Lipsky

Based on the concept that activated T cells are the key mediators of chronic autoimmune inflammation, various T cell-directed approaches have been introduced for the treatment of inflammatory rheumatic disease. Whereas attempts to down-modulate rheumatic inflammation by reducing T cell numbers have largely failed, novel treatment approaches with biologicals that specifically inhibit T cell activation are associated with considerable clinical efficiency. These compounds have clearly established the feasibility of targeted T cell-directed interventions, and the clinical benefit induced by inhibiting T cell activation supports the dominant role of T cells in rheumatic inflammation even at advanced stages of the diseases. Some interesting novel treatment approaches have been tested in animal models of autoimmune disease, but their value for clinical use in humans needs to be established. Here, we discuss current data on T cell-directed interventions in rheumatoid arthritis.

## 1   Introduction

Rheumatoid arthritis (RA) is one of the most common human autoimmune diseases. The hallmarks of RA are chronic systemic inflammation, synovial infiltrates, and a progressive cell-mediated destruction of the joints and their adjacent structures (Firestein, 2003). The pathogenetic basis of RA is a sustained specific

TABLE 18.1. Evidence for a Contribution of CD4 T Cells to Rheumatoid Inflammation.

Presence of activated CD4 memory T cells in peripheral blood, synovial fluid, and synovial infiltrates
Transfer of disease in several animal models of inflammatory arthritis
Clinical efficacy of appropriate T cell-directed therapies
Strong association of rheumatoid arthritis with HLA-DR4 and DR1 subtypes (shared epitope)

immune response against yet unknown self antigens. It is believed that in RA the persistent autoimmune response mediates local synovial inflammation and cellular infiltration, which ultimately result in tissue damage. Whereas the specific autoantigen(s) eliciting the detrimental immune reaction in RA have not been defined, it has become clear that the mechanisms culminating in the destruction of tissue and the loss of organ function during the course of the disease are essentially the same as in protective immunity against invasive microorganisms.

Of critical importance in initiating, controlling, and driving a specific immune response are CD4+ T cells. CD4+ T cells are activated by an antigen (i.e., peptide) recognized specifically by their T cell receptor (TCR) if presented in the context of a syngeneic major histocompatibility complex class II (MHC II) molecule on the surface of a professional antigen-presenting cell (APC) such as a dendritic cell. Once activated, CD4+ T cells become the master regulators of specific immune responses and determine to a large extent the outcome of immune reactions by activating different effector functions of the immune system. It is of no surprise, therefore, that activated CD4+ T cells can be found in the inflammatory infiltrates in the rheumatoid synovium, where they propel local inflammation via their effector functions. Further evidence for an important role of CD4+ T cells in autoimmune diseases derive from animal models of autoimmune diseases, in which tissue damaging autoimmunity can be initiated by transfer of CD4+ T cells from sick animals into healthy syngeneic recipients (Banerjee et al., 1992; Breedveld et al., 1989). The most compelling data implying a central role for CD4 T cells in propagating rheumatoid inflammation remain the association of aggressive forms of the disease with particular MHC II alleles, such as subtypes of HLA-DR4, that contain similar amino acid motifs in the CDR3 region of the DRβ-chain (Calin et al., 1989; Winchester, 1994). Although the exact meaning of this association has not been resolved, all interpretations imply that CD4 T cells orchestrate local inflammation and cellular infiltration, following which a large number of subsequent inflammatory events are unleashed. Thus, T cells and in particular CD4+ T cells are central for both the induction and effector phases of specific immune responses in RA and therefore represent an ideal target for immunotherapy (Table 18.1).

## 2   T Cell-Targeted Therapy by Depletion

Initial T cell-directed therapies were designed to control disease progression by means of reducing the number or activation of T cells—for example, by total lymphoid irradiation or thoracic duct drainage (for review see Emmrich et al., 1994;

Schulze-Koops and Kalden, 2003; Schulze-Koops and Lipsky, 2004). However, these approaches have provided only modest and inconsistent clinical benefit and have been associated with a number of side effects. It became obvious that the generation of T cell lymphopenia is insufficient to combat established autoimmune responses. Numerous studies have in fact shown that manipulations that generate functional T cell lymphopenia in animals result in the development of a variety of organ-specific autoimmune disease in these models (reviewed in Gleeson et al., 1996). Examples of such manipulations include the TCR α-chain-deficient mice, which develop inflammatory bowel disease associated with an array of autoantibodies (Mizoguchi et al., 1997; Mombaerts et al., 1993); the interleukin (IL)-2 knockout (KO) mouse, which develops prominent autoimmune colitis (Sadlack et al., 1993); TCR α-chain transgenic mice (Sakaguchi et al., 1994b); neonatal application of cytotoxic intervention protocols, such as cyclosporine (Sakaguchi and Sakaguchi, 1989); total lymphoid irradiation (Sakaguchi et al., 1994a) or thymectomy (Kojima and Prehn, 1981); and lymphotoxic treatment of adult animals (Barrett et al., 1995). Further studies revealed that the development of autoimmunity in these models was critically dependent on α/β TCR-expressing CD4+ T cells, indicating that lymphopenia promotes the induction of autoimmune inflammation by self-reactive peripheral blood CD4 T cells in these animals. In fact, it could be demonstrated that the peripheral T cell population that emerged in mice in which lymphopenia was induced by cytotoxic treatment with cyclophosphamide or streptozotocin, preferentially consisted of interferon-γ (IFNγ)–secreting proinflammatory T helper (Th)1-like cells (Ablamunits et al., 1999). Although in humans lymphopenia is not sufficient for the development of autoimmune diseases (Schulze-Koops, 2004), it is conceivable that lymphopenia in patients with existing autoimmune diseases exacerbates the condition by permitting the homeostatic expansion of autoreactive T cells, thereby resulting in the reappearance of autoimmune inflammation and thus the recurrence of clinically overt autoimmune phenomena.

## 3   T Cell-Targeted Therapy With Disease-Modifying Antirheumatic Drugs

Significant advances in the understanding of T cell biology in recent years has led to the specific development of compounds designed to interfere exclusively with T cell activation without reducing T cell numbers. Cyclosporine and FK506 (tacrolimus), for example, inhibit T cell activation by interfering with calcineurin-mediated transcriptional activation of a number of cytokine genes, such as IL-2, IL-3, IL-4, IL-8, and IFNγ. Leflunomide, a potent noncytotoxic inhibitor of the key enzyme of the de novo synthesis of uridine monophosphate (Bruneau et al., 1998), dihydroorotate dehydrogenase, blocks clonal expansion and terminal differentiation of T cells as activated T cells critically depend on the de novo pyrimidine synthesis to fulfill their metabolic needs.

Cyclosporine, FK506, and leflunomide belong to the group of disease-modifying antirheumatic drugs (DMARDs), which are clinically effective in ameliorating autoimmune inflammation and are important components of the current therapeutic repertoire in autoimmune diseases. Whereas the molecular mechanisms by which cyclosporine, FK506, or leflunomide inhibit T cell activation are well defined, several studies have indicated that a number of DMARDs might be effective in rheumatic diseases at least in part because of their immunomodulatory effects on T cell subsets. In this regard, current evidence suggests that RA may reflect ongoing inflammation largely mediated by activated proinflammatory Th1 cells without the sufficient differentiation of immunoregulatory Th2 cells to down-modulate inflammation (Miltenburg et al., 1992; Schulze-Koops et al., 1995; Simon et al., 1994; Skapenko et al., 1999). Importantly, Th1 and Th2 cells antagonize each other by blocking the generation of the other and by blocking each other's effector functions. Thus, a shift in the balance of Th1/Th2 effector cells toward antiinflammatory Th2 cells can be expected to induce clinical benefit in RA.

The concept of modulating the Th1/Th2 balance as a treatment for chronic autoimmunity has been successfully applied in a number of animal models of autoimmune diseases (Bessis et al., 1996; Joosten et al., 1999). It is therefore of great interest to note that DMARDs appear to be able to modulate the Th1/Th2 balance (Table 18.2). For example, leflunomide selectively decreases the activation of proinflammatory Th1 cells while promoting Th2 cell differentiation from naive precursors (Dimitrova et al., 2002). Clinical efficacy of cyclosporine is associated with decreased serum levels of IFNγ, IL-2, and IL-12 and with significant increases in IL-10 (de Groot and Gross, 1998). Likewise, methotrexate (MTX) significantly decreases the production of IFNγ and IL-2 in in vitro stimulated peripheral blood mononuclear cells (PBMCs) while increasing the concentration of IL-4 and IL-10 (Constantin et al., 1998). Sulfasalazine potently inhibits the production of IL-12 in a dose-dependent manner in mouse macrophages stimulated with lipopolysaccharides (LPS). Importantly, pretreatment of macrophages with sulfasalazine in vitro or in vivo reduces their ability to induce the Th1 cytokine IFNγ and increases the ability to induce the Th2 cytokine IL-4 in antigen-primed CD4+ T cells (Kang et al., 1999). Bucillamine decreases the frequency of IFNγ–producing CD4+ T cells among CD4+ T cells generated after priming

TABLE 18.2. Effect of Immunosuppressive Drugs on Th1 and Th2 Cytokines.

| Immunosuppressive drug | Interferon-γ | Interleukins |
|---|---|---|
| Leflunomide | IFNγ ↓ | IL-4 ↑ |
| MTX | IFNγ ↓ | IL-4 ↑, IL-10 ↑ |
| Sulfasalazine | IFNγ ↓ | IL-4 ↑ |
| Cyclosporine | IFNγ ↓ | IL-4 ±, IL-10 ↑ |
| Gold | IFNγ ↓ | IL-4 ↑ |
| D-Penicillamine | IFNγ ↓ | IL-4 ↑ |
| Bucillamine | IFNγ ↓ | |
| Chloroquine | IFNγ ↓ | |
| Corticosteroids | | IL-4 ↑ |

cultures of PBMCs (Morinobu et al., 2000). Finally, reports have suggested that glucocorticoids inhibit cytokine expression indirectly through promotion of a Th2 cytokine secretion profile, presumably through their action on monocyte activation (Almawi et al., 1999). Together, the data suggest that clinical benefit following treatment with a number of DMARDs is associated with a significant shift in the Th1/Th2 balance in favor of the later. Moreover, the cumulative data further suggest that some of the current treatment modalities in RA exert their antiinflammatory effects by directly inhibiting Th1 cell activation and/or differentiation and by favoring Th2 differentiation, thereby shifting the Th1/Th2 balance toward the Th2 direction.

# 4    T Cell-Directed Therapies With Biologicals

Despite the progress that has been made in the treatment of rheumatic diseases, standard immunosuppressive therapy, even if T cell-directed, is still clinically ineffective in many patients and is associated with a number of side effects related to toxicity and immunosuppression. Moreover, standard therapy with DMARDs and corticosteroids has failed to interrupt and permanently halt autoimmune inflammation. The substantial progress in our understanding of molecular and cellular biology in recent years has permitted the design of therapeutic tools with defined targets and effector functions ("biologicals") that were hoped to meet the demands of an optimal therapy. Based on the increased knowledge of molecular mechanisms involved in the pathogenesis of rheumatic diseases, biologicals were developed to target selectively only those cells and/or pathways driving the disease while maintaining the integrity of the remainder of the immune system. Based on the concept that activated T cells are the key mediators of chronic autoimmune inflammation, a number of approaches have been designed to target mature circulating T cells specifically. They include the use of monoclonal antibodies (mAbs) to T cell surface receptors, such as CD2, CD3, CD4, CD7, CD18, CD25, and CD52 as well as mAbs to surface molecules on non-T cells that interact with T cells during their recruitment into inflamed tissue, such as CD54, or during T cell co-stimulation, such as CD80 or CD86 (Table 18.3).

Detailed reviews of the experience with in vivo use of mAbs to these individual surface receptors and the outcome of clinical trials with such mAbs have been published elsewhere. Although some of the mAbs employed were clearly associated with convincing and prolonged clinical benefit, the concept arose from these trials that targeting surface receptors of CD4 T cells by mAbs, in contrast to blocking co-stimulatory ligands, was generally not sufficient to ameliorate established autoimmune inflammation (Schulze-Koops and Kalden, 2003; Schulze-Koops and Lipsky, 2000). Of importance, the induction of permanent unresponsiveness of autoreactive T cells that would have resulted in sustained clinical improvement without the need for continuous immunosuppressive therapy was never achieved in any of the studies, regardless of the targeted cell population. With the exception of a limited number of trials with biologicals blocking CD2 (Gottlieb, 2006),

TABLE 18.3. Target Structures for T Cell-Directed Therapies for Rheumatoid Arthritis.

| Intended target cell | Target molecule | Main cellular distribution | Biological | Ligand |
|---|---|---|---|---|
| T cell | CD2 | T cells, NK cells | sCD58-IgG1 | CD58 |
| | CD3 | T cells | mAb | MHC II/peptide |
| | CD4 | T helper cells, monocytes | mAb | MHC class II |
| | CD5 | T cells, memory B cells | mAb | Unknown |
| | CD7 | T cells, NK cells | mAb | Unknown |
| | CD25 | Activated T cells, activated B cells, activated monocytes | mAb | IL-2 |
| | CD52 | Lymphocytes, monocytes | mAb | Unknown |
| Professional APC | CD80 | Activated APCs, activated T cells, activated B cells | mAb CTLA4Ig | CD28 CTLA4 |
| | CD86 | Activated APCs, activated T cells, activated B cells | mAb CTLA4Ig | CD28 CTLA4 |
| Endothelial cell | CD54 | Activated endothelial cells, leukocytes | mAb | CD11a-c/CD18 |

[1]NK, natural killer; APCs, antigen-presenting cells; mAb, monoclonal antibody.

CD3 (Utset et al., 2002), or CD4 (Hepburn et al., 2003), clinical studies with mAbs to T cell surface receptors in rheumatic diseases have largely been discontinued for several years.

A number of reasons, such as selection of the targeted molecules, the design of the biologicals, and the selection of patients at advanced stages of their disease during the initial clinical trials with biologicals might have contributed to the unfavorable results in humans. A further problem in targeting specifically the disease-promoting T cells in human autoimmune rheumatic diseases is, as outlined above, the fact that neither the eliciting (auto)antigens nor the specific disease initiating or perpetuating T cells are known. Therefore, the most rational approach to treating human autoimmune diseases has been interference with the activation of CD4+ T cells in a rather broad, non-antigen-specific manner. Moreover, CD4 T cells comprise functionally different subpopulations that include inflammatory effector T cells as well as T cells with a regulatory capacity, such as CD4+CD25+ regulatory T cells (Tregs). As depletion or functional inactivation of CD4+CD25+ Tregs results in the development of several autoimmune phenomena in humans and animals (Fontenot et al., 2003; Sakaguchi et al., 1995; Wildin et al., 2002), it is conceivable that the blockade of CD4 T cells by mAbs indiscriminately directed against all T cells (mAbs to CD2, CD3, CD52), all CD4 T cells (mAbs to CD4) or CD25-expressing T cells (mAbs to CD25) resulted in the functional inhibition of CD4+CD25+ Tregs, thereby promoting reactivation and/or development of autoimmune inflammation.

An alternative approach to blocking T cell activation directly with mAbs to T cell surface receptors is to inhibit T cell activation by interrupting the interaction between T cells and accessory cells required for T cell stimulation. By blocking the ligand for a T cell surface molecule on the surface of the cells interacting with

T cells, receptor/counter-receptor interaction can be prevented. This approach has been successfully employed in an attempt to block CD28-mediated co-stimulation of T cells (Genovese et al., 2005; Kremer et al., 2003, 2005; Moreland et al., 2002). Co-stimulation is an absolute requirement for the activation of naive T cells. Therefore, co-stimulation controls the initiation of specific primary immune responses. In fact, activation of a naive T cell through its TCR without providing appropriate co-stimulation can render the T cell anergic, which essentially restricts the initiation of a specific immune response to professional APCs that are able to engage co-stimulatory molecules on naive T cells. CD28-mediated co-stimulation can be blocked by coating the binding partners of CD28 on APCs, CD80 and CD86, with a soluble immunoglobulin fusion protein of the extracellular domain of CD152 (cytotoxic T lymphocyte antigen 4, or CTLA4). CTLA4 is a homologue to CD28 and is expressed by activated T cells. It can bind both CD80 and CD86 with higher affinity than CD28. Because CD152 has a high affinity for CD80 and CD86, soluble forms of CTLA4 inhibit the interaction of CD28 with its ligands.

## 5    Clinical Experience With CTLA4Ig (Abatacept)

The safety and preliminary efficacy of CTLA4Ig was assessed in a multicenter multinational study in 214 patients with active RA who had failed to respond to at least one DMARD (Moreland et al., 2002). In this placebo-controlled study, the patients received placebo or different doses of either CTLA4Ig (abatacept) or a modified CTLA4Ig fusion protein (LEA29Y) on days 1, 15, 29, and 57. Both inhibitors of co-stimulation were well tolerated at all dose levels. The most frequent adverse event that could be related to the study medication was headache. Although the primary endpoint of this initial study was safety, the clinical efficacy of the study medication was apparent, as 23%, 44%, and 53% of the CTLA4Ig-treated patients and 34%, 45%, and 61% of the LEA29Y-treated patients achieved the ACR20 response criteria at day 85 in response to four infusions of 0.5, 2, or 10 mg of the study medication, respectively. In a subsequent randomized, double-blind, placebo-controlled study, 339 patients with active RA despite treatment with MTX were treated either with placebo or with 2 mg or 10 mg CTLA4Ig on days 1, 15, and 30 and monthly thereafter for a total of 6 months (Kremer et al., 2003). All patients remained on weekly treatment with MTX. Whereas the ACR20 response was similar in the placebo and the 2-mg treatment group (35.3% and 41.9%, respectively, after 6 months), significantly more patients achieved the ACR20 response criteria as early as day 60 in the 10-mg group (60% after 6 months). Remarkably, both verum-treated groups had significantly higher ACR50 and ACR70 responses at 6 months than the placebo-treated group (ACR50: 22.9% and 36.5% vs. 11.8% for the 2-mg, 10-mg, and placebo groups, respectively; ACR70: 10.5% and 16.5% vs. 1.7% for the 2-mg, 10-mg, and placebo groups, respectively). Again, CTLA4Ig was well tolerated, and no deaths, malignancies, or opportunistic infections were reported after the end of the 6-month treatment. After an additional 6 months of treatment, the clinical response rates to CTLA4Ig were similar to the response rates after 6 months (Kremer et al., 2005). In that study, significantly more patients achieved the ACR20 response criteria after 12 months in the 10-mg group but not in the 2-mg group compared to the placebo group. Higher frequencies of patients treated with 10 mg CTLA4Ig also achieved ACR50 and ACR70 responses at 12 months, compared to the placebo group (ACR50: 41.7% vs. 20.2%, for the 10-mg and placebo groups, respectively; ACR70: 20.9% vs. 7.6%, for the 10-mg and placebo groups, respectively). The incidence of adverse effects was comparable for all groups. These studies suggested that CTLA4Ig in combination with MTX had the potential to play an important role in the future therapy of RA.

In a subsequent study, CTLA4Ig was tested in patients who had failed to respond to at least 3 months of anti-tumor necrosis factor (TNF) therapy (Genovese et al., 2005). A total of 391 RA patients were randomly assigned in a 2:1 ratio to receive abatacept 10 mg/kg body weight or placebo on days 1, 15, and 29 and every 28 days thereafter up to and including day 141. As early as day 15, significantly more patients who were treated with abatacept achieved ACR20 response criteria compared to placebo-treated patients (45 vs. 7 in the abatacept and placebo groups, respectively); and at 6 months, ACR20, ACR50, and ACR70 response rates were significantly higher in the verum group compared to the placebo group (50.4%, 20.3%, and 10.2% for ACR20, ACR50, and ACR70 vs. 19.5%, 3.8%, and 1.5% in the abatacept and placebo groups, respectively). Likewise, at 6 months, significantly more patients in the abatacept group compared to the placebo group had a clinically meaningful improvement in physical function. The incidences of adverse effects and periinfusional adverse events were 79.5% and 5.0%, respectively, in the abatacept group and 71.4% and 3.0%, respectively, in the placebo group (Genovese et al., 2005). Together, these studies show that inhibition of co-stimulation is feasible, safe, and clinically effective in patients with severe RA and even in those clinically difficult situations in which anti-TNF therapy has failed to ameliorate sustained systemic inflammation.

## 6   Experience With Alefacept

An alternative co-stimulatory pathway involved in T cell activation is the CD2/CD58 pathway. Following the promising results from an open-label study with alefacept — a soluble fully human recombinant fusion protein comprising the first extracellular domain of CD58 and the hinge, CH2 and CH3 sequences of human immunoglobulin $G_1$ ($IgG_1$) — in patients with psoriatic arthritis (Patel et al., 2001), a Phase II study of alefacept in combination with MTX for psoriatic arthritis has recently been presented (reviewed in Gottlieb, 2006). Three months after a 12-week period of weekly intramuscular application of 15 mg alefacept, 54% of the verum-treated patients (compared to 23% of the placebo-treated control) achieved an ACR20 response. The data suggest that prevention of T cell activation by targeting CD2/CD58 interactions is possible and might result in reduction of autoimmune joint inflammation. Further studies are required to substantiate these observations.

Together, the successful therapy of clinically active rheumatic diseases with biologicals interrupting T cell co-stimulatory pathways clearly emphasize the important role of T cells in the pathogenesis even at advanced stages of these diseases. Importantly, in contrast to naive T cells, memory and effector T cells are independent of co-stimulation (Skapenko et al., 2001). The data therefore also strongly suggest that inflammatory joint activity in RA and psoriasis depends on the continuous activation and recruitment of naive T cells.

# 7   Other Approaches to Inhibiting T Cell Activation

T cell recruitment to sites of inflammation was successfully prevented with a murine mAb to CD54 (ICAM-1), which is critical for transendothelial migration of T cells and their subsequent activation (Kavanaugh et al., 1994). Because of the immunogenicity of this mAb, however, retreatment with the agent was associated with immune complex-mediated side effects, including urticaria, angioedema, and serum complement protein consumption (Kavanaugh et al., 1997). Thus, further studies were not conducted. The concept of modulating autoimmune inflammation by selectively interfering with T cell migration, however, was tested again in a more recent randomized placebo-controlled trial of an antisense oligodeoxynucleotide to ICAM-1 in patients with severe RA (Maksymowych et al., 2002). In this study, clinical efficacy was not noted, presumably because of insufficient dosage, as suggested by a subsequent study of Crohn's disease, in which the dose required for therapeutic efficacy was higher than the dose employed in the RA trial (Yacyshyn et al., 2000). The clinical value of an antisense oligodeoxynucloetide approach to CD54 in RA thus remains to be shown.

Apart from the treatment principles described above, other innovative T cell-directed therapeutic strategies have been defined in experimental inflammatory arthritides, some of which have already entered preliminary clinical trials. For example, the antiinflammatory role of 3-hydroxy-3-methylglutaryl-coenzyme A reductase inhibitors (statins) has been documented in a murine model of inflammatory arthritis. Simvastatin not only markedly inhibited developing, but also clinically established, collagen-induced arthritis in doses that were unable to alter cholesterol concentrations significantly in vivo (Leung et al., 2003). Importantly, simvastatin reduced anti-CD3/anti-CD28 induced T cell proliferation and IFNγ production and, moreover, demonstrated significant suppression of collagen-specific Th1 humoral and cellular immune responses. Studies in humans, though, have not been reported to date.

# 8   Conclusion

Because of the central role that CD4+ T cells play in the pathogenesis of RA, various T cell-directed therapies were introduced for the treatment of this chronic, destructive autoimmune disease. The initial approaches that aimed to ameliorate inflammatory activity by reducing T cell numbers, however, provided only modest and inconsistent clinical benefit. Compounds that specifically interfere with T cell activation, such as some of the DMARDs currently used as standard therapy in rheumatoid inflammation are clinically effective in most patients but are still associated with a number of side effects related to toxicity and immuno-suppression. Owing to the substantially increased knowledge of cellular and molecular mechanisms of the pathogenesis of RA and the increased understanding of molecular and cellular biology, molecules (biologicals) have specifically been designed to target only those cells perpetuating the chronic inflammation, with minimal effects on other aspects of the immune or inflammatory systems. Various T cell-directed biologicals have been employed in rheumatic diseases with varying clinical success. The inhibition of T cell activation by interfering with co-stimulatory signals has been shown to be an effective means to ameliorate rheumatoid inflammation. Together, the recent clinical studies have clearly established the feasibility of targeted T cell-directed interventions, and the clinical benefit induced by inhibiting T cell activation supports the dominant role of T cells in rheumatic inflammation even at advanced stages of the diseases.

*Acknowledgment.* The work of H.S.-K. is supported in part by the Deutsche Forschungsgemeinschaft (Schu 786/2-4) and the Interdisciplinary Center for Clinical Research in Erlangen (Projects B3 and A18)

# References

Ablamunits, V., Quintana, F., Reshef, T., et al. (1999) Acceleration of autoimmune diabetes by cyclophosphamide is associated with an enhanced IFN-gamma secretion pathway. *J. Autoimmun.* 13:383-392.

Almawi, W. Y., Melemedjian, O. K., Rieder, M. J. (1999) An alternate mechanism of glucocorticoid anti-proliferative effect: promotion of a Th2 cytokine-secreting profile. *Clin. Transplant.* 13:365-374.

Banerjee, S., Webber, C., Poole, A. R. (1992) The induction of arthritis in mice by the cartilage proteoglycan aggrecan: roles of CD4+ and CD8+ T cells. *Cell. Immunol.* 144:347-357.

Barrett, S. P., Toh, B. H., Alderuccio, F., et al. (1995) Organ-specific autoimmunity induced by adult thymectomy and cyclophosphamide-induced lymphopenia. *Eur. J. Immunol.* 25:238-244.

Bessis, N., Boissier, M. C., Ferrara, P., et al. (1996) Attenuation of collagen-induced arthritis in mice by treatment with vector cells engineered to secrete interleukin-13. *Eur. J. Immunol.* 26:2399-2403.

Breedveld, F. C., Dynesius-Trentham, R., de Sousa, M., Trentham, D. E. (1989) Collagen arthritis in the rat is initiated by CD4+ T cells and can be amplified by iron. *Cell. Immunol.* 121:1-12.

Bruneau, J. M., Yea, C. M., Spinella-Jaegle, S., et al. (1998) Purification of human dihydro-orotate dehydrogenase and its inhibition by A77 1726, the active metabolite of leflunomide. *Biochem. J.* 336:299-303.

Calin, A., Elswood, J., Klouda, P. T. (1989) Destructive arthritis, rheumatoid factor, and HLA-DR4: susceptibility versus severity, a case-control study. *Arthritis Rheum.* 32:1221-1225.

Constantin, A., Loubet-Lescoulie, P., Lambert, N., et al. (1998) Antiinflammatory and immunoregulatory action of methotrexate in the treatment of rheumatoid arthritis: evidence of increased interleukin-4 and interleukin-10 gene expression demonstrated in vitro by competitive reverse transcriptase-polymerase chain reaction. *Arthritis Rheum.* 41:48-57.

De Groot, K., Gross, W. L. (1998) Wegener's granulomatosis: disease course, assessment of activity and extent and treatment. *Lupus* 7:285-291.

Dimitrova, P., Skapenko, A., Herrmann, M. L., et al. (2002) Restriction of de novo pyrimidine biosynthesis inhibits Th1 cell activation and promotes Th2 cell differentiation. *J. Immunol.* 169:3392-3399.

Emmrich, F., Schulze-Koops, H., Burmester, G. (1994) Anti-CD4 and other antibodies to cell surface antigens for therapy. In: *Immunopharmacology of Joints and Connective Tissue.* London, Academic Press, pp 87-117.

Firestein, G. S. (2003) Evolving concepts of rheumatoid arthritis. *Nature* 423:356-361.

Fontenot, J. D., Gavin, M. A., Rudensky, A. Y. (2003) Foxp3 programs the development and function of CD4+CD25+ regulatory T cells. *Nat. Immunol.* 4:330-336.

Genovese, M. C., Becker, J. C., Schiff, M., et al. (2005) Abatacept for rheumatoid arthritis refractory to tumor necrosis factor alpha inhibition. *N. Engl. J. Med.* 353:1114-1123.

Gleeson, P. A., Toh, B. H., van Driel, I. R. (1996) Organ-specific autoimmunity induced by lymphopenia. *Immunol. Rev.* 149:97-125.

Gottlieb, A. B. (2006) Alefacept for psoriasis and psoriatic arthritis. *Ann. Rheum. Dis.* 64:58-60.

Hepburn, T. W., Totoritis, M. C., Davis, C. B. (2003) Antibody-mediated stripping of CD4 from lymphocyte cell surface in patients with rheumatoid arthritis. *Rheumatology (Oxford)* 42:54-61.

Joosten, L. A., Lubberts, E., Helsen, M. M., et al. (1999) Protection against cartilage and bone destruction by systemic interleukin-4 treatment in established murine type II collagen-induced arthritis. *Arthritis Res.* 1:81-91.

Kang, B. Y., Chung, S. W., Im, S. Y., et al. (1999) Sulfasalazine prevents T-helper 1 immune response by suppressing interleukin-12 production in macrophages. *Immunology* 98: 98-103.

Kavanaugh, A. F., Davis, L. S., Nichols, L. A., et al. (1994) Treatment of refractory rheumatoid arthritis with a monoclonal antibody to intercellular adhesion molecule 1. *Arthritis Rheum.* 37:992-999.

Kavanaugh, A. F., Schulze-Koops, H., Davis, L. S., Lipsky, P. E. (1997). Repeat treatment of rheumatoid arthritis patients with a murine anti-intercellular adhesion molecule 1 monoclonal antibody. *Arthritis Rheum.* 40:849-853.

Kojima, A., Prehn, R. T. (1981) Genetic susceptibility to post-thymectomy autoimmune diseases in mice. *Immunogenetics* 14:15-27.

Kremer, J. M., Westhovens, R., Leon, M., et al. (2003) Treatment of rheumatoid arthritis by selective inhibition of T-cell activation with fusion protein CTLA4Ig. *N. Engl. J. Med.* 349:1907-1915.

Kremer, J. M., Dougados, M., Emery, P., et al. (2005) Treatment of rheumatoid arthritis with the selective costimulation modulator abatacept: twelve-month results of a phase iib, double-blind, randomized, placebo-controlled trial. *Arthritis Rheum.* 52:2263-2271.

Leung, B. P., Sattar, N., Crilly, A., et al. (2003) A novel anti-inflammatory role for simvastatin in inflammatory arthritis. *J. Immunol.* 170:1524-1530.

Maksymowych, W. P., Blackburn, W. D., Jr., Tami, J. A., Shanahan, W. R., Jr. (2002) A randomized, placebo controlled trial of an antisense oligodeoxynucleotide to intercellular adhesion molecule-1 in the treatment of severe rheumatoid arthritis. *J. Rheumatol.* 29:447-453.

Miltenburg, A. M., van Laar, J. M., de Kuiper, R., et al. (1992) T cells cloned from human rheumatoid synovial membrane functionally represent the Th1 subset. *Scand. J. Immunol.* 35:603-610.

Mizoguchi, A., Mizoguchi, E., Smith, R. N., et al. (1997) Suppressive role of B cells in chronic colitis of T cell receptor alpha mutant mice. *J. Exp. Med.* 186:1749-1756.

Mombaerts, P., Mizoguchi, E., Grusby, M. J., et al. (1993) Spontaneous development of inflammatory bowel disease in T cell receptor mutant mice. *Cell* 75:274-282.

Moreland, L. W., Alten, R., Van den Bosch, F., et al. (2002) Costimulatory blockade in patients with rheumatoid arthritis: a pilot, dose-finding, double-blind, placebo-controlled clinical trial evaluating CTLA-4Ig and LEA29Y eighty-five days after the first infusion. *Arthritis Rheum.* 46:1470-1479.

Morinobu, A., Wang, Z., Kumagai, S. (2000) Bucillamine suppresses human Th1 cell development by a hydrogen peroxide-independent mechanism. *J. Rheumatol.* 27:851-858.

Patel, S., Veale, D., FitzGerald, O., McHugh, N. J. (2001) Psoriatic arthritis—emerging concepts. *Rheumatology* 40:243-246.

Sadlack, B., Merz, H., Schorle, H., et al. (1993) Ulcerative colitis-like disease in mice with a disrupted interleukin-2 gene. *Cell* 75:253-261.

Sakaguchi, N., Miyai, K., Sakaguchi, S. (1994a) Ionizing radiation and autoimmunity. Induction of autoimmune disease in mice by high dose fractionated total lymphoid irradiation and its prevention by inoculating normal T cells. *J. Immunol.* 152:2586-2595.

Sakaguchi, S., Ermak, T. H., Toda, M., et al. (1994b) Induction of autoimmune disease in mice by germline alteration of the T cell receptor gene expression. *J. Immunol.* 152:1471-1484.

Sakaguchi, S., Sakaguchi, N. (1989) Organ-specific autoimmune disease induced in mice by elimination of T cell subsets. V. Neonatal administration of cyclosporin A causes autoimmune disease. *J. Immunol.* 142:471-480.

Sakaguchi, S., Sakaguchi, N., Asano, M., et al. (1995) Immunologic self-tolerance maintained by activated T cells expressing IL-2 receptor alpha-chains (CD25): breakdown of a single mechanism of self-tolerance causes various autoimmune diseases. *J. Immunol.* 155:1151-1164.

Schulze-Koops, H. (2004). Lymphopenia and autoimmune diseases. *Arthritis Res. Ther.* 6:178-180.

Schulze-Koops, H., Kalden, J. R. (2003) Targeting T Cells in rheumatic diseases. In: Smolen, J. S., Lipsky, P. E. (eds) *Biological Therapy in Rheumatology.* London, Martin Dunitz, pp 3-24.

Schulze-Koops, H., Lipsky, P. E. (2000) Anti-CD4 monoclonal antibody therapy in human autoimmune diseases. *Curr. Dir. Autoimmun.* 2:24-49.

Schulze-Koops, H., Lipsky, P. E. (2004) T cells in the pathogenesis of rheumatoid arthritis. In: Clair, W. W., Pisetsky, D. S., Haynes A. (eds) *Rheumatoid Arthritis.* Philadelphia, Lippincott Williams & Wilkins, pp 184-196.

Schulze-Koops, H., Lipsky, P. E., Kavanaugh, A. F., Davis, L. S. (1995). Elevated Th1- or Th0-like cytokine mRNA in peripheral circulation of patients with rheumatoid arthritis: modulation by treatment with anti-ICAM-1 correlates with clinical benefit. *J. Immunol.* 155:5029-5037.

Simon, A. K., Seipelt, E., Sieper, J. (1994) Divergent T-cell cytokine patterns in inflammatory arthritis. *Proc. Natl. Acad. Sci. USA* 91:8562-8566.

Skapenko, A., Wendler, J., Lipsky, P. E., et al. (1999) Altered memory T cell differentiation in patients with early rheumatoid arthritis. *J. Immunol.* 163:491-499.

Skapenko, A., Lipsky, P. E., Kraetsch, H. G., et al. (2001) Antigen-independent Th2 cell differentiation by stimulation of CD28: regulation via IL-4 gene expression and mitogen-activated protein kinase activation. *J. Immunol.* 166:4283-4292.

Utset, T. O., Auger, J. A., Peace, D., et al. (2002) Modified anti-CD3 therapy in psoriatic arthritis: a phase I/II clinical trial. *J. Rheumatol.* 29:1907-1913.

Wildin, R. S., Smyk-Pearson, S., Filipovich, A. H. (2002) Clinical and molecular features of the immunodysregulation, polyendocrinopathy, enteropathy, X linked (IPEX) syndrome. *J. Med. Genet.* 39:537-545.

Winchester, R. (1994) The molecular basis of susceptibility to rheumatoid arthritis. *Adv. Immunol.* 56:389-466.

Yacyshyn, B. W., Chey, W., Salzberg, G. B., et al. (2000) Double-blinded, randomized, placebo-controlled trial of the remission inducing and steroid sparing properties of two schedules of ISIS 2302 (ICAM-1 antisense) in active, steroid-dependent Crohn's disease. *Gastroenterology* 118:S2:2977.

# 19
# Estrogens in the Treatment of Multiple Sclerosis

Rhonda R. Voskuhl

## 1  Introduction

It has been appreciated for decades that symptoms of patients with autoimmune diseases are affected by pregnancy and the postpartum period. The most well characterized observations include those on multiple sclerosis (MS), rheumatoid arthritis (RA), and psoriasis. These patients experience clinical improvement during pregnancy with a temporary "rebound" exacerbation postpartum (Abramsky 1994; Birk et al. 1990; Confavreux et al. 1998; Damek and Shuster 1997; Da Silva and Spector 1992; Nelson et al. 1992; Runmarker and Andersen 1995; Whitacre et al. 1999). This phenomenon of alleviation of the disease during pregnancy is a unique opportunity to gain insight into MS disease pathogenesis and to capitalize on a naturally occurring situation in which the disease is downregulated. Here we review the preclinical and clinical data suggesting that high levels of estrogens, which occur during pregnancy, may contribute to some of the protective effect of pregnancy. Safety issues with respect to estrogen use in patients with disease as well as future goals aiming to develop selective estrogen receptor modifiers are discussed. Finally, with regard to the timing of estrogen treatment relative to disease stage, what has been learned from the use of estrogens as treatment of neurodegenerative diseases is highlighted. Enhanced understanding of the above issues of estrogen treatment may be useful not only in MS but also in other autoimmune diseases characterized by significant improvement during pregnancy.

# 2   Estrogen Treatment: Attempting to Recapitulate Pregnancy

## 2.1   Effect of Pregnancy on Clinical MS

Most MS patients have either relapsing remitting (RR) or secondary progressive (SP) MS. The RR phase is characterized by a higher incidence of gadolinium-enhancing lesions on cerebral magnetic resonance imaging (MRI) and relapses clinically. Many RRMS patients transition to SPMS, which is a less inflammatory disease with a much lower incidence of enhancing lesions and gradual neurological decline (Trapp et al., 1999). All of the currently available therapies for MS were designed to act primarily through antiinflammatory mechanisms (Aharoni et al., 1998; Duda et al., 2000; Kozovska et al., 1999; Rudick et al., 1998). Therefore, it is not surprising that all of these therapies have been shown to be more effective in RRMS than in SPMS. Indeed, they each are of proven benefit in RRMS whereby a significant reduction in gadolinium-enhancing lesions and a reduction in relapse rates have been shown compared to placebo control. On the other hand, they each remain of questionable benefit in SPMS (Li et al., 2001). This difference in therapeutic efficacy between the two phases of the disease is likely due to differences in immunopathogenesis during the two phases. Two hypotheses that are not mutually exclusive are that inflammation is more important during the RR phase and axonal pathology is more important during the SP phase (Barkhof et al., 2001; Trapp et al., 1999)—and that the nature of the immune dysregulation differs between the two phases (Balashov et al., 2000; Jensen et al., 2001; Soldan et al., 2004; van Boxel-Dezairc ct al., 1999).

### 2.1.1   Effect of Pregnancy on Relapses

What is the precise effect of pregnancy on MS? During decades of observations that MS improved during late pregnancy, the early studies did not separate the MS patients into RR and SP groups (Abramsky, 1994; Birk et al. 1988, 1990). However, what was generally described was that there was a period of relative "safety" with regard relapses during pregnancy followed by a period of increased relapses postpartum. These clinical observations were supported by a small study of two patients who underwent serial cerebral MRI scans during pregnancy and postpartum. In both women there was a decrease in MR disease activity (T2 lesion number) during the second half of pregnancy and a return of MR disease activity to prepregnancy levels during the first months postpartum (van Walderveen et al., 1994). Other studies found that in addition to having a decrease in disease activity in patients with established MS the risk of developing the first episode of MS was decreased during pregnancy compared to nonpregnant states (Runmarker and Andersen 1995). The most definitive study of the effect of pregnancy on MS came in 1998 by the Pregnancy in Multiple Sclerosis (PRIMS) Group (Confavreux et al., 1998). Relapse rates were determined in 254 women with MS during 269 pregnancies and for up to 1 year after delivery. Relapse rates were significantly reduced from 0.7 per woman-year during the year before pregnancy to 0.2 during the third trimester. Rates then increased to 1.2 during the first 3 months postpartum before returning to prepregnancy rates. No significant changes were observed between relapse rates during the first and second trimesters compared to the year prior to pregnancy. Together these data clearly demonstrated that the latter part of pregnancy is associated with a significant reduction in relapses, and there is a rebound increase in relapses postpartum.

In a 2-year follow-up report by the PRIMS group, clinical factors that predicted postpartum flares were examined. Neither breast-feeding nor epidural anesthesia affected the likelihood of relapsing postpartum. The best predictor of which subjects would relapse postpartum was their prepregnancy relapse rates. Those with the most active disease before pregnancy were the most likely to relapse postpartum (Vukusic et al., 2004).

### 2.1.2   Effect of Pregnancy on Disability

Because late pregnancy is associated with a reduction in relapses and the postpartum period with a transient increase in relapses, what is the net effect of pregnancy on the accumulation of disability? In a short-term 2-year follow-up, no net effect of a single pregnancy on disability accumulation was observed (Vukusic et al., 2004). However, long-term follow-up studies suggested that disability accumulation may be reduced significantly in those with pregnancies after the onset of MS (Verdru et al., 1994). A study by Damek and Shuster indicated that a full-term pregnancy increased the time interval to reach a common disability endpoint (walking with the aid of a cane or crutch). In essence, pregnancy increased the time interval to having a secondary progressive course (Damek and Shuster, 1997). Runmarker compared the risk of transition from an RR to an SP course in women who were pregnant after MS onset with that in women who were not pregnant after MS onset. Importantly, the two groups were matched for neurological deficit, disease duration, and age. There was a significantly decreased risk of a progressive course in women who were pregnant after MS onset compared to those who were not pregnant (Runmarker and Andersen, 1995). The fact that the patients were matched for neurological deficit, disease duration, and age is extremely important in the latter study, as one might predict that there might be a selection bias such that women with less disability would be more likely to get pregnant and a difference in baseline disability could explain the longer time interval to reach an

SP course. Careful matching of the groups made this explanation unlikely, and therefore the study indeed provided support for a net beneficial effect of pregnancy on the accumulation of disability in MS.

Although there is clearly a short-term effect of pregnancy on decreasing relapse rates and possibly a long-term effect on increasing the time interval to reach a given level of disability, there appear to be no conclusive data supporting a long-term effect of pregnancy in healthy individuals and their subsequent risk to develop MS. One study reported that women of parity zero to two developed MS twice as often as women with a parity of three or more, thereby implying a protective effect of multiple pregnancies, but the difference did not reach statistical significance (Villard-Mackintosh and Vessey, 1993). Another found no association between parity and the subsequent risk of developing MS (Hernan et al., 2000). Together these data indicate that pregnancy in healthy women has no long-lasting effects with regard to reducing their risk of developing MS in the future, and hence pregnancy does not have a permanent effect on the immunopathogenesis of MS. However, if women with MS get pregnant, there is indeed a temporary reduction in relapses during the pregnancy. The effect of pregnancy appears to be similar to what is observed when patients take the approved antiinflammatory therapies for MS: Relapses are reduced temporarily when patients are on the treatments, but when they are discontinued the relapses return.

Given that late pregnancy is a state of temporary immunomodulation lasting 4 to 5 months, one then returns to the question of whether multiple pregnancies would be expected to have permanent effects on disability. Because it is known that up to 5 years of continuous treatment with immunomodulatory treatments has only a modest impact on disability in MS, a temporary antiinflammatory effect of the third trimester of pregnancy would not be expected to affect long-term disability. However, an as yet unidentified pregnancy-associated neuroprotective effect combined with the temporary antiinflammatory effect could reconcile the finding of a beneficial effect of pregnancy on disability.

## 2.2   Effect of Pregnancy on the Immune System

Because mechanisms of action of the approved injectable therapies for MS involve antiinflammatory effects and these treatments result primarily in a reduction in relapse rates, it is logical to hypothesize that mechanisms underlying the protective effect of pregnancy on MS relapses involve antiinflammatory effects. Indeed, pregnancy has been shown to have significant effects on the immune system.

Pregnancy is a challenge for the immune system. From the mother's standpoint, the fetus is an allograft, as it harbors antigens inherited by the father. It is thus evolutionarily advantageous for the mother to suppress transiently the cytotoxic, cell-mediated, Th1 type immune responses involved in fetal rejection during pregnancy. However, not all immune responses should be suppressed because humoral, Th2 type immunity is needed for passive transfer of antibodies to the fetus. Thus, a shift in immune responses with downregulation of Th1 and upregulation of Th2 is thought to be necessary for fetal survival (Formby, 1995; Hill et al. 1995; Raghupathy, 1997; Wegmann et al., 1993). Indeed, it has been shown in both mice and humans that failure to shift immune responses in this manner results in an increase in spontaneous abortion (Hill et al., 1995; Krishnan et al., 1996; Marzi et al., 1996). This shift in immune responses from Th1 to Th2 occurs both locally at the maternal–fetal interface (Lin et al., 1993; Sacks et al., 2001; Wegmann et al., 1993) and systemically (Dudley et al., 1993; Elenkov et al., 2001; Fabris et al., 1977; Hill et al., 1995; Krishnan et al., 1996; Marzi et al., 1996). The systemic shift away from Th1 and toward Th2 was initially shown in murine systems by a decrease in mixed lymphocyte reactions of splenocytes and an increase in antibody production during pregnancy (Fabris et al., 1977). Antigen-stimulated splenocytes were then shown to produce less Th1 cytokines and more Th2 cytokines when derived from pregnant mice (Dudley et al., 1993; Krishnan et al., 1996). In humans, peripheral blood mononuclear cells (PBMCs) in women with successful pregnancies produced interleukin-10 (IL-10) but no interferon-$\gamma$ (IFN$\gamma$) upon stimulation with trophoblast antigens (Hill et al., 1995). In another study, antigen- and mitogen-stimulated peripheral blood mononuclear cells (PBMC) derived from patients with normal pregnancies demonstrated a decrease in the production of IL-2 and interferon-$\gamma$ (IFN$\gamma$) and an increase in production of IL-4 and IL-10, with the lowest quantities of IL-2 and IFN$\gamma$ and the highest quantities of IL-4 and IL-10 present during the third trimester of pregnancy (Marzi et al., 1996). During the third trimester, ex vivo monocytic IL-12 production was also found to be about threefold lower and tumor necrosis factor-$\alpha$ (TNF$\alpha$) production was approximately 40% lower than the postpartum values (Elenkov et al., 2001). Two other studies have been completed wherein MS subjects were followed longitudinally for immune responses during pregnancy and postpartum. Gilmore et al. (2004) demonstrated that ex vivo stimulated PBMCs had increased IFN$\gamma$ production postpartum compared to the third trimester, and myelin protein-specific T cell lines derived from subjects during the third trimester produced more IL-10. Al-Shammri et al. (2004) found that six of the eight MS patients' ex vivo stimulated PBMCs showed a distinct shift from a Th2 cytokine bias (IL-4 and IL-10) during pregnancy toward a Th1 cytokine bias (IFN and TNF) after delivery. In light of these data demonstrating a relative shift to Th2 systemically during pregnancy with a rebound to Th1 postpartum, it becomes highly plausible that these alterations in the immune response could underlie the improvement in putative Th1-mediated autoimmune diseases during pregnancy as well as the exacerbation postpartum.

# 3    Estrogen Treatment in Experimental Autoimmune Encephalomyelitis

## 3.1    Estrogen Treatment in EAE: Effects on Clinical Disease

### 3.1.1    Estrogen Treatment Ameliorates EAE

Experimental autoimmune encephalomyelitis (EAE) is a widely used animal model to study immune mechanisms in MS. EAE models vary depending on the species, the strain, and the method of disease induction, with some models being

RR, others chronic progressive, and still other monophasic with full recovery (Voskuhl, 1996). The most appropriate EAE model is generally selected to answer the question being examined. It was shown more than a decade ago that EAE in guinea pigs, rats, and rabbits improved during pregnancy (Abramsky, 1994). Furthermore, it was shown that RR EAE in SJL mice was alleviated during late pregnancy (Langer-Gould et al., 2002; Voskuhl and Palaszynski, 2001). The EAE model was then used to determine if an increase in levels of a certain hormone during pregnancy might be responsible for disease improvement. Because estrogens and progesterone increase progressively during pregnancy to the highest levels during the third trimester, these hormones were candidates for possibly mediating a protective effect. Two estrogens, estradiol and estriol, each increase progressively during pregnancy. Estradiol is otherwise present at much lower fluctuating levels during the menstrual cycle in nonpregnant women and female mice. Estriol, in contrast, is produced by the fetoplacental unit and is thus not otherwise present in nonpregnant states, being an estrogen unique to pregnancy. Over the last decade it has been shown in numerous studies that estrogen treatment (both estriol and estradiol) can ameliorate both active and adoptive EAE in several strains of mice (SJL, C57BL/6, B10.PL, B10.RIII) (Bebo et al., 2001; Ito et al., 2001; Jansson et al., 1994; Kim et al., 1999; Liu et al., 2002, 2003; Matejuk et al., 2001; Polanczyk et al., 2003; Subramanian et al., 2003). Estriol treatment has also been shown to be effective in reducing clinical signs in EAE when administered after disease onset (Kim et al., 1999). Finally, both estradiol and estriol have been shown to be efficacious in both female and male mice with EAE (Palaszynski et al., 2004).

### 3.1.2   Estrogen Type and Dose in Ameliorating EAE

Clinical amelioration of EAE occurred when estriol was used at doses to induce serum levels that were physiological with pregnancy. On the other hand, estradiol had to be used at doses severalfold higher than pregnancy levels to induce the same degree of disease protection (Jansson et al., 1994). Thus, although it is clear that high doses of estradiol are protective in EAE, it has not yet been clearly established whether low doses of estradiol are protective. Due to major differences in metabolic rates between humans and mice, it is difficult to determine what a low-dose estrogen in an oral contraceptive pill in humans would equate to in a mouse. Thus, one can only use available physiological benchmarks, such as the dose needed to induce a level in blood equal to that in pregnant mice or the dose needed to induce an estrus level in an ovariectomized mouse. Rigorous comparisons of blood levels in pregnant or estrus mice, assessed in parallel with levels in estradiol-treated mice are needed. Because ovariectomy removes physiological levels of estradiol, as well as progesterone, data on the effect of ovariectomy on EAE is somewhat informative in this regard. Some reports have found that ovariectomy in female mice makes EAE worse (Matejuk et al., 2001), whereas others have found that ovariectomy does not have a significant effect on disease (Voskuhl and Palaszynski, 2001). Thus, it is controversial whether low levels of endogenous estradiol, which fluctuate during the menstrual cycle, have a significant influence on EAE.

### 3.1.3    The Other Pregnancy Hormone, Progesterone, in EAE

What is the effect of the other major hormone of pregnancy, progesterone, in EAE? Whereas the severity of EAE was significantly reduced in estriol- or estradiol-treated mice compared to the placebo-treated mice, the results of treatment with progesterone were indistinguishable from those achieved with placebo (Kim et al., 1999). Then, a variety of progesterone doses ranging from low (physiological with menstrual cycle levels) to moderate (physiological with late pregnancy levels) to very high (supraphysiological) were used in combination with estrogen (either estradiol or estriol) in EAE: None of the progesterone doses significantly altered the protective effect of estrogen treatment during the EAE disease course (Voskuhl and Palaszynski, 2001). The lack of an effect of progesterone treatment on EAE was somewhat disappointing, as progesterone had been shown to enhance remyelination in vitro (Baulieu and Schumacher, 2000; Ghoumari et al., 2003; Schumacher et al., 2004). Interestingly, in the Lewis rat, progesterone treatment alone was shown to be deleterious, causing increased motor deficits, increased inflammation, and increased neuronal apoptosis during acute EAE, whereas estrogen treatment in combination protected against these deleterious effects of progesterone (Hoffman et al., 2001). These data in EAE are consistent with previous work by the Holmdahl group done during the mid-1990s demonstrating that treatment with estrogens could ameliorate collagen-induced arthritis (Jansson et al. 1990, 1994; Jansson and Holmdahl, 1992), whereas progesterone treatment alone had no effect. Progesterone treatment had only mild synergistic effects when used in combination with estrogen treatment (Jansson and Holmdahl, 1989). Together these data indicate that it is primarily estrogen, not progesterone, that contributes to disease protection during EAE.

### 3.1.4    Hormones and the Postpartum Period

Evidence in animal models suggests that not only high levels of estrogens are protective in Th1-mediated autoimmune diseases during pregnancy but that the precipitous drop in estrogens postpartum may lead to disease exacerbation. In type II collagen-induced arthritis (CIA) in DBA/1 mice, a characteristic feature is remission during gestation and exacerbation of the disease during the postpartum period. Two possibilities were pursued with regard to hormonal changes underlying the postpartum flare: (1) the precipitous fall in estrogens postpartum and (2) the surge of prolactin after delivery. It was shown that treatment with high-dose estrogens during a short period immediately after parturition protected mice from postpartum flares, and treatment with bromocriptine, a drug known to inhibit endogenous prolactin release, has a less marked effect. Furthermore, studies of lactating (i.e., animals with physiological stimulation of endogenous prolactin release) and nonlactating arthritic mice revealed no clear-cut differences in flares, indicating that prolactin was of minor importance in the induction of postpartum flares (Mattsson et al., 1991).

These data on arthritis in mice are consistent with data on MS in women: The PRIMS group found that whether women were or were not breast-feeding had no effect on the increase in relapse rates postpartum (Vukusic et al., 2004). Because estriol is the predominant estrogen of pregnancy, with levels increasing progressively during pregnancy, and because estriol administered at pregnancy doses has been shown to be protective in both EAE and collagen-induced arthritis, these data together suggest that a precipitous drop in the protective hormone estriol after delivery may be responsible, at least in part, for postpartum exacerbations. A trial that entails treating RRMS patients with estriol during the postpartum period in an effort to prevent postpartum flares should be considered.

## 3.2  Estrogen Treatment in EAE: Effects on the Immune System

Protective mechanisms of estrogen treatment (both estriol and estadiol) in EAE clearly involve antiinflammatory processes, with estrogen-treated mice having fewer inflammatory lesions in the central nervous system (CNS) (Kim et al., 1999). In adoptive EAE in SJL mice, an increase in IL10, with no change in IFNγ, was observed in ex vivo stimulated myelin basic protein (MBP) specific responses, which was accompanied by an increase in MBP-specific antibody of the immunoglobulin G1 (IgG1) isotype, with no change in those of the IgG2a isotype (Kim et al., 1999). In active EAE in C57BL/6 mice, a decrease in TNFα, IFNγ, and IL6 has been observed, with an increase in IL5 (Bebo et al., 2001; Ito et al., 2001; Liu et al., 2003; Palaszynski et al., 2004). Estrogen treatment has also been shown to downregulate chemokines in the CNS of mice with EAE and may affect expression of matrix matalloprotease-9 (MMP-9), each leading to impaired recruitment of cells to the CNS (Matejuk et al., 2001; Subramanian et al., 2003). In addition, estrogen treatment has been shown to impair the ability of dendritic cells to present antigen (Liu et al., 2002; Zhang et al. 2004). The effect of estrogen treatment on dendritic cell function in EAE may involve upregulated expression of indoleamine 2,3-dioxygenase (IDO) (Xiao et al., 2004). IDO is an enzyme involved in the catabolism of tryptophan, which is expressed in antigen-presenting cells (APCs) of lymphoid organs and in the placenta. It was shown that IDO prevents rejection of the fetus during pregnancy, probably by inhibiting alloreactive T cells (Munn et al., 1998), and it has been shown that IDO expression in APCs may also control autoreactive immune responses (Mellor and Munn, 2004). Finally, estrogen treatment has been shown to induce CD4+CD25+ regulatory T cells in EAE (Matejuk et al., 2004; Polanczyk et al., 2004). Thus, estrogen treatment has been shown to be antiinflammatory via a variety of mechanisms.

## 3.3  Estrogen Treatment in EAE: Effects on the Central Nervous System

An antiinflammatory effect of estrogen treatment in EAE does not preclude an additional more direct neuroprotective effect. Estrogens are lipophilic, readily traversing the blood-brain barrier, with the potential to be directly neuroprotective (Brinton, 2001; Garcia-Segura et al., 2001; Wise et al., 2001). Estrogen treatment has been shown to decrease microglial activation in vitro (Drew and Chavis, 2000; Vegeto et al., 2001), and estrogen-mediated protection of neurons has been demonstrated in a variety of in vitro models of neurodegeneration, including those induced by excitotoxicity and oxidative stress (Behl et al., 1995, 1997; Goodman et al., 1996; Harms et al., 2001). Also, estrogens have been shown to decrease glutamate-induced apoptosis and preserve electrophysiological function in primary cortical neurons (Sribnick et al. 2003, 2003). In addition, in vitro studies have demonstrated estrogen's ability to modulate the astrocytic response to injury (Azcoitia et al., 1999; Garcia-Segura et al., 1999) and protect oligodendrocytes from cytotoxicity (Cantarella et al., 2004; Sur et al., 2003; Takao et al., 2004). Estrogen treatment has previously been shown to be neuroprotective in vivo in numerous neurodegenerative disease models including Parkinson's disease, spinal cord injury, cerebellar ataxia, Down's syndrome, epilepsy, and some models of stroke and Alzheimer's disease (Dubal et al., 2001; Granholm et al., 2003; Green et al., 2005; Heikkinen et al., 2004; Jover et al., 2002; Leranth et al., 2000; Rau et al., 2003; Sierra et al., 2003; Sribnick et al., 2003, 2005; van Groen and Kadish, 2005; Veliskova et al., 2000; Wise et al., 2001). Whether a direct neuroprotective effect of estrogen treatment in EAE exists remains to be determined.

# 4    Estrogen Treatment in Multiple Sclerosis

## 4.1    Oral Contraceptive Use in MS

Estrogen levels lower than those that occur during pregnancy, such as levels induced by doses in oral contraceptives or hormone replacement therapy, may or may not be high enough to be protective in MS. Although some studies have attempted to simulate a situation of treatment with oral contraceptives in EAE mice and have shown an effect on disease (Bebo et al., 2001; Subramanian et al., 2003), doses used in mice are not readily translatable to humans. In fact, the data in humans thus far has suggested that treatment with oral contraceptives is not likely to suppress MS. The incidence of MS onset in both former and current oral contraceptive users was not different from that in never-users (Thorogood and Hannaford, 1998). It is not surprising that former use of oral contraceptives in healthy women would have no effect on subsequent risk to develop MS, as one would not anticipate that the effect of treatment on the immune system would be permanent. However, the fact that the incidence of MS in current oral contraceptive pill users was not decreased, compared to never-users, suggests that the estrogens in oral contraceptives are not of sufficient type or dose to ameliorate the immunopathogenesis of MS even temporarily during current use. This remains an unresolved issue as controversial results emerge with respect to the use of oral contraceptives and MS risk during current use (Alonso et al., 2005; Hernan et al., 2000).

Studies of hormone replacement therapy and effects on disease activity in RA can provide further clues with respect to which doses of estrogens could potentially be protective in MS (Da Silva and Hall, 1992). A randomized placebo-controlled trial of transdermal estradiol in 200 postmenopausal RA patients, who continued other antirheumatic medications, found that those who achieved a serum estradiol level > 100 pmol/liter had significant improvements in articular index, pain scores, morning stiffness, and erythrocyte sedimentation rates, whereas those with lower estradiol levels did not demonstrate improvement (Hall et al., 1994). Together these reports suggest that it is likely that a sustained level of a sufficient dose of an estrogen is necessary to ameliorate disease activity in MS and RA.

## 4.2    Pilot Trial of Estriol in MS

Because estriol is the major estrogen of pregnancy and an estriol dose that yielded a pregnancy level in mice was protective in disease (Kim et al., 1999), estriol was administered in a prospective pilot clinical trial to women with MS in an attempt to recapitulate the protective effect of pregnancy on disease (Sicotte et al., 2002). A crossover study was used whereby patients were followed for 6 months before treatment to establish baseline disease activity; it included cerebral MRI every month and a neurological examination every 3 months. The patients were then treated with oral estriol (8 mg/day) for 6 months and were observed for 6 months more during the posttreatment period. Six RRMS patients and four SPMS patients finished the 18-month study. The RRMS subjects were then retreated with oral estriol and progesterone during a 4-month extension phase. Estriol treatment resulted in serum estriol levels that approximated levels observed in untreated healthy control women who were 6 months pregnant. Interestingly, a significant decrease in a prototypic in vivo Th1 response (the delayed-type hypersensitivity response to the recall antigen tetanus) was observed at the end of the 6-month treatment period compared to the pretreatment period. Furthermore, mRNA levels of the Th1 cytokine IFNγ were decreased significantly in unstimulated PBMCs derived from subjects at the end of the 6-month treatment period compared to the baseline pretreatment period in the RRMS patients. When PBMCs were stimulated ex vivo, a favorable shift in cytokine profile (decreased TNFα, increased IL-10 and IL-5) was observed during treatment compared to baseline (Soldan et al., 2003). On serial MRIs, the RRMS patients demonstrated an 80% reduction in gadolinium-enhancing lesions within 3 months of treatment compared to pretreatment (Sicotte et al., 2002), and this improvement in enhancing lesions correlated with the favorable shift in cytokine profiles (Soldan et al., 2003) (Fig. 19.1). Importantly, gadolinium-enhancing disease activity gradually returned to baseline during the posttreatment period, and the favorable cytokine shift also returned to baseline. Furthermore, during the 4-month extension phase of the study, both the decrease in brain enhancing lesions and the favorable immune shift returned upon retreatment with estriol in combination with progesterone in the RRMS group. The latter data have important translational implications because progesterone treatment is needed in combination with estrogen treatment to prevent uterine endometrial hyperplasia when estrogens are administered for a

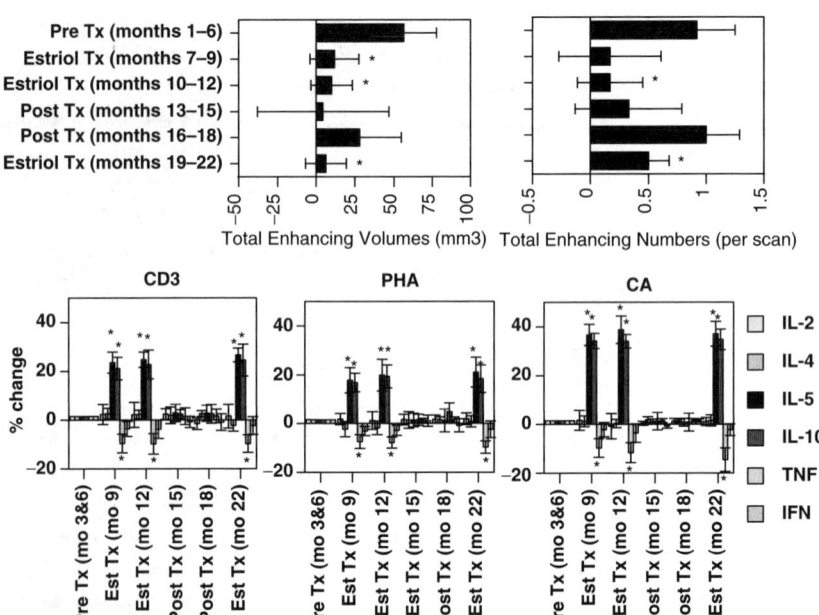

FIGURE 19.1. *Top.* Enhancing lesion volumes and numbers on serial magnetic resonance imaging (MRI) scans were significantly decreased with treatment (Tx) compared to the pretreatment baseline, increased back to baseline after treatment, then were again significantly decreased with reinstitution of treatment in relapsing-remitting multiple sclerosis (RRMS) patients. Median total volumes and numbers of gadolinium-enhanced lesions were determined during the 6-month pretreatment period as a baseline (months 1–6), during estriol treatment at study months 7–9 and 10–12, during the posttreatment period at study months 13–15 and 16–18, and during retreatment at months 19–22. Error bars indicate standard errors between patients for each median volume or number at each time. $^*p < 0.05$ (Sicotte et al., 2002). *Bottom.* In vivo estriol treatment increased secreted interleukins IL-5 and IL-10 and decreased secreted tumor necrosis factor-α (TNFα) from stimulated peripheral blood monocytic cells (PBMCs). Secreted cytokine levels—IL-2, IL-4, IL-5, IL-10, TNFα, and interferon-γ (IFNγ) were assessed in culture supernatants 48 hours after stimulation by cytometric bead array. PBMCs from RRMS patients were stimulated with αCD3, phytohemagglutinin (PHA), or *Candida albicans* lysate (CA). Cytokine levels are expressed as the mean percentage change of the level in the samples at the indicated treatment time point compared to the mean from two pretreatment baseline time points (months 3 and 6), with error bars indicating standard errors between patients. $^*p < 0.005$; Wilcoxson/Kruskal Wallis rank sum analysis). The range of detection was 20–5000 pg/ml for each cytokine (Soldan et al., 2003).

year or more. These results indicate that treatment with progesterone in combination with estriol did not neutralize the beneficial effect of estriol treatment on these biomarkers of disease. A double-blind, placebo-controlled trial of estriol in RRMS is now warranted.

# 5   Risk/Benefit Ratio of Estrogen Treatment

## 5.1   Risk/Benefit Ratio of Estrogen Treatment in Health and Disease

The risks and benefits of estrogen plus progesterone treatment in healthy post-menopausal women has been evaluated (Anonymous, 2002). A randomized, placebo-controlled prevention trial (with a planned duration of 8.5 years) included 16,608 postmenopausal women aged 50 to 79 years. Participants received either conjugated equine estrogens 0.625 mg/day (Premarin) plus medroxyprogesterone acetate 2.5 mg/day (Provera) versus placebo. The primary outcome was whether treatment was protective in coronary heart disease; while the primary adverse outcome was invasive breast cancer. Also assessed were stroke, pulmonary embolism, endometrial cancer, colorectal cancer, hip fracture, and death due to other causes. In May 2002, after a mean of 5.2 years of follow-up, the data and safety monitoring board recommended stopping the trial because the risks exceeded the benefits. Results indicated the following risks: Among 10,000 women taking the drug for a year, there were seven more coronary heart disease events, eight more invasive breast cancers, eight more strokes, and nine more pulmonary emboli; there were also six fewer colorectal cancers and five fewer hip fractures. Considering all events together over the 5.2-year duration of the trial, the excess number of all events in the active drug group was 100 per 10,000 (or 1/100 women).

Subsequently, in 2004 the arm of the trial that involved women treated with Premarin only (no progestin in subjects with no uterus) was also stopped after 6.8 years. It was stopped because no improvement in the primary outcome measure (heart disease) was observed (no effect, either good or bad). Similar to the Premarin/Provera arm, there was increased risk for stroke and decreased risk for hip fracture. However, there was a trend for reduction in breast cancer risk in women treated with Premarin (Anderson et al., 2004).

What are the ramifications of the finding that the risk/benefit ratio for Premarin as hormone replacement therapy (HRT) do not justify its use in healthy menopausal women, specifically with respect to the use of estrogens as a treatment for RRMS? There are three major considerations.

First, the risk/benefit ratio for treatment of MS is quite different from that of preventive treatment of healthy individuals. No toxicity is tolerable when treating healthy individuals, as in cases of HRT for healthy women or vaccination of children, for example. At the other extreme, significant toxicity is acceptable when treating a terminal disease such as cancer. Modest levels of toxicity are acceptable when treating individuals with a chronic disabling disease such as MS. This difference in what is acceptable toxicity is most evident when considering the risks and side effects of IFNβ, for example. The adverse effects of IFNβ treatment would be considered far in excess of what is acceptable for preventive treatment of healthy individuals, but they are considered appropriate for treatment of RRMS. Ideally, one prefers to have no risk and all benefit, even when treating patients with chronic disabling diseases such as MS. Realistically, the next best thing is to have a treatment with such a low level of toxicity that its use has been extensively debated for preventive medicine in healthy individuals (HRT in menopausal women) for decades.

Second, the age of subjects in the HRT study was 50 to 79 years, whereas the RRMS subjects were much younger, age 18 to 50 years. Younger subjects are at much less risk for cancer and vascular events. Some information concerning the risk of estrogen treatment in younger women can be found in the use of oral contraceptives. The primary concern is vascular events in those who smoke. In the future, it would therefore be judicious to pursue estrogen treatment approaches only in RRMS women who do not smoke. In addition, those RRMS subjects with a family history of breast cancer should undergo screening mammography and be treated with an estrogen only with caution.

Third, estrogens are not all alike. The three major estrogens are estradiol, estrone, and estriol. Estrogens differ in their affinity for estrogen receptor $\alpha$ (ER$\alpha$) and ER$\beta$. Quantitatively, estriol is a very weak estrogen compared to both 17$\beta$-estradiol and estrone, with the estriol binding of ER$\alpha$ and ER$\beta$ being 14% and 21%, respectively, compared to that of estradiol (100%) (Enmark and Gustafsson, 1999; Katzenellenbogen, 1984; Kuiper et al., 1997). Estriol has been accepted as the safest of the three estrogens in reviews dating from the 1970s to the 2000s (Follingstad, 1978; Head, 1998; Taylor, 2001; Utian, 1980). It has been used extensively in Europe and Asia for the treatment of menopausal symptoms (Cardozo et al., 1993; Cheng et al., 1993; Graser et al., 2000; Hayashi et al., 2000; Itoi et al., 2000; Lundstrom et al., 2001; Takahashi et al., 2000; Tzingounis et al., 1978); and, unlike 17$\beta$-estradiol, estriol causes minimal endometrial proliferation when moderate doses are used (Cardozo et al., 1993; Kirkengen et al., 1992; Lauritzen, 1976, 1987; Tzingounis et al., 1978; Utian, 1980). The fact that estrogens differ in their relative preference to bind ER$\alpha$ versus ER$\beta$ (Barkhem et al., 1998; Enmark and Gustafsson, 1999; Katzenellenbogen, 1984; Kuiper et al., 1997) is important because binding of each ER can result in opposite effects on transcription (Paech et al., 1997). In some tissues, estriol's actions have been shown to be antagonistic to those of 17$\beta$-estradiol (Bergink, 1980; 1997; Enmark and Gustafsson, 1999; Head, 1998; Melamed et al., 1997; Son et al., 2002). For example, estriol was once considered as a treatment for breast cancer because it opposed the breast cancer-promoting action of endogenous 17$\beta$-estradiol (Follingstad, 1978; Head 1998; Lemon et al., 1966; Lemon, 1973, 1975, 1980, 1987). Finally, it is known that sex hormones regulate sex hormone receptors, and again not all estrogens are alike in this regard. For example, Premarin (a conjugated equine estrogen) was shown to downregulate ER$\alpha$ in the hippocampus and cerebral cortex of ovariectomized rats with no effect on ER$\beta$ expression, whereas Progynova (estradiol valerate) upregulated ER$\beta$ and had no effect on ER$\alpha$ expression in these same brain regions (Jin et al., 2005).

In summary, when the risk/benefit ratio is weighed in the context of estrogen treatment for MS, one must consider (1) that MS is a chronic potentially disabling disease; (2) the age and other risk factors of the patient population; and (3) the type and dose of the estrogen planned for use.

## 5.2    Selective Estrogen Receptor Modifiers

Although the risk/benefit ratio in a chronic disease such as MS is clearly differ-ent than the risk/benefit ratio in healthy individuals, optimizing efficacy and min-imizing toxicity is a major goal. Hence, determining which estrogen receptor mediates the protective effect of an estrogen in disease is of central importance. The actions of estrogen are mediated primarily by nuclear estrogen receptors, ERα and ERβ, although nongenomic membrane effects have also been described (Weiss and Gurpide, 1988). Originally it was thought that ERα and ERβ would each have distinct tissue distributions, thereby providing a means through which the use of selective estrogen receptor modifiers (SERMs) could improve tissue selec-tivity. However, the relation between ERα and ERβ became complex, with most tis-sues expressing some detectable level of each of these receptors (Kuiper et al., 1998). The two receptors at times did, and at other times did not, co-localize to the same cells in a given tissue (Enmark and Gustafsson, 1999; Kuiper et al., 1997). Furthermore, the two receptors were shown to act synergistically in some tissues but antagonistically in others. These tissue-specific differences in biological out-comes are thought to be due in part to tissue-specific differences in transcription factors, which become activated upon binding of each ER by ligand (Nilsson et al., 2001; Paech et al., 1997; Shang and Brown, 2002).

Despite the fact that both ERα and ERβ are expressed in both the immune sys-tem and the CNS (Enmark and Gustafsson, 1999; Erlandsson et al., 2001; Igarashi et al., 2001 Kuiper et al., 1998), estrogen receptor knockout mice have been used to show that the protective effect of estrogen treatment (estradiol and estriol) in EAE was dependent on the presence of ERα, not ERβ (Liu et al., 2003; Polanczyk et al., 2003). However, whether targeted stimulation of ERα in developmentally normal mice was both necessary and sufficient for the estrogen-mediated protection in EAE remained unknown. ERβ could theoretically act synergistically or antagonis-tically with ERα in either the CNS or the immune system during estrogen treatment of EAE. A highly selective ERα ligand, propylpyrazole-triol (PPT) (Harrington et al., 2003), was then used to determine whether stimulation of ERα in develop-mentally normal mice would be sufficient for estrogen-mediated protection in EAE. Our group (Fig. 19.2) and another (Elloso et al., 2005) found that treatment with this ERα selective ligand was indeed sufficient for protection in EAE and was capable of mediating an antiinflammatory effect on cytokine production.

## 6    Estrogen Treatment: Early Versus Late in Disease

Now we review, and attempt to reconcile, findings of estrogen treatment in basic science with estrogen treatment in clinical trials of other diseases, mostly neurode generative diseases. Despite promising results in various in vitro and in vivo animal systems describing neuroprotective effects of estrogen treatment (reviewed above), a large study of estrogen replacement therapy in patients with an average age of 75 years and mild to moderate Alzheimer's disease yielded disappointing results

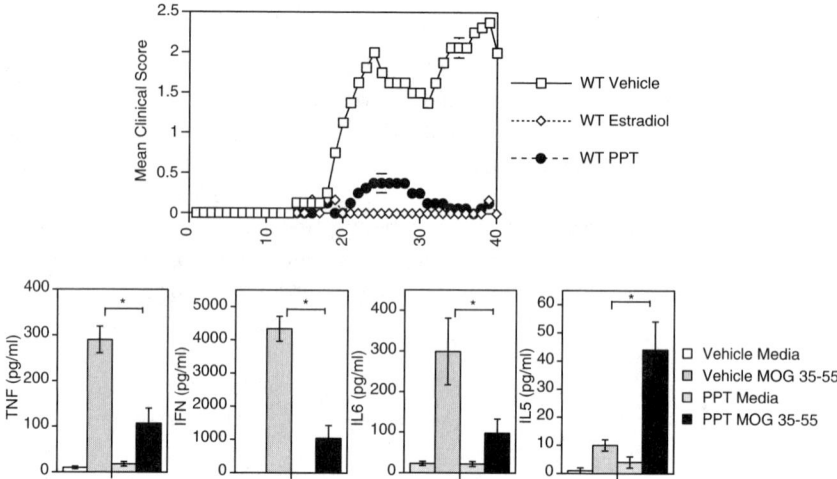

FIGURE 19.2. *Top.* Treatment with an ERα selective ligand is sufficient to reduce the clinical severity of experimental autoimmune encephalomyelitis (EAE). EAE was ameliorated in ovariectomized wild-type C57BL/6 female mice treated with the ERα-selective ligand propyl pyrazole triol (PPT). Daily treatments of ovariectomized mice with injections of vehicle (negative control), estradiol (positive control), or PPT (10 mg/kg/day) began. Seven days later, active EAE was induced with MOG 35-55 peptide. Mean clinical scores were significantly reduced in both estradiol- and PPT-treated mice compared to vehicle treated mice ($p < 0.0001$; ANOVA). Error bars indicate the variability of clinical scores between mice within a given treatment group ($n = 5$ mice per treatment group). Data are representative of experiments repeated a total of five times. *Bottom.* Treatment with an ERα ligand reduced proinflammatory cytokine production by peripheral immune cells in ovariectomized wild-type C57BL/6 female mice with EAE. At day 40 after EAE induction, mice were sacrificed, and the cytokine production by MOG 35-55-stimulated splenocytes was determined. TNFα, IFNγ, and IL-6 levels were each significantly reduced with PPT treatment, whereas IL-5 levels were increased with PPT treatment. $^{*}p < 0.05$. Error bars indicate the variability of cytokine values for splenocytes among individual mice within a given treatment group, with $n = 5$ mice for each treatment group. Data are representative of experiments repeated three times.

(Mulnard et al., 2000). It has been hypothesized that either the age of the subject population or the stage of disease may be critical in whether a neuroprotective effect of estrogen treatment can be appreciated (Mulnard et al., 2004). Adequate levels of estrogen receptors or associated intracellular cofactors may no longer be optimally expressed in either hormonally senescent or significantly diseased brains. Indeed, the effect of estrogen treatment on the blood-brain barrier has been shown to be quite different when comparing senescent versus younger rats (Bake and Sohrabji, 2004). Clinical reports also indicate differential effects of estrogen treatment depending on age and disease status. Estrogen replacement had no effect on cognition in elderly menopausal women (Binder et al., 2001), whereas it preserved cognition in younger women undergoing surgical hysterectomy (Sherwin, 1988; Verghese et al., 2000). In a large trial assessing the effects of estrogen treatment in stroke patients, treatment had no effect on cognition in women with stroke who had cognitive impairment at baseline, but it reduced the risk for cognitive decline in those with normal cognition at baseline (Viscoli et al., 2005). Finally, an improvement in cognitive function was observed with estriol treatment during early RRMS but not late SPMS (Sicotte et al., 2002). A "healthy cell bias of estrogen action" has been hypothesized to explain the disparity between promising results of estrogen treatment in animal models of neurodegenerative diseases versus disappointing results in estrogen treatment in some clinical neurodegenerative disease trials (Brinton, 2005). Efficacy of estrogen treatment appears to depend critically on it being administered early, as preventative therapy, before neurodegeneration has occurred (Mulnard et al., 2000). The question is then prompted regarding whether estrogen treatment must also be instituted early in autoimmune diseases, such as MS, RA, or psoriasis. Together these data warrant further study of treatment with estrogens, specifically ERα selective ligands, in MS and other autoimmune diseases with consideration of the age and disease duration of the subjects.

## 7  Conclusion

The reduction in relapse rates in MS during the third trimester of pregnancy provides a unique opportunity to identify naturally occurring factors that are responsible for this disease improvement. Knowledge in this area could theoretically be exploited to develop a novel therapy for RRMS. Although numerous factors capable of immunosuppression in putative Th1-mediated autoimmune diseases may be present during pregnancy, it is important to discriminate between those factors that are, or are not, increased during the third trimester and those that are, or are not, immunosuppressive when used at concentrations physiological with pregnancy. The pregnancy estrogen, estriol, meets these criteria for a possible role in the decrease in relapse rates during late pregnancy. However, in light of the complex nature of events that occur during late pregnancy, it is highly possible that estriol's effects are synergistic with effects of other pregnancy factors, ultimately creating the beneficial effect on MS that has been reported by patients and observed by clinicians for decades.

## References

Abramsky, O. (1994) Pregnancy and multiple sclerosis. *Ann. Neurol.* 36(Suppl 1):S38-41.

Aharoni, R., Teitelbaum, D., et al. (1998) Bystander suppression of experimental autoimmune encephalomyelitis by T cell lines and clones of the Th2 type induced by copolymer 1. *J. Neuroimmunol.* 91(1-2):135-146.

Alonso, A., Jick, S. S., et al. (2005) Recent use of oral contraceptives and the risk of multiple sclerosis. *Arch. Neurol.* 62(9):1362-1365.

Al-Shammri, S., Rawoot, P., et al. (2004) Th1/Th2 cytokine patterns and clinical profiles during and after pregnancy in women with multiple sclerosis. *J. Neurol. Sci.* 222(1-2):21-27.

Anderson, G. L., Limacher, M., et al. (2004) Effects of conjugated equine estrogen in post-menopausal women with hysterectomy: the Women's Health Initiative randomized controlled trial. *J.A.M.A.* 291(14):1701-1712.

Anonymous (2002) Risks and benefits of estrogen plus progestin in healthy postmenopausal women: principal results from the Women's Health Initiative randomized controlled trial. *J.A.M.A.* 288(3):321-333.

Azcoitia, I., Sierra, A., et al. (1999) Localization of estrogen receptor beta-immunoreactivity in astrocytes of the adult rat brain. *Glia* 26(3):260-267.

Bake, S., Sohrabji, F. (2004) 17Beta-estradiol differentially regulates blood-brain barrier permeability in young and aging female rats. *Endocrinology* 145(12):5471-5475.

Balashov, K. E., Comabella, M., et al. (2000) Defective regulation of IFNgamma and IL-12 by endogenous IL-10 in progressive MS. *Neurology* 55(2):192-198.

Barkhem, T., Carlsson, B., et al. (1998) Differential response of estrogen receptor alpha and estrogen receptor beta to partial estrogen agonists/antagonists. *Mol. Pharmacol.* 54(1):105-112.

Barkhof, F., van Waesberghe, J. H., et al. (2001) T(1) hypointense lesions in secondary progressive multiple sclerosis: effect of interferon beta-1b treatment. *Brain* 124(Pt 7): 1396-1402.

Baulieu, E., Schumacher, M. (2000) Progesterone as a neuroactive neurosteroid, with special reference to the effect of progesterone on myelination. *Steroids* 65(10-11):605-612.

Bebo, B. F., Jr., Fyfe-Johnson, A., et al. (2001) Low-dose estrogen therapy ameliorates experimental autoimmune encephalomyelitis in two different inbred mouse strains. *J. Immuno.* 166:2080-2089.

Behl, C., Widmann, M., et al. (1995) 17-Beta estradiol protects neurons from oxidative stress-induced cell death in vitro. *Biochem. Biophys. Res. Commun.* 216(2):473-482.

Behl, C., Skutella, T., et al. (1997) Neuroprotection against oxidative stress by estrogens: structure-activity relationship. *Mol. Pharmacol.* 51(4):535-541.

Berkink, E. W. (1980) Oestriol receptor interactions: their biological importance and therapeutic implications. *Acta Endocrinol. Suppl* 233:9-16.

Binder, E. F., Schechtman, K. B., et al. (2001) Effects of hormone replacement therapy on cognitive performance in elderly women. *Maturitas* 38(2):137-146.

Birk, K., Smeltzer, S. C., et al. (1988) Pregnancy and multiple sclerosis. *Semin. Neurol.* 8(3):205-213.

Birk, K., Ford, C., et al. (1990) The clinical course of multiple sclerosis during pregnancy and the puerperium. *Arch. Neurol.* 47(7):738-742.

Brinton, R. D. (2001) Cellular and molecular mechanisms of estrogen regulation of memory function and neuroprotection against Alzheimer's disease: recent insights and remaining challenges. *Learn. Mem.* 8(3):121-133.

Brinton, R. D. (2005) Investigative models for determining hormone therapy-induced outcomes in brain: evidence in support of a healthy cell bias of estrogen action. *Ann. N. Y. Acad. Sci.* 1052:57-74.

Cantarella, G., Risuglia, N., et al. (2004) Protective effects of estradiol on TRAIL-induced apoptosis in a human oligodendrocyte cell line: evidence for multiple sites of interactions. *Cell Death Differ.* 11:503-511.

Cardozo, L., Rekers, H., et al. (1993) Oestriol in the treatment of postmenopausal urgency: a multicentre study. *Maturitas* 18(1):47-53.

Cheng, G. J., Liu, J. L., et al. (1993) Nylestriol replacement therapy in postmenopausal women: a three-year prospective study. *Chin. Med. J.* 106(12):911-916.

Confavreux, C., Hutchinson, M., et al. (1998) Rate of pregnancy-related relapse in multiple sclerosis: Pregnancy in Multiple Sclerosis Group. *N. Engl. J. Med.* 339(5):285-291.

Da Silva, J. A., Hall, G. M. (1992) The effects of gender and sex hormones on outcome in rheumatoid arthritis. *Baillieres Clin. Rheumatol.* 6(1):196-219.

Da Silva, J. A., Spector, T. D. (1992) The role of pregnancy in the course and aetiology of rheumatoid arthritis. *Clin. Rheumatol.* 11(2):189-194.

Damek, D. M., Shuster, E. A. (1997) Pregnancy and multiple sclerosis. *Mayo Clin. Proc.* 72(10):977-989.

Drew, P. D., Chavis, J. A. (2000) Female sex steroids: effects upon microglial cell activation. *J. Neuroimmunol.* 111(1-2):77-85.

Dubal, D. B., Zhu, H., et al. (2001) Estrogen receptor alpha, not beta, is a critical link in estradiol-mediated protection against brain injury. *Proc. Natl. Acad. Sci. USA* 98(4):1952-1957.

Duda, P. W., Schmied, M. C., et al. (2000) Glatiramer acetate (Copaxone) induces degenerate, Th2-polarized immune responses in patients with multiple sclerosis. *J. Clin. Invest.* 105(7):967-976.

Dudley, D. J., Chen, C. L., et al. (1993) Adaptive immune responses during murine pregnancy: pregnancy-induced regulation of lymphokine production by activated T lymphocytes. *Am. J. Obstet. Gynecol.* 168(4):1155-1163.

Elenkov, I. J., Wilder, R. L., et al. (2001) IL-12, TNF-alpha, and hormonal changes during late pregnancy and early postpartum: implications for autoimmune disease activity during these times. *J. Clin. Endocrinol. Metab.* 86(10):4933-4938.

Elloso, M. M., Phiel, K., et al. (2005) Suppression of experimental autoimmune encephalomyelitis using estrogen receptor-selective ligands. *J. Endocrinol.* 185(2): 243-252.

Enmark, E., Gustafsson, J. A. (1999) Oestrogen receptors: an overview. *J. Intern. Med.* 246(2):133-138.

Erlandsson, M. C., Ohlsson, C., et al. (2001) Role of oestrogen receptors alpha and beta in immune organ development and in oestrogen-mediated effects on thymus. *Immunology* 103(1):17-25.

Fabris, N., Piantanelli, L., et al. (1977) Differential effect of pregnancy or gestagens on humoral and cell-mediated immunity. *Clin. Exp. Immunol.* 28(2):306-314.

Follingstad, A. H. (1978) Estriol, the forgotten estrogen? *J.A.M.A.* 239(1):29-30.

Formby, B. (1995) Immunologic response in pregnancy: its role in endocrine disorders of pregnancy and influence on the course of maternal autoimmune diseases. *Endocrinol. Metab. Clin. North Am.* 24(1):187-205.

Garcia-Segura, L. M., Naftolin, F., et al. (1999) Role of astroglia in estrogen regulation of synaptic plasticity and brain repair. *J. Neurobiol.* 40(4):574-584.

Garcia-Segura, L. M., Azcoitia, I., et al. (2001) Neuroprotection by estradiol. *Prog. Neurobiol.* 63(1):29-60.

Ghoumari, A. M., Ibanez, C., et al. (2003) Progesterone and its metabolites increase myelin basic protein expression in organotypic slice cultures of rat cerebellum. *J. Neurochem.* 86(4):848-859.

Gilmore, W., Arias, M., et al. (2004) Preliminary studies of cytokine secretion patterns associated with pregnancy in MS patients. *J. Neurol. Sci.* 224(1-2):69-76.

Goodman, Y., Bruce, A. J., et al. (1996) Estrogens attenuate and corticosterone exacerbates excitotoxicity, oxidative injury, and amyloid beta-peptide toxicity in hippocampal neurons. *J. Neurochem.* 66(5):1836-1844.

Granholm, A. C., Sanders, L., et al. (2003) Estrogen alters amyloid precursor protein as well as dendritic and cholinergic markers in a mouse model of Down syndrome. *Hippocampus* 13(8):905-914.

Graser, T., Koytchev, R., et al. (2000) Comparison of the efficacy and endometrial safety of two estradiol valerate/dienogest combinations and Kliogest for continuous combined hormone replacement therapy in postmenopausal women. *Climacteric* 3(2):109-118.

Green, P. S., Bales, K., et al. (2005) Estrogen therapy fails to alter amyloid deposition in the PDAPP model of Alzheimer's disease. *Endocrinology* 146(6):2774-2781.

Hall, G. M., Daniels, M., et al. (1994) A randomised controlled trial of the effect of hormone replacement therapy on disease activity in postmenopausal rheumatoid arthritis. *Ann. Rheum. Dis.* 53(2):112-116.

Harms, C., Lautenschlager, M., et al. (2001) Differential mechanisms of neuroprotection by 17 beta-estradiol in apoptotic versus necrotic neurodegeneration. *J. Neurosci.* 21(8):2600-2609.

Harrington, W. R., Sheng, S., et al. (2003) Activities of estrogen receptor alpha- and beta-selective ligands at diverse estrogen responsive gene sites mediating transactivation or transrepression. *Mol. Cell. Endocrinol.* 206(1-2):13-22.

Hayashi, T., Ito, I., et al. (2000) Estriol (E3) replacement improves endothelial function and bone mineral density in very elderly women. *J. Gerontol. A Biol. Sci. Med. Sci.* 55(4):B183-B190; discussion B191-B193.

Head, K. A. (1998) Estriol: safety and efficacy. *Altern. Med. Rev.* 3(2):101-113.

Heikkinen, T., Kalesnykas, G., et al. (2004) Estrogen treatment improves spatial learning in APP + PS1 mice but does not affect beta amyloid accumulation and plaque formation. *Exp. Neurol.* 187(1):105-117.

Hernan, M. A., Hohol, M. J., et al. (2000) Oral contraceptives and the incidence of multiple sclerosis. *Neurology* 55(6):848-854.

Hill, J. A., Polgar, K., et al. (1995) T-helper 1-type immunity to trophoblast in women with recurrent spontaneous abortion. *J.A.M.A.* 273(24):1933-1936.

Hoffman, G. E., Le, W. W., et al. (2001) Divergent effects of ovarian steroids on neuronal survival during experimental allergic encephalitis in Lewis rats. *Exp. Neurol.* 171(2): 272-284.

Igarashi, H., Kouro, T., et al. (2001) Age and stage dependency of estrogen receptor expression by lymphocyte precursors. *Proc. Natl. Acad. Sci. USA* 98(26):15131-15136.

Ito, A., Bebo, B. F., Jr., et al. (2001) Estrogen treatment down-regulates TNF-alpha production and reduces the severity of experimental autoimmune encephalomyelitis in cytokine knockout mice. *J. Immunol.* 167(1):542-552.

Itoi, H., Minakami, H., et al. (2000) Comparison of the long-term effects of oral estriol with the effects of conjugated estrogen on serum lipid profile in early menopausal women. *Maturitas* 36(3):217-222.

Jansson, L., Holmdahl, R. (1989) Oestrogen induced suppression of collagen arthritis. IV. Progesterone alone does not affect the course of arthritis but enhances the oestrogen-mediated therapeutic effect. *J. Reprod. Immunol.* 15(2):141-150.

Jansson, L., Holmdahl, R. (1992) Oestrogen-induced suppression of collagen arthritis; 17 beta-oestradiol is therapeutically active in normal and castrated F1 hybrid mice of both sexes. *Clin. Exp. Immunol.* 89(3):446-451.

Jansson, L., Mattsson, A., et al. (1990) Estrogen induced suppression of collagen arthritis. V. Physiological level of estrogen in DBA/1 mice is therapeutic on established arthritis, suppresses anti-type II collagen T-cell dependent immunity and stimulates polyclonal B-cell activity. *J. Autoimmun.* 3(3):257-270.

Jansson, L., Olsson, T., et al. (1994) Estrogen induces a potent suppression of experimental autoimmune encephalomyelitis and collagen-induced arthritis in mice. *J. Neuroimmunol.* 53(2):203-207.

Jensen, J., Krakauer, M., et al. (2001) Increased T cell expression of CD154 (CD40-ligand) in multiple sclerosis. *Eur. J. Neurol.* 8(4):321-328.

Jin, M., Jin, F., et al. (2005) Two estrogen replacement therapies differentially regulate expression of estrogen receptors alpha and beta in the hippocampus and cortex of ovariectomized rat. *Brain Res. Mol. Brain Res.* 142(2):107-114.

Jover, T., Tanaka, H., et al. (2002) Estrogen protects against global ischemia-induced neuronal death and prevents activation of apoptotic signaling cascades in the hippocampal CA1. *J. Neurosci.* 22(6):2115-2124.

Katzenellenbogen, B. S. (1984) Biology and receptor interactions of estriol and estriol derivatives in vitro and in vivo. *J. Steroid Biochem.* 20(4B):1033-1037.

Kim, S., Liva, S. M., et al. (1999) Estriol ameliorates autoimmune demyelinating disease: implications for multiple sclerosis. *Neurology* 52(6):1230-1238.

Kirkengen, A. L., Andersen, P., et al. (1992) Oestriol in the prophylactic treatment of recurrent urinary tract infections in postmenopausal women. *Scand. J. Primary Health Care* 10(2):139-142.

Kozovska, M. E., Hong, J., et al. (1999) Interferon beta induces T-helper 2 immune deviation in MS. *Neurology* 53(8):1692-1697.

Krishnan, L., Guilbert, L. J., et al. (1996) T helper 1 response against Leishmania major in pregnant C57BL/6 mice increases implantation failure and fetal resorptions: correlation with increased IFN-gamma and TNF and reduced IL-10 production by placental cells. *J. Immunol.* 156(2):653-662.

Kuiper, G. G., Carlsson, B., et al. (1997) Comparison of the ligand binding specificity and transcript tissue distribution of estrogen receptors alpha and beta. *Endocrinology* 138(3):863-870.

Kuiper, G. G., Shughrue, P. J., et al. (1998) The estrogen receptor beta subtype: a novel mediator of estrogen action in neuroendocrine systems. *Front. Neuroendocrinol.* 19(4):253-286.

Langer-Gould, A., Garren, H., et al. (2002) Late pregnancy suppresses relapses in experimental autoimmune encephalomyelitis: evidence for a suppressive pregnancy-related serum factor. *J. Immunol.* 169(2):1084-1091.

Lauritzen, C. (1987) Results of a 5 years prospective study of estriol succinate treatment in patients with climacteric complaints. *Horm. Metab. Res.* 19(11):579-584.

Lauritzen, C. H. (1976) The female climacteric syndrome: significance, problems, treatment. *Acta Obstet. Gynecol. Scand. Suppl.* 180(51):47-61.

Lemon, H. M. (1973) Oestriol and prevention of breast cancer. *Lancet* 1(7802):546-547.

Lemon, H. M. (1975) Estriol prevention of mammary carcinoma induced by 7,12-dimethylbenzanthracene and procarbazine. *Cancer Res.* 35(5):1341-1353.

Lemon, H. M. (1980) Pathophysiologic considerations in the treatment of menopausal patients with oestrogens; the role of oestriol in the prevention of mammary carcinoma. *Acta Endocrinol. Suppl.* 233:17-27.

Lemon, H. M. (1987) Antimammary carcinogenic activity of 17-alpha-ethinyl estriol. *Cancer* 60(12):2873-2881.

Lemon, H. M., Wotiz, H. H., et al. (1966) Reduced estriol excretion in patients with breast cancer prior to endocrine therapy. *J.A.M.A.* 196(13):1128-1136.

Leranth, C., Roth, R. H., et al. (2000) Estrogen is essential for maintaining nigrostriatal dopamine neurons in primates: implications for Parkinson's disease and memory. *J. Neurosci.* 20(23):8604-8609.

Li, D. K., Zhao, G. J., et al. (2001) Randomized controlled trial of interferon-beta-1a in secondary progressive MS: MRI results. *Neurology* 56(11):1505-1513.

Lin, H., Mosmann, T. R., et al. (1993) Synthesis of T helper 2-type cytokines at the maternal-fetal interface. *J. Immunol.* 151(9):4562-4573.

Liu, H. B., Loo, K. K., et al. (2003) Estrogen receptor alpha mediates estrogen's immune protection in autoimmune disease. *J. Immunol.* 171(12):6936-6940.

Liu, H. Y., Buenafe, A. C., et al. (2002) Estrogen inhibition of EAE involves effects on dendritic cell function. *J. Neurosci. Res.* 70(2):238-248.

Lundstrom, E., Wilczek, B., et al. (2001) Mammographic breast density during hormone replacement therapy: effects of continuous combination, unopposed transdermal and low-potency estrogen regimens. *Climacteric* 4(1):42-48.

Marzi, M., Vigano, A., et al. (1996) Characterization of type 1 and type 2 cytokine production profile in physiologic and pathologic human pregnancy. *Clin. Exp. Immunol.* 106(1):127-133.

Matejuk, A., Adlard, K., et al. (2001) 17Beta-estradiol inhibits cytokine, chemokine, and chemokine receptor mRNA expression in the central nervous system of female mice with experimental autoimmune encephalomyelitis. *J. Neurosci. Res.* 65(6):529-542.

Matejuk, A., Bakke, A. C., et al. (2004) Estrogen treatment induces a novel population of regulatory cells, which suppresses experimental autoimmune encephalomyelitis. *J. Neurosci. Res.* 77(1):119-126.

Mattsson, R., Mattsson, A., et al. (1991) Maintained pregnancy levels of oestrogen afford complete protection from post-partum exacerbation of collagen-induced arthritis. *Clin. Exp. Immunol.* 85(1):41-47.

Melamed, M., Castano, E., et al. (1997) Molecular and kinetic basis for the mixed agonist/antagonist activity of estriol. *Mol. Endocrinol.* 11(12):1868-1878.

Mellor, A. L., Munn, D. H. (2004) IDO expression by dendritic cells: tolerance and tryptophan catabolism. *Nat. Rev. Immunol.* 4(10):762-774.

Mulnard, R. A., Cotman, C. W., et al. (2000) Estrogen replacement therapy for treatment of mild to moderate Alzheimer disease: a randomized controlled trial; Alzheimer's Disease Cooperative Study. *J.A.M.A.* 283(8):1007-1015.

Mulnard, R. A., Corrada, M. M., et al. (2004) Estrogen replacement therapy, Alzheimer's disease, and mild cognitive impairment. *Curr. Neurol. Neurosci. Rep.* 4(5):368-373.

Munn, D. H., Zhou, M., et al. (1998) Prevention of allogeneic fetal rejection by tryptophan catabolism. *Science* 281(5380):1191-1193.

Nelson, J. L., Hughes, K. A., et al. (1992) Remission of rheumatoid arthritis during pregnancy and maternal-fetal class II alloantigen disparity. *Am. J. Reprod. Immunol.* 28(3-4):226-227.

Nilsson, S., Makela, S., et al. (2001) Mechanisms of estrogen action. *Physiol. Rev.* 81(4):1535-1565.

Paech, K., Webb, P., et al. (1997) Differential ligand activation of estrogen receptors ERalpha and ERbeta at AP1 sites. *Science* 277(5331):1508-1510.

Palaszynski, K. M., Liu, H., et al. (2004) Estriol treatment ameliorates disease in males with experimental autoimmune encephalomyelitis: implications for multiple sclerosis. *J. Neuroimmunol.* 149(1-2):84-89.

Polanczyk, M., Zamora, A., et al. (2003) The protective effect of 17beta-estradiol on experimental autoimmune encephalomyelitis is mediated through estrogen receptor-alpha. *Am. J. Pathol.* 163(4):1599-1605.

Polanczyk, M. J., Carson, B. D., et al. (2004) Cutting edge: estrogen drives expansion of the CD4+CD25+ regulatory T cell compartment. *J. Immunol.* 173(4):2227-2230.

Raghupathy, R. (1997) Th1-type immunity is incompatible with successful pregnancy. *Immunol. Today* 18(10):478-482.

Rau, S. W., Dubal, D. B., et al. (2003) Estradiol attenuates programmed cell death after stroke-like injury. *J. Neurosci.* 23(36):11420-11426.

Rudick, R. A., Ransohoff, R. M., et al. (1998) In vivo effects of interferon beta-1a on immunosuppressive cytokines in multiple sclerosis. *Neurology* 50(5):1294-1300. Erratum in *Neurology* 1998;51(1):332.

Runmarker, B., Andersen, O. (1995) Pregnancy is associated with a lower risk of onset and a better prognosis in multiple sclerosis. *Brain* 118(Pt 1):253-261.

Sacks, G. P., Clover, L. M., et al. (2001) Flow cytometric measurement of intracellular Th1 and Th2 cytokine production by human villous and extravillous cytotrophoblast. *Placenta* 22(6):550-559.

Schumacher, M., Guennoun, R., et al. (2004) Local synthesis and dual actions of progesterone in the nervous system: neuroprotection and myelination. *Growth Horm. I.G.F. Res.* 14(Suppl A):S18-S33.

Shang, Y., Brown, M. (2002) Molecular determinants for the tissue specificity of SERMs. *Science* 295(5564):2465-2468.

Sherwin, B. B. (1988) Estrogen and/or androgen replacement therapy and cognitive functioning in surgically menopausal women. *Psychoneuroendocrinology* 13(4):345-357.

Sicotte, N. L., Liva, S. M., et al. (2002) Treatment of multiple sclerosis with the pregnancy hormone estriol. *Ann. Neurol.* 52(4):421-428.

Sierra, A., Azcoitia, I., et al. (2003) Endogenous estrogen formation is neuroprotective in model of cerebellar ataxia. *Endocrine* 21(1):43-51.

Soldan, S. S., Retuerto, A. I., et al. (2003) Immune modulation in multiple sclerosis patients treated with the pregnancy hormone estriol. *J. Immunol.* 171(11):6267-6274.

Soldan, S. S., Alvarez Retuerto, A. I., et al. (2004) Dysregulation of IL-10 and IL-12 p40 in secondary progressive multiple sclerosis. *J. Neuroimmunol.* 146(1-2):209-215.

Son, D. S., Roby, K. F., et al. (2002) Estradiol enhances and estriol inhibits the expression of CYP1A1 induced by 2,3,7,8-tetrachlorodibenzo-p-dioxin in a mouse ovarian cancer cell line. *Toxicology* 176(3):229-243.

Sribnick, E. A., Wingrave, J. M., et al. (2003) Estrogen as a neuroprotective agent in the treatment of spinal cord injury. *Ann. N. Y. Acad. Sci.* 993:125-133; discussion 159-160.

Sribnick, E. A., Ray, S. K., et al. (2004) 17Beta-estradiol attenuates glutamate-induced apoptosis and preserves electrophysiologic function in primary cortical neurons. *J. Neurosci. Res.* 76(5):688-696.

Sribnick, E. A., Wingrave, J. M., et al. (2005) Estrogen attenuated markers of inflammation and decreased lesion volume in acute spinal cord injury in rats. *J. Neurosci. Res.* 82(2):283-293.

Subramanian, S., Matejuk, A., et al. (2003) Oral feeding with ethinyl estradiol suppresses and treats experimental autoimmune encephalomyelitis in SJL mice and inhibits the recruitment of inflammatory cells into the central nervous system. *J. Immunol.* 170(3): 1548-1555.

Sur, P., Sribnick, E. A., et al. (2003) Estrogen attenuates oxidative stress-induced apoptosis in C6 glial cells. *Brain Res.* 971(2):178-188.

Takahashi, K., Manabe, A., et al. (2000) Efficacy and safety of oral estriol for managing postmenopausal symptoms. *Maturitas* 34(2):169-177.

Takao, T., Flint, N., et al. (2004) 17Beta-estradiol protects oligodendrocytes from cytotoxicity induced cell death. *J. Neurochem.* 89(3):660-673.

Taylor, M. (2001) Unconventional estrogens: estriol, biest, and triest. *Clin. Obstet. Gynecol.* 44(4):864-879.

Thorogood, M., Hannaford, P. C. (1998) The influence of oral contraceptives on the risk of multiple sclerosis. *Br. J. Obstet. Gynaecol.* 105(12):1296-1299.

Trapp, B. D., Bö, L., et al. (1999) Pathogenesis of tissue injury in MS lesions. *J. Neuroimmunol.* 98(1):49-56.

Tzingounis, V. A., Aksu, M. F., et al. (1978) Estriol in the management of the menopause. *J.A.M.A.* 239(16):1638-1641.

Utian, W. H. (1980) The place of oestriol therapy after menopause. *Acta Endocrinol. Suppl.* 233(5):51-56.

Van Boxel-Dezaire, A. H., Hoff, S. C., et al. (1999) Decreased interleukin-10 and increased interleukin-12p40 mRNA are associated with disease activity and characterize different disease stages in multiple sclerosis. *Ann. Neurol.* 45(6):695-703.

Van Groen, T., Kadish, I. (2005) Transgenic AD model mice, effects of potential anti-AD treatments on inflammation and pathology. *Brain Res. Brain Res. Rev.* 48(2):370-378.

Van Walderveen, M. A., Tas, M. W., et al. (1994) Magnetic resonance evaluation of disease activity during pregnancy in multiple sclerosis. *Neurology* 44(2):327-329.

Vegeto, E., Bonincontro, C., et al. (2001) Estrogen prevents the lipopolysaccharide-induced inflammatory response in microglia. *J. Neurosci.* 21(6):1809-1818.

Veliskova, J., Velisek, L., et al. (2000) Neuroprotective effects of estrogens on hippocampal cells in adult female rats after status epilepticus. *Epilepsia* 41(Suppl 6):S30-S35.

Verdru, P., Theys, P., et al. (1994) Pregnancy and multiple sclerosis: the influence on long term disability. *Clin. Neurol. Neurosurg.* 96(1):38-41.

Verghese, J., Kuslansky, G., et al. (2000) Cognitive performance in surgically menopausal women on estrogen. *Neurology* 55(6):872-874.

Villard-Mackintosh, L., Vessey, M. P. (1993) Oral contraceptives and reproductive factors in multiple sclerosis incidence. *Contraception* 47(2):161-168.

Viscoli, C. M., Brass, L. M., et al. (2005) Estrogen therapy and risk of cognitive decline: results from the Women's Estrogen for Stroke Trial (WEST). *Am. J. Obstet. Gynecol.* 192(2):387-393.

Voskuhl, R. R. (1996) Chronic relapsing experimental allergic encephalomyelitis in the SJL mouse: relevant techniques. *Methods* 10(3):435-439.

Voskuhl, R. R., Palaszynski, K. (2001) Sex hormones and experimental autoimmune encephalomyelitis:implications for multiple sclerosis. *Neuroscientist* 7(3):258-270.

Vukusic, S., Hutchinson, M., et al. (2004) Pregnancy and multiple sclerosis (the PRIMS study): clinical predictors of post-partum relapse. *Brain* 127(Pt 6):1353-1360.

Wegmann, T. G., Lin, H., et al. (1993) Bidirectional cytokine interactions in the maternal-fetal relationship: is successful pregnancy a TH2 phenomenon? *Immunol. Today* 14(7):353-356.

Weiss, D. J., Gurpide, E. (1988) Non-genomic effects of estrogens and antiestrogens. *J. Steroid Biochem.* 31(4B):671-676.

Whitacre, C. C., Reingold, S. C., et al. (1999) A gender gap in autoimmunity. *Science* 283(5406):1277-1278.

Wise, P. M., Dubal, D. B., et al. (2001) Minireview: neuroprotective effects of estrogen-new insights into mechanisms of action. *Endocrinology* 142(3):969-973.

Xiao, B. G., Liu, X., et al. (2004) Antigen-specific T cell functions are suppressed over the estrogen-dendritic cell-indoleamine 2,3-dioxygenase axis. *Steroids* 69(10):653-659.

Zhang, Q. H., Hu, Y. Z., et al. (2004) Estrogen influences the differentiation, maturation and function of dendritic cells in rats with experimental autoimmune encephalomyelitis. *Acta Pharmacol. Sin.* 25(4):508-513.

# 20
# Immune Regeneration Through Hematopoietic Stem Cell Transplantation to Restore Tolerance in Autoimmune Disease

Paolo A. Muraro

## 1  Introduction

Immune depletion followed by autologous hematopoietic cell support (abbreviated as hematopoietic stem cell transplantation, or HSCT) is an emerging experimental therapy for severe autoimmune diseases (Tyndall and Saccardi, 2005). The clinical principles and application of HSCT to autoimmune disease are reviewed in Chapter 17. We focus here on discussing the mechanisms through which HSCT exerts its clinical effects. The underlying principle of HSCT is that maximal suppression or ablation of the immune system can destroy the pathogenic autoreactive cells, and a healthy immune system can be regenerated from autologous hematopoietic progenitor cells. However, this has long been an unproven assumption. Since the beginning of trials of HSCT for autoimmune diseases during the mid-1990s, a controversy has lingered as to whether amelioration of the course of the disease observed after HSCT could simply be the result of prolonged immunosuppression induced by the cytotoxic chemo- or radiotherapy rather than the outcome of more permanent regeneration of the immune system. For several years, research to address this fundamental issue had made little progress. Recently, however, there have been important advances, and studies starting to characterize the mechanism of action of HSCT in autoimmune disease have been published.

Here, we present the state of the art in the field and discuss some of the important questions that remain to be answered. This is not a purely academic matter, as thorough understanding of the mode of action can lead to optimization of safety and efficacy of HSCT regimens and to better selection of the patient population who can benefit from such treatment schemes.

The autoimmune diseases that have been treated with HSCT most frequently are multiple sclerosis (MS) among neurological disorders and systemic lupus erythematosus (SLE), systemic sclerosis (SS), rheumatoid arthritis (RA), and juvenile idiopathic rheumatoid arthritis (JRA) among the rheumatological

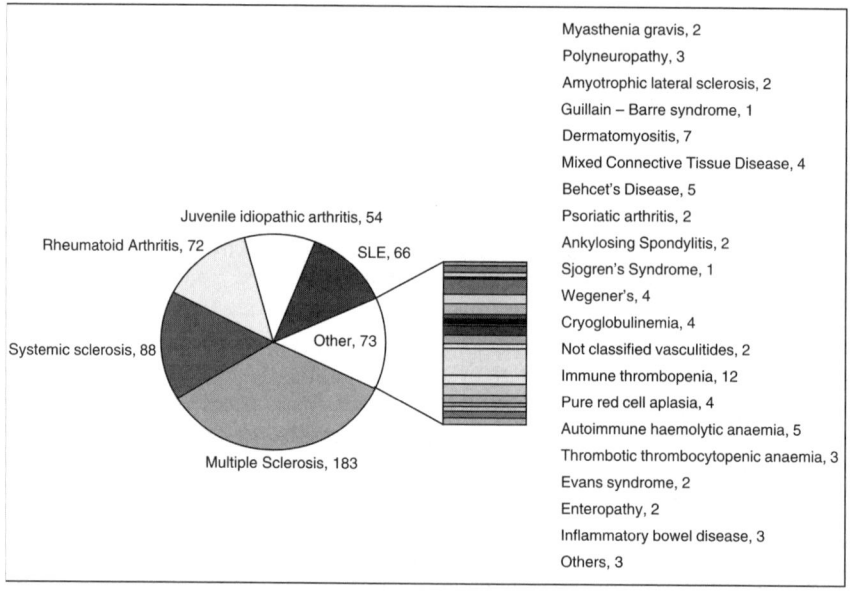

FIGURE 20.1. Autoimmune disease patients treated with autologous hematopoietic stem cell transplantation (HSCT) in the EBMT/EULAR database (October 2004). The pie chart illustrates the proportions and actual numbers of patients treated with autologous HSCT for various autoimmune diseases. Source of data is the EBMT/EULAR database updated to October 2004 as reported by Tyndall and Saccardi (2005).

diseases. Taken together, patients suffering from MS, SLE, SS, RA, and juvenile-onset RA (JRA) represented 463 (86%) of the 536 subjects treated with autologous HSCT for autoimmune disease recorded in the EBMT/EULAR database as of October 2004 (Tyndall and Saccardi, 2005) (Fig. 20.1). Although a formal meta-analysis of North American trials is not available, MS, SLE, SS, and RA are also the autoimmune disorders that have most frequently been treated with HSCT in this region. Many other autoimmune conditions, some of which are listed in Figure 20.1 ("Other") have been treated with HSCT in small pilot trials, or individual cases were treated on a compassionate basis or coincidentally, with the primary target a solid or hematological neoplasm (Marmont, 2004). However, the anecdotal reports of which we are aware do not provide much immunological or mechanistic insight. On the other hand, there is a substantial body of literature on intensive chemotherapy and HSCT for cancer that describes general features of immune reconstitution, and to mention some relevant aspects we refer to existing review articles covering that field.

# 2    Treating Autoimmune Diseases With HSCT: The Underlying Hypotheses

## 2.1    What Are We Trying to Fix in Autoimmune Disease?

For most autoimmune disorders, the etiology is unknown and the pathogenesis is incompletely understood. No direct causative relation has been indisputably demonstrated for any genes or environmental agents that have been found in association with autoimmune disease. Inappropriate recognition of constituents of self by immune cells has been a widely accepted way to schematize the basic problem in those idiopathic disorders, hence the name "autoimmune." Whether the error causing the aggression of self originates from erroneously selected "forbidden clones," as first theorized by Burnet (1959), or from the context in which self antigens are presented and recognized, as postulated in the "danger" model of autoimmunity (Matzinger, 1994), has been matter of intense debate; and various refinements of these conceptual models have been recently proposed.

Regardless of the uncertainties about the triggering mechanism, most experts agree that immune cells are central players in the effector phases of autoimmune disorders. To explain the pathogenesis of MS and SLE, for example, current models envision a crucial role for T and B cells inappropriately recognizing self antigens and initiating a cell-mediated or humoral reaction, or both, leading to inflammatory tissue and vascular damage (Shlomchik et al., 2001; Sospedra and Martin, 2005). Development of this autoreactive process is thought to require two conditions: (1) a complex, polygenic susceptibility of the individual; and (2) an encounter with one or more environmental agents, possibly microbial, capable of acting as triggering factor. Adding further complexity, there may be significant latency before the initial environmental (possibly infectious) exposure and the development of clinical expression of disease.

Such a model can reconcile several features of autoimmune diseases that are not explained by mendelian genetic transmission or by a purely infectious etiology (Marrie, 2004). During life, contacts between the individual's immunogenetic background and the environment shape the composition and qualities of the adaptive immune system in every person. In a subject genetically predisposed to autoimmune disease, some of these interactions may be abnormal and lead to autoimmunity. Until the abnormalities are fully understood, it will of course be difficult to correct or counteract them specifically. Even then, heterogeneity in the pathogenesis of each autoimmune disorder might require a highly individualized treatment approach, which may not be possible to offer to patients on a population scale. For example, the immunodominant epitopes of myelin basic protein (MBP), a constituent of myelin thought to be a major target of autoreactive T cells in MS, differ in function depending on the individual's immunogenetic background (Muraro et al., 1997; Pette et al., 1990; Vergelli et al., 1997). Although specific immunotherapy is conceptually attractive, individual variation and epitope spreading (Lehmann et al., 1993) may limit its efficacy. A broader immune intervention targeting one or more common mechanisms of autoimmunity would give the advantage of not requiring knowledge of specific molecular components of the autoimmune disorder in a particular patient to restore tolerance.

## 2.2   Why Should HSCT Work in Autoimmune Disease?

The rationale for using HSCT to treat autoimmune diseases is apparently unsophisticated. Intensive immune depletion (immune ablation) could allow elimination of virtually all mature lymphocytes, regardless of their antigen specificity. Therefore, primed lymphocytes that are mediating an autoimmune attack or have the potential to initiate a reexacerbation can also be eliminated. Reconstitution of the adaptive immune system from multipotent hematopoietic progenitors, such as CD34+ cells, is expected to regenerate a new immune system that no longer harbors disease-mediating immune cells at an elevated frequency from previous activation and expansion. In an analogy with computers that can be restarted to resolve a lockup or other software error, this concept has been known as the immune "resetting" or "rebooting" hypothesis. Another way to describe this is that autologous HSCT could reset the autoimmune disease to an earlier, latent phase of disease development (Fassas and Kimiskidis, 2004).

In this perspective, clinical efficacy of HSCT relies on the assumption that the unidentified events, possibly environmental, that originally lead to breakdown of tolerance either do not recur or at least require several years before they can reinduce active disease. The other condition that must be met for autologous HSCT to be efficacious is that the defect causing autoimmune disease must not be a genetically determined stem cell disorder. In this case, as for computers, the "hardware" error could not be corrected by "restarting" the system, and it would fail again, resulting in continuation of the autoimmune process. Studies of bone marrow transplantation (BMT) in experimental models of autoimmune disease have shown successful clinical outcomes not only after syngeneic but also after autologous or pseudoautologous BMT (reviewed in van Bekkum, 2000). The relevance of these studies to the issue of therapeutic potential in the autologous setting is limited by the artificial induction, in contrast to spontaneous onset, of disease in these models. It is nevertheless remarkable that in one study arthritic rats treated with syngeneic BMT were refractory to relapses even if reimmunized (van Bekkum et al., 1989).

In humans, theoretical reassurance on the curative potential of autologous HSCT comes from epidemiological data on concordance for autoimmune disease in twins. Because the concordance rate for MS in genetically identical twins typically does not exceed 25% (Ristori et al., 2005; Willer et al., 2003), one can infer that for approximately three-fourths of MS patients genetic factors alone are necessary but not sufficient to induce expression of clinical disease. One may further speculate that if the mature immune system is ablated and reconstituted with an autologous graft, and if environmental disease factors relevant to disease induction are not immediately recurring and active, approximately three-fourths of the patients should benefit from prolonged remission from disease activity. In line so far with this prediction, results of clinical trials in MS documented progression-free survival at 3 or 5 years in about 75% of patients (Fassas et al., 2000; Fassas and Kazis, 2003; Tyndall and Saccardi, 2005).

## 2.3  Potential Common Mechanisms of HSCT in Various Autoimmune Diseases

Is it meaningful to review the immunological studies performed in clinical trials for various autoimmune diseases, such MS, SLE, RA, and SS, and speculate on possible shared mechanisms of HSCT across these disorders? There are no strong reasons to expect that the HSCT procedure per se exert differential actions on the immune system of individuals with different autoimmune diseases. In particular, to understand the mode of action of HSCT, we should not necessarily expect to see a selective effect on organ autoantigen-specific immune responses (Muraro and Douek, 2006; Sykes and Nikolic, 2005).

Evidence of amelioration of the course of various autoimmune diseases from clinical trials of HSCT suggests that more general effects on the immune system are required to explain its clinical effectiveness. The underlying autoimmune disease, however, can and most likely does involve different abnormalities of immune function. In addition, a variety of immunosuppressive and immunomodulatory agents and schemes are employed to treat autoimmune disorders. Therefore, both the disease process and previously received treatments may introduce alterations in the baseline (i.e., pre-HSCT) composition and function of the patient's immune system. It has been suggested that these changes may determine differences in the modalities of immune recovery after HSCT (Isaacs and Thiel, 2004). Aside from this potential variance, there is consolidated evidence that the HSCT scheme, particularly the conditioning (immuno-depleting) regimen, the type of hematopoietic graft (e.g., enriched in hematopoietic stem cells or unmanipulated), and the patient age are the major determinants of the kinetics and qualities of immune reconstitution.

## 2.4    Complexity of Correlating Mode of Action of HSCT With Clinical Results

Heterogeneity is an increasingly recognized feature of the pathogenesis of systemic (e.g., SLE, SS) and even organ-specific (e.g., MS) autoimmune diseases (Croker and Kimberly, 2005; Morales et al., 2006) and can act as a confounding factor in the evaluation of efficacy based on clinical outcomes. Some of the heterogeneity is related to the existence of different phases and possibly subtypes of disease and to the different type and extent of organ damage that has been inflicted.

The importance of accumulated damage for clinical outcome has been demonstrated by experiences in an animal model of autoimmune disease. Data from experimental autoimmune encephalomyelitis (EAE), a model for MS, have shown that BMT before onset of EAE prevents occurrence of the disease. However, BMT administered after clinical onset arrests clinical progression of disease but does not reverse the neurological deficit (Burt et al., 1995, 1998a; Karussis et al., 1993; van Bekkum, 2000; van Gelder et al., 1993). Therefore, it appears likely that when the degree of tissue injury has overcome the capacity for spontaneous repair and functional recovery, the clinical status of the affected subject cannot be significantly improved by HSCT.

Furthermore, with severe autoimmune disease, organ or tissue dysfunction may start accumulating progressively even independent of an inflammatory attack owing to a variety of factors including vascular damage, cellular dystrophy, and damage and fibrosis of organs and support tissues (astrogliosis in the nervous system). For example, in MS, neurodegenerative processes are thought to play a major role in the progressive accumulation of disability observed in advanced stages of disease (Zamvil and Steinman, 2003). Progressive MS is characterized pathologically by global white matter injury, which may depend on low-grade diffuse inflammation, probably sustained by central nervous system (CNS) resident cells (microglia) rather than peripheral immune cells (autoreactive lymphocytes) such as in acute focal demyelinating lesions (Kutzelnigg et al., 2005). Whatever the mechanisms, it is highly unlikely that advanced neural degeneration involving significant axonal loss may be ameliorated by immunotherapy, including HSCT. It has been argued that toxicities from high-dose conditioning regimens, particularly if employing radiation, may even worsen the degenerative progression in MS either by inflicting direct organ damage or by preventing spontaneous repair (Burt et al., 2003). Furthermore, some studies have shown beneficial effects on the course of EAE from the administration of neurotrophic growth factors that can be produced by immune cells, leading to the hypothesis, still unproven in humans, that inflammation may have a neuroprotective role (reviewed in (Kerschensteiner et al., 2003).

Investigators should be aware of these complexities when evaluating correlations of mechanistic immunological studies with clinical outcomes in HSCT trials. Clearly, the definition of success or failure in clinical trials depends to some degree on arbitrary stipulation of the outcomes. As mentioned before, heterogeneous pathogenic mechanisms contribute to determining the course of autoimmune disease. Some of the pathogenic components may be unrelated to inflammation (hence likely unaffected by immunotherapeutic interventions) and yet could significantly influence the clinical outcomes. A possible consequent pitfall is assuming failure of HSCT in the presence of clinical progression even when it cannot be attributed to persistence or exacerbation of inflammatory disease activity. An example of such misinterpretation in MS trials is concluding that HSCT failed to control autoimmunity in patients with advanced disease at baseline who continue to progress slowly in their disability despite the absence of new inflammatory lesions in the brain or clinical exacerbations. In these circumstances, it cannot be assumed that immune abnormalities are responsible for clinical progression; therefore, no conclusion can be drawn about the effectiveness of HSCT to resolve or counteract these abnormalities. The only conclusion that could be safely made from such results is that there was no evidence of clinical benefit from HSCT in those patients.

For all these reasons, to understand the mechanisms of HSCT in autoimmune disease, we need to study patients for whom there are measurable outcomes that can be clearly related to inflammatory disease activity. This, then, allows us to determine the success or failure of the therapy.

# 3   Immunological Results from Studies on HSCT for Autoimmune Disease

## 3.1   Quantitative Reconstitution of Immune Cells

Basic cell counts and immunophenotyping data have been reported in most clinical trials of HSCT for autoimmune disorders, defining the characteristic of numerical recovery after transplant. A consistent finding across studies in adults has been the rapid and complete posttransplant reconstitution of B cell, natural killer (NK) cell, and CD8+ T cell compartments at levels comparable to those before transplant but a delayed and often incomplete recovery of CD4+ T cell numbers for up to 2 years after HSCT (Burt et al., 1998c; Carreras et al., 2003; Farge et al., 2005; Fassas et al., 1997, 2002; Kozak et al., 2000; Muraro et al., 2005; Nash et al., 2003; Passweg et al., 2004; Saccardi et al., 2005; Sun et al., 2004; Verburg et al., 2001). In adult patients, repletion of the CD4 subset was observed only when follow-up was extended to $\geq 2$ years (Muraro et al., 2005; Passweg et al., 2004) (Fig. 20.2). A study in children with refractory juvenile idiopathic arthritis has shown CD4+ recovery in 12 months (Wulffraat et al., 2004). These studies essentially replicated the previous findings in hematological and solid malignancies (Guillaume et al., 1998), indicating that the kinetics of reconstitution of major lymphocyte subsets in autoimmune disease are similar.

## 3.2   Naive and Memory T Cell Subpopulations and Thymopoiesis

A few trials of HSCT in autoimmune disease have enumerated naive and memory T cell subpopulations in patients longitudinally. A paradoxical increase of

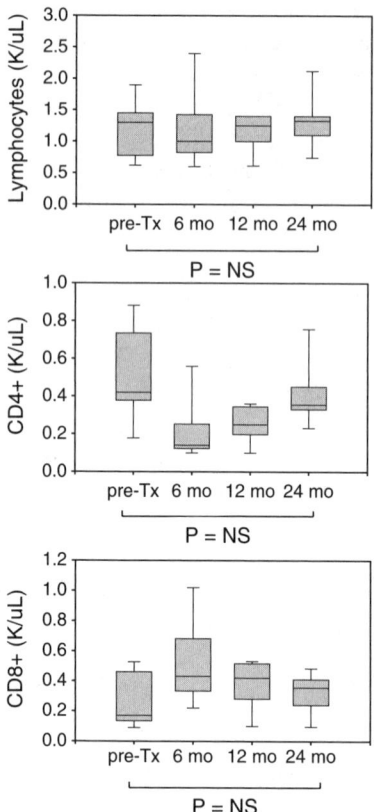

FIGURE 20.2. Reconstitution of absolute lymphocyte (top), CD4+ (middle), and CD8+ (bottom) counts in seven patients with multiple sclerosis (MS) after HSCT. The initial recovery of total lymphocyte count was mediated by CD8+ cells. Repletion of the CD4+ cell subset was slower, requiring 2 years. NS = not statistically significant.

T cells with memory (CD45RO+) phenotype has repeatedly been observed during the early phases (up to 6 months) after transplant (Muraro et al., 2005; Rosen et al., 2000; Saccardi et al., 2005; Wulffraat et al., 2004). In an inverse relation to memory cells, the naive T cell subset (CD45RA+) was depleted for up to 1 year after transplant (Carreras et al., 2003; Farge et al., 2005; Kozak et al., 2000; Muraro et al., 2005; Saccardi et al., 2005; Verburg et al., 2001).

These data are consistent with results in hematological and solid organ malignancies (Koehne et al., 1997) and with the established notion of a delay in the thymus-dependent recovery of naive CD4+ cells (Guillaume et al., 1998). The age-dependent characteristics of thymic renewal have been recently described in depth (Hakim et al., 2005). However, the thymus may not be the exclusive source of naive cell recovery. The correlation found between the numbers of naive cells infused with hematopoietic allografts and the numbers of naive CD4+ cells found during the first 6 months of immune reconstitution has suggested that even for the naive arm mature T cells may have a role in early immune reconstitution (Storek et al., 2003). A fundamental limitation in the identification of naive and memory cells based on cell surface phenotypic markers is the process of homeostatic proliferation that occurs in a host with relative lymphopenia, such as in a patient during the first months after HSCT. Douek and I have discussed elsewhere the mechanisms of homeostatic proliferation during immune reconstitution and the implications for autoimmune disease (Muraro and Douek, 2006).

In our studies of HSCT in MS, we confirmed the previous findings of expansion of T cells displaying memory/effector phenotypes early after transplant; but having extended the duration of the posttherapy follow-up, we found significantly increased numbers of naive T cells after long-term immune reconstitution (Muraro et al., 2005) (Table 20.1). Although at present comparable data from studies with equally long follow-up are not available, it is also possible that the degree of naive cell repopulation may be different in various HSCT regimens. In fact, depletion of T cells in the graft through CD34+ selection appeared to be a factor delaying T cell reconstitution but promoting thymic regeneration in adults who have undergone total body irradiation (Malphettes et al., 2003). Recent thymic emigrant cells (RTEs) can be identified by a high content of intracytoplasmic T cell receptor excision circles (TRECs) and by the co-expression of CD45RA and CD31 (on CD4+ cells) on the cell surface. In our study, the correlation of an increase of naive cells with expansion of cells with RTE phenotype and high TREC levels has demonstrated the thymic origin of those cells (Muraro et al., 2005), answering a longstanding question on the origin of the late-arriving CD4+ CD45RA+ "naive" phenotype peripheral blood T cells (Haynes et al., 2000). Interestingly, better initial recovery of TRECs was observed at 1 year after HSCT in a subgroup of patients with SS who showed sustained clinical responses compared to the subgroup who had a poor clinical response (Farge et al., 2005). It is important to determine if those differences are sustained over a longer period of time.

TABLE 20.1. Increased frequency of naive T cells before after long-term immune reconstitution following HSCT in patients with MS.

| T cell subset | %<br>Pre-HSCT<br>(mean ± SD)[1] | % At 2-year<br>follow-up<br>(mean ± SD) | Absolute<br>% change | Relative<br>% change | $p$ |
|---|---|---|---|---|---|
| CD4 T naive | 19.3 ± 15.9 | 42.0 ± 24.4 | +22.7% | +118% | 0.032 |
| CD4 T central memory | 67.1 ± 9.0 | 41.4 ± 11.1 | −25.7% | −38% | 0.008 |
| CD4 T effector memory | 13.2 ± 7.8 | 15.9 ± 16.6 | +2.7% | +21% | NS (a = 0.75) |
| CD8 T naive | 45.2 ± 21.5 | 46.5 ± 23.9 | +1.3% | +3% | NS (a = 0.26) |
| CD8 T central memory | 31.7 ± 12.1 | 21.6 ± 8.7 | −10.1% | −32% | NS (a = 0.26) |
| CD8 T effector memory | 16.7 ± 9.9 | 19.4 ± 9.4 | +2.7% | +16% | NS (a = 0.05) |

[1]The values represent the means and standard deviations (SD) of the percentages of the indicated T cell subsets as determined by flow-cytometry in 7 patients with MS before and 2 years after HSCT. For methodological details see Muraro et al. 2005. Naïve CD4+ T cells increased >2-fold at 2-year follow-up. There was a contraction of the same magnitude of the central memory T cell subsets.

## 3.3  Cytokines and Other Soluble Mediators

Dysregulation of T helper (Th)1/Th2 balance is thought to be a mechanism of autoimmune disorders. For example, on the basis of data from experimental models and observations in patients, MS has been categorized as a Th1 disease and SLE as a Th2-like disease. In general, the cytokine balance shift is much less clear in the human diseases than in the related animal models, but some studies have tried to examine the effects of HSCT on Th balance.

Cytoplasmic expression of interleukin (IL)-4, a Th2 cytokine, was detected at elevated frequency in T cells before transplant and normalized after transplant in SLE patients (Traynor et al., 2000). Conversely, production of Th1 cytokine interferon-$\gamma$ (IFN$\gamma$) by stimulated T cells was lower than normal before transplantation and restored to normal levels after transplantation—in line with normalization of an imbalance favoring Th2 responses and theoretically T cell-dependent B cell hyperactivity in SLE (Traynor et al., 2000).

Serum levels of cytokines, including, tumor necrosis factor-$\alpha$ (TNF$\alpha$), IFN$\gamma$, interleukin (IL)-4, IL-10, and IL-12 were measured in patients with MS undergoing HSCT (Sun et al., 2004). In this study, transient elevations of IFN$\gamma$, TNF$\alpha$, and IL-10 were detected between 3 and 6 months after HSCT, whereas, more interestingly, serum IL-12 production decreased persistently in all patients.

Serum levels of matrix metalloproteinase-9 (MMP-9), whose levels are increased in MS patients compared to healthy subjects, were found to be significantly decreased for up to 3 years after HSCT (Blanco et al., 2004). There was a strong correlation of MMP-9 mRNA levels with the CD4+ T cell counts, suggesting that decreased MMP-9 production was a consequence of the reduced number of CD4+ T cells after transplant. The authors concluded that it remained to be determined whether the reduction in proteolytic load contributed to the reduced disease activity observed in the patients or merely reflected hematological effects of HSCT.

Overall, the available data do not elucidate whether global effects on cytokines or other soluble factors that have been implicated in autoimmune disease can explain clinical remission after HSCT. The redundancy and complexity of the cytokine system and the incomplete adherence of human autoimmune diseases to the Th1/Th2 paradigm makes this question a challenging one to address. T cell cytokine production in response to specific antigens can be more informative and is discussed in the next section.

## 3.4  T Cell Responses to Mitogens and Autoantigens

An early report showed suppressed proliferation of peripheral blood mononuclear cells (PBMCs) to phytohemagglutinin and, to a lesser extent, to IL-2/anti-CD3 stimulation after 30, 90, and 180 days after HSCT (Burt et al., 1998b). Because CD4+ T cells sustain most of the peripheral blood proliferation response to mitogens, these results may simply reflect decreased CD4+ cell numbers during the first year after therapy. As a measure of cell activation, Traynor et al. measured the percentage of cells expressing CD69 following activation with phorbol myristate acetate (PMA) (Traynor et al., 2000). The unusually high percentages of CD69+ T cells found in resting PBMCs from the four patients with SLE studied were corrected after transplantation, and the ability of T cells to upregulate CD69 after PMA activation was preserved.

Another early report of immunological monitoring after HSCT in MS included evaluation of lymphocyte proliferation to myelin antigens (performed in three patients) and of cytokine production (assessed by ELISPOT in one patient) (Openshaw et al., 2000). Proliferative responses (primary proliferation with IL-2 addition to boost weak responses) were assessed in response to a panel of 14 MBP peptides and a set of five immunodominant myelin protein peptides (MBP 84-106, MBP 142-163, PLP 104-117, PLP 142-153, MOG 42-53) and were suppressed after HSCT. Modest positive proliferative responses were found in one patient 8 months after HSCT (to MBP 1-12) and 20 months after HSCT (to PLP 104-117) The authors hypothesized a relation of this finding with progression of clinical disability, observed at months 15 to 17 after HSCT (Openshaw et al., 2000). More recently, Sun and colleagues estimated by ELISPOT for IFNγ the precursor frequency of MBP-reactive T cells in four MS patients before and after HSCT (Sun et al., 2004). They observed prominent suppression of MBP-specific T cell frequency during the first 9 months after therapy, but the frequency returned to near, or above, baseline levels at the 12-month follow-up. The authors' interpretation suggested that the T cell repertoire that initially expanded from memory T cells does not contain sufficient MBP-reactive T cells to mount T cell responses to MBP. Further investigation into the epitope recognition patterns was carried out using MBP overlapping peptides; it showed that a more heterogeneous pattern of reactivity to MBP was reconstituted at 12 months. Preexisting dominant responses to specific peptides in individual patients were also reconstituted, albeit at reduced frequency; and new epitope specificities emerged. These changes resulted in a broader repertoire and a different hierarchy of immunodominant peptides of MBP than was present before transplantation.

Taken together, these important results indicated that the responses to MBP inducing IFNγ secretion followed the kinetics of the repletion of the CD4+ T cell repertoire, which is limited during the first year after HSCT. Recovery of MBP reactivity at the 12-month follow-up in the absence of new disease activity is not too surprising, as MBP-specific T cells can be found in healthy individuals at a frequency comparable to that in MS patients. Whether a different distribution of MBP recognition patterns or other features of the reconstituted MBP-specific repertoire (e.g., secretion of other cytokines or chemokines, avidity/affinity for antigen/MHC) can explain tolerance to myelin requires further investigation. Alternatively, the data may be seen as evidence that MBP-specific T cells are not involved in the pathogenesis of MS. An additional possibility that currently receives a great deal of attention in the field of autoimmunity is that tolerance is not maintained by an absence of autoreactive cells but by the presence of normal regulatory cell function (Shevach et al., 2001).

## 3.5   T Cell Receptor Repertoire

Although HSCT has been performed for decades in thousands of patients for hematological malignancies, the reconstitution of T cell antigen receptor (TCR) repertoire has received relatively little attention. Measurement of chimerism has covered the need to quantify the contribution of the donor's immune cells in allogeneic HSCT. Research efforts have mainly focused on measuring the diversity of the TCR repertoire as a correlate of recovery of immune competence and on identifying the possible association of clonal expansions with graft-versus-host disease after allogeneic transplantation. In autologous transplantation, the possibility of distinguishing reconstitution from mature cells "surviving" in the host or reintroduced via the graft from new immune cells relies on the ability to compare the clonal composition of the pretransplant with the posttransplant TCR repertoire. This is a difficult task given the complexity of the human TCR repertoire (Nikolich-Zugich et al., 2004).

Only recently improved availability of reagents and methods to assess the TCR repertoire and an increased interest in immune reconstitution have fueled research in this area. Traynor and colleagues (2000) first observed increased polyclonality of the TCR repertoire following HSCT in three patients with SLE, as assessed by CDR3-length spectratyping (Gorski et al., 1994). This study provided preliminary evidence of recovery of TCR repertoire diversity after HSCT. To start exploring the regeneration of the T repertoire at the single clone level, a small number of TCR clones were sequenced in patients who received who received unselected autologous HSCT for malignancies (Protheroe et al., 2000). This report described the persistence after therapy of one preexisting clonal expansion in each of the three patients examined, suggesting that mature T cells participated in regeneration of the T cell repertoire. The origin of the maintained expanded clonotypes was not addressed, but the infusion of respectable doses of unselected hematopoietic stem cells ($2 \times 10^6$/kg of unselected CD34+ cells) pointed to a possible reintroduction of clones from the T cells present in the hematopoietic graft. Yet reconstitution of the clonal repertoire after CD34+ selected autologous HSCT remains poorly understood.

More detailed studies of T cell repertoire reconstitution after high-dose treatment and HSCT for cancer and for autoimmune diseases have now been published (Hakim et al., 2005; Muraro et al., 2005; Sun et al., 2004). Sun and colleagues examined the evolution of the TCRBV repertoire in PBMCs by quantitative polymerase chain reaction (PCR) and *immunoscope* (CDR3 spectratyping) in four patients with MS who underwent autologous HSCT (Sun et al., 2004). The study identified selective expansion of T cell clones that shared BV20 as a common variable region gene during the first 3 to 9 months of follow-up. At 12 months after therapy, a broader TCRBV gene usage repertoire was restored, suggesting reconstitution of a more heterogeneous T cell repertoire.

My colleagues and I have examined reconstitution of the TCR repertoire in the CD4 and CD8 subsets using a combination of flow cytometry, CDR3 spectratyping, and sequencing-based clonotyping (Fig. 20.3). Our study confirmed the

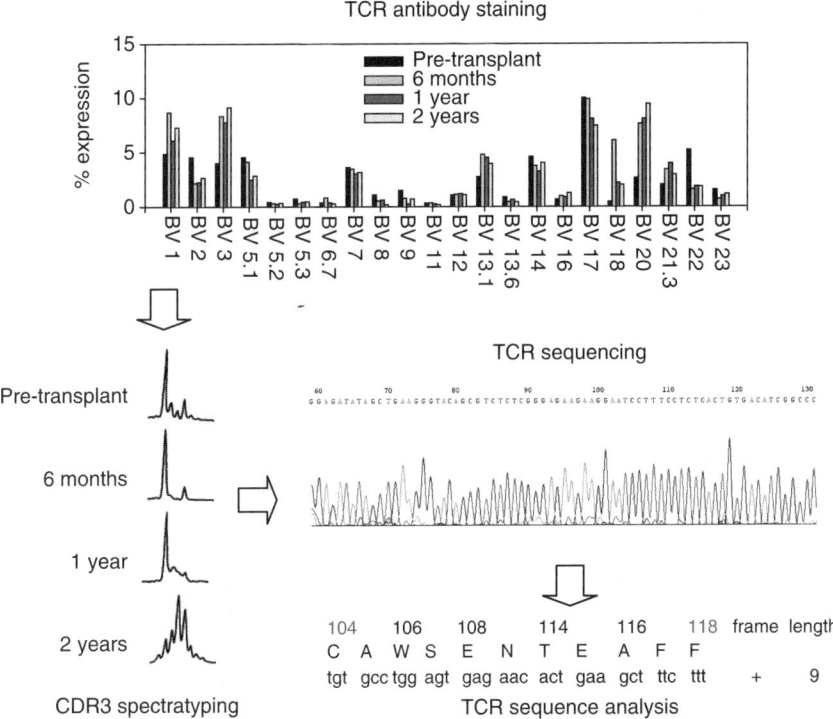

FIGURE 20.3. Analysis of the T cell receptor (TCR) repertoire in immune reconstitution after HSCT. We employed a combination of staining with TCR Vβ-specific antibodies and flow cytometry (top), CDR3 spectra typing (bottom left), and TCR sequencing analysis (bottom right). The combined use of these techniques allowed us to describe recovery of T cell repertoire diversity and substantial clonal renewal in patients with MS after HSCT (Muraro et al., 2005).

initial evidence from other studies that incomplete reconstitution of diversity and oligoclonal expansions (particularly in the CD8 subset) predominate during the first year after HSCT (Muraro et al., 2005). However, in conjunction with the recovery and increase of thymic output during the second year after transplant, we found the T cell repertoire diversity to be enhanced, leading to normalized distribution of CDR3-length spectratypes. In addition and perhaps most importantly, we showed through sequencing-based clonotyping that most (typically > 90%) T cell clones molecularly identified at baseline were replaced by new clones at 2 years after transplant. Although some preexisting clones persisted after transplant, sometimes at expanded frequency, their degree of expansion was usually reduced, and other newly appeared clones became dominant. Our non-antigen-biased read-out is in remarkable analogy with the data by the Zhang group, who focused on T cell responses to MBP peptides (Sun et al., 2004), and further suggests that the effects of HSCT on the patterns of immunodominance may not be limited to MBP-specific cells but may be a more global phenomenon.

## 3.6    B Cells, Immunoglobulins, and Autoantibodies

B cells and autoantibodies have received the most attention in HSCT trials for SLE, based on the generally accepted notion that lupus is an autoimmune disease of B cell hyperactivity (Lipsky, 2001). In the largest single-center study of non-myeloablative HSCT for SLE published so far, anti-nuclear (ANA) and anti-double stranded (ds)DNA autoantibodies were suppressed in most of the patients following HSCT, in line with previous studies utilizing more intense conditioning regimens (Burt et al., 2006). Reduced C3, an index of complement consumption in the disease process, tended to normalize after HSCT in this study. Levels of lupus anticoagulant and anti-cardiolipin antibodies also improved after HSCT (Burt et al., 2006; Statkute et al., 2005). Overall, these results indicate that HSCT may exert profound effects on abnormalities of B cell function and autoantibody production in SLE.

In the recent French study on SS, peripheral blood (CD19+ and CD20+) B cells were reduced for up to 9 months after transplant in the group of clinical responders but not in the nonresponder patient group, who showed full reconstitution as early as the 3-month follow-up (Farge et al., 2005). There was a modest but significant positive correlation between B cell counts and the presence of high levels of anti-topoisomerase I (Anti-Scl-70) antibodies, which in fact were suppressed to some extent only in the responder patient group.

Serum levels of immunoglobulins (Igs) G, IgM, IgA, and IgE examined in MS patients before HSCT and at 3-month intervals after therapy did not change significantly from baseline (Sun et al., 2004). Because MS is a CNS disorder, however, immunoglobulins produced intrathecally (i.e., inside the blood-brain barrier) may be more significant. Cerebrospinal fluid (CSF) oligoclonal IgG bands (OBs) are a nonspecific but characteristic finding in MS. A decrease in the CSF IgG synthesis rate has been observed after HSCT in one report (Healey et al., 2004). However, in most studies, including the one by Healey et al., oligoclonal banding has been shown to persist at the 1-year follow-up after HSCT (Carreras et al., 2003; Healey et al., 2004; Openshaw et al., 2000; Saiz et al., 2001). In the study by Sun and colleagues, CSF OBs decreased in number in two of four patients, disappeared in one patient, and persisted in one patient (Sun et al., 2004). The study by Saiz et al. provides an interesting comparison of pre- and posttransplant CSF OB in four patients (Saiz et al., 2001). In three of the patients new OBs appeared in the serum at month 3 after HSCT, many of which were also present in the CSF. The same CSF OBs that were present at baseline were found at 1 year after transplant. In addition, some of the OBs that had appeared at 3 months after transplant persisted at the 1-year follow-up. Given the long half-life of autoantibodies and plasma cells, reassessment after a longer follow-up is required to understand the evolution of CSF OBs after HSCT. Because CSF OBs are not a surrogate marker of disease activity, however, and extensive attempts have failed to establish involvement in the pathogenic process of MS, the interpretation of their disappearance or persistence even after long-term follow-up after transplant requires caution.

Overall, there is still much to be learned on B cell reconstitution and its possible role in the mechanisms of action of HSCT for autoimmune diseases. An ongoing National Insitutes of Health (NIH) protocol of reduced intensity immunodepletion in active and severe SLE that utilizes rituximab, a B cell-depleting therapeutic antibody (http://clinicalstudies.info.nih.gov/detail/A_2004-C-0095.html/; Muraro et al., 2006), offers the opportunity to further our understanding of this area.

## 4    Conclusions

There is now robust evidence that HSCT does not act in a purely quantitative way (i.e., through immunosuppression), but substantial qualitative changes in the immune system are regenerated after transplantation. Transient numerical depletion of immune cells does not explain the long-term clinical remissions that have extended up to several years after HSCT in the absence of maintenance immunotherapy (Tyndall and Saccardi, 2005). The sustained clinical effects are better explained by qualitative changes in the reconstituted immune repertoire. Substantial posttransplant modifications have been shown for the adaptive immune system. More data are available to support the notion of regeneration of T cell immunity, including changes in the proportions of T cell subsets, modification of antigen-specific T cell responses, and generation of a different and more diverse TCR repertoire after transplant. Ongoing and future work may reveal similar aspects of immune renewal for B cells, which so far has been less extensively investigated partly owing to the emphasis of T cell immunity in studies of MS, which has been the frontline of HSCT in autoimmune disease. Further work is also required in the area of regulation of T cell and B cell immunity. The substantial progress made in the last few years has attracted a strong interest in integrating immunological studies with clinical trials of HSCT in autoimmune disease, so the field is now likely to evolve more rapidly. Exciting discoveries can be anticipated in the near future.

*Acknowledgments.* This work was supported by the National Institute of Neurological Disorders and Stroke Intramural Program. The author thanks Roland Martin, Richard Burt, Danny Douek, Ron Gress, Henry McFarland, and Steve Pavletic for their contributions to and support of research in this field. In the loving memory of my father, Fortunato Pietro Muraro (1928–2006).

## *References*

Blanco, Y., Saiz, A., Carreras, E., Graus, F. (2004) Changes of matrix metalloproteinase-9 and its tissue inhibitor (TIMP-1) after autologous hematopoietic stem cell transplantation in multiple sclerosis. *J. Neuroimmunol.* 153:190-194.

Burnet, F. M. (1959) The Clonal Selection Theory of Acquired Immunity. Cambridge University Press, Cambridge.

Burt, R. K., Burns, W., Ruvolo, P., et al. (1995) Syngeneic bone marrow transplantation eliminates V beta 8.2 T lymphocytes from the spinal cord of Lewis rats with experimental allergic encephalomyelitis. *J. Neurosci. Res.* 41:526-531.

Burt, R. K., Padilla, J., Begolka, W. S., et al. (1998a) Effect of disease stage on clinical outcome after syngeneic bone marrow transplantation for relapsing experimental autoimmune encephalomyelitis. *Blood* 91:2609-2616.

Burt, R. K., Traynor, A. E., Cohen, B., et al. (1998b) T cell-depleted autologous hematopoietic stem cell transplantation for multiple sclerosis: report on the first three patients. *Bone Marrow Transplant.* 21:537-541.

Burt, R. K., Traynor, A. E., Pope, R., et al. (1998c) Treatment of autoimmune disease by intense immunosuppressive conditioning and autologous hematopoietic stem cell transplantation. *Blood* 92:3505-3514.

Burt, R. K., Cohen, B. A., Russell, E., et al. (2003) Hematopoietic stem cell transplantation for progressive multiple sclerosis: failure of a total body irradiation-based conditioning regimen to prevent disease progression in patients with high disability scores. *Blood* 102:2373-2378.

Burt, R. K., Traynor, A., Statkute, L., et al. (2006) Nonmyeloablative hematopoietic stem cell transplantation for systemic lupus erythematosus. *J.A.M.A.* 295:527-535.

Carreras, E., Saiz, A., Marin, P., et al. (2003) CD34+ selected autologous peripheral blood stem cell transplantation for multiple sclerosis: report of toxicity and treatment results at one year of follow-up in 15 patients. *Haematologica* 88:306-314.

Croker, J. A., Kimberly, R. P. (2005) SLE: challenges and candidates in human disease. *Trends Immunol.* 26:580-586.

Farge, D., Henegar, C., Carmagnat, M., et al. (2005) Analysis of immune reconstitution after autologous bone marrow transplantation in systemic sclerosis. *Arthritis Rheum.* 52:1555-1563.

Fassas, A., Kazis, A. (2003) High-dose immunosuppression and autologous hematopoietic stem cell rescue for severe multiple sclerosis. *J. Hematother. Stem Cell Res.* 12:701-711.

Fassas, A., Kimiskidis, V. K. (2004) Autologous hemopoietic stem cell transplantation in the treatment of multiple sclerosis: rationale and clinical experience. *J. Neurol. Sci.* 223:53-58.

Fassas, A., Anagnostopoulos, A., Kazis, A., et al. (1997) Peripheral blood stem cell transplantation in the treatment of progressive multiple sclerosis: first results of a pilot study. *Bone Marrow Transplant.* 20:631-638.

Fassas, A., Anagnostopoulos, A., Kazis, A., et al. (2000) Autologous stem cell transplantation in progressive multiple sclerosis—an interim analysis of efficacy. *J. Clin. Immunol.* 20:24-30.

Fassas, A., Passweg, J. R., Anagnostopoulos, A., et al. (2002) Hematopoietic stem cell transplantation for multiple sclerosis: a retrospective multicenter study. *J. Neurol.* 249:1088-1097.

Gorski, J., Yassai, M., Zhu, X., et al. (1994) Circulating T cell repertoire complexity in normal individuals and bone marrow recipients analyzed by CDR3 size spectratyping: correlation with immune status. *J. Immunol.* 152:5109-5119.

Guillaume, T., Rubinstein, D. B., Symann, M. (1998) Immune reconstitution and immunotherapy after autologous hematopoietic stem cell transplantation. *Blood* 92: 1471-1490.

Hakim, F. T., Memon, S. A., Cepeda, R., et al. (2005) Age-dependent incidence, time course, and consequences of thymic renewal in adults. *J. Clin. Invest.* 115:930-939.

Haynes, B. F., Markert, M. L., Sempowski, G. D., et al. (2000) The role of the thymus in immune reconstitution in aging, bone marrow transplantation, and HIV-1 infection. *Annu. Rev. Immunol.* 18:529-560.

Healey, K. M., Pavletic, S. Z., Al-Omaishi, J., et al. (2004) Discordant functional and inflammatory parameters in multiple sclerosis patients after autologous haematopoietic stem cell transplantation. *Mult. Scler.* 10:284-289.

Isaacs, J. D., Thiel, A. (2004) Stem cell transplantation for autoimmune disorders: immune reconstitution. *Best Pract. Res. Clin. Haematol.* 17:345-358.

Karussis, D. M., Vourka-Karussis, U., Lehmann, D., et al. (1993) Prevention and reversal of adoptively transferred, chronic relapsing experimental autoimmune encephalomyelitis with a single high dose cytoreductive treatment followed by syngeneic bone marrow transplantation. *J. Clin. Invest.* 92:765-772.

Kerschensteiner, M., Stadelmann, C., Dechant, G., et al. (2003) Neurotrophic cross-talk between the nervous and immune systems: implications for neurological diseases. *Ann. Neurol.* 53:292-304.

Koehne, G., Zeller, W., Stockschlaeder, M., Zander, A. R. (1997) Phenotype of lympho-cyte subsets after autologous peripheral blood stem cell transplantation. *Bone Marrow Transplant.* 19:149-156.

Kozak, T., Havrdova, E., Pit'ha, J., et al. (2000) High-dose immunosuppressive therapy with PBPC support in the treatment of poor risk multiple sclerosis. *Bone Marrow Transplant.* 25:525-531.

Kutzelnigg, A., Lucchinetti, C. F., Stadelmann, C., et al. (2005) Cortical demyelination and diffuse white matter injury in multiple sclerosis. *Brain* 128:2705-2712.

Lehmann, P. V., Sercarz, E. E., Forsthuber, T., et al. (1993) Determinant spreading and the dynamics of the autoimmune T-cell repertoire. *Immunol. Today* 14:203-208.

Lipsky, P. E. (2001) Systemic lupus erythematosus: an autoimmune disease of B cell hyperactivity. *Nat. Immunol.* 2:764-766.

Malphettes, M., Carcelain, G., Saint-Mezard, P., et al. (2003) Evidence for naive T-cell repopulation despite thymus irradiation after autologous transplantation in adults with multiple myeloma: role of ex vivo CD34+ selection and age. *Blood* 101:1891-1897.

Marmont, A. M. (2004) Stem cell transplantation for autoimmune disorders: coincidental autoimmune disease in patients transplanted for conventional indications. *Best Pract. Res. Clin. Haematol.* 17:223-232.

Marrie, R. A. (2004) Environmental risk factors in multiple sclerosis aetiology. *Lancet Neurol.* 3:709-718.

Matzinger, P. (1994) Tolerance, danger, and the extended family. *Annu. Rev. Immunol.* 12:991-1045.

Morales, Y., Parisi, J. E., Lucchinetti, C. F. (2006) The pathology of multiple sclerosis: evi-dence for heterogeneity. *Adv. Neurol.* 98:27-45.

Muraro P. A., Nikolov N. P., Butman J. A., et al. Granulocytic invasion of the central nerv-ous system after hematopoietic stem cell transplantation for systemic lupus erythe-matosus. *Haematologica* 2006 June; 91(6 suppl):ECR21.

Muraro, P. A., Douek, D. C. (2006) Renewing the T cell repertoire to arrest autoimmune aggression. *Trends Immunol.* 2006 Feb; 27(2):61-67.

Muraro, P. A., Vergelli, M., Kalbus, M., et al. (1997) Immunodominance of a low-affinity MHC binding myelin basic protein epitope (residues 111-129) in HLA-DR4 (B1*0401)

Feb subjects is associated with a restricted T cell receptor repertoire. *J. Clin. Invest.* 100:339-349.

Muraro, P. A., Douek, D. C., Packer, A., et al. (2005) Thymic output generates a new and diverse TCR repertoire after autologous stem cell transplantation in multiple sclerosis patients. *J. Exp. Med.* 201:805-816.

Nash, R. A., Bowen, J. D., McSweeney, P. A., et al. (2003) High-dose immunosuppressive therapy and autologous peripheral blood stem cell transplantation for severe multiple sclerosis. *Blood* 102:2364-2372.

Nikolich-Zugich, J., Slifka, M. K., Messaoudi, I. (2004) The many important facets of T-cell repertoire diversity. *Nat. Rev. Immunol.* 4:123-132.

Openshaw, H., Lund, B. T., Kashyap, A., et al. (2000) Peripheral blood stem cell transplantation in multiple sclerosis with busulfan and cyclophosphamide conditioning: report of toxicity and immunological monitoring. *Biol. Blood Marrow Transplant.* 6:563-575.

Passweg, J. R., Rabusin, M., Musso, M., et al. (2004) Haematopoetic stem cell transplantation for refractory autoimmune cytopenia. *Br. J. Haematol.* 125:749-755.

Pette, M., Fujita, K., Wilkinson, D., et al. (1990) Myelin autoreactivity in multiple sclerosis: recognition of myelin basic protein in the context of HLA-DR2 products by T lymphocytes of multiple-sclerosis patients and healthy donors. *Proc. Natl. Acad. Sci. USA* 87:7968-7972.

Protheroe, A. S., Pickard, C., Johnson, P. W., et al. (2000) Persistence of clonal T-cell expansions following high-dose chemotherapy and autologous peripheral blood progenitor cell rescue. *Br. J. Haematol.* 111:766-773.

Ristori, G., Cannoni, S., Stazi, M. A., et al. (2005) Multiple sclerosis in twins from continental Italy and Sardinia: a nationwide study. *Ann. Neurol.* 20:20.

Rosen, O., Thiel, A., Massenkeil, G., et al. (2000) Autologous stem-cell transplantation in refractory autoimmune diseases after in vivo immunoablation and ex vivo depletion of mononuclear cells. *Arthritis Res.* 2:327-336.

Saccardi, R., Mancardi, G. L., Solari, A., et al. (2005) Autologous HSCT for severe progressive multiple sclerosis in a multicenter trial: impact on disease activity and quality of life. *Blood* 105:2601-2607.

Saiz, A., Carreras, E., Berenguer, J., et al. (2001) MRI and CSF oligoclonal bands after autologous hematopoietic stem cell transplantation in MS. *Neurology* 56:1084-1089.

Shevach, E. M., McHugh, R. S., Thornton, A. M., et al. (2001) Control of autoimmunity by regulatory T cells. *Adv. Exp. Med. Biol.* 490:21-32.

Shlomchik, M. J., Craft, J. E., Mamula, M. J. (2001) From T to B and back again: positive feedback in systemic autoimmune disease. *Nat. Rev. Immunol.* 1:147-153.

Sospedra, M., Martin, R. (2005) Immunology of multiple sclerosis. *Annu. Rev. Immunol.* 23:683-747.

Statkute, L., Traynor, A., Oyama, Y., et al. (2005) Antiphospholipid syndrome in patients with systemic lupus erythematosus treated by autologous hematopoietic stem cell transplantation. *Blood* 106:2700-2709.

Storek, J., Dawson, M. A., Maloney, D. G. (2003) Correlation between the numbers of naive T cells infused with blood stem cell allografts and the counts of naive T cells after transplantation. *Biol. Blood Marrow Transplant.* 9:781-784.

Sun, W., Popat, U., Hutton, G., et al. (2004) Characteristics of T-cell receptor repertoire and myelin-reactive T cells reconstituted from autologous haematopoietic stem-cell grafts in multiple sclerosis. *Brain* 127:996-1008.

Sykes, M., Nikolic, B. (2005) Treatment of severe autoimmune disease by stem-cell transplantation. *Nature* 435:620-627.

Traynor, A. E., Schroeder, J., Rosa, R. M., et al. (2000) Treatment of severe systemic lupus erythematosus with high-dose chemotherapy and haemopoietic stem-cell transplantation: a phase I study. *Lancet* 356:701-707.

Tyndall, A., Saccardi, R. (2005) Haematopoietic stem cell transplantation in the treatment of severe autoimmune disease: results from phase I/II studies, prospective randomized trials and future directions. *Clin. Exp. Immunol.* 141:1-9.

Van Bekkum, D. W. (2000) Stem cell transplantation in experimental models of autoimmune disease. *J. Clin. Immunol.* 20:10-16.

Van Bekkum, D. W., Bohre, E. P., Houben, P. F., Knaan-Shanzer, S. (1989) Regression of adjuvant-induced arthritis in rats following bone marrow transplantation. *Proc. Natl. Acad. Sci. USA* 86:10090-10094.

Van Gelder, M., Kinwel-Bohre, E. P., van Bekkum, D. W. (1993) Treatment of experimental allergic encephalomyelitis in rats with total body irradiation and syngeneic BMT. *Bone Marrow Transplant.* 11:233-241.

Verburg, R. J., Kruize, A. A., van den Hoogen, F. H., et al. (2001) High-dose chemotherapy and autologous hematopoietic stem cell transplantation in patients with rheumatoid arthritis: results of an open study to assess feasibility, safety, and efficacy. *Arthritis Rheum.* 44:754-760.

Vergelli, M., Kalbus, M., Rojo, S. C., et al. (1997) T cell response to myelin basic protein in the context of the multiple sclerosis-associated HLA-DR15 haplotype: peptide binding, immunodominance and effector functions of T cells. *J. Neuroimmunol.* 77:195-203.

Willer, C. J., Dyment, D. A., Risch, N. J., et al. (2003) Twin concordance and sibling recurrence rates in multiple sclerosis. *Proc. Natl. Acad. Sci. USA* 100:12877-12882.

Wulffraat, N. M., de Kleer, I. M., Prakken, B. J., Kuis, W. (2004) Stem cell transplantation for autoimmune disorders: refractory juvenile idiopathic arthritis. *Best Pract. Res. Clin. Haematol.* 17:277-289.

Zamvil, S. S., Steinman, L. (2003) Diverse targets for intervention during inflammatory and neurodegenerative phases of multiple sclerosis. *Neuron* 38:685-688.

# 21
# Cell Transplantation of Peripherally Derived Adult Cells for Promoting Recovery from CNS Injury

Christine Radtke, Peter M. Vogt, and Jeffery D. Kocsis

## 1   Introduction

Cell-based therapies are being considered in clinical trials for a number of neurological diseases including multiple sclerosis, spinal cord injury, Parkinson's disease, and stroke. The rationale is that cells could serve as replacement therapy, provide neuroprotection, or stimulate neurogenesis or angiogenesis by producing chemokines and neurotrophins. Although the use of embryonic stem cells shows promise because of the totipotency of these cells and the prospect of developing large numbers of cells, a number of other cells derived from adult peripheral tissues are also being actively investigated. These cells include Schwann cells from peripheral nerve, olfactory ensheathing cells (OECs) from the olfactory system, and stromal cells from bone marrow (mesenchymal stem cells, or MSCs). In principle, these cells could be derived autologously, expanded in culture, and used for cell-based therapies. This chapter reviews experimental work demonstrating the potential of peripherally derived cells as a tool for promoting recovery in CNS injury.

## 2   Bone Marrow-Derived Cells for Cell Therapy in CNS Disorders

Bone marrow-derived mesenchymal stem cells (BMSCs) are thought to represent a very small proportion of cells in the mononuclear population of bone marrow. As originally described by Friedenstein (1976), these cells grow to confluency in appropriate culture conditions as flattened fibroblast-like cells (Majumdar et al., 1998) and have been suggested to differentiate into bone, cartilage (Kobune et al., 2003), cardiac myocytes (Toma et al., 2002), and neurons and glia (Honma et al., 2005; Iihoshi et al., 2004; Kobune et al., 2003; Nomura et al., 2005; Prockop et al., 1997; Woodbury et al., 2000 ) in vitro and in vivo. MSCs prepared from human bone marrow have been used in clinical studies for metachromatic leukodystrophy and Hurler syndrome (Koc et al., 2002), graft-versus-host disease (Aggarwal and Pittenger, 2005), and stroke (Bang et al., 2005).

Under some conditions MSCs can fuse with other cells, thereby making the distinction between cell fusion and transdifferentiation difficult (Alvarez-Dolado et al., 2003; Castro et al., 2002). MSCs have been demonstrated to differentiate into hepatic cells (Lagasse et al., 2000), myocytes (Ferrari et al., 1998), and neurons and glia (Azizi et al., 1998; Woodbury et al., 2000) in culture. They have been shown to cross the blood-brain barrier upon intravenous infusion and to enter normal brain (Brazelton et al., 2000; Mezey et al., 2000). It has been shown that intravenous delivery results in "homing" of these cells to areas of demyelination in the spinal cord (Akiyama et al., 2002a). Systemic delivery of neurosphere-derived progenitors has been shown to home to CNS lesions in experimental autoimmune encephalitis (EAE) in the mouse and to abate disease progression (Plucchino et al., 2003, 2005). Chopp et al. (2000) demonstrated that intravenous infusion of MSCs has a similar effect.

It is also important to note that ectopic cells have not been found outside the lesion zone. One possibility for this apparent homing mechanism is that disruption of the blood-brain barrier by lesion induction and local inflammatory responses in the lesion zone provide a strong signaling mechanism to direct the transplanted cells to the injury site. Autologous MSCs can be harvested in large number, and the potential for systemic delivery and homing to multiple lesion sites offers considerable advantage as a potential therapeutic approach.

Akiyama et al. (2002b) separated bone marrow by growing plastic-adherent cells for several weeks in culture (i.e., stromal cells). This population of cells is enriched in MSCs and depleted of hematopoietic stem cells. To test the hypothesis that myelin-forming precursor cells were present in this stromal fraction, identified stromal cells were injected directly into the demyelinated spinal cord. The results indicated relatively extensive remyelination; and because green fluorescent protein (GFP)-expressing donor cells were used, the clear association of transplanted cells and myelination could be made (Akiyama et al., 2002b). Figure 21.1 shows the presence of GFP-MSCs in the demyelinated dorsal columns of the rat spinal cord 3 weeks after direct microinjection. Plastic semithin sections shows that the demyelinated axons were remyelinated (Fig. 21.1D,G). Moreover, GFP cells can be seen surrounding axon profiles, indicating that the injected MSCs were able to remyelinate the spinal cord axons (Fig. 21.1F). Intravenous delivery of these cells required a much larger number of cells to achieve the same density of remyelination (Inoue et al., 2003).

Identification and expansion of specific cell types in bone marrow responsible for this repair potential could allow minimalization of the number of cells needed

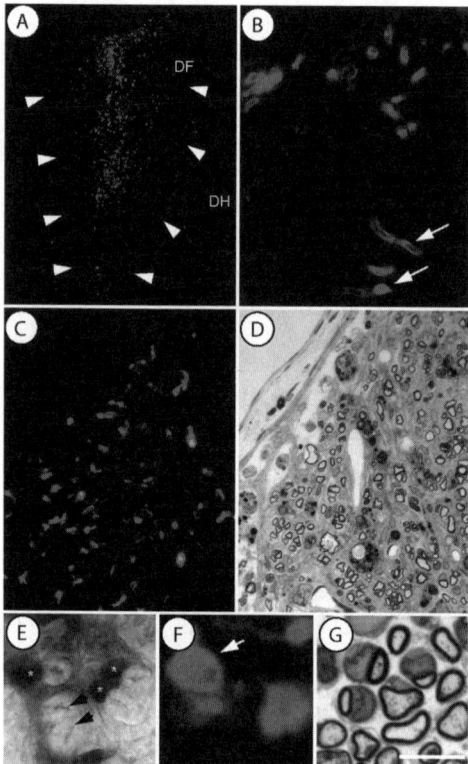

FIGURE 21.1. **A–G.** Identification of donor bone marrow cells in vivo. **A.** Dorsal funiculus of coronally cut spinal cord from a rat that was injected with bone marrow stromal cells from green fluorescent protein (GFP)-expressing mice 3 weeks after transplantation. *Arrows* indicate the lateral margins of the dorsal funiculus. Numerous GFP-positive cells are observed in the remyelinated region. **B.** Higher-power image of same field showing profiles reminiscent of myelinated axons. **C, D.** Frozen and plastic embedded sections, respectively, from the same animal showing co-localization of GFP fluorescence and more clearly defined myelination in the plastic section. **E, F.** H&E-stained frozen section and a fluorescent unstained image with GFP fluorescence at the same high power. Note that in the frozen H&E section the axon cylinder is collapsed (*arrows*) and the myelin is "puffy" as is typical with this staining technique. **G.** Comparable semithin plastic section from the same animal showing myelinated axons. The myelin is better preserved and the tissue more shrunken form dehydration protocols required for plastic embedding. Bar in **G** corresponds to: 250 μm in **A**, 50 μm in **B**, 40 μm in **C** and **D**, 12 μm in **E** and **F**, and 10 μm in **G**. (Modified from Akiyama et al., 2002b.)

for intravenous transfer and enhance the extent of remyelination. Moreover, a number of issues related to the mulitpotency of stem cells, including the possibility of ectopic cell generation, need to be further investigated. Recently a homogeneous population of defined human MSCs has been generated and immortalized using a telomerase gene (hTERT-MSCs) (Kobune et al., 2003).

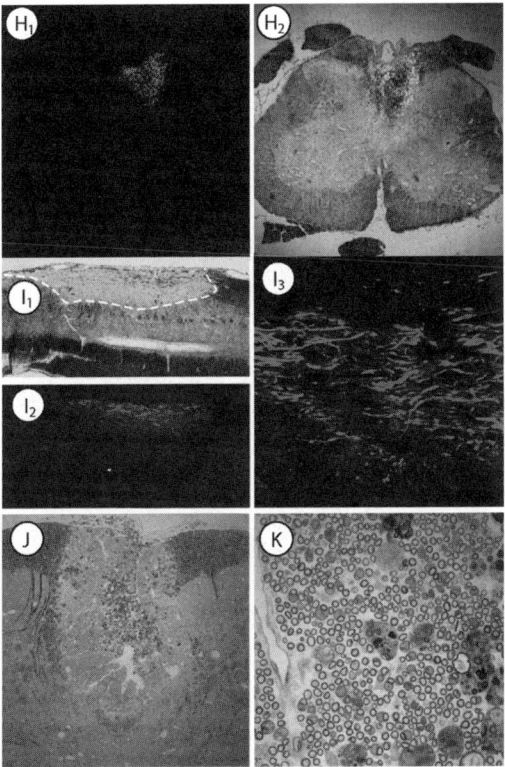

FIGURE 21.1. (*Continued*) **H–K.** Distribution of eGFP-expressing OECs transplanted into the demyelinated dorsal funiculus. **H₁.** Coronal spinal cord section from rat that was transplanted with Schwann cells from GFP-expressing mice 3 weeks after transplantation. **H₂.** Superimposition of the GFP fluorescent and DIC images. Note that donor Schwann cells expressing GFP are localized only in the dorsal funiculus. **I.** Sagittal sections through the lesion site showing the distribution of transplanted OECs. The lesion site is within the stippled area in **I₁. I₂.** The transplanted cells extend throughout the lesion site. **I₃.** Higher-power micrograph showing clusters of transplanted cells. **J, K.** Low- and high-power micrographs of plastic semithin sections of the lesion with remyelinated axons. (Modified from Akiyama et al., 2004.) (*See Color Plates between pages 256-257.*)

Although these cells have not yet been studied in demyelination models, they have been shown to reduce infarction size and improve functional outcome in rodent cerebral ischemia (Honma et al., 2005). Figure 21.2A is a 2,3,5-triphenyltetrazolium chloride (TTC)-stained brain section showing areas of infarction (white) 1 week after middle cerebral artery occlusion (MCAO) by a microfilament. A brain section 1 week after MCAO, but from an animal that was intravenously injected with hTERT-MSCs 12 hours after MCAO, is shown in Figure 21.2B; the infarction volume is reduced. Figure 21.2C shows that the

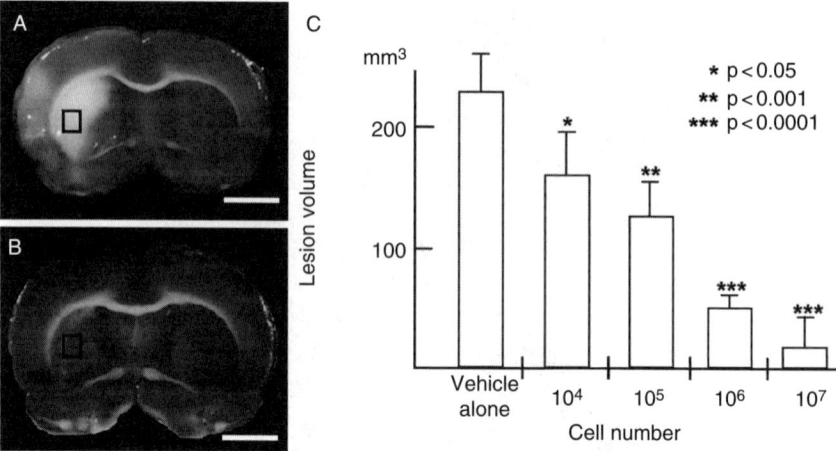

FIGURE 21.2. Summary of the 2,3,5-triphenyltetrazolium chloride (TTC)-unstained lesion volumes in each group. Brain slices stained with TTC to visualize the damaged lesions. A brain slice from the nontreated middle cerebral artery occlusion (MCAO) rats is shown for comparison **A**. The TTC-stained brain slices from rats that were intravenously transplanted with $10^7$ hTERT-mesenchyal stem cells (MSCs) 12 hours after MCAO are shown in **B**. The sections were also counterstained with hematoxylin. Although the glial scar tissue and a large number of inflammatory cells were obvious in the lesion without cell transplantation **C**, parenchymal brain tissue was greatly preserved in the treated group. Bars = 3 mm (**A, B**). (Modified from Honma et al., 2005.)

infarction volume decreases with the increasing number of intravenously delivered MSCs. Behavioral improvement was greater in the MSC-treated groups, indicating that the reduction in infarction volume is accompanied by improved functional outcome.

There is low-level basal secretion from hMSCs of several neurotrophic factors in culture, but ischemic rat brain extracts induce production of neurotrophins and angiogenic growth factors (Chen et al., 2002). One of these, brain-derived neurotrophic factor (BDNF), constitutively expressed at low levels in primary hMSC cultures (Kurozumi et al., 2004; Nomura et al., 2005), increased in the ischemic lesion following hMSC treatment in the rat MCAO model (Kurozumi et al., 2004; Nomura et al., 2005). Transplantation of BDNF gene-modified MSCs resulted in stronger therapeutic effects with increased BDNF levels in the ischemic lesion (Kurozumi et al., 2004; Nomura et al., 2005). Thus, BDNF and other trophic factors may play an important role in the therapeutic benefits on cerebral ischemia following MSC transplantation. The capacity of MSCs to release growth and trophic factors has been suggested to be the key to the beneficial effect in cerebral ischemia (Chen et al., 2001). Munoz et al. (2005) demonstrated that transplantation of human MSCs into rat hippocampus can produce an array of neurotrophic factors, including nerve growth factor, neurotrophin-4/5, and ciliary neurotrophic factor; and it promotes neurogenesis of endogenous neural stem cells. In addition to neurotrophic factors which may confer neuroprotective effects, isolated cultured bone marrow-derived stromal cells secrete angiogenic cytokines, vascular endothelial growth factor (VEGF), and placental growth factor (PlGF) (Kinnaird et al., 2004a,b). These factors promote the growth of new, stable vessels in cardiac and limb ischemia (Autiero et al., 2003; Luttun et al., 2002). Thus, the therapeutic effects of MSCs may result from direct and indirect effects on host tissues via conferring neuroprotection and promoting angiogenesis.

# 3   Peripheral Myelin-Forming Cells

An alternative approach to circumventing current concerns with the multipotency of stem cells and potential safety issues is the use of committed myelin-forming cell types, such as olfactory ensheathing cells (OECs) or Schwann cells (SCs), where the prospect of transforming into undesirable cell types is diminished. Although OECs are pluripotent and can differentiate into astrocyte-like and SC-like cells in culture, they appear to predominantly differentiate into SC-like myelin-forming cells when transplanted into chemically demyelinated lesions (Akiyama et al., 2004; Imaizumi et al., 1998; Sasaki et al., 2004). Remyelination of the dorsal columns of the spinal cord following transplantation of GFP-OECs is shown in Figure 21.3.

Both SC and OEC transplantation in rodent models of demyelination results in peripheral-like myelination with improved impulse conduction (Akiyama et al., 2004; Honmou et al., 1996; Imaizumi et al., 1998). The demyelination lesion model was established by X-irradiating the spinal cord and 3 days later focally injecting ethidium bromide (EB) directly into the dorsal columns. X-irradiation stops progenitor cells from dividing, and EB kills glial cells including oligodendrocytes. The center of the lesion is not remyelinated by endogenous cells until about 8 weeks after lesion induction (Sasaki et al., 2004). Figure 21.3 shows conduction velocity data from in vivo recordings in the spinal cord at various times following demyelination and after transplantation of either Schwann cells or OECs. Note that following EB injection to demyelinate the axons but without X-irradiation there is progressive improvement in the conduction velocity over 6 weeks. With the addition of X-irradiation, the conduction velocity remains low and the axons are not remyelinated by endogenous cells. However, the conduction velocity is significantly increased if Schwann cells or OECs are transplanted into the lesion (Fig. 21.3B).

Despite of the potential of SCs and OECs for myelin repair, collecting sufficient numbers of human OECs and SCs for transplantation procedures is difficult because obtaining sufficient biopsy tissue for harvesting adequate numbers of autologous OECs is problematic. One approach potentially to overcoming this concern is the use of porcine xenografts, where large numbers of acutely prepared OECs can be obtained from large porcine olfactory bulbs. Porcine xenografts are

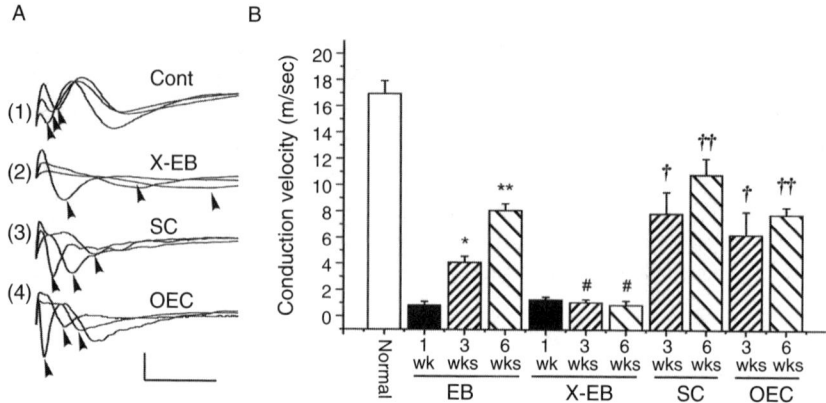

FIGURE 21.3. Conduction velocity measurements recorded in vivo of the demyelinated and remyelinated axons. **A.** Superimposed compound action potentials recorded at 1.0-mm increments longitudinally along the dorsal columns in normal (*cont; 1*), X-EB lesion (*2*) and three weeks after Schwann cell (SC) (*3*) and olfactory ensheathing cell (OEC) (*4*) transplantation. **B.** Histograms of conduction velocity (error bars indicate SEM) of dorsal column axons obtained from normal, 1, 3, and 6 weeks EB injection without prior x-irradiation, after X-EB lesion induction, and 3 and 6 weeks after Schwann cell and OEC transplantation into the X-EB lesion. $^*p < 0.1$, $^{**}p < 0.01$, $^\dagger p < 0.05$, $^{\dagger\dagger}p < 0.05$, and $^\#$not statistically significant. (Modified from Akiyama et al., 2004.)

used in a variety of forms, but the immune rejection of the donor tissue by the host creates a problem, which is usually reduced with immunosuppression of the host. An approach to reducing the immune response further is the use of transgenic pigs expressing antigens compatible with human transfer—e.g., expression of human complement inhibitors or the H-transferase (H-T) gene; H-T synthesizes the carbohydrate backbone of the cell membrane similar to that of the universal donor type O blood group (see discussion in Radtke et al., 2004).

Radtke et al. (2004) prepared OECs from transgenic pigs expressing the H-T gene and transplanted them into the demyelinated nonhuman primate spinal cord (African green monkey). These results indicated that a large number of highly pure transgenic (H-T) porcine OECs could be harvested from pig olfactory bulbs. The OECs displayed antigenic properties (p75 NGFR, S100, GFAP) similar to the well studied rodent OECs.

The transgene was expressed in the OECs, and functional assays showed a reduction in human complement activation. The in vivo transplantation studies demonstrated the ability of transgenic porcine OECs to remyelinate demyelinated monkey spinal cord. All previous studies utilizing OECs for spinal cord repair have been carried out in rodents. Virtually all experimental studies on the potential of stem cells and myelin-forming cells to repair the demyelinated CNS have been carried out on infraprimate species. However, primates present with a number of unique properties that do not ensure extrapolation from rodent work. Thus,

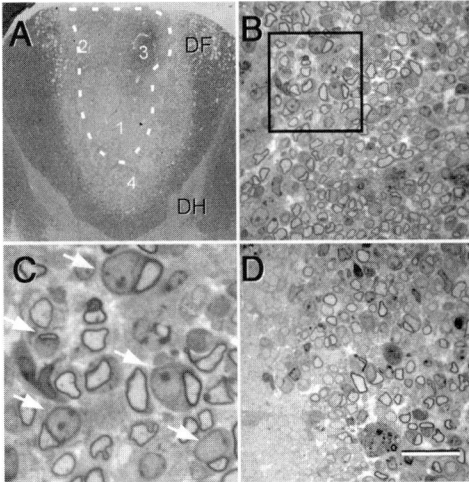

FIGURE 21.4. Transplantation of transgenic pig OECs into the demyelinated monkey spinal cord. **A.** Low power micrograph of lesion following cell transplantation in the dorsal funiculus (DF). Remyelination was observed within the white dashed lines and most of the dorsal funicular region outside of the dashed lines remained demyelination. **B.** The central core of the lesion was densely remyelinated. **C.** The boxed area in B showing myelinated axon profiles exhibiting a peripheral pattern of remyelination. **D.** The edge of the densely remyelinated zone shows a transition from demyelinated (left) to remyelinated axons. Bar in **D** = 0.5 mm in **A**, 20 μm in **C**, 125 μm in **B** and **D**. (Modified from Radtke et al., 2004.)

a limited number of studies on cell transplantation in the nonhuman primate would represent an important set of data on feasibility for initiation of human cell-based therapy studies with regard to remyelination in the CNS. Indeed, transplantation of H-T OECs into the demyelinated monkey spinal cord results in remyelination with a peripheral pattern (Fig. 21.4) (Radtke et al., 2004), thus demonstrating that the repair potential is not limited to rodents but also occurs in primates. This is as an important prerequisite for consideration of such an approach in human studies.

A large body of work supports the proposal that transplantation of OECs into traumatic spinal cord injury and models can promote axonal regeneration and functional recovery (Franklin et al., 1996; Imaizumi et al., 2000; Keyvan-Fouladi et al., 2003; Li et al., 1997, 1998; Lu et al., 2002; Plant et al., 2003; Ramon-Cueto et al., 1998, 2000; Ramon-Cueto and Nieto-Sampedro, 1994; Sasaki et al., 2004). A unique feature of presumptive axonal regeneration following OEC transplantation is the occurrence of groups of axons in the transection lesion site with a peripheral pattern of myelination surrounded by a fibroblast-like cell that forms a "tunnel" around small clusters of myelinated axons (Li et al., 1997). These tunnels have not been reported following SC transplantation into transection lesions (Imaizumi et al., 2000) or into demyelinating lesions (Lankford et al.,

2002). Li et al. (1998) referred to the surrounding fibroblast-like cells as "A" cells and the myelin-forming cells as "S" cells and suggested that both can be derived from the donor OECs. Although transplanted SCs can myelinate spinal cord axons (Blakemore and Crang, 1985; Baron-Van Evercooren et al., 1992; Honmou et al., 1996) and are associated with improved functional outcome (Takami et al., 2002), the "A" and "S" cell organization appears to be unique to OEC transplantation (Imaizumi et al., 2000; Li et al., 1997, 1998). Figure 21.5 shows spinal cord 5 weeks after dorsal hemisection and transplantation of OECs. Regenerated axons are observed across the transection site, and they organize in bundles of myelinated axons surrounded by a fibroblast-like cell; these cells may correspond to "A" cells.

Boyd et al. (2004) did not find LacZ-expressing "S" cells following transplantation of E18-derived OECs into a spinal cord compression injury model—only LacZ-expressing fibroblast-like cells. They concluded that only the fibroblast-like cell is derived from OEC transplantation and that "S" cells are exclusively

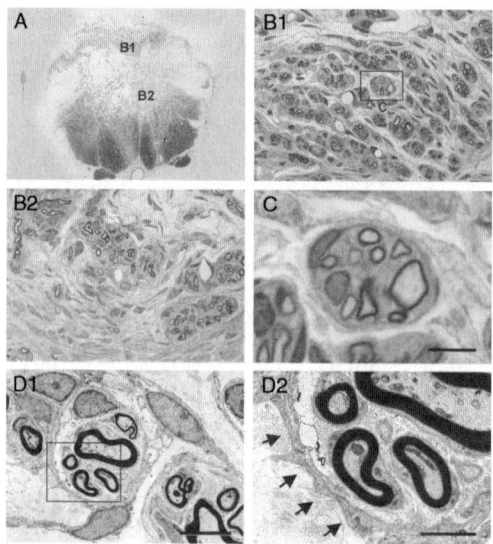

Figure 21.5. Semithin plastic sections stained with methylene blue/azure II through the transection site 5 weeks after transplantation of OECs. **A.** Low-power micrograph showing completeness of the transection through the entire dorsal funiculus and beyond. **B1, B2.** Higher-power micrographs showing groups of myelinated axons within the lesion zone obtained from regions indicated in **A. C.** Higher power micrograph from boxed region in **B1.** Note the clustering of myelinated axons surrounded by a cellular element. **D1.** Electron micrograph from same lesion showing myelinated axons surrounded by a cellular element forming a tunnel. **D2.** *Boxed* region from **D1.** *Arrows* point to cytoplasmic region of cell surrounding the bundle of myelinated axons. Bar in **C** = 0.75 mm for **A**; 30 μm for **B1, B2**; 6 μm for **C**. Bar in D1= 5 μm; bar in D2 = 2 μm. (Modified from Sasaki et al., 2004.)

derived from invading SCs. It is clear, however, that following OEC transplantation into transected spinal cord cellular tunnels are formed and regenerating axons grow through them (Sasaki et al., 2004), possibly providing a unique permissive environment for regeneration.

# 4  Challenges in Cell Therapy Approaches for Demyelinating Disorders

Experimental models of demyelination of the spinal cord indicate the feasibility of cell transplantation approaches to elicit at least some degree of functional recovery through remyelination of demyelinated axons. Challenges to develop clinical approaches include selecting the appropriate cell type and delivery method for the appropriate neurological disease. Although oligodendrocytes normally form myelin in the CNS, the prospect of using Schwann cells or OECs to remyelinate CNS lesions is intriguing because they can be harvested autologously and they may not have antigens present on oligodendrocytes, which may be immunogenic in MS because central, but not peripheral, myelin is the immune target in multiple sclerosis. Yet, there are limitations with the use of these cells. One is that Schwann cell remyelination in the CNS does not precisely recapitulate the pattern of remyelination by oligodendrocytes. The density of axonal spacing is less with Schwann cell myelination (Honmou et al., 1996), and it must be asked what potential negative effects this could have on the system. Indeed, it has been speculated that a selective force for the evolution of oligodendrocytes was to provide for maximum conduction velocity conferred by myelin deposition, with the most economic utilization of space (for review see Kocsis, 1999). The advantage this organization offers with oligodendrocyte myelination is a relatively compact CNS, but a disadvantage is that pathology to a single oligodendrocyte affects a number of myelin segments. Despite this limitation, it is reasoned that restoration of myelin and conduction by SCs or OECs in even a limited subset of axons could potentially be beneficial.

Another important concern with regard to cell transplantation therapies is, as mentioned above, the harvesting of sufficient numbers of appropriate cells for transplantation. Whereas in the laboratory a number of cell types can be expanded in culture with trophic factors and mitogens, it is not clear if such expansion alters the physiology of the cells in a negative way with regard to their ability to form functional myelin. Indeed, studies to examine the myelinogenic potential of expanded cells are required. Another concern with expanded cells is the potential risk of tumor formation. It is absolutely essential to determine if experimental in vivo transplantation of expanded cells not only allows the cells to retain their ability to carry out neural repair but that they do not form tumors. This concern emphasizes the importance of Phase I clinical studies to assess safety in small numbers of patients. It cannot be assumed that the observed safety in animal models applies to humans.

Another point is that cell-induced remyelination and axonal regeneration may not recapitulate uninjured structures but may establish new neuronal circuits and conduction channels (Bareye et al., 2005). Transplantation of glial cells may not only remyelinate axons but could also produce trophic factors that may alter axonal sprouting and synapse formation (Keyvan-Fouladi and Raisman, 2002). One concern is that the newly organized neural structures could be maladaptive and elicit neurological problems, such as pain or allodynia (Hofstetter et al., 2005). Care must be taken to ensure that new pathways and myelination elicited by interventional approaches such as cell transplantation do not result in maladaptive responses. Appropriate design of functional experiments to assess efficacy is essential for evaluating these approaches in clinical studies. Additionally, because patterns of remyelination and potentially partial restoration of axon white matter tracts and synapses may not recapitulate preinjury structures, rehabilitation strategies may be needed to understand and maximize the extent of achievable functional recovery.

Unlike animal models, where cell survival and the anatomical pattern of remyelination can be studied at the termination of the experiments with histological processing, extensive histological examination in human studies is not possible. Moreover, although magnetic resonance imaging (MRI) techniques are powerful, there are currently limitations when assaying donor cell survival in vivo after transplantation in humans. For example, while MS lesions can be identified with MRI, this technique cannot yet clearly distinguish sites of remyelination. Techniques to label donor cells for recognition with MRI are experimental and not yet routinely employed. As these techniques become compatible with routine clinical use, more powerful methods will be available to assess donor cell survival and distribution in human clinical studies. Some information of cell survival in transplanted patients could be obtained from CNS biopsy, but the amount of tissue is limited and safety issues are associated with biopsy procedures.

## 5    Prospects of Cell-Based Clinical Approaches

Experimental work in animal models indicates that a number of cell types transplanted into a variety of CNS lesion models can integrate into the host CNS, and functional benefits can result. The most extensive use of cell transplantation to address a clinical problem has been in Parkinson's disease, where extensive studies transplanting human fetal mesencephalic neurons into the caudate nucleus has been carried out in an attempt to replace dopaminergic neurons lost in this disease (see Bjorklund, 2005, for an overview). To date, results from these studies are complex; and although some positive results have been reported, there remain many issues yet to be addressed before practical implementation of this approach is possible.

In cell transplantation approaches to treat Parkinson's disease, neuronal replacement of a specific neuronal type is required (i.e., mesencephalic dopaminergic neurons). This presents many challenges in terms of cell collection and expansion in culture and directing cells to a specific neuron type that must synaptically integrate into the CNS. In many ways, the use of myelin-forming cells or precursors to remyelinate CNS axons is less daunting. Endogenous remyelination can occur in both the CNS and peripheral nervous system (PNS), and committed myelin-forming cells and precursors can be derived from a number of sources including autologous CNS, PNS, and bone marrow. There are strong signaling mechanisms between axons and appropriate glial cells for induction of myelin formation. Therefore, delivery of sufficient numbers of myelin-forming cells to areas of demyelination may readily result in new myelin formation.

Multiple sclerosis is an obvious disease in which to study the potential for a cell therapy approach using myelin-forming cell transplantation. Yet, because MS is an immunological disease with central myelin presenting antigens that elicit white matter inflammation, introduction of exogenous oligodendrocytes could exacerbate the disease. However, with recent advances in immunological thera-pies for MS, it is conceivable in the future that the basic disease process in MS could be controlled. Many patients would have residual demyelination plaques that could be associated with neurological symptoms. Under these conditions an effective cell therapy to targeted sites of demyelination could remyelinate axons, improve axonal function, and reduce neurological symptoms in these patients. Recently it was demonstrated that systemic delivery of embryonic stem cells into EAE mice abated disease progression (Plucchino et al., 2003, 2005). Moreover, intravenous delivery of MSCs can also abate disease progression in EAE (Zhang et al., 2005). Thus, cell therapeutic approaches may also be applied for immunomodulatory neuroprotective strategies.

Another approach might be to take advantage of the ability of peripheral myelin-forming cells as a donor source because peripheral myelin is not a target in MS. Although these cells may form myelin that is different than normal oligo-dendrocyte myelin, axonal conduction has been shown to improve in Schwann cell-mediated myelination of spinal cord axons following cell transplantation (Honmou et al., 1996). Moreover, human Schwann cells transplanted into the demyelinated immunosuppressed rat spinal cord also results in improved impulse conduction velocity (Kohama et al., 2001).

MS plaques contain demyelinated axons and are gliotic because of astrocytic proliferation. This astrocytic "scar" is thought to impede the migration of regen-erating axons and cells. It will be important in the future to understand and pos-sibly develop myelin-forming cell types that can migrate through these regions to achieve maximal repair. In this regard, OECs have been suggested to have greater ability to migrate through astrocytic regions (Lakatos et al., 2000).

In addition to addressing demyelinating diseases, cell therapy approaches may be useful for traumatic CNS injury. Contusive spinal cord injury results in necro-sis of spinal cord tissue; but even in humans there are often areas surrounding the central necrotic core that are demyelinated. OECs are being used in clinical cell transplantation studies in China (Huang et al., 2003) for spinal cord injury patients. Transplantation of both OECs (Imaizumi et al., 1998; Li et al., 1997; Ramon-Cueto and Nieto-Sampedro, 1994) and Schwann cells (Imaizumi et al., 2000; Takami et al., 2002) suggests that axonal regeneration with subsequent myelination can be elicited by implantation of these cells. It is thought that the peripheral myelin-forming cells can bridge a lesion zone in the spinal cord and provide a permissive environment for axonal regeneration. Significant behavioral improvement (Li et al.1997; Sasaki et al., 2004) occurs after OEC transplantation. However, whether the functional improvement is the result of axonal regeneration or preservation (i.e., neuroprotection) is an important concern. Several studies transplanting OECs and bone marrow cells into a contusive spinal cord injury model demonstrate that the extent of tissue damage is significantly reduced. This suggests that a neuroprotective influence resulting in tissue sparing is effected by the cell transplantation. Greater sparing of spinal cord was observed if the transplantation was delayed by 1 week after injury rather than if done immediately (Plant et al., 2003). This suggests that a therapeutic window may be present to allow such an approach in patients. Interestingly, beneficial effects of systemic delivery of bone marrow cells have been reported in spinal cord injury models (Chopp et al., 2000), cerebral ischemia models (Chen et al., 2001; Iihoshi et al., 2004; Honma et al., 2005), and in EAE (Zhang et al., 2005).

It is important to emphasis that a number of neural cells such as Schwann cells, olfactory ensheathing cells, astrocytes, and neural stem cells may provide trophic and antiinflammatory effects to the damaged lesions upon transplantation. Transplanted cells may produce neurotrophic factors in vivo, and several trophic factors have been reported to have therapeutic effects following CNS trauma (Hirouchi and Ukai, 2002). Exogenous application of certain factors—brain derived neurotrophic factor (BDNF), glial cell line-derived neurotrophic factor (GDNF), nerve growth factor (NGF), endothelial growth factor (EGF), and basic fibroblast growth factor (bFGF),—has been reported to limit the extent of ischemic lesion volume (Ay et al., 2001; Schabitz et al., 1997). The neural tissue rescue may through several mechanisms, such as free radical scavenging, antiapoptotic activity, antiinflammatory activity, and antiglutamate excitotoxicity (Hirouchi and Ukai, 2002). Thus, in addition to the prospect of reconstructing myelin and enhancing axonal regeneration, cell therapy approaches utilizing OECs, Schwann cells, or stem cells may provide an important neuroprotective strategy in the treatment of traumatic CNS injury, including spinal cord injury.

Although precise mechanisms of action are not fully defined and experimental work must continue, clinical studies examining safety and efficacy of cell therapy approaches in the treatment of CNS trauma must continue in parallel. Beneficial therapeutic effects may well result from both an initial neuroprotective effect mediated by donor cell production of trophic factors and later reparative effects on spared tissue such as axonal regeneration and remyelination. The potential of intravenous delivery of stem cells to home to and repair lesions in the spinal cord suggests that autologous systemic infusion of these cells may provide a novel, relatively noninterventional approach for the treatment of acute spinal cord injury or demyelinating disorders. Although much more laboratory investigation certainly is required, progress to date in this field holds the exciting prospect of novel cell therapy-based interventional approaches to treat demyelinating and traumatic CNS diseases in the not too distant future.

# References

Aggarwal, S., Pittenger, M. F. (2005) Human mesenchymal stem cells modulate allogeneic immune cell responses. *Blood* 105:1815-1822.

Akiyama, Y., Radtke, C., Honmou, O., Kocsis, J. D. (2002a) Remyelination of the spinal cord following intravenous delivery of bone marrow cells. *Glia* 39:229-236.

Akiyama, Y., Radtke, C., Kocsis, J. D. (2002b) Remyelination of the rat spinal cord by transplantation of identified bone marrow stromal cells. *J. Neurosci.* 22:6623-6630.

Akiyama, Y., Lankford, K. L., Radtke, C., et al. (2004) Remyelination of spinal cord axons by olfactory ensheathing cells and Schwann cells derived from a transgenic rat expressing alkaline phosphatase marker gene. *Neuron Glia Biol.* 1:1-9.

Alvarez-Dolado, M., Pardal, R., Garcia-Verdugo, J. M., et al. (2003) Fusion of bone-marrow-derived cells with Purkinje neurons, cardiomyocytes and hepatocytes. *Nature* 425:968-973.

Autiero, M., Luttun, A., Tjwa, M., Carmeliet, P. (2003a) Placental growth factor and its receptor, vascular endothelial growth factor receptor-1: novel targets for stimulation of ischemic tissue revascularization and inhibition of angiogenic and inflammatory disorders. *J. Thromb. Haemost.* 1:1356-1370.

Ay, I., Sugimori, H., Finklestein, S. P. (2001) Intravenous basic fibroblast growth factor (bFGF) decreases DNA fragmentation and prevents downregulation of Bcl-2 expression in the ischemic brain following middle cerebral artery occlusion in rats. *Brain Res. Mol. Brain Res.* 87:71-80.

Azizi, S. A., Stokes, D., Augelli, B. J., et al. (1998) Engraftment and migration of human bone marrow stromal cells implanted in the brains of albino rats—similarities to astrocyte grafts. *Proc. Natl. Acad. Sci. USA* 95:3908-3913.

Bang, O. Y., Lee, J. S., Lee, P. H., Lee, G. (2005). Autologous mesenchymal stem cell transplantation in stroke patients. *Ann. Neurol.* 57:874-882.

Baron-Van Evercooren, A., Clerin-Duhamel, E., Lapie, P., et al. (1992) The fate of Schwann cells transplanted in the brain during development. *Dev. Neurosci.* 14:73-84.

Bjorklund, A. (2005) Cell therapy for Parkinson's disease: problems and prospects. *Novartis Found. Symp.* 265:174-186.

Blakemore, W. F., Crang, A. J. (1985) The use of cultured autologous Schwann cells to remyelinate areas of persistent demyelination in the central nervous system. *J. Neurol. Sci.* 70:207-223.

Boyd, J. G., Lee, J., Skihar, V., et al. (2004) LacZ-expressing olfactory ensheathing cells do not associate with myelinated axons after implantation into the compressed spinal cord. *Proc. Natl. Acad. Sci. USA* 101:2162-2166.

Brazelton, T. R., Rossi, F. M., Keshet, G. I., Blau, H. M. (2000) From marrow to brain: expression of neuronal phenotypes in adult mice. *Science* 290:1775-1779.

Castro, R. F., Jackson, K. A., Goodell, M. A., et al. (2002) Failure of bone marrow cells to transdifferentiate into neural cells in vivo. *Science* 297:1299.

Chen, J., Li, Y., Wang, L., et al. (2001) Therapeutic benefit of intracerebral transplantation of bone marrow stromal cells after cerebral ischemia in rats. *J. Neurol. Sci.* 189:49-57.

Chopp, M., Zhang, X. H., Li, Y., et al. (2000) Spinal cord injury in rat: treatment with bone marrow stromal cell transplantation. *Neuroreport* 11:3001-3005.

Ferrari, G., Cusella-De Angelis, G., Coletta, M., et al. (1998) Muscle regeneration by bone marrow-derived myogenic progenitors. *Science* 279:1528-1530.

Franklin, R. J., Gilson, J. M., Franceschini, I. A., Barnett, S. C. (1996) Schwann cell-like myelination following transplantation of an olfactory bulb ensheathing cell line into areas of demyelination in the adult CNS *Glia* 17:217-224.

Friedenstein, A. J. (1976) Precursor cells of mechanocytes *Int. Rev. Cytol.* 47:327-359.

Hirouchi, M., Ukai, Y. (2002) Current state on development of neuroprotective agents for cerebral ischemia. *Nippon Yakurigaku Zasshi* 120:107-113.

Hofstetter, C. P., Holmstrom, N. A., Lilja, J. A., et al. (2005) Allodynia limits the usefulness of intraspinal neural stem cell grafts; directed differentiation improves outcome. *Nat. Neurosci.* 8:346-353.

Honma, T., Honmou, O., Iihoshi, S., et al. (2006). Intravenous infusion of immortalized human mesenchymal stem cells protects against injury in a cerebral ischemia model in adult rat. *Exp. Neurol.* 199:56-66.

Honmou, O., Felts, P. A., Waxman, S. G., Kocsis, J. D. (1996) Restoration of normal conduction properties in demyelinated spinal cord axons in the adult rat by transplantation of exogenous Schwann cells. *J. Neurosci.* 16:3199-3208.

Huang, H., Chen, L., Wang, H., et al. (2003) Influence of patient's age on functional recovery after transplantation of olfactory ensheathing cells into injured spinal cord. *Chin. Med. J.* 116:1488-1491.

Iihoshi, S., Honmou, O., Houkin, K., et al. (2004) A therapeutic window for intravenous administration of autologous bone marrow after cerebral ischemia in adult rats. *Brain Res.* 1007:1-9.

Imaizumi, T., Lankford, K. L., Waxman, S. G., et al. (1998) Transplanted olfactory ensheathing cells remyelinate and enhance axonal conduction in the demyelinated dorsal columns of the rat spinal cord. *J. Neurosci.* 18:6176-6185.

Imaizumi, T., Lankford, K. L., Burton, W. V., et al. (2000) Xenotransplantation of transgenic pig olfactory ensheathing cells promotes axonal regeneration in rat spinal cord. *Nat. Biotechnol.* 18:949-953.

Inoue, M., Honmou, O., Oka, S., et al. (2003) Comparative analysis of remyelinating potential of focal and intravenous administration of autologous bone marrow cells into the rat demyelinated spinal cord. *Glia* 44:111-118.

Keyvan-Fouladi, N., Li, Y., Raisman, G. (2002) How do transplanted olfactory ensheathing cells restore function? *Brain Res. Brain Res. Rev* 40:325-327.

Keyvan-Fouladi, N., Raisman, G., Li, Y. (2003) Functional repair of the corticospinal tract by delayed transplantation of olfactory ensheathing cells in adult rats. *J. Neurosci.* 23:9428-9434.

Kinnaird, T., Stabile, E., Burnett, M. S., et al. (2004a) Marrow-derived stromal cells express genes encoding a broad spectrum of arteriogenic cytokines and promote in vitro and in vivo arteriogenesis through paracrine mechanisms. *Circ. Res.* 94:678-685.

Kinnaird, T., Stabile, E., Burnett, M. S., et al. (2004b) Local delivery of marrow-derived stromal cells augments collateral perfusion through paracrine mechanisms. *Circulation* 109:1543-1549.

Kobune, M., Kawano, Y., Ito, Y., et al. (2003) Telomerized human multipotent mesenchymal cells can differentiate into hematopoietic and cobblestone area-supporting cells. *Exp. Hematol.* 31:715-722.

Koc, O. N., Day, J., Nieder, M., et al. (2002) Allogeneic mesenchymal stem cell infusion for treatment of metachromatic leukodystrophy (MLD) and Hurler syndrome (MPS-IH). *Bone Marrow Transplant.* 30:215-222.

Kocsis, J. D. (1999) Restoration of function by glial cell transplantation into demyelinated spinal cord. *J. Neurotrauma* 16:695-703.

Kohama, I., Lankford, K. L., Preiningerova, J., et al. (2001). Transplantation of cryopreserved adult human Schwann cells enhances axonal conduction in demyelinated spinal cord. *J. Neurosci.* 21:944-950.

Kurozumi, K., Nakamura, K., Tamiya, T., et al. (2004) BDNF gene-modified mesenchymal stem cells promote functional recovery and reduce infarct size in the rat middle cerebral artery occlusion model. *Mol. Ther.* 9:189-197.

Lagasse, E., Connors, H., Al Dhalimy, M., et al. (2000). Purified hematopoietic stem cells can differentiate into hepatocytes in vivo. *Nat. Med.* 6:1229-1234.

Lakatos, A., Franklin, R. J., Barnett, S. C. (2000) Olfactory ensheathing cells and Schwann cells differ in their in vitro interactions with astrocytes. *Glia* 32:214-225.

Lankford, K. L., Imaizumi, T., Honmou, O., Kocsis, J. D. (2002) A quantitative morphometric analysis of rat spinal cord remyelination following transplantation of allogenic Schwann cells. *J. Comp. Neurol.* 443:259-274.

Li, Y., Field, P. M., Raisman, G. (1997) Repair of adult rat corticospinal tract by transplants of olfactory ensheathing cells. *Science* 277:2000-2002.

Li, Y., Field, P. M., Raisman, G. (1998) Regeneration of adult rat corticospinal axons induced by transplanted olfactory ensheathing cells. *J. Neurosci.* 18(24):10514-10524.

Lu, J., Feron, F., Mackay-Sim, A., Waite, M. E. (2002) Olfactory ensheathing cells promote locomotor recovery after delayed transplantation into transected spinal cord. *Brain* 125:14-21.

Luttun, A., Tjwa, M., Carmeliet, P. (2002) Placental growth factor (PlGF) and its receptor Flt-1 (VEGFR-1): novel therapeutic targets for angiogenic disorders. *Ann. N. Y. Acad. Sci.* 979:80-93.

Majumdar, M. K., Thiede, M. A., Mosca, J. D., et al. (1998) Phenotypic and functional comparison of cultures of marrow-derived mesenchymal stem cells (MSCs) and stromal cells. *J. Cell. Physiol.* 176:57-66.

Mezey, E., Chandross, K. J., Harta, G., et al. (2000) Turning blood into brain: cells bearing neuronal antigens generated in vivo from bone marrow. *Science* 290:1779-1782.

Munoz, J. R., Stoutenger, B. R., Robinson, A. P., Spees, et al. (2005) Human stem/progenitor cells from bone marrow promote neurogenesis of endogenous neural stem cells in the hippocampus of mice. *Proc. Natl. Acad. Sci. USA* 102:18171-18176.

Nomura, T., Honmou, O., Harada, K., et al. (2005) I.V. infusion of brain-derived neurotrophic factor gene-modified human mesenchymal stem cells protects against injury in a cerebral ischemia model in adult rat. *Neuroscience* 136:161-169.

Plant, G. W., Christensen, C. L., Oudega, M., Bunge, M. B. (2003) Delayed transplantation of olfactory ensheathing glia promotes sparing/regeneration of supraspinal axons in the contused adult rat spinal cord. *J. Neurotrauma* 20:1-16.

Pluchino, S., Quattrini, A., Brambilla, E., et al. (2003) Injection of adult neurospheres induces recovery in a chronic model of multiple sclerosis. *Nature* 422:688-694.

Pluchino, S., Zanotti, L., Rossi, B., et al. (2005) Neurosphere-derived multipotent precursors promote neuroprotection by an immunomodulatory mechanism. *Nature* 436:266-271.

Prockop, D. J. (1997) Marrow stromal cells as stem cells for nonhematopoietic tissues. *Science* 276:71-74.

Radtke, C., Akiyama, Y., Brokaw, J., et al. (2004) Remyelination of the nonhuman primate spinal cord by transplantation of H-transferase transgenic adult pig olfactory ensheathing cells. *FASEB J.* 18:335-337.

Ramon-Cueto, A., Nieto-Sampedro, M. (1994) Regeneration into the spinal cord of transected dorsal root axons is promoted by ensheathing glia transplants. *Exp. Neurol.* 127:232-244.

Ramon-Cueto, A., Plant, G. W., Avila, J., Bunge, M. B. (1998) Long-distance axonal regeneration in the transected adult rat spinal cord is promoted by olfactory ensheathing glia transplants. *J. Neurosci.* 18:3803-3815.

Ramon-Cueto, A., Cordero, M. I., Santos-Benito, F. F., Avila, J. (2000) Functional recovery of paraplegic rats and motor axon regeneration in their spinal cords by olfactory ensheathing glia. *Neuron* 25:425-435.

Sasaki, M., Lankford, K. L., Zemedkun, M., Kocsis, J. D. (2004) Identified olfactory ensheathing cells transplanted into the transected dorsal funiculus bridge the lesion and form myelin. *J. Neurosci.* 24:8485-8493.

Schabitz, W. R., Schwab, S., Spranger, M., Hacke, W. (1997) Intraventricular brain-derived neurotrophic factor reduces infarct size after focal cerebral ischemia in rats. *J. Cereb. Blood Flow Metab.* 17:500-506.

Takami, T., Oudega, M., Bates, M. L., et al. (2002) Schwann cell but not olfactory ensheathing glia transplants improve hindlimb locomotor performance in the moderately contused adult rat thoracic spinal cord. *J. Neurosci.* 22:6670-6681.

Toma, C., Pittenger, M. F., Cahill, K. S., et al. (2002) Human mesenchymal stem cells differentiate to a cardiomyocyte phenotype in the adult murine heart. *Circulation* 105:93-98.

Woodbury, D., Schwarz, E. J., Prockop, D. J., Black, I. B. (2000) Adult rat and human bone marrow stromal cells differentiate into neurons. *J. Neurosci. Res.* 61:364-370.

Zhang, J., Li, Y., Chen, J., et al. (2005) Human bone marrow stromal cell treatment improves neurological functional recovery in EAE mice. *Exp. Neurol.* 195:16-26.

# 22
# Multiple Sclerosis: Future Directions and Prospects

Hartmut Wekerle

## 1   Is Multiple Sclerosis an Important Disease?

Multiple sclerosis (MS) is a widely known and dreaded disease, but is it really that important? The answer is clear: MS is important indeed and for several reasons. MS is of paramount socioeconomic significance. First, the disease is not rare, afflicting an estimated million of people worldwide. Second, it is a long-lasting disease: It typically affects people at a young age and then accompanies them throughout life but without significantly reducing overall life expectancy (Confavreux and Compston, 2006). Third, MS often is a debilitating disease, causing severe distress in the patient and requiring special care. Fourth, MS is becoming a global problem. Whereas it has traditionally been considered a disease restricted to the Western world, there is now evidence that the incidence and prevalence of "Western type MS" are increasing in other societies (e.g., Japan) (Kira, 2003).

Clinical complexity and an unpredictable course are the hallmarks of MS. The neurological deficits are multiple and variegated, affecting motor, sensitive, and sensory functions. Each individual patient seems to present with a unique profile of symptoms and signs. In many people the disease starts out with a relapsing/remitting course, whereas in others it progresses relentlessly from the very beginning. The disease leads to severe disability in one patient but remains benign in the next. There is no diagnostic marker available that could predict the fate of a patient on the day of diagnosis. Regrettably, there is no causative treatment of MS.

These features certainly provide sufficient incentive to try to decrypt the complicated mechanisms underlying the disease. In the following, I shall view the pathogenesis of MS from an immunological angle and list some of the questions which at present are considered among the most pressing. This list, it should be made clear, is by no means complete; rather, it reflects largely the author's personal bias.

## 2  Is MS One Disease?

The complexity of clinical MS fundamentally impedes diagnosis, which ultimately is often made through the neurologist's intuition. John Kurtzke 's clairvoyant dictum—"MS is what a good neurologist would call MS" (quoted by Compston et al., 2006)—masterly describes the protean nature of the disease and, at the same time, the state of diagnostic helplessness that has surrounded MS even beyond the advent of additional tools, such as cerebrospinal fluid (CSF) diagnostics and modern imaging.

The clinical complexity of MS has raised the question of whether what we call "MS" is indeed a single disease. This query gains additional weight considering the varied courses the disease may take and the distinct responsiveness to particular therapies. Early-onset patients with the relapsing/remitting course often respond well to immunomodulatory therapy, whereas people with primary progressive, or long-standing, MS are much poorer responders. Does this mean that the term "MS" does not label one disease entity but, rather, a collection of central nervous system (CNS) disorders that all end up producing demyelinating plaques in the CNS?

The complexity of MS goes even further: The histology of MS plaques is by no means uniform and allows the classification of different distinctive patterns, each hinting at a particular terminal pathogenesis. Lucchinetti and Lassmann proposed four main classes of active MS plaques (Table 22.1): Type I suggests a cellular inflammatory attack against myelinated tissue; type II hints at a pathogenesis implying humoral autoantibodies plus complement and/or macrophages; class III has a strong vasculitic component; and, finally, type IV may be initiated by a degenerative process in myelin-forming oligodendrocytes (Lassmann et al, 2001).

Yet there are denominators common to most cases of MS. Inflammatory, demyelinating lesions are visualized by magnetic resonance imaging (MRI); oligoclonal immunoglobulin bands are seen in the CSF of most, though not all, patients; and most active plaques contain inflammatory infiltrates. Thus, although we do not exclude distinct demyelinating CNS diseases masquerading as MS, there are good reasons to adhere to the current terminology.

TABLE 22.1. Four Subtypes of the Actively Demyelinating MS Plaques.

| Subtype | Pathology | Putative effector mechanism |
|---|---|---|
| I | Perivenous localization<br>Infiltrates: macrophages and T cells<br>Macrophages and microglia around<br>myelin debris | T cell-mediated inflammation with<br>recruitment and activation of<br>macrophages |
| II | Like pattern I but immunoglobulin<br>and complement deposits on myelin<br>debris | T cell-mediated inflammation with<br>macrophage/microglia activation<br>Complement-mediated lysis of antibody-<br>complexed myelin |
| III | Inflammation with T cells and<br>macrophages<br>Small vessel vasculitis with endothe-<br>lial damage and thrombosis;<br>degeneration of oligodendrocyte<br>processes proceeding to cell bodies | T cell-mediated vasculitis with secondary<br>ischemic damage of white matter |
| IV | Like pattern I but with prominent<br>degeneration of oligodendrocytes<br>in periplaque white matter | T cell-mediated inflammation with recruit-<br>ment and activation of macrophages<br>Demyelination by macrophage toxins<br>acting on predamaged oligodendrocytes<br>Primary oligodendrocyte defect? |

Adapted from Lassmann et al. (2001)

## 3  Is MS an Autoimmune Disease?

The concept of an autoimmune pathogenesis of MS rests on several complementary lines of evidence derived from morphological and functional studies. The active MS plaque, as described in a reductionist manner, displays several cardinal changes: The lesion is dominated by an inflammatory response, with leukocytes forming perivascular cuffs and parenchymal infiltrates. Then, there is a breach of the tight endothelial blood-brain barrier (BBB), leading to local edema formation and complement deposition. Typically, local glia cells are activated: Activated microglia cells dominate especially during the early plaque phase, and astrogliosis may prevail during chronic stages (Lassmann and Wekerle, 2006). Although this pattern of pathological change is highly suggestive of inflammatory, immunopathological lesions, it does not distinguish between immune responses against foreign agents, such as microbial pathogens, and attacks against self structures.

Indirect evidence pointing to autoimmune responses comes from animal experiments and clinical observations. First, the changes in the active MS lesions strikingly resemble the ones seen in various versions of experimental autoimmune encephalomyelitis (EAE) (Storch et al., 1998a,b), an experimental CNS inflammation caused by autoimmunization of susceptible experimental animals against protein components of its own myelin. Then, there is evidence of humoral autoantibodies binding to myelin in the human MS lesion, and at least some of these antibodies seem to be specific for myelin autoantigens (Genain et al., 1999; O'Connor et al., 2005). Indeed, patients whose lesions present prominent immunoglobulin binding, respond particularly well to the elimination of antibodies from blood by plasmapheresis, whereas others, lacking this feature, remain unresponsive (Keegan et al., 2005). Impressive support of autoimmune mechanisms in MS finally comes from antigen-targeted immunotherapy. Treatment of MS patients with altered peptide ligands (APLs), variants of the suspected main encephalitogenic target autoantigen myelin basic protein (MBP), was remarkably effective, in a good way and a bad way. In some patients, such treatment resulted in a reduction of the size and number of inflammatory brain lesions (Kappos et al, 2000). In others, however, the treatment produced hypersensitivity reactions and even worsening of the disease (Bielekova et al., 2000).

Myelin-specific, autoreactive T cell clones have been isolated from the blood of MS patients. However, they are not restricted to the immune repertoire of MS patients but also populate the immune systems of healthy people, even at a similar frequency (Martin et al., 1990; Pette et al., 1990). This fact per se does not argue against a possible active role of myelin-autoreactive T cells in the pathogenesis of MS, as is indicated by experiments using nonhuman primates. In effect, myelin-specific T cells have been isolated from the blood of untreated, healthy monkeys. The cells were expanded as pure lines in vitro and injected back into the original donor animals. Some, though not all, of these short-term lines transferred (mostly mild) EAE to the recipients, proving that the myelin self-reactive T cells seen in healthy primates were not only *autoreactive* but had pathogenic potential (Genain et al., 1994; Meinl et al., 1997). Given the close similarity of monkey and human immune systems, one may extend these experimental findings to humans and conclude that myelin-specific T cells in healthy or MS-affected persons have autoimmune potential.

The encephalitogenic potential of myelin-specific human T cells is further substantiated by *humanized* transgenic mice. Several groups have inserted the genes of paired *human* MBP-specific T cell receptor (TCR) α- and β-chains into the germline of mice along with additional, ancillary genes required for productive presentation and recognition of (auto)antigen. The TCR genes were derived from human CD4+ T cell clones recognizing MBP epitopes in the context of human major histocompatibility complex class II (MHC II) molecules, HLA DR2. Under special circumstances, especially in immunodeficient mice (due to the deletion of RAG recombinases, enzymes that control the assembly of diversified T and B cell receptors), a high proportion of the humanized mice developed clinical and histological EAE (Ellmerich et al., 2005; Madsen et al., 1999). The disease of these transgenic mice was unequivocally caused by T cells using human receptors for myelin autoantigen. These transgenic T cells attacked their body's own CNS, recognizing locally produced myelin proteins presented by local antigen-presenting cells (APCs) in the context of proper human MHC proteins.

Thus, at this point it appears reasonable to conclude that the human immune repertoire contains myelin-autoreactive T cells with an encephalitogenic potential. However, these findings do not yet formally demonstrate the actual contribution of myelin-specific T cells to the pathogenesis in an individual patient. Direct demonstration of autoimmune responses in the MS lesion is still outstanding.

# 4    What is the Role of CD8 T Cells in MS?

So far, neuroimmunologists devoted most of their attention to the study of CD4+ T cells as potential pathogenic agents in MS. Other important agents, CD8+ T cells and B cells, have been disregarded. There are methodological reasons for this bias: mainly that most EAE types are mediated by CD4+ T cells and, at least in rodents, CD4+ T cells are much easier to manipulate in culture than CD8+ T cells and B cells. In particular, it has been much more difficult to establish monospecific lines from the latter cell types.

It should be noted, however, that the human MS infiltrate, much in contrast to the classical EAE lesion, is dominated by CD8+, not CD4+, T cells (Friese and Fugger, 2005). This has been recognized in early studies using immunocytochemistry, which located CD8+ T cells especially among the parenchymal infiltrates (Booss et al., 1983; Hauser et al., 1986) (Table 22.2). Later studies examined the TCR repertoire of infiltrate cells and found evidence of clonal expansion in infiltrating CD8+ cells but much less so CD4+ T cells (Babbe et al., 2000). Indeed, some of these T cells persisted in the immune system of patients over periods of more than 5 years (Skulina et al., 2004). Unfortunately, to the present day, the target autoantigens of brain infiltrating CD8 T cells remain unknown.

How would autoimmune CD8+ T cells attack the target tissue? First, they could act, similar to their CD4+ counterparts, by secreting proinflammatory cytokines, which could affect neuronal function directly or via recruiting and activating ancillary macrophages. Second, CD8+ T cells could assault target cells directly and destroy them. Direct attack requires expression of the specific autoantigen in context of MHC I proteins in the target cell. Contrary to traditional views, which considered neurons unable to produce and express MHC of any class, several groups had shown recently that neurons are indeed inducible to MHC I under particular circumstances. MHC I induction in neurons requires interferon-$\gamma$ (IFN$\gamma$) as a positive signal as well as reduction of electric membrane activity through paralysis by neurotoxins (Neumann et al., 1995) or degeneration following peripheral axotomy (Lindå et al., 1998). MHC I-induced neurons are indeed able to process and productively present antigenic peptides to specific CD8+ T cells, which, following recognition, directly attack neuronal axons and somata (Medana et al., 2001).

TABLE 22.2. Evidence Supporting a Pathogenic Function of CD8+ T Cells in MS Plaques.

- CD8+ T cells are dominant in active human MS infiltrates (Booss et al., 1983)
- Extensive induction of MHC I within active MS lesion (Woodroofe et al., 1986)
- Clonal expansion of T cell receptors in infiltrating CD8+, less in CD4+ T cells (Babbe et al., 2000)
- Cytotoxic attack of CD8+ T cells against MHC I-induced neurons (Medana et al., 2001)
- Amplification of transferred EAE by myelin-specific CD8+ T cells (Huseby et al., 2001; Sun et al., 2001)

MHC I, major histocompatibility complex class I; EAE, experimental autoimmune encephalomyelopathy.

The studies of CD8 attacks against neurons made use of transgenic T cells recognizing artificial viral peptides presented by MHC I-induced cultured neurons, a situation remote from clinical MS, where myelin or neuronal autoantigens are thought to serve as targets. Myelin-specific CD8+ T cells were demonstrated in rodent models after autoimmunization of MBP-deficient mutant mice (Huseby, 2001) or after expansion in vitro (Sun et al., 2001); and these cells mediated EAE-like disease when transferred into immunocompromised or healthy hosts.

# 5    A Role for B Cells in MS?

There is manifold evidence arguing in favor of a B cell involvement in MS (Table 22.3). It has been know for a long time that B cells and plasma cells are quite common components of immune infiltrates of MS plaques (Esiri, 1998) and that these B cells produce antibodies of diverse specificities.

B cells infiltrating the brain parenchyma or the overlaying leptomeningeal membrane parts are responsible for the production of oligoclonal immunoglobin bands (OCBs) in the CSF. These antibodies are oligoclonal because they are produced by a limited number of local B cell clones. The target (auto)antigens of OCBs in MS are, however, unknown. The lack of specific binding may be explained by absorption of true autoantibodies by local myelin structures. Much in contrast, microbial brain infections also show OCBs, but some of these antibody bands indeed bind antigens of the causative agents.

B cells may have functions beyond the secretion of humoral autoantibodies. They can act as APCs, thereby imprinting the character of a T cell response. In particular, specific B cells are able to pick up highly diluted specific antigen, concentrate it, and present it to T cells. Finally, B cells secrete cytokines that create particular immune milieus that may support or divert autoimmune T cell responses.

It seems that B cells may persist over extensive periods of time in CNS tissues, which thus seem to offer a favorable milieu. This collides with our current concept of immune hostility of CNS microenvironments. Although it is well known that T cells rapidly undergo apoptotic cell death upon intrusion into the CNS, B cells could behave differently. In fact, recent evidence indicates that factors enabling B cells to function and survive (e.g., BAFF and CXCL13) are produced

TABLE 22.3. Evidence Favoring B Cells in the Pathogenesis of Multiple Sclerosis.

- Intrathecal production of oligoclonal immunoglobulin bands in CSF (Tourtellotte et al., 1985)
- Somatic hypermutation and dominant clonotypes in CSF and brain parenchyma B cells (Owens et al., 1998; Qin et al., 1998)
- Decoration of myelin debris by activated complement and immunoglobulin (Storch et al., 1998), including anti-MOG antibodies (Genain et al., 1999)
- Beneficial effect of plasmapheresis in some MS patients (Keegan et al., 2005)
- Beneficial effect of therapeutic B cell depletion (Stüve et al., 2005)

Adapted from Meinl et al. (2006).
CSF, cerebrospinal fluid; MOG, myelin oligodendrocyte glycoprotein.

in inflamed rodent (Magliozzi et al., 2004) and human (Krumbholz et al., 2005, 2006) CNS tissue.

Single-cell cloning techniques that allow isolation, expression, and characterization of CNS infiltrated B cells are required to better understand the pathogenic function of these persisting immune cells (Haubold et al., 2004). In fact, previous work based on cloning of CNS-associated B cells has helped establish a "pedigree" of B cells suggesting that these lymphocytes not only stay inertly in CNS parenchyma or spinal fluid but may actively respond to (unknown) local antigen (Owens et al., 1998; Qin et al., 1998).

Identification of the target (auto)antigen(s) activating CNS-persistent B cells has become a major challenge to MS research. Future work will continue studying human specimens but will also require the use of suitable animal models.

# 6   Which are the Target Autoantigens?

Formal proof of autoaggressive T cells and identification of their autoantigenic targets must satisfy "Koch's postulates," rules established by Robert Koch to identify pathogenic agents in microbial infections. Autoimmune (as microbial) candidate pathogens must be isolated from an affected tissue and propagated in pure culture. The isolated putative pathogens then must be transferred into a healthy recipient, where they would produce pathogenic changes identical to the one seen in the original donor organ. Obviously this rigorous proof is absolutely impossible in the case of human T cell-mediated autoimmune diseases. Most importantly, there are overriding ethical concerns that forbid the transfer of potentially pathogenic immune cells from a diseased to a healthy person. Second, immunological incompatibility (i.e., rejection of foreign cells) would in addition prevent successful transfer of lymphocytes between people other than identical twins.

As a consequence, our knowledge of encephalitogenic T cells and their putative target autoantigens has to rely on experimentally induced CNS autoimmune models, variants of EAE. From these models we know that many, if not all, proteins of the white substance can serve as targets of autoimmune attacks. Practically all accessible myelin proteins have been tested for their encephalitogenic potential, and they were found to be active targets for autoimmune T cells. Interestingly, different myelin proteins, or peptide segments thereof, produce EAE of different intensity and clinical patterns in various experimental animals. Furthermore, it turned out that in vivo reactivity against encephalitogenic autoantigens is not static but can fluctuate profoundly over time. Sercarz and his colleagues coined the term "determinant spreading" to describe the dynamic expansion of the number of target autoantigens during an ongoing encephalitogenic immune response. In the original report (Lehmann et al., 1992), mice were initially immunized against the MBP peptide Ac1-11, a treatment that activated T cells uniquely specific for the immunizing peptide and eventually induced EAE. Later, during the chronic phase, the T cells' immune response spread out, targeting MBP epitopes beyond the immunizing peptides, including sequences 35-47, 81-100, and 121-140.

The mechanisms underlying determinant spreading are not definitively known. Current concepts focus on professional APCs, or dendritic cells (DCs). These are not components of the normal CNS but may make their way into the EAE lesion from outside during the initial inflammation. Sitting in the CNS tissue, DCs are supposed to pick up locally available myelin as shed soluble proteins or particulate vesicles, process the material, and present an array of myelin autoantigenic peptide fragments to specific T cells. At least in theory, the APCs may act either locally, recruiting T cells spreading antigen specificity in the EAE lesion, or in lymphoid tissues, after having emigrated back into peripheral immune organs (Vanderlugt and Miller, 2002).

We presume that the human autoimmune response behaves in a way comparable to its rodent counterpart, but our information is scanty. Studies of human immune cells, though restricted to peripheral blood or CSF, identified T cells not only specific for classical myelin autoantigens (MBP; myelin oligodendrocyte glycoprotein, or MOG; and proteolipid protein, or PLP) but also for most other encephalitogenic autoantigens originally detected in rodents. It is important to note, with few exceptions, that the frequencies of CNS-specific T cell clones are similar in the MS and healthy donor immune repertoires (Sospedra and Martin, 2005).

As mentioned above, there is good evidence that in human MS CD8 T cells play a dominant role in local tissue destruction. In humans, in contrast to EAE models, early active lesions are dominated by CD8+, not CD4+, T cell infiltrates; and these infiltrating CD8+ T cells show signs of clonal expansion. Studies of MS plaque-infiltrating CD8+ T cells are thus of paramount importance, but such studies meet enormous difficulties. For example, studies of CD8+ T cells have been hampered by the difficulty of establishing antigen-specific cell lines and clones. Even more hampering are the ethical reasons that forbid access to living MS plaque tissue. Thus, we require new technology that can overcome these problems and yet allow us to study plaque-infiltrating CD8+ T cells.

One approach has been to characterize the plaque-invading TCR repertoire, which was undertaken by Oksenberg et al. (1990), who applied polymerase chain reaction (PCR) amplification to plaque-derived RNA preparations. The investigators found a limited repertoire of TCR α-chain genes, but at that stage it was impossible to ascribe their results to specified T cell subsets, let alone individual T cells. Babbe et al. (2000) went one decisive step further: They succeeded in amplifying the rearranged genes for TCR β-chains from individual CD8+-infiltrated T cells. Using immunocytochemistry, they identified CD8+ parenchyma infiltrating T cells in fixed tissue sections, and they then amplified genomic DNA for analysis of rearranged TCR genes. T cells from clones that had expanded in the CNS lesion were also detected in peripheral blood and CSF populations in the same patient (Skulina et al., 2004).

A breakthrough toward functional studies of single-infiltrate T cells from fixed specimens was achieved recently by Dornmair and colleagues (Seitz et al., submitted). These investigators succeeded in isolating from fixed human autoimmune tissues CD8+ T cells with known Vβ usage and then cloned the paired TCR α- and β-chains. These were then grafted into living lymphoma cells, which do not produce TCRs of their own but have fully functional donor TCR. A similar approach has been reported for human B cells in optic neuritis (Haubold et al., 2004).

The identification of unknown target autoantigens remains a major problem. Combination of peptide libraries with bioinformatics (Nino-Vasquez et al., 2005) is a promising approach, but it is complicated by the unlimited diversity of the peptide universe and the redundancy of peptide presentation by MHC and degenerate recognition by TCR.

Notwithstanding the considerable technical progress, so far no myelin candidate autoantigen has been identified with certainty as a target of the pathogenic effector cell in MS lesions. There is converging evidence that, as in EAE models, a multitude of myelin (and extramyelin) has the potential to act as a target autoantigen in a T and B cell attack against the CNS. Established encephalitogens, such as MBP, PLP, and MOG, are definitely prime candidates, but other antigens may have a role as well. Different autoantigens may have different functions in individual patients. Furthermore, one autoimmune attack may target more than one autoantigen at one point in time, and the autoantigenic profiles may change over time. Identification of autoantigenic profiles in patients is a major objective in human neuroimmunology. Such diagnostics are required as the basis of immune-specific T cell-targeted immunotherapy.

# 7    Which Factors Precipitate Onset and Relapses?

It is common clinical knowledge that MS typically attacks people of young age, that women are affected more frequently than men, and that the disease often starts out as bouts and remissions. However, the events that precipitate the onset and relapse of diseases remain largely obscure. We know that susceptibility to MS is governed by both intrinsic factors (genes) and environmental factors (i.e., infection). Studies of twin pairs have shown that the concordance rate of MS is more than 25% in genetically identical twins but is much lower in nonidentical twins, about 3% (Ebers et al., 1986; Hansen et al., 2005). This discrepancy underlines the importance of genetic factors; but the fact that concordance is substantially less than 100% points to additional factors, which, apart from epigenetic factors, should be mostly environmental.

Indeed, processes surrounding infections have often been invoked as contributing to the precipitation of onset and relapses of MS. Infections could influence MS prevalence in several, quite opposite ways. Bach recently pointed to a general trend in a global trend of declining infectious diseases coincident with an increase in certain autoimmune diseases, such as MS and type 1 diabetes mellitus (Bach, 2002). Accordingly, intensive and sustained immune reactions against microbial or parasitic infection would keep autoimmune diseases and allergies away—would be protective.

Under particular conditions, however, infections could have an opposite function: They could trigger autoimmune diseases rather than protect against them.

Disease precipitation by infections is based on several grounds (Gilden, 1999). First, often cited is an "MS epidemic" that occurred on the Faroe Islands during World War II. This archipelago had not seen MS patients before 1943, when British troops were stationed on some islands. These troops no doubt carried with them novel infectious agents; and according to one hypothesis, these agents might have been responsible for a wave of MS cases that appeared shortly thereafter (Kurtzke et al., 1995).

Then there is a link between relapses and sporadic, often banal infections. One study (Sibley et al., 1985) found a substantial association of disease exacerbation with common viral infections, while at the same time patients with MS seem to suffer less frequently from such infections than healthy people.

Following currently prevailing views, infections could trigger MS (and other organ-specific autoimmune diseases) by pathologically activating the self-reactive T cells that populate the healthy immune repertoire at considerable frequency but remain innocuous in most individuals. At least in theory, microbes could stimulate these T cells in several ways (Wekerle, 2006). Autoreactive T cells may be activated by microbial components that structurally resemble organ-specific autoantigens, a process called "molecular mimicry." Activated by microbial encounter, the self-reactive T cells would turn autoaggressive, enter their target tissue, and mount a pathogenic attack. Alternatively, a number of bacterial and viral agents produce "superantigens," proteins that activate *subsets* of TCRs rather than antigen-specific clones. Activation of a subset that contains a particularly high proportion of autoreactive T cells could be the trigger of an autoimmune response. Finally, microbial infection could start autoimmune disease via mechanisms of innate immunity. Infection could create a strongly inflammatory milieu that ultimately allows (unspecific?) activation of autoreactive T cells in the sense of a "bystander" reaction.

Microbial activation of autoreactive T cells as a trigger of autoimmune disease has been demonstrated in some experimental models, but in MS these mechanisms remain to be demonstrated beyond doubt. In more general terms: Although to date no single infectious agent has been identified as *the* MS agent, it is highly probable that a number of infections can affect development and course of MS. Studying these mechanisms remains a research area of top priority.

# 8     What is the Interplay of Immune Response and Neuronal Degeneration?

It should be noted that many but not all contemporary MS investigators subscribe to the autoimmune concept of pathogenesis. There are divergent views, however, that maintain that the inflammatory responses seen in active lesions are, if anything, secondary to primary degenerative processes. This view was resuscitated recently by a neuropathological study of several brains (one of them in depth) from people who had died shortly after onset of a relapse. In these tissues, signs of oligodendrocyte degeneration and microglia activation were noted but no signs of inflammatory cell infiltration (Barnett and Prineas, 2004). Although these lesions coming from highly exceptional cases may be representative for a certain subgroup of MS (or demyelinating disease masquerading as MS?), it appears dated to extend primary degeneration as the cause of all cases of MS.

Yet, there are important interconnections between CNS inflammation and neuronal degeneration. Thus, it is well established that intensive local inflammation puts the CNS tissues at serious risk. Inflammatory mediators released during such a response may have detrimental effects on the integrity of neural cells—neurons as well as glia. One telling example is the intensive inflammation underlying acute EAE models. Furthermore, transgenic mice with exaggerated production of proinflammatory cytokines (i.e., tumor necrosis factor-$\alpha$, or TNF$\alpha$) by local glia cells suffer from progressive neurodegenerative changes (Akassoglou et al., 1997).

Conversely, it is also known that neurodegenerative changes support the presence of inflammatory disease. An important examples is X-linked adrenoleukodystrophy (ALD), a primary degenerative disease of the CNS, adrenal glands, and testes, that is definitely due to mutation of a gene encoding a faulty peroxisomal membrane transporter protein (Mosser et al., 1993). In a proportion of ALD cases, the affected CNS displays lesions that stunningly resemble active MS plaques. They show, in addition to destruction of oligodendrocyte and myelin sheaths, prominent perivascular round cell infiltrates and production of proinflammatory cytokines (Powers and Moser, 1998).

It is important to note, however, that intact CNS tissue represents a strongly antiinflammatory milieu and that this status is tightly controlled by normally functioning neurons. In healthy CNS tissue, neurons actively suppress the production of "immune molecules" (e.g., MHC antigens, cytokines, chemokines, cell adhesion molecules) that are required to support a regular immune response. This deficiency is one key factor warranting the "immunoprivileged" status of the healthy CNS. Loss of neuronal activity by genetic, toxic, or degenerative processes (e.g., following peripheral axotomy) cancels neuronal suppression and thus allows de novo production of "immune molecules" (Neumann and Wekerle, 1998). Neurodegenerative CNS tissues thus turn "immune friendly" and allow the local immune responses to unfold, either against foreign agents, as in microbial infections, or against self antigens, as in CNS autoimmune disorders.

To date, the relevance of the interconnections between neuronal activity, degeneration, and inflammation in the pathogenesis of MS remain largely unknown. Again, suitable experimental models are required to fill this gap.

# 9    Immune (Dys)regulation?

Late in 1971 Gershon and Kondo reported that lymphocytes taken from mice tolerant to a foreign antigen were able to transfer this state of specific tolerance adoptively to a second, completely naive host (Gershon and Kondo, 1971). Their article and two others (Droege, 1971; Jacobson et al., 1972) opened the gates to a veritable flood of reports confirming and extending the finding—the birth of the suppressor T cell. Some of these early reports described a possible role of suppressor T cells in regulating the course and activity of MS (Arnason and Waksman, 1980). One method to identify suppressor cells in MS was by measuring their effect in vitro on mitogen-driven immunoglobulin secretion by peripheral blood lymphocytes. Suppressor T cells, often found in CD8 populations, were reduced during active MS phases, which fit well with a general deficit of CD8 T cells in the peripheral blood of affected patients.

The suppressor cell concept fell from grace for a while but reemerged recently under a new guise, regulatory T cells. "Tregs" sensu strictu are defined as CD4 T cells that express the constitutively activation markers CD25 [interleukin-2 (IL-2) receptor α-chain] and GITR (glucocorticoid induced TNF receptor-like), a member of the TNF receptor gene superfamily, on their membranes and, more important (because more specific), the transcription factor FoxP3 in their cytoplasm (Sakaguchi, 2005). Indeed, mice or people with nonfunctional, mutant FoxP3 (scurfy mice and IPEX patients) suffer from a lymphoproliferative disease characterized by autoimmune-like infiltrates of many organs, especially endocrine glands. Remarkably, however, the CNS does not seem to be affected by FoxP3 deficiency (Ulmanen et al., 2005).

One early study screened circulating blood lymphocytes for CD4 T cells expressing high levels of CD25, thus corresponding to Tregs of murine models (Viglietta et al., 2004). The number of CD4+CD25$^{hi}$ T cells retrieved in the blood from MS patients was markedly lower than in control samples from healthy donors. This was in line with a poor cloning efficiency of CD4+CD25$^{hi}$ T cells in MS-derived blood. This work was confirmed and extended by two subsequent reports. One of the reports showed that the reduced suppressive effect in MS blood lymphocytes was not due to decreased numbers of CD4+CD25$^{hi}$ Tregs but was explained by a reduced suppressive capacity per cell (Haas et al., 2005). The other report noted reduced suppressive capacity in MS T cells to suboptimal expression of FoxP3 RNA and protein (Huan et al., 2005).

Another intriguing regulatory cell derives from the CD8 T cell lineage. This regulatory cell is not targeted on the putative autoantigen but recognizes clonotypic determinants on the autoimmune effector T cells—in a loose sense qualifies as an anti-idiotypic member of a receptor-directed regulatory network. First clues for regulatory CD8 T cells came from studies of T cells transferred to EAE. At about the same time, two groups observed that transfer of encephalitogenic CD4+ T cell lines into a recipient activated a specific counterregulatory response by CD8+ T cells (Lider et al., 1988; Sun et al., 1988b). These cells specifically destroyed the original CD4 T cells in culture and neutralized their encephalitogenic potential in vivo. Depletion of CD8 T cells from Lewis rats using monoclonal antibody (mAb) OX-8 terminated the vaccination effect but did not change the monophasic course of an individual EAE episode (Sun et al., 1988a). Subsequent studies confirmed and extended these finding in mouse EAE models. They showed that, indeed, depletion of CD8 T cells by treatment with mAbs affected resistance against repeated induction but not the duration of individual bouts (Jiang et al., 1992). Transgenic mice deficient in CD8 T cells showed milder EAE but a propensity to clinical relapses (Koh et al., 1992).

Regulatory T cells affect encephalitogenic responses by distinct mechanisms (Table 22.4). Khoury et al. described CD8+ regulatory T cells with undetermined antigen specificity that limit expansion and activation of pathogenic CD4+ T cells possibly indirectly by suppressing co-stimulatory activity on APCs (Najafian et al., 2003).

Jiang et al. discovered that "anti-idiotypic" CD8 T cells recognize fragments of encephalitogenic CD4 T cell receptors in the molecular context of Qa-1 molecules (Jiang et al., 1998), class Ib products of the MHC that carry relevant receptors from the immune repertoire (Jiang et al., 1995), a process that profoundly shapes the autoantigen-specific TCR repertoire (Jiang et al., 1995, 2003). In line with these observations, mice lacking Qa determinants show exaggerated (auto)immune responses in the absence of CD8+ suppressor T cells (Hu et al., 2004).

There are immunoregulatory lymphocytes in addition to classical Tregs, including natural killer (NK) cells, NK1 T cells, and T cells with a γδ receptor; and many if not all lineages seem to exhibit abnormalities in MS (Sospedra and Martin, 2005). Whether these defects occur independently in individual patients or they are the results of a deficient master plan remains open so far. It is not yet clear to what extent these defects cause or are the consequence of MS pathogenesis or how far they indeed influence autoimmune reactivity in MS patients. There is an immense need to answer these questions. Clear answers might provide the basis for immunotherapy.

TABLE 22.4. Potentially Self-Protective Regulatory T cells in Autoimmune Diseases.

| Subset | Targets | Regulatory molecules | Induction in vivo |
|---|---|---|---|
| CD4+CD25+ Treg | Effector T cells; APCs | CTLA-4; IL-10; TGFß | Natural; activation via co-stimulatory molecules? Innate immune responses? |
| Th3 | Effector T cells | IL-10; TGFß | Oral tolerization |
| CD8, Qa-1 restricted | Activated T cells expressing Qa-1 | TCR recognizing TCR peptides from target effector T cells in Qa context | Natural; T cell vaccination |
| CD8+ Tregs | Effector T cells? | TCR (including TCR-gd) specific for autoantigen? | Vaccination with autoantigen in aerosol |
| NK1 T cells | Activated T cells | Activation of "invariant" TCR by glycolipids/CD1d complexes; IL-4; IL-10; TGFß; cytotoxicity | Natural; vaccination with glycolipids |

APCs, antigen-presenting cells; CTLA, cytotoxic T lymphocyte antigen; IL-10, interleukin-10; TGFβ, transforming growth factor-β; TCR, T cell receptor.

# References

Akassoglou, K., Probert, L., Kontogeorgos, G., Kollias, G. (1997) Astrocyte-specific but not neuron-specific transmembrane TNF triggers inflammation and degeneration in the central nervous system of transgenic mice. *J. Immunol.* 158:348-445.

Arnason, B. G. W., Waksman, B. H. (1980) Immunoregulation in multiple sclerosis. *Ann. Neurol.* 8:237-240.

Babbe, H., Roers, A., Waisman, A., et al. (2000) Clonal expansion of CD8$^+$ T cells dominate the T cell infiltrate in active multiple sclerosis lesions shown by micromanipulation and single cell polymerase chain reaction. *J. Exp. Med.* 192:393-404.

Bach, J-F. (2002) The effect of infections on susceptibility to autoimmune and allergic diseases. *N. Engl. J. Med.* 347:911-920.

Barnett, M. H., Prineas, J. W. (2004) Relapsing and remitting multiple sclerosis: pathology of the newly forming lesion (p NA). *Ann. Neurol.* 55:458-468.

Bielekova, B., Goodwin, B., Richert, N., et al. (2000) Encephalitogenic potential of the myelin basic protein peptide (amino acids 83-99) in multiple sclerosis: results of a phase II clinical trial with an altered peptide ligand. *Nat. Med.* 6:1167-1175.

Booss, J., Esiri, M. M., Tourtellotte, W. W., Mason, D. Y. (1983) Immunohistochemical analysis of T lymphocyte subsets in the central nervous system in chronic progressive multiple sclerosis. *J. Neurol. Sci.* 62:219-232.

Compston, A., Lassmann, H., McDonald, I. (2006) The story of multiple sclerosis. In: Compston, A., Confavreux, C., Lassmannn H., et al. (eds) McAlpine's Multiple Sclerosis. Churchill Livingstone Elsevier, New York, pp 3-62.

Confavreux, C., Compston, A. (2006) The natural history of multiple sclerosis. In: Compston, A., Confavreux, C., Lassman, H. (eds) McAlpine's Multiple Sclerosis. Churchill Livingstone Elsevier, New York, pp 183-272.

Droege, W. (1971) Amplifying and suppressive effect of thymus cells. *Nature* 234: 549-551.

Ebers, G. C., Bulman, D. E., Sadovnick, A. D., et al. (1986) A population-based study of multiple sclerosis in twins. *N. Engl. J. Med.* 315:1638-1642.

Ellmerich, S., Mycko, M., Takács, K., et al. (2005) High incidence of spontaneous disease in an HLA-DR15 and TCR transgenic multiple sclerosis model. *J. Immunol.* 174: 1938-1946.

Esiri, M. M. (1980) Multiple sclerosis: a quantitative and qualitative study of immunoglobulin-containing cells in the central nervous system. *Neuropathol. Appl. Neurobiol.* 6:9-21.

Friese, M. A., Fugger, L. (2005) Autoreactive CD8+ T cells in multiple sclerosis: a new target for therapy? *Brain* 128:1747-1763.

Genain, C. P., Lee-Parritz, D., Nguyen, M-H., et al. (1994) In healthy primates, circulating autoreactive T cells mediate autoimmune disease. *J. Clin. Invest.* 94:1339-1345.

Genain, C. P., Cannella, B., Hauser, S. L., Raine, C. S. (1999) Identification of autoantibodies associated with myelin damage in multiple sclerosis. *Nat. Med.* 5:170-175.

Gershon, R. K., Kondo K. (1971) Infectious immunological tolerance. *Immunology* 21:903-914.

Gilden, D. H. (1999) Chlamydia: a role for multiple sclerosis or more confusion? *Ann. Neurol.* 46:4-5.

Haas, J., Hug, A., Viehöver, A., et al. (2005) Reduced suppressive effect of CD4+CD25$^{high}$ regulatory T cells on the T cell immune response against myelin oligodendrocyte glycoprotein in patients with multiple sclerosis. *Eur. J. Immunol.* 35:3343-3352.

Hansen, T., Skytthe, A., Stenager, E., et al. (2005) Risk for multiple sclerosis in dizygotic and monozygotic twins. *Mult. Scler.* 11:500-503.

Haubold, K., Owens, G. P., Kaur, P., et al. (2004) B-lymphocyte and plasma cell clonal expansion in monosymptomatic optic neuritis cerebrospinal fluid. *Ann. Neurol.* 56:97-107.

Hauser, S. L., Bhan, A. K., Gilles, F., et al. (1986) Immunohistochemical analysis of the cellular infiltrate in multiple sclerosis lesions. *Ann. Neurol.* 19:578-587.

Hu, D., Ikizawa, K., Lu, L. R., et al. (2004) Analysis of regulatory CD8 T cells in Qa-1-deficient mice. *Nat. Immunol.* 5:516-523.

Huan, J., Culbertson, N., Spencer, L., et al. (2005) Decreased FOXP3 levels in multiple sclerosis patients. *J. Neurosci. Res.* 81:45-52.

Huseby, E. S., Liggitt, D., Brabb, T., et al. (2001) A pathogenic role for myelin specific CD8+ T cells in a model for multiple sclerosis. *J. Exp. Med.* 194:669-676.

Jacobson, E. B., Herzenberg, L. A., Riblet, R., Herzenberg, L. A. (1972) Active suppression of immunoglobulin allotype synthesis. II. Transfer of suppressing factor with spleen cells. *J. Exp. Med.* 135:1163-1176.

Jiang, H., Zhang, S-L., Pernis, B. (1992) Role of CD8+ T cells in murine experimental allergic encephalomyelitis. *Science* 256:1213-1215.

Jiang, H., Ware, R., Stall, A., et al. (1995) Murine CD8+ T cells that specifically delete autologous CD4+ T cells expressing Vβ8 TCR: a role of the Qa-1 molecule. *Immunity* 2:185-194.

Jiang, H., Kashleva, H., Xu, L-X., et al. (1998) T cell vaccination induces T cell receptor Vβ-specific Qa-1 restricted CD8+ T cells. *Proc. Natl. Acad. Sci. USA* 95:4533-4537.

Jiang, H., Curran, S., Ruiz-Vazquez, E., et al. (2003) Regulatory CD8+ T cells fine-tune the myelin basic protein-reactive T cell receptor Vβ repertoire during experimental autoimmune encephalomyelitis. *Proc. Natl. Acad. Sci. USA* 100:8378-8383.

Kappos, L., Comi, G., Panitch, H., et al. (2000) Induction of a non-encephalitogenic Th2 autoimmune response in multiple sclerosis after administration of an altered peptide ligand in a placebo controlled, randomized phase II trial. *Nat. Med.* 6:1176-1182.

Keegan, M., König, F., McClelland, R., et al. (2005) Relation between humoral pathological changes in multiple sclerosis and response to therapeutic plasma exchange. *Lancet* 366:579-582.

Kira, J-I. (2003) Multiple sclerosis in the Japanese population. *Lancet Neurol.* 2:117-127.

Koh, D-R., Fung-Leung, W-P., Ho, A., et al. (1992) Less mortality but more relapses in experimental allergic encephalomyelitis in CD8-/- mice. *Science* 256:1210-1213.

Krumbholz, M., Theil, D., Derfuss, T., et al. (2005) BAFF is produced by astrocytes and upregulated in multiple sclerosis lesions and primary central nervous system lymphoma. *J. Exp. Med.* 201:195-200.

Krumbholz, M., Theil, D., Cepok, S., et al. (2006) Chemokines in multiple sclerosis: CXCL12 and CXCL13 up-regulation is differentially linked to CNS immune cell recruitment. *Brain* 129:200-211.

Kurtzke, J. F., Hyllested, K., Heltberg, A. (1995) Multiple sclerosis in the Faroe islands: transmission across four epidemics. *Acta Neurol. Scand* 91:321-325.

Lassmann, H., Wekerle, H. (2006) The pathology of multiple sclerosis. In: Compston, A., Confavreux, C., Lassmann, C., et al (eds) McAlpine's Multiple Sclerosis. Churchill Livingstone Elsevier, New York, pp 557-600.

Lassmann, H., Brück, W., Lucchinetti, C. (2001) Heterogeneity of multiple sclerosis pathogenesis: implications for diagnosis and therapy. *Trends Mol. Med.* 7:115-121.

Lehmann, P. V., Forsthuber, T., Miller, A., Sercarz, E. E. (1992) Spreading of T-cell autoimmunity to cryptic determinants of an autoantigen. *Nature* 358:155-157.

Lider, O., Reshef, T., Béraud, E., et al. (1988) Anti-idiotypic network induced by T-cell vaccination against experimental autoimmune encephalomyelitis. *Science* 239:181-183.

Lindå, H., Hammarberg, H., Cullheim, S., et al. (1998) Expression of MHC class I and β2-microglobulin in rat spinal motoneurons: regulatory influences of IFN-gamma and axotomy. *Exp. Neurol.* 150:282-295.

Madsen, L. S., Andersson, E. C., Jansson, L., et al. (1999) A humanized model for multiple sclerosis using HLA DR2 and a human T cell receptor. *Nat. Genet.* 23:343-347.

Magliozzi, R., Columba-Cabezas, S., Serafini, B., Aloisi, F. (2004) Intracerebral expression of CXCL13 and BAFF is accompanied by formation of lymphoid follicle-like structures in the meninges of mice with relapsing experimental autoimmune encephalomyelitis. *J. Neuroimmunol.* 148:11-23.

Martin, R., Jaraquemada, D., Flerlage, M., et al. (1990) Fine specificity and HLA restriction of myelin basic protein-specific cytotoxic T cell lines from multiple sclerosis patients and healthy individuals. *J. Immunol.* 145:540-548.

Medana, I., Martinic, M. M. A., Wekerle, H., Neumann, H. (2001) Transection of MHC class I-induced neurites by cytotoxic T lymphocytes. *Am. J. Pathol.* 159:809-815.

Meinl, E., Hoch, R. M., Dornmair, K., et al. (1997) Encephalitogenic potential of myelin basic protein-specific T cells isolated from normal rhesus macaques. *Am. J. Pathol.* 150:445-453.

Meinl, E., Krumbholz, M., Hohlfeld, R. (2006) B lineage cells in the inflammatory CNS environment: migration, maintenance, local antibody production and therapeutic modulation. *Ann. Neurol.* 59:880-892.

Mosser, J., Douar, A-M., Sarde, C-O., et al. (1993) Putative X-linked adrenoleukodystrophy gene shares unexpected homology with ABC transporters. *Nature* 361:726-730.

Najafian, N., Chitnis, T., Salama, A. D., et al. (2003) Regulatory functions of CD8+CD28- T cells in an autoimmune disease model. *J. Clin. Invest.* 112:1037-1048.

Neumann, H., Wekerle, H. (1998) Neuronal control of the immune response in the central nervous system: linking brain immunity to neurodegeneration. *J. Neuropathol. Exp. Neurol.* 58:1-9.

Neumann, H., Cavalié, A., Jenne, D. E., Wekerle, H. (1995) Induction of MHC class I genes in neurons. *Science* 269:549-552.

Nino-Vasquez, J. J., Allicotti, G., Borras, E., et al. (2005) A powerful combination: the use of positional scanning libraries and biometrical analysis to identify cross-reactive T cell epitopes. *Mol. Immunol.* 40:1063-1074.

O'Connor, K. C., Appel, H., Bregoli, L., et al. 2005. Antibodies from inflamed central nervous system tissue recognize myelin oligodendrocyte glycoprotein. *J. Immunol.* 175:1974-1982.

Oksenberg, J. R., Stuart, S., Begovich, A. B., et al. (1990) Limited heterogeneity of rearranged T-cell receptor Vα transcripts in brains of multiple sclerosis patients. *Nature* 345:344-346.

Owens, G. P., Kraus, H., Burgoon, M. P., et al. (1998) Restricted use of $V_H4$ germline segments in an acute multiple sclerosis brain. *Ann. Neurol.* 43:236-243.

Pette, M., Fujita, K., Wilkinson, D., et al. (1990) Myelin autoreactivity in multiple sclerosis: recognition of myelin basic protein in the context of HLA-DR2 products by T lymphocytes of multiple sclerosis patients and healthy donors. *Proc. Natl. Acad. Sci. USA* 87:7968-7972.

Powers, J. M., Moser, H. (1998) Peroxisomal disorders: genotype, phenotype, major neuropathologic lesions, and pathogenesis. *Brain Pathol.* 8:109-120.

Qin, Y., Duquette, P., Zhang, Y., et al. (1998) Clonal expansion and somatic hypermutation of $V_H$ genes of B cells from cerebrospinal fluid in multiple sclerosis. *J. Clin. Invest.* 102:1045-1050.

Sakaguchi, S. (2005) Naturally arising Foxp3-expressing CD4+CD25+ regulatory T cells in immunological tolerance to self and non-self. *Nat. Immunol.* 6:345-352.

Sibley, W. A., Bamford, C. R., Clark, K. (1985) Clinical viral infections and multiple sclerosis. *Lancet* 1:1313-1315.

Skulina, C., Schmidt, S., Dornmair, K., et al. (2004) Multiple sclerosis: brain-infiltrating CD8+ T cells persist as clonal expansions in the cerebrospinal fluid and blood. *Proc. Natl. Acad. Sci. USA* 101:2428-2433.

Sospedra, M., Martin, R. (2005) Immunology of multiple sclerosis. *Annu. Rev. Immunol.* 23:683-747.

Storch, M. K., Piddlesden, S., Haltia, M., et al. (1998a) Multiple sclerosis: in situ evidence for antibody- and complement-mediated demyelination. *Ann. Neurol.* 43:465-471.

Storch, M. K., Stefferl, A., Brehm, U., et al. (1998b) Autoimmunity to myelin oligodendrocyte glycoprotein in rats mimics the spectrum of multiple sclerosis pathology. *Brain Pathol.* 8:681-694.

Stüve, O., Cepok, S., Eliaset B., et al. (2005) Clinical stabilization and effective B-lymphocyte depletion in the cerebrospinal fluid and peripheral blood of a patient with fulminant relapsing-remitting multiple sclerosis. *Arch. Neurol.* 62:1620-1623.

Sun, D., Ben-Nun, A., Wekerle, H. (1988a) Regulatory circuits in autoimmunity: recruitment of counter-regulatory CD8+ T cells by encephalitogenic CD4+ T line cells. *Eur. J. Immunol.* 18:1993-2000.

Sun, D., Qin, Y., Chluba, J., et al. (1988b) Suppression of experimentally induced autoimmune encephalomyelitis by cytolytic T-T cell interactions. *Nature* 332:843-845.

Sun, D. M., Whitaker, J. N., Huang, Z. G., et al. (2001) Myelin antigen-specific CD8+ T cells are encephalitogenic and produce severe disease in C57BL/6 mice. *J. Immunol.* 166:7579-7587.

Tourtellotte, W. W., Staugaitis, S. M., Walsh, M. J., et al. (1985) The basis of intra-blood-brain-barrier IgG synthesis. *Ann. Neurol.* 17:21-27.

Ulmanen, I., Halonen, M., Ilmarinen, T., Peltonen, L. (2005) Monogenic autoimmune diseases—lessons of self-tolerance. *Curr. Opin. Immunol.* 17:609-615.

Vanderlugt, C. L., Miller, S. D. (2002) Epitope spreading in immune mediated diseases: implications for immunotherapy. *Nat. Rev. Immunol.* 2:85-95.

Viglietta, V., Baecher-Allan, C., Weiner, H. L., Hafler, D. A. (2004) Loss of functional suppression by CD4+CD25+ regulatory T cells in patients with multiple sclerosis. *J. Exp. Med.* 199:971-979.

Wekerle, H. (2006) Breaking tolerance: the case of the brain. In: Radbruch, A., Lipsky, P. E. (eds) Current Concepts in Autoimmunity and Chronic Inflammation. Springer, Berlin, pp 25-50.

Woodroofe, M. N., Bellamy, A. S., Feldmann, M., et al. (1986) Immunocytochemical characterization of the immune reaction in the central nervous system in multiple sclerosis: possible role for microglia in lesion growth. *J. Neurol. Sci.* 74:135-152.

# Author Index

Huston, J. S., 329
Hutchings, P. R., 111–112
Hutchins, B., 32
Hutchinson, M., 455, 460
Hutloff, A., 19, 135–136
Hutton, G., 478
Hyllested, K., 511
Hynes, R. O., 324

Iannucci, G., 261–262, 327
Ibanez, C., 455
Ichikawa, T., 114
Igarashi, H., 451, 456
Ignatowicz, L., 169
Iihoshi, S., 481, 492, 494
Ikemizu, S., 184, 330
Ikizawa, K., 51, 511
Illes, Z., 7, 18–19, 140, 144, 146,
    148–149, 151, 169
Ilmarinen, T., 513
Ilonen, J., 115
Im, S. H., 16–17, 114
Im, S. Y., 435
Imaizumi, T., 485, 487–488, 491,
    494–495
Imberti, L., 406
Imitola, J., 18, 137
Inaba, K., 32
Indiveri, F., 57, 64
Infante-Duarte, C., 253, 260
Ingle, G. T., 250, 257
Inglese, M., 416, 422
Inobe, J., 15, 64
Inoue, M., 410, 422, 481, 494
Irmler, M., 284
Isaacs, J. D., 465, 477
Ishida, Y., 126, 135, 138
Ishikawa, T., 18
Ishizaka, K., 21, 35
Ismail, N., 53, 149
Ito, A., 443, 445, 456
Ito, I., 456
Ito, Y., 494
Itoi, H., 450, 456
Itoyama, Y., 233–234, 257
Itskovitz-Eldor, J., 424
Itsykson, P., 423
Ivars, F., 166
Iwai, H., 19, 124–125, 135, 138

Iwai, Y., 135
Iyoda, T., 408
Izikson, L., 19, 35

Jackson, K. A., 493
Jackson, R. A., 110, 117
Jacob, A., 334
Jacob, N., 330
Jacobs, F., 211
Jacobs, H., 114
Jacobs, K. M., 367
Jacobs, R., 405–406
Jacobsen, H., 184, 330
Jacobsen, M., 178, 183, 281
Jacobson, D. L., 24, 32
Jacobson, E. B., 507, 511
Jacobson, S., 188, 200, 208
Jahn, G., 207
Jahng, A. W., 140, 142–143, 149
Jakel, P., 424
James, A. G., 110
Janeway, C. A., Jr. 39–40, 51, 53, 70, 86,
    122, 136, 165, 170, 394, 402
Janson, C. H., 403
Janssens, W., 162, 167
Jansson, L., 184, 282, 366, 412, 443–444,
    456–457
Jaraquemada, D., 366, 512
Jauberteau, M. O., 84
Jeffery, D. R., 298, 329
Jeffery, E. W., 15, 50, 54
Jenkins, M. K., 10, 12, 16, 38, 51, 91,
    112, 136
Jenkins, R. N., 355, 366, 391, 403
Jenne, D. E., 184, 283, 512
Jensen, J., 274, 281, 335, 439, 457
Jeong, M. C., 81, 181, 253
Jerne, NK., 40, 51
Jersild, C., 266, 281
Jezzard, P., 254
Jiang, F., 227, 257
Jiang, H., 23, 26, 29, 32–33, 36–37,
    40–41, 44–48, 51–52, 54, 57, 64,
    267, 281, 392, 403, 509, 511
Jick, S. S., 453
Jin, F., 457
Jin, L., 107
Jin, M., 457
Jin, W., 50

# Subject Index

## A

Acetylcholine receptor (AChR), 69, 390

ACR50, 355

ACR70, 355

Activation-induced cell death (AICD), 21, 90–91

Acute disseminated encephalomyelitis (ADEM), 72, 225

Affinity/avidity model of peripheral T cell regulation, 47–49

a-Galactosylceramide (a-GalCer), 140

AIRE gene, 88

AIRE transcription factor, 4

Aletumzumab (Campath-1), 313

Alpha lipoic acid (ALA), 321

Altered peptide ligands (APLs), 100, 310

Alzheimer's disease, 446

Amyloid precursor protein (APP), 244

AndroGel, 309

Antibodies, variable regions of, 68

Antibody-dependent cell-mediated cytotoxicity (ADCC)-dependent effector mechanisms, 70

Anti-CD28 antibodies, 105

Anti-CD25 monoclonal antibody, 163

Anti-ergotypic responses
  anti-ergotypic T cells component of, 59–61
  ergotope components of, 58–59
  in immunized individuals, 60
  in nonimmunized individuals, 59–60
  overlaps and interactions with other regulatory population, 61–62
  target T cells components of, 57

  triggered by therapeutic vaccination, 60–61

Antigen-presenting cells (APCs), 10, 12, 22, 28, 38, 40, 44, 48, 121, 143, 175, 194, 267, 274, 291, 344, 385

Antigen/T cell receptor (Ag/TCR) transgenic systems, 89

Anti-idiotypic responses, 57

Anti-nuclear antibodies (ANA), 72

Anti-T cell receptor (TCR) CD8aa+ Treg, 22

Anti-TCR CD8 Treg cells, 28

Anti-thymocyte globulins, 96

Anti-TNFa treatment, 25

Anti-tumor necrosis factor-a (TNFa), 25

APECED, *see* Autoimmune polyendocrinopathy-candidiasis-ectodermaldystrophy

Apoptotic cell death, 37

AQP4, 239

Aquaporin-4 (AQP4) water channel, 73

Arachidonic acid, 227

Arthritis, 58

Athymic BALB/c mice, 22

Atorvastatin, 308

Autoimmune disease, immunotherapeutic strategies to treat
  induction of antigen-specific Treg, 25–27
  induction of non-specific Treg, 25

Autoimmune disorders (AIDs), 293

Autoimmune hemolytic anemia, 69, 80

Autoimmune polyendocrinopathy-candidiasis-ectodermal dystrophy, 5, 89

Printed in the United States of America.